On the Move

On the Move

How and Why Animals
Travel in Groups

Edited by
Sue Boinski and Paul A. Garber

The University of Chicago Press
Chicago and London

Sue Boinski is associate professor in the Department of Anthropology and faculty member in the Division of Comparative Medicine at the University of Florida. Paul A. Garber is professor in the Department of Anthropology at the University of Illinois and Director of Education and Research at the La Suerte Biological Research Center in Costa Rica and the Ometepe Biological Research Station in Nicaragua.

The University of Chicago Press, Chicago 60637
The University of Chicago Press, Ltd., London
© 2000 by The University of Chicago
All rights reserved. Published 2000

09 08 07 06 05 04 03 02 01 00   1 2 3 4 5
ISBN: 0-226-06339-9 (cloth)
ISBN: 0-226-06340-2 (paper)

Library of Congress Cataloging-in-Publication Data

On the move : how and why animals travel in groups / edited by
   Sue Boinski and Paul A. Garber.
      p.   cm.
   Includes bibliographical references. (p.   ) and index.
   ISBN 0-226-06339-9 (cloth : alk. paper). — ISBN 0-226-06340-2
   (paper)
      1. Social behavior in animals.   2. Primates—Behavior.
   I. Boinski, Sue.   II. Garber, Paul Alan.
   QL775.O6   2000
   591.56—dc21   99-13382
                                                                      CIP

⊗ The paper used in this publication meets the minimum requirements of the American National Standard for Information Sciences—Permanence of Paper for Printed Library Materials, ANSI Z39.48–1992.

CONTENTS

ACKNOWLEDGMENTS

We wish to thank the many friends and colleagues who helped us during the long editing process. All the contributors participated in the internal review process, most notably Dick Byrne, Colin Chapman, Marina Cords, and Charlie Janson. This volume also benefited greatly from the external chapter reviews kindly provided by Susan Alberts, Susan Antón, Mark Bekoff, Irwin Bernstein, Christophe Boesch, Monique Borgerhoff-Mulder, Robert Boyd, Nancy Caine, Dorothy Cheney, Anne Clark, Guy Colishaw, Scott Creel, Leslie Digby, Diane Doran, Robin Dunbar, John Eisenberg, Mark Elgar, Alejandro Frid, Doree Fragaszy, Dan Gebo, Deborah Gordon, Harold Gouzoules, Thomas Grubb, Walter Hartwig, Eckhard Heymann, Lynne Isbell, Alison Jolly, Andreas Koenig, Joanna Lambert, Steve Leigh, Bob Malina, John Mitani, Jim Moore, Dario Maestripieri, Lori Marino, Marilyn Norconk, William Oloput, Deb Overdorff, Ryne Palombit, Michael Pereira, Ann Pusey, Melissa Remis, John Robinson, Liz Rodgers, Lisa Rose, Dan Rubenstein, Michelle Sauther, Ann Savage, Carel van Schaik, Jeanne Sept, Tom Struhsaker, Elisabetta Visalberghi, Hal Whitehead, Andy Whiten, Bruce Winterholder, and additional anonymous reviewers. Alison Jolly and Andy Whiten also rendered gracious and eloquent service above and beyond the call of professional courtesy as the two external reviewers for the entire volume. Jennifer Rehg and Melanie Brandt drew the illustrations at the head of each chapter; their splendid work is appreciated. Christie Henry, Theresa Biancheri, and Charles Clifton at the University of Chicago Press and copyeditor Norma Roche provided much-needed support and advice at every stage.

Unraveling the Complexities of Group Travel

Clotted upon branches overhanging a cool stream, a group of about forty squirrel monkeys were immersed in their regular midday snooze. But for the occasional head cocking upward with watchful eyes, the adult males could have been mistaken for bromeliads sprawled over branches with pendulous limbs and tails. Females and juveniles huddled with tails tidily wrapped around their bodies such that they assumed the size and shape of golden loaves of bread. Only a group of infants and yearlings were awake, chasing and tumbling on the sturdy play area afforded by a wide, low branch of a kapok tree. Eventually, an adult female stood up, gave herself a vigorous scratch, and started to scrutinize a nearby *Cecropia* leaf for invertebrate denizens. Soon most of the females and immatures were active, searching foliage or walking slowly along foliated branches in search of promising foraging sites. The males continued to doze.

An adult female moved to the periphery of the troop and stared outward. She turned back toward the troop, emitted three loud, twittery vocalizations in short succession, and then moved several meters farther away from the other monkeys. Soon she was joined by another adult female, and both engaged in another intense bout of twittering calls. This time, however, there was an apparent rejoinder. Another female on the troop periphery, across from the first two, began twittering, and traveled 10 meters away from the periphery before returning. Within 10 minutes the twittering had escalated to the point that all fifteen adult female troop members had seemingly chosen sides, and were partitioned into two opposing groups. Was it mere happenstance that the two trajectories indicated by the females' positions relative to the troop center both intersected with favored foraging areas? In the midst of this cacophony, the immatures continued foraging and playing and the males dozing.

Suddenly the females clustered around the second calling point stopped twittering and returned to the main body of the troop, while the females at the first calling position started moving farther away from the troop. A few of those females, those at the vanguard edge, continued twittering at sporadic intervals. In their wake followed the remaining females and a number of juveniles. As the last of this adult female and immature cortege was leaving the streamside, the adult males finally stirred and followed languidly. The troop traveled in a directed fashion, in a column about 20 meters wide by 50 meters long, in a more or less straight line. Deviations from a straight trajectory were caused only by gaps in the interconnected branches that formed their arboreal highway. After traveling in this manner for nearly a kilometer, the troop reached a patch of feral guava trees. Twittering by the females completely ceased, and all the females and juveniles quickly became engrossed in eating. The adult males resumed their naps, and the infants and yearlings again started to play.

Field studies of primates, as well as those of social groups of birds, mammals, and invertebrates, are rife with similar intriguing reports of group travel that appears to require complex thinking and planning. As individuals, social animals must make ego-based behavioral decisions that serve to reduce vulnerability to predation, minimize the time and energy spent in traveling to and exploiting feeding sites, and increase reproductive success. As members of a group, however, individuals act within a wider social network in which cooperation, competition, and alliance formation require more complex processes of communication and coordination. These challenges and opportunities of group living are handled in diverse and fascinating ways, and social animals rely on different sets of cognitive, sensory, and social processes to coordinate group movement. The following descriptions of African elephants, hamadryas baboons, and common chimpanzees serve to highlight the complex and dynamic social drama that is an integral part of both individual and group-level decision making.

> Often when members of [an African elephant *(Loxodonta africana)*] family are involved in a stationary group activity such as feeding, mud-splashing or resting, one member moves to the periphery of the group and stands facing away from the rest. Then, typically, with one leg lifted she gives a long unmodulated, and to our ears soft rumble accom-

panied by steady ear flapping. This call may be repeated every few minutes until, one by one, other members of the family join her and the group moves off in unison. Long unmodulated calls that are associated with this behavior are referred to as the *"let's go"* rumble. (Poole et al. 1988, 389)

In the case of hamadryas baboons, Kummer (1971, 24–25) reports,

As a consequence of the ever more frequent shifts, the troop changes its shape like an ameba. Nobody has departed yet, but here and there the troop's periphery protrudes in a kind of pseudopod which may persist or withdraw again. After perhaps half an hour, a number of male baboons in the center of the troop finally get up in quick succession and walk towards one of the pseudopods. Their straightforward walk is quite different from the previous hesitant shifts at the periphery; many animals begin to move in a parallel direction, and suddenly the troop is on the move.

Chimpanzees are our closest living relatives. They hunt, use and manufacture tools in the wild, and in captive settings have been taught to communicate symbolically. Chimpanzees live in fission-fusion social communities in which individuals feed, forage, rest, travel, and socialize in small temporary parties. Membership in these parties changes daily, sometimes hourly, as individuals separate from their present social partners, seek out and join other parties, travel alone, or use the loud, raucous pant-hoots of distant chimpanzees as a beacon to locate new feeding sites. Given the social fluidity of these communities and the relative freedom each individual has to associate with particular individuals or subgroups, how do chimpanzees coordinate their movement patterns within this complex and changing social environment? The following description by Jane Goodall (1986, 140–41) illustrates the intricate set of coercive and persuasive signals used by these primates to coordinate and at times negotiate group movement.

The chimpanzees who follow often monitor the movements of a leader and, when he (or she) starts to move on, may quickly stop feeding or grooming and hurry after. The leader usually glances toward the others as he leaves and sometimes waits if they do not immediately follow, occasionally giving a soft or extended grunt. Scratching can be a clear-cut signal in this context, indicating that he is about to go. A mother, having moved to a low branch prior to descending a tree, stops, glances at her infant, and gives slow but vigorous scratches down her side. An obedient infant responds rapidly, hurrying to cling to her for the descent. The branch-shaking "Follow me!" signal, used by a male at the start of a consortship, occasionally occurs in other travel contexts. Sometimes the higher ranking of a pair of males, impatient to move on, shakes branches when a companion does not immediately follow. Once Figan shook branches at Jomeo, apparently signaling that he should follow him on a hunt. And Humphrey, as described, used this signal when trying to persuade Mr. McGregor, crippled by polio, to follow him.

Why did Figan encourage Jomeo to follow him on a hunt, and why did Humphrey signal a crippled Mr. McGregor to travel with him? Why do subordinate male baboons attempt to initiate group movement using a slow, hesitant gait, whereas more dominant individuals wait until several pseudopods form and then walk quickly and directly to influence the direction of travel? How does one elephant decide that it is time for the group to move on?

Why is coordinated group travel and cohesion so important to these large mammals? *Who* leads group travel, decides *where* the group will travel, using *what* information, and *how* do they accomplish these tasks? This book represents our quest for answers to these questions, focusing on the complex social, cognitive, and ecological processes that underlie patterns and strategies of group travel in primates and other animals. Drawing on the literature for human and nonhuman primates, its chapters examine how factors such as group size, resource distribution and availability, the costs of travel, predation, social cohesion, and cognitive skills affect the manner in which individuals as well as social groups exploit their environment. Comparisons with nonprimates, including arthropods, birds, carnivores, and cetaceans, provide a broad taxonomic perspective and offer new insights into questions concerning whether primates are unique in their ability to coordinate group-level activities, or whether there exist a set of socioecological principles that underlie travel and ranging patterns and group cohesion in all social foragers.

Specialists in anthropology, physiology, psychology, cognitive science, ecology, and animal behavior have contributed chapters to this volume, but most contributors found that they could not limit themselves to their nominal fields of specialty. Issues of group travel often demanded unusually broad and ambitious syntheses. As a consequence, some chapters do overlap in scope, and certain chapters could easily fit into one or more sections of the volume. In general, however, the chapters can comfortably be assigned to five topical sections that each encompass a set of closely related questions:

Part 1, Ecological Costs and Benefits: Can travel costs limit group size? How heavily does predation risk influence travel decisions, and how can this be quantified? Why do some taxa form stable mixed-species groups, and how does their behavior differ from that of single-species groups? When is territorial behavior advantageous, and how much time and energy must be allocated to territorial defense?

Part 2, Cognitive Abilities, Possibilities, and Constraints: What are the mechanisms underlying group travel in social insects? How can we observe or experimentally manipulate the environment of social groups so as to evaluate the objectives and cognitive processes underlying travel decisions? Are there inherent limits to the capacity for processing travel decisions? What does comparative neuroanatomy tell us about the selective regimes influencing brain evolution? Is sociality or foraging more important? Has group-level selection influenced the process by which travel decisions are made? What is the best way to operationalize the relationship of the individual to travel decisions expressed by social groups?

Part 3, Travel Decisions and Outcomes: Do animals use coordinate-based or route-based navigation? Can there be both public and private information in group movement? Over what scale of space do individuals and groups make travel decisions? How important is learning in efficient selection of travel routes? Do the travel decisions of one group affect the travel decisions of nearby groups? How would we know if the ecological and social parameters affecting group travel differed between taxa? How sensitive is group movement to food availability? What is the long-term variation in habitat use? Is there feedback between habitat use and habitat quality? Do groups have randomizing components to group movement decisions? Relative to species in stable social groups, do fission-fusion social organizations entail greater or lesser communication to achieve travel decisions?

Part 4, Social Processes: What social interactions contribute to group travel? What are the ultimate and proximate factors underlying individual differences in travel route preferences? How important is leadership versus consensus in achieving travel decisions? How do group members communicate travel goals—or do they?

Do prosimians differ from monkeys and apes in the social and spatial cognition underlying their travel decisions?

Part 5, Group Movement from a Wider Taxonomic Perspective: How do mixed-species bird flocks coordinate travel? Are dolphins the aquatic equivalent of monkeys in terms of travel decisions? Does group travel by social carnivores differ from that of primates in cognitive demands? Did hominids have different ranging patterns than modern humans? Are the physiological and energetic bases of habitat use by human and nonhuman primates different? Does the social and cognitive basis of travel decisions differ between human and nonhuman primates?

Concluding Remarks: Are primates achieving group travel with different cognitive mechanisms or decision processes than other social animals? If so, what makes primates different? Which is more important in explaining taxonomic differences in group travel, ecological adaptations or cognitive abilities?

Last, mention must be made of what is pervasively implicit in this volume: the conservation implications of group movement. The lack of a specific "conservation" chapter is not because we understand these implications to be few or unimportant. A thorough appreciation of ranging needs and processes for social species is crucial to their successful management and protection. We felt that to compartmentalize the conservation implications in a separate chapter would almost trivialize the fundamental relationship of group movement to the conservation of primates, ungulates, cetaceans, birds, and other taxa. Conservation biologists reading these chapters should have little difficulty in identifying biological and behavioral features of group movement of key relevance.

PART ONE

Ecological Costs and Benefits

Cost-benefit analysis is a technique commonly used in behavioral ecology to deconstruct a complex selective regime into its component parts. Once contributing factors are detailed, predictions and alternative hypotheses are usually more readily constructed and tested. Taken together, these chapters present group movement in its daunting ecological intricacy, at the intersection of foraging and travel efficiency, predation avoidance, territorial defense, mixed-species association, and group size effects. Taken separately, however, the chapters comfort the reader by demonstrating that these ecological factors can be made amenable to investigation once dissected into testable hypotheses. An added payoff from these focused analyses of the ecology of group movement is that their results provide important insights into the fundamental costs and benefits of group living, an issue central to behavioral science. Inadequate theory, however, is plainly not the factor limiting further developments. Rather, more field data appropriate to test the large body of existing theory are desperately needed. In the future, as now, the power of the various hypotheses on the ecological basis of group movement will be determined by the extent to which the distribution of key resources and risks can be quantified or controlled and individual behaviors measured.

CHAPTER ONE

The Physiology and Energetics of Movement: Effects on Individuals and Groups

KAREN STEUDEL

The distance that an individual or group travels in a day, season, or year is determined, on the one hand, by the various benefits to be gained by having access to greater range and, on the other, by the costs of travel. This chapter focuses on the energetic cost of travel, discussing how this cost can be assessed and applying it to individuals and groups. While there are other possible costs involved in locomotion, such as exposure to increased risk of predation (Sih 1992), increased water loss (e.g., Eckert 1988), and increased heat load (e.g., Baudinette 1980), these ancillary costs will not be considered. (Key terms used in this chapter can be found in table 1.1.)

Definitions of Travel Costs

How much does it cost an animal to travel for a given time or distance? Energetic cost is usually measured as the rate of oxygen consumed per unit time or distance during "steady-state" locomotion on a treadmill. "Steady-state" locomotion means that an animal is fueling its movement by aerobic means via oxidative phosphoryla-

Table 1.1 Summary of key terms

Cost of locomotion	Energy expended in locomotion per unit time
Cost of transport	Energy expended in locomotion per unit distance
Ecological cost of transport (ECT)	Percentage of the total daily energy expenditure accounted for by locomotion
Daily energy expenditure (DEE)	Total daily energy expenditure
Incremental cost of locomotion (ICL)	The minimum cost of locomotion (the slope of the relationship between energy expended and speed)
Postural cost	The difference between the y-intercept of the relationship between the cost of locomotion and speed and BMR

tion. Thus, the rate of oxygen consumption becomes an excellent estimator of the rate of energy production (e.g., McArdle, Katch, and Katch 1991). Anaerobic energy stores are used to fuel very intense exercise and the early stages of moderate exercise, so the costs of very intense (e.g., a sprint) or brief activities are more difficult to measure. Measuring the cost of steady-state locomotion involves either collecting all expired gases and then measuring the volume of the expired air and level of depletion of oxygen compared with room air (Douglas bag: e.g., McArdle, Katch, and Katch 1991) or pulling room air past the face of the subject at a known rate and measuring the depletion of oxygen (open circuit: e.g., Fedak, Rome, and Seeherman 1981).

Most studies, on a wide variety of species, have found that the cost of locomotion increases linearly with speed (Taylor, Schmidt-Nielsen, and Raab 1970; Taylor, Heglund, and Maloiy 1982). To facilitate comparisons among animals of different sizes, physiologists often report the mass-specific cost of locomotion, that is, the total cost divided by body mass. Figure 1.1 demonstrates this pattern for a representative series of species. A very few species, with unusual locomotor behaviors, show other patterns. For example, the cost of locomotion in the red kangaroo *(Macropus rufus)* actually decreases with increasing speed once the animal begins hopping (Dawson and Taylor 1973). Locomotor costs increase curvilinearity with speed in at least some gaits and in some species. In the presence of such curvilinearity, there are optimal speeds at which cost is lowest for each gait. Cotes and Meade (1960) and Margaria et al. (1963) demonstrated that the cost of human walking shows such a pattern. Hoyt and Taylor (1981), in a study run under highly controlled conditions, demonstrated a slight curvilinearity of cost in all three gaits that they tested in ponies. Their results were partic-

Karen Steudel

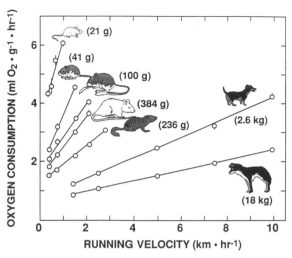

Figure 1.1 The relationship between the energy required to move a gram of their mass at a variety of speeds for animals ranging in size from 21g to 18 kg.

ularly interesting because the deviation from linearity was so slight. Fitting a linear regression to the data resulted in $r = .99$. This finding suggests that cost may generally increase with speed in a slightly curvilinear fashion in nearly all species, but that the usual levels of error variance mask the effect. Nevertheless, a simple linear model provides a good fit to most cost-speed data.

It is also clear that cost is very highly correlated with body size. The rate at which cost increases with speed decreases systematically with increasing body size (Taylor, Schmidt-Nielsen, and Raab 1970; Taylor, Heglund, and Maloiy 1982). This means that while the total cost incurred by a large animal to move a given distance is greater than that for a small animal, it costs a larger animal less per kg body mass to travel that distance. Economy refers to the thrifty use of resources, while efficiency is the ratio of work done to the cost of that work. Large animal species are, in general, less economical than small species; that is, it requires more total energy for them to travel from place to place. At the same time, however, they are more efficient than small species because their energy expended per unit mass is lower. Which of these variables is more crucial to a species depends on the nature of its particular adaptation to its environment. Since the focus of this chapter is the contribution of locomotion to the total energy budget of an individual or species, the remainder of the chapter will focus chiefly on total, rather than mass-specific, locomotion costs (see also Altmann 1998).

The Energetic Costs of Different Locomotor Behaviors

We can make some additional generalizations on the costs of various forms of locomotion. Terrestrial running is the most expensive means of transport. Flying is consistently more economical, and swimming (for natural swimmers) is the cheapest (Tucker 1975). Within the general category of terrestrial mammals, however, one can make only a few generalizations. The bipedal hopping of small rodents is about as costly as the quadrupedal locomotion of other mammals of equivalent size (Thompson et al. 1980). (This is not true for large hoppers such as kangaroos: Dawson and Taylor 1973.) Long-legged mammals are not more efficient than those with shorter limbs (Steudel and Beattie 1995).

Modes of locomotion that do more work against gravity are, as one might suppose, more costly. As we all know, walking or running uphill feels much more energetically intensive. That this is in fact the case has been demonstrated in humans (*Homo sapiens:* Dill 1965), red squirrels (*Tamiasciurus hudsonicus:* Wunder and Morrison 1974), rats (*Rattus norvegicus:* Armstrong et al. 1983), mice *(Mus musculus),* and chimpanzees (*Pan troglodytes:* Taylor, Heglund, and Maloiy 1972). The amount of additional energy required increases with increasing speed and inclination. Brachiation has also been shown to be more costly than walking in spider monkeys (*Ateles geoffroyi:* Parsons and Taylor 1977). The cost to brachiate at 4 km/hr is about 30% greater than the cost to walk at the same speed. As Parsons and Taylor point out, however, brachiation may nevertheless be energetically advantageous since it allows an animal to travel more directly in an arboreal environment than would be possible by walking on branches. Similarly, climbing is energetically costly. Norton, Jones, and Armstrong (1990) measured the cost of locomotion in rats climbing a laddermill at a 75% incline. They reported that climbing at 5 and 10 m/min was metabolically equivalent to level running at 45 and 60 m/min! Possibly this cost would be less in an animal for which climbing constitutes a higher percentage of its total locomotor activities than in rats. Interestingly, the increment in mass-specific energetic cost involved in running uphill is less for small animals than for large ones, chiefly because small animals have such high mass-specific costs for horizontal locomotion (Taylor, Heglund, and Maloiy 1972).

More detailed comparisons of the costs of different modes of locomotion are extremely problematic. Individual variation can be great. Parsons and Taylor (1977) measured the cost of locomotion in the slow loris *(Nycticebus coucang)* using two individuals. One had a cost of locomotion about what would be expected for an animal of that body size. The other had a cost 25% less than predicted (Taylor, Heglund, and Malory 1982). Of two goats whose costs of locomotion were reported by Taylor, Heglund, and Maloiy (1982), one was 7% below its predicted cost, the other 20% above. Similar discrepancies characterize other species in which more than one individual has been studied (see table 1 in Taylor, Heglund, and Maloiy 1982). Furthermore, one should bear in mind that the measurement of steady-state locomotion does not lend itself to measurement of the costs to animals of traveling through trees or on uneven ground.

What Is the Causal Basis of the Energetic Cost of Locomotion?

The physiological, morphological, and kinematic determinants of the cost of locomotion are widely debated. One of the difficulties in determining the causal basis of its cost is that a wide variety of locomotor parameters (stride length, stride frequency, step length, contact time, rate of muscular contraction) are highly correlated with both speed and body size, making them plausible candidates as cost determinants. Workers in human exercise physiology have long supposed that it is the mechanical work of locomotion that determines its energetic cost. In contrast, Taylor, Heglund, and Maloiy (1982) found that while mass-specific energetic cost decreased with increasing size (as noted above), the mass-specific mechanical work of locomotion did not change in any regular way with size. Thus, the most obvious explanation for variation in cost is not viable.

Because animals use higher stride frequencies at higher speeds and at smaller body sizes, many workers believe that the rate at which muscles generate force is an important component of energetic cost (Heglund and Taylor 1988; Kram and Taylor 1990). At higher stride frequencies, the propulsive muscles must contract more rapidly, requiring recruitment of muscle fibers with higher intrinsic contraction rates. Such faster fibers appear to be more energetically expensive due to higher rates of formation and breakage of the cross-bridges between the actin and myosin filaments (Rall 1986). In addition, there is evidence that muscle activation costs may represent as much as 30% of the cost of an individual muscle contraction (Homscher et al. 1972).

Stride frequency, however, is an imperfect determinant of muscle contraction rate. Strang and Steudel (1990) point out that increases in stride length also influence muscle contraction rate by requiring that the limb be moved through a greater arc in a given amount of time. The ongoing search for an explanation for the cost of locomotion in terms of muscle contraction rate is complicated by the fact that the time interval in which a foot exerts force on the ground (contact time) influences cost in two ways. Myers, Steudel, and White (1993) reported the results of a factorial experimental design in which stride frequency was held constant while speed was varied and speed held constant while stride frequency was varied. They found that contact time was not significantly correlated with overall variation in cost, and suggested the dual influences of contact time on cost as the explanation. A shorter contact time requires higher muscle contraction rates, which tend to increase cost, as described above. On the other hand, contact time is also an estimate of the time interval in which muscle-tendon springs are loaded and unloaded in the storage and recovery of elastic strain energy. While measurement of the magnitude of elastic effects on cost is notoriously difficult, it is generally supposed that this mechanism is more effective with shorter contact times (Blickhan 1989; Ettema et al. 1990; Farley, Glasheen, and McMahon 1993). Thus, the rate of doing work, the mechanical work done, the efficiency of energy transfers among body segments, and the storage and recovery of elastic strain energy all influence the energetic cost of locomotion. The relative contributions of each are not well understood at this time.

What Is the Most Ecologically Relevant Way of Estimating Locomotor Cost?

The rate of oxygen consumption by an organism can be measured either per unit time (the cost of locomotion) or per unit distance (the cost of transport). Which of these is most useful in an ecological/behavioral context? As indicated above, the cost of locomotion increases linearly with increasing speed in most gaits and most species. This does not mean, however, that to travel cheaply, one should travel slowly. If we consider the cost of transport (per unit distance), a different picture appears. The cost to travel a given distance actually *decreases* with increasing speed. This counterintuitive result stems from the fact that both measures of locomotor cost include the cost of maintaining the basal metabolic rate (BMR) during that time interval. If one travels faster, one arrives sooner, and a smaller amount of the daily BMR is included in the measurement of locomotor cost. Since an animal, unlike a car, cannot simply turn itself off when it arrives at its destination, it is probably preferable to subtract out the effects of BMR when applying the results in a behavioral/ecological context. There remains a small but consistent decrease in the cost of transport with increasing speed. This decrease results from the fact that the line describing the relationship of either the cost of locomotion or the cost of transport to speed does not have its y-intercept at BMR, but at some greater value. This difference between BMR and the y-intercept of the cost of locomotion has been called the "postural cost" (Taylor 1977). If one subtracts out the postural cost, the cost to travel a given distance is essentially independent of speed (table 1.2). This is equivalent to the incremental cost of locomotion (ICL) defined by Garland (1983). This subtraction seems entirely appropriate because BMR is measured as a minimal cost that would typically be lower than that observed in an animal engaged in any activity. My focus in this chapter is on the energetic consequences for an animal of deciding to move from point A to point B. Postural cost is typically about 1.7 times basal metabolic rate (Taylor 1977).

James and Schofield (1990) summarize estimates of physical activity ratios (PAR) for a wide range of activities in humans:

$$\text{PAR} = \frac{\text{the energy cost of an activity per minute}}{\text{the energy cost of basal metabolism per minute}}.$$

Quite light exercises such as playing darts, playing the piano, carpet weaving, or cooking have PARs similar to that of the postural cost of locomotion. It seems likely that animals that are awake are usually engaging in activities at least as strenuous as these, and would approach the basal metabolic rate only during sleeping. Thus, using the ICL as an estimator of the cost to travel a given distance, as

Table 1.2 Mass-specific locomotor costs to travel for a given time (cost of locomotion), for a given distance (cost of transport), and for a given distance subtracting out the postural cost

	Cost of locomotion (ml O$_2$/sec/kg)		Cost of transport (ml O$_2$/km/kg)		Cost to travel 1 km, omitting postural cost (ml O$_2$/km/kg)	
	1 kg	10 kg	1 kg	10 kg	1 kg	10 kg
0.5 m/s	.632	.278	1,264	556	534	258
1.0 m/s	.833	.407	833	407	533	257.5

opposed to being active in the same location, seems appropriate. Furthermore, including postural costs removes the velocity independence of the ecological cost of transport (ECT). If one were engaged in estimating an animal's total energy budget over the course of a day, the postural cost would have to be added as a separate term. Thus, the best estimator of the locomotor cost incurred will be distance, rather than time, traveled. The estimation of locomotor costs through measures of time spent traveling requires that one combine that information with the rate of travel. Distance traveled estimates the total cost of that travel.

How Important Are Locomotor Costs in the Total Energy Budget of an Individual?

Is the cost of locomotion for a typical mammal of sufficient magnitude that it really must be considered as a significant cost in the cost-benefit analysis of travel by individuals in groups? Garland (1983) calculated the ecological cost of transport (ECT) as the percentage of the total daily energy expenditure accounted for by locomotion. Garland defines ECT as

$$ECT(\%DEE) = 100 \times \frac{DMD \text{ (km/day)} \times ICL \text{ (J/km)}}{DEE \text{ (J/day)}}.$$

where DMD is daily movement distance, ICL is the incremental cost of locomotion, and DEE is the total daily energy expenditure. Estimates of DMD are taken from the literature. Here I have used the equation for estimating DEE in eutherian mammals given by Nagy (1994):

$$DEE \text{ (kJ/day)} = 894.5M^{0.762}.$$

To estimate ICL, I have used the first term in Equation 9 of Taylor, Heglund, and Maloiy (1982) because it is based on a wider array of data than available to Garland (1983). Thus,

$$ICL \text{ (J/km)} = 10,700M^{0.684}.$$

As one might expect, ECT is larger in larger animals, both because of the greater distances traveled by large animals and because of the higher values for BMR in small mammals. Carnivores have substantially greater ECTs than other mammals because of their typically greater daily movement distances. The actual values reported for ECT by Garland (1983), however, were quite modest, "unbelievably low" according to Altmann (1987). Predicted costs ranged from 1.2% of total daily energy expenditure for a typical 1 kg mammal to 5% for a typical 1,000 kg mammal. Are costs of this magnitude, in fact, negligible? Perhaps.

Altmann (1987) questions the adequacy of these estimates. Daily movement distance is an important component of ECT. Altmann argues that published values for daily movement distance may be "off by a factor of two, three, or even more." If he is correct, the estimates for ECT would increase by two- or threefold. Although in some cases the data in Garland (1983) are from monitoring a marked individual for a day or more, or from snow tracks, others are estimated from periodic sightings or radiotelemetry fixes. Thus some of the estimates may well be underestimates (see also Harvey and Clutton-Brock 1981). Altmann (1998) notes that in Amboseli, an average female baboon walks 6.1 km per day, spending 2.63 hours per day in the process. Thus approximately one-fourth of the animal's active hours are engaged in locomotion. Nevertheless, locomotion accounts for only 6% of a female's total energy budget. Altmann regards this as "unreasonably low" in that the remainder of the baboon's active hours are spent feeding, resting, and socializing, and these activities seem to him to be, on average, even less energetic than locomotion. One way to resolve this issue would be to carefully document the nonlocomotor activities an animal engages in over the course of a day and estimate their energetic cost, either using the human PAR values cited above or some more accurate measure. Either Altmann is right, or feeding, resting, and socializing are more energetically expensive than one might suppose. Thus ECT values for "typical" mammals may in fact be much higher than suggested by existing data.

Whether or not these estimates are generally low, ECT is a consequential percentage of daily costs for many species whose daily movement distance is substantial. Carnivores generally travel farther than do most noncarnivorous mammals. Consequently, their ECTs are higher, ranging from 8.4% for a typical 10 kg carnivore to 14% for a 100 kg carnivore (Garland 1983). Calculating the ECT for a 23 kg wolf based on cost data in Taylor, Heglund, and Maloiy (1982) and daily movement distance values in Garland (1983) yields a value of 16.3%. Using the same sources yields a value of 9.0% for hamadryas baboons *(Papio hamadryas)*. Thus a notable percentage of total daily energy expenditure is allocated to locomotion in some species. The significance of locomotor costs in determining group movement will generally increase in larger species and ones that travel longer distances.

If the energetic consequences of time spent in locomotion are an important agent of selection, one might expect that, when possible, animals would select gaits and speeds that allow a lower energetic cost. While the measured cost of transport for most animals is virtually constant across speeds, there is evidence (see above) in some species for a curvilinearity of cost of locomotion that results in an optimal travel speed(s). Is there any evidence that such speeds are used preferentially? Hoyt and Taylor (1981) noted that measured speeds of a horse moving freely tended to cluster at the energetically optimal speeds within each gait. As indicated above, the data revealing this curvilinearity were recorded under highly controlled experimental conditions, substantially exceeding the conditions under which such data are usually acquired. This suggests that a similar relationship might obtain in other species, but that the standards of experimental control required to demonstrate the curvilinearity are rarely achieved. If this is the case, there may also be optimal speeds within a gait for other quadrupeds. For example, C. J. Pennycuick (1975) reported that freely moving wildebeests (*Connochaetes taurinus*) tended to walk and canter within narrow ranges of speeds near the center of the possible speeds for each gait. Likewise, walking humans select the stride length and stride frequency that minimizes energetic costs at any given speed (Zarrugh and Radcliffe 1978; Minetti et al. 1995). Efficiency is not, however, the only determinant of gait choice. Hreljac (1993) has shown that human subjects chose to switch from a walk to a run at a speed for which walking actually requires less energy. Interestingly, the subjects reported that their perceived level of exertion was less for running than for walking at this speed, even though measurement of expired gases demonstrated the reverse.

How Do Locomotor Costs Vary among Group Members?

Because the energetic cost of locomotion is highly correlated with body size, size variation among group members results in variation in their locomotor costs. To assess the magnitude of this effect, I have calculated the average daily energy expenditure (DEE) and the ecological cost of transport (ECT) for the two sexes and for juveniles of hamadryas baboons (*Papio hamadryas*) and for Costa Ri-

Karen Steudel

can squirrel monkeys *(Saimiri oerstedii)*. The former species was chosen because of its notable daily movement distance, large body size, and high level of sexual dimorphism; the latter because it is much smaller, less dimorphic, and has a much shorter daily movement distance. Data on daily movement distance are from Siggs and Stolba (1981) and Mitani and Rodman (1979); locomotor costs are calculated from the general mammal equation in Taylor, Heglund, and Maloiy (1982); DEE estimates are from Nagy (1994); and ECT is calculated as in Garland (1983). Adult body weights are from Phillips-Conroy and Jolly (1981) and Mitchell, Boinski, and van Schaik (1991); juvenile body weights are from Coelho (1985) and Kaack, Walker, and Brizee (1979). DEE and ECT were also calculated for the closely related, but less wide-ranging, *P. anubis* based on weights from Phillips-Conroy and Jolly (1981) and DMD from Harvey and Clutton-Brock (1981). Estimates of daily movement distance are assumed to be the same for both sexes.

As can be seen in table 1.3, the total daily energy requirements of baboon males greatly exceed those of females due to their larger body size. Females expend only 66–68% as much energy as do conspecific males. In the less dimorphic squirrel monkeys, the difference in DEE is commensurately less. The effect of size differences on ECT as a percentage of DEE, however, is very small (see table 1.2). Males and females of the same species must allocate a similar proportion of their daily energy requirements to locomotion in spite of any size differences. This situation, of course, changes when a female becomes pregnant. For example, female hamsters *(Mesocricetus auratus)* show a 21% increase in their daily energy expenditure in the last third of their pregnancy (Quek and Trayhurn 1990). Female brown bats *(Myotis lucifugus)* when pregnant or lactating showed an even greater increase in metabolic rate, over 40% above that expected for a nonreproductive mammal of their size (Kurta et al. 1989). This extremely high value is thought to result from the very high contribution of foraging flight to the energy budget of these animals. Similarly, female primates have higher locomotor costs when they are pregnant. Coelho (1985) gives 0.95 kg as the average birth weight in baboons. Taking 11.5 kg as the weight of the female baboon at term results in an incremental cost of transport of 56,872 J/km as compared with 51,333 J/km for a nonreproductive female, a 10% increase. This is certainly not a trivial amount. However, given that the ecological cost of transport is only 12% of DEE

Table 1.3 Daily movement distance (DMD), daily energy expenditure (DEE), and ecological cost of transport (ECT) for different-sized individuals of three primate species

	Average body mass (kg)	(DMD) (km)	DEE[a] (J/day)	ECT
Hamadryas baboon *(Papio hamadryas)*				
Female	9.9	9.5	5,131,618	9.5%
Male	16.9	9.5	7,713,118	9.1%
Juvenile	2.5	9.5	1,798,087	10.6%
Juvenile	4.0	9.5	2,572,468	10.2%
Yellow baboon *(Papio anubis)*				
Female	11.7	3.6	5,828,247	3.6%
Male	21.2	3.6	9,167,444	3.4%
Squirrel monkey *(Saimiri oerstedii)*				
Female	0.60	3.5	606,082	4.4%
Male	0.75	3.5	718,417	4.3%
Juvenile	0.30	3.5	357,394	4.6%
Juvenile	0.20	3.5	262,401	4.7%

Note: data sources are given in the text.

[a]DEE values for juvenile individuals are based on equations developed for adults and, therefore, neglect the cost the growth.

in these animals, this is a relatively small increment in locomotor energy requirements. The hamster and bat data suggest that the increased metabolic rate would have a much more substantial effect.

What about the contribution of locomotor costs to the energy budget of much smaller juvenile individuals, whose energy intake must also include an allocation to growth? Table 1.3 contains estimates of DEE and ECT for juvenile troop members, assuming that they are traveling independently. These estimates are based on equations developed for adult animals of that body size and hence neglect the energy allocated by the juvenile to growth. DEE values for juveniles are, therefore, underestimates, while ECT values are overestimates. Nevertheless, it is interesting that ECT is similar in very different sized adults and offspring. This similarity results from the fact that, while the mass-specific locomotor costs are considerably higher for the juveniles, so are their mass-specific basal metabolic costs.

These ECT estimates assume that a juvenile is moving with the troop under its own power. In most primates, infants and young juveniles are carried by their mothers or by other group members for varying periods of time. Thus, at the times when an infant's

Karen Steudel

cost of locomotion would otherwise be very high, transport by the mother allows it to allocate more energy toward growth. In baboons, for example, mothers meet some of their offspring's nutritional needs with milk throughout the first year of life and into the second (Altmann 1980). Even as a juvenile becomes partially self-supporting nutritionally, less energy spent on locomotion allows more energy to be allocated to growth and other activities. Consequently, maternal behavior that minimizes her offspring's locomotor costs may be selected. This is particularly true because of the high per gram cost of locomotion in small animals.

The cost of carrying loads is well documented (Taylor et al. 1980; Soule, Pandolf, and Goldman 1978; Steudel 1990). How does the cost to carry an infant baboon compare to the cost of allowing it to move independently? Altmann (1980) reports that by a juvenile's fourth month, it uses its mother for transportation only if group progression is quite rapid or the journey is long (see also Kummer 1968a). Presumably by the sixth month, the juvenile is largely weaned from riding. Does this weaning occur at the juvenile body size at which it becomes energetically cheaper for the mother and infant to travel independently? Coelho (1985) reports data on body size for baboon juveniles at birth and for each year of age up to 8 years. To estimate weight at 6 months, I assumed that the rate of growth was steady so that half of the year's weight gain occurred in the first 6 months. At this age there is little sexual dimorphism in mass. Both sexes should weigh about 2.5 kg at 6 months and about 4.0 kg at 1 year, an age by which maternal transport presumably has ceased completely.

Figure 1.2 shows the combined cost for a hamadryas baboon female and a dependent offspring to travel 1 km. The cost for the two individuals to travel separately was calculated by summing the costs of transport for a 2.5 kg or a 4 kg animal and a 9.9 kg animal based on Taylor, Hegelund, and Maloiy (1982). Taylor et al. (1980) showed that the increase in locomotor cost incurred when carrying loads of between 7% and 27% of body mass was directly proportional to the increase in mass. The cost for a 9.9 kg female to carry juveniles of 2.5 kg and 4.0 kg were calculated based on Taylor et al.'s equation for the cost to carry loads. The values reported are the sum of the female's cost to travel 1 km plus the cost to carry the 2.5 kg or 4.0 kg load for the same distance. It is apparent that the cost for a female to carry her offspring is about 10% lower, even at 1 year of age, than the combined cost for them to travel

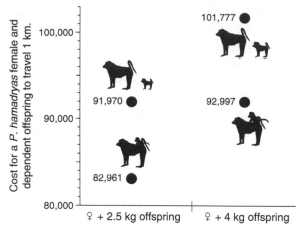

Figure 1.2 Energetic cost for a 9.9 kg hamadryas baboon and her dependent offspring to travel 1 km. Costs have been calculated under two conditions: with the offspring traveling independently and with the offspring being carried by the mother.

independently. Thus weaning juvenile baboons from being carried does not appear to result from considerations of locomotor energetics.

The data are a bit different for the very small Costa Rican squirrel monkey, in which offspring are born at a large size relative to their mothers. By the end of its fourth month, a juvenile is rarely in contact with its mother (Boinski and Fragaszy 1989) and, consequently, must be traveling under its own power. At this age, the juvenile weighs approximately 200–225 g (Brizzee and Dunlap 1986). The cost for a typical 600 g female (Mitchell, Boinski, and van Schaik 1991) and a 200 g offspring to travel independently for 1 km would total 11,104 J. The cost to travel the same distance with the mother carrying the offspring is 10,060 J, 9% less. Even at 250 g, the cost to a female to carry a juvenile would be 4% less than the combined costs of their independent progression. Therefore, other factors, such as foraging opportunities (Boinski and Fragaszy 1989), thermal considerations, or muscle/joint strain are more important in producing the juvenile's locomotor independence from its mother (see Altmann and Samuels 1992 for a more complete discussion).

These results are similar to those of Altmann and Samuels (1992) for savanna baboons, in spite of some considerable differences in the techniques used to compare the cost of carrying offspring with that of their independent transport. There are two main differences in our approaches. First, Altmann and Samuels estimated the cost of transport using the equation for primates in Taylor, Heglund,

Karen Steudel

and Maloiy (1982), while I used the general mammals equation in the same paper. I prefer the latter because the values in the primate equation are very heavily determined by the fact that the only two large primate species studied (chimpanzees and humans) have disparate cost values. There is no indication that this is the case in other primates. Second, Altmann and Samuels calculated locomotor cost including the postural cost, while I omitted postural cost. As I discussed above, it seems unlikely that primates return to basal metabolic rate in the diurnal intervals between locomotor bouts; often they would be foraging or engaging in social activities, which seem likely to result in a activity cost similar to that of the postural cost of locomotion, based on James and Schofield's (1990) data on humans. I, therefore, believe that the incremental cost of locomotion is a better estimator of locomotor costs, although either approach has merit. In spite of these methodological differences, our results are completely consistent.

Conclusions
The energetic cost of locomotion can best be estimated in the field by documenting the distance traveled rather than the time spent traveling. The extent to which locomotor costs influence the daily energy budget of an individual or group is highly variable and is determined chiefly by the daily movement distance. In an animal that travels as much as wolves or some baboons, it can be significant. Smaller individuals require substantially less energy per day than larger individuals, but body size has a minimal effect on the ECT of individuals. It is very energetically advantageous for females to carry their juvenile offspring. Females, however, wean their offspring from riding at a body size at which it would still be advantageous for them to be carried, suggesting that some other factor must be driving age at weaning.

There remains a conspicuous need for a better understanding of an individual's daily energy budget. The ECT is surprisingly small for most animals. Perhaps this is due to underestimates of daily movement distances. On the other hand, no one has yet attempted to estimate the energetic costs of other activities such as foraging and social behavior. An initial attempt might usefully be based on the PAR values developed for humans, as discussed above. It would also be illuminating to know what happens to metabolic rates at night. Do many species reach BMR during sleep, or must they run higher metabolic rates for thermoregulation?

CHAPTER TWO

Determinants of Group Size in Primates: The Importance of Travel Costs

COLIN A. CHAPMAN AND
LAUREN J. CHAPMAN

Identifying the ecological factors underlying group size and social organization has been a central theme of primate behavioral ecology (Gartlan and Brain 1968; van Schaik 1983; van Schaik and van Hooff 1983; Terborgh 1983; Butynski 1990). This interest has stemmed from the fact that group size in primates is highly varied across the order (1–320 members) and even within species (e.g., patas monkey, *Erythrocebus patas:* 21–36 members; table 2.1). Determinants of group size have been extensively evaluated in terms of costs and benefits. While disagreement exists as to the relative importance of the different potential advantages of grouping, various authors have suggested that it confers such predictable benefits that differences in group size can be explained by its disadvantages

Table 2.1 Group size, day range, body size, and social system for primates

Species	Group size[a]	Day range (km)	Body size (kg)[b]	Social system[c]	Reference[d]
Lemuirdae					
Lemur catta	18.0	0.95	2.5	MF	1
Lemur fulvus	9.5	0.14	1.9	MF	1
Lemur mongoz	2.6	0.61	1.8	SF	1
Indriidae					
Indri indri	3.0	0.25	10.5	SF	1
Propithecus verreauxi	6.5	0.85	3.5	MF	1
Callitrichinae					
Saguinus oedipus	7.4	2.06	0.5	SF	2
Saguinus fuscicollis	4.7	1.37	0.4	SF	2
Saguinus fuscicollis	6.5	1.22	0.4	SF	2
Saguinus imperator	4.0	1.42	0.5	SF	2, 3
Pitheciinae					
Callicebus moloch	3.2	0.57	1.1	SF	4
Callicebus moloch	4.2	0.67	1.1	SF	4
Callicebus torguatus	3.9	0.82	1.1	SF	4
Chiropotes albinasus	25.0	3.75	2.5	MF	4
Chiropotes satanas	19.0	2.50	2.7	MF	4
Atelinae					
Alouatta palliata	9.1	0.12	5.7	MF	5
Alouatta palliata	12.2	0.60	5.7	MF	5
Alouatta palliata	15.5	0.44	5.7	MF	5
Alouatta seniculus	9.5	0.39	6.4	MF	5
Alouatta seniculus	7.1	0.54	6.4	MF	5
Alouatta seniculus	9.0	0.71	6.4	MF	5
Ateles paniscus	18.0 (3.0)	2.70	5.8	F-F	6
Ateles belzebuth	18.0 (3.0)	2.30	5.8	F-F	6
Brachyteles arachnoides	24.5	1.28	9.8	MF	8
Brachyteles arachnoides	45.0 (5.0)	0.63	9.8	F-F	9
Lagothrix lagotricha	33.0	1.00	5.8	MF	6
Cebinae					
Saimiri oerstedii	23.0	3.35	0.6	MF	6
Saimiri sciureus	42.0	1.50	0.6	MF	6
Cebus albifrons	15.0	1.85	2.6	MF	3, 6
Cebus capucinus	17.5	2.00	2.7	MF	6
Cebus olivaceus	20.0	2.30	2.3	MF	6, 7
Cebus apella	10.0	2.00	2.1	MF	6
Colobinae					
Procolobus badius	34.0	0.56	5.8	MF	10
Procolobus badius	20.0	0.60	5.8	MF	10
Colobus guereza	12.0	0.54	9.3	MF	10
Colobus satanas	15.5	0.46	9.5	MF	10
Semnopithecus entellus	19.0	0.36	11.4	MF	10
Presbytis obscura	10.3	0.95	6.5	MF	10
Presbytis melalophus	9.3	1.15	6.6	MF	10
Presbytis melalophus	14.0	0.61	6.6	MF	10
Cercopithecinae					
Cercopithecus ascanius	26.3	1.54	2.9	MF	11

Table 2.1 (*continued*)

Species	Group size[a]	Day range (km)	Body size (kg)[b]	Social system[c]	Reference[d]
Cercopithecus ascanius	32.5	1.45	2.9	MF	11
Cercopithecus cephus	10.0	0.90	2.9	MF	1
Cercopithecus mitis	32.6	1.14	4.4	MF	11
Cercopithecus mitis	18.7	1.30	4.4	MF	11
Cercopithecus nictitans	20.0	1.50	4.2	MF	1
Cercopithecus pogonias	15.0	1.75	3.0	MF	1
Cercopithecus neglectus	4.0	0.53	4.0	SF	11
Chlorocebus aethiops	24.0	0.95	3.6	MF	11
Miopithecus talapoin	112.0	2.32	1.1	MF	11
Erythrocebus patas	35.5	4.33	5.6	MF	11
Erythrocebus patas	20.6	2.25	5.6	MF	11
Lophocebus albigena	14.4	1.27	6.4	MF	12
Cercocebus galeritus	19.0	1.29	5.5	MF	1
Papio c. cynocephalus	80.0	6.40	15.0	MF	12
Papio c. ursinus	47.2	10.46	16.8	MF	12
Papio c. ursinus	45.0	4.67	16.8	MF	12
Papio c. hamadryas	68.0 (7.3)	8.60	9.4	F-F	13, 14
Macaca fascicularis	27.0	1.90	4.1	MF	12
Macaca nemestrina	35.0	2.00	7.8	MF	12
Theropithecus gelada	320.0 (113)	2.50	13.6	F-F	1
Hylobatidae					
Hylobates agilis	4.4	1.22	5.7	SF	15
Hylobates klossii	3.7	1.51	5.9	SF	15
Hylobates lar	3.5	1.49	5.3	SF	15
Hylobates syndactylus	4.0	0.79	10.6	SF	15
Hylobates syndactylus	5.0	0.97	10.6	SF	15
Hylobates syndactylus	3.0	0.74	10.6	SF	15
Hylobates syndactylus	3.8	0.93	10.6	SF	15
Pongidae					
Pongo pygmaeus	1.8	0.50	37.0	F-F	16
Pan troglodytes	28.0 (4.0)	3.90	31.1	F-F	1
Gorilla gorilla	9.0	0.70	93.0	MF	17

[a]The number in parentheses is the average subgroup size for species with a fission-fusion social organization.

[b]Primate body weights are for adult females from Clutton-Brock and Harvey 1977b.

[c]MF, multifemale groups; SF, single-female groups; F-F, fission-fusion societies.

[d]1, Clutton-Brock and Harvey 1977b; 2, Goldizen 1986; 3, Terborgh 1983; 4, Robinson, Wright, and Kinzey 1986; 5, Crockett and Eisenberg 1986; 6, Robinson and Janson 1986; 7, Robinson 1988; 8, Strier 1987; 9, Milton 1984; 10, Struhsaker and Leland 1986; 11, Cords 1986b; 12, Melnick and Pearl 1986; 13, Sigg and Stolba 1989; 14, Stammbach 1986; 15, Leighton 1986; 16, Rodman and Mitani 1986; 17, Stewart and Harcourt 1986.

(Terborgh and Janson 1986; Wrangham, Gittleman, and Chapman 1993; Janson 1992).

Benefits of grouping can be considered to fall within three broad categories: predator avoidance, foraging advantages, and avoidance of conspecific threat. Predator avoidance hypotheses suggest that group living facilitates (1) increased probability of predator detection (Rodman 1973b; Struhsaker 1981; Gautier-Hion, Quris, and Gautier 1983; Boinski 1987a, 1989; van Schaik and van Noordwijk 1985, 1989; Cords 1990b; Norconk 1990; Terborgh 1990; Chapman and Chapman 1996), (2) greater confusion of a predator trying to focus on an individual prey (Morse 1977), (3) a decreased probability of each individual being captured by predators (Hamilton 1971; Wolf 1985), and (4) increased defense against predators (Struhsaker 1981; van Schaik and van Noordwijk 1985; Boinski 1987a; Gautier-Hion and Tutin 1988; van Schaik and van Noordwijk 1989; van Schaik and Hörstermann 1994). Foraging benefits may include (1) access to foods otherwise not available (e.g., adult males opening large fruits that immatures cannot open: Struhsaker 1981; Gautier-Hion, Quris, and Gautier 1983; Waser 1984a), (2) efficient use of shared resources (e.g., not returning to areas just depleted by conspecifics: Cody 1971; Terborgh 1983; Cords 1986a, 1987; Whitesides 1989; Oates and Whitesides 1990; Podolsky 1990), (3) increased feeding rates when in a group, possibly associated with a decreased need for vigilance (Klein and Klein 1973; Munn and Terborgh 1979; Podolsky 1990), (4) increased resource detection (Gartlan and Struhsaker 1972; Struhsaker 1981), and (5) cooperative resource defense (Wrangham 1980; Garber 1988a). Recently, several authors have cautiously suggested that the risk of conspecific attack (e.g., infanticide) from nongroup members may also favor group living (Watts 1989; van Schaik and Dunbar 1990; Smuts and Smuts 1993; van Schaik and Kappeler 1993; Janson and Goldsmith 1995; Treves and Chapman 1996). If conspecifics outside of a group create situations in which group members have to bear some costs, such the risk of infanticide, group members should adopt strategies to minimize this cost, such as cooperative defense. The number and age/sex composition of the animals cooperating against outsiders would influence the success of deterring nongroup members.

While the relative importance of these different advantages of group living remains controversial, there has been little disagreement when it comes to considering the disadvantages. The most widely acknowledged cost of group living is within-group feeding competition. Such competition has clear fitness effects, including increased mortality (Dittus 1979) and lower female reproductive rates (Whitten 1983). Within-group competition can reduce foraging efficiency in two ways: direct contests over food resources (interference competition: Nicholson 1954; Janson 1985, 1988a,b; van Schaik 1989) or reduction of resources merely by competitors using the resource, independent of any direct interaction (exploitation competition: Terborgh 1983; Janson 1988b; van Schaik and van Noordwijk 1988). The relative frequency of occurrence of these two types of competition has rarely been quantified. This probably stems from the fact that while contest competition is obvious (e.g., two animals engaged in a fight over a food source), exploitation competition is difficult to verify. If one animal simply beats a second animal to a food source, when the second animal approaches the place where that food source was, it is difficult to say whether there is no food left, or whether the animal does not wish to eat in the area.

This agreement on the costs of group living has led to the development of a model suggesting that exploitation competition can limit group size whenever an individual moves more while in a group than alone. This ecological constraints model, which is detailed in the next section, highlights the importance of understanding determinants of group movement in order to comprehend what constrains the size of primate groups. The foundations of this model have been well established through studies on a variety of vertebrates (Bradbury and Vehrencamp 1976; Pulliam and Caraco 1984; Clark and Mangel 1986; Elgar 1986). However, primates are a particularly well-suited group for testing the generality of the model, since a great deal of descriptive data is available on the order and there is great variation in foraging strategies and group size (see table 2.1).

Currently, we understand little about the mechanisms underlying the ecological constraints model. In this chapter, we propose that an increase in group size leads to increased travel costs through two mechanisms: (1) patch depletion and (2) avoidance of overlap of search field. We review and evaluate support for these two mechanisms, identify gaps in our knowledge, and outline directions for future research.

Evidence suggests that perceived predation risk alters animals' selection of habitats, length of time spent in a patch, and thus probably the size of the group of which an animal chooses to be a

C. A. Chapman and L. J. Chapman

member. For example, desert baboons *(Papio cynocephalus ursinus)* spend less time than expected feeding in high-risk, food-rich habitats, but more time than expected feeding in low-risk, relatively food-poor habitats (Cowlishaw 1997). Unfortunately, there is little quantification of how animals respond to perceived predation risk. As a result, we concentrate on how ecological conditions can constrain group size, but recognize the effect that predation risk could have on how animals weigh the costs and benefits of groups of different sizes (see Boinski, Treves, and Chapman, chap. 3, this volume).

Conceptual Framework of Ecological Constraints on Group Size

Animals must forage over an area that can meet their energetic and nutritional requirements. Therefore, an increase in group size may be expected to increase the area that must be searched to find adequate food supplies (Eisenberg, Muckenhirn, and Rudran 1972; S. A. Altmann 1974; Bradbury and Vehrencamp 1976). Thus, individuals must travel farther and expend more energy if they are in a large group than if they forage in a smaller group or alone (Wrangham, Gittleman, and Chapman 1993; Steudel, chap. 1, this volume). With an increase in the time and energy spent traveling, a point will be approached at which energy spent in travel exceeds the energy obtained from the environment, and a smaller group size should become advantageous. In this way ecological factors can influence movement patterns and foraging efficiency and thereby constrain the size of groups that can efficiently exploit available food resources.

This model assumes that an increase in group size will lead to an increase in within-group feeding competition, which may be expressed as increased day range. The nature of this relationship will vary depending on the nature of the resources used by particular species. With frugivorous and possibly folivorous primates, large groups may deplete patches faster than smaller groups, resulting in longer day ranges. For insectivorous species, resources may not occur in patches, or the patches may not be divisible, so additional group members may lead to an increase in the overlap of individual search fields, reducing per capita encounter rates with food and increasing the area that must be searched. (A search field is the area over which a foraging animal is visually exploring for food items.) As a result, the ecological constraints model has one or two key assumptions: (1) food items are assumed to occur in discrete depleting patches, and an increase in group size leads to more rapid patch depletion, necessitating increased travel between patches and/or (2) it is assumed that as group size increases, individual search fields overlap, reducing per capita encounter rates with food, and consequently the size of the search area increases.

The Patch Depletion Process

For species that use resources that occur in discrete depleting patches, an increase in group size increases the rate of patch depletion, simply because there are more mouths to feed. Once a patch is depleted, animals will have to travel on in search of other feeding sites. Thus, an increase in group size increases the time and energy invested in travel. For this process to operate, resources must occur in patches and patches must be depleting.

Examining the assumption that food items occur in discrete depleting patches has proved difficult. The first challenging step has simply been to define a patch. A number of theoretical or conceptual definitions have been proposed (S. A. Altmann 1974; Hassell and Southwood 1978; Addicott et al. 1987). However, most field studies have simply considered a patch to be an aggregation of food items structured so that animals can use the area without interrupting their feeding. For forest-dwelling primates this definition is often operationalized as an isolated tree (e.g., a fruiting fig tree; Chapman 1988a, 1989a, 1990a,b; Symington 1988b; White and Wrangham 1988; Strier 1989; Chapman, White, and Wrangham 1994).

While this definition may apply to large-bodied primates that specialize on high-quality fruit resources, it is difficult to see how the resources of other species could be considered to occur in patches. For insectivorous primates, resources may be more or less uniformly dispersed (or they may occur in indivisible patches). For example, redtail monkeys *(Cercopithecus ascanius),* for which insects constitute 22–28% of the diet (Cords 1987), typically feed in groups that are dispersed over a 50 m swath. Individuals catch isolated insects, and rarely does an animal stay at an insect capture site for more than a few seconds. These animals tend to congregate at fruiting trees, yet it is rare to find all group members feeding in one tree. For such species, insects may represent a dispersed food resource, and even when they are feeding on patchy food resources (e.g., a fruiting tree), they may have the option of reducing competi-

C. A. Chapman and L. J. Chapman

tion by shifting from feeding in the fruit patch to searching nearby for insects. The decision to shift between fruit patch feeding and searching for insects probably depends on an interaction between competition, which is a function of the size and richness of a fruit patch, and the density of dispersed insects. With species such as the redtail monkey, the question then becomes, does the mechanism involving avoidance of overlap of search fields operate, or, since they rely on both fruit and insect resources, to what degree are they constrained by one process over the other, or does the model apply at all?

If we follow the simplifying assumption that for many primates a patch is equal to a tree, the question becomes whether or not primates typically deplete the patches they use. Theoretically, a patch may be considered depleted when the feeding activity of the consumer has led to the disappearance of all food items. However, as food items become rare within a tree, they become progressively harder to obtain. Thus, a patch will be functionally depleted before all of the food items are eaten. From this perspective, patches can be considered depleted when the rate of food intake drops to a level equal to the average intake in the environment (Charnov 1976; Stephens and Krebs 1986).

Although the concept is fundamental to several models of primate social organization (van Schaik 1989; Isbell 1991; Cheney 1992), there are few data on primate patch depletion (Janson 1988b). We have examined the assumption that primates deplete patches using four different species (cebus monkeys, *Cebus capucinus;* spider monkeys, *Ateles geoffroyi;* howler monkeys *Alouatta palliata;* and chimpanzees, *Pan troglodytes:* Chapman 1988a; Chapman, Wrangham, and Chapman 1995). Four lines of evidence suggest that patch depletion occurs. First, all species were commonly observed feeding in a number of individual trees of the same species in direct succession, rarely revisiting the same tree on the same day (see also Garber 1988a). Second, for most trees, the rate of food intake was higher at the start of the feeding bout than later. (For some tree species, feeding rate did not change over the feeding bout, suggesting that this relationship is not caused simply by satiation.) Third, for species with variable group sizes (e.g., spider monkeys and chimpanzees), members spent more time traveling as subgroup size increased, suggesting that large subgroups deplete patches of equal size faster than smaller subgroups. Finally, for species with variable group sizes, the amount of time spent feeding in a patch was generally a function of the size of the patch and the number of animals using the patch. Preliminary data on the folivorous red colobus *(Procolobus badius)* provides similar evidence of patch depletion for some trees (C. A. Chapman and L. J. Chapman, unpub.).

These data, and studies reviewed by Janson (1988b, 1992), suggest that some primate species frequently deplete their food patches. However, for specific food types or during some seasons, food resources may be so abundant that within-group competition is relaxed. For spider monkeys, for example, the rate of intake of food items during a feeding bout in a single patch is typically higher at the start of the feeding bout than later in the same feeding session (Chapman 1988a). However, when spider monkeys feed in large fruiting fig trees *(Ficus* sp.), fruit intake rate does not change over the feeding bout, suggesting that fruit is so abundant in these trees that spider monkeys are unable to deplete them. Similarly, in brown capuchins *(Cebus apella),* Janson (1988a) found that per capita feeding time decreased with group size in small patches but was independent of group size in large patches. In primate populations that may have been reduced below carrying capacity by disease (Collins and Southwick 1952; Work et al. 1957) or hunting (Peres 1990; Chapman and Onderdonk 1998), patch depletion may be less common than in high-density populations.

Three aspects of depleting resources can affect movement patterns and in turn affect group size: patch size, density, and distribution (fig. 2.1). Patch size may influence day range directly by determining the amount of food available in a given patch. A larger group will spend less time in a depleting patch (e.g., a fruit tree with a finite supply of ripe fruit) of a given size than a smaller group because it depletes the patch more quickly. Once a patch is depleted, animals will have to travel in search of other feeding sites. At this point, the density of patches directly influences travel costs and group size. When resource patches occur at a high density, the distance to the next patch is short, travel costs are low, and animals can afford to be in large groups. In this situation, additional costs associated with being a member of a large group, such as the need to visit many patches, can be easily recovered. For example, when chimpanzees of Kibale National Park, Uganda, are feeding on ripe fruits of *Pseudospondias microcarpa,* the distance to the next fruit-

C. A. Chapman and L. J. Chapman

ing tree is short because *P. microcarpa* is found in groves. In 1993 one small grove contained twenty-eight large fruiting trees with an average interpatch distance of only 35 m (C. A. Chapman and L. J. Chapman, unpub.). At this time chimpanzees were found in large subgroups (Chapman, Wrangham, and Chapman 1995), and although these large subgroups had to travel to many *P. microcarpa* trees in a single day, the distance between feeding trees was small. While the density of trees is often a good index of presumed travel costs, travel costs are more appropriately evaluated relative to intake. Thus, if the patches are small or unproductive, travel costs could be high relative to intake, even if the patches are close together. Resources such as corms and insects may occur in small or unproductive patches, but for frugivorous primates feeding in large canopy trees, or even smaller understory trees, when trees are at high densities, travel costs are low, permitting large groups.

The distribution of patches is a critical parameter that is often ignored. When large or small food patches are clumped, the distance to the next patch is short, travel costs are low, and animals can form large groups as long as the clump of patches can support the group's foraging activity (fig. 2.1B). At such times, any additional cost associated with being a member of a large group, such as the need to visit many patches, can be easily recovered. If food patches are clumped, scarce, and found in either large or small patches, animals may not be constrained from being in large groups in the short term (e.g., spider monkeys: Chapman, Wrangham, and Chapman 1995), but may be forced to live in small groups if those resource conditions persist on a longer temporal scale. When food patches are uniformly distributed, regardless of their size, we expect food density to be the key determinant of group size for the following reasons: when patches are dense, animals can congregate because the distribution and density of their food resources do not impose high travel costs. When depleting patches are rare, small groups are favored. Individuals minimize travel costs by being in small groups that can feed in a single patch for long periods, since there are only a few mouths to feed, and patches are depleted slowly. Similarly, when depleting patches are uniformly distributed, large, and rare, small groups will be advantageous.

While the size, density, and distribution of patches may be key variables determining travel costs and group size, the situation can be assessed in a simpler fashion. When animals generally deplete

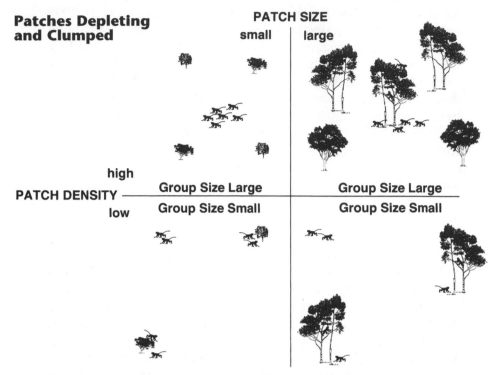

Patches Depleting and Clumped

PATCH SIZE

small | large

PATCH DENSITY — high / low

| Group Size Large | Group Size Large |
| Group Size Small | Group Size Small |

Figure 2.1 Hypothesized associations between food states (patch size, density, and distribution) and primate group size. Two classes of depleting patches are considered: *(this page)* clumped and *(opposite page)* uniformly distributed. Within these classes, each "quadrat" represents a different patch size/patch density combination. The predicted group size is indicated in each quadrat. This figure represents extremes of the parameters and should not be considered to represent all situations an animal could experience. Its purpose is to visualize how the size, density, and distribution of patches could influence travel costs and thereby constrain group size. A hypothetical travel route is illustrated for each ecological condition.

the patches they use, measures of habitat-wide food availability will probably adequately reflect the size, density, and distribution of patches. Thus, group size may be a function of food availability. An illustration of this is provided in figure 2.2 for the spider monkeys of Santa Rosa National Park, Costa Rica. This species has a very flexible fission-fusion type of social organization. In Santa Rosa, subgroup size can range from one to thirty-five individuals, but on average about five individuals are found traveling together. For this population, 50% of the variance in mean monthly subgroup size can be predicted from relatively crude measures of the size, density, and distribution of food patches (Chapman, Wrangham, and Chapman 1995).

C. A. Chapman and L. J. Chapman

Patches Depleting and Uniform

PATCH SIZE

small | large

high

PATCH DENSITY

	Group Size Large	Group Size Large
low	Group Size Large	Group Size Large

Testing the Model: Patch Depletion Process

The relationships between group size, factors thought to influence exploitation competition (patch size, density, and distribution), and day range have been examined in detail for only a few primate species (e.g., spider monkeys: Chapman 1988a, 1990a; Symington 1987; Chapman, Wrangham, and Chapman 1995; woolly spider monkeys *(Brachyteles arachnoides):* Milton 1984; Strier 1989; howler monkeys: Leighton and Leighton 1982; Chapman 1988a, 1990b; chimpanzees: White and Wrangham 1988; Chapman, White, and Wrangham 1994; Chapman, Wrangham, and Chapman 1995; bonobos *(Pan paniscus):* White and Wrangham 1988; Chapman, White, and Wrangham 1994), and rarely have all of the components (size, density, and distribution of patches) been examined in one study. Tests of the ecological constraints model are difficult to conduct because one must relate changes in group size to a set of ecological conditions, and for most species group size tends to be only slowly modifiable through births and deaths. To examine this model with such species would require either a long-term research program or a correlative approach requiring the habituation of many

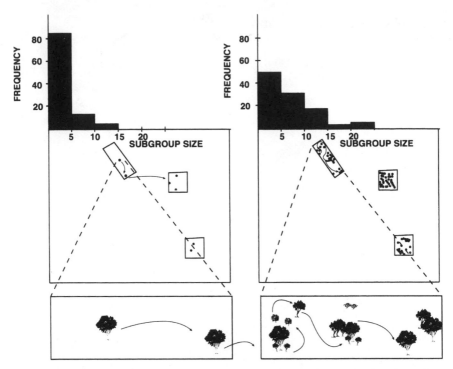

Figure 2.2 The ecological conditions corresponding to the sample periods when the spider monkeys in Santa Rosa National Park, Costa Rica, were observed to occur in *(left)* the smallest and *(right)* the largest subgroups. The frequency histogram of subgroup size is depicted at the top of the figure. The large squares represent the home range of the spider monkeys. Within their home range were three 4 ha ecological sampling grids (represented by the smaller squares or rectangles). Each dot within a sampling grid represents a food patch available to the spider monkeys. One of these sampling grids is expanded to illustrate the actual size, density, and distribution of food trees in the sampling grid. (Adapted from Chapman 1990b.)

groups. Consequently, researchers have often used fission-fusion societies (e.g., chimpanzees, spider monkeys, woolly spider monkeys) to examine the ecological constraints model (Milton 1984; Symington 1987; Chapman 1990a; Chapman, Wrangham, and Chapman 1995). In fission-fusion societies, animals from a single community are found in small subgroups that change size and composition frequently (i.e., two or three times a day). Because subgroup size is flexible, animals can respond to ecological changes that occur over short temporal and spatial scales. It is then possible to relate short-term variation in food resources to changes in subgroup size (Klein and Klein 1977; Milton 1984; Symington 1987; Chapman 1990a,b).

Studies of fission-fusion societies have provided support for

C. A. Chapman and L. J. Chapman

many of the components of the ecological constraints model. Relationships between patch size and feeding group size has been documented for woolly spider monkeys (Strier 1989), howler monkeys (Leighton and Leighton 1982; Chapman 1988a), and chimpanzees (White and Wrangham 1988; Ghiglieri 1984). In spider monkeys *(Ateles paniscus)* in Peru, 62% of the variance in monthly subgroup size was accounted for by habitat-wide food availability (Symington 1987). Symington also found that individuals in larger subgroups travel farther each day and spend less time feeding than individuals in smaller subgroups. Based on two 6-year studies, Chapman (1990a,b) and Chapman, Wrangham, and Chapman (1995) were able to explain 50% of the variance in spider monkey subgroup size and 22% of the variance in chimpanzee subgroup size using measures of patch size, density, and distribution. Furthermore, during specific periods when the chimpanzees were feeding almost exclusively on one species of fruiting tree and it was possible to measure food abundance and travel costs directly, 77% of the variance in subgroup size could be explained by the density and distribution of resources (Chapman, Wrangham, and Chapman 1995). Although this chimpanzee study community had a diverse diet, changes in the abundance of only three key fruiting tree species *(Mimusops bagshawei, Pseudospondias microcarpa, Uvariopsis congensis)* correlated with changes in chimpanzee party size (C. A. Chapman and R. W. Wrangham, unpub.).

Wrangham, Gittleman, and Chapman (1993) conducted an empirical review to determine whether variation in primate and carnivore group size relates to exploitation competition. They suggested that two factors directly affect the intensity of exploitation competition: density of food resources and travel efficiency. These variables are important because they influence the relationship between group size and per capita energy balance. Thus, as food density rises (with travel efficiency constant), more individuals can feed within a given travel distance. If selection favors large groups, increases in food density will therefore result in increased group size. Similarly, these authors argue that as travel efficiency rises, the intensity of competition falls, because by traveling farther, each individual encounters more food. A multiple regression analysis using indices of the density of food resources and travel efficiency explained up to 46% of the variance in primate group size and 57% of the variance in carnivore group size (see Janson and Goldsmith 1995 for an alternative analysis).

Avoidance of Overlap of Search Fields

Several studies have provided evidence to suggest that patch depletion may not occur in some species (e.g., folivores) or may occur only sometimes in others (e.g., when species eating fruits and insects congregate to feed in a fruiting tree). An alternative explanation to account for group size constraints in these situations is the avoidance of search field overlap. Larger groups travel farther than smaller groups in some species (*Lophocebus albigina:* Waser 1974, 1977a; Olupot et al. 1994; *Macaca fascicularis:* van Schaik et al. 1983; van Schaik and van Noordwijk 1988; *Cebus olivaceus:* de Ruiter 1986; *Cebus apella:* Janson 1988a; *Ateles paniscus:* Symington 1988a,b), but not all (*Procolobus badius:* Isbell 1983; *Papio anubis:* Bronikowski and Altmann 1996). Isbell (1983) documented a red colobus group of nine individuals that had a day range of 578 m, while a group of sixty-eight red colobus had a day range of 593 m (Isbell 1983, 1991; Struhsaker and Leland 1979, 1986). One could question the generality of this finding with respect to red colobus: it could simply be that the group of sixty-eight individuals had significantly more resources in its home range than the group of nine individuals. Alternatively, it may be that folivores like the red colobus do not deplete leaf resources (Isbell 1991).

A second situation in which the patch depletion process may not be operating to constrain group size involves species that rely on dispersed food items (or small nondivisible patched ones), such as some insects. For such species, additional group members may not increase the rate of patch depletion and thus may not lead to increased day ranges.

For species that either do not deplete the patches in which they feed or which feed on dispersed foods, an alternative process may be operating. As group size increases, individual search fields may overlap, reducing per capita encounter rates with food, thereby increasing the area that must be searched to find food. Van Schaik et al. (1983) described such a process in their study of long-tailed macaques *(Macaca fascicularis)*. In this species time spent traveling increased monotonically with group size (n = 7 groups). Long-tailed macaques have a mixed diet that includes not only a significant proportion of fruit, but also more dispersed foods including insects; young leaves, and mushrooms (van Schaik and van Noordwijk 1988). Van Schaik et al. (1983) suggested that foraging animals tend to move away when approached by others, presumably since their conspecifics reduce the availability of dispersed food items. They termed this behavioral mechanism "pushing forward."

C. A. Chapman and L. J. Chapman

Chapman (1990b) conducted a field study on a group of white-faced capuchin monkeys *(Cebus capucinus)* to examine the ecological constraints model. Since this species has cohesive groups, it was hypothesized that group members would spread out to reduce within-group competition. Contrary to what was predicted, interindividual distance was not related to the density and distribution of tree resources. However, when fruiting trees occurred at low densities and patch depletion was more likely, capuchin monkeys spent more time eating insects. Similarly, when the monkeys were using small trees that could hold only a few individuals at a time, again insect feeding was prevalent. This finding suggests that within-group competition may be reduced by shifting from feeding at a fruiting tree to searching nearby for insects when the fruit patch is occupied by others. Identifying mechanisms whereby species reduce within-group competition will be a profitable avenue for future research. A detailed quantification of intake and movement rates in relation to nearest neighbors among insectivorous primates or species whose diets are composed of corms and bulbs of grasses (Whiten, Byrne, and Henzi 1987) is needed to understand the interactive and independent effects of patch depletion and search field overlap on travel costs.

Pushing Forward Mechanism
For animals that are feeding on dispersed food items, it may be costly to have to share the area that is being searched for food because such sharing will lead to a reduction in per capita encounter rates with food. To avoid overlap of search fields, animals may move on to new areas. If increasing group size increases the tendency for search fields to overlap, then larger groups will tend to travel farther than smaller groups to avoid the overlap of search fields. The "pushing forward" mechanism has not been fully explored and may be applicable to species that forage extensively on insects or species that feed on grasses, corms, or bulbs. This pushing forward process could operate for purely ecological reasons, whereby animals avoid others entering their foraging space and depressing their rate of food intake (van Schaik and van Noordwijk 1988). Alternatively, ecological factors may interact with social factors to produce the observed patterns (e.g., animals avoiding dominants approaching from behind). Observations on chimpanzees suggest that displacements can occur for social reasons. When subadult male chimpanzees are feeding in large fruiting trees, they are often displaced from the tree by adult males, even when there appear to be sufficient

feeding locations in the tree for all members of the subgroup and when adjacent trees may also be fruiting. Such displacements may ensure the maintenance of the dominance hierarchy, but they also directly influence the immediate movement patterns of the animals involved. An intriguing avenue for future research is the relative importance of exploitation competition versus social factors in determining the rate and extent of pushing forward. If some proportion of the pushing forward effect is a function of social avoidance, this may represent a hidden cost of sociality that has not been fully investigated (Smuts and Smuts 1993). When groups are feeding on very small fruit patches (e.g., shrubs) or slowly depleting one-animal insect patches (e.g., bark insects), investigations integrating local food resource availability, the tendency to be pushed forward, and the dominance rank of the patch holder may provide a means to explore social versus ecological influences on the pushing forward mechanism.

It seems likely that the pushing forward process may be affected differently by food types that engender aggressive (contest) food competition (e.g., small patches in which only one animal can feed) than by foods that engender purely exploitation (scramble) competition (e.g., dispersed food items). Resources that engender contests may not result in increased travel distances in all situations. For example, if a subordinate animal is feeding in a small patch in which only one animal can feed, it may be aggressively displaced by a dominant animal, leading to increased travel, but if the dominant has access to the patch, it may not give it up, and no increased travel will occur. When feeding on dispersed items that engender exploitation competition, both dominant and subordinate animals may avoid overlap of search field by increasing travel. Investigations of the degree to which displacements occur with respect to food type may provide insights into primate grouping patterns.

Conclusions: What Next?

The current body of evidence supports the notion that exploitation competition can limit primate group size whenever a group must travel farther per day than a solitary forager to satisfy its food requirements. Animals must forage over an area that can meet their energetic and nutritional requirements. Therefore, an increase in group size may be expected to increase the area that must be searched to find adequate food supplies. Thus, individuals must travel farther and expend more energy if they are in a large group

C. A. Chapman and L. J. Chapman

than if they forage in a smaller group or alone. With an increase in the time and energy spent traveling, a point will be approached at which energy spent in travel exceeds the energy obtained from the environment, and a smaller group size should become advantageous. In this way ecological factors can influence movement patterns and foraging efficiency and thereby constrain the sizes of groups that can efficiently exploit available food resources.

To date, detailed tests of the ecological constraints model are limited to a relatively few studies of species that have similar ecological requirements and similar social systems. In addition, the model rests on a number of assumptions for which there are reasons to question their widespread acceptance. For example, it is unclear whether the ecological constraints model is useful in accounting for variation in group size in insectivorous species that rely on dispersed food items or in folivores that have been suggested to feed in nondepleting patches. Furthermore, tests of the model have been primarily restricted to species with fission-fusion social organizations. Thus, it may be useful to ask how well a model developed through assessment of the short-term costs and benefits of being in a "subgroup" that changes in size and composition a number of times throughout the day will apply to "group" sizes that are relatively stable, modified only by births and deaths. With questions like these remaining to be answered, it is clear that investigations attempting to understand what determines animal group size will be an exciting area for future research.

Areas of future research include:

1. Do strict folivores, such as red colobus or black-and-white colobus (*Colobus guereza*), deplete leaf resources (paying attention to the fact that not all leaves are the same)? Do increases in folivore group size lead to increased day range?
2. How does competition over dispersed foods, such as possibly insects, corms, or bulbs, increase with group size? Is the shape of the relationship between group size and per capita foraging loss the same for species feeding on clumped depleting patches as for species feeding on dispersed food resources?
3. Is there a social component to the pushing forward effect that is independent of an ecological component?
4. Is the pushing forward process affected differently by food types that engender aggressive (contest) food competition than by foods that engender purely exploitation (scramble) competition?
5. How do primate populations reduced by disease or hunting respond to a reduction in within-group competition, and how does this influence group movement?

6. Are there viable strategies to reduce the effects of within-group competition, and if so, what are their effects on group size?
7. For species with mixed diets (e.g., insects and tree fruits), what are the interactive and independent effects of patch depletion and search field overlap on travel costs?
8. How are the ecological constraints model's predictions affected by individual differences in competitive abilities (e.g., costs and benefits of traveling for lactating adult females vs. adult males)?
9. How does variation in perceived predation risk affect animals' decisions to be in groups of different sizes?

Acknowledgments
Our project in Kibale has been supported by the Wildlife Conservation Society, National Geographic grants, NSF funding, local USAID funding, a Lindbergh grant, and USAID PSTC funding. We thank the government of Uganda, the National Parks Service, and Makerere University for permission to work in the Kibale National Park. We thank the following individuals for fruitful discussion of topics in this chapter: Sue Boinski, Dick Byrne, Joanna Lambert, Adrian Treves, Richard Wrangham, and particularly Charlie Janson. Sophia Balcomb, Sue Boinski, Dick Byrne, Mark Elgar, Paul Garber, Charlie Janson, Joanna Lambert, Daphne Onderdonk, and four anonymous reviewers provided helpful comments on this manuscript.

C. A. Chapman and L. J. Chapman

A Critical Evaluation of the Influence of Predators on Primates: Effects on Group Travel

SUE BOINSKI,

ADRIAN TREVES, AND

COLIN A. CHAPMAN

A troop of Costa Rican squirrel monkeys *(Saimiri oerstedii)* is in a patch of early second-growth forest, eating fruits of *Cecropia* (Moraceae) and *Piper* (Piperaceae) and the occasional caterpillar. The sudden onset of raucous and frenetic alarm calls disturbs the tranquil scene, yanking the observer's attention to a voracious-looking crested eagle *(Morphnus guianensis)* perched 4 meters above the ground in what had been the center of the troop's dispersion. Adult females and immatures immediately coalesce into a writhing ball of more than forty squirrel monkeys within the protective confines of a dense vine tangle. Meanwhile, twenty foolishly valiant adult and subadult males are mobbing the raptor, literally throwing themselves at the bird. The squirrel monkeys slide off the raptor and thump to the ground below without ruffling a feather, much less the predator's composure. After several minutes of barrage by the small-bodied squirrel monkeys, the crested eagle spreads its wings and launches into a flight that skims the shrubby growth. It makes a brief dip to grab a young adult male with its talons, and is last seen landing with its prey in the canopy of a distant tree. The remaining troop members are immobilized for nearly 15 minutes, so still and quiet that they become almost invisible. Then, ever so quietly and slowly, they creep to the ground, following the lead of four adult males. The troop walks single-file into the grass, and continues walking on the ground for more than a kilometer, until it reaches the portion of its range with the densest and lowest ground cover, too impenetrable for the human observer to follow. The squirrel monkey troop remains in this thicket for the next 3 days (S. Boinski, pers. obs.).

There is good evidence that even some of the largest species in the primate order have strategies designed to decrease the risk of predation. For example, Tutin, McGrew, and Baldwin (1981, 1983) report a complex of antipredator behaviors used by savanna-dwelling chimpanzees *(Pan troglodytes)* at Mt. Assirik, Senegal, a habitat containing abundant predators, including leopards, lions, spotted hyenas, and wild dogs. Chimpanzees, when moving long distances, congregate in large parties and move in a rapid, directed fashion, while remaining unusually silent. These traveling parties are also described as being intensely alert when moving through open areas, frequently standing bipedally to scan their surroundings.

These and many other anecdotes imply that groups, and the individuals within groups, behave so as to limit predation risk. Moreover, predation has traditionally received the widest attention as a factor making group living advantageous for primates (van Schaik 1983). A group-living animal is thought to obtain antipredation benefits in the form of (1) decreased individual vulnerability (the more group members, the smaller the likelihood that any single individual will be killed during a predator attack), (2) increased effectiveness in detecting and deterring potential predators, (3) increased opportunities to confuse the predator, and (4) greater information on the presence of predators from alarm calls emitted by other group members (Hamilton 1971; Pulliam 1973; Powell 1974; Bertram 1978; van Schaik 1983; Terborgh and Janson 1986). Nevertheless, predation risk as an influence on primate behavior in general, and group movement in particular, remains controversial. Quantitative and particularly experimental documentation are scanty. Correspondingly, Janson (1992) notes that theoretical models and indirect tests provide the bulk of the evidence indicating that predation avoidance is a beneficial consequence of primate grouping.

In this chapter we consider the proposition that group movement is influenced by the risk of predation, both in the absence of attack and subsequent to an attack, presumably a period of more certain and elevated risk. The focus of our inquiry is a set of traits thought to affect predation risk—group travel, spatial structure, and habitat use—and the arsenal of presumptive predator avoidance, detection, and deterrence techniques. We critically examine the methods currently employed to evaluate the effects of predation on primate sociality. Then we propose specific methodological approaches useful in future research on predation and group movement. Last, potential interactions between predators and the many components of group movement in primates are considered in detail. We do not consider how group movement is influenced by primates when they themselves act as predators of vertebrates (Stanford 1995b; Rose 1997).

Evidence that Predation Is a Finite Risk to Primates

Views on the impact of predation on primate behavior within the discipline of primatology have swung like a pendulum. Early fieldworkers presumed that susceptibility to predation was a stringent constraint on the behavior and morphology of primates, although predation attempts on primates were rarely documented in these early studies (Carpenter 1934; Chance 1955; DeVore and Hall 1965; Nishida 1968). By the 1970s the dearth of predation data had become so conspicuous that a groundswell of researchers expressed doubts as to its significance (Aldrich-Blake 1970; Rodman 1973b). Most of the classic edited volumes in primatology from this period did not even include a predation entry in the index (e.g., DeVore 1965; Jay 1968). This skepticism continued into the early 1980s. Wrangham (1979, 1980), Fittinghoff and Lindberg (1980), and Collins (1984) all concluded that predation on primates was so seldom observed that its impact as a selective pressure was best considered weak. Yet Stuart Altmann (1974) countered that predation's biological importance cannot be evaluated on the basis of frequency of occurrence. He noted that although births, like deaths, are only rarely observed among wild primates, they represent a biological event of immense consequence.

By the mid-1980s, however, a substantial number of publications reporting successful and unsuccessful predation attempts on primates were collated by Anderson (1986) and Cheney and Wrangham (1986). Sufficient data had accrued to allow preliminary quantitative analyses relating interspecific differences in predation rate to group size, body size, arboreal versus terrestrial habits, and diurnal versus nocturnal activity (Anderson 1986; Cheney and Wrangham 1986; Isbell 1994). In general, higher predation rates were significantly predicted by smaller body and group sizes, but terrestrial primates did not have higher rates of predation per capita than arboreal primates. These studies relied on small sample sizes, explained little variance, and sometimes produced conflicting results (Boinski and Chapman 1995). Nevertheless, predation could no longer be easily dismissed as a negligible ecological factor on the basis of rarity. Despite these advances, primatology has remained locked in a tradition of "bean counting" predation events. What is really needed to evaluate the impact of predation are data that compare victims with survivors and track behavioral and genotypic changes over time.

We are not going to attempt quantitative analyses incorporating the more recent predation data (see our concerns expressed below regarding currently attempting cross-species analyses). Instead, readers are encouraged to examine documented cases of successful predation reported in the literature (tables 3.1 and 3.2). Mammals, birds, and reptiles are the major primate predators. Only in Africa are primates a significant predator of other primates. Both terrestrial and arboreal primates are taken by every type of predator, although there seems to be a tendency toward greater vulnerability of arboreal primates to raptors and terrestrial primates to mammalian carnivores. It is also evident that for most genera the cumulative total of the published number of successful predation events in our tally remains relatively small (mode = 1), and reporting and/or actual predation is highly skewed (mean = 32.6, median = 6, SD = 122, skew = 5). Even if this figure were doubled to compensate for an incomplete literature search or incomplete reporting of observed predation events, the cumulative total of documented predation events would be minuscule compared with the cumulative number of hours invested by field-workers in observation of primates. For example, for squirrel monkeys observed at two sites, the rates were 0.0016 and 0.0024 predation deaths per hour of observation respectively (Mitchell, Boinski, and van Schaik 1991).

Quantitative Analyses Based on Multiple Species
There is great temptation to incorporate every scrap of predation data, such as those presented in table 3.2, into a single analytic model. However, many difficulties arise in testing predictions concerning the evolutionary, ecological, and behavioral consequences

S. Boinski, A. Treves, and C. A. Chapman

Table 3.1 Types of predators reported to kill primates and their wild primate prey by genus and region

Region	Class	Predator type	Primate genera observed as prey or found in the remains of kills	References[a]
Asia				
	Reptiles	Snakes	Macaca, Semnopithecus	1, 2, 43
		Crocodiles	Homo, Macaca	3, 4
	Birds	Hawks, eagles	Macaca	5
	Mammals	Carnivores	Homo, Hylobates, Loris, Macaca, Nycticebus, Pongo, Semnopithecus	1, 2, 6–9
		Primates*	Nycticebus	10
Africa				
	Reptiles	Snakes	Cercopithecus, Hapalemur, Microcebus, Papio	1, 11, 12
		Crocodiles	Homo, Papio	12, 13
	Birds	Hawks, eagles	Avahi, Cercopithecus, Cheirogaleus, Colobus, Eulemur, Homo, Lemur, Lepilemur, Lophocebus, Microcebus, Mirza, Papio, Procolobus, Propithecus	11, 12, 14–16
		Owls	Lepilemur, Microcebus	11
	Mammals	Carnivores	Cercopithecus, Colobus, Eulemur, Gorilla, Homo, Lemur, Microcebus, Pan, Papio, Procolobus, Propithecus	11, 12, 17–22
		Primates	Cercopithecus, Colobus, Galago, Homo, Lophocebus, Microcebus, Papio, Perodicticus, Procolobus	1, 12, 23–28
Americas				
	Reptiles	Snakes	Cebus, Saguinus	29, 30
	Birds	Hawks, eagles	Alouatta, Ateles, Callithrix, Cebus, Chiropotes, Saguinus, Saimiri, Pithecia	31–36
	Mammals	Carnivores	Alouatta, Aotus, Ateles, Homo, Saguinus, Saimiri	37–41
		Primates	Callicebus	42

Note: Primate predators are restricted to nonhuman primates because humans probably hunt most primates. Attacks on humans are restricted to those that involve predation (i.e., no elephant trampling or snakebites).

[a]1, Cheney and Wrangham 1986; 2, Rajpurohit and Sommer 1991; 3, Galdikas and Yeager 1984; 4, M. Leighton, pers. comm.; 5, Rodman 1988; 6, Rijksen 1978; 7, Stanford, 1989; 8, Seidensticker 1983; 9, Sunquist 1981; 10, Utami and van Hooff 1997; 11, Goodman, O'Connor, and Langrand 1993; 12, Cowlishaw 1994; 13, *Uganda News* 1997; 14, Boshoff et al. 1991; 15, Struhsaker and Leakey 1990; 16, Steyn 1982; 17, Boesch 1991a; 18, Fay et al. 1995; 19, Cheney, Lee, and Seyfarth 1981; 20, Busse 1980; 21, Struhsaker 1975; 22, Treves and Naughton-Treves 1999; 23, Hladik, Charles-Dominique, and Petter 1980; 24, Butynski 1982b; 25, Wrangham and Riss 1990; 26, Hausfater 1976; 27, Uehara et al. 1992; 28, Treves and Naughton-Treves 1999; 29, Chapman 1986; 30, Heymann 1987; 31, Izor 1985; 32, Mitchell, Boinski, and van Schaik 1991; 33, Juillot 1994; 34, Ferrari and Lopes-Ferrari 1990; 35, Goldizen 1986; 36, Stafford and Ferreira 1995; 37, Peetz, Norconk, and Kinzey 1992; 38, Emmons 1987; 39, Galef, Mittermeier, and Bailey 1976; 40, Schaller 1983; 41, Hill and Hurtado 1995; 42, Freese and Oppenheimer 1981; 43, Shine, Harlow, Keogh, and Boeadi 1998; 44, Wright and Martin 1995; 45, Maisels et al. 1994; 46, Stanford et al. 1994; 47, Goodall 1986.

Table 3.2 Reported killings of primates

Taxonomic group	Genus	Number of kills	References[a]
Family Cheirogaleidae	*Microcebus*	7	11
	Cheirogaleus	3	44
	Eulemur	2	11
	Lemur	1	11
	Hapalemur	1	11
Family Lepilemuridae	*Mirza*	1	11
	Lepilemur	1	11
Family Indridae	*Avahi*	1	11
	Propithecus	6	11
Family Lorisidae	*Galago*	5	1, 24
	Nycticebus	7	10
Subfamily Atelinae	*Alouatta*	6	31, 37
	Ateles	1	33
Subfamily Callitrichinae	*Callithrix*	2	34, 36
	Saguinus	11	1, 30
Subfamily Cebinae	*Cebus*	7	1, 29
	Saimiri	12	32
Subfamily Cercopithecinae	*Cercopithecus*	33	1, 15, 19, 26, 45
	Macaca	18	3, 6, 43
	Papio	59	12, 14, 20
Subfamily Colobinae	*Colobus*	1	15
	Procolobus	629	21, 25, 27, 46, 47
	Semnopithecus	14	2, 43
	Trachypithecus	1	7
Family Hominidae	*Pan*	3	17
	Pongo	9	6
	Homo	33	22, 41

Note: This table is a selected review of reported killings of primates. Caution should be used when interpreting these values since, as noted in the text, reporting is biased to certain taxa, the number of hours of observations on different taxa vary greatly, and many reports stem from observations of primates killing other primates. To exclude the possibility of including observations of scavenging events, this listing does not include observations of remains of prey found in feces or at nest sites, or of prey being consumed by predators unless the freshly killed body was seen.

[a]As given in Table 3.1.

of predation when using between-species comparisons. Foremost among these difficulties is that inferential data structure our current knowledge of how predation shaped or is shaping primate behavior (i.e., predator risk is inferred from logical deductions by human observers regarding how predators and prey should behave). A second difficulty is that a panoply of parameters potentially contribute to the influence of predation risk on group movement. It is desirable in comparative studies to statistically control the interaction be-

S. Boinski, A. Treves, and C. A. Chapman

tween variables, in effect statistically removing the effects of various parameters when considering the effect of a single variable. An obviously desirable control, for example, would be removal of the effect of body size when considering the relationship between predation risk and terrestriality. At the present time, however, such controlled comparisons are limited given the relatively restricted number of primate species for which data are available. As a consequence, the amount of variance explained in comparative analyses is low, usually less than 15% (Isbell 1994; Boinski and Chapman 1995). Third, when attempting to extract comparable predation data from many primate species, one is limited to extremely crude estimators of predation pressure, such as predation events per unit time. The following are some of the more obvious factors that also reduce the utility of comparative analyses on the influence of predation.

1. Published reports of the effects of predation are often based on disappearance data, not events (e.g., Boinski 1987a; Isbell 1990; Peetz, Norconk, and Kinzey 1992; Boinski and Chapman 1995). Disappearance can result from a variety of causes (e.g., mortality due to disease, dispersal), of which predation is only one. Thus, disappearance data should be used with caution and only when the researcher has a detailed understanding of the pattern and rate of dispersal in the species.
2. Reporting of predation events is biased to the spectacular. Little incentive exists for publishing a report stating that in 2,000 hours of observation no predation attempts were observed. Yet this is exactly the data needed for an accurate representation of predation pressure.
3. The present predation rate in any population may not reflect the former predation regime that selected the current antipredation behavior (Cheney and Wrangham 1986; Byers 1997).
4. The appropriate taxonomic level at which to conduct comparative analyses may not be obvious. Species are not independent events, but are nested within phylogenies. There are statistical procedures that can be used to deal with such difficulties (Cheverud, Dow, and Leutenegger 1985; Martins and Hansen 1996).
5. Group size counts for a primate species in comparative data sets are usually based on many groups of widely varying sizes, but predation rates come from only one or a few groups (Boinski and Chapman 1995).
6. If the relationships between predation rate and group size are simply compared across species, we ignore species and individual differences in the costs of group living. For example, some folivorous primates are thought to experience reduced between- and within-group feeding competition compared with many frugivorous

primates (Isbell 1991). Furthermore, males and females may experience different costs associated with group membership (Chapman 1990a; Chapman, Wrangham, and Chapman 1988; Chapman 1995; Treves 1998).

7. Observed predation rates may seriously underestimate actual rates because the presence of the observer deters many predators. The impact of this effect is extremely difficult to quantify and no doubt varies among predator species and sites according to the degree of local development, size and type of research station, and human activity in the area. Anecdotal data could be useful in evaluating to what degree observers deter predators. Thus, field-workers might consider recording information on relevant observations (e.g., the number of predators seen avoiding the observer, the number of times a group mobs a predator even when the identity of the predator is unknown). For example, more than half of the fifty successful and unsuccessful predation attempts reported for Costa Rican squirrel monkeys occurred when the observer was still and obscured by foliage or an umbrella (Boinski 1987a).

8. On an evolutionary time scale increased predation pressure may favor large groups, but on a shorter ecological time scale high predation levels may decrease group size directly, through increased mortality rates due to predation. In Stanford's (1995b) study of chimpanzee predation on red colobus *(Procolobus badius),* colobus troops within the core of the chimpanzee hunting area averaged 46% smaller than troops on the periphery of the chimpanzee range, where hunting pressure on red colobus troops seemed much reduced.

9. Tremendous variation exists across sites in the densities of alternative mammalian prey species available to predators. This fact suggests that the predation rate for any single species is dependent on the current availability of other potential prey in its community (Wright, Gompper, and DeLeon 1994).

10. Different predators within a primate community present different risks and may evoke widely divergent antipredator responses. For example, large mammalian carnivores and raptors are characterized by markedly different senses to detect prey, morphological adaptations to pursue and capture prey, times of hunting activity (cats are usually nocturnal and raptors diurnal), and microhabitats searched for prey (van Schaik and Kappeler 1996). Primates often distinguish between aerial and terrestrial predators in their alarm calling and evasive responses (Struhsaker 1967a; Cheney and Seyfarth 1990; Macedonia and Evans 1993; Wright 1998).

11. Individual variation in response to predation risk and attack is evident even within a primate species. Scanning patterns vary within groups by age, sex, and dominance rank (Rose and Fedigan 1995; Gould, Fedigan, and Rose 1997; Treves 1997a,b, 1998). On the other hand, group behavior can mask diverse antipredator responses among the troop members. Group movement decisions in

the troops of some, but not all, primate species are determined by one or a few group members (Boinski 1996; Boinski, chap. 15, this volume). Plausibly, the antipredator tactics of subordinate individuals may be compromised by the actions of leaders.

12. Predation risk to infants probably does not have the same influence on individual and group decisions with regard to group movement as would, say, risk to an old male past his prime. Similarly, predation risk to infants in a species with a very long interbirth interval (e.g., black-handed spider monkey *Ateles geoffroyi,* 34 months: Chapman and Chapman 1990) may not have the same impact as predation risk to infants in a species with a short interbirth interval (e.g., spectral tarsier, *Tarsius spectrum,* 152 days: Harvey, Martin, and Clutton-Brock 1986).

13. Many fishes (Christenson and Persson 1993) and birds (Lima and Valone 1991) rely upon structurally complex microhabitats as refuges from predators. Casual inspection of the physical environments used by primates also suggests that habitats are selected to decrease the likelihood of predator detection or attack. Cords (1990b) found that increasing density of foliage was associated with decreased vigilance in redtail monkeys *(Cercopithecus ascanius)* and blue monkeys *(C. mitis)*. This finding suggests that individual movement patterns and microsite preferences may influence investment in self-protection.

14. Circumstantial data, unlike metric measures such as distances, rates, and group size, are difficult to incorporate into quantitative models of predation pressure. Yet the anecdotal literature on predation on primates is often sufficiently rich to provide convincing evidence for the importance of predation to specific taxa.

15. Researchers have warned that the rate and pattern of predation observed by field-workers is that occurring despite the array of antipredation behaviors primates exhibit (Boesch 1991a; Cowlishaw 1994; Dunbar 1997). In effect, the actual predation rate is not a reflection of the total risk to which group members are exposed, but the net predation risk after all precautions have been taken.

16. The relationship between predation rate and the antipredator behavior expressed by primate groups is unlikely to be linear (Sih 1987; Lima 1993). Although a sudden increase in predation pressure is predicted to instigate a surge in antipredator behavior, a decrease or even absence of predators in a community is unlikely to extinguish antipredator behaviors, especially if these behaviors are not costly in terms of time and energy. The degree to which the relationship between predation rate and the antipredator behavior is linear could be assessed, but only under situations in which the majority of predator attempts are known.

A review of the above list suggests that caution should be used when evaluating the influence of predation using comparative data that contrast many divergent taxonomic groups. Comparisons of a simple index of predation risk and primate behavioral patterns are unlikely to be robust.

Adaptive Story Telling: The Difficulty of Providing Functional Interpretations

A significant proportion of the antipredator behaviors ascribed to specific primate populations are probably best regarded as plausible, yet unsubstantiated, adaptive hypotheses. Quantitative data supporting claims of the functional significance of particular behaviors are frequently not provided, and alternative hypotheses are not considered. Two aspects of primate lifestyles often assumed to be under stringent antipredator selection, activity schedules and sleeping site selection, are prime examples of such conflicting scenarios. In regard to activity schedules, Moynihan (1976) claims that most primates have become diurnally active so as to be able to sleep in comparative safety at night. Similarly, Wright (1989) argues that by being nocturnal, the owl monkey *(Aotus trivirgatus)* avoids predation from a multitude of diurnal raptors. Yet, the anomalous extended bouts of foraging activity observed in Paraguay during bright daylight by owl monkeys are explained as a response to intense nocturnal predation pressure by great horned owls combined with the local extinction of many diurnal raptors (Wright 1994). Garcia and Braza (1987) argue instead that the frequently cold temperate-latitude nights in Paraguay provide a better explanation for the activity cycle shift in this population. The second example concerns sleeping site selection. Following the logic of many previous researchers, van Schaik, van Amerongen, and van Noordwijk (1996) concluded that persistent selection of sleeping sites on river banks by long-tailed macaques *(Macaca fascicularis)* is consistent with a functional hypothesis of predator avoidance. Nevertheless, the authors also clearly warn readers that their field data are unable to exclude the alternative explanations that these macaques are merely avoiding mosquitoes or conspecific troops.

Predation and behavioral data from many primate populations come from sites where predator populations have been reduced (Bishop et al. 1981; Seidensticker 1983; Rajpurohit and Sommer 1991). Attempts have been made to exploit these populations at artificially reduced predation risk as "natural" experiments. Yet too many alternative interpretations remain in these situations to allow robust conclusions to be drawn. A good illustration is Goodman's (1994) proposal that a large Malagasy eagle of the genus *Aquila* went extinct sometime between 500 and 4,000 years ago. He reasons

S. Boinski, A. Treves, and C. A. Chapman

that the stereotyped, strong antipredator response of large Lemuridae is a relic response to this raptor because extant raptors pose little threat to adults and subadults. As Csermely (1996) has pointed out, however, there are several problems with this argument, and multiple alternative hypotheses cannot be excluded. First, *Aquila* are not forest-hunting eagles presently, so the forest-living *Lemur* and *Propithecus* considered by Goodman may never have faced predation from this extinct raptor. Second, the argument that present-day antipredator responses are too strong and stereotyped is unjustified. Third, occasional predation by the small extant raptors may be sufficient to maintain the antipredator response described by Goodman. Fourth, the antipredator response of (putatively immune) adults could be designed to help immatures to recognize danger, as they are not immune to extant raptors.

Even the presence or absence of significant predation risk is a contested premise of some predation scenarios. Most notably, Southeast Asia is the focus of conflicting accounts of predation risk from raptors and felids. In regard to raptors, Southeast Asia is described by some as devoid of monkey-killing raptors (Bennett and Davies 1994), although the Philippine monkey eagle is reported to be an efficient hunter of primates (Kennedy 1977). Rodman (1988) recounts the case of a raptor killing a long-tailed macaque and uses this occurrence and others to dispute the long-standing contention about felid predators put forth by van Schaik and van Noordwijk (1985). The latter authors suggest that the Simeulue Islands lack felid predators and thus propose a direct comparison of group size with that at Ketambe, a Sumatran site with a full complement of predators. Unfortunately, the monkeys watched by van Schaik and van Noordwijk on Simeulue were not habituated, while the Ketambe macaques had been studied for 6 years. In this situation, censuses of subgroups might be expected to vary simply because of differences in observer avoidance by the two study populations. Food abundance, habitat differences, trail cutting, and temporal differences also were uncontrolled. Finally, Rodman also notes that raptors may have been present on Simeulue, so the absence of felids does not mean the absence of predators.

Recommended Strategies: Hypothesis Testing through Experimentation and Detailed Field Observations

In the previous sections we have outlined the obstacles to obtaining useful comparative data on the influence of predation on group movement and other aspects of social behavior and the difficulty of testing alternative adaptive scenarios. What can be done? Isbell (1994) advocated a quest for better data on predation rates, in the form of studying predators or conducting experiments. Both approaches should certainly be explored. Since researchers interested in the influence of predation on primate group movement are not likely to launch detailed studies of the predators themselves, we focus on the second approach: conducting experiments.

While providing useful observations, studies that draw conclusions about predation in a post hoc manner (Boinski 1987a; Chapman 1986) seldom allow for alternative hypotheses to be distinguished. For more powerful studies, explicit hypotheses should be laid out prior to the collection of data. We emphasize two useful methodologies: field experiments and detailed field observations, both of which must be finely tailored to the study species' apparent antipredator adaptations. We illustrate the utility of this approach with a number of examples in table 3.3. The hypothesis that sleeping site selection is made so as to maximize concealment and minimize accessibility to predators, for example, has at least two possible approaches. Not only can observations of the vegetation surrounding the selected site versus alternative sites be compared quantitatively, but the availability and the structure of sleeping sites can also be altered in a manipulative protocol.

The value of field experiments is well illustrated by studies of vervet monkeys *(Chlorocebus aethiops)*. Building on the original field observations by Struhsaker (1967a,b) of predator avoidance by vervets, Cheney and Seyfarth published a now classic series of field experiments examining the production of alarm calls in response to perceived predators as well as responses to playbacks of those calls (Cheney and Seyfarth 1981, 1985, 1990). Field experiments on other primate systems using stuffed predator models (Kortlandt 1963; van Schaik and van Noordwijk 1989; van Schaik and Mitrasetia 1990; Macedonia and Young 1991) and sound playbacks of alarm or predator calls (van Schaik and van Noordwijk 1989; van Schaik and Mitrasetia 1990; Chapman and Chapman 1996; Treves 1997a [thesis]) have also yielded persuasive corroboration of antipredator hypotheses based solely on field observations.

Precise sampling and quantification of individual and group-level behavior can provide powerful tests of hypotheses regarding the effects of predation on travel routes and the spacing, activity, and positioning of individuals within groups (Janson 1990a,b; Cords

S. Boinski, A. Treves, and C. A. Chapman

1990b; Rose and Fedigan 1995; Hall and Fedigan 1997; Cowlishaw 1994, 1997; Treves 1997a). The field situation that Rose and Fedigan (1995) exploited for their observational study of the function of vigilance in white-faced capuchins *(Cebus capucinus)* was particularly propitious. Four capuchin troops at the Costa Rican study site were well habituated and individually recognized, thus allowing four replicate tests of the research questions, among which were, first, are males more vigilant than females? and second, does the number of adult males in a group affect individual investment in vigilance? These questions were formulated after years of previous fieldwork at this site. Not only did males in all four groups invest more time in vigilance than females, but the overall mean rate of vigilance in each group was negatively related to the number of males and independent of group size.

Deconstruction of Predation Risk

> The space-specific risk of predation for baboons results from the fact that the baboons' predators tend to concentrate their hunting to particular habitats within the home range. For example, leopards stalk from cover and are seldom seen in areas of low open grassland. The baboon's problem then is to avoid areas of high risk and yet still get at areas with concentrated resources. (S. A. Altmann 1974, 245)

A successful predation episode reflects a dynamic sequence of events involving both predator and prey, and prey greatly benefit from obstructing this interaction whenever and however possible (Lima and Dill 1990; Endler 1991). Also, survivors presumably benefit by observing and learning from successful and unsuccessful predation attempts. From the perspective of the predator, a predation episode can be summarized as prey location, pursuit, attack, and retention (Kerfoot and Sih 1987). Two basic counterstrategies are available to prey: avoidance and deterrence (van Schaik and van Hooff 1983, 1996; Pulliam and Caraco 1984; Sih 1987; Brodie, Formanowicz, and Brodie 1991). In the course of predator avoidance, prey employ concealment, crypticity, and avoidance of habitats with predators to reduce the opportunities for prey detection. In contrast, deterrence by prey attempts to foil pursuit, attack, and retention by a predator once the prey has been detected. Of course, some behaviors, such as selection of propitious travel routes, might under most circumstances be best described as predator avoidance measures, but switch into predation deterrence in the aftermath of a predator attack.

Table 3.3 Suggested tests of hypotheses about group movement and predation

Formal hypothesis	Quantification	Specific predictions	Observational tests	Experiments
Sleeping sites maximize concealment and inaccessibility so as to reduce detection and approach of predators, respectively.	Tag sleeping sites and compare with structure of unused nearby sites: e.g., the last tree fed in or every tree passed in the previous 2–3 hours.	(1) Where raptors hunt by night, sleeping primates should avoid terminal branch tips and open canopy.	Sleeping trees should have heavier vegetation and fewer sectors of broken canopy than other trees.	(a) Modify sleeping tree structure or access and assess use over time.
		(2) Where carnivores hunt by night, sleeping primates should avoid tree trunks and branches within reach of neighboring trees' major branches.	Sleeping trees should have fewer major branches reaching the ground or touching neighboring trees; tree trunks may be smoother or thornier.	(b) Play back leopard roar or wingbeats to influence site choice and individual positioning.
		(3) Where snakes hunt by night, sleeping primates should choose trees with few access points (e.g., overhanging water).	Sleeping tree canopies should have fewer connections to neighboring tree canopies	(c) Model predator presentation to influence site choice.

Group movement, speed, and cohesion should vary with habitat and the predator types present to reduce ambush and detection by predators.[a]

Measure visibility and background noise in different areas or habitats. Measure behavior (individual vigilance and vocalizations; group movement, speed, and cohesion) as the group moves through its range.

Prediction	Expected result	Method
(1) Sit-and-wait predators using concealment to launch surprise attacks may be more dangerous when a group enters an area with impediments to vigilance.	Group movement speed and spread should decrease within habitats of low visibility or high background noise, while individual vigilance and contact calling should increase.	(a) In captive settings, modify visibility or background noise and measure behavioral changes.
(2) Hidden predators pose a greater risk when a group moves to a new area or an area it has not visited for some time.	The amount of time spent in an area should correlate positively with time spent playing, socializing, and feeding, negatively with vigilance, contact calling, and cohesion.	(b) Hide model predator in core vs. periphery; study discovery latency, identity/position of discoverer.
(3) Searching predators travel far and stalk or chase prey they have detected. Prey groups that move and/or use exposed habitats may be detectable at a distance.	Group movement, speed, and individual vigilance should increase and group spread should decrease when moving in habitats offering long-distance visibility.	(c) Measure vigilance and cohesion in sites that vary: grass, height, deciduous trees, croplands, etc.

continues

Table 3.3 (*continued*)

Formal hypothesis	Quantification	Specific predictions	Observational tests	Experiments[c]
Group movement produces audible and visible signs that may increase detection by predators.	Measure the sensory traces produced by groups varying in size, movement speed, cohesion, and arboreality. Sensory traces may be visual, auditory, or olfactory (e.g., feces).	(1) Groups that vary in size, group spread, biomass, movement speed, loud call production, and arboreality will differ in detectability.	Observers should have more difficulty finding groups with low detectability. Predator encounter rates should be lower also.[b]	(a) Naive observer[c] searches for groups; records species, traces (sound, sight, smell), height, size, spread.
		(2) Conspicuousness will be reduced by behavioral tactics and timing. Such reductions should be associated with predator encounter or dangerous times of day.	The timing of conspicuous displays will coincide with periods of predator inactivity or will profit from concealment and synchrony, or will be followed by rapid movement.	(b) Measure conspicuous activity (signaling, playing, moving) before and after predator playbacks or real attempts.
		(3) Running water, dense vegetation, and windy habitats will provide more background "noise" and reduce the conspicuousness of primate groups.	Groups in these habitats will be noisier, more colorful, use more elaborate visual displays, and move less cohesively.	(c) Within habitats, species differing in conspicuousness will respond differently to predator or alarm call.

[a]Note that type ≠ species, i.e., the same predator may use different hunting styles in different contexts.

[b]Accurate determination of predator encounter rates requires habituation of predators and prey.

[c]Must be absolutely unfamiliar with ranges, habitat features, etc.

Discussion of the evolutionary impact of predation can be framed in terms of costs and benefits (Ydenberg and Dill 1986; Smith and Winterhalder 1992). A primary trade-off in behavioral options confronting the typical prospective prey item, such as a primate, is thought to exist between minimizing predation risk and maximizing foraging success (Schoener 1971; Stephens and Krebs 1986; Ferguson, Bergerud, and Ferguson 1988; Kennedy et al. 1994). Although the cost-benefit ratio to be optimized can be described tidily as a trade-off between two endpoints on a continuum, this does not mean that the underlying decision-making process is likely to be simple. Instead, predation risk is probably incorporated into each step of a cascade of hierarchical decisions that precede a foraging decision, including when, where, and what food to search for, and how to search for, capture, and process it (Lima and Dill 1990). Travel figures significantly in the resolution of the predation-foraging trade-off. Nonforaging factors that can engender group movement are probably weighted by predation risk, including defense of food resources and mates and travel to safe sleeping sites.

A still more realistic perspective is that multiple factors, not just predation risk versus foraging success, contribute to movement decisions. The travel pattern of a small troop of a given primate species might be predicted to differ from that of a large troop given varying predation risk and anticipated foraging success (see Chapman and Chapman, chap. 2, this volume). Further complexity is added when additional factors are incorporated into the algorithm, such as the probability of both attack and escape in alternative habitats (Lima 1992), the availability of effective predator detection and deterrence (Cheney and Seyfarth 1981), and pregnancy-enhanced vulnerability to predation (Berger 1991). Despite the probable multifactorial nature underlying most travel decisions, simple cost-benefit models do work. Trade-offs between foraging success and predation risk, for example, are documented in the movements of social groups of baboons *(Papio cynocephalus)* (Cowlishaw 1997) and mountain sheep *(Ovis dalli)* (Berger 1991; Frid 1997).

Do primates overestimate predation risk? Clearly the primary effect of predation on primate groups is not the absolute mortality inflicted. The successful predation of a primate is at best an uncommon event (see table 3.1; Anderson 1986; Cheney and Wrangham 1986). About the only situation for which all the information needed to determine the optimal solution is available is theoretical models presented in academic texts (Stephen and Krebs 1986;

Smith and Winterhalder 1992). In the daily reality faced by primate troops, the quality of data on predation risk probably ranges from moderately to grossly inaccurate. Obtaining accurate information would also be expensive; "testing the waters" by having one or a few troop members act as scouts to ascertain the presence or absence of predators in alternative foraging areas would probably be quite hazardous, at least for the scouts. In fact there is negligible evidence that primates expend much effort beyond vigilance in updating knowledge of predation risk. Optimization models suggest that in situations of imperfect knowledge, if the total cost of overestimating predation risk is less than the cost of underestimating predation risk, overestimation should be the favored strategy (Bouskila and Blumstein 1992; Abrams 1994); mortality rates are likely to be higher in primates who underestimate than in those who overestimate. Accurate estimates, even if sought, would be difficult to garner because predation risk is a phenomenon resulting from the complex behavioral interactions of predator and prey (Lima and Dill 1990; Abrams 1994).

Features of Group Travel Suggested to Reduce Susceptibility to Predation

In this section we survey the many components of group travel that may be manipulated by primates to reduce predation risk. They vary widely in both the number of primate species in which they are evident and the extent to which their purported antipredation function has been critically evaluated. This discussion highlights parameters and concepts meriting careful examination in future research. Our intent is not to exhaustively review the literature, but to indicate the manifold predator avoidance and predation deterrence tactics available to primates.

At the outset we admit that setting firm limits on the influence of predation on primate social behavior is difficult. Although it is often suggested as an ultimate factor making group living advantageous, little evidence suggests that predation directly affects the internal social structure of primate groups (van Schaik 1983; Terborgh 1983; Terborgh and Janson 1986; Treves and Chapman 1996; but see Stanford 1998). Predation rate does not broadly predict interspecific differences in social relationships, such as dominance and affiliation, or sex differences in dispersal patterns (Cheney and Wrangham 1986). With reference to group movement issues, predation risk is also unlikely to be a factor of broad influence on territo-

S. Boinski, A. Treves, and C. A. Chapman

riality (see Peres, chap. 5, this volume) or other situations of intertroop aggression (see Boinski, chap. 15, this volume). Yet these broad exclusions must be qualified because in specific instances predation might well be a contributing factor in territorial behavior and intertroop aggression, as when the resources being defended are safe sleeping sites or refuges.

For ease of presentation, the following list of antipredation behaviors is divided into spatial and social tactics. These categories are not mutually exclusive. *Spatial tactics* encompass those behaviors that result in changes in the position of animals in three-dimensional space. *Social tactics* are frequency-dependent behavioral adaptations whose expression by an individual hinges on the behavioral responses of other group members. As in mate selection or foraging behavior (Hinde 1983), the expression of individual behavioral strategies to reduce predation risk may well be affected by kin selection and reciprocal altruism (Trivers 1971; Seyfarth and Cheney 1984).

Spatial Tactics

Familiarity. Perhaps the most prominent commonality evident in primate group movement is the existence of core areas, heavily used portions of the home range. Great advantages appear to follow from using familiar habitat for which the best, most often updated information on the whereabouts of potential predators is available (Lima and Dill 1990; van Schaik, van Amerongen, and van Noordwijk 1996; Isbell and van Vuren 1996). There is some evidence that mortality increases when primate groups are forced out of familiar ranging areas (Isbell 1990; Isbell, Cheney, and Seyfarth 1990; Peetz, Norconk, and Kinzey 1992). An alternative interpretation is that familiarity with the distribution of food sources is the primary advantage responsible for the existence of core areas (Waser 1977a; Chapman 1990b; Newton 1992).

It is also interesting to note that no nomadic groups of nonhuman primates exist comparable to the nomadic human *(Homo sapiens)* pastoralists described by McCabe (chap. 22, this volume). The most wide-ranging nonhuman primates, including solitary gorillas *(Gorilla gorilla)* and orangutans *(Pongo pygmaeus)* (Watts 1994; van Schaik and van Hooff 1996), bachelor herds of langurs *(Semnopithecus entellus)* (Vogel and Loch 1984), and, possibly, seasonal range shifts in gorillas (Vedder 1984), have annual ranges of probably less than 40 km² Yet the annual range of Turkana pastoralists in northwestern Kenya can exceed 2,400 km^2 (McCabe, chap. 22, this volume).

Travel Route Selection. Numerous researchers working with terrestrial primates note that travel routes appear to minimize exposure to "risky" habitats, those with high-density vegetation offering cover for predators, such as lions and leopards (Rasmussen 1983; Isbell 1994; Cowlishaw 1997). For forest primates there is evidence of at least transitory avoidance of, and precautions to carefully inspect, areas in which a predation attempt has occurred (Boinski 1989). A typical example is the behavior of a moustached tamarin *(Saguinus mystax)* group after the successful predation of a group member on a fallen tree trunk (Heymann 1987). The site was part of a traditional, frequently used travel route, but 8 days passed without the group coming near the predation site. For an additional 4 days the group cautiously approached the tree trunk in a markedly excited manner while emitting alarm calls at a high rate, and still did not venture across the tree trunk.

Sleeping Site Selection. Circumstantial evidence suggests that sleeping site preferences in at least some primates may be an adaptation to reduce predation risk. First, arboreal sleeping sites, nest holes, steep cliffs, or other sites limiting access by terrestrial predators are typical sleeping sites for primates, even large terrestrial species (Gautier-Hion 1973; Anderson 1984; Chapman 1989b; Chapman, Chapman, and McLaughlan 1989; Ferrari and Lopes-Ferrari 1990). Second, in those species switching between a number of alternative sites, effort is apparently expended to avoid signaling the location of that evening's sleeping site to predators (Tutin, McGrew, and Baldwin 1981; Heymann 1995). Third, primates spend a large portion of their lives at sleeping sites (tamarins often more than 60%: Heymann 1995); therefore circumspect selection of a sleeping site to reduce vulnerability to predators would appear to be important.

Caution should be used, however, when considering the argument that sleeping site selection by primate groups reflects strategies to reduce predation risk. First, aside from baboons (Hamilton, Buskirk, and Buskirk 1975; Busse 1980; Hamilton 1982), there are not many records of predation attempts at sleeping sites (Galef, Mittermeier, and Bailey 1976; Wright, Heckscher, and Durham 1997). This is what would be expected if primates were successful

S. Boinski, A. Treves, and C. A. Chapman

at choosing nonrisky sleeping sites. It would also be expected because most researchers do not watch their study groups at night. This fact results in there being few studies of predation at sleeping sites and thus limits our ability to consider whether sleeping site selection is influenced by predation risk. Second, both repeated and nonrepeated use of sleeping sites (e.g., consistently using one or a few sleeping locations versus using many different locations) are argued to be strategies to elude predators. Repeated use of the same sleeping site is often presumed to result from the specific location being particularly effective in deterring surprise attacks or permitting rapid escape (Wright 1981; Snowdon and Soini 1988). On the other hand, changing sites frequently is said to make location of the group more difficult for the local predators (Ferrari and Lopes-Ferrari 1990; Wright 1998).

Whether or not sleeping sites that curtail or minimize exposure to predators are a limited commodity to primate troops has two important implications. First, the distribution of species locally and regionally might depend on secure sleeping sites or other environmental features that allow predator avoidance (Anderson 1984; Lima 1993). Dietz, Peres, and Pinder (1997) suggest that golden lion tamarins *(Leontopithecus rosalia)* are limited to primary forest and prevented from exploiting secondary forest because of the dearth of suitable tree cavities for sleeping sites in the latter habitat. Second, if sleeping sites are merely rare within a troop's ranging area, their distribution may well constrain the troop's ranging behavior (Sigg and Stolba 1981; Chapman 1989b; Chapman, Chapman and McLaughlan 1989) If a troop uses one of only a few sites each night, the likelihood of troop movement to a specific foraging area decreases as distance from the refuge(s) increases (Hamilton and Watt 1970; Schoener 1971; Chapman, Chapman, and McLaughlan 1989). In at least some primate species, sleeping sites appear to be limited in number, and troops select those sites closest to their current feeding area, thereby minimizing travel costs (spider monkeys, *Ateles geoffroyi:* Chapman 1989b; Chapman, Chapman, and McLaughlan 1989; Chapman 1990a; baboons: Rasmussen 1979). In other primates, however, sleeping sites appear not to affect travel distances or routes significantly (wedge-capped capuchin, *Cebus olivaceus:* Robinson 1986).

Height Above Ground. In most forests exploited by primates, predation risk from terrestrial canids, felids, and snakes is argued to decrease as height above ground increases (Emmons 1987; Peres 1993a). Yet once a primate is so high in the forest as to approach the top of the canopy, or is on the edge of a forest gap or clearing, its vulnerability to raptor predation may increase (Eason 1989). Unfortunately, exceptions can easily be found to such tidy risk scenarios. For example, Maisels et al. (1994) observed a crowned hawk eagle *(Stephanoactus coronatus)* kill that occurred at a height of only 15 m, suggesting that there is danger even at lower levels. In practice, constructing broad generalizations of how predation influences height selection is difficult. The outcome will depend on the both the local abundance and diversity of predators and each species' hunting strategies, and is still further confounded by biomechanical and foraging requirements. Thus, the influence of predation risk on height selection should be determined separately for each species within each locality.

Comparisons between species within one habitat may be informative, particularly if predation risk differs between similarly sized species as a result of predator preference. For example, in Kibale National Park, Uganda crowned hawk eagles *(Stephanoaetus coronatus)* have been documented to take more blue monkeys than redtail monkeys or red colobus monkeys (Skorupa 1989; Struhsaker and Leakey 1990). Blue monkeys respond to real encounters with raptors or playback of their calls by decreasing their height in the trees, whereas redtails do not change their height, and red colobus actually increase their height (Treves 1997a). In the same way, chimpanzee encounters or playbacks prompt no response in redtails (rarely preyed on by chimpanzees), but prompt increases in height by red colobus (frequently preyed on by chimpanzees). These patterns of microhabitat avoidance can translate into group-level changes in movement or preferences for certain strata in the forest.

Foliage Cover. Many animal taxa rely on vegetation and other complex environmental features to avoid detection by predators and to evade attacking predators (Lima 1992). One of the supposed benefits of the use of lower canopy levels by primates, avoidance of raptors, in great part reflects the dense and continuous foliage cover that is often characteristic of this stratum (Brown 1966; Crandlemire-Sacco 1988; Oates and Whitesides 1990). For example, Cords (1990b) demonstrated that foliage density influenced vigilance in redtail monkeys in the Kakamega Forest, Kenya. Yet the same measure used by Treves (1997a,b) failed to correlate with vigilance in

S. Boinski, A. Treves, and C. A. Chapman

redtail monkeys in Kibale National Park, Uganda. If such intersite differences (in the same type of habitat) in the effectiveness of foliage cover as a refuge from predators are found to be common, simple categorical distinctions between forest, woodland, and savanna may be misleading in evaluations of predation risk.

Cryptic, Conspicuous, and Cautious Styles of Travel. A cryptic mode of group travel is often interpreted as a mechanism to elude detection by predators (Tutin, McGrew, and Baldwin 1983). Species described as cryptic often, but not always, have small body sizes and small group sizes and are well suited to wend their way unnoticed through the interstices of dense vegetation (Clutton-Brock and Harvey 1977a; Watanabe 1981; Terborgh and Stern 1987). Creeping, stealthy travel are ascribed to the buffy-headed marmoset (*Callithrix flaviceps:* Ferrari and Lopes-Ferrari 1990), the de Brazza's monkey (*Cercopithecus neglectus:* Wahome, Rowell, and Tsinglia 1993) the spot-nosed monkey (*Cercopithecus petaurista:* Oates and Whitesides 1990), and moustached and saddleback tamarins (*Saguinus mystax, S. fuscicollis:* Peres 1993a), particularly as they approach that evening's sleeping site. Moreover, individuals of typically social primate species, such as blue monkeys, move in a stealthy and cautious manner when solitary: these solitary animals may be stealthy to avoid predators or to avoid detection by resident males (Tsinglia and Rowell 1984; Treves 1997b).

In contrast, for taxa living in large groups, such as red colobus (Struhsaker 1975), gray woolly monkeys (*Lagothrix lagotricha:* Peres 1996b), Costa Rican squirrel monkeys (Boinski 1987b), and white-faced saki monkeys (*Pithecia pithecia:* Walker 1996), typical group movement can be conspicuous to the point of flamboyance: branches sway so exuberantly and crash so noisily as a troop travels that even the most naive human field-worker can detect the passing troop from a distance. In these easily detected groups the costs of high visibility to predators are thought to be counteracted by the benefits of large group size (Chapman and Chapman, chap. 2, this volume). Yet some larger-bodied species, such as white-faced sakis (Walker 1996) and the diademed sifaka (*Propithecus diadema:* Wright, 1998), presumably less susceptible to predation than smaller species, are described as cryptic because little noise or substrate movement is engendered during group movement. Extreme examples of the exploitation of stasis for crypticity may include taxa (*Aotus, Callicebus,* callitrichines) that are active at night or spend significant portions of the day still in nest holes or other concealed refuges (Terborgh 1983).

Escape Responses.   Changes in the demeanor of a troop's movement after a predator attack are often striking. Dramatic increases and decreases in the apparency of group members to the predator can occur, and both are described as antipredation responses (Goeldi's marmoset, *Callimico goeldi:* Pook and Pook 1979; golden iron tamarin; common marmoset, *Callithrix jacchus:* Stafford and Ferreira 1995; tamarins, *Saguinus* spp.: Heymann 1990b; blue monkey: Cords 1987; redtail monkey: Cords 1987, Treves 1997a). For example, after the successful predation of a troop member by a raptor, the five surviving members of a tamarin *(Saguinus nigricollis)* troop clung motionless to tree trunks for 37 minutes, and once the troop started moving, they traveled slowly and close to the ground (Izawa 1978). Similarly, one of the common reactions by female and young de Brazza's monkeys to a dangerous situation is complete immobilization and absolute silence (Gautier-Hion 1973; Wahome, Rowell, and Tsinglia 1993). This "freezing" can be maintained for periods exceeding 10 minutes (Gautier-Hion 1973). In contrast, when a primate troop flees from a predator as soon as it is detected, great commotion may occur; red colobus monkeys respond to predators by extremely rapid locomotion characterized by a high frequency of leaping and vertical bounding (Gebo et al. 1994; Chapman and Chapman 1996).

It is debatable how socially coordinated either rapid movements or cryptic evasions truly are. A more conservative interpretation is that these group responses are best explained as synchronous individual responses. The basis of these evasions in separate individual reactions is most obvious among smaller primates, in which the response to alarming stimuli is to either immediately drop to the ground or scatter in different directions to protected positions (Pook and Pook 1979; Heymann 1990b; Stafford and Ferreira 1995).

Social Tactics
Group Dispersion.   In a famous simile Kummer (1967) likens group dispersion to an amoeba, constantly changing its shape in space. Group dispersion commonly refers both to the expanse and shape of the area encompassed by a group and to the density of group members within that area. The dispersion of a group reflects the

S. Boinski, A. Treves, and C. A. Chapman

results of spacing decisions by individual group members. Usually dispersion is greatest in foraging contexts and clumped in nonforaging contexts (Stolz and Saayman 1970; Busse 1984; Boinski 1987b). This continuum is thought to result from a trade-off between foraging competition and predation risk. Individuals are predicted to distance themselves from others to enhance foraging efficiency, but at a cost of greater susceptibility to predation as the separation from neighbors increases (Janson 1990a). Near neighbors potentially offer antipredation advantages in the form of (1) increased probability of predator detection (Rodman 1973b; Struhsaker 1981; Gautier-Hion, Quris, and Gautier 1983; Boinski 1987a, 1989; van Schaik and van Noordwijk 1985, 1989; Cords 1990; Norconk 1990a,b; Terborgh 1990; Chapman and Chapman 1996), (2) greater confusion of a predator trying to focus on an individual prey (Morse 1977), (3) a decreased probability of each individual being captured by predators (Hamilton 1971), and (4) increased defense against predators (Struhsaker 1981; van Schaik and van Noordwijk 1985, 1989; Boinski 1987a; Gautier-Hion and Tutin 1988; van Schaik and Hörstermann 1994).

Three additional sets of observations reinforce the generalization that predation risk curtails the tolerable extent of group dispersion. First, groups often become most clumped and cohesive during those periods when the perceived predation risk is apparently greatest (Tutin, McGrew, and Baldwin 1983; Boinski 1987a) or subsequent to some predator attacks (van Schaik and Mitrasetia 1990). Baboons, for example, often cluster tightly when traveling through portions of their range where predators are often encountered (Altmann and Altmann 1970; Harding 1977).

Second, in many species, group members seldom wander far from the periphery of a troop. Field studies commonly report that group members avoid separation from the main body of the group. Anecdotal evidence suggests that the smaller the species, the shorter the tolerated distance from the group, a response consistent with the expectation that small body size is linked with enhanced susceptibility to predation. While the limit for a woolly monkey, 8.25 kg, is about 100 m (Peres 1993c, 1996b), that for adult brown capuchins *(Cebus apella)* and white-faced capuchins, approximately 3–4 kg, is much smaller, usually less than 50 m (Janson 1990a; Hall and Fedigan 1997). For Costa Rican and Peruvian squirrel monkeys (Boinski 1991; Boinski and Mitchell 1992) and golden lion tamarins (Boinski et al. 1994), all less than 1 kg, the maximum dispersion of an individual from the main body of the troop seldom exceeds 10 m.

Third, the extreme of interanimal dispersion is seen in large-bodied taxa such as spider monkeys (6 kg) and chimpanzees (37 kg), whose fission-fusion social organization involves animals often traveling apart from others for long periods (Terborgh and Janson 1986; Chapman 1990a; Chapman, Wrangham, and Chapman 1995). Adults of both of these species spend their time in small subgroups that change size and composition frequently as animals join existing subgroups or as subgroups split. Thus, unlike most primate species, which have cohesive social groups, spider monkeys and chimpanzees are not always in the same group. Each individual has the option of associating with subgroups of different sizes and compositions, and individuals are frequently solitary. One inference here is that large body size affords a reduced susceptibility to predation that permits adjustment of group size to food patch size (see Chapman and Chapman, chap. 3, this volume), a strategy unavailable to smaller primate species. However, long-tailed macaques, baboons, and howler monkeys *(Alouatta palliata)* also form subgroups (Fittinghoff and Lindburg 1980; van Schaik et al. 1983; Anderson 1981, 1983; Chapman 1988a). Moreover, mothers with infants and juveniles would logically be suspected to face high levels of predation risk, yet they often travel alone in several species (Chapman, Wrangham, and Chapman 1995). These counter-examples suggest that the connection between the tendency of a species to form subgroups and that species' risk of predation warrants further study.

Position Within the Dispersion of a Troop.   The configuration of troop members within a troop perimeter does not result in equal exposure to predation risk (Collins, Henzi, and Motro 1984; Ron 1996). In a stationary group, predation risk is often viewed to be greatest on the periphery of the group and to decrease toward the center; peripheral group members have the fewest neighbors to assist in the detection and deterrence of predators and will be the first to encounter approaching predators (DeVore 1965; Hamilton 1971; Vine 1971). Of course, primate troops are not perpetually stationary, but allocate a significant proportion of time to travel. Advantages of different positions vary in a moving troop in that there is a food depletion cost to troop members foraging in the wake of previous troop members (Whitten 1983; Watts 1985, 1992; Chapman

1988a; Janson 1990a,b; Hall and Fedigan 1997). Furthermore, it has been suggested that group members within the vanguard, or leading edge, of a moving group may incur enhanced predation risk compared with subsequent group members (Boinski, chap. 15, this volume). Not only will these individuals be the first to enter the capture radius of "sit and wait" predators (Rhine 1975), but the group benefits of vigilance will be reduced because these areas will have been less thoroughly scanned for predators (Hall and Fedigan 1997).

Vigilance. Visual scanning with the apparent goal of detecting potential predators is a behavior commonly described in primates (Cords 1990b; Janson 1990a; Treves 1997a, 1998). Furthermore, vigilance rates are often exploited by researchers as a measure of predation risk perceived by individual group members. The proportion of time individuals typically allocate to vigilance varies within and between species. In some primate species, adult males are most vigilant (e.g., Costa Rican squirrel monkey: Boinski 1987b), yet females predominate in others (e.g., *Lemur catta:* Gould 1996). Juveniles and immatures, although often the most susceptible to predation, typically exhibit negligible amounts of vigilance behavior (Janson and van Schaik 1988).

While vigilance appears to afford antipredation benefits, these benefits are usually argued to be costly in terms of the reduction of time available for foraging (Pulliam and Caraco 1984). An individual cannot efficiently search for food and concurrently to survey its surroundings in search of predators. Direct tests of this purported trade-off between feeding and vigilance in primates remain to be performed.

Vigilance must be documented carefully and interpreted cautiously, as both social and antipredator functions may underlie the expression of this behavior in some primate species. Much of the time allocated to vigilance by adult male Peruvian squirrel monkeys (Mitchell 1990), vervets (Balldellou and Henzi 1992), white-faced capuchins (Rose and Fedigan 1995), and red colobus and redtail monkeys (Treves 1997a) is aimed at conspecifics, not at potential predators.

Alarm Calls. Alarm calls may act to increase the cohesion of groups, warn predators away, and teach naive individuals about potential threats (Klump and Shalter 1984; Srivastava 1991; Ross

1993; Zuberbühler, Noë, and Seyfarth 1997). Variation in alarm call production by individual group members in at least some species is well predicted by inclusive fitness and reciprocal altruism (Sherman 1977; Owings and Leger 1980; Seyfarth and Cheney 1984). Alarm calls often appear extremely efficient, effective, and important in mobilizing appropriate antipredation responses among group members. Information on the source of the threat (i.e., what type of predator, aerial or terrestrial) can be conveyed by features of the vocalization's acoustic structure (Cheney and Seyfarth 1990; Macedonia and Evans 1993). Peres (1993a) credits the rapid response to alarm calls produced by moustached tamarin group members with foiling all nine predation attempts by raptors that he observed during the study.

Alarm calls are not produced automatically when an individual detects a predator, and recipients of the signal do not necessarily exhibit unvarying, automatic responses (Cheney and Seyfarth 1990). Solitary primates do not emit aerial predator alarm calls in response to frightening stimuli (S. Boinski, pers. obs; C. A. Chapman, pers. obs.), although solitary individuals of many species of primates and other social animals commonly produce terrestrial predator alarm vocalizations on encountering terrestrial stimuli perceived as threatening (Marler and Evans 1996; Boinski, Gross, and Davis in press). Responses to both aerial and terrestrial predator alarm calls by recipients in social groups may range from completely ignoring the vocalization, to looking about, to flight and evasion. Troop members appear to weight their response to the alarm calls produced by an individual by that individual's recent history of accuracy (Cheney and Seyfarth 1990) and the recent history of predator attacks (Heymann 1990b). Response to mistaken alarm calls can be costly in terms of lost foraging opportunities and risk of injury (Cheney and Seyfarth 1990; Macedonia and Young 1991; Peres 1993a). Production of alarm calls may also be costly, as some predators may target callers (Ivins and Smith 1983).

Mobbing and Other Forms of Active Predator Deterrence. Mobbing by primates usually entails vigorous vocal threats and displays by multiple group members at close proximity to a predator. It is sometimes accompanied by physical aggression, including punches, kicks, and bites (Teleki 1973; Kortlandt 1980; Ferrari and Lopes-

S. Boinski, A. Treves, and C. A. Chapman

Ferrari 1990; Passamani 1995; Iwamoto et al. 1996) and blows with branches (Boinski 1988b; Chapman 1986) and stones (Hamilton, Buskirk, and Buskirk 1975). A poorly habituated white-faced capuchin once grabbed and threw a nearby Costa Rican squirrel monkey at a field-worker (Boinski 1988b). Chimpanzees (Haraiwa-Hasegawa et al. 1986) and baboons also chase predators away (Altmann and Altmann 1970; Baenninger et al. 1977; Iwamoto et al. 1996). Mobbing undoubtedly entails risk in the form of enhanced exposure to the predator and the possibility of injury. In many species adult males are the age-sex class most reliably participating in mobbing episodes; this pattern is suggested to represent kin selection, especially male parental investment in some taxa (Boinski and Mitchell 1994; van Schaik and van Noordwijk 1989b). Cowlishaw (1994) also notes that adult male baboons face lower risk during predator mobbing.

"Contact" Calls.   The vocal repertoire of many primate species contains vocalizations commonly labeled "contact" or "separation" calls (Marler 1965; Byrne 1981; Caine and Stevens 1990; Hohmann 1991; Snowdon and Hodun 1981; Boinski et al. 1994; Peres 1996b). These vocalizations appear to permit the exchange of information among group members so as to facilitate group cohesion, especially in contexts where foliage or extensive interanimal distances impede visual contact among group members. Contact calls, unlike alarm calls, are often produced at high rates (Caine 1993). This is especially the case in those species in which group members forage while traveling, resulting in numerous opportunities for scattered individuals to stray and lose contact with the main body of the troop (Byrne 1981; Boinski 1991). For example, adult female squirrel monkeys in Costa Rica and Peru typically emit over a thousand contact calls a day (Boinski 1991; Boinski and Mitchell 1992, 1995, 1997). One probable consequence of the barrage of contact calls audible within a troop at any moment is that group members obtain reasonably accurate, continually updated information on the dispersion, rate of travel, and even activity of other group members (Boinski 1991; Boinski and Mitchell 1992, 1997). As such, contact calls may well be a crucial coordination mechanism allowing group members to disperse from one another sufficiently to accommodate foraging demands while retaining swift access to the other social antipredator tactics listed above.

Conclusion

Does predation have a pervasive effect on group movement? The answer is probably yes. But this answer is derived largely from indirect observations and circumstantial correlations. When studying any particular species, observers are invariably impressed by a varied arsenal of spatial and social strategies that appear to counter predation risk. Yet it is apparent that in our understanding of how predation risk influences group movement, there is a large gap between the rich descriptions of antipredator behaviors and what can truly be demonstrated. It is difficult to get beyond the presentation of evolutionary just-so stories to obtain adequate evolutionary and ecological data to distinguish between alternative hypotheses. As a result, we advocate species-level studies that test explicit hypotheses through both experiments and field observations as a useful and complementary strategy to the comparative approach.

Acknowledgments

During the preparation of this chapter S. B.'s research was supported by a NSF grant (SBR 972284) and a grant from the National Geographic Society. C. A. C.'s research was supported by the Wildlife Conservation Society, USAID internal support grants, a Conservation Food and Health Grant, a PSTC USAID grant, NSF grants (INT 93-08276, SBR 9617664), National Geographic Society grants, and the Lindbergh Foundation. A. T. was supported by an NIH-MH grant (35,215) to C. T. Snowdon. We would like to thank Paul Garber and Marina Cords for offering very constructive comments on this work.

S. Boinski, A. Treves, and C. A. Chapman

CHAPTER FOUR

Mixed Species Association
and Group Movement

MARINA CORDS

Most people who consider a "group" of primates probably imagine at least three individuals of the same species who are either physically or socially more cohesive with each other than they are with members of other groups. At any moment, most group-living primates belong to just one such group. There are, however, several primate species whose members belong to more than one group at a time. Their dual membership arises from the fact that their group of conspecifics associates with one or more groups of other species, so that each individual belongs to a smaller group of conspecifics as well as to a larger mixed-species group. (Mixed-species groups may also involve single members of one species associating with a group of heterospecifics—e.g., Fleury and Gautier-Hion 1997—but such associations will not be considered in this chapter.) While some of these mixed-species associations may result from brief chance encounters, others seem to occur too often, or last too long, to be explained as random events (Cords 1987, 1990a; Gautier-Hion 1988; Whitesides 1989; Buchanan-Smith 1990; Holenweg, Noë, and Schabel 1996). These latter associations appear instead to reflect an evolved propensity for interspecific socializing by one or more of the species involved, for which a variety of adaptive benefits have been documented (table 4.1).

In the context of group movement, the existence of such nonrandom mixed-species associations raises several questions. First, we might ask whether and how the movements of a group of conspecifics are influenced by association with heterospecifics. A full understanding of the determinants of group movement in those species whose groups do associate regularly requires an answer to this question. Second, we might ask whether and how changes in movement patterns caused by association with other species are costly or beneficial. An answer to this question should enlighten our understanding of the functional interpretation of mixed-species association. Finally, we might ask how coordination of group movement is achieved in mixed-species groups. By virtue of their different species identities, the members of these groups are even less likely than the members of a single-species group to share a set of ecological priorities and a communicative repertoire. The potential for conflict in determining group movement patterns is thus increased, while the means to assure coordination among individuals would seem to be more limited.

In this chapter, I will review evidence from the published literature relevant to these three questions. The reader should be forewarned, however, that complete answers will not be forthcoming, and my efforts should be taken only as a first attempt, for two reasons. First, and most importantly, the number of detailed and comprehensive studies on mixed-species groups of primates is still small, despite a growing interest in this aspect of social life (Norconk 1990a). Even for species that live associated with others for much of the time and over much of their geographic range, often only one population (and perhaps only one group) has been intensively studied. Generalizations would therefore be premature, and we are limited to describing variation, as far as we know it.

A second reason for caution is that even many of the detailed studies of mixed-species association do not provide truly conclusive evidence for the various effects on group movement they try to document. Some of the mixed-species groups that have been studied occur essentially permanently, so that it is not possible to compare the behavior of a study group when associated and when alone. At best, groups associating permanently are compared with *different* groups that associate less often, but which also use different areas,

Marina Cords

and have different compositions and different histories. It is therefore difficult to be confident that differences in the degree of association are the most important determinants of observed differences in group movement patterns. This is especially true in light of the fact that a typical study focuses on one or perhaps two groups, which is all that a field-worker can practically manage. Another problem, at least for present purposes, is that the predominant focus of most studies of mixed-species association has been the functional question of why association between species is adaptive, so that descriptions of the mechanisms of association are often secondary and unquantified, if present at all. More explicit attention to such mechanisms, and the application of experimental methods such as sound playbacks to their study, would undoubtedly greatly improve both the quality and the quantity of available information, but few such studies have been undertaken to date.

How Group Movements Are Affected by Mixed-Species Group Partners

To document precisely how mixed-species association affects various parameters of group movement, it is best to compare associated and unassociated groups. Such comparisons have been made for several species of African guenons and for squirrel monkeys and capuchins in Peru. Although there is a considerable literature on mixed-species association in tamarins, data from most sites have not been reported in a way that allows comparison between single- and mixed-species groups. (Terborgh's 1983 study of *Saguinus fuscicollis* and *S. imperator* is the sole exception). In some tamarin communities, such comparisons would be impossible in principle because the monkeys spend essentially all of their time associated.

Still, many studies of tamarins have documented remarkable coordination of the travel paths and patterns of associated groups (Pook and Pook 1982; Terborgh 1983; Yoneda 1984; Garber 1988a; Buchanan-Smith 1990; Norconk 1990b; Peres 1992b, 1996a). At one site in Peru, for example, members of the associating species fed simultaneously and in the very same tree in the vast majority of feeding bouts (Garber 1993b; P. A. Garber, pers. comm.). Spatial coordination on a scale of individual tree crowns may be exceptional (e.g., see Terborgh 1983; Yoneda 1984), but even larger-scale coordination strongly suggests that adjustments in group movement by associating species have occurred, although they cannot be measured directly.

Table 4.1 Evidence for various adaptive benefits to participants in primate mixed-species associations

Benefit	Species deriving benefit	Association partners	Reference
Protection from predators, including reduction in vigilance	*Procolobus badius* *Cercopithecus diana*	*Cercopithecus diana* *Procolobus badius*	Noë and Bshary 1997; Bshary and Noë 1997
	Cercopithecus ascanius *Cercopithecus mitis*	*C. mitis* *C. ascanius*	Cords 1987
	Cercopithecus cephus *Cercopithecus nictitans* *Cercopithecus pogonias*	*C. nictitans, C. pogonias* *C. cephus, C. pogonias* *C. cephus, C. nictitans*	Gautier-Hion, Quris, and Gautier 1983; Gautier-Hion and Tutin 1988
	Lophocebus albigena	*Cercopithecus ascanius,* *Procolobus badius*	Chapman and Chapman 1996; Waser 1982b
	Saguinus fuscicollis *Saguinus mystax*	*S. mystax* *S. fuscicollis*	Peres 1993a
	Saguinus fuscicollis *Sanguinus labiatus*	*S. labiatus* *S. fuscicollis*	Hardie and Buchanan-Smith 1997
	Saguinus fuscicollis *Saguinus imperator*	*S. imperator* *S. fuscicollis*	Windfelder 1997
	Saimiri sciureus	*Cebus apella* or *C. albifrons*	Terborgh 1983

Increased access to plant food or feeding sites	*Cercopithecus ascanius*	*C. mitis*	Cords 1987
	Cercopithecus cephus *Cercopithecus nictitans* *Cercopithecus pogonias*	*C. nictitans, C. pogonias* *C. cephus, C. pogonias* *C. cephus, C. nictitans*	Gautier-Hion, Quris, and Gautier 1983
	Cercopithecus ascanius	*Lophocebus albigena*	Chapman and Chapman 1996; Waser 1982b
	Saimiri sciureus	*Cebus apella* or *C. albifrons*	Terborgh 1983
Increased access to invertebrate prey	*Saguinus fuscicollis*	*S. mystax*	Peres 1992b
Guiding to more profitable feeding areas	*Cercopithecus ascanius*	*C. mitis*	Cords 1987
	Saimiri sciureus	*Cebus apella* or *C. Albifrons*	Terborgh 1983
Joint defense of feeding territories	*Saguinus fuscicollis* *Saguinus mystax*	*S. mystax* *S. fuscicollis*	Peres 1992a; Garber 1988a; Garber and Teaford 1986

Note: This table includes only selected studies that provide considerable quantitative data to support the hypothesized advantages, and in which the null hypothesis of random association was either formally rejected or is unreasonable (given the permanence of associations).

Data that directly relate changes in group movement to polyspecific association are thus somewhat limited, but nevertheless indicate that such association can affect many parameters of group movement (table 4.2; the reader is asked to consult this table for details of and full references to the research findings discussed in the remainder of this section).

Rate of Movement
One such parameter is the rate of movement, which may be speeded up or slowed down when members of one species associate with others. In some cases, all the groups that associate together adjust their pace to some value that is intermediate to those of monospecific groups traveling on their own. In other cases, adjustments appear to be unilateral, with groups of only one species changing their pace in association. The one study that reported no change concerned *Cercopithecus diana,* which associated with several congeneric species at levels that did not exceed chance expectations; however, even in associations with *Procolobus verus,* with which the dianas did associate nonrandomly, there was no evidence of compromise with respect to travel rates, perhaps because these two species had similar travel rates to begin with (Whitesides 1989).

Use of Space
There are also several reports of mixed-species association affecting the area used over a given period. Data are available for range use over periods from several months to single days, as well as for the width of the swath covered by the group as it moves through the forest. In light of the above results on rate of movement, and because the area covered (at least on a daily basis) is likely to be influenced by the rate of movement, it is not surprising to find that mixed-species grouping can increase, decrease, or not change the area covered. Some of the changes are quite dramatic, as in the case of *Cercopithecus cephus,* which uses an area at least twice as large when in association with *C. nictitans* and *C. pogonias.* In several cases, however, there is insufficient quantitative information to assess the magnitude of the changes. These observations still contribute to the general finding that the nature of changes in the area covered by a particular species may not be the same from one spatial scale to the next. For example, *Cebus apella* confines its activities to a roughly 80 ha area regardless of association state, although the group diameter swells almost twofold when *Saimiri* are in asso-

ciation. *Cercopithecus nictitans* and *C. pogonias* use a smaller total area when associated with *C. cephus,* although the area used per day is greater in the mixed-species group.

Changes in the intensity with which monkeys use a particular area also appear to be variable. Whereas *Cercopithecus ascanius* in Kenya used parts of their home range more intensively when they associated often with *C. mitis,* three congeneric species in Gabon showed a less intensive use of their range when in mixed-species groups. Unfortunately the methods used to assess intensity of use of a given area were very different in these two studies: Gautier-Hion, Quris, and Gautier (1983) looked at backtracking from one 30-minute interval to the next, whereas Cords (1987) considered any redundancy of use of particular areas at any time on the same day, or on subsequent days. Perhaps there would be more consistency in the results if the same temporal scale were used to monitor the degree to which areas were used more than once.

Trajectory
Descriptions of the way in which members of one group approach or return to a heterospecific group from which they have become separated suggest that the particular route a group takes may be influenced by the location of heterospecifics with whom association is sought. Observations like these also come closest to indicating the potential for conflict between constituent groups in deciding on a joint route for the mixed party. It sometimes appears that one of the constituent groups goes conspicuously out of its way (quite literally) to maintain or re-form an association with the other. For example, Pook and Pook (1982) described several occasions when, in mixed *Saguinus* groups, *S. fuscicollis,* which usually trailed behind the leading *S. labiatus* in their joint progression, veered off the common path; *S. labiatus* were then detected circling around to rejoin their partners.

There are, however, other reports of heterospecific groups simply flowing together (Cords 1987), in which case the continuity of both the speed and direction of movement of both parties does not suggest so strongly that one is influencing the other. Such influences may occur, though, even though they are not obvious.

Much more work is needed in this area. The only quantitative comparisons of trajectory shape in and out of association come from a study of guenons in Gabon. Gautier-Hion, Quris, and Gautier (1983) found that monospecific groups of *Cercopithecus cephus*

Table 4.2 Effects of association in mixed-species groups (MSG) on group movement

Type of influence	Species and particular effect	Statistical test?	Reference
Change in rate of movement	*Cercopithecus mitis* increase mean hourly movement rate by 12% in MSG. *Cercopithecus ascanius* show no change.	Yes	Cords 1987, M. Cords, unpub.
	Cercopithecus cephus in MSG have mean daily path length 53% longer than *cephus* alone. *Cercopithecus nictitans* and *C. pogonias* increase mean daily path length by 8% when not associated with *C. cephus*.	No	Gautier-Hion, Quris, and Gautier 1983
	Cercopithecus diana do not change mean rate of movement (per 10 min sample) in MSG with other species, including *Procolobus verus*, *Colobus polykomos*, *Cercopithecus petaurista*, *Cercopithecus campbelli*, *Procolobus badius*, and *Cercocebus atys*.	Yes	Whitesides 1989
	Procolobus badius increase rate of movement by about 25% when together with *Cercopithecus diana*. *C. diana* do not change rate of movement.	Yes Yes	Holenweg, Noë, and Schabel 1996
	Cebus apella increase hourly movement rate by ~40% when in MSG. *Saimiri sciureus* decrease hourly movement rate by 30–40% when in MSG.	Yes No	Terborgh 1983
	Saguinus imperator travel about 30% faster when in MSG with *Saguinus fuscicollis*. *Saguinus fuscicollis* travel 17–20% slower when in MSG with *Saguinus imperator* (but see reference for alternative interpretations).	No No	Terborgh 1983
Change in amount of area used	*Cercopithecus mitis* use a greater area on days when association lasts longer. *Cercopithecus ascanius*: no such correlation exists.	Yes Yes	Cords 1987
	Cercopithecus cephus group's total 4-month home range is 2 times larger (119 vs. 60 ha) when associated; *C. cephus* group that associates regularly has 4-month home range 2.3 times the size of a *C. cephus* group that rarely associates (119 vs. 52 ha); mean daily area used is 2.4 times greater for *cephus* group that associates often.	No	Gautier-Hion, Quris, and Gautier 1983

		MSG	Reference
	C. nictitans and C. pogonias decrease the 4-month home range by 20% when associated; mean daily area used is increased by 27%	No	Starin 1993
	Procolobus badius adolescents travel over a larger area in MSG with Erythrocebus patas and Chlorocebus aethiops. No changes mentioned for latter species.	No	
	Saimiri sciureus in MSG remain in 80 ha home ranges of Cebus apella; alone, Saimiri may move through several Cebus home ranges.	No	Terborgh 1983
	Cebus apella does not change the area it uses in MSG.	No	
	Cebus apella increase group diameter 1.9 times when in MSG.	Yes	Podolsky 1990
	Saimiri sciureus do not change group diameter.	Yes	
Change in intensity with which area is used	Cercopithecus ascanius show greater between-day overlap in areas used in months with more time in MSG.	Yes	Cords 1987
	Cercopithecus mitis show no relationship between time spent in MSG and between-day overlap in quadrat use.	Yes	
	In neither C. ascanius nor C. mitis is the amount of time in MSG correlated with redundancy of area used on a daily basis.	Yes	
	Cercopithecus cephus move on to new areas every 30 min more often when in MSG.	Yes	Gautier-Hion, Quris, and Gautier 1983
	Cercopithecus nictitans and C. pogonias move on to new areas every 30 minutes or more often when in MSG with C. cephus.	Yes	
Change in trajectory	Cercopithecus cephus, C. nictitans, and C. pogonias do less backtracking when in MSG. C. cephus move toward C. nictitans and C. pogonias if they call from 300–500 m away.	Yes	Gautier-Hion, Quris, and Gautier 1983; Gautier and Gautier-Hion 1983
		No	
	Saguinus imperator and Saguinus fuscicollis move toward the heterospecific group's resting site after periods of rest.	No	Terborgh 1983
	Saguinus labiatus circle around to join the Saguinus fuscicollis members of their MSG who had suddenly veered off from a common trajectory.	No	Pook and Pook 1982
	Saguinus labiatus return to S. fuscicollis after the S. labiatus had set off in one direction, and S. fuscicollis did not follow.	No	Buchanan-Smith 1990

continues

Table 4.2 (*continued*)

Type of influence	Species and particular effect	Statistical test?	Reference
Change in types of habitat used	*Saimiri sciureus*, when associating with *Cebus apella* who are temporarily stationary, make closed-loop forays and return to *Cebus* every 30 min or so.	No	Terborgh 1983
	Cercopithecus ascanius use more open habitat on days when MSG last a longer time;	Yes	Cords 1987
	Cercopithecus mitis do not change the types of habitat used.	Yes	Cords 1987
	Cercopithecus cephus use more high, open forest when in MSG than when alone;	No	Gautier-Hion, Quris, and Gautier 1983
	Cercopithecus nictitans and *C. pogonias* reduce their use of high, open forest and increase their use of intermediate high and dense forest when in MSG.	No	Gautier-Hion, Quris, and Gautier 1983
Change in height of canopy used	*Cercopithecus mitis* and *Cercopithecus ascanius* show no change.	Yes	Cords 1987
	Cercopithecus cephus increase use of higher strata when in MSG.	Yes	Gautier-Hion, Quris, and Gautier 1983
	Cercopithecus nictitans and *C. pogonias* do not change the canopy strata used.	Yes	Gautier-Hion, Quris, and Gautier 1983
Changes in timing of activity	*Cercopithecus nictitans* wait for *Cercopithecus pogonias* to begin moving in the early morning, or to finish feeding on fruits, before beginning group movement.	No	Gautier-Hion and Gautier 1974
	Cercopithecus cephus in MSG flee rapidly when alarmed, while they don't move and hide when alone.	No	
	Procolobus verus have periods of movement and stasis that coincide with those of *Cercopithecus diana.*	No	Oates and Whitesides 1990
	Saguinus imperator and *Saguinus fuscicollis* start and stop different activities at the same time, especially when moving through a series of trees or vines	No	Terborgh 1983
	Saguinus mystax and *Saguinus fuscicollis* coordinate their activities in time, especially rapid travel.	Yes	Norconk 1990b
	Saguinus fuscicollis follow lead of *S. labiatus* in resuming travel after rest periods.	No	Yoneda 1984

were more likely to backtrack and less likely to continue proceeding directly forward than were mixed-species groups of *C. cephus, C. nictitans,* and *C. pogonias.*

Habitat and Microhabitat Used

More quantitative information is available documenting how association in mixed-species groups influences the type of forest habitat that is used. In two African forest communities, one guenon species increases its use of less dense forests when it associates with its congeners. The congeners may adjust their ranging to include a greater proportion of denser forests when in mixed-species groups, so that the species in a mixed group appear to compromise on the degree to which different habitat types are visited. In other cases, however, the adjustment seems to be unilateral.

Changes in vertical location in the canopy as a function of mixed-species association have been investigated in African guenons. In one Kenyan forest, the use of different canopy levels was unaffected by association with other species, while in Gabon, *Cercopithecus cephus* in a monospecific group occupied heights that were on average 5 m lower than *C. cephus* in a mixed-species group. Unfortunately, this latter comparison concerns two groups of *C. cephus* that differed in the amount of time spent associated with other guenons, and so it may be confounded by differences in forest structure in their respective home ranges: the *C. cephus* group that rarely associated had a proportionately and absolutely lower amount of high, open forest available in their range, which bordered on a river.

Scheduling of Movement

Changes in spatial aspects of group movement, like those discussed above, are not the only kinds of changes that association with other species can bring about. The scheduling of movement and rest periods can also be influenced by heterospecific associates. While many studies of mixed-species association imply that some kind of temporal coordination of movement occurs, relatively few discuss it explicitly, and almost none have analyzed it quantitatively. In tamarins, coordination of activity seems to be especially close during periods of directed travel, which makes sense if this is when it is easiest to lose heterospecific associates. Guenons have been reported to wait for their associates before beginning to move. Changes in the scheduling of activities may also result more indirectly; for example, the different alarm responses that *Cercopithecus cephus* shows when alone or with its congeners in a mixed group influence how rapidly it begins to move again after an alarming stimulus (Gautier-Hion and Gautier 1974).

Although the data in table 4.2 begin to make a case for the many effects that mixed-species association can have on group movement patterns, there is still much work to be done. Most effects have been described only qualitatively, and even quantitative data are often not analyzed statistically. Furthermore, the data often fall short of showing conclusively that association in mixed-species groups is causing adjustments in movement patterns. This is either because comparisons involve different groups, with different association tendencies but also different habitats, or because associations are so permanent that one cannot observe any of the constituent species on their own, and so be sure that a change has occurred when they join other species.

Are the Changes in Group Movement Related to Mixed-Species Association Costly or Beneficial?

As noted previously, the major focus of studies of mixed-species associations in primates has generally been to discern the costs and benefits of participating in them for each of the species involved, and so to understand them from a functional or evolutionary perspective. Data relating to group movement have been important in evaluating alternative hypotheses. The changes in group movement that result from associating with other species sometimes appear on the credit and sometimes on the debit side of the ledger.

On the credit side, changes in the amount of area that is covered, or the type of habitat included, or both, figure most prominently. An increase in area covered means that the animals are using a larger supplying area, and so should have access to more food sources, or a greater variety of them. Combining information on the larger area covered per day by *Cercopithecus cephus* in mixed groups and spatial distribution of food plants, Gautier-Hion, Quris, and Gautier (1983) calculated minimum increases of 25–37% in the number of food plant species the monkeys were likely to encounter. As these authors pointed out, the actual increase is likely to be even larger because the animals do not move randomly, but can exploit one another's ability to find foods in the larger area being searched. Increases in the types of habitat used may also increase the diversity of food plants visited, or make a greater number of feeding sites

Marina Cords

accessible. For example, among guenons, it seems that species that typically use dense vegetation can increase their use of more open sites or areas when they are associated with other species, and their increased daringness has been attributed to increased levels of vigilance for predators in mixed-species groups (Gautier-Hion, Quris, and Gautier 1983; Cords 1987; Gautier-Hion 1988).

A larger home range can also lead to more varied social opportunities, as in the case of red colobus juveniles who travel with groups of vervets or patas monkeys. Starin (1993) has proposed that the red colobus, while still under the protective cover of the guenon groups, get a chance to evaluate different groups of their own species into which they might later transfer.

Finally, changes in the intensity with which an area is used have been seen as benefits to animals in mixed-species groups because these changes are presumably associated with foraging efficiency. On the one hand, increases in intensity of use may be associated with the advantages of being guided by heterospecifics who know a smaller area better, and so travel efficiently between closely spaced feeding areas. On the other hand, decreases in intensity of use may be associated with the advantages mentioned above, namely, greater diversification of the types of habitat used and larger supplying areas.

Changes in movement patterns that have been scored as costs are of two general kinds. First, increases in rate and distance of movement in mixed-species groups suggest that there may be travel costs (i.e., energetic costs) associated with the increased membership in these groups relative to smaller single-species groups. These costs need not accrue to members of all species that associate, however, and groups of some species even slow down when in association with others. Energetic costs of covering greater distances may also be recouped if an increased foraging area increases access to foraging sites, as suggested above.

Less easily documented are the opportunity costs associated with adjusting activity schedules and movement patterns to fit a partner group. For example, if one group has to wait for another before moving off from a sleeping or resting area, the amount of time it has for feeding, or the distance it can ultimately travel, may be reduced. Also, it is possible, at least in principle, that association constrains the actual area used, so that animals forgo visiting certain areas to maintain their association. Relevant evidence is utterly lacking, however, although it seems that there are limits to the degree to which animals will modify their movement patterns to remain with a partner group. For example, when a mixed-species group comes to the range or territory boundary of only one of its member groups, that group may stop and turn around. Some researchers have noted that the associating groups that are not at their intraspecific range boundaries may keep on moving in these situations, instead of staying with their partners and backtracking (*Cercopithecus ascanius:* Cords 1987; *Cercopithecus diana:* Wachter, Schabel, and Noë 1997). This strategy would be predicted by the hypothesis that association with other species prevents duplication of ranging effort.

In general, data on how mixed-species association affects group movement have been central to documenting the costs and benefits of these associations. Many of the observed changes in movement patterns are associated with presumed changes in foraging efficiency. Yet detailed studies of foraging efficiency, and how it is affected by changes in movement patterns that are related to mixed-species grouping, have not been carried out. Chapman and Chapman (1996) present some data on feeding rate as a function of association status for five monkey species in the Kibale Forest in Uganda. Because of limited sample sizes, they were not able to control for the type of food consumed in their comparisons of a given species that was or was not in association with another species, and a close inspection of their results suggests that there must have been considerable variation in the rates measured in any particular circumstance. They report a complex pattern, in which feeding rates are sometimes higher, sometimes lower, and sometimes unchanged in single-species versus mixed-species groups, depending on which particular combination of species is considered. Most mixed-species associations in this forest seem to occur randomly, however (Struhsaker 1981; Waser 1982b, 1984a), so increased feeding rates in association might not be expected. The one species pair that may associate more often than expected by chance includes the two guenon species *Cercopithecus ascanius* and *C. mitis,* although even in this case it is not clear whether the monkeys are attracted to each other per se, or just to common feeding sites which each species finds independently (Struhsaker 1981; Cords 1990a). *C. ascanius* individuals showed a 64% increase in mean feeding rate when associated with *C. mitis; C. mitis* individuals showed a 73% increase when associated with *C. ascanius.* Only the former difference was

Marina Cords

statistically significant. While the direction of these changes in feeding rates is as expected for two species that seem to associate more often than expected by chance, data are not available for this community that would allow one to attribute gains in feeding efficiency to changes in group movement patterns. On the other hand, Chapman and Chapman do present data that refute an alternative explanation for increased feeding efficiency, namely, that shared vigilance in a larger group allows each individual to spend more time, or more uninterrupted time, feeding, and so increases feeding efficiency. Both species showed increased rather than decreased vigilance rates when in association with each other (with the *ascanius* again showing significant changes).

Clearly, more exhaustive study is required to relate changes in group movement to changes in foraging efficiency. Not only must data be collected simultaneously on foraging and movement, but various scales of efficiency also need to be considered, and perhaps applied to various components of the diet. It may not be mouthfuls per minute that matter, but rather the variety of plant species that are visited over longer periods. It is also possible that changes in movement patterns are related more significantly to the efficiency with which invertebrate prey are captured than to plant feeding (Cords 1987). Such study, if carried out in the future, will serve to confirm, or perhaps refute, the preliminary conclusions based on observations of changes in group movement patterns as a function of mixed-species association.

How Is Group Movement Coordinated in Mixed-Species Groups?
Just like the members of a single-species group (Boinski, chap. 15, this volume), members of mixed-species groups must coordinate a common travel route to maintain the cohesion that is the reason for their existence. Studies of travel coordination in single-species groups focus on the roles of individual animals in initiating and directing group movement. Those who have considered mixed-species groups typically focus on the roles of one or both of the associating *groups,* as if the group were the entity that forms, directs, maintains, and ends a period of coordinated travel. In view of this focus in published reports, I will proceed with a discussion mainly at this group level, but we should realize that the group consists of individuals, whose priorities may differ with respect to where they want to be, and how much they want to be near mem-

bers of a heterospecific group. The influence of one group on another is really the result of the negotiations and decisions of individuals within each of the groups.

Another point to keep in mind is that the task of coordinating the movement of heterospecific groups may not fall equally to the species concerned, especially if they do not experience equal net benefits from associating with others. Although many well-studied nonrandom associations have documented benefits for all participating species (see table 4.1), there are cases in which it is not clear that all species benefit equally, or at all. For example, Terborgh (1983) suggested that the net effect of associating with *Saimiri sciureus* may be slightly negative for *Cebus apella* in Peru. When the benefits of associations are one-sided, it is unlikely that coordinating behavior will be performed equally by all parties.

In yet other cases, in which mixed-species groups include at least three species, coordination of movement between any two of them may be accomplished indirectly, with the third species acting as a mutual point of attraction. For example, Höner, Leumann, and Noë (1997) attribute associations between *Procolobus badius* and several other monkeys in an African forest to the mutual attraction that all these species have to highly vigilant groups of *Cercopithecus diana*. Coordination of movement between two non-diana groups may occur essentially as a side effect of the fact that each of these groups coordinates its movement with the dianas. When mixed-species associations include more than two species, such indirect coordination becomes possible.

Coordinating the Formation of Mixed-Species Groups
The task of coordinating movement in a mixed-species group begins with getting the two partner groups together. It seems that in many species that form mixed-species associations, partner groups often sleep separately to some degree (Gautier-Hion and Gautier 1974; Pook and Pook 1982; Gautier and Gautier-Hion 1983; Terborgh 1983; Yoneda 1984; Cords 1987; Heymann 1990a; Norconk 1990b; Peres 1992b). Even those species that spend essentially all of their time together during the day may be tens or even hundreds of meters apart overnight, and must come together to intermingle before setting off in the morning.

Loud vocalizations seem to play an important role in orchestrating the formation of mixed-species groups, at least in certain taxa. Among tamarins, for example, loud calling typically precedes di-

Marina Cords

rected movements away from the sleeping area (Pook and Pook 1982; Terborgh 1983; Yoneda 1984; Heymann 1990a; Norconk 1990b; Peres 1992b). Although Norconk reported that one species *(S. mystax)* called first on all but one occasion, other authors have reported that loud calling can be initiated approximately equally often by members of either species that forms associations (Terborgh 1983; Heymann 1990a; Peres 1992b). According to Peres (1992b), adults do the calling. Often, calls from one species are answered by calls from the other (Pook and Pook 1982; Terborgh 1983; Heymann 1990a; Peres 1992b), though not necessarily immediately (Norconk 1990b). Calls of different species are easily distinguished by ear and carry over long distances.

The function of these vocal exchanges in facilitating the formation of mixed-species groups is suggested mainly by the facts that they are often antiphonal (but see Norconk 1990b) and that they are closely associated in time with the formation of mixed-species groups. While these patterns are consistent with the hypothesis that loud calls serve to communicate between species and thus coordinate mixed-species group formation, it remains possible that these calls are intended only for a conspecific audience. In fact, these calls (or at least calls that sound identical to humans) are known to be used for within-species communication to neighboring groups and in situations in which one or more individuals became separated from the rest of their own group (Pook and Pook 1982; Terborgh 1983; Heymann 1990a). In such contexts, these loud calls have been described as "rallying" calls, and it is easy to imagine the necessity of rallying the group to initiate travel in the morning, whether a heterospecific group is involved or not. While members of partner groups could gather information from these rallying calls, even if they were not the intended recipients, there may be other acoustic or visual cues that they use to find each other. Early morning playbacks of partner group calls, in the absence of partner groups, might provide one means of assessing whether heterospecific groups actually do respond to the calls they hear, as opposed to any alternative cues.

Heymann (1990a) is the only researcher who distinguished events that occurred when early morning loud calls were or were not given. He found that it took longer for partner groups of tamarins to coalesce on mornings when calls were given. While at first glance these results seem to refute the idea that loud calls facilitate the formation of mixed-species groups, they actually end up supporting the idea of an interspecific communicative function when one notes that calls are more likely to be given when the groups have slept at greater distances, just when the monkeys most need additional cues to be able to find their partners. Because they have slept farther apart, it also takes them longer to get together in the morning. Terborgh's (1983) description of coalescence between *S. imperator* and *S. fuscicollis* on one particular morning is also quite suggestive of an interspecific communicative function. A *fuscicollis* group, whose home range overlapped that of two *imperator* groups, arrived in the home range of one of them, at which point the *fuscicollis* called repeatedly until they got a response from the resident *imperator;* shortly thereafter, the two groups joined. Additional observations of this sort, relating the occurrence or intensity of early morning calling to spatial relations between groups, would be very useful in clarifying the role of vocalizations in mixed-species group formation.

The coordinating role of loud calls has also been described extensively among West African guenons by Gautier and Gautier-Hion (1983; see also Gautier-Hion and Gautier 1974). Like some of the tamarins just described, *Cercopithecus pogonias* and *C. nictitans* form essentially permanent and stable associations, but sleep at least slightly segregated, and so must reunite in the morning. Loud calls occur early in the morning, usually within about half an hour of waking, and before the monkeys begin any directed movement as a mixed party (although they may start feeding). In guenons, only adult males make loud calls, and countercalling between the males of different species is common. The fact that the calls of individual males follow one another so closely in time, even though those males may not be very close together, suggests that the males are calling in response to one another, and not to some other stimulus that all of them perceive more or less simultaneously. Relative to the tamarins, the guenons seem to show a clearer asymmetry in terms of initiation and participation in early morning calling bouts, with *C. pogonias* calling first, and calling in more bouts, than *C. nictitans.* Similar asymmetries have been reported as characterizing other pairs of west African guenons that form mixed-species groups (Gautier-Hion 1988). In guenons, these bouts of loud calling by males occur not only in the early morning, but also at other times when there are changes in major activities and the monkeys have to reestablish coordinated movement. Such contexts include the resumption of activity after midday resting periods, after alarm epi-

Marina Cords

sodes, and after territorial battles, which typically involve just one of the associating species fighting with a neighboring group of conspecifics, while partner groups seem to wait in the general vicinity (Gautier-Hion 1988; M. Cords, unpub.).

Some guenon species do not form mixed-species associations that endure as long as those of *C. pogonias* and *C. nictitans. C. cephus* associates with these latter species for variable amounts of time, and associations do not necessarily begin first thing in the morning. Nevertheless, when *cephus* joins a mixed *pogonias* and *nictitans* group, a bout of male loud calls is likely to occur, and may include males of all three species. As Gautier and Gautier-Hion (1983) describe, *pogonias* and *nictitans* males typically call when they come to a part of their range in which *cephus* occur, and *cephus* groups approach from distances as far off as 500 m. Presumably such distances in dense habitat would preclude detection of the *cephus* before the other males call, and one is left with the impression that the *pogonias* and *nictitans* males are advertising their presence, like Terborgh's *S. fuscicollis.* On other occasions, however, something different may be going on. Gautier-Hion and Gautier (1974) suggest that sometimes male loud calls punctuate, but do not precede, the coalescence of a *cephus* group with a *pogonias-nictitans* group. They note a general increase in activity as well as "contact" calls as *cephus* individuals begin to intermingle with *pogonias-nictitans* individuals. The loud calls by males may occur in response to the perturbation indicated by these quieter "contact calls" given by adult females and juveniles, and as such be intended primarily for a conspecific audience. Still, since males of all three species are likely to participate with repeated calls, information is available to members of each species as to the source of perturbation, namely, the proximity of heterospecific individuals. During the adult males' calling bout and afterward, other group members continue giving contact calls, and members of all three species become fully intermingled without any aggression.

While the guenon groups studied by Gautier and Gautier-Hion seem to use male loud calls extensively to coordinate the formation of polyspecific groups, reports on other guenons either fail to mention (Struhsaker 1981) or failed to find (Cords 1987; Oates and Whitesides 1990) a coordinating role for such vocalizations. Similarly, Terborgh (1983) notes that *Cebus apella* lacks any pronounced vocal response to the comings (and goings) of *Saimiri sciureus* in Peru. Whether these monkeys use alternative communicative signals to mediate the coalescing of their groups is not known, but alternatives are certainly not obvious.

It is tempting to relate the development of vocal exchanges during mixed-species group formation to the permanence and stability of the resultant mixed-species groups. All the cases in which vocalizations seem to be less important in orchestrating mixed group formation involve impermanent (though nonrandom) associations, and typically a given group has more than one heterospecific partner with whom associations occur. In many of the tamarins and West African guenons, whose vocalizations conspicuously precede or accompany mixed-species group formation, associations are nearly or essentially permanent, or involve one particular heterospecific group, or both. In light of this distinction, it is interesting to note that *Cercopithecus cephus* in Gabon formed far less permanent associations than its partner species did, and also participated less often in bouts of loud calls.

Not all data may fit the pattern, however. For example, although Pook and Pook's (1982) description stresses the importance of vocalizations in coordinating mixed-species group formation among *Saguinus* species and *Callimico* in Bolivia, the associations of these monkeys at this site are apparently not as long-lasting as those among callitrichid species in some other communities. More conclusive evidence awaits more extensive and more quantitative information on both vocalization behavior and the permanence and stability of mixed-species groups. If such further study confirms a relationship between calling and the permanence and stability of mixed-species groups, we will have to explain it. At present, however, there is no indication that permanent associations, or those that occur regularly with one particular heterospecific group, are any less cohesive or coordinated than others, or any more valuable, so the more overt communication that accompanies mixed-species group formation remains puzzling.

Keeping Mixed-Species Groups Together
Once mixed-species groups have formed, the behavior of their constituents needs to remain coordinated if they are to persist. Just like members of single-species groups (Rowell and Olson 1983), individuals in mixed-species groups can use both overt forms of communication and monitor-and-adjust regulation of their own activities to ensure that they stay together with their groupmates. Adjustments in the scheduling of activities, especially rapid movement, have al-

ready been mentioned (see also table 4.2). Vocalizations are an example of overt communication, and several researchers allude to the long-distance calls, or bouts of countercalling, that occur periodically as heterospecific groups move around together (Pook and Pook 1982; Gautier and Gautier-Hion 1983; Terborgh 1983; Cords 1987; Gautier-Hion 1988; Norconk 1990b; Windfelder 1997). Evidence that these calls actually function to maintain the group's integrity are essentially nonexistent for most mixed-species groups. A recent study that focused on tamarin long calls (Windfelder 1997) provides a partial exception. *Saguinus fuscicollis* and *S. imperator* in Peru form associations in which each group associates with only one group of the other species. When playbacks were made to unassociated tamarin groups, they responded as strongly to the calls of their heterospecific associates as to the calls of their own groupmates, but responded far less to the calls of howler monkeys, with which associations are very rare. Measured responses included both immediate approaches to the playback source and vocalizations. These results are consistent with the hypothesis that long calls can be used both to locate heterospecific associates (which will countercall in response) and to bring them nearer. Careful analyses of the contexts in which such calls are produced would provide additional critical evidence.

There are differences between studies in the degree to which vocalizations or adjustments in activities are made equally by all partner species in a mixed-species group. Among tamarins, it seems that vocalizations generally come from both partner groups, and in the one case in which changes in movement patterns could be evaluated, both groups made adjustments and traveled at an intermediate rate when together (Terborgh 1983). Podolsky (1990) also reports reciprocal adjustments in travel rates made by *Cebus apella* and *Saimiri sciureus* in their mixed groups. With guenons, greater asymmetries have been reported, both in terms of frequency of vocalizations and in terms of adjustment of movement patterns. One might expect a greater investment in keeping a mixed-species group together by members of that species which has most to gain from the association. At present, it is hard to evaluate this idea carefully because in many cases, all members of the association have something to gain, but their gains may be in different currencies (see table 4.1), and so are hard to compare quantitatively. Turning this proposition on its head, however, several researchers have used asymmetries in the behavior involved in maintaining mixed-species associations as guides to the relative magnitude of the net benefits that different species achieve as a result of associating in mixed-species groups (Cords 1987; Whitesides 1989; Oates and Whitesides 1990; Holenweg, Noë, and Schabel 1996; Windfelder 1997).

Determining the Speed and Direction of Travel

It would be natural to assume that whichever animals are at the leading edge of a moving group, including a mixed-species group, are the ones that set the speed and direction of travel. This assumption has been made by several researchers, allowing them to identify one species in a mixed group that typically takes on the primary role with regard to travel decisions (e.g., Gautier-Hion and Gautier 1974; Oates and Whitesides 1990; Podolsky 1990; Peres 1992b, 1996a). In some cases, there is additional evidence that a leading species is controlling travel, as, for example, when it uses loud vocalizations earlier and more often than the associated species that tend to follow. It was based on both vocalization and spatial positioning data, for example, that Gautier and Gautier-Hion (1983) argued for the predominant role of *Cercopithecus pogonias* in organizing mixed-species group travel with *C. nictitans* and *C. cephus*. Sometimes careful consideration of the context in which a particular partner group leads seems necessary. For example, Terborgh (1983) found that *Saimiri sciureus* were usually out in front in mixed parties with *Cebus apella,* but when the monkeys were moving rapidly to a fruiting tree or a new foraging area, *Cebus* were usually out in front, and so seemed to be determining the direction and pace of the major steps in the daily trajectory.

It is also possible, however, for non-leading animals to influence and even determine travel pace and direction. Groups that slow down when associating with other species (see table 4.2) seem to be responding to the slower pace set by their straggling associates. Non-leading groups can influence the direction of travel by refusing to follow the leading group. For example, Pook and Pook (1982) described how the trailing *Saguinus fuscicollis* occasionally veered away from the path taken by the leading *S. labiatus,* who would circle around to rejoin their partners, perhaps taking the lead again, but moving in a different direction than the one they initially chose. In one such case, a returning *S. labiatus* lunged at a *S. fuscicollis,* but it is of course impossible to be confident that this conflict was related to changes in the trajectory.

Detecting such a voice from behind is certainly more difficult than detecting which animals are at the leading edge of the group, but some of the most interesting communication between species probably occurs through these voices. It would probably help us to find out more about such strategies if researchers looked for them explicitly. They are probably least likely to occur in those species that use their allospecific partner groups as guides to profitable feeding sites, or perhaps as beaters for mobile prey, since in those cases the followers benefit directly from their second position, and refusing to follow might be costly. They are probably most likely to occur when the benefits of participating in mixed-species groups concern reduction of predation risk or joint defense of a common territorial boundary, so that the partner groups gain in similar, symmetrical ways from their association.

What Happens when Mixed-Species Groups Disband?
There is some evidence that the disbanding of mixed-species groups into single-species groups, whether for the night, for a midday resting period, or at other times, is accompanied by loud vocalizations, which may rally conspecifics, indicate the end of association to heterospecifics, or both. Gautier-Hion and Gautier (1974) described how loud calls precede a midday resting period, in which guenon species separate partially into monospecific units. At night, there is greater separation between groups when males have engaged in loud calls prior to settling down. Terborgh (1983) described how *Saimiri sciureus* sometimes produce an unusual number of loud calls when *Cebus apella* become stationary and the *Saimiri* move on, and away, without them.

These observations suggest that a change to single-species travel requires communication. That one function of that communication is intraspecific, presumably coordination of the conspecific members of the caller's group, is suggested by two other observations relating calling frequency to mixed-species association. Boinski and Mitchell (1992) compared the vocal repertoires of two *Saimiri* species, and found one of them, *S. sciureus,* to lack intraspecific orientation calls found in the other. They proposed that *S. sciureus,* which spends much of its time in association with *C. apella* in the population where it was studied, may depend on its partners for direction so completely that it totally lacks this class of vocalizations. Study of additional populations would be needed to confirm this hypothesis. It is quite similar, however, to propositions made by Gautier-Hion and Gautier (1974) regarding guenons in Gabon.

In calling bouts of adult males, the males of the three species they studied are not equally represented. They described the males who call least often as acoustically parasitizing the males who call more, and whose calls serve to organize the travel of groups of different species, as well as the spacing between groups of different species. Interestingly, the species identity of the male who calls most often may not be the same in neighboring mixed-species groups (Gautier-Hion 1988).

Conclusions and Future Directions

The behavior of your average primate is enveloped in several layers of social influence. It has especially strong bonds with its immediate family. It usually also belongs to a group or network. Its group may interact with neighboring groups in the population. Its group may also be part of a larger group that includes individuals of other species. All of these social units can affect group movement, and a thorough understanding of group movement cannot be achieved without considering all of these influences and the interactions among them. Clearly we are not there yet, but we are moving in the right direction.

This chapter has considered only one of these layers of influence, namely, the heterospecific partner groups with which some primates form long-lasting associations. Although much of the evidence is as yet anecdotal, it seems clear that associating in mixed-species groups does influence various parameters of group movement for the species involved, and that the changes it brings about are sometimes costly and sometimes advantageous. The mechanisms by which these changes are brought about are less clear. In some cases overt forms of communication, especially vocalizations, seem key, but conclusive evidence that such communication is designed to be received by heterospecifics, or even *that* it is received by them, is scanty. Furthermore, there are species pairs that do not communicate in any obvious way when beginning or maintaining an association. Monitor-and-adjust strategies of group coordination may also be occurring, but they are much harder to study.

Perhaps more than it has answered the three questions posed at the outset, this review has made clear that there is much left to find out. We might begin by recognizing the need for studies that focus systematically on these questions; to date, available answers have been abstracted from studies with different aims. Studies that focus on the function of associations do not necessarily address the subject of group movement, and very rarely address the behavioral mechanisms that coordinate the movement of heterospecific groups. Careful documentation of the timing and direction of group movements is the first step, but to evaluate both the significance of changes that result from mixed-species association and the behavioral mechanisms that ensure coordination between heterospecific groups, data on group movement must be combined with equally detailed records of other types of behavior. These include vocalizations, whose production schedule has been successfully related to changes in group movement in single-species groups (see Boinski, chap. 15, this volume). When two groups are involved, the job will require more manpower, as both groups should be monitored simultaneously. The behavior of individuals within mixed-species groups must also be examined, with an eye open to the possibility that refusing to follow could be a way of leading from behind. Careful documentation of feeding and alarm behavior and predator encounter rates is needed to demonstrate thoroughly some of the presumed costs and benefits of changes in group movement that relate to feeding efficiency, diet, and reduction of risk of predation.

There is also a general need for more information from more populations and species, as a comparative perspective can both test and sharpen existing hypotheses. Across the primate order, there is variation in the nature of mixed-species associations along three important dimensions that are likely to be related to group movement. One concerns the amount of time spent associated, which can vary substantially, even among animals who associate more often than expected by chance. The second concerns the degree to which *all* members of a mixed-species association actively seek out such association, which is probably related to the degree to which all members benefit from such association. The third concerns the degree to which associating heterospecific groups have one-to-one relationships, so that each group of species A associates with only one group of species B. These parameters may influence the kinds of adjustments in group movement that are made as a result of association, whether (and for whom) these adjustments are primarily costly or beneficial, the elaboration of signaling to ensure coordination, and the bi- or unilateral nature of such coordination.

Perhaps the biggest gap in our understanding of how mixed-species association relates to group movement comes from the fact that nearly all studies to date have been observational, so that manipulation of the variable of interest—namely, whether a group is associated with heterospecifics—is impossible. As noted previously, some associations are so permanent that it is impossible to study the constituent groups when they are not associated with heterospecifics. But even in those communities in which association partners can be studied both in and out of association, an investigator relies on natural variation in association status, which is probably not random, and so a comparison of the animals' behavior when associated and apart may be confounded by other factors that are poorly understood. Experimental manipulations would be a means to ameliorate this situation, and a few pioneering studies suggest they should be feasible.

Hardie and Buchanan-Smith (1997; Buchanan-Smith and Hardie 1997) have pioneered studies of mixed-species tamarin groups in captivity. Studying vigilance levels in single- and mixed-species *Saguinus* groups, they documented that individual vigilance levels are reduced in mixed-species groups, even though there is an increase in the amount of time when at least one group member is vigilant. They also documented differences in the kinds of vigilance (looking up vs. scanning in other directions) that members of different species undertake, and how association influences the performance of these different kinds of vigilance. While these investigators worked within the confines of small zoo cages, the extension of this technique to larger enclosures could allow a study of group movement with controlled variation in important parameters, such as the distribution of resources and shelter or the number and identity of the animals, as well as the presence of heterospecifics.

Field experiments in which one partner group was removed temporarily would be a more daring solution, but probably more revealing. Although such selective removals have not been undertaken in the context of studying mixed-species groups, cropping of wild tamarin populations has been undertaken on a larger scale (to stock captive breeding programs), and resultant changes in the ecology and demography of the remaining animals have been studied (Glander, Tapia, and Augusto 1984; Ramirez 1984). With small groups that are relatively easily trapped, tamarins would seem to be prime candidates for removal and reintroduction experiments.

Finally, playback experiments may prove useful in studying both the function of vocalizations in coordinating movement of heterospecific groups and the way in which heterospecific groups influence one another's movement patterns. The attractive function of loud calls can be tested even when heterospecific groups are not nearby, as Windfelder (1997) showed with *Saguinus fuscicollis* and *S. imperator* in Peru. Playbacks that mimic potential coordinating vocalizations of one or both species could be played from particular locations in the mixed-species group to see whether they predict changes in travel speed or direction. Even if there are no vocal signals designed to communicate with heterospecifics, vocalizations could be used to mimic the presence of heterospecific groups, possibly even moving heterospecific groups. Playbacks made in various locations and contexts could be used to delimit constraints on the degree to which association with other species influences group movement patterns.

Acknowledgments

I thank Sue Boinski and Paul Garber for inviting me to write this chapter. I am grateful to them and to Colin Chapman for comments on earlier versions of the manuscript. For help in accessing and clarifying parts of the literature on tamarins, I am especially grateful to Paul Garber, John Terborgh, Eckhard Heymann, and Hannah Buchanan-Smith. My field research was carried out with permission from the Government of Kenya, and sponsorship of the Zoology Department, University of Nairobi.

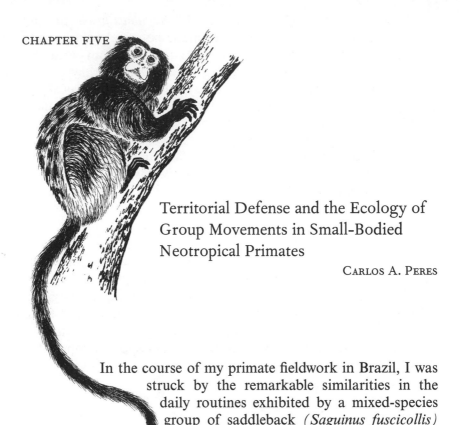

CHAPTER FIVE

Territorial Defense and the Ecology of
Group Movements in Small-Bodied
Neotropical Primates

CARLOS A. PERES

In the course of my primate fieldwork in Brazil, I was
struck by the remarkable similarities in the
daily routines exhibited by a mixed-species
group of saddleback *(Saguinus fuscicollis)*
and moustached tamarins *(Saguinus mys-*
tax) at a central Amazonian site and a group of golden lion tamarins *(Leontopithecus rosalia)* at a coastal Atlantic forest over
3,200 km away. Within minutes of emerging from their nocturnal
shelters, groups of both of these primate genera frequently emitted
their loud long calls toward neighboring groups, while remaining
motionless in the upper levels of the forest. There was a certain
tension in the air during those early morning periods as the animals
apparently listened carefully to any countercalls coming from outside the group. This behavior was all the more remarkable because
the immediate urge to find food—as would be expected for these
small-bodied mammals (0.4–0.6 kg) after their unusually long overnight fasting period of over 14 hours—was not apparent. On most
mornings, the deceptive composure of the animals was suddenly broken as each group rapidly traveled in a straight trajectory toward a
specific portion of their home range boundary, where frantic and
prolonged intergroup encounters ensued.

This patent intolerance of and spatial scrutiny for neighboring troops is typical of small-bodied primates living in small groups, which allocate a substantial portion of their time and energy to intergroup resource defense. In the Neotropics these species are rather uniquely represented by the callitrichines—including the pygmy marmosets (*Cebuella:* Soini 1982), marmosets (*Callithrix:* Hubrecht 1985), tamarins (*Saguinus:* Garber 1988a; Garber et al. 1993; Peres 1992a), and lion tamarins (*Leontopithecus:* Peres 1989; 1991b)—the owl monkeys (*Aotus:* Wright 1978),[1] and the titi monkeys (*Callicebus:* Robinson 1979; Robinson, Wright, and Kinzey 1986). The poorly studied and elusive Goeldi's monkey (*Callimico goeldii*) may be an exception to this generalization. The low rates of intergroup interaction in this species probably result from low densities of breeding groups, which may be separated by distances of up to several kilometers (Pook and Pook 1981; Christen and Geissmann 1994; Peres 1997).

Although there is a possibility that intergroup interactions in some species may also serve as a mechanism of mate defense (van Schaik and Dunbar 1990; Cowlishaw 1992), territoriality in small-bodied platyrrhines living in small groups is best defined as a form of site-dependent priority of access to resources reinforced by considerable energetic investments and frequent boundary contests (see Waser and Wiley 1979; Kaufmann 1983). However, exclusive use of space is not a necessary consequence of territoriality, despite the substantial foraging benefits derived from exclusive access to resources (Peres 1989, 1992a,b). In fact, the extent to which home ranges overlap is highly variable, from virtually none in pygmy marmosets (Soini 1982), common marmosets (Hubrecht 1985; Alonso and Langguth 1989), and titi monkeys (Mason 1968; Robinson 1979; Kinzey and Becker 1983) to 76–88% in buffy-headed marmosets (Ferrari 1988) and tamarins of the genus *Saguinus* (Izawa and Yoneda 1981; Peres 1991a; de la Torre, Campos, and de Vries 1995).

Forest primates living in large groups, on the other hand, are often unable to exhibit territoriality because of their large spatial requirements (Clutton-Brock and Harvey 1977a; Mitani and Rodman 1979; Mace, Harvey, and Clutton-Brock 1983; Lowen and Dunbar 1994), particularly when the effects of diet on group mobility are considered (Grant, Chapman, and Richardson 1992). The costs of territorial defense of increasingly larger home ranges should therefore become prohibitive beyond a certain point (Brown 1964; Brown and Orians 1970; Schoener 1983; Davies and Houston

1984). Costs of defense should increase with territory size if increasing size forces residents to spend more time and energy patrolling boundaries and to travel farther each day to expel intruders (Schoener 1987). In large territories, the costs of regular intergroup encounters at range boundaries may thus become prohibitive if the distances that must be covered on a daily basis are far longer than necessary to satisfy immediate foraging requirements, especially if few long-term benefits can be gained from exclusive access to resources.

Group movement and use of space in territorial primates are thus expected to be strongly influenced by mechanisms of resource competition (*sensu* Park 1954), including direct depletion of resources shared by other groups (a form of exploitative or scramble competition) and agonistic interactions between groups at range boundaries (a form of interference or contest competition). While displays of interference competition in territorial species can hardly go unnoticed by even the beginning field primatologist (see Cheney 1986 for a review), the occurrence of exploitative competition can be easily overlooked, and has received very little attention in the primate literature (but see Janson and van Schaik 1988; Peres 1989; Lucas and Waser 1989; Isbell 1991). In small-bodied platyrrhines living in small groups, interactions between neighboring groups are initially mediated by different forms of long-range signals that tend to result in intergroup approaches prior to the escalation of frequently reinforced and somewhat ritualized boundary contests (Robinson 1981b; Robinson, Wright, and Kinzey 1986; Peres 1989, 1992a; Boinski, chap. 15 this volume). The space and resource requirements of these species, however, can be drastically different, ranging over three orders of magnitude from 0.1 ha in pygmy marmosets (*Cebuella pygmaea:* Soini 1982) to well over 150 ha in several populations of tamarins and lion tamarins (Peres 1991a; Valadares-Padua 1993; Passos 1997). This variation should have significant consequences for the costs and benefits of resource defense, as illustrated by the wide variation in home range overlap in different populations across different species and genera.

Here I examine how the ecology of group movements in territorial platyrrhines might be constrained by the everyday demands imposed by intergroup defense of home range boundaries. A more general review of the ecology of these species is beyond the scope of this chapter, and can be found elsewhere (callitrichines: Sussman and Kinzey 1984; Stevenson and Rylands 1988; Snowdon and Soini 1988; Ferrari and Lopes-Ferrari 1989; Garber 1993a; Rylands 1993;

Callicebus and *Aotus:* Robinson, Wright, and Kinzey 1986; Wright 1994). Instead, I compare the patterns of group displacements and the mechanics of intergroup interactions in two closely related genera of large-bodied callitrichines, *Saguinus* and *Leontopithecus* (Nagamachi et al. 1997), studied in closed-canopy habitats in Brazilian Amazonia (Peres 1991a, 1992a,b) and the Atlantic forest (Peres 1989, 1991b). Daily movements in both of these genera are markedly constrained by systematic forays into specific portions of their range boundaries where intergroup encounters take place. I draw attention to several other similarities in the patterns of group movement and intergroup spacing exhibited by these genera, despite large differences in habitat structure and topography, floristic composition, and seasonality in resource production at each forest site (Peres 1994). Finally, I evaluate the relationship between group mobility, spatial requirements, and the energetic costs of range defense across the territorial platyrrhines by compiling available data on group movement, home range size, and intergroup spacing in all species, populations, and groups studied to date. In particular, I examine how the relationship between group mobility and range size affects home range overlap between neighboring groups, which may be used to gauge the extent to which territoriality can effectively secure exclusive use of resources. Home range overlap in highly territorial primates should be low where resource abundance is high, home ranges are defensible, and the overall costs of defense are low.

Previous comparative studies of the relative costs of resource defense in primates (Mitani and Rodman 1979; R. D. Martin 1981a; Isbell 1991; Grant, Chapman, and Richardson 1992; Lowen and Dunbar 1994) have targeted a larger and far more divergent set of species in terms of phylogeny, body size, ecology, group size, and social organization, all of which can profoundly affect ranging behavior (Milton and May 1976; Clutton-Brock and Harvey 1977a; Wrangham, Gittleman, and Chapman 1993; Chapman, Wrangham, and Chapman 1995; Kinnaird and O'Brien, chap. 12, this volume; Watts, chap. 13, this volume). Restricting this analysis to a phylogenetically and socioecologically cohesive set of species will sidestep the potential confounding effects of these factors.

Comparative Field Studies
The behavioral and spatial data reported here were obtained from comparable studies of a habituated group of five golden lion tamarins and a mixed-species group of five to eight saddleback tamarins

and eight to ten moustached tamarins. The sizes and compositions of these groups, and the forest types they used, were representative of their populations based on observations of at least five neighboring groups followed less systematically during each study. The study groups of *Saguinus* and *Leontopithecus* were followed in 1988–1989 (Peres 1991a, 1992a,b, 1993b, 1996a) and in 1984–1985 (Peres 1986a, 1989, 1991b; Dietz, Peres, and Pinder 1997), respectively, at two Neotropical forest sites that I shall now briefly describe.

Poço d'Antas Reserve. Golden lion tamarins were studied at a 5,900 ha remnant of coastal Atlantic forest in the state of Rio de Janeiro, southeastern Brazil. Following a long history of small-scale deforestation and fires, this reserve consists of a mosaic of selectively logged primary forest and regenerating secondary forest (60%), and nonforest patches (40%) not used by lion tamarins. Forest structure thus results from the interaction of habitat topography (20–180 m above sea level) and human disturbance, which largely defines the boundaries of six habitat types (see Peres 1986a,b; Dietz, Peres, and Pinder 1997). Total annual rainfall at Poço d'Antas averages 1,760 mm per year ($N = 10$ years, 1983–1992), with a strongly demarcated dry season from June to August.

The Urucu Forest. The mixed-species group of saddleback and moustached tamarins was studied at an upland (= *terra firme*) forest located 4 km inland from the upper Urucu River, central Amazonas, Brazil (Peres 1992b; 1993b,c, 1994, 1996a). The almost entirely undisturbed 900 ha study plot (55–70 m above sea level) consisted of tall high-ground forest in moderately undulating terrain (93%), creekside forest abutting a perennial forest stream (6%), palm swamps in poorly drained, low-lying soils (0.5%), and forest edge and second growth adjacent to two 3 ha man-made clearings (0.5%). Access to this remote study site was facilitated by helicopters subcontracted to an oil company. Total annual rainfall at the Urucu Forest averages 2,900 mm per year ($N = 9$ years, 1988–1996), with a fairly demarcated dry season from July to September.

Methods and Definitions
Ranging Ecology. The study groups of *Saguinus* and *Leontopithecus* were located every 15 minutes over 111 and 106 days of observations, respectively ($N = 2,596$ quarter-hour locations for *Saguinus;* 3,655 locations for *Leontopithecus*). This was done to examine

how the energetic costs of range defense might be incorporated into the ranging ecology of these species. The mixed tamarin group was treated as a single spatial unit despite its heterospecific composition because overall group cohesion was extremely high; distances between the centers of the saddleback and the moustached tamarin groups rarely exceeded 25 m, and were usually shorter than monospecific group spreads (Peres 1992b; see also Cords, chap. 4, this volume).

Quarter-hour steps are defined as the straight-line distances moved by each group between any two consecutive 15-minute locations. Day range (or daily path length) is defined as the sum of 15-minute displacements of the group center over a full day of observations. Quarter-hour steps were calculated by a FORTRAN program using the Pythagorean theorem applied to x,y coordinates defining consecutive locations. This method produced surprisingly accurate estimates despite a resolution error of up to 12.5 m determined by quadrats of 25×25 m. For example, computer-generated estimates of step distances for the mixed tamarin group (mean = 53 m, $N = 2,121$ steps) provided an overall accuracy of 96.5% compared with that of more time-consuming manual measurements of group movements per hour of observation obtained using group locations plotted on daily range maps on a scale of 1:5,000 (mean = 55 m, $N = 662$ h).

Group centrality was computed by a FORTRAN program, and is defined as the straight-line distance between any given quarter-hour location and the geometric center of the home range. This analysis was done to examine the study group's position at any one time in relation to its range boundaries. Because home range geometry may affect group proximity to boundaries, the average distance from the home range center to territorial boundaries (c) was evaluated as a function of range eccentricity (E), which ranges from 0 in a perfect circle to 1 in an infinite straight line. Home range shape has essentially no effect on c unless it is highly eccentric ($E > 0.9$: Mitani and Rodman 1979). Home range eccentricity for both study groups was low ($E < 0.5$ in both cases), suggesting that this analysis is reasonably robust with respect to home range shape.

Intergroup Spacing. An exclusive area is defined as the combination of contiguous quadrats used exclusively by each study group, whereas overlapping areas refer to the sum of quadrats shared by neighboring groups. Encounter zones are defined as those areas occupied during direct intergroup encounters, when groups were within sighting distance (< 40 m) of one another (Peres 1991b, 1992a). In both studies these quadrats were largely located in the periphery of each group's home range, were not necessarily centered around key food sources, and were consistently used during encounters with the same neighboring groups throughout each study.

On the basis of audible long calls, I estimated intergroup distances at least once every 15 minutes when groups appeared to move toward or withdraw from one another within up to 90 minutes before and 120 minutes after the beginning of each encounter. I also tested whether the forays of each study group into encounter zones on consecutive days of group follows were independent of one another in order to assess whether movements to different portions of the range boundary were randomly allocated across different encounter zones. This was done to examine whether the directions of movements toward range boundaries were affected by those of previous days. Forays into encounter zones were recorded whether or not encounters actually took place. If independent boundary forays were randomly allocated to different encounter zones, the number of consecutive returns to a given zone would be predicted by its daily probability of single forays to the power two, times the number of independent events observed on consecutive days. Daily probabilities were based on the number of times each zone was visited on different days of observation, and whether or not these visits actually fell within a sequence of consecutive days of observations (i.e., nonconsecutive days of observation were assumed to yield independent events). Differences between observed and expected return times to each encounter zone were then assessed for each study group using a nonparametric test following the binomial distribution (Sign test: Zar 1996). For the mixed tamarin group, I obtained a total of 72 pairwise consecutive day boundary forays into any of six encounter zones over 111 days of observations. For the lion tamarin group, a total of 52 consecutive day forays into any of seven encounter zones were recorded over 106 days of observations. Further details on the mechanics, and costs and benefits, of intergroup interactions in *Saguinus* and *Leontopithecus* can be found elsewhere (Peres 1986a, 1989, 1991a,b; 1992a).

Use of Space and Group Movements
The ranging behavior of *Saguinus* at Urucu and *Leontopithecus* at Poço d'Antas is summarized in table 5.1. The large area require-

Table 5.1 Summary of the ranging ecology and intergroup interactions in a mixed-species group of tamarins *(Saguinus fuscicollis* and *S. mystax)* at the Urucu Forest and in a group of golden lion tamarins *(Leontopithecus rosalia)* at Poço d'Antas

Parameter	*Saguinus* spp.	*Leontopithecus*
Group size	5–8 (*f*) + 8–10 (*m*)	5 (both years)
Total group biomass (g)	5,632	2,779
Total home range size (ha)	149	56
Total home range overlap (%)	76	61
Mean ± SD day range (m)	1,991 ± 431	1,480 ± 322
Min.–max. day range (m)	1,150–2,700	955–2,405
Quarter-hour step length (m)	53 ± 51	45 ± 53
Duration of encounters (min)	74 ± 64	83 ± 76
Frequency of encounters (per day)	0.55 (42/76)	0.57 (48/84)
Time spent in direct encounters (%)	9.0	9.9
Number of neighboring groups	5	6–7
Number of encounter zones	6	7
Number of complete days of observation	76	84
Total number of days of observation	111	106

Sources: Peres 1986a, 1989, 1991a,b, 1992a.

ments of both of these populations are consistent with data on two neighboring mixed-species groups of tamarins (home ranges > 120 ha: Peres 1991a) and five other groups of lion tamarins (36–73 ha: Dietz, Peres, and Pinder 1997). Patterns of home range use and investments in territorial defense were remarkably similar between the two study groups. On complete days of observation, the *Saguinus* group traveled an average of 1,991 ± 431 m, whereas the *Leontopithecus* group covered shorter distances of 1,480 ± 322 m, which is proportional to the smaller home range used by the latter. Quarter-hour step distances averaged 53.0 ± 51.1 m ($N = 2,121$) for the mixed tamarin group and 43.9 ± 53.0 m ($N = 2,553$) for the lion tamarin group. Group movements were faster early and late in the day, but tended to slow down at midday (fig. 5.1). Both groups rarely used the same sleeping site on consecutive nights (Peres 1986a, 1991b), and distances between sleeping sites used on consecutive nights were considerably variable (*Saguinus:* 529 ± 242 m, N = 63 nights; *Leontopithecus:* 199 ± 177 m, $N = 75$ nights). However, both groups tended to retire to sleeping sites close to their home range centers, which is consistent with their shorter distances from the range center early and late in the day and longer distances at midday (fig. 5.1). Daily peaks in group velocity thus coincided with the time when neighboring groups approached or withdrew from the range boundaries where the bulk of intergroup interac-

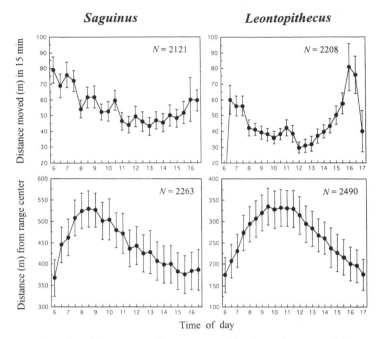

Figure 5.1 Diurnal variation in step distances covered by the study groups of *Saguinus* and *Leontopithecus (above)* and in group centrality *(below)*. The total number of step distances and quarter-hour group locations obtained for each study group is indicated on each plot.

tions took place. In these peripheral parts of their territories, step distances were comparatively short, and the groups usually ground to a halt and rested during the midday hours not far from one of the intergroup encounter zones visited earlier in the day.

Patterns of Intergroup Interactions

Both *Saguinus* and *Leontopithecus* tended to cover a large proportion of the quadrats in their home ranges on any given day, following rapid movements toward their range boundaries early in the morning (Peres 1986a, 1991a). The numbers and locations of encounter zones along range boundaries were highly stable for each study group, and proportional to the number of neighboring groups overlapping each home range (see table 5.1). In both genera, the direction of movement toward range boundaries was usually determined by long-range vocal interactions (i.e., long calls) between neighboring groups. These almost always escalated into direct encounters, which accounted for roughly a tenth of the waking hours of each study group. Encounters usually occurred during the early

Carlos A. Peres

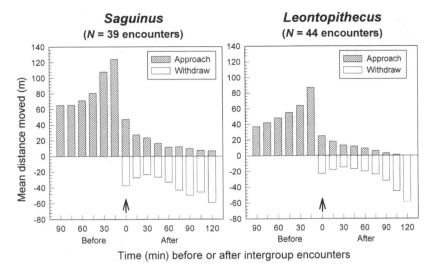

Figure 5.2 Intergroup approaches and withdrawals by a mixed-species group of *Saguinus fuscicollis* and *Saguinus mystax (left)* and a group of *Leontopithecus rosalia (right)* as measured by distances moved during 15-minute intervals up to 1 hour 30 min before and 2 hours after intergroup encounters took place. Arrows indicate the time at which encounters began.

hours of the day, and often waxed and waned until mid-morning, when neighboring groups withdrew from encounter sites and drifted apart.

The relationship between rates of group movement and intergroup spacing up to 90 minutes before and 120 minutes after intergroup encounters followed a similar pattern for both *Saguinus* and *Leontopithecus* (fig. 5.2). The longer step lengths in *Saguinus* merely reflect their higher group velocity, as expected from a three-fold larger home range size. Neighboring groups in both genera countercalled to one another and rapidly approached a common "collision point" in a given encounter zone. Encounters then usually lasted between 1 and 2 hours until the groups eventually withdrew from one another. Vocal activity was particularly intense prior to and during encounters, which were typically characterized by frequent long calls and vigorous staccatos of short whistles interspersed throughout these contests. This was confirmed by experimental playbacks of long calls of moustached tamarins and golden lion tamarins targeted to each study group, which invariably resulted in rapid approaches toward the tape recorder, whether playbacks were conducted in central or peripheral portions of each home range (C. Peres, unpub.).

Movements to Encounter Zones
Independent forays into different portions of the range boundaries did not follow a random pattern in either *Saguinus* or *Leontopithecus.* Single encounter zones were visited either more often or less often than expected by chance (*Saguinus:* $\chi^2 = 18.03$, 6 d.f., $p < .02$; *Leontopithecus:* $\chi^2 = 16.94$, 5 d.f., $p < .01$). However, the observed probabilities of forays into the same encounter zone on consecutive days were consistently lower than those expected by chance for both study groups (table 5.2). In other words, once an encounter zone had been visited on a given day, it was highly unlikely to be visited again on the following day (*Saguinus:* 6 of 6 cases: sign test; $P = .016$; *Leontopithecus* 7 of 7 cases: $P = .008$). This finding suggests that reinforcement of boundary contests was nonrandom with respect to the identity of neighboring groups. Both *Saguinus* and *Leontopithecus* thus appeared to maximize return times to specific sections of home range boundaries to at least a certain extent, although sequential forays into different encounter zones did not match a perfect rotation cycle. The directions of daily movement in both genera thus appear to be affected by those of previous days and by the recent history of boundary interactions with other groups. Maximizing return times would increase the efficiency of range defense by both exploitative or interference mechanisms.

Ranging Ecology of Territorial Platyrrhines
A compilation of data on body mass, group size, group biomass, mean day ranges, and home range size and overlap was carried out for all small-bodied Neotropical primates (< 1.2 kg) living in small groups based on a comprehensive survey of 68 published and several unpublished studies. The final data set incorporated comparable information on 101 study groups of 27 species of pygmy marmosets, marmosets, tamarins in monospecific groups, tamarins in mixed-species groups, lion tamarins, Goeldi's monkeys, owl monkeys, and titi monkeys occurring at 60 Neotropical forest sites (appendix 5.1). In a few cases, data on more than one study group in the same study were assumed to be independent since differences in habitat types used by different groups sharing the same study area were often substantial (e.g., Dawson 1979; Valladares-Padua 1993; Dietz, Peres, and Pinder 1997). Relationships among some of these variables may be based on a reduced number of study groups

Carlos A. Peres

Table 5.2 Observed number of forays into encounter zones, and expected versus observed numbers of consecutive day returns to different encounter zones, by the mixed tamarin (*Saguinus fuscicollis* + *Saguinus mystax*) and lion tamarin (*Leontopithecus rosalia*) study groups

Saguinus fuscicollis + *Saguinus mystax*				*Leontopithecus rosalia*			
		Consecutive day returns				Consecutive day returns	
Encounter zone[a]	Obs. forays	Obs.	Exp.	Encounter zone[a]	Obs. forays	Obs.	Exp.
N	23	1	3.09	Ban	10	0	1.23
NE	26	0	3.95	Cal	20	2	4.92
NW	21	0	2.58	Cap	5	0	0.31
S	8	0	0.37	ES	6	0	0.44
SW	24	3	3.37	Por	11	0	1.49
W	9	0	0.47	RN	8	0	0.71
				VM	5	0	0.31
Total	111	4.00	13.83[b]		65	2.00	9.49[b]

[a]*Saguinus*: N, North; NE, Northeast; NW, Northwest; S, South; SW, Southwest; W, West; *Leontopithecus*: Ban, Banana ouro; Cal, Calcareo; Cap, Capim; ES, Eastern slopes; Por, Portuense; RN, Reflorestamento Norte; VM, Vale Médio.

[b]Sign tests: *Saguinus*: $P = .0156$; *Leontopithecus*: $P = .0078$.

since data on day ranges and range overlap were unavailable for several studies.

For each population for which data on day ranges and home range size were available, I calculated two indices that can successfully discriminate between territorial and nonterritorial primate species (or populations) on the basis of simple ranging data (Mitani and Rodman 1979; Lowen and Dunbar 1994). The Mitani-Rodman defensibility index *(DI)* is defined as $DI = DR/HR\varnothing$, where DR is the mean day range in km, and $HR\varnothing = (4A/\pi)^{0.5}$ and describes the diameter of a circle that is equivalent in area to the size of the home range *(A)* in km^2. A *DI* value of 1.0 or greater thus describes primate groups moving at least one average home range diameter per day.

The Lowen-Dunbar monitoring index *(MI)* for species in cohesive foraging groups (i.e., one independent foraging party) is defined as $MI = s * DR/HR\varnothing^2$, where s is the mean detection distance at which intruders can be detected, which was arbitrarily set at 50 m (Lowen and Dunbar 1994). Despite the lack of any immediate biological significance associated with this index, these authors have noted that an *MI* value of 0.08 best differentiates between territorial ($MI > 0.08$) and nonterritorial species ($MI < 0.08$). The discriminating power of the Lowen-Dunbar *MI* index has been shown to be marginally greater than that of the Mitani-Rodman *DI* index because it takes into account both the length of the boundary to be defended and the distance at which neighboring groups are detected, in addition to linear measures of range size (Lowen and Dunbar 1994).

Comparative tests of spatial requirements (e.g., home range size per unit group biomass) based on compilations of this kind can be confounded by differences in soil fertility, forest types, and resource productivity in what amounts to a highly heterogeneous set of study sites. My primary aim here, however, was to examine how the relationship between daily distance moved and range size in territorial platyrrhines under different ecological scenarios might affect their ability to defend space. In addition, differences in habitat quality are canceled out to some extent because the variation in group density within genera appears to be comparable to that between genera. No attempt was made to control for phylogeny in the comparative data set considered here. While this decision is clearly open to criticism (e.g., Pagel and Harvey 1989), phylogenetic contrasts are most relevant when comparing life history parameters across an

Carlos A. Peres

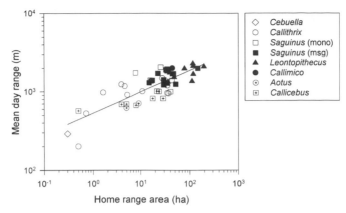

Figure 5.3 Relationship between the home range area and the average day range for small-bodied Neotropical primates living in small groups. Monospecific and mixed-species groups of tamarins (*Saguinus* spp.) are coded separately.

unbalanced distribution of data points among a wide range of taxa sharing no immediate common ancestry. Despite differences in body size and group size, the genera considered in this study are reasonably cohesive in terms of socioecology and intergroup spacing system, and belong to only two major clades of platyrrhine primates (Schneider et al. 1993). Moreover, I found no clear differences in these relationships when analyzing the callitrichine and the cebid data sets separately, suggesting weak effects of phylogeny. The underlying assumption is that differences between the taxa are largely a function of ecology and environmental gradients rather than phylogenetic differences between them.

Range Use and Group Movement Territorial Platyrrhines
As might be expected, daily distances moved are a reasonably good predictor of home range area in small-bodied Neotropical primates, despite considerable interspecific differences in ecology and habitat use (on $\log_{10}$-transformed data; $r = .805$, $P < .001$, $N = 57$; fig. 5.3). This relationship appears to be independent of differences in absolute metabolic requirements, as a greater group biomass did not necessarily result in commensurately larger home ranges (on $\log_{10}$-transformed data; $r = .255$, $P = .012$, $N = 97$).

In general, groups of large-bodied callitrichines (*Saguinus, Leontopithecus,* and *Callimico*) move the longest distances over the largest home ranges of all territorial platyrrhines (fig. 5.3). These genera also have the greatest spatial requirements per unit group biomass and the lowest ratios of daily distance moved per unit home range area (table 5.3). The mechanics of territorial defense in *Saguinus* and *Leontopithecus* may therefore involve very different constraints compared with territorial platyrrhine groups accommodated by smaller home ranges. First, directional movement toward boundaries in a large range is not necessarily informed by vocalizations giving away the locations of neighboring groups. While range areas can be drastically variable across different genera (table 5.3), the maximum distance at which intergroup vocalizations can be effectively transmitted does not appear to increase with range size. For example, the duets produced by titi monkeys and hylobatids can intersect the home ranges of several neighboring groups and be heard by human observers up to 1 km away (Mason 1968; Chivers, Wright, and Kinzey 1986). Although the auditory sensitivities of callitrichines and humans are considerably different (Aitkin et al. 1986), a similar case could be made for *Cebuella* and *Callithrix,* which typically occupy small home ranges (see Boinski, chap. 15, this volume). In contrast, long calls of at least some *Saguinus* and *Leontopithecus* populations may not reach beyond even one radius of a typical home range size (Peres 1986a, 1991b). Groups near the range center soon after daybreak may thus be positioned beyond the audible range of a neighboring group's long calls, as animals "sit tight" at high levels of the forest, apparently striving to locate their neighbors. Moreover, if territorial boundaries are to be regularly monitored, increasingly larger home ranges will result in far greater energetic costs associated with longer travel distances, since the range perimeter increases at a rate over three times faster than the range diameter (see also Steudel, chap. 1, this volume). This is all the more relevant for at least some group members in small-bodied primates such as callitrichines, which are forced to commute to range boundaries burdened with a relatively high total litter weight for several months of the year (Leutenegger 1973; Tardif, Harrison, and Simele 1993).

Home Range Defensibility and Overlap
I evaluated the relationship between the daily distance moved *(DR)* and the approximate home range diameter (*HR∅*) in territorial platyrrhines, assuming no extreme deviation from a circular home range shape. Although these species generally exhibited high group mobility in relation to range size (Mitani-Rodman *DI* index = 2.7 ± 1.4, range = 1.1–7.1, $N = 57$; Lowen-Dunbar *MI* index = 0.51 ± 0.88, range = 0.04–4.46, $N = 57$), the best fit between day range

Carlos A. Peres

Table 5.3. Summary (mean ± SD) of the ranging ecology of territorial platyrrhines studied to date

Genus	Home range size (ha)	HRS/GB[a] (ha/kg)	N[b]	Mean day range (m)	DR/HRS[c] (m/ha)	DI[d]	MI[e]	N[b]
Cebuella	1.1 ± 1.3	1.5 ± 1.4	(4)	290	967	4.7	3.8	(1)
Callithrix	10.9 ± 12.3	4.6 ± 5.4	(21)	968 ± 351	248 ± 234	3.6 ± 1.9	1.0 ± 0.9	(12)
Saguinus								
Monospecific	23.9 ± 16.1	10.2 ± 7.9	(16)	1,368 ± 521	85 ± 62	2.9 ± 1.4	0.3 ± 0.2	(6)
Mixed-species	48.9 ± 40.8	22.2 ± 20.9	(24)	1,582 ± 297	47 ± 22	2.3 ± 0.6	0.2 ± 0.1	(18)
Leontopithecus	92.5 ± 61.8	38.2 ± 26.2	(19)	1,794 ± 345	23 ± 12	1.7 ± 0.4	0.1 ± 0.1	(9)
Callimico	65.0 ± 28.3	28.8 ± 22.6	(2)	2,000	44	2.6	0.2	(1)
Aotus	10.6 ± 6.4	2.6 ± 1.6	(3)	555 ± 194	100 ± 33	2.3 ± 0.3	0.4 ± 0.1	(3)
Callicebus	12.8 ± 9.9	3.5 ± 2.7	(10)	786 ± 161	212 ± 377	2.8 ± 1.9	0.8 ± 1.5	(8)

Note. See appendix 5.1 for sources of data.

[a] Home range size (ha) per unit group biomass (kg).

[b] Number of distinct study groups for which data are available.

[c] Mean day range (m) per unit home range size (ha), calculated for individual populations.

[d] Mitani and Rodman's (1979) defensibility index.

[e] Lowen and Dunbar's (1994) monitoring index.

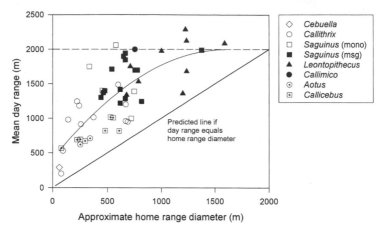

Figure 5.4 Relationship between the approximate home range diameter and mean day ranges in callitrichenes, titi monkeys, and night monkeys. All values above the diagonal line amount to defensibility values (*sensu* Mitani and Rodman 1979) greater than 1 (*DI* > 1).

and home range diameter was provided by a power curve ($r = .814$, $N = 52$, $DR = 42.15 * HR\emptyset^{0.54}$; fig. 5.4). Indeed, relative group mobility was proportionately lower within increasingly larger home ranges. This finding is confirmed by the strong negative correlation between $\log_{10}$-transformed values of range size and either defensibility indices (*DI:* $r = -.75$, $p < .001$, $N = 57$; *MI:* $r = -.83$, $p < .001$, $N = 57$; fig. 5.5). In other words, increasingly larger ranges appear to be less defensible despite the fact that, in absolute terms, groups moved farther each day within larger ranges. Day range to range size ratios were thus particularly low in the case of some groups of *Saguinus* and *Leontopithecus* that occupied extremely large range areas.

If daily movements of territorial platyrrhines could be described in terms of (1) an outward displacement to the range boundary, (2) a portion of the range boundary, and (3) an inward displacement back to the range center, day ranges would consist of two home range radii (equivalent to 1 *HR*∅) and an angular section of the range periphery. Boundary patrols could then be defined as arched, peripheral movements covering a given proportion of the range perimeter. I thus examined the relationship between observed day ranges and those predicted by home range size if groups were to exhibit this model of centrifugal movements, covering increasingly longer arched segments of the range perimeter (0°, 60°, 120°, 180°, and 240°). Observed data on the relative group mobility of territorial platyrrhines fitted to this hypothetical model suggest that these

Carlos A. Peres

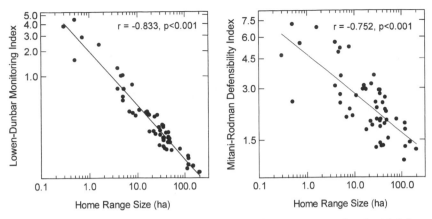

Figure 5.5 Relationship between home range size and two measures of territorial defensibility for small-bodied platyrrhines living in small groups. The axes representing the Mitani-Rodman and the Lowen-Dunbar defensibility indices have been scaled to the power 0.1.

species would at most cover a boundary segment of roughly one-third (120°) of the range perimeter (fig. 5.6). In an idealized world of quasi-circular home ranges and a radially symmetrical intergroup spacing, this would result in all boundary areas being monitored every third day, should defense of range boundaries be evenly spread over time. While unrealistically idealistic, this simple model roughly portrays the daily routine of *Saguinus* and *Leontopithecus,* and perhaps other New World primates in small groups (e.g., *Pithecia*) for which ranging data may yet become available.

Home range overlap is expected to be high if investments in territorial defense are insufficient to exclude neighbors, either because group mobility is too low or home ranges are too large to be monitored effectively. This prediction is consistent with the fact that the enormous variation in home range overlap with neighboring groups in territorial platyrrhine populations (mean = 33 ± 25%, range = 0–88%, $N = 32$) was positively correlated with home range size ($r = .575$, $P < .001$, $N = 32$) and negatively correlated with the defensibility indices (Mitani-Rodman DI: $r = -.558$, $P = .002$; Lowen-Dunbar MI: $r = -.448$, $P = .01$, $N = 27$). While low defensibility values did not necessarily translate into high overlaps, high values ($DI > 4$) were invariably associated with low overlaps. These relationships, however, were clearly nonlinear (range overlap vs. range size: $r^2 = .39$, 30 d.f., $F = 9.5$, $P = .001$; range overlap vs. DI value: $r^2 = .33$, 24 d.f., $F = 5.8$, $P = .009$; fig. 5.7), again indicating

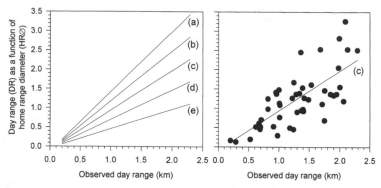

Figure 5.6 *(Left)* Hypothetical relationships between observed day ranges in territorial platyrrhines and those predicted if daily movements were equivalent to one home range diameter (two radii) plus a gradually shorter section of the range perimeter equating to boundary arches of *(a)* 240°, *(b)* 180°, *(c)* 120°, *(d)* 60°, and *(e)* 0°. *(Right)* Scatter plot showing the actual distribution of available data points. Day ranges in these taxa are thus best predicted by the distance corresponding to two home range radii plus one-third of the range perimeter.

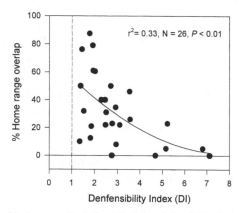

Figure 5.7 Relationship between the Mitani-Rodman defensibility index *(DI)* and the proportion of a study group's home range overlapping those of all neighboring groups in twenty-seven populations of territorial platyrrhines.

that substantial increases in range overlap were primarily a function of excessively large range sizes. Moreover, the degree of overlap showed a nearly identical relationship with home range size per unit group biomass ($r^2 = .39$, 30 d.f., $F = 9.5$, $P = .001$), which provides a more meaningful measure of spatial requirements across varying levels of resource density.

Intergroup Spacing in Large-Bodied Callitrichines

Tamarins at Urucu and lion tamarins at Poço d'Antas used some of the largest home ranges reported for any territorial platyrrhine

Carlos A. Peres

(see reviews in Sussman and Kinzey 1984; Snowdon and Soini 1988; Ferrari and Lopes-Ferrari 1989), but comparable to those of other large-bodied callitrichine populations (Terborgh and Stern 1987; Rylands 1989; Valladares-Padua 1993; Dietz, Sousa, and Silva 1994; Christen and Geissmann 1994; Passos 1997). Both study groups covered long distances per unit time in their relatively large home ranges. Directional early morning travel was generally led by one or two adults as they made a "beeline" to the location of neighboring groups' long calls or large food patches (Peres 1996a). On the other hand, use of space was affected to only a minor extent by the location of sleeping sites because these were relatively numerous and widely distributed across the home range of each group (Peres 1986a, 1991b).

Despite their widely disjunct distributions, there were also a number of parallels between *Saguinus* and *Leontopithecus* in the way territorial investment affected group movement and use of space. These parallels included similar patterns of intergroup interactions and spacing and a diurnal pattern of group centrality matching a bimodal distribution in group velocity early and late in the day, coinciding with movements to and from the range boundaries. When tested against Waser's (1977b) two-dimensional ideal gas model,[2] both genera also showed frequencies of intergroup encounters that far exceeded those predicted by chance alone (Peres 1991b, 1992a). They also both clearly adopted a "doughnut-shaped" activity field (*sensu* Waser and Wiley 1979) in that the amount of time spent within encounter zones was far greater than predicted by the proportion of those areas in their home ranges. Moreover, their peripheral bias in range use involved similar foraging costs, as both genera incurred increasingly lower rates of prey capture success as they moved away from central home range areas held to the exclusion of neighboring groups (Peres 1989, 1992a). Yet peripheral foraging facilitates detection of intruders, allows resource defense by exploitation (*sensu* Davies and Houston 1981), and depletes resources that would otherwise be lost to neighbors in the large peripheral areas of range overlap (for other vertebrates that also defend territories via active depletion of peripheral resources, see Paton and Carpenter 1984; Gittins 1984; Gill 1988). Persistent resource harvesting in the range periphery by both genera thus appears to complement energetic investments allocated to defense by interference, which alone was clearly unable to exclude competitors from large feeding territories. It could be suggested, therefore, that for territorial primates harvesting constantly renewable resources (see Waser 1981), such as the highly insectivorous callitrichines, defense through interference is most effective within small home ranges, whereas defense through exploitation becomes increasingly important within large home ranges.

Territoriality and Group Movement in Platyrrhines
Other small-bodied species of New World monkeys also allocate considerable investment to territorial defense, and regularly engage in intergroup encounters at range boundaries (see appendix 5.1). Although well within the range of body mass of territorial platyrrhines, squirrel monkeys (*Saimiri* spp.) live in extremely large groups (Peres 1997), target relatively large food patches (Terborgh 1983; but see Mitchell, Boinski, and van Schaik 1991), and occupy large overlapping home ranges that are not defended against neighboring groups (Mitchell 1990; Boinski and Mitchell 1994). While *DI* values for two Amazonian groups of squirrel monkeys are greater than unity (*DI* = 1.12–1.64; calculated from Mitchell 1990), this is probably a function of high group mobility resulting from greater foraging demands in a large group (Clutton-Brock and Harvey 1977a; Isbell 1991; Janson and Goldsmith 1995). Intergroup encounters, as defined here, are thus restricted to platyrrhine species living in small groups, and tend to occur at conventional and relatively stable boundary zones, which might be reinforced throughout the entire life span of a given group (e.g., Goldizen et al. 1996). One group of owl monkeys studied by a number of workers in Peru, for example, maintained nearly identical boundaries for 9 years (Terborgh 1983; Wright 1986). In at least some populations of pygmy marmosets, Goeldi's monkeys, and collared titis *(Callicebus torquatus)*, however, there are few opportunities for stable intergroup interactions because territories are either not contiguous or shift quite frequently (Soini 1982; Pook and Pook 1981; Easley and Kinzey 1986; Peres 1993c). Intergroup spacing in most territorial platyrrhines is thus mediated by site-dependent aggression, often following movement toward range boundaries prior to the escalation of intergroup encounters (Mason 1968; Hubrecht 1985; Wright 1986; Robinson, Wright, and Kinzey 1986; Campos, de la Torre, and de Vries 1992; Peres 1989, 1991b, 1992a).

In territorial platyrrhines, the relationship between day range and home range size suggests an upper threshold of group mobility where spatial requirements are particularly high (or resource abundance particularly low). A reduction in relative group mobility

Carlos A. Peres

within exceptionally large ranges suggests that these species would incur an increasingly prohibitive cost in defending even larger range areas. At low resource density, the time required to patrol a large territory containing sufficient resources should render the costs of defense increasingly prohibitive (Davies and Houston 1984). It is therefore not surprising that primates using ranges even larger than those compiled here invest little in intergroup resource defense (Mitani and Rodman 1979; Lowen and Dunbar 1994). This conclusion is confirmed by a comprehensive data set on both territorial and nonterritorial Neotropical primates, including all species studied to date (C. Peres, unpub.).

Conclusions

In this chapter I have presented data supporting the following points: (1) the ranging patterns of *Saguinus* and *Leontopithecus* living at sites over 3,200 km apart converge in being constrained by similar strategies of intergroup resource defense; (2) resource defense in these large-bodied callitrichines—which travel the longest distances over the largest home ranges of all territorial platyrrhines—operates via multiple mechanisms of intergroup competition; (3) interference and exploitative resource defense in peripheral home range areas can be seen as complementing each other, the relative importance of the latter becoming increasingly greater in larger home ranges, which cannot be defended solely by direct intergroup interactions; (4) in small-bodied New World monkeys, there appears to be an upper threshold in home range size beyond which the cost of territoriality becomes prohibitive, largely because of greater travel costs incurred by range defense; and (5) the degree of exclusive access to resources can be largely explained by the relationship between home range size (or perimeter) and travel costs associated with boundary forays.

This analysis largely supports Mitani and Rodman's (1979) economic defensibility hypothesis as applied to small-bodied, largely insectivorous species across a wide range of spatial requirements. The same cannot be said for other studies addressing similar issues and based on a broader sample of primates (Grant, Chapman, and Richardson 1992), or on a single highly folivorous genus (*Presbytis:* van Schaik, Assink, and Salafsky 1992). These sets of species have a greater variation in ranging strategies and intergroup spacing systems, and do not necessarily defend space for the same reasons as do the taxa addressed here. Given the range use constraints resulting from the ecology of different species in different habitats, the efficiency with which territories can be defended by callitrichines, titis, and owl monkeys does appear to decrease with increasing home range size. This decrease results in a substantially greater degree of home range overlap between neighboring groups, especially where spatial requirements are atypically high. In these cases, the importance of interference mechanisms of resource defense, in the form of highly conspicuous intergroup contests, may be outweighed by that of the more subtle defense by exploitation, which is more compatible with short-term metabolic needs. Both mechanisms, nevertheless, should complement each other in the long-term maintenance of a feeding territory. It remains to be seen, however, whether the behavioral features constituting such an important part of the ranging ecology and intergroup spacing of most small-bodied New World monkeys are shared by other territorial mammals.

Notes

1. Although the genus *Aotus* has been traditionally aligned with *Callicebus* and other pitheciines on the basis of morphological characters (Rosenberger 1984), the most recent phylogenies based on genetic evidence no longer support this hypothesis (Schneider et al. 1993).

2. This null model offers a framework for calculating the expected frequency of group collisions—derived on the assumption of random movements from known values of group density and mean group velocity—with which actual rates of intergroup encounters can be compared.

Carlos A. Peres

Appendix 5.1 Sources of data on the ranging ecology of small-bodied Neotropical primates living in small groups

Species	Sources
Callitrichinae	
Cebuella pygmaea	Castro and Soini 1977; Ramirez, Freese, and Revilla 1977; Soni 1982; Terborgh 1983; Peres 1993c
Callithrix humeralifer	Rylands 1986
Callithrix argentata	Albernaz 1993; Veracini 1996
Callithrix jacchus	Maier, Alonso and Langguth 1982; Hubrecht 1985; Stevenson and Rylands 1988; Alonso and Langguth 1989; Scanlon, Chalmers, and Monteiro da Cruz 1989; Digby and Barreto 1996
Callithrix penicillata	Fonseca and Lacher 1984; Faria 1989
Callithrix kuhli	Rylands 1982, 1989
Callithrix flaviceps	Ferrari 1988
Callithrix aurita	Torres de Assumpção 1983; Muskin 1984; Ferrari, Corrêa, and Coutinho 1996; L. Brandão, in litt.
Saguinus fuscicollis	Terborgh 1983; Terborgh and Stern 1987; Soini 1987; Soini and Cóppula 1981; Pook and Pook 1982; Garber 1988a; Heymann 1990a; Castro 1991; Norconk 1986; Peres 1991a; Crandlemire-Sacco 1986; Izawa and Yoneda 1981; Lopes and Ferrari 1994
Saguinus nigricollis	Izawa 1978; de la Torre, Campos, and de Vries 1995
Saguinus mystax	Ramirez 1986; Norconk 1986; Garber 1988; Garber et. al. 1993; Heymann 1990a; Castro 1991; Peres 1991a
Saguinis labiatus	Pook and Pook 1982; Buchanan-Smith 1990, 1991; Izawa and Yoneda 1981
Saguinus imperator	Terborgh 1983; Terborgh and Stern 1987
Saguinus oedipus	Dawson 1979; Neyman 1977; Garber 1984
Saguinus bicolor	Egler 1992
Saguinus midas	Mittermeier 1977; Kessler 1995; Oliveira 1996
Callimico goeldii	Pook and Pook 1982; Christen and Geissman 1994
Leontopithecus rosalia	Peres 1986b; Dietz, Peres, and Pinder 1997
Leontopithecus chrysomelas	Rylands 1989; Dietz, Sousa, and Silva 1994
Leontopithecus chrysopygus	Carvalho and Carvalho 1989; Keuroghlian 1990; Valladares-Padua 1993; Passos 1992, 1997
Leontopithecus caissara	C. Valladares-Padua, pers. comm.
Pitheciinae	
Aotus nigriceps	Wright 1986, 1989
Aotus brumbacki	Solano 1996
Aotus azarae	Garcia and Braza 1987; Wright 1994
Callicebus moloch	Mason 1968; Kinzey 1981; Wright 1986; Robinson 1979; Robinson, Wright, and Kinzey 1986
Callicebus personatus	Kinzey and Becker 1983; Müller 1995; S. Heiduck, in litt.
Callicebus torquatus	Kinzey 1981; Easley 1982; Kinzey and Robinson 1983; Defler 1983; Campos, de la Torre, and de Vries 1992; Peres 1993c

PART TWO

Cognitive Abilities, Possibilities, and Constraints

All invertebrate and vertebrate social foragers can be safely characterized as encountering multiple social and ecological problems in locating resources and organizing group movements. From that point on, however, as these chapters demonstrate, firm generalizations across major taxonomic boundaries regarding the cognitive processes underlying group travel become sparse. Group-living animals may differ widely in the information available to them on the distribution of food and predators in their environment, in their sensory and neurological capabilities, and in the role of conspecifics in learning and decision making. The evolution of the order Primata, for example, is associated with several significant changes in the visual system, enhanced eye-hand coordination, grasping feet and hands, increased brain size, distinct configurations of food distribution and presentation, and social skills. Compared with most other groups of mammals, monkeys and apes have large brains, and especially large neocortices, for their body size. But what a large neocortex signifies for how primates store, categorize, recall, and integrate spatial, temporal, sensory, and social information is unclear. Another factor confounding easy descriptions of the cognition entailed in group movement is that success at the group level may enhance success at the individual level. Attention will be focused profitably on decision-making processes and learning through social facilitation, among other phenomena, in efforts to document such synergistic individual- and group-based behavioral strategies.

CHAPTER SIX

Group Movement
and Individual
Cognition: Lessons
from Social Insects
FRED C. DYER

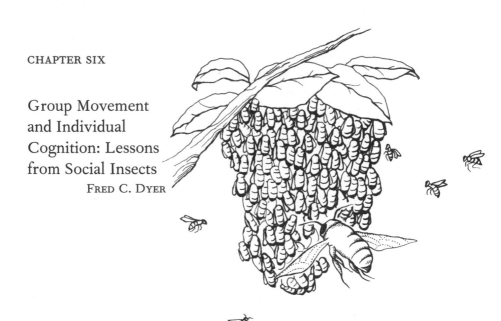

In spite of the inclusive title of this volume, many readers will find it somewhat surprising that a chapter on insects has been included. My task is to show that social insects can provide a useful model for examining how animals solve the problems that they face in moving as a part of a group. Because of the high degree of integration among the members of social insect colonies, the causes and consequences of group movement are relatively easy to characterize. Furthermore, a vast literature has developed concerning the sensory systems, navigation, and learning abilities of social insects, as well as the mechanisms underlying social integration and collective action. This literature provides a framework for addressing questions about mechanisms for initiating group movement, maintaining cohesion during a move, and learning the locations of resources after the group has moved. No doubt, many primates and other mammals use much more sophisticated (or at least divergent) individual and social strategies during group movement. Still, because of the relative ease with which they can be studied, social insects offer the opportunity to develop a "worked example" of group movement that could serve as a guide for building integrated models of group movement in other species.

This chapter has three main parts. First, I briefly summarize the natural history of social insects, including the circumstances under which colonies of various species need to move. This section will serve to orient readers not familiar with the biology of this group of animals. Next, I offer my summary of the behavioral challenges associated with group movement by social insect colonies. Although presented in reference to the social insects, the ideas in this section overlap with those presented in other chapters. Finally, I describe in some detail studies of individual and group movement in one social insect taxon, the honeybees (genus *Apis*), which has been far better studied than other taxa. The focus in this last section is on the navigational challenges that arise as a consequence moving with a colony and the cognitive abilities that have evolved to meet these challenges.

Natural History of Group Movement in Social Insects

An Introduction to the Social Insects
As any reader of this volume is well aware, some of the most complex and highly integrated animal societies are found among the social insects. Many insect orders show at least mild forms of sociality, such as care of the young by the mother or the father (or both) (Scott 1996). The term "social insect" is conventionally reserved, however, for insects in two orders: the Hymenoptera (bees, wasps, and ants) and the Isoptera (termites) (reviews: Hölldobler and Wilson 1990; Michener 1974; Seger 1991; Wilson 1971).

In the Hymenoptera we can find various degrees of sociality (Michener 1974; Seger 1991). All ants are eusocial, but most species of bees and wasps are in fact not social at all. Instead, females either rear young in a solitary nest or lay their eggs parasitically on other organisms (usually other insects, but sometimes, as in the gall-forming wasps, on plants). Familiar examples of solitary Hymenoptera include carpenter bees (e.g., genus *Xylocopa*), which bore out nest tunnels in wooden buildings, and digger wasps (e.g., *Philanthus, Bembix*), which dig nests deep into sandy soil. In a smaller number of species, females cooperate in the care and provisioning of young. Some species exhibit cooperative associations consisting of a set of reproductively competent females that share a nest and the responsibilities of parental care. Other species, including ants, paper wasps, hornets, and honeybees, are "eusocial," exhibiting not only cooperative care of young, but also overlapping generations and a division of reproductive labor between fertile queens and sterile female workers. Males are never part of the workforce in the eusocial Hymenoptera, but instead have only a reproductive function. Eusociality is exhibited by such familiar species as honeybees, bumblebees, and wasps of the hornet family (including paper wasps and yellow jackets). All ants are eusocial, and it is thought that this family of Hymenoptera descended from a eusocial species of wasp (Hölldobler and Wilson 1990).

The life cycles of eusocial hymenopterans fall into two very general patterns related to the mode of colony reproduction. One mode involves independent founding of new nests by solitary females, or by small groups of reproductive females in coalition. At first the newly founded colony grows through the production of sterile workers, in the so-called ergonomic stage (Oster and Wilson 1978). Eventually the colony enters the reproductive stage and begins producing fertile females and males, which leave the nest to mate and found new nests. The parental colony then either disbands (in temperate climates) or reenters the ergonomic stage (Gadagkar 1991). The other mode of reproduction involves fission of a parent colony into two separate units, each with its own queen. One unit, often called the "swarm" in flying insects, disperses to establish a new nest. Most eusocial hymenopterans reproduce by independent colony founding, but swarm-founding species occur in each of the main hymenopteran groups.

I provide only a brief summary of eusociality in the Isoptera, to contrast them with the Hymenoptera and to explain why I will not deal with them in my discussion of group movement. All termites are eusocial, implying that, as in the ants, all members of this diverse insect order descended from a eusocial ancestor (Wilson 1971). Although similar to the eusocial Hymenoptera in the organization of their colonies and the degree of social integration, termites differ from the Hymenoptera in certain important respects. For one thing, the sterile workforce is composed about equally of males and females. This difference between the two social insect groups probably stems from the strikingly different genetic systems related to sex determination. The Hymenoptera are haplodiploid, with females arising from fertilized eggs and males arising from unfertilized eggs. This results in an asymmetry in relatedness that potentially favors reproductive cooperation in females, but not in males. Termites, by contrast, are diploid, so that any selective pressure in favor of cooperation applies equally to males and females. Another trait distinguishing termite societies from those of the Hymenoptera is that they are always founded by solitary reproductives rather than through colony fission. Finally, perhaps because of the nature of their food resource (cellulose), which is available in vast quantities close to the nest (indeed, the nest often occupies the food resource), termite colonies generally do not change nesting sites once formed. These last two features—absence of colony fission and the permanence of nests—mean that group movement plays a negligible role in the biology of these organisms; hence they will not play a part in this chapter.

Based on phylogenetic evidence, eusociality is thought to have arisen at least a dozen times in the insects alone (once in the Isoptera, and the remainder in the Hymenoptera). Among the few species outside the insects to have evolved eusociality is the naked mole-rat (Sherman et al. 1991), a mammal whose social structure (a workforce composed of sterile male and female workers) and dietary niche (plant matter that is indigestible to most other animals) qualify it as an honorary termite.

Eusociality presents tantalizing evolutionary puzzles, as Charles Darwin (1859) himself was the first to recognize. Current research on eusociality focuses on two general problems. One is how the traits of sterile workers, including sterility itself, could have arisen by natural selection. If natural selection works by favoring heritable phenotypes that contribute disproportionately to the next generation, then how could it result in the spread of a phenotype expressed by nonreproducing individuals? Darwin first posed this question, and also first outlined the correct answer: eusocial colonies are invariably family groups, so that heritability traits that favor helping behavior by sterile workers are passed along by kin in whose reproduction they assist. These traits include not only sterility itself, but also the specialized behavioral and morphological traits that equip workers to carry out their tasks of nourishing and defending the colony.

The second general area of research into social insect biology concerns how colonies are organized to meet the challenges of surviving and reproducing. The most conspicuous examples of social organization are the elaborate systems of division of labor seen in many species. Division of labor among workers enables the colony to feed and defend itself with great efficiency because multiple tasks can be carried out in parallel and workers can be specialized for particular tasks rather than being jacks of all trades (Oster and Wilson 1978; Hölldobler and Wilson 1990; Seeley 1995). Another prominent example of social organization is the teamwork that is

Fred C. Dyer

exhibited in carrying out many tasks. This too contributes to colony success because teams can perform tasks that individual workers could not perform on their own (Franks 1985; Bourke and Franks 1995).

To obtain the benefits of division of labor and teamwork, the colony must have mechanisms to coordinate the activities of a large number of workers. A striking phenomenon that becomes apparent to any observer of a social insect colony is that no one seems to be in charge. Unlike the human organizations with which they are often compared, social insect colonies defend themselves without generals, build their nests without architects or foremen, and regulate the nest climate and the flow of materials into and out of the nest without engineers or building supervisors. There is no one individual or subgroup within the colony to which information about the colony's needs flows and from which instructions issue. Instead, work is organized in a decentralized fashion, with coherent group behavior emerging from the interactions of individuals responding in specific ways to local information (Franks 1989; Seeley 1989, 1995; Hölldobler and Wilson 1990; Bourke and Franks 1995). This principle is well worth keeping in mind as we consider how social insect colonies move.

Benefits and Costs of Group Movement
This chapter concerns the mechanisms of group movement in social insects, but to set the stage further we need to consider the functional question of why they move. In general, colonies move either to change nesting sites when the current site deteriorates in quality, or to disperse from the mother colony after reproductive fission (fig. 6.1). Let us look at each of these in turn.

Deterioration of nest quality can occur in a variety of ways (Hölldobler and Wilson 1990; Jeanne 1991). Perhaps the most common is for the nest to come under attack by predator, forcing the colony to abandon the nest on short notice and reassemble elsewhere. Emergency moves may also be elicited by physical disasters such as flood or fire. The technique of pacifying honeybees with smoke, which has long been practiced by beekeepers as well as honey hunters, is effective because the smoke induces the workers to begin preparations for moving to escape fire, and to become less likely to sting those disturbing the nest.

Colony movement may also be precipitated by more gradual deterioration of nest quality. For example, an increase in populations

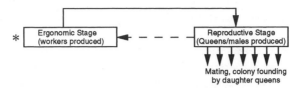

a. Independent-founding species

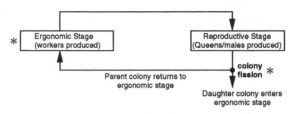

b. Swarm-founding species

Figure 6.1 Generalized life cycles of eusocial insects. Asterisks indicate where in the life cycle colony movement may occur.

of parasites or microorganisms affecting the brood may induce the colony to establish a clean nest elsewhere. Also, changes in the nest that increase the colony's vulnerability to predators may lead the colony to search for a more secure nest. A response to elevated predation risk has been documented in one of the Asian honeybee species, *Apis florea* (Seeley, Seeley, and Akratanakul 1982). Colonies of this species construct their nests in the open, suspending a wax comb the size of a dinner plate from a thin branch in dense vegetation and then protecting the comb with a curtain of interlinked worker bees. Because the workers and the colonies are very small, and can mount only a relatively weak stinging defense when attacked, the cover of vegetation plays a crucial role in their defense against predation. Loss of leaf cover as seasons change (or through experimental pruning) is sufficient to trigger a move.

In some species, colonies may change their nesting sites in response to a deterioration of food resources or climatic conditions in the habitat. Benefits might arise even through short-range movements that bring the colony nearer to rich food resources or allow it to occupy a nest with a more favorable microclimate. In more extreme cases, the colony's movement may carry it well out of its original foraging range. One well-documented case is the Neotropical army ants (*Eciton* spp.), in which colonies do not actually occupy nests, but instead form "bivouacs," the workers massing in a

Fred C. Dyer

cluster that is protected from predators and climatic fluctuations only by an outer mantle of interlinked workers (fig. 6.2). These colonies alternate between a "statary phase," in which the bivouac remains in the same location for several weeks while foragers raid the surrounding habitat, and a "nomadic phase," in which the bivouac is moved each night until the colony arrives in a new, undepleted part of the rainforest (Schneirla 1971; Franks and Fletcher 1983). Hölldobler and Wilson (1990) review this "legionary" behavior in several other ant species. Even more dramatic examples of colonial migration are seen in various species of honeybees. For example, in the Asian "rock bee", *Apis dorsata,* whose colonies nest on huge (1 diameter) exposed combs, colonies undertake seasonal migrations of more than 100 km between habitats with different flowering seasons (Koeniger and Koeniger 1982). A closely related species of rock bee, *Apis laboriosa,* which spends summers at high elevations

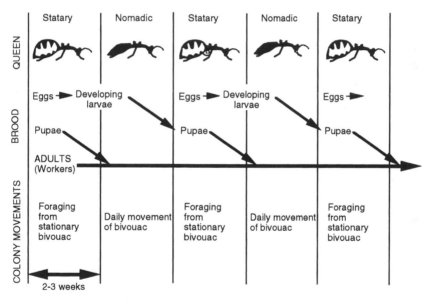

Figure 6.2 Alternation of statary and nomadic phases by a colony of army ants *(Eciton burchelli).* During the statary phase, the queen lays a large batch of eggs, which begin larval development. Also during the statary phase, the generation of larvae produced in the previous statary phase undergoes pupation. The statary colony sends raiding columns several tens of meters into the surrounding habitat, but the ants return each night to the same bivouac site. The nomadic phase begins on emergence of the pupae. Each day the bivouac changes location, usually choosing a site along one of the raiding paths. The queen slims down for the nomadic phase. The larvae of her most recent brood are carried singly by the migrating workers. The next statary phase begins when these offspring pupate. (From Wilson 1971; based on the work of Schneirla 1971).

(>2,000 m above sea level) in the Himalayas, migrates to lower elevations in the winter, waiting for spring in combless clusters (Underwood 1990).

I have described several possible reasons why social insect colonies might move to seek improved nesting conditions. Any benefits that arise from such movements must, of course, be balanced against potential costs. Costs of colony movement might include, for example, the energetic and time costs of traveling and transporting brood and nest materials; attrition of the workforce as individuals are disoriented or lost to predators; loss of any food stores that have been left behind; the energetic costs of constructing the new nest; and interruption of colony growth during the time required to move, establish a new nest, and find food sources in the new location. As we might expect, many social insects have evolved strategies that have the effect of mitigating these costs, thus enhancing the net benefit of the move.

Even with minimal time for preparation, colonies are often able to take steps that increase the probability of re-forming the colony and returning to the tasks of acquiring resources, growing in size, and reproducing. Assembly pheromones and colony-specific odors allow individuals to find nestmates quickly. In ant colonies that are forced into an emergency move (easily observed if one simulates a drastic predatory attack by exposing a nest inside a rotten log or under a flagstone), workers can often be seen carrying brood (larvae and pupae) as they escape. In species, such as honeybees, that hoard food in the nest, workers try to load themselves to capacity with food before they escape the nest (Winston 1987). In the Asian honeybee *Apis florea,* foragers from colonies forced to abandon a nest return to it in the first few days after the move to retrieve honey that may have been left behind and to collect wax for constructing the new comb (Seeley, Seeley, and Akratanakul 1982).

When the deterioration of conditions is more gradual, or in the context of migration in search of improved conditions, the colony may be able to prepare more effectively for its departure. For example, in both *Apis florea* and *A. dorsata,* both Asian honeybees that occupy exposed nests, colonies start to prepare for a local or long-distance move several days in advance by ceasing brood rearing and allowing developing brood to reach the adult stage (Koeniger and Koeniger 1980). In this way, brood in which the colony has already invested can join the workforce for the move. Analogously, in the army ant *Eciton burchelli,* the alternation between the

Fred C. Dyer

nomadic and statary phases is synchronized with the cycle of brood production (see fig. 2). During the statary phase, the queen produces tens of thousands or even hundreds of thousands of eggs, which then hatch and begin larval development. Before the nomadic phase begins, the queen stops laying eggs and slims down, making it easier for her to travel with the colony on its repeated treks through the rainforest. Also, the previous generation of brood undergoes pupal development during the statary phase, and their emergence as callow workers precedes the onset of the nomadic phase. During the nomadic phase the current larval brood are carried from site to site each day. The next statary phase begins when these larvae pupate. It remains unclear how this elaborate system is synchronized. Presumably it is partly influenced by local depletion of resources during the statary phase, but there may be feedback processes endogenous to the colony, for example, as a result of the synchrony of egg laying by the queen (reviewed by Hölldobler and Wilson 1990).

So far in this section I have focused on movement of nonreproducing colonies in search of better nesting conditions. Now I turn my attention to movement in the context of reproduction in species that undergo colony fission. This mode of reproduction has evolved several times independently in the Hymenoptera. Among the best-studied examples are the army ants, *Eciton* spp. (Schneirla 1971), certain wasps in the family Vespidae (especially in the subfamily Polistinae, the paper wasps) (Jeanne 1991), the honeybees, *Apis* spp. (Seeley 1985; Winston 1987), and the stingless bees, *Melipona* spp. and *Trigona* spp. (reviews by Michener 1974; Wille 1983; Roubik 1989). Honeybees and stingless bees are closely related and may have inherited this reproductive strategy from a common ancestor.

Dispersal of daughter colonies after colony fission is functionally equivalent to the reproductive dispersal that is seen in solitary animal species. In both cases, the newly produced individuals (or colonies) move away from the natal territory (or nest) to establish reproductive independence. In both cases, the dispersal entails travel and start-up costs associated with the expenditure of time and energy in moving, establishing a new territory (or nest), and discovering new sources of food. Colonial reproductive dispersal obviously involves specialized behavioral traits, however; specifically, those required to coordinate the activities of hundreds or thousands of workers as the colony initiates movement and searches for a new nesting site.

In general, reproductive and nonreproductive colonial movements involve many of the same challenges and, presumably in most species, the same underlying behavioral processes. At least two important differences can be identified, however. First, the factors precipitating reproductive colony movement would include some cues specific to the reproductive state of the colony (e.g., colony size, presence of a new queen). Second, during colony fission, each worker faces a decision about whether to go or to stay behind. For theoretical reasons (e.g., Seeley 1985), there is reason to believe that the expected fitness returns associated with these alternatives will be different for different workers, depending upon their relatedness to the dispersing and nondispersing queens. Thus, one might expect to find specialized mechanisms that allow workers in fissioning colonies to make the appropriate choice. So far, however, the evidence for such mechanisms is limited.

In the next section, when I review these challenges and the mechanisms involved in meeting them, I will for the most part not distinguish between reproductive and nonreproductive colonial movement.

Behavioral Challenges Associated with Colony Movement
There is wide variation among taxa in the complexity of the behavioral repertoires associated with colony movement. In the simplest case, the colony merely abandons the original nest and wanders (walking if in ants, otherwise flying) until it comes upon a new nesting site with appropriate features. At the opposite extreme, as in the honeybees and stingless bees, colony movement may be preceded by a search for a new nesting site by scouts, which then communicate the location of the new nest to nestmates.

Despite this diversity, group movement in all social insects entails certain common challenges. This section reviews these challenges and the mechanisms by which social insects meet them. Note that not all of these challenges are unique to group movement; some would also be faced even by solitarily nesting animals that disperse to a new nest site. Note also that most of the challenges faced by social insect groups would also be faced by dispersing groups in other species. Finally, keep in mind that, when we speak of the challenges facing the group, it is really the individuals composing the group that face these challenges. Thus, we are interested in the behavioral capacities of individuals, including those capacities in-

Fred C. Dyer

volved in social coordination, that allow them to move as part of an integrated social unit.

It is most convenient to organize this list of challenges in the order in which they would arise for a dispersing colony.

Assessing the Need for a Move

Given that the colony might benefit by moving, the first challenge it faces is deciding that a move is appropriate. In the case of emergency moves, this decision may be forced by circumstances such as a sudden deluge, a fire, or an attack by predators. More interesting is colony movement in response to more gradual deterioration of local conditions, or in the context of reproductive fission. In these situations, all members of the dispersing colony must somehow come to agree that it is time to prepare for a move. How might proximate stimuli that provide information about the need for a move produce such a change in behavioral state in the large collection of workers composing a colony?

One possibility is that a subset of the workers experience relevant stimuli indicating a need for a move and then somehow convey the need for a move to the rest of the colony. This mechanism would require, first, that these "forecasters" be in a position to acquire better (or earlier) information about changes in factors affecting the colony, and second, that they have a way of communicating their assessment to their nestmates. For example, if the colony is afflicted by a disease of the brood so serious that the colony would do best to move to a new nest and leave the brood (and the population of disease organisms) behind, perhaps the workers engaged in brood care could communicate the state of their charges to the rest of the workforce.

An alternative possibility, which would not entail a specialized mechanism for detecting and communicating the need for a move, would be for workers to respond to environmental cues experienced in parallel. For example, decreases in the availability of food, which lead to colony movement in tropical honeybees (Koeniger and Koeniger 1980) and in army ants (Schneirla 1971), would affect all the workers in the colony simultaneously. The same would be true of climatic fluctuations, as when the Asian rock bee *Apis laboriosa* moves to lower elevations with the arrival of winter (Underwood 1990), or deterioration of the nest site, as in the open-nesting Asian dwarf honeybees such as *A. florea* (Seeley, Seeley, and Akratanakul

1982). Finally, evidence from honeybees suggests that proximate cues that cause workers to prepare for colony fission come from intranest environmental factors associated with crowding, such as elevated CO_2 levels, which would be available to all of the workers simultaneously (reviews by Seeley 1985; Winston 1987).

I suggested that emergency moves may often be forced by the circumstances of the emergency, and may not require specialized mechanisms for preparing the colony for the move. Even in the case of an emergency, however, there may sometimes be a rather subtle decision-making process involved in assessing whether the threat to the colony is great enough to warrant a move. For example, in India I once observed (unpub.) an attack by weaver ants *(Oecophylla smaragdina)* on a colony of *Apis cerana,* the Asian hive bee, that was nesting in a fallen tree trunk. The attack was under way when I discovered the colony, and the worker bees were engaging in various defensive tactics, such as hovering above groups of ant foragers as if to distract them, smoothing the wood around the nest entrance, and buzzing their wings to blow ants away from the entrance (see also Seeley, Seeley, and Akratanakul 1982). The bees held the ants at bay for at least a day, but sometime on the next afternoon (unfortunately, I did not see the event), the colony gave up and abandoned their nest (and its rich bounty of brood and stored food) to the marauding ants. This move surely qualifies as an emergency move. It also illustrates that colonies must continuously assess their need to move, and then somehow decide collectively that the costs of moving would be outweighed by the costs of staying.

Choosing a Destination
In some species, an important step in preparation for colony movement is for the members of the dispersing colony to agree upon a destination. This process involves a division of labor in which a subset of the workers decide upon the location, or perhaps only the direction, where the colony should move, and then communicate this to other workers.

The social mechanisms by which the dispersing colony agrees on a new nest location need not be very complicated. For example, in the case of swarm-founding wasps (Jeanne 1991) and stingless bees (Wille 1983), the departing swarm moves gradually to a new nest site. The move seems to be initiated by a few pioneer individuals starting construction on a new nest and then, as a result of their activities, presenting the other workers and a dispersing queen from

Fred C. Dyer

the mother colony with a visual and olfactory choice between the old nest and the new nest. This may be all there is to the recruitment process in some species of swarm-founding wasps. In stingless bees, however (Michener 1974, Wille 1983), and in some wasps (Jeanne 1991), the pioneers facilitate the recruitment of new nestmates by laying a series of odor marks along the path from the natal nest. (Such odor trails are also used in most stingless bee species when foragers recruit their nestmates to rich sources of food.) In addition, stingless bees obtain some of the construction materials for the new nest from the old nest, as the workers shuttle back and forth, bringing bits of wax and the plant resins that are used to seal the nest against the elements. After construction is well under way and a substantial number of recruits and a queen have been mustered, the new colony stops exchanging workers with the mother colony. The connection with the mother colony may last several weeks or months (Wille 1983).

Honeybees exhibit the most elaborate set of mechanisms for selecting the destination to which the swarm will move. These mechanisms have been best studied in the European honeybee *Apis mellifera*, beginning with the classic work of Martin Lindauer (1955; reviews by Seeley 1985; Winston 1987). The basic chronology of swarm movement is as follows: First, the departing swarm, consisting of a queen and about half of the original workforce, leaves the mother colony in a sudden rush and clusters on a structure (usually a tree branch) nearby. Although this initial move is obviously coordinated, there is no evidence that the clustering site is decided in advance. Instead, the queen or a group of workers probably chooses a site with appropriate features, and then attracts the rest of the bees to the site by emitting a recruitment pheromone.

From this temporary clustering site, the dispersing colony seeks out new nesting sites and then chooses one to move to. Prospective nests are discovered by scouts, which then perform waggle dances to inform other bees about the direction and distance of the sites they have found (reviewed by Lindauer 1961; Seeley 1985). These dances are essentially identical to those that foragers use to indicate the location of food (von Frisch 1967). The dancer repeatedly runs for a few moments in a straight direction while buzzing her wings and shaking her body from side to side. The direction in which she runs indicates the direction of the goal relative to the sun. The duration of each repeated waggling run indicates the distance. Other bees follow the dancer, then fly from the colony to search for the goal. Many of those that do discover the site return to the colony to perform their own dances. In this way the fraction of the colony knowledgeable about a particular nesting site gradually increases. When this fraction is large enough, the colony takes to the air and flies directly to the chosen nest.

One of the most intriguing features of this process occurs when scouts have discovered more than one prospective nesting site (fig. 6.3). Each scout is able to evaluate the nest that it has found according to a number of criteria—the volume of the cavity, the size of its entrance, its height above the ground, presence of comb remnants from previous colonies (Seeley and Morse 1978; Seeley 1985). The scout then probably decides for herself whether the nest is good enough to indicate in a dance. But if different scouts are indicating different nests, the colony in effect has to decide which one to move to. It is clear that colonies do decide. Even if a dozen prospective nests are discovered initially, eventually dances are performed to only one of them, and the swarm will usually not move until such a consensus is reached (Lindauer 1955; Seeley and Buhrman 1999).

It remains unclear just how a consensus is reached. A major question is how scouts that have initially discovered good nest sites come to abandon them in favor of even better sites being indicated in dances by other scouts. One possibility is that bees directly compare alternative nests by flying out to inspect them and then switching if they find one better than the one they have discovered. This hypothesis implies that the decision is made at the individual level, and is made in parallel over all of the hundreds of house-hunting scouts. An alternative possibility is that something about the dynamics of recruitment via dances leads to an increase in the number of workers visiting good nests and a decrease in the number visiting poor nests, without any direct comparison by individual workers. This hypothesis implies that the decision occurs at the level of the group, akin to the way a colony decides among feeding sites, so that at any particular time it preferentially sends recruits to the best feeding sites available. The evidence is clear that such foraging decisions emerge without any individual bee comparing feeding sites, or even dances to different feeding sites (reviewed by Seeley 1995). Instead, bees detect that they are visiting a relatively poor feeding site, and hence should not perform the dances that would produce more recruitment to that site, by integrating information about the intrinsic profitability of their food source with information about the current needs of the colony. When a colony is flush with stored

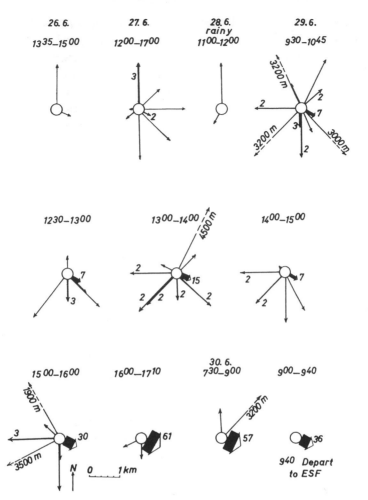

Figure 6.3 Directions and distances indicated by dancers on a honeybee swarm in its temporary clustering site. The arrows point in the compass directions indicated by dancers. The length of each arrow represents the distance indicated. The width of each arrow indicates the number of dancers seen indicating that direction and distance. The diagrams illustrate a series of observations over consecutive days in June (dates, 26–30). They show the initial discovery of several possible nest sites and eventual agreement on one site (in the direction in which the swarm actually departed). (From von Frisch 1967; after Lindauer 1955.)

food or is experiencing a strong flow of food into the nest, a given food source has to be intrinsically more profitable for foragers to reach the threshold of dancing. The proximate cue by which returning foragers obtain information about colony needs is the time that it takes for them to unload their food: the latency to unload is rapid in a hungry colony, slow in one that is well fed.

Recent studies suggest that a decision-making process akin to that seen in foraging patch selection is at work in nest site selection (Seeley and Buhrman 1999; Visscher and Camazine 1999). To begin with, it seems clear that most scouts do not visit multiple nest sites, and hence do not have the opportunity to compare alternative sites directly. This observation points to a process of decision making on the group level and not on the individual level. One important component of the process appears to be a tendency of bees to dance more intensively to better-quality nests (where nest quality has been determined by the dancer). This would lead to a greater accumulation of recruits at good sites. A second component of the process is the tendency of bees to stop dancing to the presumably poorer-quality nests. Working together, these two components could lead to the building of a consensus on the group level, even if no bees directly compared alternative nesting sites.

It is worth keeping in mind that not all of the bees in the swarm follow dances and visit the new nest. Perhaps only a minority do. Certainly the queen does not. Thus, although the colony has reached a consensus on which of a set of available nests to choose, only part of the colony participates in this consensus. This raises the question of how the other bees find their way there. I return to this question in the next section.

An interesting variation on the house-hunting process described in European honeybees is seen in the migratory dispersal of some tropical honeybees. In the Asian rock bee *Apis dorsata* (Koeniger and Koeniger 1982; Dyer and Seeley 1994), and the African hive bee *A. mellifera scutellata* (Schneider and McNally 1994), colonies migrate at the end of the flowering season, apparently in search of better conditions. In both of these cases, as in reproductive swarming, dances play a role in organizing the colony's departure, but migratory dispersal differs from reproductive swarming in a number of respects. First, the whole colony abandons the natal nest, leaving behind a comb barren of brood and stored food. Furthermore, there is no intermediate clustering stage, but instead the colony appears to choose its destination while still in the nest, for that is where the predispersal dances take place. (This raises the question of why reproductive swarms do not choose the new nest cavity while still in the natal nest. One possibility is that a temporal separation between the division of the colony and the selection of a new nest site allows for a more efficient spread of information among the subset of workers that are dispersing. There would be no such advantage when the entire colony is dispersing.)

Fred C. Dyer

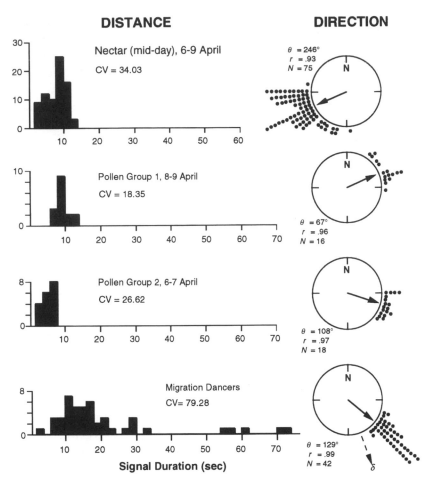

DISTANCE

DIRECTION

Nectar (mid-day), 6-9 April

CV = 34.03

θ = 246°
r = .93
N = 75

Pollen Group 1, 8-9 April

CV = 18.35

θ = 67°
r = .96
N = 16

Pollen Group 2, 6-7 April

CV = 26.62

θ = 108°
r = .97
N = 18

Migration Dancers

CV= 79.28

Signal Duration (sec)

θ = 129°
r = .99
N = 42

Figure 6.4 Comparisons of dances performed by the Asian honeybee *Apis dorsata* when indicating the direction and distance of feeding sites and the direction of colony migration. Both migratory and foraging dances are highly consistent in orientation. Migratory dances indicate distances (indicated by signal duration) with far more variability, and also indicate distances well beyond the typical foraging range, implying that they are not actually signaling specific locations, but only the migratory direction. (From Dyer and Seeley 1994.)

One of the most interesting differences between migratory dispersal and reproductive swarming is that the dancers in a colony preparing to migrate do not appear to signal a particular location, but instead indicate the migratory direction only (fig. 6.4). The evidence for this, from both *Apis dorsata* (Dyer and Seeley 1994) and *A. mellifera scutellata* (Schneider and McNally 1994), is circumstantial, but fairly compelling. Dances in a colony about to migrate are highly consistent in the direction indicated, and this direction corresponds to the direction in which the colony departs. These dances are highly variable, however, in the distances indicated. Furthermore, the distances indicated tend to be much greater than the normal flight range of foragers from the colony, implying that dancers are not signaling paths through regions that they have actually visited.

One final story will further sharpen the contrast between dances to food and dances in the migratory direction. In one colony that I observed in Thailand (Dyer and Seeley 1994), bees began migratory dances in the morning before any bees flew from the nest. Initially only a few bees began dancing, all signaling the same direction and all closely attended by other bees. Within an hour, still before any bees had left the nest, scores of simultaneous dances could be seen indicating this direction. Thus, unlike dances to feeding sites and nesting sites, the migratory dances do not appear to report information acquired during a flight just preceding the dance. Furthermore, this observation raises the intriguing possibility that information about the migratory direction, after being decided somehow by a small number of workers, spreads through the colony via dances only—in "word-of-mouth" fashion—so that new dancers are recruited to the effort without personally experiencing the flight direction being signaled.

Initiation of Move
Since the workers in most social insect species are utterly dependent on the success of the colony for their genetic fitness, the risk of getting lost must be one of the major costs associated with colony movement. Agreeing on a destination is one way to ensure cohesion of the workforce during colony movement. Another way is for everyone to start moving at the same time. For emergency moves, synchrony could be easily achieved if the workers all exhibited a similar threshold response to an external cue. In the absence of such a stimulus, however, the colony must possess some mechanism for mobilizing the workforce simultaneously. In honeybees, the mechanism involves some bees running among their sisters buzzing their wings vigorously. Workers performing these "buzz-runs" (or *Schwirrlaufen:* Lindauer 1955) seem to try to run through the most tightly clustered parts of the colony as if trying to force it to disintegrate. Indeed, buzz-runners rapidly transform the swarm cluster from a relatively placid structure, with most workers hanging head upward in an interlinked three-dimensional network, to a chaotic, quickly dissolving jumble. The time elapsed from the beginning of

Fred C. Dyer

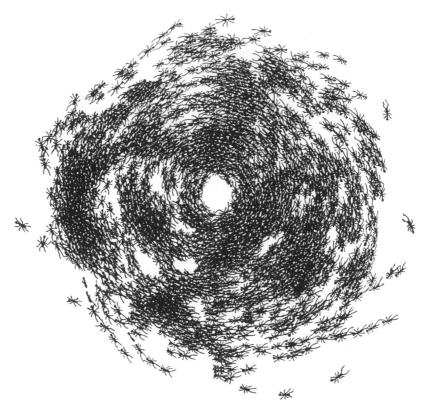

Figure 6.5 A group of workers of the army ant *Labidus praedator* that were separated from the rest of their colony during a rainstorm. Workers normally follow odor trails laid by their fellow workers. When small groups are isolated from the main colony, they become attracted to each other and perform "circular milling" until they die. (From Hölldobler and Wilson 1990; after Schneirla 1971.)

buzz-running to the liftoff of the swarm is usually less than a minute.

The mechanisms mediating the initiation of the move have not been studied in other social insect species, but it is likely that they involve similar physical stimuli, or perhaps chemical cues.

Maintaining Social Cohesion during the Move

Even if the colony chooses its destination (or at least its direction) in advance and the workers and queen all take off together, there still remains a major risk of getting lost in transit to the new site. This is especially true for flying insects, although even ants, despite the possibility of following odor cues applied to the substrate, can easily go astray (fig. 6.5). Again, I shall discuss this problem in reference to honeybees, for which our understanding is most complete, albeit still fairly spotty.

As I mentioned earlier, many members of a reproductive or migratory swarm are likely to be ignorant about the destination chosen by the house-hunting scouts and recruits. The queen, in particular, has no way of obtaining this information, since she never follows dances. So how do the queen and ignorant workers manage to go the right way? One possibility is that knowledgeable scouts lead their colony to the new nest by emitting recruitment pheromone. It is hard to imagine that this would allow much cohesion or specify the flight direction very precisely. Furthermore, if the ignorant bees follow recruitment pheromone during flight, then it should be possible to lead a flying swarm anywhere using an artificial source of pheromone. A swarm will follow an artificial lure, but only if it is moved in the direction that the dances have indicated (Avitabile et al. 1975), which implies that the bees are responding to some other source of information besides chemical cues.

One clue has been provided by observations of the early stages of the swarm's flight to its new nest. As the swarm takes to the air, it initially seems quite disorganized, and moves slowly away from its clustering site. A swarm of ten thousand or so bees, which would form a cluster about 15 cm in diameter, spreads out in the air over a diameter of 50 m or more, and most of the bees seem to soar around aimlessly. At this stage, however, one can see some bees seeming to shoot through the group in the direction that has been indicated in the dances, and then circle back to repeat this maneuver (Lindauer 1955; F. C. Dyer, pers. obs.). The swarm gradually picks up speed in this direction, and as it does, contracts to a diameter of roughly 10 m. Over the next few minutes, the swarm accelerates to its full speed of about 10–15 km/h, and maintains this speed all the way to the new nest.

It is tempting to speculate that the initial stage of the swarm's flight serves to "teach" the departure angle to those bees, including the queen, that have not followed dances or visited the new nesting site, and that the teachers are the bees that shoot through the swarm toward the new nest. If the bees that shoot through the swarm are ones that are already informed about the location of the goal, and if these bees provide information to the naive bees in the swarm, then this behavior might regarded as a large-scale version of the waggle dance. By visually following the bees shooting through,

Fred C. Dyer

naive bees would be able to learn the direction (relative to a compass reference such as the sun) in which they should continue their flight.

The role of the queen in this process is very interesting. Despite her ignorance of the swarm's destination, her presence (or her odor) is necessary for the swarm to make progress in the initial stages of its trip. If one cages the queen to prevent her from taking off, the swarm will take off (proving that she does not participate in initiating the move), but will travel no more than about 50 m away before returning and reassembling around her (Avitabile, Morse, and Boch 1975). If one carries her (or a dummy soaked in queen pheromone) with the swarm as it accelerates in its departure direction, then it will proceed to the new nest. Occasionally, a swarm that has gathered speed will leave a caged queen behind and not return to her (F. C. Dyer, pers. obs.). Thus, it is as if the workers, having checked that the queen is present, simply assume that she will stay with them. Clearly the queen is not being herded toward the new nest by the knowledgeable bees, but instead must have her own way of figuring out which way to go.

This line of speculation may seem far-fetched at first glance. On the other hand, one must keep in mind that the phenomenon it seeks to explain is quite real: naive members of the swarm somehow learn the direction to the new nest only after the swarm takes flight, and this information must be provided by knowledgeable members of the swarm. Only olfactory and visual cues seem remotely possible as explanations for this phenomenon, and there are strong reasons to doubt that olfactory cues play a role. The challenge now is to understand exactly how visual information could do the job.

Assembly in a New Location

At best, only some of the workers in a dispersing colony are informed about its destination. In the case of colonies forced to move in an emergency, or colonies undertaking a long-distance migration, none of the members of the colony know in advance where the colony is going to stop. Thus, the colony must have some way of calling an end to the trip and assembling its members in the new nest or a resting place. Surely chemical signals play the most important role in the assembly process itself; assembly pheromones have been described in numerous species. But this implies that some individuals play a leading role in initiating settling, and that others are ready at any moment to respond to a signal to settle. The queen might play this role so long as the workers monitor her presence in the moving colony. Alternatively, if a large enough subpopulation of workers slowed and stopped, then an accumulation of recruitment pheromone could induce settling by the rest of the colony, including the queen.

The assembly process can sometimes be a messy affair. In honeybees, where again our knowledge is best, the assembly pheromone is produced by the Nasanov gland near the tip of the worker's abdomen. When a migratory or reproductive swarm begins to settle, the first workers to land emit Nasanov pheromone while fanning their wings to disperse it, which attracts more workers, who then emit pheromone (Lindauer 1955; Seeley, Morse, and Visscher 1979). The sweet Nasanov scent is easily detectable to a human observer standing within a few meters downwind of a settling swarm. The vast majority of workers are drawn into the main swarm by this stimulus, but often clusters of bees form on branches and leaves up to 2 meters away. To complete the reassembly of the colony, workers from the cluster containing the queen (usually the main swarm, but sometimes one of the satellite clusters) fly to other groups of bees and burrow through them, performing buzz-runs. This disperses these clusters and seems to cause the bees in them to seek out the main cluster.

The power of the Nasanov stimulus is seen most dramatically in the behavior of Africanized honeybees (the infamous "killer bees") in Central and South America (O. Taylor, pers. comm.). Like the African bees from which they descended, these colonies are migratory (see Ratnieks 1991). During the migration season, swarms move several kilometers each day, resting at night. Sometimes a swarm that has settled to rest manages (presumably accidentally) to attract passing swarms to the same site with its Nasanov pheromone. This often results in aggression among the workers, and workers of each colony mob the other colony's queen. This behavior makes it easy to find queens in such swarms: they are at the center of tight balls of aggressive workers. Taylor has observed megaswarms consisting of hundreds of thousands of workers and more than a dozen queens.

Establishing a New Nest

Once the colony has moved to a new nest, it must fortify the nest and prepare it for brood rearing. For army ants, these are negligible challenges, since the "nest" is composed of a mass of interlinked

Fred C. Dyer

workers clustered at the base of a tree or rock (Schneirla 1971). Most other mobile social insects, however, must seal cracks in cavity walls, lay down sticky bands to defend against ants, or build wax or paper cells for rearing brood and storing food. During this period the colony may be highly vulnerable. Predators may more easily breach the walls of the nest. Rainy weather may prevent the collection of food, leading to starvation if not enough has been hoarded. The lag between the move to the nest and the production of the first new workers delays the resumption of colony growth and the replacement of any workers lost to predators or starvation.

One way to ensure success is to choose a nest site in advance that is readily defended against predators, rain, or cold. This strategy is employed by honeybees, stingless bees, and swarm-founding paper wasps. An additional strategy, employed by stingless bees (Wille 1983) and possibly by paper wasps (Jeanne 1991), is to prepare the new nest for occupation before the main move of the swarm, by transferring nesting materials and food to serve as a grubstake. Finally, in the dwarf honeybee *Apis florea,* a colony forced to abandon its old nest by predators or loss of cover will recycle wax from the abandoned comb (Seeley, Seeley, and Akratanakul 1982). Workers return to the comb and carry bits of wax on their hind legs like pollen balls. One colony that I studied in Thailand harvested about 50% of the mass of its old comb. This strategy results in a substantial savings of energy and time because new wax, which is secreted from glands on the abdomens of young workers, would otherwise have to be synthesized from sugars collected in the form of floral nectar.

Spatial Cognition and Group Movement

Like other animals whose fitness is partly a function of the efficiency with which resources are extracted from a home range, social insects have evolved an array of strategies for finding and exploiting rich sources of food. Central to the success of these strategies are sophisticated navigational abilities based on the ability to encode and use visual information about spatial relationships in the environment (see recent reviews by Dyer 1994, 1996, 1998; Wehner, Michel, and Antonsen 1996). We have seen in the previous section that among the challenges that arise in changing nest locations, the workers in a social insect colony must find a new nesting site, avoid returning to the old site, and find new feeding sites. For insects that renest within their original foraging range, meeting these challenges may mean adopting new responses to environmental features that have been used for navigation while foraging from the old nesting site. Thus, previously learned spatial relationships would have to be replaced or modified by new information. To understand how an animal's knowledge of spatial relationships could be affected by a change in nest location, it is useful to understand what animals know about spatial relationships and how they come to know it. Here I discuss these issues in reference to honeybees, in which our understanding is best developed.

First, a word about applying the term "cognition" to insects. I favor using the term in a broad sense to refer to any mechanisms involved in the acquisition and storage of sensory information and the use of that information to produce adaptive behavior. Some authors restrict the term to processes mediated by internal representations of objects, events, and relationships in the outside world. Representations are neural records of external patterns experienced by the animal. As we shall see, behavioral evidence from honeybees and other insects justifies the use of "cognition" even by the most stringent definition. Indeed, there is little evidence of a true discontinuity, either taxonomically or in the underlying mechanisms, separating "cognitive" and "noncognitive" behavioral phenomena. Thus, I will not make a special effort to justify using the term for insects. (Interested readers should refer to Gallistel 1990, Dyer 1994, and, for an alternative point of view, Kennedy 1992.)

Basics of Insect Spatial Cognition
The basic navigational task that social insect workers face is to find their way back to the nest after a long foraging trip in search of food. In the case of species that exploit resources distributed in rich patches, workers may also have to find their way repeatedly to sources of food discovered previously. In either case, the animal's task is to find a specific familiar point in its environment. For insects that rely primarily on chemical cues applied to the substrate (most ants and all termites), this task is relatively straightforward. The animal just has to locate the trail and determine which direction it needs to travel along the trail (reviewed by Hölldobler and Wilson 1990). Chemical cues are useful primarily to walking insects, although stingless bee foragers recruit nestmates to food using a series of chemical marks applied to vegetation en route to the food.

For most flying insects, chemical cues are useful only when the animal has approached to within a few meters or tens of meters the goal. At greater distances, the spread of chemical cues through diffusion and air movement makes it extremely difficult to obtain information about the location of the goal. For such species, vision is the chief sensory modality for navigation (see Wehner 1981). Some ants also have abandoned the use of chemical trails in finding familiar food sources, and instead do so visually. The literature suggests that ants that use visual cues for navigation are those living in habitats (e.g., deserts) where chemical trails would be costly to maintain, or those whose foraging strategy involves the exploitation of small, ephemeral prey items (Wehner, Harkness, and Schmid-Hempel 1983).

Since the task of using visual cues for long-distance navigation is so familiar to human beings, it is easy to underestimate the challenges that it entails, especially when we consider the limited neural resources that most insects bring to it. For a foraging bee preparing to head home, the goal is in all likelihood not directly in view, and it will not be in view until she has traveled several hundred meters. (One must bear in mind that bees and other flying insects tend to stay within a few meters of the ground during their foraging flights, and thus have a vantage point not very different from that of a terrestrial animal.) Thus, the bee must use visual cues that she can see at her starting point and along the way. Most insects use celestial cues and landmarks for this task. But to use the sky and landmarks for navigation, the bee must be informed of the spatial relationships between the specific features visible at her starting point and the location of her goal. These spatial relationships must be learned by the animal, since there is no way that they could be innately specified.

To find the way to a distant point, a navigator (of any species) needs to obtain two kinds of information from the environment (Dyer 1998). First, it needs to be able to discriminate different directions from one another; thus it needs the equivalent of a compass. Second, it needs to determine which of the directions that it can discriminate is the one that will lead it to its goal; hence it needs a sense of position. A sense of position may include information not only about the direction but also about the distance of the goal. We now know a great deal about how insects obtain directional and positional information from the environment.

A major source of directional information, for insects as well as for many vertebrates, is the sun (reviews by von Frisch 1967;

Wehner 1984; Schmidt-Koenig, Ganzhorn, and Ranvaud, 1991; Dyer 1996, 1998). Insects can also use sun-linked patterns of polarized light in blue sky if the sun is obscured by clouds. The sun is a highly reliable directional reference, but a major difficulty is that its direction relative to earth-based features changes over the day as a result of the earth's rotation. Thus, to use the sun to travel a straight course over the earth's surface, an animal needs to compensate for the shift in the sun's direction relative to the line of travel. Many animals can do this, and in most species in which sun compass orientation has been studied, the knowledge of the sun's course is based on an internal representation that develops through experience early in life (reviews by Gallistel 1990; Dyer and Dickinson 1996). This internal representation, which encodes the sun's position as a function of time as indicated by the animal's internal clock, allows animals to update their estimate of the sun's direction even during intervals when they cannot see the sun.

Landmarks, obviously, can also allow an animal to discriminate directions at a particular location. Most landmarks are not, however, very useful for compass orientation; that is, as references for maintaining a straight direction of travel over a long distance. As the animal moves along a straight path, the directions of individual landmarks shift relative to its body axes, and eventually the landmarks are lost from view. Some animals can, however, maintain a straight course relative to features of the terrain that are aligned with the direction of travel. For insects, such features include forest edges and shorelines (von Frisch 1967). Some migratory birds are thought to use shorelines or mountain ranges to steer themselves during long-distance migrations.

Some of the most intriguing puzzles in animal navigation concern the sources of positional information that allow animals to determine which direction is the one that will lead toward their goal. For nesting insects, this problem arises most often when they need to set a homeward course from a feeding site. Two sources of positional information are available. First, during the outward trip in search of food, a forager measures the directions (relative to her celestial compass) and distances traveled over successive segments of the path, and then, even if the path has been quite circuitous, she computes her net direction and distance of displacement from the nest (von Frisch 1967; Wehner 1982, 1991). This process is known as path integration, and is closely analogous to the process of "dead reckoning" performed by human navigators. What is striking is that

Fred C. Dyer

the outcome of path integration is a piece of knowledge—the insect's estimate of the direction and distance she should travel to reach home—that may not correspond to any portion of the outward path, although it is derived from information acquired during the outward path. This is an example of a capacity of insects that meets a fairly stringent of definition of cognition. Path integration allows animals to determine their position relative to home when they have wandered into unfamiliar terrain, and thus is an essential component of the ability to explore a new environment and discover new sources of food.

The second major source of positional information—landmarks—allows insects to find their way home from locations within familiar terrain. Normally, the information provided by landmarks reinforces the information provided by the path integration system. Landmarks can also serve as a backup system for navigation when path integration is impossible—for example, when clouds obscure the celestial compass cues that provide the directional reference for path integration, or when the animal has been displaced passively by the wind (or by the hand of an experimenter). Landmark learning, like path integration, raises fascinating questions about the cognitive processes underlying behavior, and has received a great deal of attention in vertebrates as well as invertebrates.

Studies of landmark learning in insects have drawn a distinction between the use of large-scale features of the terrain for navigation over long distances and the use of nearby landmarks to pinpoint the location of a familiar goal. These tasks entail distinct challenges. The large-scale task, in essence, requires knowing how to respond to a given set of landmarks in order to leave them behind; that is, to move toward a goal that is beyond the current horizon. The small-scale task involves finding a specific location (e.g., a nesting or feeding site) relative to a currently visible array of landmarks. The vast majority of studies of landmark learning in vertebrates (including all the classic laboratory research paradigms involving cue-controlled arenas) present animals with the small-scale task. A truly integrative study of landmark learning requires understanding both of these tasks. As we shall see, the navigational challenges presented to insects as a result of changing nesting sites may also require analysis on both small and large spatial scales.

To a first approximation, insects solve the small-scale task of pinpointing the location of a goal by finding the spot where their current view of surrounding landmarks matches the view remembered

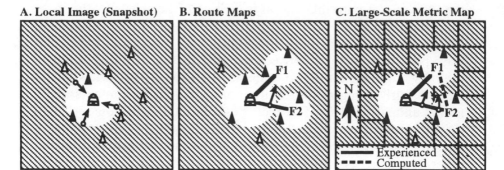

A. Local Image (Snapshot)   B. Route Maps   C. Large-Scale Metric Map

Figure 6.6 Three ways in which an animal may learn about spatial relationships in its foraging range. Shaded regions are *terra incognita;* solid triangles are known landmarks; open
triangles are unknown landmarks. Arrows originating at small circles indicate paths bees
can select toward home *(A)* or toward feeding site F1 *(B and C)* from various starting
points. *(A)* Local image, or "snapshot," of an array of landmarks surrounding a given goal;
bees can use such snapshots to approach the nest from a variety of directions (Cartwright
and Collett 1983). *(B)* Route maps for two feeding sites, F1 and F2; each route map consists
of two local images encountered in sequence along the path to the food. Bees can use such
maps to head for an unseen goal, even if they have been displaced so that they see the
landmarks associated with the route to their goal from new vantage points (Dyer, 1991; Dyer
et al. 1993). (C) Large-scale metric map, in which bees have encoded the directions and
distances separating familiar sites in reference to a common coordinate system. A metric
map would allow a bee to set a novel course to F1 even if she found herself near F2, from
which she could not see any landmarks associated with the route from the nest to F1.

from previous visits (Tinbergen and van Kruyt 1938; reviews by
Wehner 1981; Gallistel 1990; Dyer 1994). What is encoded in memory is a panoramic snapshot that contains information about the
size, shape, and distance of surrounding landmarks as well as their
overall geometric configuration (fig. 6.6A). This basic finding is
quite similar to what has been learned from studies of rodents in
the Morris water pool or the radial arm maze (see Gallistel 1990;
Dyer 1994 for review). The big riddle has been how animals use
these memory images to guide their search for the goal. If the use
of familiar landmarks were simply a matter of recognizing a familiar scene once you had arrived at the vantage point from which you
saw it before, then it would not require a very complicated cognitive
process. In various species, however, the evidence suggests that animals can set a course for a familiar goal even when starting at a
different location that offers a different view of the landmarks (Collett 1996; McNaughton et al. 1996). In the case of insects, this flexibility has been cited as evidence that the insect brain encodes a
generalized "map" of the landmarks relative to an external coordinate system, in much the same way as we can learn locations in a

Fred C. Dyer

room relative to room-based coordinates, and not only relative to our own body axes (Gallistel 1990). To form a representation of this sort, the insect's brain would have to perform some sort of calculation to transform the initially egocentric view of the landmarks as detected by the visual system into an allocentric map. As tantalizing as the behavioral evidence is for such processes, careful experimental investigations have only reinforced the notion that insects encode visual images in an egocentric reference frame (i.e., relative to retinal coordinates). Their ability to use these egocentric representations to set a course from a novel vantage point is based on various simple strategies that ease the task of comparing the current view of landmarks to the egocentric image stored in memory (Collett 1996).

The question of whether insects encode simple or complex maps of familiar landmarks also arises when we consider the large-scale task of setting a course for a goal that is too distant to be seen from the animal's starting point. This task would seem to require animals to develop an integrated map encoding the spatial relationships among landmarks experienced in different parts of the terrain. The nature of the large-scale spatial maps memorized by insects has generated considerable interest, and no small amount of controversy (see reviews by Gallistel 1990; Dyer 1996; Wehner, Michel, and Antonsen 1996).

The simplest sort of large-scale map of the terrain is one in which the animal encodes a sequence of visual images corresponding to the landmarks seen in successive stages of a familiar route through the terrain (fig. 6.6B). For animals with a large foraging range, this would require enough memory capacity to encode multiple images, but would be computationally very simple. Conceivably, each image could be encoded along with information about the direction to be traveled relative to the landmarks encoded in the image. Once in view of the landmarks encoded in one of the images, the animal would be led from scene to scene along the route to the goal. The map would not need to encode information about sequence per se, or about the distance from the goal at which each image was encoded. Such a simple representation would allow an animal to navigate accurately over a long distance. Route maps formed for multiple paths radiating from the nest would allow an animal to behave as if it had a more generalized map of the terrain.

The route map hypothesis was first outlined more than 50 years ago by Baerends (1941) to account for the homing abilities of digger wasps; it long served as the standard model of large-scale landmark learning in insects (see Wehner 1981 for review). More recently, Gould (1986) suggested that honeybees, at least, form a more sophisticated representation of spatial relationships in familiar terrain. Specifically, Gould argued that honeybees can integrate their experience on two different, separately traveled routes into a topographically accurate cognitive map resembling those that human beings can form (Byrne 1982). To form such a map, the animal must encode the directions and distances traveled on each route in a common reference frame (for example, that provided by the celestial compass). The map then allows the user to compute novel routes connecting familiar locations in different parts of the terrain (fig. 6.6C). Following Gallistel (1990), I shall refer to this sort of large-scale map as a "metric map," since it entails encoding the metric relationships (distance and direction) among sites charted in the map.

Gould's evidence for such a map came from experiments in which he trained bees to an artificial flower at a particular location in the terrain, and then observed that the bees could set a direct (i.e., shortcut) course for the feeding station when he captured them at the hive and displaced them to a third location off the foraging route. Unfortunately, Gould's results proved hard to replicate (e.g., Wehner et al. 1990; Menzel et al. 1990). I was able to replicate them, but I also showed that the bees' ability to set the shortcut course after displacement vanished under certain conditions (Dyer 1991; Dyer, Berry, and Richard 1993). Specifically, if bees could not see from the release site landmarks that had previously led them to the food, then they flew in other directions and not toward the food. My results, as well as Gould's, are most easily explained by the hypothesis that the displaced bees were guided by a route map encoding landmarks previously seen on the way to the food. An important provision of this hypothesis must be that the bees can compensate for displacements from the route that present them with an altered view of the landmarks (see fig. 6.6B). As discussed, such flexibility is well known in connection with the use of landmarks on a small spatial scale (i.e., to pinpoint the goal during the final approach to it), and it is not hard to imagine that bees use similar mechanisms for responding to larger-scale terrain features. The crucial point is that there remains no decisive evidence that bees, or any other insects, can compute a truly novel course over unfamiliar terrain. Instead, all evidence suggests that, in using landmarks for

Fred C. Dyer

navigation, insects are constrained to follow routes that offer a view of landmarks previously learned (for evidence from ants, see Collett et al. 1992; Wehner, Michel, and Antonsen 1996).

Navigational Challenges Presented by Colony Movement
The preceding section provides a context for evaluating how workers deal with the navigational challenges that they face as a result of their colony changing nesting sites. The challenges that I am considering here are faced only by individuals that have had experience as foragers prior to the move. Such workers have already learned the location of the new nest relative to landmarks in the environment. Many of them have also already learned the locations of feeding places to which they have traveled from the original nesting site. Thus, a change in nesting site requires them to update or replace previously learned spatial information in order to find feeding sites and then return safely home.

We can consider this challenge in two situations. First, the colony may move far enough that its new nesting site is outside its original foraging range. This situation would arise for colonies of migratory species such as the African bee *Apis mellifera scutellata* (Schneider and McNally 1994) or the Asian rock bee *Apis dorsata* (Dyer and Seeley 1994). After the arrival of a migratory colony at a new nesting site, experienced foragers are confronted with a completely new terrain. They must learn the location of the nest in reference to a new set of landmarks and learn the routes to new feeding places. With respect to the landmarks that they need to learn for navigation, these bees would be in much the same position as naive bees making their first flights outside the nest. A major difference is that bees with prior flight experience would have a knowledge of the sun's course (provided that the migratory flight did not carry them over too many degrees of latitude). This knowledge could enable them to undertake long foraging flights and, by using the celestial compass as the reference for path integration, find their way with little risk of getting lost. However, new landmarks would need to be learned from scratch.

Second, the colony might renest within its original foraging range, confronting experienced foragers with the challenge of adopting new responses to landmarks that they have previously seen while foraging from the old nesting site. This is the challenge faced by honeybee workers that depart with a reproductive swarm. Swarms typically select nests within a few hundred meters of the natal nest, well within the several-kilometer foraging range that workers might have covered from the natal nest (Seeley, Morse, and Visscher 1979). Similarly, most swarm-founding wasps probably also disperse relatively short distances from the natal nest (Jeanne 1991). Finally, most emergency moves, whether in social or solitary species, probably result in renesting well within the original foraging range. Thus, the challenge of adopting new responses to familiar terrain features is probably common enough that a wide variety of animals have evolved mechanisms to deal with it.

We can examine the problems faced by short-dispersing individuals for both homing animals and animals searching for a familiar feeding site. Both situations have been studied experimentally in honeybees. Consider first the task faced by a foraging bee seeking to return to the new nest cavity occupied by the swarm with which she has dispersed. In all likelihood, some foragers may find themselves on or near routes that they have previously used to fly back to the natal nest. To find the swarm's new nest, such a bee would therefore have to suppress or alter her learned response to these landmarks. How good are swarm bees at doing this, and what mechanisms allow them to do so? Robinson and Dyer (1993) studied this problem with artificial swarms, each of which consisted of a cluster of bees discharged from the natal nest and allowed to cluster around an encaged queen. Such swarms behave in every discernible way like natural swarms. We studied bees in these artificial swarms after hiving them a very short distance (about 20 m) away the natal nest.

The first question we addressed was how quickly swarm bees reoriented to a new nest: were they immediately faithful to the new nest, or did they initially drift back to the natal nest? We hived the swarm at dusk, when it was too dark for the bees to fly, and then observed the first departures from the new nest at dawn the following day, which was the bees' first opportunity to fly. We found that the bees were faithful to the new nest even at this early stage: over the first few hours; just as many bees returned to the swarm nest as departed from it (fig. 6.7A). By contrast, as a control, we displaced some colonies directly over a similar distance, without letting them form a swarm prior to the move, and observed a high rate of attrition after the colony was established at the new nest site. We also observed that the first bees departing from the swarm nests, but not those departing from the control nests, performed "orientation flights," the characteristic hovering maneuvers that have been im-

Fred C. Dyer

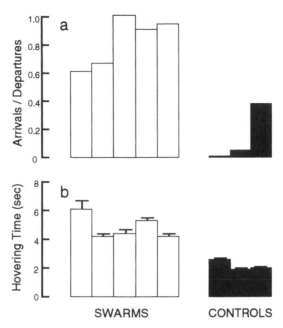

Figure 6.7 Evidence that bees reorient quickly to a new nest site after moving into it with a swarm. *(A)* With the very first departures from the new nest, the ratio of arrivals to departures is very high, even though the natal colony is only 20 m away in plain view. In control colonies, which were displaced a similar distance but without intervening experience in a swarm, the ratio of arrivals to departures is low, indicating that the workers are drifting away (back to the natal nest). *(B)* Bees departing swarm nests perform longer orientation flights than bees departing control nests. (Data from Robinson and Dyer 1993.)

plicated in learning the visual appearance of the nest and the surrounding landmarks (Becker 1958; Collett and Lehrer 1993; Capaldi & Dyer 1999). Control bees spent significantly less time hovering in front of the new nest before flying away (fig. 6.7B). We concluded that something about the experience of being in a swarm had induced bees to perform orientation flights on their first departure from the new nest, and that this had allowed them to learn the location of the new nest and avoid the natal nest, even though it stood in plain view only 20 m away.

The next question that we addressed was whether the reoriented bees avoided the natal nest because they had forgotten where it was, or because they retained a knowledge of its location but had suppressed this information in favor of the newly formed memory of the swarm nest. We captured a large number of the reoriented bees (which had been labeled with paint dots or numbered tags so that we could keep track of them), removed the swarm's nest box, and

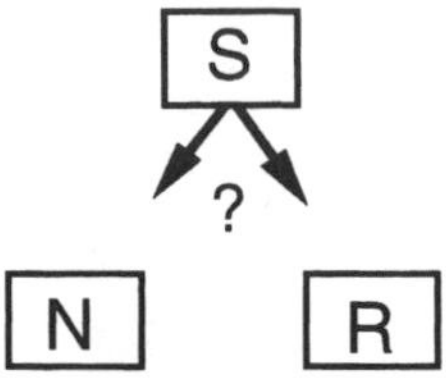

	Number of swarm bees recovered at:	
	Reference hive	Natal hive
	3	65 (95.6%)
	2	45 (95.7%)
	1	39 (97.5%)
	13	325 (96.2%)
	3	84 (96.6%)

Figure 6.8 When bees that have reoriented to a new nest as part of a swarm are then deprived of the new nest, they preferentially arrive at the site of the natal nest rather than at a reference nest presenting an equivalent stimulus in a different location. Data show the results from five different swarm colonies. (Data from Robinson and Dyer 1993.)

then released the captured bees (fig. 6.8). We observed that the bees gathered predominantly at the location of the natal nest and not at a control nest that offered an equivalent visual and olfactory stimulus the same distance away. This result proved that the reoriented bees had retained a memory of the spatial location of the natal nest. This memory evidently had been completely suppressed so long as the swarm nest remained in place, but remained available to guide the bees to the location of their former home when the swarm nest was removed.

This ability to remember the location of the natal nest is perhaps not surprising when we consider its likely selective advantage. Establishing a new nest is a risky proposition; in northern temperate habitats, newly established colonies have a much lower probability of surviving the winter than mature colonies do (see Seeley 1985 for review). The swarm starts off with only a limited grubstake—the honey that the workers can carry in their stomachs—and cannot grow in size until a comb is built and the first generation of newly laid eggs undergo the two-and-a-half-week development period. Thus bees must occasionally find themselves in swarms that, because of bad weather or an ill-chosen nest, are doomed from the start. In this situation it may pay a worker to be able to return to the natal nest so that the rest of her working life span will not be wasted. Furthermore, we can find a parallel to this experimental result in the tendency of workers in the Asian dwarf honeybee *Apis florea* to harvest wax from their old comb when they have been forced to change nesting sites (Seeley, Seeley, and Akratanakul 1982). In this case, the workers clearly need to remember the location of their old nest, if only because it has now become an invaluable resource.

Fred C. Dyer

Although it is easy to make adaptive sense of this ability, to me it is still impressive that the individual bee is able to reorient to the new nest while remembering the location of the old one. This phenomenon is especially striking in the case of our experiments (Robinson and Dyer 1993), in which the old and new nest sites were close enough to each other to be surrounded by the same panorama of landmarks. This implies that the reorientation process does not involve replacing or overwriting information about how to respond to these landmarks to find the nest, but instead learning a second response to the landmarks in addition to the first.

Now I shall consider the implications of colony movement for the task of finding feeding sites. If a colony relocates to a new nesting site within its original foraging range, then clearly experienced foragers in the swarm are within striking distance of feeding sites that they have already discovered on foraging trips from the natal nest. It would enhance their contribution to the new colony to be able to return to those feeding sites rather than seeking out new sites. But this, in turn, would require the bees to approach those feeding sites from a different starting point in the terrain.

To mimic this problem experimentally, I used artificial swarms containing bees that had been trained to artificial flowers (lightly scented sugar water) prior to the swarm's move (Dyer 1993). In the example shown in figure 6.9, one of several versions of this experiment that I have performed, the foraging route from the natal nesting site was more than 100 m long, and the swarm was allowed to move more than 100 m to a new nesting site, so that the foragers would indeed have to approach to the food from a different direction. As soon as the swarm settled in its new nest, I placed a feeder in the original feeding site, and also distributed three or four control feeders in other locations relative to the swarm nest. If the experienced foragers (which had been individually labeled) retained no memory of the original feeding site, then they should show no preference for this site in searching for food. In fact, I found a very strong preference, which suggests that they remembered it. One possibility is that they did not remember its location, but only its odor. However, all of the control feeders were scented in the same way. Furthermore, in the experiment shown in Fig. 6.9, observed the vanishing bearings of experienced foragers as they made their first departures from the new nest. The bees were strongly oriented toward the original feeding site, which was more than 150 m away, and not upwind. Thus, the bees most likely were able to recognize

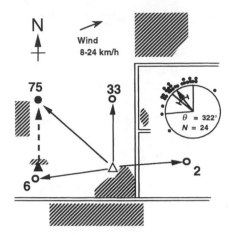

Figure 6.9 How bees find familiar feeding sites after changing nesting sites with a swarm. The triangles show the old (solid) and new (open) nesting sites, which were about 100 m apart. The circles show the locations of feeders offering sugar water. The solid circle is the feeder to which the bees had been trained before the swarm moved. The open circles show feeders placed out after the swarm moved. The numbers indicate how many bees arrived at each feeder after the swarm occupied the new nest. Most arrived at the original feeding site. Some, however, found the food placed in the original compass direction to which the bees had been trained. The inset shows the vanishing bearings of a subset of the previously trained bees, which were observed as they left the swarm's new nest. The bearings indicate that the bees set a course directly for the feeding places where they in fact ended up. (From Dyer 1993.)

from their new starting point in the terrain the direction that would lead to the familiar feeding site.

I have just described two lines of experimental evidence that reveal considerable plasticity in the responses of bees to familiar landmarks. When experienced bees change nesting sites as members of a reproductive swarm, they rapidly reorient to the new nesting site, but do not forget either the location of their original nest or feeding sites that they visited prior to the move. This evidence suggests strongly that spatial information governing bees' responses to landmarks is encoded in a very flexible fashion. Although this does not require us to discard the basic hypothesis that bees are guided by route maps encoding sequences of visual images encountered en route to specific goals in the environment, we must recognize that bees are not rigidly constrained to follow a specific set of paths radiating from a nesting site.

Conclusions
For investigating the mechanisms and adaptive design of social behavior, insect colonies are profitably viewed as "superorganisms,"

Fred C. Dyer

exhibiting on the social level many phenomena that individual organisms also exhibit (Seeley 1989, 1995). In both cases we have a collection of parts (the workers of an insect colony; the cells of an organism) whose fitness is dependent on the fitness of the whole. Like an organism, a colony faces the challenges of surviving and propagating itself, and can deal with multiple challenges simultaneously by coordinating the activities of its multiple parts. As in organisms, we find elaborate mechanisms that serve to coordinate the functioning of the parts in support of the whole.

A major disanalogy between insect colonies and organisms, of course, is that the workers in an insect colony, unlike the tissues of an organism, are physically separate entities. Thus the parts of the whole are held together, and their labors coordinated, not through biochemical interactions but through behavioral interactions. Each worker must be able to obtain information about the environment and colony's needs and then respond in an appropriate way, which may involve moving physically through the colony or in the environment outside. Thus, whereas the individual cells composing an organism typically have a greatly reduced "behavioral repertoire," social insect workers in many species can respond to a wide variety of stimuli with considerable flexibility. Indeed, this behavioral sophistication may be essential given the special challenges that arise in coordinating the activities of entities that are physically, if not biologically, autonomous.

The biology of group movement in social insects provides a particularly clear illustration of these issues. In effect, the group is the entity that must decide whether to move, when to move, and where to go, but for this to happen, the individual workers composing the group must detect the need for a move, coordinate the timing of the move, agree on the destination, maintain social cohesion during the move, and then learn the new location of the colony in the environment. Whereas all the tissues of a dispersing organism get carried along for the ride (an animal's liver or gut has little chance of being left behind in a thicket), each member of a dispersing colony must actively participate in the move, or it may go astray and thus lose its only opportunity to contribute to its own fitness. Thus, as we have seen in this chapter, each colony member must be equipped with the sensory and cognitive mechanisms that allow it to participate.

Most of what I have said in the preceding paragraphs applies with almost equal force to the coordination of social behavior, including group movement, in species with less intense social cohesion. Indeed, although the focus of this chapter has been on insects, I have tried to make clear that the challenges associated with group movement in social insects would arise in any social animal, whether invertebrate or vertebrate. Some mammals face challenges not encountered by social insects—for example, the fission-fusion social organization, in which individuals move repeatedly among different patterns of association, has no exact parallel in insects. Furthermore, precisely because most social vertebrates retain the ability to reproduce on their own, they face a decision—whether to remain with the natal group or to leave it—that does not arise in the highly eusocial species I have been focusing on (although it does arise in some primitively eusocial wasps; see Gadagkar 1991). Finally, the decisions that group-living vertebrates face about where and when to move, and whom to move with, often take into account the identities of other individuals in the group, which requires capacities of social cognition not required in insect societies. In spite of these differences, however, an individual social primate or carnivore would face many, even most, of the same challenges that I have outlined for social insects.

To the degree that social insects resemble other social animals in the problems they encounter in group movement, the study of social insects offers an example of how to categorize and study the solutions that animals employ to solve these problems. The convergence of problems, and the ability to solve similar problems, certainly does not mean that the mechanisms underlying the solutions are similar. Indeed, my view, which presumably would stir little controversy among readers of this book, is that insects are unlikely to provide a useful model of the underlying mechanisms in other animals. On the other hand, the literature on social coordination, spatial cognition, and the biology of group movement in social insects is more mature than the corresponding literature on primates and other animals. This literature could provide a valuable guide to primate researchers by outlining a set of concepts regarding possible mechanisms underlying spatial orientation, and by providing a model research program for investigating these mechanisms.

Acknowledgments
This chapter was written during a sabbatical at the University of California, San Diego. I thank Jack Bradbury and Sandy Vehrencamp for their hospitality and friendship. My research described here has been supported by several grants from the National Science Foundation Animal Behavior Program.

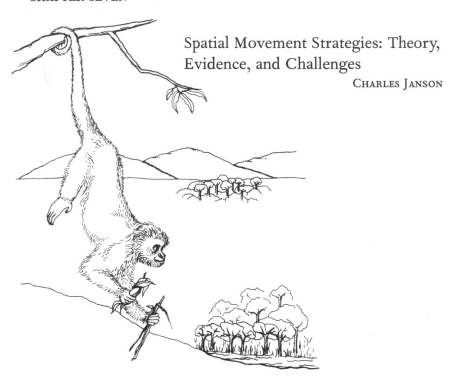

CHAPTER SEVEN

Spatial Movement Strategies: Theory, Evidence, and Challenges

CHARLES JANSON

Why is it that pedestrians on college campuses consistently ignore paved walkways and create shortcuts across once-green lawns to save at best a few meters of walking between buildings? Apparently, natural selection has favored a nearly irresistible urge in humans to minimize travel distances between goals. This fanatical drive is by no means restricted to humankind. Studies of several primate species suggest that they also choose highly efficient routes (e.g., Boesch and Boesch 1984; Garber 1989; Menzel 1973a). Insects integrate complex outbound foraging paths and then return to their nests using the proverbial "beeline" (Gallistel 1989), and "as the crow flies" is a metaphor for the most direct route between two points.

Despite this seemingly universal tendency to reduce travel time and effort, it is not at all obvious how this result is achieved. Despite decades of work, mathematicians have failed to find a solution to the famous "traveling salesman problem," or TSP—that is, an efficient direct algorithm guaranteed to produce the shortest travel circuit among many points in two dimensions (Johnson and Papa-dimitriou 1985). What, then, are the best search strategies to discover unknown resources? How do you decide in what order to harvest a set of previously known resources? When should you travel farther to obtain greater rewards? How do constraints of knowledge, food storage, or spatial nonuniformity change your decisions?

This chapter is devoted to reviewing the variety of approaches taken by distinct sciences in tackling these questions. First, I briefly review current applications of spatial routing problems in economics. Then, I lay out the basic structure of spatial movement problems for animals and humans, and discuss whether or not the problems typically encountered in human applications can easily be adapted to understand animal spatial movements. Next, I consider systematically the assumptions and constraints of various versions of spatial movement problems, before presenting an overview of common methods suggested for solving them. Finally, I address the evidence that animals or humans (without computers) apply any of these methods in their spatial movements, and point out some difficulties with this evidence and directions for future research.

Economic Applications of the Theory of Spatial Movements
Because of their economic applications, problems of efficient spatial routing have been studied for a long time (Hoffman and Wolfe 1985). On the one hand, human economic geographers wish to solve practical problems such as locating a new shopping area, predicting large-scale population movements, and understanding why Aunt Lizzie prefers to shop at the Grand Union rather than Joe's Market closer to home (e.g., Borgers and Timmermans 1986). This research tends to be dominated by theoretical studies of how humans should behave and empirical studies of the average behavior of large numbers of travelers; there is remarkably little study of how individual humans actually choose spatial routes. On the other hand, a number of distinct business applications involving optimal scheduling or routing of sequential tasks have attracted mathematical interest because they are variants of the TSP (Hoffman and Wolfe 1985). Classic examples of such problems include determining the best allocation of a set of delivery vehicles across a group of customers plus the actual delivery routes of each vehicle (Christophides 1985), and job sequencing in assembly or chemical plants, in which different jobs use the same equipment but it costs different amounts of time or money to convert the equipment from one job to the next (Tandon, Cummings, and Le Van 1995).

Charles Janson

A more recent problem is designing computer circuits; the demand for ever faster computers has placed a premium on efficient routing of instructions and connections on computer chips, which has in turn generated a large theoretical and applied literature. Many of the problems considered by computer scientists are a simplification of the TSP called the *shortest path problem* (SPP): what is the shortest route that starts at A and ends at B, using *any* combination of existing (often not straight-line) paths between intermediate sites? In contrast to TSP, the shortest path need not visit all points, or even any intermediate points if a straight-line direct path exists between A and B. Despite its similarity to the TSP, the SPP is relatively easy, and many efficient algorithms exist to solve it completely (Mehmet Ali and Kamoun 1993), including variations that include obstacles in the plane around which the path must detour (e.g., Storer and Reif 1994); dividing a task among several parallel processors (e.g., Topkis 1988); inability to pass through a particular site (Lee and Chang 1993); and scheduling multiple users of single computer servers (Mehmet Ali and Kamoun 1993). I will not discuss the latter papers in much detail as they do not generally deal with constraints of biological interest—few animals are likely to be forced to travel along predetermined paths rather than just heading directly from one goal to the next. Nevertheless, these algorithms might be biologically relevant if most animals code spatial knowledge not in a map (allowing straight-line movement between goals), but only as lists of instructions of how to move from one landmark to the next (see Bennett 1996).

In contrast to the large theoretical and applied literature on solutions to TSP problems in economics, the zoological literature contains many studies of how animals use space, both in the wild and in captivity, yet remarkably few biologists have tried to create models of adaptive animal movement. It appears that biologists and human geographers are unaware of each other's research, as they almost never cite each other's journals. My hope in this chapter is to begin to bridge this gap and point out areas where additional modeling or data would be useful. I provide a glossary of specialized terms that may be unfamiliar to readers in table 7.1.

Spatial Movement Strategies: Defining the Problem

Fitness Measures
A basic requirement for any adaptive model is a measure of success. For animals, success is usually defined as evolutionary fitness, but

Table 7.1 Glossary of key terms

Accessibility	An integrative measure of clumping of resources weighted by their benefits
Algorithm	A set of rules or procedures to derive a result
Analog	Concerning properties that vary continuously, such as gravity or distance
Center of mass	The mean x (horizontal) and y (vertical) position of a set of points, where each point's position is weighted in proportion to its mass
Cognitive map	A mental representation of space that allows direct movement between any pair of points
Convex	Bowed outward (see fig. 7.5)
Euclidean space	Space that is isometric in all directions
Exponential	Proportional to e^k, where k can be positive or negative (see fig. 7.2)
Feedback	The output of a process that acts either to speed up or slow down that process
Gradient	A gradual change in a variable across space
Heuristics	Efficient but approximate permutational methods to solve GTSP
Inertia	A tendency to keep moving in the same direction and at the same velocity
Isometric	Scaled the same way (regardless of direction)
Likelihood	Probability
Look-ahead rule	A method that considers all possible permutations of k unvisited sites out of a total pool of N sites and chooses the sequence that minimizes distance
NP-complete problems	Problems, such as TSP, for which no efficient rules are known that guarantee the best possible solution
Optimal	The best, within constraints (e.g., of having a route end up where it started)
Path	A route through a set of points with distinct starting and ending sites
Permutational algorithms	Methods for solving problems that scramble sequences of discrete units, such as separate segments of a multisegment path
Polynomial	Equal to $\Sigma c_i k^i$, for $i = \{0,1,2, \ldots\}$; some of the c_i may equal 0 (see fig. 7.2)
Rules of thumb	Approximate and simple methods to solve GTSP, usually based on one-by-one comparisons of analog properties of potential choices
Satisficing	A fitness criterion in which fitness increases with a variable up to some value V^*, beyond which higher values of the variable yield the same fitness
SPP	Shortest Path Problem: to find the shortest path through N points
Tour	A route through a set of points starting and ending at the same point
Trade-off	A case in which increased fitness via one attribute comes at the cost of decreased fitness via another attribute
TSP	Traveling Salesman Problem: to find the shortest tour through N points
Utility	A measure of perceived human preference

in foraging models, more easily measured proximate correlates of fitness, such as long-term average net energy intake, are often used instead (Stephens and Krebs 1986). An alternative criterion of fitness used in spatial movement problems is time minimizing, or satisficing (see Stephens and Krebs 1986), in which an animal is assumed to forage only as much as needed to satisfy its minimal metabolic needs. Still other measures of fitness exist (e.g., probability of survival to the end of a specified period, as used in dynamic programming: McNamara and Houston 1986), but these have not been employed in the spatial movement literature. For humans, success is often defined as "utility," a rather vague concept that is perhaps best understood as a measure of psychological satisfaction, which may or may not correlate with financial or reproductive gains (Deaton and Muellbauer 1980). Despite the vagueness of the utility concept, it is still useful if it predicts human preferences accurately. In some studies of human spatial behavior, utility is replaced by explicit functions of gains and costs (e.g., Ghosh and McLafferty 1984). Whichever way success is measured, it is important to consider whether short-term effects or long-term average success is the most relevant.

Many spatial movement models agree that foragers should prefer to obtain a given level of benefit at the least possible cost in travel distance. For the TSP and related problems, all the sites to be visited (hence total foraging benefits) are usually dictated in advance, so that fitness can be maximized only by minimizing travel costs across the whole set of points. Because they specify which sites have to be visited, TSP-like problems are not perfect models for most animal foraging problems. Real animals do not need to visit all sites, may revisit some destinations more often than others, and may not regularly return to a starting point unless they have a nest. Because of this added flexibility in decision making, it is necessary to define a new measure of foraging success other than minimizing total distance. Based on my analysis of foraging constraints in brown capuchin monkeys (*Cebus apella:* Janson 1988a), I propose that what real animals do every day is to choose from among the available resources the subset that (1) can be visited within a day or fulfills a relatively fixed daily food requirement *and* (2) minimizes the distance traveled that day. This fitness goal could be called "shortest-path satisficing." My rationale for this idea is that many animals can find and ingest food faster than they can digest it, so that the fitness returns of increasing food intake level out once intake exceeds what the animal can digest per day. In addition, renewal and deterioration of resources make it difficult to predict the spatial distribution and benefits of resources more than a day or two in advance, and may limit the usefulness of planning a detailed route more than one day in advance. Human geographers (Keller and Goodchild 1988) have tackled the similar problem of choosing the particular subset of N total sites that, when used in an optimal route, maximizes benefits at a fixed cost; an example would to design a delivery route that maximizes profit within a workday, given that the truck has to return to a central location at night. For such problems, animals and humans may ignore nearby poor resources in favor of more distant but more rewarding ones, or may prefer to visit spatially clustered resources over isolated ones.

Organismal Constraints

Before devising a model of how an individual should move, one has to define how much it knows about its environment. Typically, human geographers assume that humans possess rather complete memory of the locations and rewards of potential goals, whereas biologists often have to assume that animals have little or no knowledge of where to find their next resource or how good it will be. Nevertheless, some animals do have considerable prior knowledge about the locations and expected rewards of resources. For instance, food-storing birds can recover cached seeds from dozens to hundreds of locations at rates far above chance levels (Balda and Kamil 1988; Hilton and Krebs 1990). Primates similarly appear to recall the locations of specific resources (Menzel 1991) and show preferences for more rewarding sites (Garber 1989; Janson, 1998). Conversely, empirical studies of humans have shown that their spatial memory about potential destinations is often far from perfect, or at least spatially biased and non-Euclidean (e.g., Taylor and Tversky 1992b).

Storage volume (trunk size of cars, freezer space at home, gut size in animals) and costs are often ignored in spatial movement problems. However, a few models have investigated the effects of limited or expensive food storage on the frequency of shopping trips (Lentnek, Harwitz, and Narula 1981), and limited capacity can be used as a constraint in vehicle-routing versions of the TSP (Christophides 1985). An analogous application to animal movement has yet to be published, although limits on digestive rates are

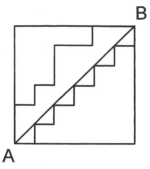

B

A

Figure 7.1 Although most distances between points in geometric problems are calculated using Euclidean (shortest straight-line) distances, at times other metrics are useful. In routing problems involving cities, the Manhattan metric is often used: the shortest distance is traced along the edges of city blocks and equals the sum of the differences in the x coordinates and the differences in the y coordinates. Manhattan distances between points do not vary according the route taken, as long as the route stays within the "box" determined by the starting and ending points, and no back-tracking in either x or y directions occurs. In this example, A and B are at the corners of a seven-by-seven array of city streets, enclosing six square blocks on a side. The Euclidean distance from A to B is $\sqrt{72} = 8.5$ blocks, whereas the Manhattan distance equals 12 blocks.

known to influence diet choice (e.g., Belovsky 1981) and food competition in social animals (Janson 1988b).

Model Assumptions

Several common assumptions are made in spatial routing problems. First, space is often considered to be isometric—that is, the cost of movement is the same in all directions. There is one important class of exceptions to this assumption, namely, the use of a so-called Manhattan metric (in which the only allowable movements are on the edges of rectangles forming a grid). Compared with a Euclidean metric (the distance between two points is the length of the straight line joining them), the Manhattan metric has the peculiar property, familiar to any city dweller, that any path between two points is as long as any other, provided that no edge is chosen that increases the previous horizontal or vertical distance to the target point (fig. 7.1). This special property of the Manhattan metric makes the particular choice of path between any two points unimportant, but it does not alter the fundamental problem of how to choose the best sequence of N points to visit—the TSP can assign real "roadway" distances as the distance between any two points rather than the Euclidean distance. More complex spatial transformations allow the cost or speed of movement to vary in diverse ways over space. The most common variant is the nonsymmetric TSP, in which the costs of moving from A to B are not the same as from B to A, which finds its application in the scheduling of tasks in manufacturing, where the time and money needed to change a machine from doing task A to doing task B may not be the same as the reverse. Such nonisometric properties of space change the shape of the most economical route between points, but again do not alter the fundamental difficulty of the TSP problem.

Another habitual assumption of TSP-like movement models is that the decision maker has complete prior knowledge of at least the spatial positions and "utilities" of the points visited. Thus, the total reward or utility of different routes can be calculated with certainty in advance. Although the possibility of chance discovery of resources along the way is not incorporated into the planning of optimal routes in the TSP, some studies of shopping behavior investigate the likelihood that a shopper will stop off casually at nonplanned points near planned stops (Borgers and Timmermans 1986; Lentnek, Harwitz, and Narula 1981). Elegant field experiments with natural food sources placed in novel locations have shown that fruit-eating bats *(Carolia perspicillara)* readily discover and use new resources while traveling between previously known foraging areas (Fleming, Heithaus, and Sawyer 1977). Distinguishing between chance discovery and planned visits to points may be difficult if an animal's detection field is greater than or equal to the mean distance between resources (Janson 1998). Unfortunately, it is relatively rare to know what an animal's detection field is for a given resource (but see Janson and Di Bitetti 1997 and references therein).

Human geographers have considered several variants of the TSP with assumptions tailored to applied human problems; collectively, these expanded TSPs are referred to as the generalized TSP, or GTSP. Perhaps the most flexible variant of the GTSP, and the most like biological foraging problems involving fixed renewing resources, is the Vehicle Routing Problem. In this variant, n customers need to be "serviced" by k trucks starting from m depots. Each customer has an associated amount of material that needs to be picked up or dropped off by the truck, and the truck has a maximum capacity for such material. The customers may be associated with a time window for deliveries (e.g., deliveries to customers on Main Street have to happen in the morning) or a time horizon (oil deliveries must be made within 60 days of the previous delivery). All trucks must return to a depot when empty (or full), but the ending depot need not be the starting depot. To translate this problem to a hummingbird foraging on flower patches or a primate moving between fruit trees, substitute "resources" for "customers," "foragers" for "trucks," "sleeping or resting sites" for "depots," "food available" for "material to be picked up," "stomach capacity" for "maximum truck capacity," "diel production patterns" for "customer time window," and "renewal interval" for "customer time horizon."

Strategy Sets

A strategy set specifies the traits that natural selection is supposed to act on to maximize fitness. For spatial foraging problems, the physical traits (neurons, hormones, endocrine organs) on which natural selection could act to produce optimal movement behaviors are rarely specified, nor is it clear how particular neural structures contribute to spatial competence (but see Shettleworth and Krebs 1986 and related articles on hippocampal volume in birds). Instead, models of spatial foraging solve for optimal behaviors or sequences of behaviors, whether or not there exist any known neural mechanisms to implement such a solution. In models that assume no spatial knowledge, behavioral decisions are assumed to depend only on the animal's immediate past (such as its prior direction) and current conditions (such as benefit from the current resource). Models that assume extensive or complete spatial memory phrase behavioral decisions as choices of a set of sites to visit and the many associated possible sequences of moves among the sites in a set; in most human applications, travelers are constrained to visit all available sites.

Solutions

No Prior Knowledge Assumed

The simplest models assume no spatial memory of resources. If an animal has no notion of where to find the next resource, and resources are depletable, it can do no better than to move in a straight line because any turns are likely to bring it back upon previously searched areas (Cody 1971). This fact should provide a cautionary note when we are interpreting animal movement patterns—although animals moving between two known resources should clearly prefer to go in a straight line, the use of straight-line movements is not by itself convincing evidence of goal-oriented movement (see also Janson 1998). If resources are not easily depleted, or if they recover quickly, then a random walk—a sequence of short moves in random directions—will be as efficient as any systematic search behavior (Stillman and Sutherland 1990; Zimmerman 1979).

Once an animal finds a resource, it has several options to increase its *short-term* success. As most resources in nature appear to be at least somewhat clumped (e.g., Hubbell 1979), finding one resource means that others are likely to be nearby. If so, then after an animal leaves a resource, it should make short movements and turn fairly sharply at each new movement so as to stay within the vicinity of the previous resource (Pyke 1983; Smith 1974). If resources are uniformly dispersed, then local discovery of a resource implies that it is unlikely that any more will be found nearby, in which case the animal should use long movements with narrow turning angles. The same basic rules apply even when individuals are foraging in a group: narrow turning angles yield more resources found, but this effect is strongest at small group sizes (Gordon 1995).

Animals may detect a gradient in the density of a resource. In this case, they should move up the resource density gradient until competition from the increasing density of conspecifics (all following the same gradient) cancels out any added benefit of higher resource availability (Kiester and Slatkin 1974). In one dimension, such local "hill-climbing" leads to the aggregation of individuals at local food abundances (e.g., Krebs 1978, 38–39). In two dimensions, climbing a density gradient may require many local exploratory movements to track the steepest slope of the gradient, as is true in models of pheromone detection (see Garber and Hannon 1993). A more complex, social variation of this problem has individuals taking cues directly from one another about the location of food (e.g., Kiester and Slatkin 1974; Krebs, MacRoberts, and Cullen 1972). Such "conspecific cueing" does not require the animal that discovers a new resource to signal its location deliberately or consciously, although many species of primates do give distinct food-associated calls that have the effect of attracting other group members (e.g., Dittus 1984). In many social animals, some individuals in a group may "scrounge" the discoveries of "producer" individuals at the latter's expense (e.g., Clifton 1991; Vickery et al. 1991).

If animals locally deplete resources but cannot remember exactly when they did so, then their best long-term strategy is to delay returning to a given place for as long as possible. Without explicit spatial memory, the only way they can delay returns is to move directly away from used areas, a goal they can achieve by always moving in a straight line, even after an interruption such as feeding at a resource (Pyke 1983). Such straight-line movements might be achieved by individual foragers that have a sense of absolute direction, but even without this capacity, high directional "inertia" could emerge in large groups or flocks because individual tendencies to deviate from a straight path will cancel each other out (Cody 1971). Although most foragers will eventually be forced by home range or territory boundaries to make sharp turns that return them toward previously used areas, maintaining straight-line movements will generally maximize their return times to a given location. Note that this strategy conflicts directly with the best strategy for increasing short-term success on clumped resources (small movements, sharp turning angles), although not for uniformly dispersed ones.

Spatial Memory Assumed
Problem Complexity. Things become considerably more convoluted when a forager has a mental map of its food sources, including prior knowledge of at least how to get from any one site to any other (possibly along habitual travel routes, like human roads), as well as, perhaps, reward amounts and rates of renewal. Relatively simple solutions are available only for models in which the animal is forced to move along a line, so that the spatial component is removed or greatly reduced, as might occur in linear habitats such as beaches or gallery forests (e.g., Arditi and Dacorogna 1988; Kiester and Slatkin 1974). In two dimensions, the classic problem in human geography is the Traveling Salesman Problem, in which a salesman needs to find the shortest route through N stops, beginning and ending at the same location but not revisiting any other stop. Such a closed route is called a *tour*, in contrast to the shortest route beginning and ending at different points, which is called a *path*. In the TSP, all points have equal value or "utility," or if they do not, it still does not pay to visit more rewarding sites first, as all points have to be visited once and only once. The distances between points may either be the Euclidean (straight-line) distances between them or be based on actual highway mileage—the important thing is that all the distance information can be condensed into a matrix of distances between all pairs of points. This problem is one of a

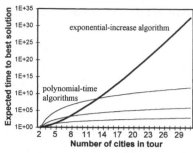

Figure 7.2 Comparison of the computing time needed to produce the best solution to a Traveling Salesman Problem with increasing numbers (N) of cities. The heavy line represents algorithms with computing times to the absolute best solution that increase exponentially with N (note the logarithmic scale of the y-axis). The light lines illustrate the time savings that could be achieved if there existed algorithms that could give the best solution in a time proportional to a polynomial (power function) of N. The three light lines of increasing height are for algorithms that grow in proportion to N^2, N^4, and N^8. If the scale on the y-axis represented seconds, and a computer could calculate and compare 1 billion tours per second, the top of the scale would still represent 3,200,000,000,000,000,000 years for a solution, whereas the longest polynomial-time solution shown here would take only about 10 minutes!

class of similar, difficult mathematical problems termed Nondeterministic-Polynomial (NP)-complete problems (Johnson and Papadimitriou 1985); the precise mathematical definition of this term is subtle and is not needed to understand the remainder of this chapter. Its practical meaning is that NP-complete problems are very difficult to solve completely—all known algorithms for the *guaranteed* best solution of NP-complete problems require computation times that increase exponentially with the size (N) of the problem (i.e., proportional to e_N, where e is the base of natural logarithms: fig. 7.2). For instance, being certain of having found the best answer to the TSP with N points may require examining all possible orderings of the $N - 1$ target sites, or $(N - 1)!$ routes, a number that increases exponentially with N. The Holy Grail of TSP research is to find an algorithm that solves an NP-complete problem with computation time that increases only as a polynomial function of N. Such a polynomial-time algorithm would prove a solution far more rapidly than any exponential-time one as N becomes large. For this reason, I shall call polynomial-increase solution rules "efficient" and exponential-increase ones "inefficient." To date, nobody has worked out a polynomial-time algorithm to solve completely any NP-complete problem.

Finding the best possible solution to a TSP-like problem requires inefficient solutions and would clearly be prohibitively difficult for most animals and humans without computers. How do you solve a

Charles Janson

problem when it is exceedingly hard to arrive at the exact answer? A rich literature has emerged that examines approximate algorithms (called heuristics or rules of thumb) for solving the TSP and similar problems. Heuristics and rules of thumb are similar in that they are efficient in giving an answer, but are not guaranteed to give the best possible answer. They differ in that most heuristics involve refining candidate travel routes by switching around, or permuting, pairs or larger sets of points; I shall refer to such methods as "permutational" algorithms. In contrast, rules of thumb typically solve a complex problem with "analog" methods, by comparing some continuous parameter of each choice either directly with other options or with a fixed standard (see Janetos and Cole 1981). Nearly all decision rules in standard foraging theory are of the analog type (e.g., leave a patch when the gain rate inside the patch declines to the value G, the long-term gain rate for that environment and forager: Stephens and Krebs 1986).

I review below a number of methods designed to come close to the best answer without being guaranteed to reach it. I start with simple analog rules of thumb that provide rough guesses as to the optimal solution, but use mechanisms that may be realistic for animal brains. I progress through more complex, more nearly optimal, but arguably less realistic algorithms. I finish with a brief treatment of computer-intensive permutational methods that can achieve near-optimal or optimal solutions, but which appear at present unlikely to be used by animals. I provide these here to help biologists who may need to use computer programs to find near-optimal solutions to compare with observed behaviors of animals; an up-to-date and thorough review of these algorithms (including programs) is available (Skiena 1997).

Rules of Thumb Involving One-by-One Comparisons of Goals. The rules of thumb offered as models of how animals or humans actually make movement decisions generally compare each potential destination with others based on a continuous "attraction" function, which depends on distance. For instance, a common method assigns an attractive force, or "gravity" value, to each potential goal i according to the formula

$$(7.1) \qquad G_i(r) = M_i/r_i^k,$$

where G_i is the gravity value, r_i is the distance from the current position to the goal, M_i is the "mass" or value of the goal, and k is a constant, often assumed to be 1.0 (Haynes and Fotheringham

1984; Sheppard 1978). Under the usual applications of this hypothesis, humans are assumed to visit different sites in proportion to each site's gravity value relative to all other choices, even though this outcome explicitly requires that many individuals make suboptimal choices by moving to sites other than the most attractive one (Haynes and Fotheringham 1984). For instance, if there are two possible goals with $G_1 = 3$ and $G_2 = 7$, it is predicted that 70% of humans will go to goal 2 and 30% to goal 1, even though it is less attractive than goal 2. It would appear more rational for individuals consistently to prefer the site of greatest gravity (fig. 7.3). If all the M_i are equal, then individuals should always use the "nearest neighbor" rule: move to the nearest (unvisited) site (Saisa and Garling 1987). A variation on the gravity rule is offered by Golden, Levy, and Vohra (1987), who advocate building up a travel route from the starting site by adding goals in decreasing order of M_i/d_i where d_i is the Euclidean distance of goal i from the spatial center of mass of the set of sites already included in the route. This rule keeps newly added sites close to the entire set of visited goals, not only to the last-visited goal.

A spatial attractiveness function applied to animals is Baker's (1978) model of animal migration, which postulates that an individual should move toward whichever of several possible goals would yield the highest fitness after deducting travel costs. This can be written as

(7.2) $$F_i(r) = M_i - cr_i,$$

with F_i as the fitness return of the ith goal as a function of distance, and c the travel costs in fitness change per unit distance. Both equations (7.1) and (7.2) specify movements only one goal at a time and, in general, are not guaranteed to return to the starting site, as required in many GTSP.

A shortcoming of both the gravity and Baker models is that they usually ignore the positions of potential goals relative to one another. For a fixed G or F value, a resource in the middle of a cluster of other resources is no more likely to be visited than if it were by itself. A few attempts have been made to account for the spatial distribution of potential destinations by summing attraction functions over multiple resources. For instance, one modification of the gravity rule assigns each potential destination j an "accessibility" measure, which is the sum of the gravities of all (unused or available) sites as calculated from site j (Haynes and Fotheringham

Charles Janson

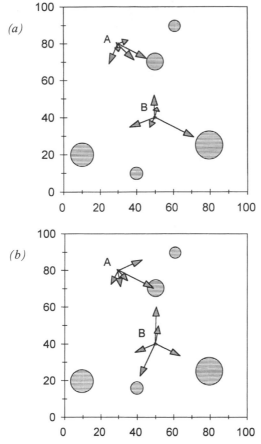

Figure 7.3 *(a)* Illustration of gravity of each of five possible destinations (solid circles) from two points in a home range, A and B. The length of the arrow in the direction of each destination is proportional to its gravity, calculated as the area of the destination circle divided by the distance from the point to the center of the destination circle. *(b)* Illustration of accessibility measures for the same five destinations as in *(a)* relative to the starting points A and B. Note that points in the center of a cluster can end up with high accessibility (and thus attractiveness) despite offering relatively small rewards. In this case, an animal at A would choose the same destination as in *(a)*, but an animal at B would move first to the smaller resource between the two large ones.

1984). The forager is assumed to prefer the destination with the largest product of accessibility and gravity attraction from the forager's present location (see Haynes and Fotheringham 1984 for details and extensions). A similar summation model is the "subgravity" measure of Golden, Wang, and Liu (1988), which differs from the accessibility measure only in that resources are devalued not as a reciprocal or linear function of distance, as in equations (7.1) and

(7.2), but as an exponentially declining function. In these accessibility models, the forager heads directly for a specific destination, although it may possibly reevaluate on the way and detour to other intermediate goals as it passes close to them.

A different summation model is called a "place utility field" (R. G. V. Baker 1982). In this model, the gravity, or *utility*, of each potential goal is evaluated at every *place* in the forager's *field* (foraging area), and then the individual utility fields are added up across all goals (fig. 7.4a), which amounts to calculating the accessibility of every point in the field. In contrast to the previous accessibility model, here the forager is expected to climb up the gradient of steepest increase in summed utility until it cannot increase utility any more, at which point it should be at a resource. As the resource is used, its M_i decreases until the local utility value becomes equal to that of the surrounding landscape, at which point $M_i = 0$ (relative to other goals), and the forager should leave to use other sites. A distinctive outcome of the place utility model is that a forager may initially head off at a bearing between those of two attractive destinations, because the summed attractiveness is higher between them than on a direct line to either one alone (fig. 7.4b); a similar behavior can occur when an animal follows multiple odor plumes (see Garber and Hannon 1993). In practice, place utility fields built up from individual gravity hypercones (see fig. 7.4A) rarely achieve much summation over space because the inverse effect of distance soon overwhelms even large differences in M. For instance, if two potential targets have "masses" $M_1 = 2$ and $M_2 = 10$, then the gravity for M_2 increases by 0.001 between 100 and 99 feet, which is slightly less than the increase in gravity for M_1 between 45 and 44 feet. Thus, an animal 100 feet from M_2 and 45 feet from M_1 will move toward M_1, even though reaching M_2 would benefit the forager five times more. If utility fields are built up from linearly declining fitness functions, as in equation (7.2), the "cones" of declining attraction fall off at a constant rate with distance so that resource values tend to add up more evenly over distance. However, this linear model has the peculiar property that if two resources have exactly the same M, then the straight line connecting them will have the exact same fitness return along its entire length, so that the forager will have no gradient to follow.

In either the attraction or place utility field models, the extent of spatial summation is determined by how quickly attractiveness declines with distance (as determined by exponent k in equation

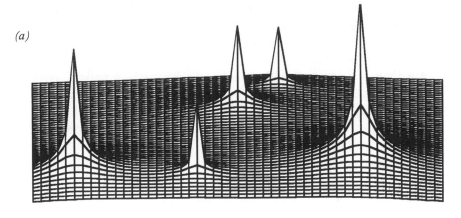

(a)

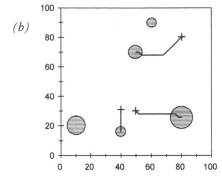

(b)

Figure 7.4 *(a)* Example of place utility field representation of the destinations in figure 7.3*a*. The height of the surface at any location is the sum of the gravity values of the five resources evaluated at that position. In theory, the utility at the exact location of the resource would be infinite (as the distance to it is zero), but this result applies only when the resource is a point (has no area). The more realistic finite peak utilities shown here assume that the resource occupies an area 1 unit in diameter. *(b)* Travel route predicted by the "hill-climbing" rule (move in the direction of the steepest increase in gravity from the current location) for the place utility field in figure 7.4*a*, starting from the three locations marked by + signs.

[7.1] or the slope c in equation [7.2]). Large scaling values lead to very steep-sided attraction cones with little summation across sites, thus approximating the simple gravity or migration models of equations (7.1) and (7.2). Low scaling values yield wide attraction cones, resulting in extensive summation of attractiveness across sites and the possibility that an animal might move toward a cluster of more distant sites instead of a single closer target. It seems likely that animals might wish to vary the extent of spatial summation according to the spatial distribution of resources or the fitness goals of the forager (maximizing intake; minimizing risk of starvation).

For instance, a very hungry animal may wish to minimize the delay to *any* resource and thus use only local summation, whereas a less desperate forager may prefer to consider more distant goals in its choice of routes.

Whether or not models that sum attractiveness functions over multiple resources yield optimal solutions to GTSP of the kind faced by animals, they are consistent with known behavioral mechanisms. It is known that spatial relationships in the environment are at least sometimes coded in brain tissue in a roughly maplike fashion (e.g., for sounds: Takahashi and Keller 1994), and local summation of signals is a standard feature of real neurons in the brain. Neurons associated with a particular location in an animal's home range could easily vary spike frequency in proportion to the benefit of visiting that site (M_i in equations [7.1] and [7.2]). If the number of axons leading from a location-signaling neuron decreases with distance from the cell body, then the neuron's effective signal strength would likewise decrease with distance and mimic some form of gravity function. A neuron that codes for motivation to move to a site could integrate such gravity-like signals coming from multiple location-signaling neurons and thereby produce an output proportional to accessibility.

At least one method of exploring the application of different continuous spatial summation models is the use of neural networks in computer science, intended to mimic brain functioning (Iyengar and Kashyap 1991). A neural network is a series of connected nodes (created either in computer memory or via parallel processing chips), each of which performs simple data transformations and is connected in simple feedback relationships to other nodes, much like real neurons. Although it would be too complex to explain their use here, neural networks have been used in many recent papers to tackle the TSP, and have produced solutions that come within 10% of the absolute best one (e.g., Bhide, John, and Kabuka 1993; Hopfield and Tank 1985).

Methods that Consider Combinations of Goals. An extension of the nearest-neighbor rule, suggested by a biologist, is the class of "look-ahead" rules (S. A. Altmann 1974; J. D. Anderson 1983). The idea is that an animal may be able to obtain an approximate answer to the TSP by examining the shortest route among only a subset of the N resources. For example, according to the one-step look-ahead model, it would examine the distances between the current location

Charles Janson

and each other available resource one at a time, the shortest of these distances would then be chosen, so that the animal would always move from each resource to its closest unused neighbor. According to the two-step look-ahead model, it would examine the distance from the current location through all possible pairs of available resources; again the shortest such route would be chosen, and so on. Obtaining the shortest possible overall route requires solving the $(N - 1)$-step look-ahead problem. You might intuitively believe that the more steps used in this look-ahead calculation, the closer you would get to the shortest possible tour, but you would be wrong! J. D. Anderson (1983) showed that in most cases, travel routes were shorter for the nearest-neighbor (one-step) rule than for larger numbers of look-ahead steps up to about $N/2$, above which the calculated routes began to approach the shortest possible tour. This paradoxical result comes about because rules that look ahead more than one resource often skip over the nearest-neighbor resource, which then becomes isolated and later has to be visited from some distant point, at a considerable cost in distance. Because of this property, it is hard to imagine how natural selection would ever favor the gradual development of the capacity to solve the TSP using many-step look-ahead rules—any tendency to consider the shortest distance among sets of two or three resources would generally reduce foraging efficiency compared with the simpler nearest-neighbor rule. Because the nearest-neighbor rule provides tours often only slightly longer than the absolute shortest solution to the TSP for modest numbers of points, it might be a simple and quite adaptive rule of thumb for use by animals and humans.

The geometry of the entire set of points can be used to construct candidate TSP tours in what I think of as the "rubber-band" algorithm (actually called the "convex-hull, minimum-cost-insertion" rule: Golden and Stewart 1985). We begin by drawing a convex polygon using the outermost points of the set, like stretching a rubber band around them (fig. 7.5). We then successively include one more "inside" point at a time by choosing it so that the angle of the two new legs formed between this point and existing points on the polygon is the most shallow available, much as though we were trying to stretch the imaginary rubber band as little as possible.

Permutational Heuristics for Improving Initial Solutions of GTSP. Several methods exist for improving an initial guess at a solution to the TSP. One common algorithm somewhat similar to the two- or

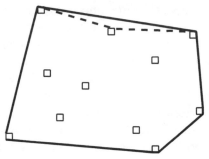

Figure 7.5 The "rubber band" (convex-hull) heuristic for generating a candidate TSP tour starts with a convex polygon drawn around the set of points to be visited. Then, the "rubber band" is stretched inward to loop around the interior point that requires the least force to reach (and hence represents the shortest added distance) among all the untouched points. New points are added until all points are included in the tour.

three-step look-ahead models, systematically reconnects all pairs or trios of points (two- or three-optimal switching: see Golden and Stewart 1985). Another method, called "simulated annealing," has achieved efficient solutions extremely close to the shortest overall path by using an analogy with the process of slowly "cooling" a metal to allow it find a minimal energy state (e.g., Yip and Pao 1995). Applied to the TSP, this method works by starting from some initial solution and randomly reconnecting points until it has a solution that remains the best for some criterion number of trials. Then the distance across which the points are allowed to reconnect is reduced (the "temperature" of the system is reduced) and the process repeated, with several cycles of distance reduction and locally best solutions, until the criterion distance is too short to allow additional local improvements by reshuffling connections. There is a large engineering literature on exactly how the "temperature" is to be reduced, as well as how local searches should be performed. Because of the random element in determining sites to reconnect, two runs of a program on identical problems may easily yield different sequences (but with similar total costs). It is not clear how such an algorithm would be implemented by the brain. Another improvement method that uses a random component is the class of so-called genetic algorithms, in which candidate solutions are subject to random "mutations" in route sequence and "recombination" of portions of routes. Each resulting new route ("genome") is compared ("competed") against existing solutions; better solutions are allowed to increase at the expense of worse ones; but replacement may take some time, thereby allowing further mutations to arise in

Charles Janson

"older" solutions. Eventually, a single tour or path comes to dominate the system, for which no small mutation will improve performance, although there is no guarantee that this solution is the absolute best one.

Permutational methods are unlikely to be realistic models of animal decision processes. First, the number of permutations increases rapidly even for only pairs of points; for instance, with ten possible destinations, there are ninety possible sequences of pairs of points. As list memory, even in primates, probably does not exceed two dozen items (Gallistel and Cramer 1996), it seems hard to believe that they could generate and remember the large number of permutations one has to sort through to solve a GTSP with even modest numbers of destinations. Second, the nervous system is primarily a graded signal system, with information coded by the frequency of impulses, rather than by combinations of on-versus-off neurons corresponding to the individual transistors of a computer chip. It is not easy to imagine how combinations of destinations could be coded easily in a graded signal system.

Exhaustive Search Methods. Even though finding the best possible solution to a TSP is guaranteed to take a relatively long time, there are cases in which it may be worth searching all possible routes; for instance, if N is not too large and/or the economic return will be large, as in mass production manufacturing. The most common method of reducing the time needed to arrive at the best possible answer to a GTSP is called "branch and bound" (Balas and Toth 1985). Imagine the set of possible sequences as a branching bush, with each fork corresponding to a choice of next sites to visit. This technique "prunes" the bush by eliminating candidate "branches" that lead to routes longer than some criterion, the "bound." There are several ways of establishing the bound. Clearly, the bound should be as close as possible to the best solution of the TSP. Approximate bounds can be placed on the best answer using either analytic (Beardwood, Halton, and Hammersley 1959) or computer-generated results (e.g., Golden and Stewart 1985), at least for randomly distributed points. For instance, Beardwood, Halton, and Hammersley (1959) showed analytically that as N becomes large, the shortest possible tour through those points will nearly always be less than a constant, k, multiplied by N/D, where D is the density of points. By using a few near-optimal instances of the TSP, they estimated k to be about 0.75. Their solution is an example of an *upper* bound criterion—all partial tours longer than this value can be discarded without calculating the length of the complete tour. A *lower* bound criterion is useful in estimating the worst-case answer for a given problem, as all real solutions must be greater than this criterion. It is usually derived by solving a simplified version of the TSP with the same constraints (e.g., distances between points); an example is to search for the shortest *set* of routes that connect N points, without requiring that the routes form a single tour (starting from one site and returning to it, visiting each site only once). The best solutions to the simplified problem (which can often be found relatively quickly) are often no more than a few percent different in total cost from the best solution to the original full problem. The "branch" part of a "branch and bound" algorithm dictates efficient methods for searching portions of the full "tree" of possibilities, given the current best lower and/or upper bounds on the answer. A simple example would be to stop examining any sequence of points as soon as the cumulative distance for the route up to the ith point surpasses the upper bound criterion, thereby "pruning" off the subsequent $(N - i)!$ permutations of the unvisited points from the bush of possible routes. A useful shareware program obtainable from <http://www.sportscoach.com.au/router.html> uses the branch-and-bound method as well as genetic and three-optimal heuristics to calculate tours for TSPs with up to 65 points.

Evidence

Foraging Movements Without Explicit Spatial Knowledge
For animals assumed to have little explicit spatial memory, traditional models of foraging movements appear to predict observed data relatively well, although alternative explanations are rarely considered. For instance, Kiester and Slatkin's (1974) model of local aggregation predicts that individuals should cluster together at rich resources because movements of individuals toward a rich resource patch will vacate foraging areas close to, but not in, the patch. Because of reduced competition, these vacated areas will become attractive foraging sites for animals farther from the resource patch. These animals will move into the vacated areas, and thus toward the resource patch, even though they may not have been able to perceive the patch from their original positions. This process of "conspecific cueing" has been documented in lizards (Stamps 1986). However, individuals could also aggregate at rich resource

Charles Janson

patches because of independent assessment of the quality of various patches followed by a choice of the richest patch to exploit (e.g., Krebs, Kacelnik, and Taylor 1978).

Several studies have evaluated the extent to which foragers move in straight lines, which is the predicted strategy when resources are renewed slowly, are not patchy, and spatial memory is absent or poor. Cody (1971) found that finch flocks tended to move forward rather than turn to the side, and almost never turned backward. To understand the significance of this pattern, he modeled movements of a finch flock on a rectangular grid superimposed on its home range. At each time step, the flock could move forward, backward, right, or left into adjacent grid squares with direction-specific fixed probabilities. For each combination of probabilities, Cody calculated the long-term intensity of use of an average grid square by the simulated flock; because the seed resources of the flock are not renewed during the winter, the lower the average intensity of use per grid square, the more seeds should be encountered by the flock when it returns to a given location. By searching the space of probability parameters, Cody found that the low turning values and the lack of backtracking observed in his real flocks minimized the intensity of grid use by the computer finches. However, the match between Cody's theory and data may be fortuitous, as the predictions of his computer model change when the grid size is modified (Pyke 1978). A similar tendency to maintain a consistent foraging direction is shown by individual bumblebees (*Bombus* sp.) when foraging on moderately clumped but slowly replenishing flowers (Soltz 1986).

Pyke (1978) predicted that when resources are patchy in distribution, foragers should turn more sharply so as to stay in the vicinity of a patch once it is located. Pyke found that bumblebees tend to turn more sharply, relative to their previous direction, after visiting flower spikes with many flowers than with few flowers. Consistent with his predictions, observed turning angles, averaged across bees, are larger when using high-density (Heinrich 1979b) or very clumped (Soltz 1986) flower distributions. At the level of individual bees, however, Soltz (1986) did not find a correlation between mean turning angle and number of flowers probed per inflorescence. Other experiments show that bumblebees may tend to turn more sharply after visiting richer flower heads simply because they have more trouble remembering from which direction they arrived as they visit more flowers and make more internal turns within an inflorescence (Taneyhill 1994). Considering the movements of foragers *within* a group, Gordon (1995) predicted that ants *(Linepithema humile)* should tend to turn more tightly when foraging in larger groups, so as to maximize the area covered by workers. Her experimental colonies conformed to this prediction, but the same trend would be obtained simply by having ants turn in a random direction in a fixed fraction of encounters with foraging nestmates; as foraging group size increased, so would the frequency of encounters between ants, and the average amount of turning observed.

Entire flocks might show high degrees of inertia despite foraging on patchy resources, which should favor large turning angles. First, if group movement direction results from compromise among many individual tendencies (see D. S. Wilson, chap. 9, this volume), then a group may tend to move straight ahead as individual preferences to turn cancel out. Second, it is difficult for any one group member to be certain that resources behind the group's present position have not been discovered and used by some other group member; this uncertainty may reduce the benefit of staying in the same patch. Third, even if an unused resource were known to lie behind the group's present position, the individuals at the front edge of a group are frequently subordinate in rank (Janson 1990b; Krause 1993) and may prefer not to backtrack so as to avoid aggression from dominants arriving behind them in the group progression. This effect in long-tailed macaques was termed "pushing forward" by van Schaik et al. (1983). If group movements are strongly influenced by inertia, it may be simplistic to assume that animals in groups will make movement decisions based solely on the distribution of available unused resources.

Evidence for Spatial Memory from Descriptive Studies of Wild Animals

Obtaining convincing evidence of spatial memory from unmanipulated wild animals is difficult, to put it mildly. The null hypothesis for any spatially structured foraging pattern should be that the animal moves at random until it detects a new foraging site or resource, after which it moves directly to exploit or pursue the resource (Garber and Hannon 1993). However, there is no universal null model of random foraging in two dimensions—any realistic model needs to incorporate species-specific parameters of detection likelihood for "prey" as a function of distance (e.g., Getty and Pulliam 1993) as well as known or assumed distributions of movement

Charles Janson

distances ("steps") and turning angles relative to the previous step (Janson 1998; Pyke 1978). For instance, Janson (1998) simulated the process of discovery by capuchin monkeys of experimental food patches by assuming (1) that they moved in straight-line segments of 100, 150, or 200 meters; (2) that they chose the next movement direction at random within 90° right or left of the direction of the previous movement segment; and (3) that they would detect any food patch within a fixed distance of the center of the group. From these simulations, Janson calculated sequences of food patch visits, the ratio of "observed" movement distances between patches to the straight-line distance between them, and the difference between the initial random departure direction from a food patch and the direction from that patch directly to the next resource visited. To compare the results of such simulations with real data, the researcher must know what the animals know about previously used resources, as well as essentially everything about the set of resources that could be discovered (without prior use) by the animals as they search. There are few descriptive studies that can come even close to meeting this standard of knowledge (but see Garber 1989; Garber and Hannon 1993). Even in the best of such descriptive studies of wild animals, the radius of detection of novel resources has not been measured (but see Janson and Di Bitetti 1997), so unambiguous conclusions about the prevalence of remembered versus newly detected resource use are rarely possible.

Captive experiments clearly can control what the animal can detect, what it knows from prior experience, the usefulness of such prior knowledge, and the availability of 'new' resources to detect (e.g., Shettleworth and Krebs 1986). However, the skills involved in foraging across restricted spaces (where local landmarks are nearly always readily available and the locations of most resource targets can be viewed directly from other targets) may be quite different from those needed for large-scale foraging across entire landscapes, where animals cannot see landmarks close to the intended target. Although you might think that efficient navigation to a goal would be easier if the goal were visible, this assumption is not always correct—seeing the goal distracts some animals from using stored route information, so that they may choose longer but visually more direct routes (Poucet, Thinus-Blanc, and Chapuis 1983). Humans asked to plan routes to several locations within a familiar city chose distance-minimizing routes more reliably when they possessed a map than when they did not, across smaller than across larger areas, and in the laboratory (where they were able to study the locations of all points on a map simultaneously) than in the field (where locations were generally not visible from each other) (Saisa and Garling 1987). In addition, it is not clear whether captive-born animals even possess or can develop the ability to navigate across large spatial scales, even if their wild-born kin do so regularly. Menzel and Beck (chap. 11, this volume) found that captive-reared golden lion tamarins *(Leontopithecus rosalia)* translocated into large outdoor enclosures frequently became lost less than 50 m from their nest box and often could not figure out a spatial route to return to it, even when it was in direct view of the animals. Their apparent ineptitude contrasts markedly with Garber's (1989) evidence in ecologically similar wild tamarins (*Saguinus* spp.) for well-developed spatial memory of even temporary resources across hundreds of meters. That spatial abilities may depend on training or experience has also been shown by training-induced and seasonal variation in hippocampal size in birds (Sherry and Duff 1996). Thus, captive-born animals may demonstrate lower spatial abilities than their wild conspecifics that use these abilities on a daily basis, or they may be trained to solve spatial problems that their wild brethren habitually do not encounter and would not know how to solve.

The strongest evidence for strategic spatial foraging comes from "traplining" species, which use a fixed set of discrete resources in predictable repeating sequences that remain stable over at least several days (e.g., Gill 1988; Janson 1996; Thomson, Slatkin, and Thomson 1997). Documenting such long-term sequential repetition has not been easy, given even modest levels of variation in sequences, as appropriate statistical techniques have been lacking until recently (Thomson, Slatkin, and Thomson 1997). One powerful method is to document that successive visits to a given site occur at relatively long intervals with a distinct modal value—if movement sequences are random and do not incorporate memory of past visits, the distribution of revisit intervals peaks at the shortest values and drops off rapidly for longer ones (Kamil 1978; Janson, unpub.). Such regularity in revisit times has been documented for several species (Gill 1988; Kamil 1978; Thomson, Slatkin, and Thomson 1997). However, even for these species, it is not known whether the observed time interval between visits optimizes the animal's fitness. Unfortunately, there are no theoretical studies that predict the optimal return time through a trapline for animal harvesters, although similar problems for humans have received some attention (e.g., fre-

Charles Janson

quency of shopping trips: Lentnek, Harwitz, and Narula 1981; periodic marketing: Ghosh 1982). Apparently, foraging theorists have been discouraged by the challenge of incorporating spatial constraints into foraging problems with renewing resources (for spatially unstructured versions of this problem, see Possingham 1989; Stephens and Krebs 1986).

Experimental Evidence for Spatial Memory
There is strong experimental evidence that many species have more or less detailed spatial memory of their foraging sites, and that this knowledge increases their foraging success relative to having no prior knowledge (see review by Gallistel 1990). Yet surprisingly little attention has been focused on how well observed spatially structured foraging decisions perform relative to the best possible behavior (or even on what the best possible behavior might be). For instance, despite a variety of convincing demonstrations of excellent spatial memory for previously hidden seeds by foraging titmice and corvids (e.g., Balda and Kamil 1988; Krebs, Healy, and Shettleworth 1990), these studies do not discuss the actual route taken by the foraging birds in terms of efficiency of harvesting relative to distance traveled. Several experimental studies on movements of primates between fixed food locations show that they strongly prefer the closest available resource (e.g., Garber 1989; Janson 1996; Menzel 1973a), but this is not evidence that they use a one-step look-ahead rule, as most two-or-more-step shortest routes will include the nearest-neighbor resource as the first step (Gallistel and Cramer 1996; Janson, unpub.). Captive vervet monkeys *(Chlorocebus aethiops)* appear to behave as though they can "look ahead" at least three steps and thus might use a permutational heuristic to solve spatial foraging problems (Gallistel and Cramer 1996). However, alternative analog models (e.g., place utility fields: R. G. V. Baker 1982) might provide equally valid and computationally much simpler explanations for the observed movements; critical tests of look-ahead rules versus other rules of thumb have yet to be performed for any animals or humans. It seems extremely doubtful that the capacity to use a three-step look-ahead rule would ever be adaptive in the wild, as look-ahead rules provide approximately optimal solutions only when resources are of equal value. This situation is probably never encountered in the wild. In fact, every published study on wild primates that demonstrates their preference for nearest-neighbor resources also notes a significant minority of

exceptions in which closer resources were passed by to obtain better but more distant ones (e.g., Boesch and Boesch 1984; Garber 1989; Janson 1996).

I have studied spatial decisions of capuchin monkeys in northern Argentina, using controlled amounts of fruit presented on platforms suspended from trees during the subtropical winter (when virtually no other fruit is available). For one study group, I was able to estimate the magnitude of the trade-off they made in swapping greater rewards for greater distance moved—the group would pass by the closest unvisited platform to use the second closest platform (roughly twice as far away as the closest one) only if the second closest contained eight times as much fruit as the closest (Janson 1998)! Clearly, these capuchins are not choosing which platform to visit next by calculating a simple reward/distance ratio (as suggested by a simple gravity model), unless they perceive the costs of increasing distance as rising very rapidly, as the cube of distance. Such a biased perception is compatible with psychological experiments on animals showing that delayed rewards are frequently strongly devalued compared with more immediate rewards (e.g., Fantino and Abarca 1985), but leaves open the question of why such steep depreciation in value should occur and why various species differ in the degree of discounting of delayed rewards (e.g., Tobin and Logue 1994; Tobin et al. 1996).

There do appear to be quantitative differences in spatial memory among species, although these differences do not relate obviously to absolute or relative brain size. Chimpanzees *(Pan troglodytes)* are able to recall upwards of sixteen spatial sites where rewards are located (Menzel 1973a), and humans can do at least as well when asked to memorize a map (Taylor and Tversky 1992a), whereas two independent tests of macaques showed them to be unable to remember more than six novel locations at a time (Gallistel and Cramer 1996). It is interesting that tamarins, with their relatively small brains (for primates), appeared able to remember up to thirty locations when their knowledge of these sites was built up and reinforced over 2 weeks (Menzel and Juno 1985). In captivity, squirrel monkeys *(Saimiri sciureus)* were less willing to repeat recently used routes than were titi monkeys, *(Callicebus moloch)* so that the titi monkeys chose established shortest foraging paths more often than did the squirrel monkeys, despite the titi's smaller brain size (Fragazsy 1980). Even 10–15-gram titmice (*Parus* spp.) can recall the locations of dozens of stored seeds at levels far above chance; tit-

Charles Janson

mice that habitually store food show better retention of cache sites than do nonstoring species, although the differences are not striking (Shettleworth and Krebs 1986). It is not yet clear whether the mental algorithms used to decide spatial movements differ between chimps and macaques (or any other species), but primates may surpass other animals in the extent to which they can plan ahead for delayed rewards. Comparative tests of "self-control" across species show that pigeons and rats discount delayed rewards so strongly that they will almost never accept a delayed reward of any size over a more immediate one, whereas macaques and humans show much stronger abilities to weigh the size of future rewards against the delay to receive them (Tobin and Logue 1994; Tobin et al. 1996). Except for these studies on "self-control," the data produced on various species (even by the same researcher) are not sufficiently comparable to allow a relative scaling of spatial abilities for more than a few closely related species.

How Efficient Are Animals in Spatial Routing?

At least some nonhuman primates achieve foraging routes that come close to minimizing total distance across a set of sites visited. In captivity, chimpanzees (Menzel 1973a), yellow-nosed monkeys (*Cercopithecus ascanius:* MacDonald and Wilkie 1990), and macaques (Gallistel and Cramer 1996) all chose efficient foraging routes. The routes actually used by yellow-nosed monkeys averaged only about 10% longer than the minimum possible route length, based on results presented in MacDonald and Wilkie (1990). In the wild, tamarins typically moved to the closest available food tree by routes that were only slightly longer than a perfect straight line between the origin and destination sites (Garber and Hannon 1993), but existing analyses do not test for more complex spatial movement rules or gauge how close the animals came to using a shortest-path solution across the entire set of chosen food trees.

For humans, the number and complexity of models of how we *should* behave are quite impressive, yet data on how humans *actually* perform are comparatively rare. There is general agreement between broad predictions for aggregates of humans and models of spatial attractiveness, but these models typically ignore the details of decision making. Thus, models of multiple-purpose shopping behavior predict that humans will often travel farther to go to a mall or large grocery store because they can buy many distinct goods in a single trip instead of making many shorter trips (e.g., Ghosh and

McLafferty 1984). While this may be true, other rules would also generate a preference for large shopping complexes, such as lower commodity prices because of competition among multiple clustered vendors. A few studies explicitly measure the trade-off between distance and cost or reward—for instance, how much farther humans are willing to go to buy gasoline at X¢ versus Y¢ a gallon—but most studies of spatial behavior focus on the summed responses of aggregates of humans rather than the decision processes of individuals (e.g., Haynes and Fotheringham 1984). Several studies have shown that humans are remarkably good at obtaining approximate solutions to GTSP on paper (e.g., Beardwood, Halton, and Hammersley 1959), but there has been little study of what algorithms humans use to achieve such good results, or how well humans perform in large-scale orienteering tasks without paper maps relative to the same tasks on paper (but see Saisa and Garling 1987; Taylor and Tversky 1992b). The only studies that explicitly test possible spatial rules used by individual humans focus on simple nearest-neighbor rules, which are nearly always rejected (Saisa and Garling 1987). Whereas these studies provide convincing evidence of what humans do not do, they do not tell us what rule humans do use to make spatial decisions.

Prospects and Unsolved Problems
A major unresolved issue in spatial foraging studies is how real animals use spatial information to devise efficient foraging routes. Do they use permutational algorithms that require explicit comparison of exact net costs and benefits among a potentially very large number of distinct sequences of points, or do they instead apply rules of thumb that integrate attractiveness functions across some area of space? I have ventured my opinion that permutational methods are not likely to be useful for most real-world foraging problems, which do not conform to the assumptions of most TSPs. If any animals do use permutational rules to solve spatial foraging problems, how is this achieved neuronally? For rules of thumb, is the area over which attractiveness functions are summed constant, or does it vary according to the density of resources, the hunger level of the animal, or other external conditions?

Difficulty of Distinguishing among Movement Rules
To test for different decision rules, it is necessary to carefully devise tests that can critically distinguish the use of one rule from other

Charles Janson

alternatives; inferring rules from the spatial choices made by animals or humans across randomly placed sites is not sufficient. For instance, the clever experiments by Menzel (1973a) on spatial choice by chimpanzees convincingly demonstrated that they could remember the locations of many foraging sites, but were less informative as to what rules they used to choose routes among the sites. In this study, a researcher carried a chimpanzee into an outdoor arena, where another researcher showed it the locations of hidden bananas. After a brief return to the indoor enclosure, the chimpanzee was released back into the arena. Typically, the chimpanzee moved rapidly from one hiding place to another, often in a sequence quite different from the original one used to show it the bananas. Despite Menzel's claim that the chimpanzees came close to distance-minimizing routes, my reanalysis of his published data shows that the actual routes chosen among the food sites visited were on average 54.4% longer than the shortest path connecting those points. Menzel suggested that the chimps mostly performed local, short-term optimizations, which would generally produce movements consistent with a nearest-neighbor movement rule. However, in the published sequences, chimpanzees moved to the nearest unused resource only 60.3% of the time across the sites they actually used in a given experiment (often not all food caches were used). Perhaps the short distances between resources in Menzel's experimental arena made minimizing travel costs a low priority for the chimpanzees.

Studies can be designed specifically to assess the importance of particular spatial decision rules. The basic strategy is to offer individuals a set of sites to visit that are arrayed in such a way that the use of one rule will generate one pattern of decisions, whereas other rules will produce different patterns (Saisa and Garling 1987). For instance, the three sets of destinations shown in figure 7.6 are intended to test which degree of look-ahead rule an animal uses (if indeed it uses any of them) to move among foraging sites. In each of these designs, an animal starting at position 1 would visit the remaining destinations in different sequences depending on what degree of look-ahead rule it was using. To control for the possibility that the animal is somehow always minimizing overall distance, the shortest route across destinations in each of the three different designs is associated with a different degree of look-ahead rule. In addition to producing different sequences, these patterns have an appreciable (>10%) difference in (1) the distances associated with

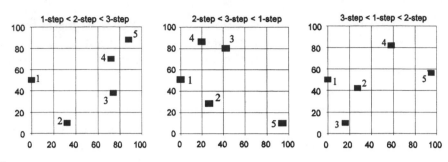

Expected
sequences for

1-step look-ahead:	1-2-3-4-5	1-2-3-4-5	1-2-3-4-5
2-step look-ahead:	1-4-5-3-2	1-4-3-2-5	1-2-4-5-3
3-step look-ahead:	1-3-5-4-2	1-2-4-3-5	1-3-2-4-5

Figure 7.6 These diagrams present three arrays of foraging sites designed to discriminate among several possible movement rules animals might use. For each array, the expected sequence of sites to be visited, listed below the array, would differ according to whether the animal used a one-, two-, or three-step look-ahead rule (see table 7.1). The arrays differ in the degree of look-ahead rule that produces the shortest overall path (see ordering of total path lengths given above each array). These varied arrays help to reduce the chance that the researcher might incorrectly assign the animal's movement pattern to a particular degree of look-ahead rule when in fact it employs a quite distinct distance-optimizing rule that happens to give the same sequence in a given array.

the shortest and longest overall paths in each design, and (2) the distance to the nearest and second-nearest targets from each site. These restrictions were imposed to reduce the chance that animals would make inconsistent choices merely because they could not perceive the difference in distances between alternative routes. The task of finding such designs to distinguish different movement rules is far from trivial—my computer program sorted through 10,000 sets of randomly placed points to find them. In the vast majority of cases, randomly placed points yielded movement sequences that were the same for many different degrees of look-ahead rule (Janson, unpub.). This fact in itself suggests that animals may be under weak selection to use two-or-more-step look-ahead strategies, as in a large fraction of real-life cases the resulting sequences are the same as those predicted by a simple nearest-neighbor movement rule.

It is important to remember that testing animals with distinctive spatial arrays can only reveal what rules the animals do not use, not what they do use. For instance, if an animal were found to use the sequence appropriate to a two-step look-ahead rule in the middle diagram figure 7.6, this would indeed imply that it does not

Charles Janson

use a one- or three-step look-ahead rule (cf. Gallistel and Cramer 1996). However, the movement sequence dictated by the two-step look-ahead rule is also that predicted by a utility field rule of thumb, as well as the overall shortest route. Experiments would have to be designed specifically to distinguish among the latter alternatives; such experiments have yet to be reported.

An added challenge is to integrate into movement models behaviors that may not be adaptive in "pure" spatial foraging problems. A good example may be the phenomenon of flock directional inertia as documented by Cody (1971; see above). Although Cody interpreted the observed degree of inertia as adaptive for his study organisms, my experience with capuchin monkeys suggests that they show group inertia even when this leads to longer short-term and long-term travel routes (Janson, unpub.). Furthermore, conflict over which of several known resources to use next can be a proximate cause of temporary group splitting (Janson and Di Bitetti, unpub.). Other examples of behaviors that may conflict with shortest-distance solutions are avoidance of high-predation areas (e.g., Cowlishaw 1997) and preferences for effort- or time-minimizing routes when spatial costs are not isometric.

Measuring the Trade-Off of Benefit versus Distance
As suggested by gravity models, wild animals seem to trade off distance to versus reward from a resource (e.g., Garber 1989; Janson 1996). What dictates the precise form of the trade-off function? Is the steep depreciation of distant food rewards shown by capuchin monkeys (Janson 1996) easily explained by existing theories of discounted delayed rewards in operant psychology? Is it caused by some perceptual constraint, or does it optimize fitness, or both? Temporal discounting of rewards in operant conditioning has been argued to be consistent with several predictions of optimal foraging theory, including preference for prey with shorter handling times and increasing preference for low-ranking foraging alternatives as average time to encounter prey increases (Fantino and Abarca 1985). Discounting of more distant rewards in space may also prove adaptive, but we currently lack an intuitive way to convert cost-minimizing solutions to fixed spatial problems directly into the fitness measures habitually used in optimal foraging problems, such as long-term energy maximization (cf. Bateson and Kacelnik 1996). Most studies on GTSP are of limited help, as they specify in advance either the set of sites that must be visited or the final destina-

tion, whereas for most animals these constraints are decisions in themselves, which will affect both the benefits and costs of the routes chosen.

There is a human sport called "rogaining" that closely mimics the animal foraging problem of choosing which points to visit in what order. Competitors are provided a map of an area with marked stops, each of which is worth a different, known number of points. The objective is to gather the maximum number of points within a specified time limit, starting and ending at predetermined sites. Because it is usually not possible to visit all the sites, competitors must decide approximately how many sites they can visit and which combination and sequence of sites will provide the highest point total. This particular human geographic problem has been modeled mathematically (Golden, Levy, and Vohra 1987; Golden, Wang, and Liu 1988). Although a potential gold mine of information about "natural" human spatial choice exists in the records of route choices for rogaining competitions, to my knowledge these have never been analyzed.

One attempt to permit multiple criteria of fitness ("objectives") in GTSP problems has been made (Keller and Goodchild 1988) by plotting the benefits (e.g., cumulative food intake) and costs (e.g., distance traveled) of particular travel routes against each other. All the points will lie below a curve that defines the best possible total benefits for routes with different total costs (fig. 7.7). To predict a

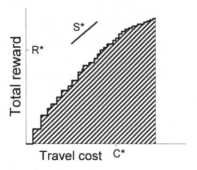

Figure 7.7 Trade-off function of total reward against total cost of travel for all possible travel routes among a large number of points. Over any small interval of total travel cost, there will be a single route that maximizes total reward, indicated by the 'corner' in the heavy line; all other travel routes will produce lower total rewards and lie within the shaded region. Given the set of maximal values of reward at each possible cost, it is possible to specify the corresponding optimal travel route for any given constraint of minimal total reward (R*), maximal total cost (C*), or minimum critical rate of reward per unit cost (slope S*). (Adapted from Keller and Goodchild 1988.)

Charles Janson

unique optimal route requires specifying additional information—
a critical maximum cost, a critical minimum total benefit, or a critical trade-off (slope) between costs and benefits. For instance, if the
costs and benefits refer to resources available only on a given day
(no renewal included), then one could compare observed behaviors
against energy-maximizing (total time cost equals the maximum activity period) and time-minimizing (total benefit equals the minimum food intake) solutions. If the costs and benefits refer to routes
that include constraints on the renewal rate of the resources, then
the slope of the outer envelope would correspond to the long-term
energy gain rate, and one could search for solutions of maximal
slope (energy-maximizing) or minimum critical slope to satisfy
daily energy needs (time-minimizing).

Integrating Space and Time in Foraging Decisions
The temporal aspect of spatial foraging remains virtually unexplored. Theoretical models of optimal return times on a trapline
have to date ignored spatial constraints (Possingham 1989). In this
case, foragers pay no cost of moving between resources, and the
only factor that affects energy yield is the rate of accumulation of
energy E as a function of time since the forager's last visit to a
resource, T. If energy increases proportionately (linearly) with time
(that is, $E(T) = cT$), then the rate of energy gain per time spent
absent by the forager is a constant ($E(T)/T = c$). This conclusion is
biologically puzzling, as can be seen in a simple counterexample.
Imagine two resources spaced far enough apart that the forager
wastes more energy traveling between them at maximal speed than
it gains once it arrives at each resource after a short period of renewal. Surely in this case it would pay the forager to slow down
and wait for more energy to accumulate before visiting each of the
resources. In fact, imposing spatial and digestive constraints leads
to predictions of unique fitness-maximizing return intervals, at least
for resources with a single, constant rate of renewal (Janson, unpub.). In essence, the forager has to balance moving too fast (not
letting enough energy accumulate in each resource) against moving
too slowly (allowing the accumulated energy in each resource to
surpass the forager's ability to harvest it). Even without spatial
movement costs included in the model, a unique optimal return
time is predicted if the rate of energy accumulation in a resource
declines over time (Possingham 1989).

The complex case in which distinct harvestable resources differ in their rates of renewal or in the kinds of benefits they offer the forager remains almost unexplored, except for limited work on frequency of human shopping trips (Lentnek Harwitz, and Narula 1981) and vending circuits (Ghosh 1982). None of these theories explicitly models the interaction of spatial and temporal choices, but observed behaviors of traplining species may provide some clues about optimal behaviors. For both primates (Janson, Stiles, and White 1986; Terborgh and Stern 1987) and bees (Heinrich 1979a), there is evidence that traplining foragers concentrate their attentions on one or two "major" food species, with other "minor" species added in opportunistically. These animals appear to resolve the problem of trading off distinct spatial distributions and renewal rates across resource types by focusing their attention on a set of resources that possesses a common rate of energy accumulation per resource and is in the aggregate productive enough to satisfy their daily foraging needs. Such an outcome may be favored especially if some learning constraint (e.g., learning to "handle" a particular flower: Lewis 1986) makes it adaptive to specialize on one resource at a time. There are few data on long-term patterns of resource use in the wild, yet these may be the most interesting aspects of foraging on spatially complex, predictably renewing resources.

Conclusions: The Broader Implications of Spatial Foraging
Simple rules of thumb, such as the nearest-neighbor rule, could allow animals to achieve nearly optimal overall travel routes. With respect to the debate on the importance of social versus ecological factors in the evolution of primate intelligence, this finding may lead some to suggest that fruit eating may not require the complex degree of intelligence apparently shown by many primates (e.g., Barton, chap. 8, this volume). As I hope I have demonstrated above, such simple algorithms may not be those actually used by primate frugivores. In separate studies of wild tamarins, capuchin monkeys, and captive chimpanzees, 30–40% of the resources chosen were not the nearest neighbors of the previously used food tree (Garber 1989; Janson 1996; Menzel 1973a), and experiments on captive macaques (Gallistel and Cramer 1996) and chimps (E. Menzel, pers. comm.) reveal an apparent ability to think beyond the nearest resource. The comparative evidence from psychological studies of "self-control" suggests that primates do have quantitatively more ability to delay gratification, and thus to incorporate more complex

Charles Janson

decision criteria in their foraging, than rats or pigeons do (Tobin and Logue 1994; Tobin et al. 1996).

If foraging efficiency has selected in primates for the ability to "think ahead," why might primates have more need for such foresight than most other animals? Fruit trees, which many primate species depend on, might favor route planning because they represent a very clumped yet sparse resource base (Milton 1981a). Fruit-eating primates are brainier than their leaf- or insect-eating relatives (Clutton-Brock and Harvey 1980), and this contrast remains even when the important effects of social group size on relative brain size are accounted for (Barton 1996b; see also Barton, chap. 8, this volume). Fruit-eating species are also brainier among bats and rodents (reviewed in Harvey and Krebs 1990), but primates have much larger relative brain sizes than these other frugivores. Why?

Two factors distinguish primates from nearly all other fruit-eating animals: first, primates do not fly, and second, most primates are both large and social. The costs per unit distance for terrestrial locomotion are three or four times greater than for flying animals (Schmidt-Nielsen 1972), and the difference is probably further exaggerated for arboreal quadrupeds, which must climb up and down trees instead of being able to travel at a single height. Furthermore, food patches worth retaining in the memory of a social primate ought to be relatively large and productive, as feeding competition from other group members would render the benefits of visiting unproductive trees small to negligible. Large and productive fruiting trees must form a small subset of the far greater density of small trees and shrubs that would attract a solitary bird or bat, or even a small social group of those small-bodied frugivores. As the density of feeding sites declines relative to ability of the forager to detect previously unknown resources, the benefits of spatial memory increase, and can grow very large. For instance, the ability of capuchin monkeys to remember the locations of experimental feeding platforms placed about 200 m apart increased their foraging efficiency nearly threefold relative to having to detect each one anew (Janson and Di Bitetti 1997); at a distance of 80 m between platforms, spatial memory would not have increased foraging success. Thus, primates may be under far greater selection pressure for spatial efficiency than many other foragers: First, they cannot use only a single food tree all day, as might a solitary forager of similar size; second, large fruit sources they require are likely to be more widely spaced apart than the smaller trees and shrubs used by smaller-bodied frugivores; and finally, the cost of a longer foraging route (relative to potential foraging gains) is far greater for an arboreal quadruped than for flying animals. If these assumptions are correct, the spatial routes used by fruit-eating birds and bats might be considerably less efficient, relative to the density of available resources, than those used by primates. Data on foraging path length for small fruit-eating Neotropical bats (Fleming 1988) show that they travel much greater distances per day than any known Neotropical primates (Janson and Goldsmith 1995), suggesting that the bats might be less concerned about route efficiency. Clearly, more data specifically documenting spatial routes of nonprimate frugivores are essential to testing the preceding scenario.

The greatest challenge of understanding animal spatial movements is incorporating resource renewal. As shown by models of periodic marketing and shoppers (Ghosh 1982; Lentnek, Harwitz, and Narula 1981), calculating an optimal interval between visits to a site (independent of the complicated question of which sequence of sites to visit) requires prior knowledge of (1) the monetary (energetic) and time costs of travel; (2) the prices (costs), depletion or renewal rates, and requirements or utilities (fitness values) per unit of different commodities; and (3) time constraints, such as (for animals) an enforced half-day resting period. Although there may exist rules of thumb to combine these many distinct variables into a nearly optimal solution for the optimal policy dictating both return times and spatial routes, at present these have neither been modeled nor measured.

The implications of considering resource renewal in primate foraging studies are profound. Primate ecologists have focused on static measurements of resource "size" (crown volume or DBH) for over 30 years. Instead, we need to consider how the spatiotemporal behavior of the primates themselves determines the available biomass of ripe fruit in tree crowns by determining the time between visits—how much primates get to eat in a tree is often a function of their own past behavior. If spatial foraging decisions in primates depend on group size, then the correlation between group size and patch size found in some studies (e.g., Terborgh 1983) might occur not only because larger patches can feed more monkeys, but also because larger groups may regulate return times to be longer and so encounter more food at each patch upon returning to it. Although much progress has been made in understanding the rela-

Charles Janson

tionship among primate group size, food competition, and resource characteristics (e.g., Janson 1992), I believe that the most exciting and interesting developments await us as we come to appreciate the important role of spatial foraging decisions in foraging success and feeding competition.

Acknowledgments
This is contribution number 1044 from the Graduate Program in Ecology and Evolution at SUNY-Stony Brook. I am grateful to many students and faculty in that program for their patience, encouragement, and advice; especially helpful were James Thomson, Mario Di Bitetti, and Neal Williams. Some of my work summarized here was supported by NSF grant IBN-9511642.

CHAPTER EIGHT

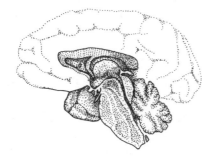 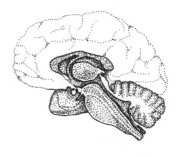

Primate Brain Evolution: Cognitive Demands of
Foraging or of Social Life?

ROBERT A. BARTON

What explains the evolution of large brains and cognitive skills in
primates? For many years, primatologists have sought factors in the
lifestyles of particular species that might place special demands on
information processing, capacities and hence that might have selected for intelligence. Some authors have drawn attention to the
possible cognitive demands of foraging, while others have emphasized social complexity. This volume posits a reconciliation between
the two by suggesting that gregarious animals may face particular
challenges in optimizing their ranging behavior under social constraints. Here, I examine the foraging and social cognition hypotheses using comparative data on the size of the brain and its relevant
component parts. I will look at several issues: Why do primates as
an order tend to have large brains? What explains variation in brain
size and structure among primate species? Which specific brain
structures, and hence what aspects of neural processing, have been
the focus of selection? How do cognitive attributes correspond to
neural specializations? I will argue that a range of comparative evidence supports the idea that brain evolution involves quite specific
sensory-cognitive adaptations (Harvey and Krebs 1990; Preuss
1993), and that these adaptations provide a basis for understanding
(1) the evolution of brain size, (2) the comparative links between

neuroanatomy and cognition, and (3) why so much debate within comparative psychology has been unproductive. More specifically, I will argue for the importance of visual specialization in the evolution of brain and cognition in primates, and will relate visual specialization to ecological and social variables such as frugivory and group size. I will suggest that cognitive mechanisms for route finding and spatial mapping have not been distinguishing features of primate brain evolution, but that the interaction between social and foraging demands is currently little understood, and may turn out to have important cognitive implications.

Foraging Cognition versus Social Cognition

In the past, proponents of the foraging cognition hypothesis have pointed to the patchy and ephemeral distribution of some primate food sources, such as fruit, arguing that large brains are needed to memorize and integrate information on the location of such resources in space and time. Milton (1988) for example, compared the large-brained spider monkey (*Ateles* spp.) with the relatively small-brained howler monkey (*Alouatta* spp.). Spider monkeys are frugivorous, ranging over a large area to find scattered fruiting trees. Howler monkeys also eat fruit, but rely much more on leaves than do spider monkeys, and have gut specializations allowing them to digest fibrous leafy material. Milton argued that foliage is more densely and evenly spread than fruit, thus requiring smaller home ranges and a less powerful spatial memory for its efficient exploitation (see Milton, chap. 14, this volume). A very similar comparison can be made between two great ape species, gorillas *(Gorilla gorilla)* and chimpanzees *(Pan troglodytes)*. Gorillas are more folivorous, have smaller home ranges, and have smaller brains than do chimpanzees. Perhaps the larger home ranges of spider monkeys and chimpanzees reflect more complex foraging strategies. The foraging hypothesis would predict that the large-brained frugivores construct efficient travel routes between widely dispersed resource patches (fruit trees), while the folivores tend to potter about in a more random fashion. Social and ecological factors may interact, however. As well as ranging over a larger area than their more folivorous relatives, spider monkey and chimpanzee groups are also larger, and they split up into temporary foraging parties in a way that howler monkey and gorilla groups do not. Not all frugivores show this fission-fusion pattern, suggesting that it is unlikely in itself to explain the general brain size-frugivory correlation. We nevertheless ought to consider the cognitive implications of social foraging.

An additional aspect that may complicate the picture is extractive foraging: the removal of foods from tough or awkward substrates (e.g., Parker and Gibson 1977). Capuchins (*Cebus* spp.) are adept at removing a variety of items from tree cavities and hard substrates; aye-ayes *(Daubentonia madagascariensis)* are specialized for extracting the grubs of wood-boring arthropods; chimps are renowned for "fishing" termites out of their mounds; and all are relatively large brained: Byrne (1995b), however, suggested that gorillas also use complex, hierarchically organized food processing techniques to "extract" the edible parts of plants from their tough, thorny, or otherwise protective exteriors, and that these techniques may be just as cognitively demanding as extractive foraging in the narrow sense.

The social cognition hypothesis was proposed by several people independently, perhaps most notably by Humphrey (see Byrne and Whiten 1988 for a historical review). Proponents of this idea claim that the most cognitively demanding aspect of the lives of many primate species is not their physical environment, but the behavior of the other members of their social group. Primate social groups are based on a delicate balance between competition and cooperation. Reproductive success in such groups probably depends partly on the skill with which an individual manages this delicate balance. Intelligence, therefore, could have evolved via a spiraling arms race for more and more sophisticated social sensitivity, knowledge, and prediction.

Each of these ideas has a degree of plausibility—and that is the problem. How can we move beyond plausible adaptive storytelling to scientific hypothesis testing? Field studies of behavior are in themselves of limited value. Inferences about the evolution of cognitive processes based on the apparent complexity of behavior in its natural setting may be alluring, but they are suspect for two reasons. First, apparent complexity may often be generated by simple underlying decision rules. Janson's chapter in this volume (chap. 7) makes this very clear: nonrandom use of a home range is far from being adequate evidence for cognitive maps in animals, since simple decision rules can generate many nonrandom patterns and solve apparently complex problems (see also Dyer, chap. 6, this volume). Even where it can be shown that memory, rather than just random

Robert A. Barton

searching and direct sensory detection of resources, is involved, this still leaves open a number of possibilities for what might be represented in memory, ranging from simple to complex. In practice, it is probably impossible to distinguish between models based on simple and on complex decision processes using observational data alone, which is why researchers are now adopting experimental methods (see Garber, chap. 10; Janson, chap. 7, this volume). This, however, brings me to the second problem. Behavioral data alone, whether based on observations or experiments, cannot tell us how decision processes are instantiated in neural processes, nor, therefore, how brains would have had to evolve to accommodate such cognition. In particular, behavioral data cannot in themselves tell us whether foraging niche, social complexity, or both have been factors in the evolution of brain and cognition.

Comparative Brain Studies
In order to test adaptive hypotheses directly, we need to turn to the comparative method—in particular, comparative studies of brain size and structure. As Harvey and Pagel (1991) pointed out, "adaptation is an inherently comparative concept." That is, adaptation implies that different species have evolved different solutions for surviving and reproducing in their particular ecological niches. Hence, when we hypothesize that large brains are an adaptation to life in complex social groups, we predict (either explicitly or implicitly) that solitary species should have smaller brains than gregarious species.

Encephalization, Frugivory, and the Evolution of Spatial Memory
The idea that comparative studies of the brain can indicate what selection pressures have driven the evolution of intelligence is an old one. Most interest has focused on species differences in brain size, based on the assumption that brain size provides a good overall index of behavioral flexibility and the capacity to learn. Brain size is, however, highly correlated with body size, and it is usually assumed that valid comparative measures of "braininess" must in some way remove the effects of body size (Jerison 1973; Harvey and Krebs 1990). Jerison (1973), in particular, developed the idea of "encephalization," the size of a species' brain relative to that expected for its body size. Various measures of encephalization or relative brain size have been used, all based on the following general procedure. Brain size is plotted against body size for a sample of

species, and some kind of regression line fitted. For each species, the observed brain size is compared with the brain size predicted for that species by the regression curve, and a difference score or ratio between the two is calculated. Species falling above the curve have high scores, those below have low scores.

Numerous studies in a wide range of taxonomic groups have looked for evidence that encephalization is related to the information processing demands of different ecological niches. Early studies found correlations between brain size and diet. Milton's interpretation of howler and spider monkey foraging strategies and brain size finds support in broader comparative studies. In both bats and primates, frugivores tend to have particularly large brains for their body size (Eisenberg and Wilson 1978; Clutton-Brock and Harvey 1980). These results have generally been interpreted as evidence for selection on spatial memory, because frugivore-omnivores have larger home ranges and a more widely scattered and ephemeral food supply than do folivores, and therefore would require bigger brains to memorize the distribution of their resources and to forage efficiently.

Objections to the Foraging Hypothesis
A number of objections have been made to this interpretation of brain size correlations. Some have pointed out that the brain is metabolically expensive (Armstrong 1983; R. D. Martin 1981b). R. D. Martin (1981b, 1996) has suggested that the size of an infant's brain is determined by maternal metabolic turnover during fetal development and lactation, and that adult brain size is directly related to infant brain size. If a frugivorous diet has a relatively high energy content, hence allowing a higher maternal metabolic rate, energetic constraints, rather than cognitive selection pressures, might explain why frugivores have bigger brains than do folivores (Martin 1996). Aiello and Wheeler (1995) have suggested that the encephalization-diet correlation reflects a trade-off between the size and energetic costs of guts (large in folivores) and brains (small in folivores), while not denying the importance of selection on cognitive traits in determining a species' optimal position on the trade-off curve. Others have suggested that social group size is a more important determinant of brain evolution (Sawaguchi 1992; Dunbar 1992), an argument that I develop below. More worryingly for allometric studies, the appropriateness of body weight as a reference variable has been called into question on both statistical and evolutionary grounds.

Robert A. Barton

The statistical problem arises because body weight is a relatively "messy" variable for the purposes of comparative analysis, showing great intraspecific variation associated with age, nutritional and reproductive status, and other factors. The error variance associated with species averages, then, potentially introduces systematic biases when the effects of body size are partialed out of brain size (Economos 1980; Harvey and Krebs 1990; Barton, in press). This is particularly true when two or more variables of interest (such as brain size and home range size) are correlated after correcting each for body size. The body size error variance will affect both sets of size-corrected indices in a similar way, potentially creating a spurious positive correlation between them. The results of all allometric studies that use such a procedure should therefore be treated with caution (for a more detailed explanation, see Barton, in press, and for possible solutions see Mace and Eisenberg 1982; Harvey and Krebs 1990).

The evolutionary objection to encephalization as a measure of comparative brain size is that it confounds selection on brain size and selection on body size. As explained above, encephalization is a composite measure—a ratio, or difference score, between observed brain size and that expected according to body size. It has been suggested, however, that brain size can lag behind rapid evolutionary changes in body size (Shea 1983; Willner and Martin 1985), the extreme cases being phyletic dwarfism and gigantism. Because of this evolutionary lag, species with high encephalization quotients might simply be those that have in recent evolutionary time decreased in body size, whereas low quotients might reflect recent increases in body size. Dietary correlates of encephalization might then be attributable to increased body size in folivores (Deacon 1990). It has been suggested, for example, that the gorilla, which has a low encephalization quotient compared with its great ape relatives, has not been selected to be smaller brained, but simply larger bodied (Byrne 1994). Encephalization, therefore, is not easily distinguished from what one might call "somatization."

All of these objections in some way potentially undermine the credibility of explanations for brain size correlations that invoke selection on cognition, including spatial cognition. Unfortunately, few decisive tests of these competing hypotheses have been made (but see Barton, in press, for tests between the energetic and cognitive hypotheses). The significance of correlations between brain size and ecology therefore remains obscure.

Testing for Correlated Evolution
A rather different problem with many of the comparative studies carried out in the past is that species were treated as independent data points in statistical analyses. This problem has been dealt with extensively elsewhere (see, for example, Felsenstein 1985; Harvey and Pagel 1991, Purvis and Rambaut 1995). Briefly, hypothesizing an adaptive association between two character states, such as brain size and lifestyle, implies that they have tended to evolve together. Valid statistical tests for such correlated evolution must be based on *independent evolutionary events.* It is not sufficient to calculate correlations across species values, because species share traits through common descent as well as through independent evolution. A hypothetical example is given in figure 8.1. Here, a statistical test for an association between activity timing and brain size might give a significant result, because the average brain size in the diurnal species is greater than in the nocturnal species. A closer look, however, would reveal no convincing evidence for an adaptive association. The phylogeny of these species indicates that there have been only three evolutionary transitions in activity timing, and that in only one of the three comparisons has the diurnal lineage evolved a larger brain size. It is these evolutionarily independent comparisons, or contrasts, that must provide the data for valid statistical tests of adaptive associations. Within primates, the two suborders, the Strepsirhinii and Haplorhinii, differ in an array of features. For example, strepsirhines have smaller brains, lower metabolic rates, shorter gestations, a different type of placentation, smaller neonates, and tend to be more nocturnal and to live in smaller groups (Martin 1990). A nonphylogenetic analysis of any of these traits would risk finding spurious correlations between them as a result of the gross differences between the two clades.

Brain Specialization
The focus on brain size also ignores the organization of the brain into functionally specialized neural systems, or modules (Fodor 1983). Could it be that selection has acted on specific neural systems, rather than just brain size and general intelligence? If so, lifestyle should be correlated with the relative size, and perhaps complexity, of specific brain regions. Emerging evidence suggests that this is indeed the case. Studies on the avian hippocampus (e.g., Krebs 1990), olfactory bulb (Healy and Guilford 1990), and song control nuclei (e.g., De Voogd et al. 1993) and on mammalian sen-

Robert A. Barton

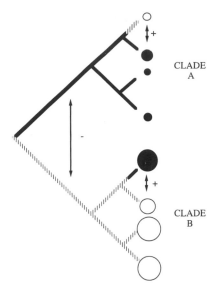

Figure 8.1 *Hypothetical comparative analysis of brain size in relation to activity timing.* Nocturnal species are represented by solid circles, diurnal species by open circles, and the size of the circle represents brain size. There are two major clades (A and B), one of which has large brain size and the other small brain size. Average brain size is less in nocturnal species than in diurnal species. However, this difference is based on just one evolutionary event, the divergence between clades A and B. There is no evidence in this hypothetical case that brain size and activity timing have evolved together. The shading on the branches of the phylogeny indicates the inferred state at each point in the evolutionary radiation of the group: black for nocturnal, hatched for diurnal. In this data set we can infer three evolutionary transitions giving rise to separate nocturnal and diurnal taxa (shown by the arrows) where we can compare brain size: larger brains in nocturnal lineages are indicated by a plus sign, and smaller brains in nocturnal lineages by a minus sign. In only one of these three cases (the comparison between the fully nocturnal parts of clade A and the fully diurnal parts of clade B) is mean brain size greater in the diurnal clade. Two of the three phylogenetic comparisons show the reverse. Hence, an analysis that did not take phylogeny into account could give a statistical relationship between two variables that was the opposite of a real evolutionary relationship.

sory systems (Barton, Purvis, and Harvey 1995) and neocortex size (Sawaguchi 1992; Dunbar 1992; Barton 1996b) have all shown that the size of specific brain structures correlates with behavioral ecology when differences in overall brain size, or the size of the rest of the brain, are taken into account. This means that, owing to specific selection pressures, certain brain structures have changed in size relative to the rest of the brain. Entirely new cortical areas may have emerged (Allman and McGuinness 1988; Preuss 1993). These evolutionary differences in particular structures do not always correlate with overall brain size, either because the relevant structures are too small to have a significant effect on overall brain size (e.g.,

Sherry et al. 1989), or because other structures show compensatory changes in size (Harvey and Krebs 1990; Barton, Purvis, and Harvey 1995). Either way, understanding the evolution of neural and cognitive specialization demands that we look beyond gross brain size. Conversely, there may be other cases in which differences in overall brain size do reflect selection on specific systems. I will argue that selection on certain neocortical systems has had a significant effect on overall brain size in primates. In such cases, understanding the evolutionary significance of brain size depends on identifying the underlying neural specializations.

Evolutionary Radiation of the Primate Brain

What Explains Large Primate Brains?

The question of why primates have large brains is based on an assumption that is only partially warranted: that primates have larger brains than other mammals. Whether this is true depends on which primate species are being compared with which other mammal species. Cetaceans and carnivores, for example, overlap the primate range of variation in brain size relative to body size. Strepsirhine primates (lemurs and lorises) as a group are generally much smaller brained than are haplorhine primates (tarsiers, monkeys, and apes). When primatologists talk about the cognitive sophistication and large brains of "primates," they usually mean monkeys and apes. Here I will look at the neuro-volumetric data of Stephan and co-workers (Stephan, Frahm, and Baron 1981) to establish key differences between the two suborders of primates, the Strepsirhinii and Haplorhinii. Stephan and other workers published comparable neuro-volumetric data on two other placental mammal groups, bats and insectivores. Because the adaptive specializations of bats are in many ways unusual and divergent from those of the other groups, and because the ancestors of contemporary primates were probably similar in some respects to living insectivores, I will make my comparisons with the latter group. This should not, of course, be taken to mean that insectivores are the ancestors of primates, or that they are evolutionarily unspecialized.

Because, as outlined above, the comparative method works by assessing correlated evolution among character states across multiple evolutionary events, the adaptive significance of single evolutionary events, such as the evolution of a large brain in the Haplorhinii, by definition, cannot be directly tested. It is, however, possible to gain some broad insights by looking at what aspects of neuro-anatomy differ between taxa. This can tell us, for example, whether large brain size is likely to be simply a side effect of some other biological change, such as increased maternal metabolic expenditure (R. D. Martin 1981b, 1996), or a result of selection on specific behavioral capacities. If the latter were true, we would expect brain size differences to be a result of the relative enlargement or contraction of those brain regions mediating the selected-for behavioral capacities. This would not altogether rule out constraints theories, because it would still be possible to argue that, when constraints on brain size are eased, investment in extra brain tissue will be more advantageous in certain brain regions than in others. It would, however, indicate that the different taxa have different information-processing specializations, and would tell us something about their general nature.

Differences in the Size of Specific Brain Regions
In fact, it is clear that the primate brain, and the haplorhine brain in particular, is not just globally large, but shows evidence of selection on specific regions, and hence specific capacities (table 8.1) Stephan, Baron, and Frahm (1988), for example, analyzed the size of individual brain regions relative to body size in insectivores and primate suborders. Some regions varied considerably in size between taxa, while other regions differed little or not at all. Some examples, using the data of Stephan, Frahm, and Baron (1981), are given in figure 8.2. The extent and direction of volumetric differences among taxa depend on which structure is compared. The neocortex is strikingly bigger in primates than in insectivores, relative to body weight, and also differs between haplorhine and strepsirhine primates. Differences in other structures are generally less striking. Indeed, some regions, such as the medulla, which controls "vegetative" functions such as respiration, and the piriform lobe, involved in olfaction, are not expanded at all in primates. The hippocampus, a structure strongly implicated in spatial memory (see below), does not differ markedly among taxa, undermining the idea that selection on cognitive mapping helps to explain the large haplorhine brain (see also Kappeler, chap. 16, this volume). The olfactory bulb is actually smaller, relative to body size, in haplorhine primates than in strepsirhines and insectivores, probably because of adaptation to diurnal versus nocturnal niches respectively (Barton, Purvis, and Harvey 1995). It is true, however, that while a

Table 8.1 Major mammalian brain regions, subregions, and functions, with the average percentage of total brain volume that each region or subregion constitutes in primates

Region	Subregion	Function
FOREBRAIN (mean volume=81%)		
Telencephalon (mean volume = 74%)	**Neocortex**	Sensory (70%), motor (10%), prefrontal (8%) and "other" areas (12%)[a]
	Basal ganglia	Control of movement
	Limbic system	Emotion, motivation, and learning, including spatial memory (**hippocampus**)[b]
	Olfactory bulbs	Olfaction
Diencephalon (mean volume = 7%)	Thalamus	Precortical nuclei, including the visual **lateral geniculate nucleus (LGN)**[c], containing separate parvocellular and magnocellular layers projecting to the visual cortex
	Hypothalamus	Autonomic and endocrine control; control of appetite behavior
MIDBRAIN (mean volume = 3%)		
	Tectum	Orienting/reactions to sudden stimuli[d]
	Tegementum	Functionally diverse nuclei, including aspects of motor control and arousal
HINDBRAIN (mean volume = 16%)		
	Cerebellum	Balance, motor control, and learning[e]
	Pons	Sleep and arousal
	Medulla	Vegetative functions (cardiovascular, respiration, etc.)

Source: Calculated from Stephan, Frahm, and Baron 1981.

Note: Structures directly referred to in the text are in boldface. References are given for selected comparative volumetric analyses that link these structures to specific ecological factors in primates and other taxa. See also Stephan, Baron, and Frahm 1988; Jerison 1991 for general reviews of allometric studies of individual brain parts. Note that the studies cited vary in the extent to which they analyzed correlated evolutionary change, rather than general phylogenetic trends (see figure 8.1).

[a]Percentages each subarea makes up of the whole neocortical surface area in macaques (data from Drury et al. 1996). The neocortex constitutes an average of 60% of total brain volume. Whole neocortex size correlated with social group size: Dunbar 1992; Sawaguchi 1992; correlated with activity timing, social group size, and frugivory: Barton 1996b; primary visual cortex size correlated with activity timing, social group size, and frugivory: Barton, Purvis, and Harvey 1995; R. A. Barton, unpub.

[b]Hippocampus size correlated with food storing in birds: see Krebs 1990 for review.

[c]LGN larger in diurnal species: Stephan, Baron, and Frahm 1988; LGN size correlated with activity timing and frugivory: Barton, Purvis, and Harvey 1995; parvocellular layers of LGN correlated with activity timing, social group size, and frugivory: Barton 1998.

[d]Tectospinal tract correlated with predatory habits: Barton and Dean 1993.

[e]Cerebellar nuclei related to locomotion: Matano 1985.

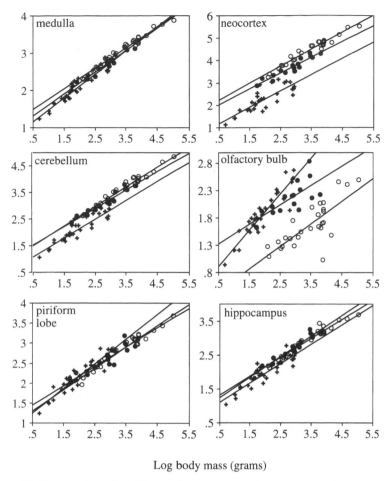

Log body mass (grams)

Figure 8.2 Volume of some individual brain structures relative to body mass in primates and insectivores. Open circles, haplorhine primates; solid circles, strepsirhine primates; crosses, insectivores. (Data from Stephan, Frahm, and Baron 1981.)

range of structures show no clear taxonomic differences, the olfactory bulb is the only one that is actually smaller in haplorhine primates than in the other groups. It is important to bear in mind that different brain structures are anatomically and functionally highly interconnected. When natural selection acts on specific brain functions, certain structures may be changed more than others, but it will often also be the case that changes will occur in a range of interconnected structures. The cerebellum, for example, is, like the neocortex, significantly larger in primates than in insectivores. This may reflect the cerebellum's important anatomical connections with the neocortex and its functional role in fine motor control and

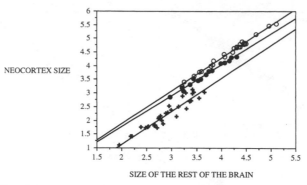

NEOCORTEX SIZE

SIZE OF THE REST OF THE BRAIN

Figure 8.3 Size of the neocortex relative to the size of the rest of the brain in primates and insectivores. Symbols as for figure 8.2.

learning. The conjunction of size differences in the neocortex and cerebellum is likely to relate to the evolution of hand-eye coordination in primates, which is to a large extent a joint function of these two structures. This idea is supported by evidence that a cortical area involved in visually guided reaching and prehension—the ventral premotor area—is unique to primates (Preuss 1993). Hence, the proposal that selection acts on specific brain *modules* must be balanced against the fact that these modules usually involve a range of *structures*.

The relative importance of neocortical evolution in primates can be clearly seen by plotting neocortex size against the size of the rest of the brain (fig. 8.3). The primate neocortex is bigger not only relative to body size, but also relative to the rest of the brain. Thus, primates' large brains result to a great extent from a disproportionately expanded neocortex. Within primates, the difference between suborders, though smaller than the difference between primates and insectivores, is statistically significant, with haplorhines having larger neocortices relative to the size of the rest of the brain than do strepsirhines (Barton 1996b). Again, the relative expansion of the neocortex in haplorhines is associated with larger brains overall because, although olfactory bulbs are reduced, there has not been a more general compensatory reduction in subcortical regions. These grade shifts in neocortex size relative to the size of the rest of the brain are clear evidence that selection can work on specific brain structures, rather than just the whole brain (cf. Finlay and Darlington 1995).

The neocortex dominates the primate brain, making up, on average, 60% of total brain volume in nonhuman species. As we have seen, this percentage varies among species. For example, the tala-

Robert A. Barton

poin *(Miopithecus talapoin)* and potto *(Perodictus potto)* both weigh roughly 1,100 grams; the talapoin, however, has a much larger brain and a neocortex that makes up about three-fourths of its total brain volume, while the potto's neocortex is only about half of its total brain volume. Among insectivores, the biggest neocortices are only about a third of total brain volume.

What is the behavioral and cognitive significance of such differences in neocortex size? At least some of the differences between orders and suborders are very likely to reflect visual specialization. Primates are highly visual animals whose early evolution has been described in terms of adaptation to visually guided predation in a fine-branch niche (Cartmill 1974). Fossil endocasts indicate relative expansion of visual cortical areas (Jerison 1973; Allman 1987; Preuss 1993), and up to about half of the neocortex of haplorhine primates is dedicated to visual processing (van Essen, Anderson, and Felleman 1992). The lateral geniculate nucleus (LGN), a relay station between retina and cortex, is larger in primates than in insectivores, and this is reflected in several unusual features of the primate visual system, including a high degree of binocular overlap, facilitating stereopsis; high visual acuity, particularly in diurnal haplorhines; a unique arrangement of the retino-tectal projections; and a lateral geniculate nucleus with up to six distinct layers, including both the two magnocellular layers common to all mammals and two to four parvocellular layers not found in other mammals (see Allman and McGuinness 1988 for more details). The separate layers of the LGN project to different regions of the neocortex, corresponding to functionally distinct visual processes (fig. 8.4): the parvocellular processing stream primarily analyzes fine detail and color, whereas the magnocellular system is primarily involved in movement detection and the analysis of dynamic form (Zeki and Shipp 1988; Livingstone and Hubel 1988; van Essen, Anderson, and Felleman 1992). Interestingly, the size of the magnocellular layers in primates is similar to the size of the whole LGN in insectivores (fig. 8.5). Perhaps the older visual subsystem has remained relatively unchanged during primate evolution, while the addition of the new—parvocellular—subsystem has led to a net expansion of subcortical visual pathways, visual neocortex, and total brain size. These developments would have underpinned the evolution of the distinctive visual abilities of haplorhine primates, including a high degree of visual acuity and trichromatic colour vision (the latter is found in apes, Old World, and some New World monkeys). Indeed, the haplorhine-strepsirhine difference in neocortex size rel-

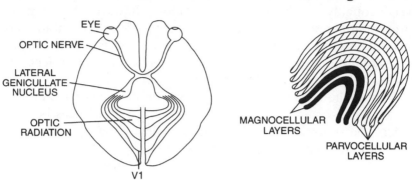

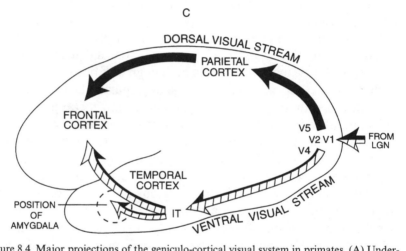

Figure 8.4 Major projections of the geniculo-cortical visual system in primates. (A) Underside of the brain, showing projections from eye to primary visual cortex via lateral geniculate nucleus. (B) Section through the right lateral geniculate nucleus, showing parvocellular and magnocellular layers. Neurons in parvocellular layers have high spatial resolution and low sensitivity to contrast, and are sensitive to wavelength. Magnocellular neurons have low spatial resolution and high sensitivity to contrast, and are insensitive to wavelength. (C) Side view of the left cortical hemisphere, with a simplified representation of dorsal and ventral visual processing streams (see van Essen et al. 1992 and Young et al. 1994 for details). IT, inferotemporal cortex. Input to the dorsal stream is magnocellular, whereas input to the ventral stream is primarily parvocellular. The dorsal stream mediates spatial vision and fast visuo-motor responses, while the ventral stream mediates detailed object vision with color and, in interaction with the amygdala (which lies beneath the cortex), processing of facial stimuli. (Redrawn with permission from Tovée 1996.)

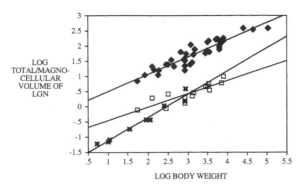

Figure 8.5 Volume of the LGN in insectivores (crosses) and primates (solid diamonds), with the volume of magnocellular layers alone (open squares) also shown for primates. The lack of a clear size difference between primates' magnocellular layers and insectivores' whole LGN may indicate that the older magnocellular pathway has evolved little since the divergence of these groups, whereas the addition of the parvocellular pathway in primates has resulted in a net increase in whole LGN size.

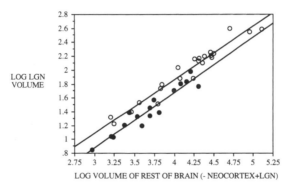

Figure 8.6 Difference between haplorhines and strepsirhines in the size of the LGN relative to the rest of the brain. Open circles, haplorhines; solid circles, strepsirhines.

ative to the size of the rest of the brain is accompanied by a very similar grade shift in both the primary visual area (V1) and the LGN (fig. 8.6). Hence, much of primate brain evolution may be associated with visual specialization, in terms of both the differentiation of primates from other orders and the differentiation of primate species from one another. I shall now deal with the latter in more detail.

Ecological Correlates of Brain Size and Structure among Primates
As with the comparisons between orders, it is instructive to examine variation in individual brain structures among primates, in this case to determine the extent to which this variation (1) correlates with

lifestyle and (2) underlies variation in overall brain size. We have already seen that the haplorhine brain is more "neocorticalized"; it has a neocortex that is more expanded relative to the rest of the brain than is the case in strepsirhines. Why? Is it related to adaptation for nocturnal versus diurnal habits? Using a version (Purvis and Rambaut 1995) of the general type of phylogenetic analysis outlined in figure 8.1, I found that neocortex size is indeed positively correlated with diurnality, based on four independent contrasts between nocturnal and diurnal primate lineages (Barton 1996b). Total brain size does not vary consistently with activity period, probably because of reciprocal changes in olfactory structures (Barton, Purvis, and Harvey 1995; Barton, in press). The association between diurnal habits and large neocortices reflects the visual functions of the neocortex already mentioned above.

The next question is whether social or dietary variables explain additional variation in neocortex size. I approached this question by partitioning the data set into clades that are monotypic with respect to activity timing and using multiple regression to tease apart the effects of social group size and foraging habits (measured as the percentage of fruit in the diet) within these groups. The only group for which there were enough species and data to do a meaningful comparative analysis was the diurnal haplorhines (see Barton, in press, for data and sources in this and subsequent analyses). The results showed that both social group size and degree of frugivory are important, each being independently correlated with relative neocortex size (Barton 1996b).

Frugivory and Neocortex Size: Spatial Memory or Color Vision?
My analysis suggests that both social life and foraging habits are linked to brain evolution in diurnal haplorhine primates. On the face of it, the correlation between neocortex size and frugivory could be taken as support for the argument that the spatial memory demands of frugivory have selected for enlarged neocortices. I do not believe this to be the correct explanation. Degree of frugivory remains significantly correlated with relative neocortex size even when home range size has been partialled out (partial $r = .49, p < .05$), whereas the converse is not true; home range size is not correlated with neocortex size when frugivory has been partialled out (partial $r = .33, p > .05$). It therefore seems to be frugivory per se, rather than the size of the ranging area that the animal must map, that is associated with neocortex size and brain size.

Robert A. Barton

There is a plausible biological explanation for the association between frugivory and neocortex size: visual specialization. As I have emphasized, much of the neocortex of monkeys and apes is dedicated to visual processing. Furthermore, the correlates of relative neocortex size (diurnality, frugivory, and social group size) are the same as those of the relative size of the primary visual cortex alone and of the LGN (Barton, Purvis, and Harvey 1995; Barton 1996b; Barton, in press), suggesting that variation in neocortex size is largely attributable to variation in the size of the visual processing areas. Differences in neocortex size therefore seem to reflect selection on visual mechanisms. Significantly, researchers in visual psychophysics, coming from quite a different angle, have suggested that color vision in primates is an adaptation for locating and selecting palatable fruit (Mollon 1989; Jacobs 1993; Osorio and Vorobyev 1996). Dichromatic color vision, present in diurnal lemurs and many New World monkeys, enables animals to distinguish between fruits of different color, while trichromatic vision (present in some New World monkeys and all Old World monkeys and apes) aids in the detection of fruits against a background of green leaves (Osorio and Vorobyev 1996). My analyses of neocortex and visual cortex size in relation to diurnal frugivory was restricted to haplorhines because, while the diurnal lemurs vary in how frugivorous they are, insufficient data are available for statistical analysis. Intriguingly, however, the primary visual cortex of the largely frugivorous ruffed lemur *(Varecia variegatus)* is about 20% larger than that of the more folivorous indri *(Indri indri),* although the indri weighs nearly twice as much, and both are diurnal. While the indri has a slightly larger brain overall, in accordance with its larger body size, the ruffed lemur is more encephalized, as a consequence, apparently, of its relatively enlarged visual structures.

I am arguing that variation in neocortex size is at least partly a product of selection on specific visual mechanisms, such as color vision. Can this hypothesis be tested? Up to a point: we can look at the broad subdivision between magnocellular and parvocellular pathways within the geniculo-cortical visual system (described above) by analyzing their corresponding layers early in the processing hierarchy, in the lateral geniculate nucleus. Selection for the ability to process fine details and color should correspond with enlargements of the parvocellular, but not the magnocellular, subsystem. Fortunately, comparative data on the number of cells and volume of these separate layers exist (Shulz 1967). Looking just at

Table 8.2 Analysis of associations between parvocellular and magnocellular LGN size (numbers of neurons) and behavioral ecology among diurnal primates, controlling for subcortical brain size

	Magnocellular layers		Parvocellular layers	
	Standardized coefficient	Partial F	Standardized coefficient	Partial F
Group size	—	0.01	0.48	20.58*
% fruit	—	0.01	0.53	13.97*

Note: These results, based on an independent contrasts analysis of the data of Shulz (1967) by Barton (1998) show that the number of neurons in parvocellular, but not magnocellular, layers of the LGN is correlated with group size and degree of frugivory. The same is true when the volume of the layers is considered (Barton, 1998). Partial F-ratios and standardized coefficients (given only where partial F-ratios reached significance) are from stepwise multiple regressions.

**$P < .01$; *$P < .05$.

diurnal species, the degree of frugivory is, as predicted, correlated with the relative size of parvocellular, but not magnocellular, layers (table 8.2). The extent to which color vision, as opposed to other aspects of parvocellular function, has been selected for in frugivores remains unclear. Further tests, looking at color vision abilities directly and at the size of V4, the primate cortical area specialized for color processing, would be helpful.

A Role for Vision in Social Cognition?
Table 8.2 shows that the size of the parvocellular LGN layers is independently correlated with social group size as well as with degree of frugivory. Recall that these two variables—group size and frugivory—are also correlates of neocortex size. And parvocellular layers, also like the neocortex (Barton 1996), are larger in diurnal than in nocturnal lineages. Hence, the neocortex and parvocellular LGN appear to have evolved together. This finding again emphasises the connection between neocortical evolution and visual specialization. But why should visual specialization—and hence neocortical evolution—be linked with social group size? I suggest that the correlation reflects the visual basis of social information processing. Recently, an authority on primate color vision suggested that its utility extends beyond fruit selection, drawing attention particularly to the "elaborate variation in facial colour patterns of many cercopithecine monkeys . . . and how such colour might be used to transmit or suppress the transmission of visual informa-

Robert A. Barton

tion" (Jacobs 1995). The parvocellular subsystem, however, does a lot more than just analyze color, and it is implausible to suggest that enhanced color processing alone underlies advanced social skills. The parvocellular system mediates a range of visual processes, particularly those that involve the perception of fine details (Livingstone and Hubel 1988; Zeki and Shipp 1988). This kind of processing is critically involved in facial recognition, perception of gaze direction, and facial expression. Complex visual cues such as these must be processed and integrated to achieve what Brothers calls the "accurate perception of the dispositions and intentions of other individuals" (1990, 28). Allman states that "as complex systems of social organisation evolved in haplorhine primates, social communication was increasingly mediated by the visual channel" (1987, 639).

Of course, magnocellular-mediated analysis of motion, as well as auditory analysis of vocalizations, must play some role in social information processing. Perhaps, however, the critical evolutionary developments for upgrading such processing were increases in the ability to analyze the fine details of socio-visual stimuli, such as facial expression, and to integrate these with memory and with emotional responses mediated by the limbic system, where cells responsive to social stimuli have also been found (Brothers 1990). The idea that processing such stimuli is unusually costly in computational and neural terms is discussed below.

Recent evidence suggests that neocortical adaptations for social information processing in primates extend beyond modifications of primary visual mechanisms. Joffe and Dunbar (1997) find that group size and neocortex size remain correlated when the size of the primary area of visual cortex (V1) is partialled out. This finding does not undermine the idea that visual specialization underlies neocortical evolution in primates. It suggests that, in addition to evolutionary changes at lower processing levels (the lateral geniculate nucleus and primary visual cortex), there have also been modifications higher up the processing hierarchy, in integrative processing areas (including higher visual areas). For example, entirely new cortical visual areas have been added at various stages of primate evolution (Allman and McGuinness 1988). It is very likely that the evolution of new higher integrative areas goes hand in hand with enhancements of the supporting lower-level architecture. In fact, it may not be valid to think of any of these integrative areas as being nonvisual, since even those not traditionally considered to

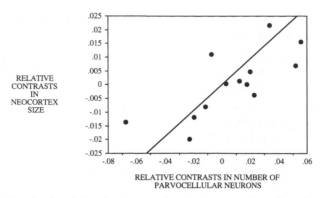

Figure 8.7 Correlated evolution in primates of neocortex size and number of parvocellular neurons, both relative to the size of the rest of the brain [brain volume − (neocortex volume + LGN volume)]. The method of independent contrasts was used. There is a significant correlation (r^2 = .46, p = .004).

be part of the visual system (e.g., the prefrontal cortex) nevertheless receive important visual inputs (see fig. 8.4).

Vision, Neocortex and Brain Size
I noted above that the pattern of ecological correlations for the size of the parvocellular LGN is identical to that for the size of the whole neocortex, suggesting that parvocellular specialization may underly variation in neocortex size. Indeed, figure 8.7 shows that about half of the variance in relative neocortex size can be attributed to parvocellular specialization (the relative number of neurons in parvocellular LGN layers): the size of the neocortex and the size of the parvocellular LGN layers have evolved together (this is true whether one uses the number of parvocellular neurons or parvocellular volume; see Barton, in press). This is true for both the primary visual area and the rest of the neocortex (which includes secondary and tertiary visual areas) when analyzed separately (R. A. Barton, unpub.). In summary, the size of the neocortex as a whole, of the primary visual cortex, and of the parvocellular LGN all correlate with the same ecological factors, and with each other, controlling for the size of the rest of the brain. Hence, there appears to be a general relationship between visual specialization and neocortical evolution in primates.

What about encephalization (overall brain size relative to body size)? The results presented above raise the possibility that frugivores' large brains have to a great extent been achieved by relative expansion of the visual neocortex. If this is true, relative neocortex size should be positively correlated with encephalization. It is (fig.

Robert A. Barton

8.8): in general, species with large brains for their body size have neocortices that are expanded relative to the rest of the brain. Crucially, the same kind of relationship with encephalization holds for the size of the parvocellular, but not the magnocellular, layers of the LGN (Barton, 1998). Thus, encephalization is associated with visual specialization.

Can ecological correlates of encephalization in other taxa, such as bats (Eisenberg and Wilson 1978) and rodents (see Harvey and Krebs 1990) be attributed to sensory specializations? Barton, Purvis, and Harvey (1995) found that, among bats, frugivores have larger visual and olfactory structures—relative to body size and to the size of the rest of the brain—than do predatory species. We suggested that these sensory specializations for frugivory underlie the correlation between diet and brain size found by Eisenberg and Wilson (1978). Similarly, among small mammals, burrowing species tend to have reduced visual systems (Barton, Purvis, and Harvey 1995; Cooper, Herbin, and Nevo 1993) and small brains (Mace, Harvey, and Clutton-Brock 1980). Further research is needed to establish the extent to which sensory specializations related to variables like diet and habitat underlie brain size variations among mammals.

In summary, selection on visual mechanisms appears to be at

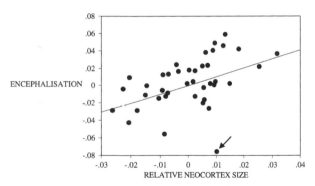

Figure 8.8 Correlated evolution of encephalization and relative neocortex size. Once again, the independent contrasts method was used, this time to examine the relationship between encephalization (brain size relative to body size) and neocortex size relative to the size of the rest of the brain. There is a significant relationship ($r^2 = .22$, $p = 0.002$), indicating that large brains evolved by disproportionate expansions of the neocortex. The arrow indicates an anomalously large outlier, the contrast between the aye-aye *(Daubentonia madagascariensis)* and the indriids. Exclusion of this outlier substantially improves the statistical relationship ($r^2 = .46$, $p < .001$). *Daubentonia*'s high encephalization score may be a result of recent evolutionary reduction in body size (Stephan, Baron, and Frahm 1988) or selection on subcortical (e.g., olfactory) mechanisms for locating concealed prey (R.=tA. Barton, unpub.).

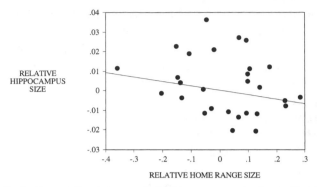

RELATIVE HIPPOCAMPUS SIZE

RELATIVE HOME RANGE SIZE

Figure 8.9 Lack of correlation between relative hippocampus size (controlling for the size of the rest of the brain) and relative home range size (controlling for body size) ($r^2 = .05$, $p = .25$). This graph represents just one of several analyses tried, based on different assumptions about the appropriate control variable for hippocampus size (e.g., the size of the rest of the brain, the size of the medulla, body weight), and using different taxa and ecological groups. None of these showed the predicted positive correlation, suggesting that range size has not been a factor in the evolution of spatial memory in primates.

least partly responsible for differences in neocortex size and overall brain size. The case for selection on spatial memory, in contrast, remains more speculative, and indeed, is weakened by the support for an alternative explanation of the correlation that originally prompted it. However, we have not yet looked at the brain structure most closely associated with spatial memory, the hippocampus.

Hippocampus Size
Experimental work on rodents has discovered hippocampal "place cells" that are responsive to spatial location (O'Keefe and Nadel 1978). Lesions of the hippocampus impair spatial memory (e.g., Rasmussen, Barnes, and McNaughton 1989), as does pharmacological blockade of hippocampal NMDA receptors (Morris et al. 1986). Comparative work has now shown that hippocampus size in birds is correlated with spatial memory demands; species that store food and have to remember the locations of hundreds of caches have larger hippocampi than those that do not (Sherry et al. 1989; Krebs 1990). In rodents, sex differences in ranging patterns are correlated with differences in hippocampus size (e.g., Jacobs et al. 1990). Is there any evidence for hippocampal specialization in primates with large home ranges? Despite an earlier report of a weak association between hippocampus size and home range size, dependent on excluding terrestrial species from analysis (Barton and Purvis 1994), new analyses with an improved data set indicate no significant relationship (e.g., fig. 8.9). Perhaps home range size is a

Robert A. Barton

poor index of the complexity of ranging patterns and a better index might correlate with hippocampus size. The hippocampus is, however, a small structure relative to overall brain size, and even where there is clear evidence of hippocampal specialization (in birds) differences in hippocampus size are not associated with differences in overall brain size (Sherry et al. 1989). While brain size differs markedly between strepsirhine and haplorhine primates, neither ranging patterns (Kappeler, chap. 16, this volume) nor hippocampus size do. Hence, ecological correlates of brain size are unlikely to reflect selection on hippocampus size.

Spatial, Visual, and Social Information Processing in Primate Brain Evolution

Inferences from Comparative Neuroanatomy
The foraging cognition hypothesis was put forward largely to accommodate the empirical correlations found between brain size and diet (Milton 1988; Clutton-Brock and Harvey 1980; Harvey and Krebs 1990). More detailed neuroanatomical analysis, reviewed above, suggests an alternative explanation for this correlation in primates: the large brains of frugivores reflect the evolutionary elaboration of parvocellular visual mechanisms. The comparative data show an evolutionary dissociation between parvocellular and magnocellular systems, just as experimental data indicate a functional dissociation (the perception of fine detail with color versus the perception of movement and dynamic form). Evolutionary dissociations like this run counter to the idea that developmental constraints prevent brain modules from evolving independently of one another, as argued by Finlay and Darlington (1995).

A paradox that arises from linking parvocellular evolution with diurnal frugivory and sociality is that it seems that the parvocellular system, together with other distinctive visual adaptations, evolved in the first primates, which, consensus has it, were almost certainly nocturnal, predominantly solitary insectivores. Cartmill (1974) argued that the basic primate adaptations were the result of an adaptive shift to visually directed predation in a fine-branch niche. This could explain the suite of visual adaptations shown by primates discussed above. Other visually guided predators, such as birds of prey and carnivores, also show frontally directed eyes and a high degree of binocularity, for example. The evolutionary origin of the primate visual system therefore may not be explained by the same factors that seem to explain its subsequent evolutionary radiation. More recently, however, Sussman (1991b) has questioned Cartmill's interpretation, arguing that primates and other mammals coevolved with flowering plants. Sussman suggests that early primates fed on and manipulated small fruits and seeds, which "might require acute powers of discrimination and precise coordination" (Sussman 1991b, 219). These powers of discrimination and coordination would then have been further elaborated as the primate order radiated into new niches, diurnal ones in particular. Sussman's model may provide a more convincing context for the evolution of the primate visual system because it emphasizes the need for acuity (whereas carnivores manage with relatively poor acuity, but excellent movement detection) as well as for a cortico-cerebellar system supporting precise, dexterous, visually guided manipulation.

Inferences from Foraging and Social Behavior
Perhaps one reason for the popularity of the spatial or foraging cognition hypothesis is that primate behavioral ecologists have spent so much time and effort collecting vast amounts of information about the ranging behavior and foraging strategies of their subjects—largely because these are among the easiest data to collect. Not to put too fine a point on it, there may be a tendency to assume that primates are especially good at the behaviors that primatologists happen to study (though this applies equally to foraging and to social behavior). Much significance has been attached to the fact that primates seem to navigate around their home ranges effortlessly and optimally, traveling in straight lines to fruit trees or other resources. Because of this, it is often stated that primates form "mental" or "cognitive" maps.

There are two problems with such statements. First, the cognitive map concept is sometimes used in a very vague way, based on observational studies rather than on experiments, so that it is not clear exactly what cognitive claims are being made or how they could be tested. Second, the assumption that cognitive mapping ability is an advanced primate feature not shared by other animals is wholly unwarranted. Other species travel in straight lines to resources, and there is a vast experimental literature on cognitive maps and their neural basis in rodents (e.g., Leonard and McNaughton 1990). Cognitive maps have also been proposed for birds (e.g., Sherry, Krebs, and Cowie 1981), fishes (Noda, Gushima, and Kakuda 1994), and even insects (Gould 1986). Recently, the attribution of

Robert A. Barton

cognitive maps to rodents has been called into question (Benhamou 1996), despite the existence of a far more detailed body of experimental work than exists for primates. Benhamou (1996) suggests that experimental results that look like evidence for cognitive mapping may actually result from simpler mechanisms. This literature has largely bypassed primatology, but it must be taken on board before we can even begin to address cognitive mapping in primates. There is some experimental work on primates (e.g., see Garber, chap. 10, and Janson, chap. 7, this volume), but at present, it contains nothing to support the view that primate spatial cognition is special in any way or better than spatial cognition in rodents or any other mammalian order. Furthermore, it is rather hard to think of any reason why it should be: while much is made of the large home ranges and spatially dispersed and ephemeral food supplies of some primate species, this must apply equally to many nonprimates—food-storing species, for example (Krebs 1990). The argument that primate ranging habits, especially those of frugivores, place exceptional demands on cognitive abilities remains essentially a hand-waving exercise, and there is little evidence to support it.

To be fair, hand-waving has been far from absent from the literature on social cognition and complexity, and the same criticism can be made that proper comparative tests have not been carried out. This is partly because testing for species differences in cognitive abilities is fraught with difficulty (e.g., Macphail 1982). We tend to be prejudiced favorably toward primates' social intelligence because primatologists have made very detailed observations that reveal, at least on the surface, social complexity. Many primates, anthropoid primates in particular, do have objectively different social lives than most nonprimate mammals in that they live in social groups that are often large and have long-term stability in membership. The suspicion that this causes them to have more complex social interactions and relationships than nonprimates really remains just that—a suspicion, rather than a scientific fact. Interestingly, evidence is emerging that where large and stable social groups are found in other mammalian orders, this also correlates with increased neocortex and brain size (Barton and Dunbar 1997). At present, the best way of augmenting comparative brain studies so as to evaluate the competing claims may be to return to first principles and ask what, in theory, would be required of a cognitive system dealing with a particular type of problem, and what its likely computational and neural costs would be.

The Neural Costs of Cognition
We know that primates do not forage randomly over their home range, or use only direct sensory cues (see Boinski, chap. 15; Garber, chap. 10; Janson, chap. 7, this volume). They must therefore hold in their heads some kind of spatial representations of their resources. However, we do not know what kinds of representations these are, how they are used, or what their costs are in terms of computation and neural network size. A highly sophisticated system for optimizing foraging routes not only would need a detailed three-dimensional spatiotemporal map of resources, but also would need to be able to hold this in working memory while calculating the consequences of the possible alternatives at each point. These two aspects of the problem, the mapping and the on-line processing of map information as the animal moves through its home range, have quite different computational and neural implications. The mapping part is basically a storage problem, probably solved by the hippocampus (O'Keefe and Nadel 1978), which, as I noted, does not need to be very large to do this. The on-line processing of the map information as the animal moves through its home range is potentially demanding because of the need to simultaneously process large amounts of information in parallel—inevitably involving the neocortex, especially the prefrontal cortex, which has an important role in working memory (Goldman-Rakic 1996). The key to discovering whether primates, or particular primate species, are specialized for handling such information lies in understanding what information is actually processed by the animals: do they take into account an array of simultaneous, spatially removed options, or do they operate according to simple rules, such as "visit the nearest tree that had fruit last time I looked"? Do they do something that rats don't? Dyer's reminder (chap. 6, this volume) that apparently difficult problems can be solved with surprisingly simple mechanisms should be borne in mind. We cannot at present rule out the possibility that foraging frugivorous primates are doing something particularly sophisticated, but equally, there seems no compelling reason to think that they are.

In contrast, I would argue, there is more reason to believe that social information processing in group-living anthropoid primates is computationally and neurally highly demanding, and that that is why larger neocortices are required to deal with life in larger groups (see also Barton and Dunbar 1997). Social information is multi-faceted, consisting of a diverse array of sensory input: auditory,

Robert A. Barton

tactile, olfactory, and, in diurnal species, especially visual (Brothers 1990). During a polyadic (multi-individual) social interaction, a group-living anthropoid must, as a minimum requirement for appropriate action, monitor the identity and social attributes (e.g., dominance rank) of other individuals, who may be numerous, and their current but dynamically and rapidly varying "dispositions and intentions" (Brothers 1990), both toward itself and toward the other animals involved. Cues to dispositions and intentions include a variety of facial expressions, vocalizations, postures, and movements. Individuals also seem to take into account the actions of particular conspecifics in the past (Cheney and Seyfarth 1990). This list is not exhaustive, but it serves to illustrate the idea of a stream of information (Barton and Dunbar 1997) that ebbs and flows according to the immediate context. There is no hard evidence that, rather than processing such a stream of information, the animals' brains do not actually use some computationally simple rules of thumb to get through their social lives, but it is difficult to conceive what these could be, given what we know about what information the animals seem to take into account (Cheney and Seyfarth 1990) and how the brain processes that information (Brothers 1990; van Essen, Anderson, and Felleman 1992; Perrett et al. 1992; Young et al. 1994). It is the massively parallel load imposed by social information processing that requires large neural networks and a large neocortex.

How does the brain bring together the disparate sensory, emotional, and memoric elements of social information? The cortical areas, such as inferotemporal cortex, specialized in primates for processing visual social information constitute part of a larger integrative system. Young (1993; Young et al. 1994) carried out a statistical analysis of neural connectivity between cortical and other brain areas that confirmed earlier suggestions of a fundamental division of the visual system into a dorsal and a ventral processing stream. The dorsal stream is almost entirely magnocellular, whereas the ventral stream has both parvocellular and magnocellular components and includes the areas associated with social information processing (van Essen, Anderson, and Felleman 1992; Merigan and Maunsell 1993). Thus, the results of the comparative analysis, linking neural (parvocellular) specialization and social evolution, are in accord with what is known about where and how social information is processed within the brain. The sensory areas culminate in a fronto-limbic complex (Young 1993), including the prefrontal cor-

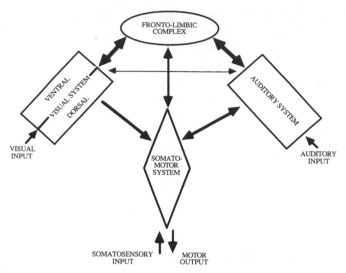

Figure 8.10 Young's general scheme of connectivity between forebrain areas in the macaque, illustrating the confluence of sensory information on a fronto-limbic complex, which includes the limbic system and prefrontal cortex. Note the division of the visual system into a dorsal and a ventral stream: it is the latter that is most associated with the processing of social stimuli, and which has major parvocellular components. The dorsal stream is relatively highly connected with the somato-motor system, reflecting the dominance of magnocellular components and its role in visually guided movements. (Redrawn with permission from Young 1993.)

tex and the whole of the limbic system, that allows immediate sensory information to be held in working memory and integrated with long-term memory and emotional significance. The prefrontal cortex must play an important role in holding social information in working memory during complex sequences of interactions (Goldman-Rakic 1996), so we would predict that its size, especially the size of its parvocellular components, should covary with the relevant visual areas and with group size (we do not yet have the data to test this). The general scheme suggested by the statistical patterns of connectivity is presented in figure 8.10.

Some caveats need to be kept in mind. As with foraging cognition, we need to examine the assumption that anthropoid social cognition is special and neurally demanding. We also need to develop models of social cognition that specify the relationship between social group size—a proxy for social complexity—and information processing load. This relationship is currently unclear. However, although group size varies markedly within species, it seems reasonable to assume that relatively large-scale evolutionary

Robert A. Barton

changes in typical group size are associated with changes in the importance of polyadic interactions, wherein individuals must process in parallel the dispositions, intentions, and qualities of numerous conspecifics simultaneously. The idea needs further development before it can be accepted. The contrast I have made between the demands of social and foraging cognition is speculative. I hope that in the future, research on functional neuroanatomy, comparative studies of primate and nonprimate foraging and social skills, and simulation modeling of cognitive processes required to solve particular tasks will establish whether it has any foundation.

Perception, Social Cognition, and Comparative Psychology
I have suggested here and elsewhere (Barton 1996b; Barton and Dunbar 1997) that social cognition is most usefully conceived as a large array of perceptual-cognitive operations occurring in parallel, and that its evolution has been based on the elaboration of functionally specific pathways such as the parvocellular visual pathway. Vision may be a crucial component of primate social cognition. This might seem surprising to those who think of sensory mechanisms as merely the basic input devices, while higher "cognitive" mechanisms do the fancy stuff. There are certainly hierarchies of processing in the brain (e.g., van Essen, Anderson, and Felleman 1992; Young et al. 1994). But I think that any clear separation between perception and "higher" cognition, between the input system and the system for analyzing the input, is fallacious. We tend to take visual processing for granted because we seem to do it effortlessly, compared with solving differential equations, for example. Primate vision is, however, extraordinarily complex and computationally demanding: in the macaque brain there are at least 305 visual pathways connecting 32 cortical visual areas, occupying more than half of the neocortex (van Essen, Anderson, and Felleman 1992). Neocortical evolution in primates has involved both the appearance of entirely new visual areas and the expansion of existing ones (Allman and McGuinness 1988; Kaas 1995). Each visual area is computationally specialized, topographically representing particular aspects of the visual scene, such as movement, form, and color. As described above, visual and other sensory information is held "on-line" and processed emotionally within a fronto-limbic complex. There is no point in this processing hierarchy that can be identified as the transition between perception and cognition. Conceiving of the neocortex as the "thinking" part of the brain, as

is popular, glosses over the fact that sensory maps are a general and fundamental feature of its organization, and that there are no areas that are "cognitive" as distinct from sensory-motor. Specific sensory areas make up about 70% of the macaque neocortex, and motor areas a further 10% (calculated from table 2 in Drury et al. 1996). Its parallel processing of topographically organized sensory information allows the neocortex to "resolve(s) the structure within the noisy sensory array" (Allman 1990).

I suggest that social information is just as much a part of the noisy sensory array as, for example, the shape and color of fruits. Neocortical specializations allow group-living primates to efficiently process the perpetual stream of social information. We do not yet have a clear understanding of how memory and emotional responses, via the limbic system, participate in resolving an animal's total social picture, as they must do, thereby generating appropriate action. In this context, new comparative studies are investigating the coevolutionary relationships between neocortex and amygdala (Joffe and Dunbar 1997; Emory et al., unpub.).

The emphasis on the perceptual subprocesses involved moves us closer to an account of social cognition rooted in an understanding of how the brain—in particular, the neocortex—works. In turn, this development moves us away from the "black box" approach to social cognition, which postulates enigmatic cognitive modules, such as "theory of mind." Such black box models have had little or no connection with knowledge about information processing in the primate brain. That is not to say that the theory of mind concept is necessarily wrong, but it needs to be given a realistic, albeit initially general, basis in neural processing terms—something that Baron-Cohen (1994) has now attempted for humans. Then we can relate the evolution of a theory of mind to the rest of the story of brain evolution. A similar point can be made about the vagueness of the attribution to primates of "cognitive maps." To be meaningful, such an attribution would have to be based on an explicit model that specifies what spatial information is processed and how (as in the model of rodent hippocampal function by McNaughton and Morris [1987]), and which generates testable predictions about foraging behavior.

The idea of a perception-cognition continuum implies, then, that cognitive evolution is not to be thought of as simply the addition of a few "black boxes" high up the processing hierarchy. The ability to process facial information, for example, requires a high degree

Robert A. Barton

of visual acuity and the ability to differentiate a wide range of expressions. So while we do not expect to find neurons that fire selectively to faces in the lateral geniculate nucleus, it is no surprise that lateral geniculate pathways projecting to areas in the inferotemporal cortex that do have such cells (Perrett et al. 1992) are modified to support that type of processing. Brain modules cut across the arbitrary distinction between perception and cognition, more in line with Jackendoff's (1992) conception of modularity than with Fodor's (1983). This has implications for the field of comparative psychology, which has for decades striven to establish cognitive differences between species as diverse as insects and apes, with remarkably little success (e.g., Macphail 1982; Bitterman 1996). To the extent that sensory and cognitive adaptations are inextricably linked, the whole debate within comparative psychology about the existence of general differences in intelligence or learning ability, as distinct from differences in perceptual apparatus, is a red herring. The vaunted "null hypothesis" that, when all "contextual variables," including differences in sensory adaptations, are controlled for, there are no differences in intelligence between a goldfish and a chimpanzee (see Macphail 1982) is not so much false as irrelevant.

Group Movement

Two sexually receptive female baboons and their male consorts and harassing followers get separated from the rest of the troop, and spend the night at a different sleeping site. In the morning, they scan the savanna from a high point as they head off to forage, making a few hopeful "wahoo" long-distance contact calls (see Byrne, chap. 17, this volume), but do not see the rest of the troop. As they approach the area where they lost contact with the others the day before, they start to make "wahoo" calls again. The frequency of calling peaks within the ¼ km quadrat where the troop was last recorded as being together (figure 8.11). Clearly, these baboons have coordinated information about spatial location and the presence of troop members and remembered it overnight. The nature and complexity of the neural algorithm involved can only be a matter for speculation, but perhaps the anecdote highlights an area for future study: socio-spatial cognition.

Group movement is an area of study that potentially reconciles foraging and social cognition hypotheses. Like competition between group members for food, group movement may require individuals to integrate information about resources and social dynam-

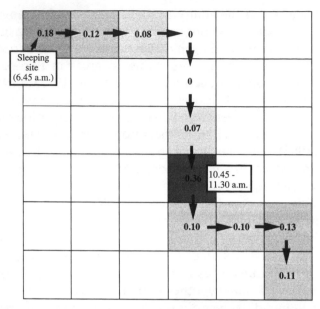

Figure 8.11 An example of primate socio-spatial memory? The diagram shows the frequency of loud "wahoo" contact vocalizations made by members of a subgroup of six olive baboons *(Papio anubis)* during the morning following their separation from the main troop (6:45 A.M.–1:00 P.M.). Figures are rates of calling per minute, averaged over 15-minute time windows, within each successively occupied quadrat. The calling rates are represented visually by the density of shading. Direction of travel is shown by the arrows. As the subgroup moved off from the sleeping rocks, some members visually scanned the area below and called, without response. As they approached the area where they had lost contact with the main troop the day before, calling was resumed, reaching a marked peak within the quadrat where they had last been observed with the troop, then declining again as they moved on. There was no direct contact with other baboons at this point, and the main troop was 2 km away. The observations were made by the author at Chololo in Laikipia, Kenya, on 31 July 1994.

ics. This would be the case, for example, if individuals had to base optimal foraging decisions partly on the location and direction of travel of other individuals, as squirrel monkeys seem to do, and if there were some kind of negotiation between group members about ranging decisions (see Boinski, chap. 15, this volume). Monitoring other individuals' foraging decisions and intentions with respect to particular resources, through visual and auditory cues, could in itself constitute a large information-processing load. A fission-fusion social organization, as in chimpanzees and spider monkeys, might present particular problems, as implied by the baboon anecdote above. According to these ideas, it is still social structure and group size that drives the evolution of brain and cognition, because the

Robert A. Barton

extra information processing load derives from the addition of a social dimension to existing foraging problems. But the interaction of social and ecological domains may turn out to be an important aspect of primate cognition, one that demands a new generation of observational, experimental, and comparative studies.

Conclusions

The comparative method is a powerful tool for understanding the selection processes that have shaped organisms, including their brains and their behavior. Analysis of correlated change in the volume of specific brain structures and in ecological factors has shown that large brains in primates evolved, at least in part, under selection on visual mechanisms. Selection on visual mechanisms, such as high acuity and color vision, rather than on spatial cognition, seems to explain why frugivores have large brains. Further tests, especially of the color vision hypothesis, need to be made. The elaboration of social skills also seems to have involved enhancements of visual mechanisms, for example, those involved in processing facial information about the identity, internal state, and intentions of other individuals. Much more work needs to be done to establish the exact nature of these mechanisms, how they are integrated with memory and emotion, and the extent to which neuro-cognitive systems have been modified at higher versus lower processing levels within the brain. There is, however, no justification for an absolute demarcation between the processes that are deemed "perceptual" and those that are deemed "cognitive." Recognition of the continuum between perception and cognition will help us to unravel the evolution of the brain and the nature of primate intelligence.

CHAPTER NINE

Animal Movement as a Group-Level Adaptation

David Sloan Wilson

Animals that live in social groups frequently move as a single unit. Group movement can appear haphazard, as when a primate troop forages in scattered subgroups through a forest canopy, or spectacularly coordinated, as when a fish school performs a flash display to evade a predator. Early biologists often described moving groups as being like single organisms in the coordination of their parts. However, this view was largely rejected as a form of naive group selectionism during the 1960s. The following quote from Pitcher and Parrish (1993, 363) on fish schools summarizes both the history and current interpretation of animal movement in general: "Travelling fish schools display impressive coordination, and were once viewed as egalitarian, leaderless societies in which cooperation preserved the species (Breder 1954, Shaw 1962, Radakov 1973). In contrast to such classical group-selectionist views, behavioural ecology reveals social behaviour to be no more glamorous than animals cooperating only when it pays."

This view of social behavior, which portrays group selection as a ghost from the past, needs to be reconsidered. To borrow Mark Twain's famous quip, rumors of group selection's death during the

1960s were greatly exaggerated. Not only were some of the original criticisms less damning than they first appeared (e.g., Maynard Smith 1964; reevaluated in Wilson 1987), but subsequent work has led to a robust theory of natural selection as a process that operates on a hierarchy of units (reviewed by Wilson and Sober 1994; Sober and Wilson 1998; see also the special issue of volume 150 of *The American Naturalist* devoted to multilevel selection). In this chapter I will attempt to show how multilevel selection theory can be used to study animal movement. I will begin by summarizing the theory and relating it to the concept of self-interest that has developed within behavioral ecology over the past 30 years.

Theoretical Background

Multilevel Selection Theory

Biological systems are often a nested hierarchy of units; genes reside within individuals, which reside within social groups, which reside within a population of groups (a metapopulation). The process of natural selection can potentially occur at any level of the biological hierarchy. For example, a single group is a population of individuals within which natural selection can occur. Within-group selection promotes traits that maximize the relative fitness of individuals in the group. To discover what these traits are, we must compare the fitness of individuals possessing some trait with the fitness of other individuals in the same group possessing alternative traits. If the traits that evolve do not have the highest relative fitness within the group, then something else is required for the model to become predictive. That "something" may be natural selection operating at a different level of the biological hierarchy, either among genes within individuals or among groups within the metapopulation. In each case we can employ the same reasoning process, comparing the relative fitness of units within the next higher unit. What actually evolves reflects the selective forces acting at different levels of the biological hierarchy.

The concept of *relative* fitness plays a central role in multilevel selection theory. Natural selection within a single group favors individuals that have more offspring than anyone else and is insensitive to the absolute number of offspring. Within-group selection sees only the relative size of the slice and not the size of the pie. A behavior that causes the individual to have one additional offspring and everyone else in the group to have two additional offspring will be selected against because it decreases relative fitness. Conversely, a behavior that causes the individual to have one fewer offspring and everyone else in the group to have two fewer offspring will be favored because it increases relative fitness. For this reason, adaptations that evolve at one level of the biological hierarchy can be maladaptive at higher levels. Genes that succeed at the expense of other genes within the same individual are often classified as diseases. Individuals who succeed at the expense of other individuals are often classified as antisocial, in everyday terms. Multilevel selection theorists have emphasized the importance of relative fitness and its maladaptive consequences at higher levels from the very beginning (e.g., Darwin 1871; Wright 1945, 1948; Williams 1966).

In one view of multilevel selection, Wright (1945) imagined groups as geographically isolated populations connected by a trickle of dispersers. This is often portrayed as the *only* view of multilevel selection, but a true reading of the literature reveals a much greater diversity of population structures. Darwin (1871) imagined groups as tribes that compete by direct warfare. Haldane (1932) imagined mobile groups that compete by fissioning at different rates. Social insect colonies are paradigmatic "superorganisms" of multilevel selection theory whose population structure bears little resemblance to Wright's view. Williams and Williams (1957) imagined groups as collections of siblings that interact for only a portion of their life cycle. Hamilton (1975) and Wilson (1975) imagined a group as any sets of individuals that interact with one another with respect to a given trait that influences fitness (termed "trait groups" in Wilson 1975). The essential feature that unites all these conceptions into a single theory is that the composition of the group determines the relative fitness of its members and the fitness of the group relative to that of other groups in the metapopulation (see chapter 2 of Sober and Wilson 1998 for a more extensive discussion). It is important to keep this definition in mind when considering group movement from the standpoint of multilevel selection theory. The early biologists who compared moving groups to single organisms had no problem identifying groups; they were the sets of individuals that moved through space as cohesive units. This conception of groups is perfectly compatible with multilevel selection theory, even though it departs from what Wright had in mind.

Identifying appropriate groups and comparing the relative fitness of units within the next higher unit are often simple tasks that can be accomplished by anyone with biological knowledge of the spe-

David Sloan Wilson

cies and its environment. Thus, multilevel selection theory provides a useful way to think about evolution even for those who have difficulty understanding formal theoretical models. However, it is important to distinguish multilevel selection theory from other conceptual frameworks that use the same terms in different ways.

The Concept of Self-Interest in Behavioral Ecology

Many readers may have found themselves nodding their heads in agreement while reading my synopsis of multilevel selection theory. It is close to Williams's (1966) portrayal and the brief descriptions of group selection that are found in textbooks. The issue, I imagine the reader thinking, is not whether group selection exists, but how strong it is relative to individual selection.

Despite this superficial agreement, behavioral ecologists often think in a way that departs from multilevel selection theory. For example, it is common to reason informally about natural selection by imagining an individual deciding whether or not to express a given trait. If expressing the trait results in higher fitness than not expressing the trait, then the trait is expected to evolve by "individual selection" and to be an example of "self-interest." This way of thinking assumes that individual selection maximizes the *absolute* fitness of individuals, which is not the same as maximizing relative fitness within single groups. If the individual increases its own absolute fitness and *no one else's,* it will indeed increase in relative fitness and be favored by within-group selection. However, if the individual increases its own absolute fitness and the fitness of everyone else in the group *as much or more,* the trait will be neutral or selected against within the group, and something else is required to explain its evolution. This nuance is lost on those who use absolute fitness to define the concepts of individual selection and self-interest. To make matters worse, absolute fitness is only one of several definitions of self-interest used by behavioral ecologists. The boundary that defines self-interest can be stretched in many ways: relative fitness within groups, absolute fitness, inclusive fitness, fitness averaged across groups, fitness calculated at the gene level (Dawkins 1976, 1982). All of these definitions have been employed by behavioral ecologists as if the term had a single well-understood meaning.

There is room for more than one theoretical framework in evolutionary biology. It may be useful to define self-interest and individual selection in a way that departs from multilevel selection theory. However, the definition of group selection should be correspondingly changed or dropped altogether. With a few exceptions (e.g., Nunney 1985) this has not happened in the field of behavioral ecology. The definition of group selection has remained constant, while definitions of self-interest have become variable. This is because group selection is not entertained as a serious hypothesis by most behavioral ecologists. Group selection is usually mentioned only briefly as an example of how not to think, without saying what group selection is or how it relates to the version of self-interest that is being employed. All of these problems are illustrated by Pitcher and Parrish's statement about fish schools quoted at the beginning of this chapter. The views of early biologists are dismissed because they invoke "classical group selection," and behavioral ecology is praised for revealing that social behavior evolves only when it "pays," as if both of these terms have well-specified meanings that articulate with each other.

Multilevel Selection Theory versus Selfish Gene Theory
A comparison of multilevel selection theory and selfish gene theory will illustrate the problems that arise when two theoretical frameworks are not properly related to each other. According to multilevel selection theory, natural selection occurs at level x of the biological hierarchy precisely when units at level x differ in fitness within the next higher unit. Gene-level selection occurs when genes differ in fitness within individuals, individual-level selection occurs when individuals differ in fitness within groups, and group-level selection occurs when groups differ in fitness within the metapopulation. When natural selection occurs exclusively at the individual level, individuals evolve into adaptive units that are designed to maximize their relative fitness in the group. We call these adaptive units organisms.

Selfish gene theory defines units of selection in terms of persistence through evolutionary time. Groups and sexually reproducing individuals fail to satisfy this criterion, leading to the conclusion that genes are "the fundamental unit of selection (replicators)." Does this mean that individual selection is a myth and individuals should never be regarded as organisms? No. An individual must be a unit of selection to become an organism in multilevel selection theory, but not in selfish gene theory. If selfish gene theory changes one major definition, it must change or create others to remain coherent. According to selfish gene theory, individuals evolve into organisms when they are vehicles of selection, which means that all

David Sloan Wilson

of the genes in the individual are "in the same boat" with respect to fitness. This is the same as saying that natural selection operates at the individual level in multilevel selection theory. The two conceptual frameworks use different languages to reach the same conclusion.

Why do we need two (or more) conceptual frameworks that reach the same conclusion? Perhaps we don't, and all but one framework should be abandoned. However, multiple frameworks can be useful if they examine a complex subject from different angles, each achieving insights that are obvious only in retrospect to others. For this advantage of pluralism to be realized, it is essential to distinguish among the frameworks and make the appropriate translations when the same terms are used in different ways. For example, what should we make of the claim that traits evolve only when they "pay" at the gene level? It is true as far as it goes, but it would be false to regard it as an argument against individuals as adaptive units, since it often "pays" for genes to "pull together" with the other genes in the same boat. Calculating fitness at the level of genes is not an argument against individuals as adaptive units.

These issues may seem clear enough when we talk about genes in individuals, but they become more confused when we talk about individuals in groups. Those bad early biologists who described groups as organismal units were comparing groups to individuals, not genes. If individuals can evolve into adaptive units without being replicators, perhaps groups can as well. According to multilevel selection theory, between-group selection is required for groups to evolve into adaptive units. For selfish gene theory to address the same question, we need to ask whether groups can be vehicles of selection (Wilson and Sober 1994). If genes can succeed not only by coordinating with other genes in the same individual, but also by causing individuals to coordinate with one another, then groups can become the same kind of "lumbering robots" as individuals, to use Dawkins's (1976) picturesque language.

This is the correct translation between conceptual frameworks, but it is obvious only in retrospect. The gene as the "fundamental unit of selection" was originally regarded as a fatal argument against group selection, without any reference to vehicles (Williams 1966; Dawkins 1976). Only gradually did it emerge that calculating fitness at the gene level is no more an argument against group selection than it is against individual selection. Both Williams (1992) and Dawkins (1982) have acknowledged this point, but the confusion remains widespread. Consider the following recent account of selfish gene theory published in the *New Yorker* (Parker 1996, 42):

> [*The Selfish Gene*] was designed to banish an infuriatingly widespread popular misconception about evolution. The misconception was that Darwinian selection worked at the level of the group or the species, that it had something to do with the balance of nature. How else can one understand, for example, the evolution of apparent "altruism" in nature? . . . Once one understands that evolution works at the level of the gene—a process of gene survival, taking place (as Dawkins developed it) in bodies that the gene occupies and then discards—the problem of altruism begins to disappear.

This passage reflects Dawkins's original position that selfish genes by themselves constitute an argument against group selection without any reference to vehicles. Like many of Sigmund Freud's ideas in psychology and Margaret Mead's ideas in anthropology, this particular implication of selfish gene theory seems to thrive at the margins even after it has been authoritatively rejected at the center of its own discipline. I would like to think that the margins are confined to journalists and their lay readers, but even Pitcher and Parrish (1993) share the confusion in their passage on fish schooling behavior quoted at the beginning of this chapter. Their statement that social behavior evolves only when it "pays" is no argument at all against group selection. The question is whether the behavior pays by increasing the individual's relative fitness within groups or by increasing group's ability to function as a coordinated unit relative to other groups. Pitcher and Parrish's failure to make this crucial distinction extends beyond the quoted passage to their entire review of fish schooling behavior.

New scientific developments take time to spread, but the developments outlined above have spread so slowly that an explanation is required. Dugatkin and Reeve (1994, 113) provide part of the answer:

> By erecting the concept of a "vehicle," Dawkins in effect began to build a bridge between his gene-selectionism and the new group-selectionism. When the bridge was half-built, however, he effectively abandoned it, apparently sensing that a completed bridge would lead him to a land inhabited by the dreaded group selection dragons.

The proper translation between theoretical frameworks makes group selection (or group-level vehicles) a legitimate possibility that must be considered for all forms of social behavior. The question of whether groups can evolve into adaptive units, which seemed

David Sloan Wilson

so decisively settled and which made selfish gene theory seem so important in the first place, is back. Let us now cross the bridge and explore the specific subject of animal movement from the standpoint of multilevel selection theory.

Group Movement and Multilevel Selection Theory

The basic elements of multilevel selection theory are easy to understand without complicated mathematical equations. First and foremost, it is important to compare fitnesses the right way. If a trait does not maximize relative fitness within groups, then something else is required to explain its evolution by natural selection. Using relative fitness as the criterion for evaluating traits is no more difficult than using other criteria (such as absolute fitness), but merely involves a different frame of comparison.

It is also important to freely speculate about the possibility of groups as adaptive units. Group selection favors traits that maximize the fitness of a group relative to other groups in a metapopulation. To know what these traits are, we must think about the design features that are required for groups to function as adaptive units. Thinking this way does not commit us to the belief that groups *are* adaptive units, but merely tells us what to look for. Behavioral ecologists are accustomed to thinking this way at the individual level, but have been taught to avoid adaptationism at the group level, almost as if it were one of the Ten Commandments. As multilevel selectionists, we must think about adaptation at all levels of the biological hierarchy and then examine the balance between levels of selection that determines what actually evolves (Sober and Wilson 1998; Wilson 1997b).

Now let's apply these general principles to the specific subject of moving groups. Of particular interest are the cognitive processes that determine *where* a group moves. Deciding where to move requires obtaining and remembering information about the alternative possibilities, evaluating them with respect to survival and reproduction, and finally choosing the best option. It is possible that a group of individuals interacting in a coordinated fashion can perform this cognitive task better than any single individual. In general, the advantages of cooperation are as relevant to cognitive activities such as memory, learning, and decision making as they are to physical activities such as hunting, defense, and aggression. At the extreme, cognition might become such a distributed process, and the role of any particular individual might become so partial, that the group could literally be said to have a mind of its own that is not located in any single individual.

Group-Level Cognition and Foraging Movements in Social Insects
The concept of a group mind may sound like science fiction, but it has been documented in honeybees *(Apis mellifera)* and other social insects (Seeley 1995, 1997). To function adaptively, a bee colony must make decisions about which flower patches to visit and which to ignore over an area of several square miles; whether to gather nectar, pollen, or water; the allocation of workers to foraging versus hive maintenance; and so on. Seeley and his colleagues have worked out in impressive detail how these decisions are actually made. In one experiment, a colony in which every bee was individually marked was taken deep into a forest where natural resources were scarce. The colony was then provided with artificial nectar sources whose quality could be experimentally manipulated. Workers ceased visiting sources when their quality was lowered below that of other sources—proof that the hive can quickly perceive and adaptively respond to changes in its environment. However, individual workers visited only one source, and therefore had no frame of comparison. Instead, individuals contributed one link to a chain of events that allowed comparisons to be made at the colony level. Bees returning from the inferior source danced less, and were themselves less likely to revisit it. With fewer bees returning from the inferior source, bees from better sources were able to unload their nectar faster, which they used as a cue to dance more. Newly recruited bees were therefore directed to the best patches.

The mechanisms of group-level cognition that direct movement in honeybees go beyond the famous symbolic dance that allows bees to communicate the location of resources to one another. In fact, many aspects of the group mind in honeybees are remarkable in their simplicity. The individuals respond to environmental cues and to one another in a simple fashion, but the interactions have emergent properties that result in complex and adaptive behaviors at the hive level. For example, the colony acts as if it is hungry when its honey supplies are low, sending more workers to collect nectar, yet no individual bee is physically hungry. Instead, the state of the colony is communicated by the amount of time that returning foragers must wait to regurgitate their load of nectar to other workers, who carry it to empty cells. When resources are scarce and many cells are empty, returning foragers can immediately unload their

David Sloan Wilson

nectar, which serves as a cue for increasing foraging effort. Even the physical architecture of the hive, including the location and dimensions of the dance floor, honeycomb, and brood chambers, plays a role in the cognitive architecture of honeybee colonies (see Seeley 1995 for details). A group mind truly exists that cannot be located in any single individual.

Let's examine one of these traits in more detail: the worker's response to having nectar unloaded within a few seconds of entering the hive. Genetic variation presumably exists for this trait, with some individuals responding by foraging more, others by foraging less, while others perhaps ignore the cue altogether. Genetic variation for the trait should exist both within and between colonies. As multilevel selectionists, we would first examine the relative fitness of individuals within single colonies that differ in their response. Then we would examine the relative fitness of colonies in the metapopulation that differ in their response. In this case, it is fairly obvious that responding the right way will increase the performance of the colony as a unit, without increasing the relative fitness of individuals within the colony, allowing us to identify the behavior as a group-level adaptation. The facts that honeybee workers are highly related to one another and reproduce primarily through a single queen are important, but they do not constitute an argument against group selection. Instead, their importance needs to be appreciated from within the framework of multilevel selection theory. High relatedness means lots of genetic variation among groups and relatively little variation within groups. Reproduction through a single queen means that workers cannot increase their relative fitness within the group by becoming freeloaders and avoiding foraging. These and other factors stack the deck in favor of group selection for honeybee colonies.

Group-Level Cognition Does Not Require a High Degree of Relatedness

Group selection was rejected so strongly during the 1960s and '70s that it was even avoided by many social insect biologists (e.g., West-Eberhard 1978). Fortunately, it has again become respectable to view social insect colonies as largely a product of group selection—or, if you prefer, as group-level vehicles of selection. However, there is a tendency to assume that group-level adaptations are *restricted* to social insect colonies and other groups whose members are highly related to one another. By this reasoning, we should avoid thinking about group-level adaptations in human groups, nonhuman primate troops, lion prides, ungulate herds, bird flocks, and fish schools, even if we accept them in the case of social insect colonies.

Nothing could be further from the truth. Natural selection at all levels requires variation among units, but the *amount* of variation depends on the balance between levels of selection. The genes in an individual don't cooperate with one another because they are related, but because they are "in the same boat" with respect to fitness. No fitness differences exist within individuals, which means that random variation between individuals is sufficient for individual-level adaptations to evolve. Whenever individuals in social groups are "in the same boat" with respect to fitness, it follows that random variation among groups can be sufficient for group-level adaptations to evolve. Group selection and group-level adaptations are a possibility for all group-living animals, regardless of genealogical relatedness. This conclusion is obvious from the standpoint of multilevel selection theory, even though it may cause kin selectionists to rub their eyes in disbelief (Sober and Wilson 1998; Wilson 1990, 1997a,b; Wilson and Dugatkin 1997).

Groups that move as a unit are especially likely to be in the same boat with respect to fitness, regardless of genealogical relatedness. If members of a group must stay in close proximity (to avoid predators, for example), then everyone stands to win or lose by the quality of the decision about where to move. I do not mean to underestimate the conflicts of interest that may exist with respect to where to move (see McCabe, chap. 22, this volume, for examples in humans), but the shared interest should not be underestimated either. To the extent that group-level cognition benefits everyone in the group and does not involve large fitness differences within the group, it should extend far more widely than the social insects.

Movement as a Group-Level Adaptation in African Buffalo Herds
African buffalo *(Syncerus caffer)* provide one possible example of movement governed by a communal decision-making process in groups whose members are not closely related. The home range of an African buffalo herd is a complex mosaic of patches whose quality depends on previous grazing history by the herd, depletion by competing species, regrowth speed, soil fertility, and distance from the current position of the herd. Prins (1996, 222) observed African buffalo for 2 years before realizing that what appeared to be a mun-

David Sloan Wilson

dane stretching behavior was actually a group-level decision making process.

> Some buffalo cows arise, shuffle around a bit and bed down again. At first I interpreted this as "stretching the legs," but one day I noticed that the cows adopt a particular stance after the shuffling and before lying down again. They seem to gaze in one direction and keep their head higher than the normal resting position but lower than the alert. . . . This standing up, gazing and lying down behaviour continues for about an hour, but the overall impression remains that of a herd totally at rest. Then at about 18.00 hours there is a sudden energizing of the herd. . . . A few moments later, everywhere in the herd buffalo start trekking. The exciting thing is that they start trekking, at the beginning and independently of each other, **in the same direction** [author's boldface]. Within seconds, the animals that initiate these movements are followed by other individuals, clusters of movement arise, and within about 3–5 minutes the whole herd of hundreds of individuals moves as if conducted by one master. They totally give the impression that they know where they are going to: apparently, some decision has been taken by the group.

Prins calls this "voting behavior," and has documented its effects on herd movement in impressive detail. Only adult females vote, and females participate regardless of their social status within the herd. When the average direction of gaze is compared with the subsequent destination of the herd, the average deviation is only 3°, which is well within measurement error. On days in which cows differ sharply in their direction of gaze, the herd tends to split and graze in separate patches for the night. In addition to this evidence for communal decision making, there is no evidence for individual leadership. For example, no individual cow or bull stays in the vanguard of the herd for more than a few minutes. Similar forms of "voting behavior" have been observed—although with less conclusive evidence—in baboon troops (see Byrne, chap. 17, this volume) and fish schools (Kils 1986; discussed by Pitcher and Parrish 1993).

A multilevel analysis of communal decision making in African buffalo would proceed as follows. First, we focus on a particular trait, such as stretching in the direction of a preferred resource patch. The energetic cost of this trait is so small that it can be ignored. Do individuals that stretch in the direction of a preferred patch increase their fitness, relative to other individuals in the same herd that behave differently? It is not obvious what this relative fitness advantage would be, if the entire herd moves as a unit. Does stretching in the direction of a preferred resource increase the fitness of the entire herd, relative to herds that behave differently? This would be the obvious advantage of the stretching behavior, which would qualify as a group-level adaptation. I do not claim that stretching has been proved to be a group-level adaptation in African buffalo, but merely wish to show how it would be studied from the multilevel perspective. Future research might involve searching for a within-group advantage of stretching that had been previously overlooked. It would also be important to confirm that herds vary in their stretching behavior, and perhaps even to document the process of smart herds expanding at the expense of other herds. The point is to determine where the fitness differences occur in the biological hierarchy (within-group vs. between-group) and to avoid classifying the behavior as an individual-level adaptation simply because it "pays" in some loose sense.

Female African buffalo remain within their natal herd for life, but this kind of stability is not required for groups to function as adaptive units with respect to movement. Imagine a fish species in which individuals always swim in schools by day but form larger aggregations by night. Group formation is a daily event, and the composition of each school is a random sample of the larger aggregation. Even for these extremely ephemeral groups, there will be genetic and phenotypic variation within and between groups. The social processes that determine how a group moves to forage or escape predators will be confined to the members of that group. The fitness consequences of these social processes can be partitioned into within- and between-group components. In fact, the daily random shuffling of individuals into groups is much like the intergenerational random shuffling of genes into individuals. In the former case, individuals play the role of replicators, but the groups remain potential vehicles of selection. Multilevel selection theory incorporates a wide range of population structures, from groups that last a fraction of a lifetime to groups that last many generations. Movement as a group-level adaptation is a legitimate possibility for virtually every species that moves in groups.

Group-Level Cognition in Humans and Its Relevance to Other Species

Humans have lived in nomadic social groups for most of their evolutionary history (see McCabe, chap. 22, this volume). The human mind is increasingly being regarded as a collection of cognitive adaptations for solving the important adaptive problems experienced

David Sloan Wilson

in our ancestral environments (Barkow, Cosmides, and Tooby, 1992). If so, then the ability to decide where to move must be chief among these adaptations. Understanding the psychology of group movement in humans is important in its own right and might also shed light on similar mechanisms in nonhuman species.

The Advantages of Thinking in Groups

Earlier I stated that the advantages of cooperation are as relevant to cognitive activities as to physical activities. The voluminous psychological literature allows some of the specific advantages of thinking in groups to be identified (reviewed by Wilson 1997b). Decision making usually begins with a phase in which alternative solutions are imagined, followed by an evaluation phase that leads to the final decision. For many problems, a single individual can imagine only a small subset of possible solutions, and the solutions imagined by different individuals only partially overlap. Thus, groups whose members freely imagine and then pool their potential solutions have an advantage over single individuals or groups that inhibit free thought and the sharing of ideas. Groups also can surpass individuals during the evaluation phase. Michaelsen, Watson, and Black (1989) had college students take a test as individuals and then take the same test in groups of three people. The group score almost invariably exceeded the individual score of the best member, for the simple reason that the best member was not right all the time, and the group was able to decide which member was most likely to be right on a question-by-question basis. These and other advantages of thinking in groups are likely to be as relevant to movement as to other activities.

Coordinating the Decision-Making Process

Because decision making is a sequential process, the behaviors that are adaptive during one phase become inappropriate during other phases. A study by Kruglanski and Webster (1991) shows how these changes can be coordinated in human social groups. Scout troops were asked to decide between sites for a work camp, a situation similar to a band of hunter-gatherers deciding where to forage. Individuals Scouts had previously filled out a sociometric scale, rating other members of their group for liking, appreciation, and respect. These three measures correlated highly with each other and were averaged to yield a single index of social status. For each group, a member whose social status score was at the median of the distribution was asked to become a confederate of the experimenters and was instructed to advocate a clearly better (conformist) or worse (deviant) site, either early or late in the decision-making process. After the decision was made, the Scouts were told that their previous sociometric ratings had been lost and were asked to again fill out the same scale, enabling changes in social status as a result of the decision-making event to be measured. The only change in social status was a decrease that occurred when the confederate expressed the deviant position late in the decision-making process. Expressing the same position early in the decision-making process had no effect on social status.

This study illustrates the importance of social norms in regulating the decision-making process in human groups. Furthermore, it shows that the norms are sufficiently flexible to allow both diversity of opinion and conformity at the appropriate times. It might seem that participating in the decision-making process is selfish because it prevents a loss of status within the group. However, we still must explain the norms that cause social status within the group to become aligned with group welfare. If I reward someone for benefiting the group by according them high status, then I am benefiting the group. If my effort to reward another involves a cost, then I am benefiting the group at the expense of my relative fitness within the group. Even if I could induce others to benefit the group at no cost to myself, my behavior would be merely neutral within the group and would require group selection to evolve. The evolution of social norms is a multilevel selection problem, just like the evolution of behaviors that are performed voluntarily. (See Boyd and Richerson 1990, 1992 and Sober and Wilson 1998 for a more general account of rewards and punishments from a multilevel perspective.)

Leadership and Social Dominance
In an early experiment, Guetzkow and Simon (1955) had members of five-person problem-solving groups communicate with one another in a variety of ways. They found that the most efficient communication network was a hierarchical one, in which four members communicated directly to a central fifth member. More richly connected groups, in which all members could communicate with one another, were less efficient and actually reverted to the hierarchical structure over time by breaking some of the connections (see Collins and Raven 1969 for a review of this research tradition). This study illustrates an important principle: that well-designed infor-

David Sloan Wilson

mation structures often (although not always) are hierarchically organized (Simon 1981). Purely from the standpoint of group fitness, we should often expect to find some individuals that we would call "leaders," others that we would call "followers," and so on.

Of course, social hierarchies can also be explained purely on the basis of relative fitness within groups. If individuals differ in their fighting ability, then the strongest will simply subjugate the weakest and take most of the resources. If the weakest are also the youngest, they might optimally bide their time as subordinate individuals until they grow large enough to challenge the dominants, and so on. Thus, multilevel selection theory provides two very different pictures of hierarchical social organization. Which picture best describes a social group, and how they interact with each other when both apply, depends on the balance between levels of selection.

The psychological literature on leadership in humans (reviewed by Bass 1990; Hogan Curphy, and Hogan, 1994) leads to two general conclusions. First, when social hierarchies are required from the information processing standpoint, the conflict between levels of selection is likely to be severe, because leaders, by definition, are in a position of power that allows them to exploit other members of their own group. For this reason, the social norms that control the behavior of leaders must be especially strong if the group is to function as an adaptive unit. There is abundant evidence that leaders in human social groups are often controlled at least as much as they do the controlling (e.g., Boehm 1982, 1993). They are selected on the basis of their ability to lead the group and are required to prove their leadership abilities by deed as well as word. They are often expected to take the same risks as other group members and are subject to exceptionally high moral standards. Often their power is confined to a single decision-making domain beyond which they have no special authority. None of these features of human leadership are expected on the basis of within-group selection, and they are easily interpreted as group-level adaptations. Of course, there is also abundant evidence that leaders frequently escape social control to exploit members of their own group. The point is that *both* within- and between-group selection are probably required to explain the hierarchical nature of human social groups.

Second, when a social hierarchy is required from the standpoint of information processing, the leader usually does *not* function as an autonomous decision maker who acts as the "brains" for the group. Decision making remains a group-level process in which the leader functions as a component. A study by Anderson and Balzer (1991) provides a good example. Leaders of decision-making groups were instructed to announce their opinion either early or late in the decision-making process. The groups generated more ideas when the leader's opinion was delayed, and the final decision was often based on ideas that were not generated by the leader. In fact, a general rule for effective leadership from the group standpoint might be "act as an organizer and moderator of group-level processes and refrain from exercising too much personal control" (Hogan, Curphy, and Hogan 1994). The concept of leaders as moderators may also apply to nonhuman primate species. According to Boinski (chap. 15, this volume), "some of the most dominance-ranked species appear to have typically long 'deliberation' periods among group members prior to travel onset, especially baboons."

Division of Labor

One of the basic advantages of cooperation is division of labor, which allows individuals to specialize on a subset of the activities required to perform a given task. Hutchins (1995) provides a fascinating example of division of labor in deciding where to move aboard large ships. When a ship is within sight of land, its position on a chart can be estimated by finding the compass directions of three landmarks, drawing lines from the landmarks on the chart at the appropriate angles, and noting where they intersect. The intersection of two lines is theoretically sufficient, but does not provide an estimate of error. The intersection of three lines creates a triangle whose size indicates the likelihood of error.

On modern ships that are comfortably away from the shore, the entire task of estimating position is performed by a single person. When the ship enters a harbor, however, this person is replaced by a six-person team whose social organization has evolved over a period of centuries. Hutchins describes in detail how the team functions as an integrated cognitive unit, swiftly gathering and transforming information, allowing spatial position to be noted at a glance at 3-minute intervals. The functional organization of the team includes a number of subtle but important design features. The members are spatially positioned and communicate in a way that allows one person's mistakes to be caught by at least one other person. There is redundancy of function so that the team can perform smoothly if one of its members is called away. The sequence in which the tasks are learned allows the team to function when its

David Sloan Wilson

experienced members leave permanently and are replaced by new initiates. Even the architecture of the chart room contributes to the adaptive decision-making process, much as the architecture of the hive contributes to group-level cognition in bees. Hutchins argues that this is only one of many examples of cognition as an adaptive distributed group-level process in humans.

The Group Mind in Humans and Its Relevance to Other Species
The concept of human social groups as information processing devices may seem radical, but it becomes plausible in the light of multilevel selection theory. The benefits of thinking as a group are potentially enormous and do not necessarily entail large fitness differences within groups. Social norms can also turn within-group selection into a tool of between-group selection by conferring status and other social benefits onto members who contribute to the welfare of their group. Some aspects of the group mind in humans might be so distributed that we play our roles as individuals without any awareness of the bigger adaptive picture, just as neurons and honeybees are presumably unaware of their roles in higher-level cognitive processes. For example, gossip and other forms of talk might prove to be a sophisticated system for gathering, transmitting, and processing information at the group level (Wilson et al., in press), but as individuals, we know only that we are dying to transmit the latest juicy bits of information to the appropriate people.

Some aspects of the group mind in humans require spoken language, a facility that is lacking in other species. Other aspects require the enforcement of social norms, which in turn may require sophisticated cognitive abilities that are lacking in other species (see Boinski, chap. 15; Byrne, chap. 17, this volume). These are questions for the future, but they cannot even be asked until behavioral ecologists learn to partition relative fitness into within- and between-group components. For example, de Waal (1995) described an event that took place in a captive chimpanzee colony, in which two juvenile females remained outside at the end of the day, preventing the other chimps from receiving their food. The zookeepers expected trouble and kept the females away from the other chimps that night, but the following morning they were severely punished and were the first to go inside the following night. De Waal uses this event to introduce the more general subject of morality, which fundamentally involves a shared conception of appropriate behavior that is maintained by rewards and punishments. He provides a compelling case for the roots of moral behavior in chimps and other primates, but his analysis is based on absolute fitness. Thus, a dominant chimp who "keeps the peace" within his group is described as self-interested, even though the benefits of keeping the peace are shared and may have their ultimate advantage in allowing the group to outcompete less harmonious groups. Similarly, Clutton-Brock and Parker's (1995) review of punishment in animal societies is based on absolute benefits to self and kin, without any reference to relative fitness within and between groups. Their review shows that the use of rewards and punishments to enforce social norms might extend beyond primates and could coordinate group-level cognition and movement in many species. However, group-level adaptations will never be recognized until the relative fitness advantages are partitioned into within- and between-group components.

The Glamour of Studying Group-Level Adaptations
I will end this chapter by pointing out a curious double standard that exists in the field of behavioral ecology, which is revealed by Pitcher and Parrish's (1993) statement that fish schooling behavior is "no more glamorous than animals cooperating only when it pays." What makes the study of individual-level adaptations exciting, even glamorous? We expect them to evolve, yet we are still fascinated to discover them, to study their underlying mechanisms, and to marvel at their often astonishing degree of sophistication. When the subject shifts to cooperation and altruism, interest shifts away from the actual adaptations to the conditions that allow them to evolve. It is obvious—therefore unglamorous—that animals can evolve to cooperate when everyone stands to gain. It is obvious—therefore unglamorous—that altruism can evolve among genetic relatives. To be glamorous, the evolution of a trait must be difficult to explain from the standpoint of current theory, such as the evolution of strong altruism among nonrelatives. Multilevel selection theory can meet this challenge (e.g., Wilson and Dugatkin 1997). In addition, however, we need to realize that group-level adaptations, like individual-level adaptations, remain exciting and glamorous after we have explained the raw fact that they can evolve. Returning to group movement, it is marvelous to contemplate that groups have evolved into computational machines that collectively gather, store, and integrate information to make adaptive decisions about where to move. Group minds and their effects on group

David Sloan Wilson

movement have only recently been documented for the social insects, and we know virtually nothing about their existence or degree of sophistication in other taxa. We are similarly ignorant about the neural, behavioral, and social mechanisms that allow groups to think and move as adaptive units. If we discover that group minds can easily evolve in many kinds of social groups because everyone gains and there is little need for self-sacrifice within the group, then group minds should become more fascinating to study, not less. For this and other subjects, multilevel selection theory transforms what appears to be mundane (animals behaving only when it pays) into glamorous and unexplored possibilities.

This chapter is intended to present a general theoretical framework (multilevel selection theory) to an audience of readers that is primarily interested in a particular biological subject (group movement). I have tried to relate the theory to group movement to the best of my ability, but I also have tried to avoid undue speculation. Group minds and their effects on group movement have been demonstrated beyond a reasonable doubt in the social insects. The case for African buffalo appears surprisingly strong, and the literature on humans provides mechanistic details that are lacking for other vertebrate species. The other chapters in this volume contain many suggestive possibilities for movement as a group-level adaptation, but more work is required to take them beyond the realm of speculation. I hope that this introduction to multilevel selection theory, combined with the biological expertise of the reader, will lead to additional well-documented examples of movement as a group-level adaptation in the future.

Table 9.1 Key terms

Absolute fitness	The fitness of an individual, without reference to the fitness of other individuals in the population (compare with relative fitness).
Between-group selection	The process of natural selection that operates between groups, favoring traits that maximize the fitness of groups, relative to other groups in the metapopulation
Group mind	A cognitive process that is distributed among members of a group and cannot be located within any single individual
Group selection	See Between-group selection

(continued)

Table 9.1 (*continued*)

Multilevel selection theory	A theoretical framework that views biological systems as a nested hierarchy of units. Genes reside within individuals, which reside within groups, which reside within populations of groups, or metapopulations. The process of natural selection can occur at all levels of the hierarchy
Organism	A biological unit whose parts work in close harmony to survive and reproduce. Individuals are organisms to the extent that they are units of selection. Other units, such as groups, can also acquire the properties of organisms to the extent that they are units of selection
Replicator	A unit that persists through evolutionary time. Genes are the most common example of replicators
Selfish gene theory	A theoretical framework that calculates fitness at the gene level, often erroneously regarded as an argument against group selection
Trait group	The set of individuals that are influenced by the expression of a trait
Vehicle	A unit that shares the same fate with respect to fitness. Individuals are often (but not always) vehicles. The group selection controversy hinges on the question of whether groups can also be vehicles
Within-group selection	The process of natural selection that operates within single groups, favoring traits that maximize the fitness of individuals, relative to other individuals in the same group

PART THREE

Travel Decisions

Observations that primate troops take what appear to be direct routes to distant feeding sites were first quantified in baboons, chimpanzees, and tamarins, and are ubiquitous in field anecdotes for virtually all primate species. Perhaps what has fascinated us most is that these travel decisions seem so reasonable and efficient, even humanlike, given what is known about the distribution of food and other resources in the ranging areas of these troops. As this section illustrates, monkeys and apes incorporate not only ecological, but also social information into their travel decisions.

Two major unresolved components of travel decision processes in primate groups are also considered here. The first is whether primates represent spatial information as a coordinate-based or a route-based system in traveling between patches. In effect, do primates rely upon the equivalent of a two-dimensional Cartesian map, or on something more akin to the circuitry of a computer chip, a network of familiar, intersecting travel paths? The second concerns the role of rule-based foraging—the ability to store, recall, and compare social and ecological information and use it to generate a set of "learning rules" or expectations that may enhance foraging success. Such rules might be general—for example, "select softer fruits over harder fruits within a food patch"—or they might be quite specific—"return to foraging areas visited yesterday only if they contained a clumped distribution of productive feeding sites." We suspect that 20 years from now, the cognitive systems underlying the "spatial maps" of most primate species will be found to be similar. On the other hand, we would not be surprised if the "foraging rules" employed by species were more varied, each set reflecting closer adaptations to the diverse natural histories of primate taxa.

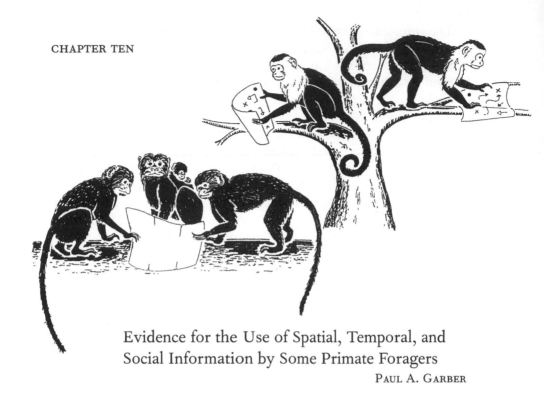

CHAPTER TEN

Evidence for the Use of Spatial, Temporal, and
Social Information by Some Primate Foragers

PAUL A. GARBER

Searching for food in a rainforest is like trying to locate a particular restaurant that you once visited among the thousands of buildings, skyscrapers, and streets of a large city. Some parts of the city are familiar to you because you have recently been there. Other streets you have never visited, and so you have either no knowledge or incomplete knowledge of how those streets might connect, whether the restaurant you are looking for is open for lunch, closed for repairs, or requires a reservation to accommodate you and your ten hungry friends. Someone in your group remembers that the restaurant was not far from the shopping mall you just passed, but is uncertain of which cross street to take. Should you follow this friend? Another member of your party smells food coming from a small fast-food restaurant located in the opposite direction. Someone suggests calling a friend who knows the restaurant's precise location. Should members of your group split up to increase the chances of finding the restaurant, opt for a greasy hamburger, walk back to the mall and find a public phone, or continue traveling in the presumed general direction of the restaurant?

The situation described above is analogous to the complex social and spatial problems that free-ranging primates face daily in locating rare or ephemeral foods. Given that sequential feeding sites exploited by arboreal primates may lie hundreds of meters apart, are visually obscured by dense foliage, and may be scattered in any direction, the ability to reduce the travel time and energy required to relocate productive food patches and to encounter new feeding sites is likely to have a significant effect on foraging success (Milton 1981a, 1988; Terborgh 1983; Robinson 1986; Garber 1988, 1989; Janson and Boinski 1992; Janson 1996).

In this chapter I address a series of questions regarding the kinds of information monkeys use in selecting feeding sites and the challenges that arboreal group-living primates face in locating and exploiting food resources that exhibit a scattered and patchy distribution across a forested landscape. These challenges involve the ability to integrate *spatial* (where to forage), *temporal* (when to return to feeding sites), and *ecological* (expectations regarding the amount of food at a feeding site) information concerning the availability of productive feeding sites as well as *social* information on dominance relationships, status, and individual access to resources also sought by other group members (Krebs and Inman 1992; Real 1994). A major objective of this discussion is to examine how primates use such information to affect patterns of group and individual movement. In order to accomplish this I will begin by reviewing recent theoretical ideas regarding how foragers might represent spatial information. Too often, researchers have either assumed that animals have no knowledge of where to look for food and forage randomly, or that animals have perfect knowledge of the location and availability of feeding sites and are like traveling salespersons, always selecting the best feeding sites in the most efficient order. Neither of these extremes is probably common. I also examine the concept of rule-guided foraging and describe how the social environment can influence the types of information available to animals that forage in dispersed versus cohesive groups.

A second major goal of this chapter is to present evidence obtained from experimental field studies on wild primates to determine how monkeys use spatial, temporal, ecological, and social information to select feeding sites. Experimental field studies offer the opportunity to examine species differences in problem-solving skills and perceptual abilities under natural conditions in which the information available to the forager is controlled and systematically

Paul A. Garber

varied (Garber and Dolins 1996). Knowledge of the sensory cues and spatial information animals use in decision making (e.g., an individual's ability to identify and select a particular behavioral response from a range of possible alternatives) is an important step in understanding patterns of individual and group movements. In addition, I review a recent model of spatial mapping presented by Poucet (1993) and test it against evidence of foraging and ranging patterns in wild tamarin monkeys. Poucet has argued that foragers are likely to represent information in "large-scale" space very differently from information in "small-scale" space. This model offers a productive framework for understanding differences in spatial strategies used by animals when navigating between distant feeding and resting sites (large-scale space) and when locating feeding trees within a locally defined area (small-scale space). I end the chapter by indicating the importance of combining experimental field studies and natural field studies in examining questions of primate cognition, information processing, and the role of the social environment in feeding ecology.

Mental Representations of Spatial Information

Frequently, primatologists have argued that monkeys and apes maintain detailed knowledge of the spatial location and phenology of resources in their home range and, from virtually any point in the forest, can assess the distance and direction to potential feeding, resting, and drinking sites (S. A. Altmann 1974; MacKinnon 1974; Janson, Terborgh, and Emmons 1981; Milton 1981a, 1988; Sigg and Stolba 1981; Estrada and Coates-Estrada 1986; Robinson 1986; Garber 1988b, 1989, 1993b; Chapman, Chapman, and McLaughlan 1989; Menzel 1991; Leighton 1993). MacKinnon (1974, 31), for example, provides the following description of ranging patterns in orangutans:

> In view of the unpredictable nature of fruit availability, orang-utans seemed to have an uncanny ability of arriving at the right place at the right time. . . . They often proceeded to good fruit trees by very direct routes, although I believed they had not fed there previously that year.

Similarly, in the case of spider monkeys, Milton (1981a, 540–41) suggests that individuals or an entire group may travel

> a considerable distance between fruiting trees without stopping to eat along the way. The following day one or two of the same trees may be revisited by various spider monkeys and from four to eight new individ-

uals of the same fruiting species added to the day route by moving over a route that minimizes travel distances between such fruiting trees. These data show that spider monkeys *know* the locations of particular trees. . . . It further suggests that spider monkeys are capable of formulating a travel route in advance that takes them to a number of different fruiting individuals over the course of a normal day's travel such that they do not double back and cover areas already visited.

In most instances, however, the evidence for spatial learning in wild primates is either anecdotal or correlational and is based principally on observations of relatively direct or "straight-line" travel to a distant set of previously visited food patches. Such behavior is reminiscent of *traplining*, a feeding pattern also found in insects, birds, and other mammals, in which several trees of a single species are visited in succession or become the focus of feeding activities over a period of days or weeks (Janson, Terborgh, and Emmons 1981). Tree species exploited in this manner are often characterized by intraspecific floral or fruiting synchrony (all or most trees of the same species flower or fruit at the same time). Phenological synchrony adds an element of temporal predictability that may offset problems associated with exploiting patchily distributed resources and may reduce the time and energy spent in exploratory foraging. Traplining is a form of *goal-directed travel* and has been reported in several species of prosimians (Sussman and Raven 1978; Charles-Dominique and Petter 1980; Overdorff 1993b), New World monkeys (Janson, Terborgh, and Emmons 1981; Milton 1981a; Terborgh 1983; Garber 1988b, 1989; Norconk and Kinzey 1994), Old World monkeys (S. A. Altmann 1974; Wheatley 1980; Sigg and Stolba 1981; Menzel 1991; Henzi, Byrne, and Whitten 1992; Gautier-Hion and Maisels 1994), and apes (Menzel 1973a; MacKinnon 1974; Chivers 1977; Boesch and Boesch 1984; Goodall 1986). Goal-directed travel is consistent with the ideas that an animal retains an expectation of where it is heading and a memory of and sense of direction from where it has traveled, and that it uses spatial or sensory information (i.e., olfaction, angle of the sun, landmarks) to navigate to targeted feeding sites (Dolins 1993). By itself, however, goal-directed travel is not an indication of the types of spatial and sensory cues an animal is using to locate feeding sites.

The ability to internally represent and organize spatial and sensory information has been referred to as a "mental map" or a "cognitive map" (Tolman 1948; O'Keefe and Nadel 1978). This term has generated great controversy in the behavioral and psychological

Paul A. Garber

literature because it has been used to describe different types of internal representations that encode information of varying complexity (Nadel 1990; Dyer 1991, 1994, 1996; Gallistel and Cramer 1996). In a general sense, these internal representations reflect the ways in which information obtained by the senses is translated, processed, and reorganized in the brain into a format that constructs a view of salient features of the environment (Real 1994). Gallistel (1990) has used the term *cognitive map* to describe any form of spatial representation maintained by an individual. In contrast, Tolman (1948) and O'Keefe and Nadel (1978) restrict the use of this term to an ability to mentally construct a geometric or Euclidean map of large-scale space in which landmarks are represented within a common coordinate system. In this type of cognitive map, the actual positions of landmarks and other features of the environment are either geometrically represented as true angles and distance relationships or can be derived computationally from other coordinate information already present on the map (Bennett 1993, 1996; Dolins 1993). This would require an ability to reorient, integrate, and reconstruct large numbers of individual local views of places and landmarks in the environment into a single geometric "global-view-from-above" of large-scale space (Benhamou 1996). Such a representation is analogous to having an exact map of your university campus in your head so that, from any starting location on campus, you are able to chart and compute shortcuts and novel routes of travel to visit each of your professors, several friends, and all campus libraries on the same day.

Recently, Poucet (1993), Benhamou (1996), Byrne (chap. 17, this volume), and others have questioned whether animals can truly represent spatial information in such a detailed and accurate way, and have offered alternative explanations for animal navigation. Benhamou (1996, 202) states, "There is considerable behavioral and neural evidence that mammals process and store large amounts of spatial information about the environment, [however,] their ability to perform 'global' place navigation has never been demonstrated nor even properly tested." A model of spatial navigation (Poucet 1993) that may better represent how primates and certain other animals use landmarks as cues to travel through the environment is presented at the end of this chapter. This model highlights an important distinction between movement in "small-scale" space and movement in "large-scale" space. Most captive research has focused on experimental studies of how animals represent and use landmarks in spatially limited areas in which they have the opportunity to see and encode several views of the same landmarks from different directions. In contrast, field studies, especially of primates, have examined group ranging patterns over areas of up to several square kilometers (Byrne, chap. 17, this volume; Watts, chap. 13, this volume). Given limited opportunities to view the spatial relationships between sets of landmarks in large-scale space, the exploitation of large ranges may require very different types of navigational skills and mental representations of space than the exploitation of small ranges.

Spatial Maps

One focus of the debate on animal cognition has involved strategies of spatial memory and the manner in which landmarks and other informational cues are remembered and used during travel. Landmarks are visual images or olfactory cues that can be used as guides or beacons, and may be of two general types. Near-to-target landmarks may function primarily as associative or reinforcement cues (Brown and Gass 1993). That is, they are not used for purposes of charting a route, but rather are a good predictor of achieving a goal such as finding nearby food. For example, the presence of several large bromeliads or other epiphytic plants on a tree might be used by an insectivorous/faunivorous forager as a cue to search the crown for prey that might feed on or find refuge in these structures. In contrast, distant landmarks, such as tall trees, edges between visually distinct habitats, natural boundaries such as rivers, treefall gaps, or perhaps particular stands or formations of trees, may be used for purposes of navigation and orientation. An animal may learn to associate a single distant landmark, or a set of sequential landmarks, with a particular goal and use a behavioral rule such as "head toward the tallest tree near the river" to locate food (see Dyer, chap. 6, this volume). In general, such orienting landmarks are likely to be more permanent and predictable features of the habitat (e.g., forest edge) rather than more ephemeral guides such as a tree that is currently in flower or fruit.

It also has been suggested that primate species from saddleback tamarins to common chimpanzees may be capable of learning the relative spatial relationships of several individual landmarks to one another and using this information to chart accurate distance and direction vectors and novel shortcuts to resources that are presently out of view (Boesch and Boesch 1984; Garber 1989). For example,

Paul A. Garber

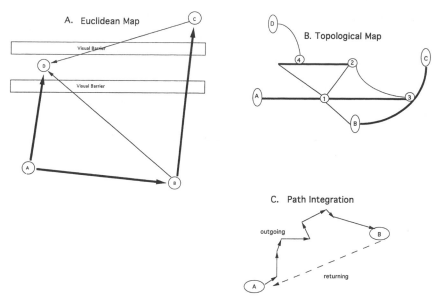

Figure 10.1 Ways in which animals might represent spatial information. *(A)* A Euclidean map, in which the true distances and directions between feeding sites are computed by the forager. Circles A, B, C, and D represent feeding sites. Heavy lines represent previously traveled routes; lighter lines represent new or novel routes of travel. If a forager is located at point B, it computes a novel route of travel to point D, which is presently out of view, by representing spatial information as a coordinate system. *(B)* A topological spatial representation. Ovals A, B, C, and D represent feeding sites. Heavy lines represent commonly traveled routes; lighter lines represent less commonly traveled routes. Circles 1, 2, 3, and 4 represent nodes of intersection (landmarks) used by the forager to reorient its line of travel. The forager cannot go directly from point A to point D, but is constrained by a network of remembered travel routes and must pass through several other points. *(C)* Path integration. The forager, in moving from A to B, tracks its own movements (distance, direction, angle, speed) in small-scale space, and then computes a reasonably direct short-distance or homing route to return to its place of origin.

as illustrated in figure 10.1A, a forager that has previously traveled from point A to D and from point A to B can compute a novel direct route (BD) from point B to a targeted but out-of-view feeding site at point D. This capability is consistent with a coordinate-based internal representation of space, and is associated with what was referred to earlier as a *geometric* or *Euclidean map* (Bennett 1996; Dyer, 1996).

Spatial information also may be encoded in a *route-based* or *topological* framework. In a topological spatial system, the forager maintains a mental representation of places and landmarks as a set of intersecting travel routes (figure 10.1B). Spatial relationships are nongeometric, however, in that the information encoded does not represent the actual distances and angles between landmarks, but rather is an exaggerated view of the landscape in which targeted goals such as feeding trees and refuge sites are located in proximity to landmarks. Travel using such a spatial framework can occur over long distances, can occur in a relatively straight line, and may involve the use of several known landmarks, but these must be visited in sequence and along the same route or set of routes in order to reach the goal (Bennett 1996). This type of internal spatial representation is similar to traveling on a Greyhound bus. You can eventually get from city A to city D, but in order to do so, you are constrained by a system of roads and highways and have to pass through cities B and C, even if those take you out of your way.

A topological representation of spatial information can be extremely complex. The forager may be able to learn "spatial relationships between different routes that are related through common points" (Dolins 1993, 26) or intersections, and in doing so may be able to recombine route sequences to reach the same and different feeding sites. MacKinnon's (1974) description of orangutans following permanent features of the landscape like streams and ridges during travel, and the existence of what he termed "arboreal highways" or paths used repeatedly despite changes in resource availability, may be indicative of a reliance on topological information. Although the idea of a route-based map implies reuse of common paths to reach feeding and resting sites, depending on the number of route segments, their relative proximity to feeding sites, and the forager's ability to encode the location of a large number of landmarks, an individual might be capable of using several different route combinations to reach its target.

Animals also can rely on a form of navigation called *path integration* (also referred to as dead reckoning) to return to a nest, sleeping site, or central foraging location. Path integration (figure 10.1C) is a computational process in which the individual tracks its own movements (distance, direction, angle, speed) in small-scale space from some fixed point in the environment and then can compute a reasonably direct short-distance or homing route to return to its place of origin (Maurer and Séguinot 1995; Etienne, Maurer, and Séguinot 1996). This type of cognitve map is describes as part of an "egocentric" spatial framework in that the animal navigates by retaining a sense of its current position relative to a landmark in the environment. Path integration, which is reported to be widespread among vertebrates and invertebrates, does not require spatial

Paul A. Garber

knowledge of the locations of landmarks or of the spatial relationships between landmarks in order to reach a target or goal. Studies have shown, for example, that in the absence of any visual cues, blindfolded humans guided through two legs of a triangle retain an internal record of their route and can compute an angle of return that takes them back to the vicinity of the original starting point (Loomis et al. 1993). Navigational errors are common, however, and human subjects tend to overestimate short-distance returns and underestimate long-distance returns (Loomis et al. 1993; see Dyer 1994 for a review). Thus, in the absence of access to visual cues to correct or reorient their route, even humans experience difficulty in accurately tracking their movements in small-scale space. Path integration is considered by Gallistel (1990) to be a prerequisite for developing a cognitive map, and may be used in conjunction with forms of topological and Euclidean information to navigate in space.

Rule-Guided Foraging
The manner in which individuals within a group represent and use spatial, temporal, perceptual, and social information must play a central role in patterns of group movement. The effectiveness of different solutions to problems of locating food, coordinating group foraging and hunting (Holecamp, Boydston, and Smale, chap. 20, this volume), avoiding predators (Boinski, Treves, and Chapman, chap. 3, this volume), minimizing within-group feeding competition (Chapman and Chapman, chap. 2, this volume), and cooperating in resource defense (Peres, chap. 5, this volume) is dependent on the ability of members of a social group to integrate, process, exchange, and on occasion, conceal disparate types of information.

The resources available to most foragers are highly ephemeral. For example, the amount of food available at a feeding site can vary on a time scale from hours in the same day (nectar renewal in a flower or gum production on the trunk of a tree) to days (patterns of fruit ripening), weeks (the flush of new leaves), or years (the flowering of bamboo), depending on the fruiting and reproductive cycles of a particular plant species. Similarly, the spatial distribution of food patches changes seasonally as previously visited feeding sites are no longer productive and new sites become the focus of current feeding activities. In order to find and relocate food and track food availability in the forest in some systematic or nonrandom way, all animals must be capable of associating certain forms of spatial, temporal, and sensory (visual, olfactory, tactile) information with specific feeding events and storing that information. Animal species are likely to vary in the precise manner in which they internally represent their environment in terms of the amount, type, and hierarchy of information stored, as well as in their ability to derive effective behavioral solutions to newly encountered ecological and social problems.

Although our understanding of how the external world is neurologically represented in the brain remains limited, it is clear that the ability to use information from previous feeding events to solve present problems of food acquisition enables animals to generate effective solutions "without re-learning cause and effect relationships involved in every new situation" (Garber and Dolins 1996, 202). This ability, which has been referred to as "hypothesis" (Harlow 1949) or "rule-guided" foraging (Menzel 1996), enables an individual to apply or generalize a set of preexisting expectations learned in one foraging context to other foraging contexts (Krebs 1978; Menzel and Juno 1982; Gibson 1990; Garber 1993b; Kamil 1994; Real 1994; Garber and Dolins 1996; Menzel 1996; De Lillo, Visalberghi, and Aversano 1997; Garber and Paciulli 1997). Such "rules" might include when to leave a patch, when to return to a patch, whether or not to enter a patch already occupied by a conspecific, where to search next, and which individuals to follow or monitor in order to locate new food items. Rules might differ between species, or between different individuals of the same species, depending on their assessments of present and future feeding opportunities, which might take into account whether the forager did (goal-oriented foraging) or did not (exploratory foraging) have knowledge of the presence of food patches in the general area, the behavior and social relationships of other group members, and species-specific patterns of information storage, communication, and food sharing (Kamil 1994; see Boinski, chap. 15, this volume).

In many cases, the use of the same rule or small set of rules may be effective in exploiting several different types of resources. For example, a forager might return to the same large-crowned, highly productive, fruit-laden tree day after day for a period of weeks. This strategy represents a commonly applied foraging rule, known as Win-Return. Another such rule might be, "continue searching for large bromeliads because over the past few days several have contained easily caught small vertebrate prey" (Occasionally Win-Continue to Search Same Type of Object). Another rule might be,

Paul A. Garber

"if red-flowered trees visited last week did not contain nectar, do not travel to red-flowered trees this week" (Lose-Shift). Or, "when uncertain of food availability in the part of your range you are in, travel to the gallery forest near the river because it often contains some trees in fruit" (bet hedging, or a fruit in your mouth is worth two on some unknown tree). Or, in social foragers, "after searching unsuccessfully for food, follow the same adult female who yesterday allowed you to enter a small food patch that she discovered" (Win-Continue to Follow).

Cohesive versus Dispersed Foragers

A major advantage of group or social foraging is the opportunity to share or parasitize information concerning individual foraging success and the location of food. The information available to social foragers includes both private (self-generated) and public (group-based) information, and is central to understanding patterns of group movement (Clark and Mangel 1984; Giraldeau 1984; Valone 1989; Chapman and Lefebvre 1990; Vickery et al. 1991; Ranta, Hannu, and Lindström 1993; Valone and Giraldeau 1993; Avery 1994; Benhamou 1994; Templeton and Giraldeau 1995; Tiebout 1996). *Public information* is obtained by monitoring the behavior, ranging patterns, feeding activities, vocalizations, and other forms of communication of conspecifics. Such information can be used to identify the location of a food patch, to determine the productivity of a food patch, to identify which patches or areas to avoid, or to collect ecological information that may aid in predicting what other types of resource patches or feeding sites are likely to contain food (Valone 1989; Templeton and Giraldeau 1995). In contrast, *private information* refers to knowledge generated by an individual's own foraging experience and its self-generated expectations regarding resource availability in patches or areas previously visited. Among groups that forage cohesively or are highly coordinated in their group movement (see Peres, chap. 5; Boinski, chap. 15; Greenberg, chap. 18, this volume), such as many species of tamarins (Garber 1993a), titi monkeys (Kinzey 1977), common brown lemurs (*Eulemur fulvus:* Sussman, 1977a), and black-and-white colobus (*Colobus guereza:* Oates 1977), personal knowledge and public information are largely the same. All group members have access to nearly identical information regarding which feeding sites were previously visited, which were depleted, and which are likely to continue to contain food. This is especially true of species that are territorial or defend feeding sites from conspecific groups, as has been suggested by Wrangham (1980) to occur in female-bonded primates, as well as species in which observational learning of dietary information, food calls, and other exchanges of foraging information are common. It is important to point out that the concept of public information does not require that all group members have access to the same information. The concept of public information does, however, represent a process by which social foragers obtain ecological and resource information by observing the behaviors and activities of conspecifics. In contrast, in primates that travel in more dispersed foraging parties, such as chimpanzees and bonobos (Chapman, White, and Wrangham 1994), spider monkeys (Chapman, Chapman, and McLaughlan 1989), and gray-cheeked mangabeys (Waser 1984b), many individuals are likely to have different personal knowledge regarding the productivity and location of feeding sites.

It also has been suggested that traveling in a dispersed front or in scattered subgroups can increase the likelihood of locating new or widely spaced food patches; especially in species that use vocalizations to signal the location of productive feeding sites (Ghiglieri 1984; Chapman and Lefebvre 1990; Vickery et al. 1991; Boinski, chap. 15, this volume). In the case of first-time feeding sites, discovery rates are likely to vary positively with day range, the size and visibility of the food patch, the number of individuals foraging together, the spread of the foraging unit, and the size of the detection field of each forager (Janson and Di Bitetti 1997). Foraging in a dispersed or wide front may increase the rate at which prey or feeding sites are encountered. White-faced capuchins *(Cebus capucinus)*, for example, are known to consume larvae from colonial insect nests (Hymenoptera, wasps) and to capture infants from squirrel *(Sciurus* sp.), coati *(Nasua narica),* and avian nests (Oppenheimer 1982; Fedigan 1990). Finding these resources may be analogous to locating first-time feeding sites. Group spread may increase the likelihood that at least some individuals will locate these foods. In this setting, monitoring the foraging activities of nearby conspecifics (public information) may play an important role in increasing a scrounger's potential access to hard-to-find foods (Vickery et al. 1991).

A potential cost of social foraging, however, is a reduction in feeding rates due to a greater number of individuals exploiting the same patch. These costs are expected to increase with decreasing

patch productivity and increasing group size, and to have a disproportionately negative effect on subordinate animals (Janson 1988a;
van Schaik and van Noordwijk 1988). Under conditions in which a
small set of dominant animals can control access to feeding sites
and in which food patches are small and widely scattered, dispersed
foragers may have an advantage over cohesive foragers in their ability to locate and exploit low-density resources. This advantage is
likely to increase if each forager or subgroup can keep private resource information private. In species such as common chimpanzees *(Pan troglodytes)*, blue monkeys *(Cercopithecus mitis)*, spider
monkeys *(Ateles)*, and gray woolly monkeys (*Lagothrix lagotricha*),
members of different subgroups may be hundreds of meters apart
(Cords 1987; Chapman and Lefebvre 1990; Chapman, White, and
Wrangham 1994; Peres 1996). As foragers distribute themselves
more widely across the landscape, ego-based decisions are expected
to be more common and to occur among all or most troop members
regardless of sex and social position. Given that each individual or
foraging subgroup has different resource information and may be
unaware of the recent feeding activities of other individuals or subgroups, information regarding the availability and distribution of
potential feeding sites is less reliable. Unknown to the forager, a
feeding site it encountered yesterday may have been encountered
and exhausted by conspecifics later that same day. For example,
Wheatley (1980) reports that when exploiting small-crowned trees,
individual crab-eating macaques *(Macaca fascicularis)* feed in the
tree alone and, on average, at a distance of 20 meters from the nearest group member. When foraging in these scattered subunits, each
group member has only limited, and presumably different, information concerning the feeding activities of conspecifics, as well as im-
precise knowledge regarding which of the trees will continue to bear
fruit. In contrast, when exploiting certain large-crowned trees, all
thirty group members feed together in the same tree, and "the troop
repeatedly exploited them for several consecutive weeks" (Wheatley
1980, 213). In these large-crowned trees, all group members have
access to the same public foraging information.

Experimental Studies of Rule-Guided Foraging
Based on captive studies, there is evidence that primates can rely on
a number of different foraging rules and apply each under a particular set of resource conditions. Menzel (1996) has clearly shown this to
be the case for long-tailed macaques *(Macaca fascicularis)*. Menzel

tested the ability of captive Japanese macaques to predict the location of new feeding sites. These monkeys were presented with visible food items in controlled spatial arrangements in an outdoor enclosure, and then observed to see if they would generalize from these distributions to predict the locations of hidden food items. The results indicated that the captive macaques applied a number of search strategies to locate hidden food, including using conspecifics as guides or beacons, random search, structure-guided search (e.g., search along the border), matching-type search (i.e., search near objects of similar type), and search at regular intervals along a straight line (distance-based search) (Menzel 1996). Evidence of the use of complex spatial or social rule-guided foraging has also been reported in recent captive studies of cotton-top tamarins (*Saguinus oedipus:* Dolins 1993), brown capuchins (*Cebus apella:* De Lillo, Visalberghi, and Aversano 1997), and vervet monkeys (*Chlorocebus aethiops:* Cramer 1995); in experimental field studies of brown capuchins (*Cebus apella:* Janson 1996; Janson and Di Bitetti 1997), white-faced capuchins (*Cebus capucinus:* Garber and Paciulli 1997), and moustached tamarins (*Saguinus mystax:* Garber and Dolins 1996); and in naturalistic field observations (Chapman, Chapman, and McLaughlan 1989; Garber 1993b; Garber and Hannon 1993). Similarities in the types of foraging rules (e.g., selection of distance-minimizing travel routes) used by captive and wild primates suggest that experimental field studies can provide an important tool with which to identify species differences in the use of spatial and temporal information and rule-guided foraging (Brown and Gass 1993; Menzel 1996). Although field experiments using playbacks of recorded vocalizations have, for two decades, proved extremely useful in studying the behavior of wild primates (see, for example, Cheney and Seyfarth 1990), it is only in the past two to three years that field experiments designed to examine the kinds of information primates use in locating and selecting feeding sites have been published.

Experimental Field Studies of Rule-Guided Foraging in Tamarins and Capuchins
Experimental field studies offer the opportunity to systematically control the information available to the forager in its natural habitat and determine the hierarchy of informational cues used by group members during foraging (Garber and Dolins 1996; Janson 1996). In particular, these studies can address questions regarding the kinds of information primates use in making foraging decisions, as

Paul A. Garber

well as how factors such as access to public and private information affect individual and group movement and foraging success.

Below I detail the results of a series of field experiments in which wild groups of moustached tamarins and white-faced capuchins were presented with a schedule of food availability analogous to patterns of resource availability they encounter naturally in the wild. The information available to the foragers was controlled, and this allowed a series of hypotheses to be tested regarding the use of spatial relationships, visual cues, olfactory cues, and foraging rules in selecting feeding sites. Both primate species were presented with a similar set of foraging problems and a similar set of information that could be used to locate food rewards. No attempt, however, was made to manipulate the type of social information available to the foragers (i.e., presence or absence of conspecifics). In these studies, species differences in group spread, dominance interactions, and social tolerance played an important role in tamarin and capuchin ranging patterns and foraging decisions.

Moustached tamarins are small-bodied New World primates (adult body weight 525–600 g) that exploit a diet composed principally of insects, ripe fruits, floral nectar, and plant exudates. Group size ranges from five to thirteen individuals (Garber et al. 1993), and individuals forage as a cohesive social unit, with all or most group members entering and leaving feeding trees at the same time. Group spread in these tamarins was generally less than 15 meters (Peres, chap. 5, this volume). High levels of group cohesion and social cooperation in callitrichines may be reinforced by kinship ties as well as by the evolution of a mating and infant care system in which all adult group members assist in transporting, guarding, and exchanging food with the young (Garber 1994, 1997).

Like that of moustached tamarins, the diet of white-faced capuchins also includes ripe fruits and insect and vertebrate prey. Group size averages ten to twenty individuals, and groups typically contain more than one adult male (Fedigan 1993; Rose 1994a,b; Fedigan, Rose, and Avila 1996). Capuchins frequently forage in a dispersed fashion, with troop members often separated by distances of 50–100 meters or more (Chapman and Fedigan 1990; Phillips 1995). In large-crowned trees, many or most individuals forage simultaneously. In medium-sized or small food patches, individuals forage in a more solitary manner (one to three freely locomoting individuals per patch). Troop integrity is maintained using a varied repertoire of vocal signals to coordinate and change the direction of group

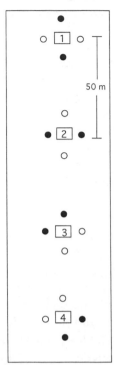

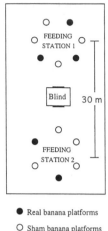

Figure 10.2 Arrangement of experimental feeding stations and feeding platforms. Platforms measured approximately 1,300 cm² in area and were raised 1.5 meters above the ground. The capuchins were presented with one group of seven platforms (station 1) and one group of six platforms (station 2), each arranged in an oval configuration. On average, platforms were positioned 3.2 meters apart, and the feeding stations were located 30 meters apart. The tamarins were presented with four feeding stations spaced 50 meters apart. Each feeding station consisted of four feeding platforms located 10 meters north, south, east, and west from a center point. In all experiments, a feeding station is analogous to a food patch, each with its own number of feeding platforms, pattern of food distribution, and rate of resource renewal.

movement and to signal the location of feeding sites (Boinski 1993; Boinski and Campbell 1995; Boinski, chap. 15, this volume).

The field experiments conducted on white-faced capuchins and moustached tamarins were designed to examine the ability of these primates to (1) use spatial information and nearby landmarks as cues to predict the location of baited feeding sites, (2) solve foraging problems using temporal and spatial rules to locate baited (banana: reward) versus sham (plastic banana: no reward) feeding sites, and (3) exploit public and private information to solve problems of food acquisition (Garber and Dolins 1996; Garber and Paciulli 1997).

In these experiments, feeding platforms were constructed in the home range of the study groups (fig. 10.2). Platforms were baited with either sham (plastic) or real bananas according to the particular protocol. In the capuchin study there were two baited feeding stations, one with seven feeding platforms and one with six (a total of thirteen platforms). In the tamarin study there were four feeding stations, each with four feeding platforms (a total of sixteen platforms). A feeding station is analogous to a food patch (e.g., crown of a tree), each with its own number of feeding platforms, pattern of food distribution, and rate of resource renewal.

Experiment 1: Spatial Information. The first set of experiments was designed to examine the ability of capuchins and tamarins to use information in small-scale space to predict the locations of baited (real banana) and sham (plastic banana) feeding platforms. The spatial positions of real and sham feeding sites were constant over the course of several days (place predictable), and then rotated (changed) on day 1 of the next test session (table 10.1). This new spatial arrangement remained constant over the course of several days, and then a second rotation was performed. This experiment offered an opportunity to identify the degree to which spatial information was dominant to visual or olfactory information in selecting a potential feeding site.

The tamarins were extremely attentive to even minor changes in conditions in the vicinity of feeding sites (Garber and Dolins 1996). Typically, the entire group arrived at a feeding station at the same time. As the group passed above the platforms, several individuals could be heard vocalizing, and a single platform became an initial focus of group foraging activity. Often two to four animals would land on the same platform and either feed or visually inspect, touch, and sniff the plastic bananas. When place was constant, group members quickly learned to discriminate between the spatial positions of reward and sham feeding platforms. During a 6-day experimental period (fig. 10.3), 78% (39/50) of all visits were to reward platforms. This is significantly above chance levels (chance level was 50%), and indicates that the tamarins were able to relocate and discriminate among sixteen feeding sites distributed across 200 meters of forest. In six of the eleven cases in which tamarins explored sham platforms, these sites were never visited again.

The tamarins were next presented with a change in the experimental conditions in which the positions of real and sham bananas were rotated, so that platforms that had previously contained real bananas now contained sham bananas. This was done to determine which cues were most salient to the tamarins in selecting feeding

Table 10.1 Summary of field experiments

Experiment	Test	Conditions
1	Associate spatial information with presence/absence of food rewards at multiple feeding sites	Place kept constant Visual cues and olfactory cues available On day 6 the positions of real and plastic bananas were rotated
	Locate real banana versus plastic banana feeding sites using only spatial information (capuchins only)	Place kept constant Bait covered, direct visual cues unavailable Olfactory cues unavailable
2	Use spatial-temporal rules to locate food rewards	Place random in morning and predictable in afternoon Bait covered, direct visual cues unavailable Olfactory cues unavailable

Source: Garber and Dolins 1996; Garber and Paciulli 1997.

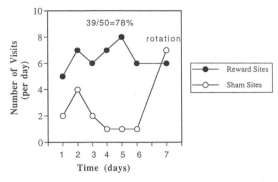

Figure 10.3 Results of field experiments on the use of spatial information by moustached tamarins. Note the decrease in visits to sham sites over time, and the marked increase in incorrect responses on the day of the rotation, indicating that the tamarins were using spatial cues to discriminate between real and sham feeding platforms.

sites. If visual or olfactory differences between real and sham bananas were the primary cues used by the tamarins to locate reward feeding sites, then group members should continue to select platforms with real bananas. If spatial information was the primary cue used to relocate reward sites, then the tamarins would be expected to return to sites that previously contained real bananas. The results (fig. 10.3) clearly indicate that the initial response of the tamarins on the first day of the rotation was to return to platforms that were currently sham sites, but previously had contained food rewards (Win-Return foraging strategy). Thus, when the tamarins were given conflicting information (i.e., platforms that had not previously contained bananas now looked and smelled like platforms that did contain bananas), spatial information was more salient than visual and olfactory information in their selection of feeding sites (Garber and Dolins 1996).

Although this experiment highlights the tamarins' ability to associate spatial information with recent feeding experiences, it does not indicate whether such information must be updated continuously or is retained over longer periods of time. This question was examined by comparing platform choices during morning and afternoon feeding bouts on the same day. For example, if foragers did significantly better in the afternoon trial of day 1 than in the morning trial of day 2, this would suggest that information is relearned or reinforced each morning and not retained overnight from the previous day's activity. In contrast, the ability to select feeding sites with a high and relatively equal level of accuracy in afternoon trails and next-day morning trials would suggest that information is retained at least from one day to the next. Although 75% of morning platform visits were to real banana sites and 90% of afternoon platform visits were to real banana sites, a comparison of morning and afternoon foraging choices indicated no statistically significant differences in accuracy. Thus, the tamarins did retain spatial information regarding the specific location of productive and unproductive feeding sites across days, and used this information in deciding where to forage. Given the cohesive nature of the tamarin foraging group and the fact that several individuals fed on the same platform at the same time, information regarding the location and productivity of individual feeding platforms is probably available to all group members (e.g., through individual experience and observational learning), and is best considered public information.

Compared with the tamarins, the capuchins exhibited a very different set of behaviors when approaching the platforms. Typically a single individual or a small number of individuals, rather than all or most of the group, arrived at the feeding stations. In some cases, the first individual to arrive would sit within 5 meters of a platform and vocalize softly for 10–20 minutes before descending to feed. Other individuals would explore the platforms immediately upon arrival. There appeared to be a hierarchy of access to resources on the platforms. Only one individual was ever observed to be on a platform at a time, and unlike the tamarins, in virtually all cases the capuchins transported the bananas from the platform to the crown of a nearby tree to feed. On several occasions aggressive vocalizations and fighting occurred between individuals who were in possession of bananas and individuals who had not yet fed. This was never observed in the tamarins. In most trials, however, capuchins that came late to the feeding platforms had no opportunity to learn the spatiotemporal schedule of resource availability. Given differences in the times when individual capuchins arrived at the feeding stations and in their access to resources, only certain group members had accurate private information regarding the location and predictability of sham and reward feeding sites.

After only a single exposure to the feeding stations, the capuchins learned to discriminate between reward and sham platforms (Win-Return/Lose-Shift). These data are presented in figure 10.4A. During the 5 days of the experiment, real banana platforms were selected 85.3% of the time (chance levels were 38.4%). If one considers the first day as a period of exploratory foraging and information gathering, then during days 2 through 5, 95.8% (23/24) of platform

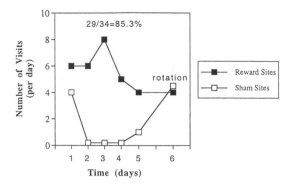

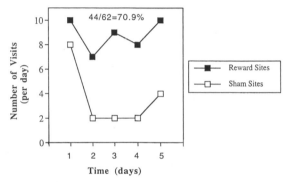

Figure 10.4 Results of field experiments on the use of spatial information by white-faced capuchin monkeys. *(A)* Conditions in which visual and olfactory cues were provided. Note the decrease in visits to sham sites over time and the marked increase in incorrect responses on the day of the rotation. *(B)* Conditions in which only spatial information was available to the capuchins (platforms were covered and banana skins were placed with the plastic bananas).

visits were to banana sites. On day 6, the locations of reward and sham platforms werc rotated. As illustrated in figure 10.4A, on the morning of the rotation, the capuchins typically returned to those locations that had previously contained food rewards, despite the availability of visual and olfactory cues indicating the new locations of reward platforms. In fact, as in the case of the tamarins, the performance of the capuchins on the day of the rotation was less efficient than on day 1 of the initial experiment, when they had no previous spatial information. A comparison between morning and afternoon foraging activities indicated no significant differences in accuracy. This finding supports the contention that individual capuchins rapidly encoded spatial information and retained this information across days (afternoon of day 1 to morning of day 2).

The capuchins (but not the tamarins) were also presented with a second condition in which visual cues were eliminated by covering real and plastic bananas with a large leaf, olfactory cues were minimized by placing banana skins with the plastic bananas, and the location of reward and sham platforms was held constant. That is, the only strategy available to the forager in order to successfully locate a food reward was to use naturally occurring near-to-site landmark cues to remember the spatial positions of real and sham feeding sites. The results are presented in figure 10.4B. Under these conditions, the capuchins were able to select reward sites 70.9% of the time, a value significantly greater than chance. The first day of this experiment was analogous to a rotation day, on which the foragers may have been relying on expectations based on their previous experience rather than on the new conditions of this experiment. If the results of day 1 are eliminated, the capuchins, relying solely on spatial information, visited reward sites 77.3% and sham sites 22.7% of the time. As in the previous experiments, there was no significant difference in accuracy between morning visits and afternoon visits.

Experiment 2: Additional Evidence for Rule-Guided Foraging. A foragers' ability to use spatiotemporal information to locate feeding sites is a critical factor in understanding patterns of individual and group movement. If foragers are unable to track or predict resource renewal rates effectively, then the benefits of following conspecifics as "knowledgeable" guides to feeding sites will be significantly reduced.

In the second experiment, the ability of capuchins and tamarins to apply a set of foraging "rules" to solve problems associated with spatial and temporal changes in resource predictability was examined (see table 10.1). The experiment added a pattern of *resource renewal* to the experimental protocol. During the morning baiting session, platforms containing real and sham bananas were assigned randomly (place not predictable). However, in afternoon baiting sessions, the positions of real and sham bananas were determined by the following rule: if a platform had real bananas in the morning, it would have plastic bananas in the afternoon, whereas if a platform had sham bananas in the morning, it would have real bananas in the afternoon. Banana skins were placed with the plastic bananas to minimize differences in olfactory cues, and any visual differences between plastic and real bananas were eliminated by covering the platforms with large leaves. This experiment presented a set of conditions in which information obtained each morning could be used to predict the location of food rewards in the afternoon, and is analogous to natural situations in which monitoring rates of resource renewal can play an important role in foraging success (e.g., when to return to flowers that have replenished their nectar supply).

During the first several days of this experiment, moustached tamarins were unable to locate the reward platforms above chance or random levels. By day 7, however, patterns of platform visits indicated evidence of the use of a set of spatiotemporal rules; namely, a Win-Shift and Lose-Return foraging strategy for locating feeding sites. During days 7–18, 53.8% of platforms visited in the morning contained real bananas, whereas in the afternoon, 85% of the platforms visited contained real bananas (Garber and Dolins 1996). A comparison of morning and afternoon trials indicates that, whereas the location of reward platforms in the morning did not deviate from chance (location of bananas in the morning was not predictable), performance in the afternoon was significantly above chance levels. These data clearly indicate that the tamarins used information regarding the location of reward and sham platforms each morning to predict the location of reward and sham sites in the afternoon. This information (place, time, food/nonfood) had to be updated daily. In contrast, once the set of behavioral rules required to solve this foraging problem was tried and proved effective, the tamarins applied it over and over again during the remainder of the experiment. In addition, this behavioral pattern, "Win-Shift and Lose-Return," was different than the foraging pattern the tamarins used in experiment 1, which was "Win-Return and Lose-Shift." Given the relatively rapid rate at which they solved this particular spatiotemporal problem, it is likely that a Win-Shift and Lose-Return rule was already part of the tamarins' foraging pattern. Moreover, rather than having to continually invent new and efficient foraging rules to accommodate each new resource condition, these animals probably tested, applied, and reapplied a series of behavioral "hypotheses" that had proved successful in previous foraging contexts.

The capuchins were exposed to an analogous set of experimental conditions (see table 10.1) in which the locations of real and plastic bananas were random in the morning and predictable in the afternoon according to a Win-Shift and Lose-Return foraging rule. Initial examination of the data indicated that over the 15-day experimental period, there was no significant difference in the ability of the capuchins to locate baited sites during morning and during afternoon trials. In both instances, the percentage of correct responses did not deviate from chance levels. Given that capuchins are noteworthy for their large brain size, use of tools in both captive and field settings, and complex problem-solving skills (Fragaszy and Visalberghi 1989; Anderson 1990; Parker and Poti 1990; Visalberghi and Fragaszy 1995), their inability to locate real bananas at levels greater than chance was unexpected. I reanalyzed the data, however, examining only the first three platform visits at each feeding station. This was done because only three platforms at each station contained food rewards, and it became apparent that social factors associated with within-group feeding competition and the dispersed nature of capuchin foraging activities were affecting platform choice. Individuals traveling at the vanguard of the troop consistently arrived at the platforms several minutes earlier than other troop members, and these foragers had access to resource information (presence/absence/location/predictability) not available to other troop members. Although individuals were not marked, based on differences in facial features, body size, pelage, and in the case of adult females, whether or not they were carrying an infant, it appeared that the set of early-arriving animals was relatively consistent from day to day. Animals arriving later typically explored undisturbed platforms that did not contain a food reward.

Reanalyzing the data in this way, it became apparent that by day 4 of this field experiment, the first animals to visit feeding platforms were applying a Win-Shift and Lose-Return foraging rule to locate baited feeding sites in the afternoon. The frequency of correct responses in the afternoon trials went from 44% during days 1–3 to 76% during days 4–15 (50% correct was chance level). In contrast, in the absence of information from which to predict the location of real and sham sites in morning trials, capuchin performance during days 4–5 did not deviate from chance levels. Given that some individuals solved this foraging problem over the course of only 3 days (it took the tamarins 6 days), it is likely that a Win-Shift and Lose-Return foraging rule was already part of the capuchin behavioral repertoire and was simply applied in this new feeding context.

It was also apparent that social dominance played a critical role in the foraging strategies used by subordinate troop members, many of which never applied the correct foraging rule in the experimental setting. There were several cases in which subordinate white-faced

Paul A. Garber

capuchins adopted a behavioral strategy of always going to a particular platform while dominant animals were visiting other platforms. Such a pattern, for example, was common for one female capuchin carrying a very young infant. She consistently traveled directly to the same platform during both morning and afternoon trials. Such a tactic may be analogous to a form of scramble competition enabling her to obtain access to a platform not occupied by other individuals. The pattern used by this female resulted in her obtaining a food reward in approximately 50% of trials. Two other capuchins relied on a similar tactic, and by doing so made it difficult for dominant animals to closely track or monopolize food availability on other platforms. Thus, patterns of group movement and individual decisions regarding which platforms to visit were strongly influenced by social dominance and differences in resource information among group members.

Experimental Field Studies of Brown Capuchins in Large-Scale Space

Brown capuchins *(Cebus apella)* were the subject of some highly innovative field experiments carried out in Argentina by Charlie Janson and his colleagues (Janson 1996; Janson and Di Bitetti 1997). These studies were designed to examine how factors such as group spread, patch size, and patch distribution affect the ability of capuchins to encounter new and unknown feeding sites in large-scale space. In one study, Janson (1996) tested how factors such as the amount of food at a feeding site, distance between feeding sites, and opportunities for a dominant animal to monopolize access to a feeding site influenced group movement and individual feeding success. In this experiment, fifteen platform sites (each site had one or two platforms separated by a distance of between 2 and 50 meters) baited with varying amounts of fruit were positioned throughout the group's home range. Sites with baited platforms were located at least 180 meters apart, so that it was not possible for a forager to sight directly from one set of platform sites to another. The distance between platform sites was considerably greater than group spread (40–50 meters), which meant that members of the group visited only one platform site at a time.

Given that the spatial distribution, number of platforms, and amount of food at baited feeding sites were controlled, there are several conclusions that can be offered regarding brown capuchin foraging abilities. The results indicated that the capuchins consistently traveled to nearest-neighbor feeding sites, and did so using a variety of different travel routes and starting points. This finding suggests that group movement was nonrandom, and that these primates rapidly (over the course of a 2-week learning period) encoded detailed knowledge of the relative spatial locations of the feeding sites in large-scale space (an area totaling tens of hectares). In addition, these primates were able to retain information regarding differences in the amount of food at these fifteen feeding sites, and to use temporal information in regulating return times to previously visited sites. In terms of group movement patterns, platforms were rarely revisited in the same day, platforms with more food were selected over platforms with less food, and nearer platform sites were selected over more distant platform sites. Janson (1996, 320) reports that the foraging strategy of these brown capuchins was characterized by a preference for "closer platforms even when they are less rewarding [i.e. have less fruit] per unit distance traveled."

There was also evidence that, as in the case of white-faced capuchins, social factors had a profound influence on individual foraging success. Janson reports that dominant animals were able to monopolize access to feeding sites, especially when adjacent platforms were less than 10 meters apart. The quantity of food on a platform had a significant influence on rates of aggression as well. Levels of within-group feeding competition were greatest at platforms containing an intermediate amount of fruit (10–16 fruits).

In a second set of field experiments, Janson and Di Bitetti (1997) examined how factors such as group spread, patch size, and speed of travel affected the ability of a brown capuchin foraging group to detect previously unknown feeding sites. The results indicated that as the speed of travel increased, the likelihood that any member of a brown capuchin group would encounter a feeding platform decreased. When traveling slowly, the capuchins detected feeding platforms from an average distance of approximately 30 m. When traveling rapidly, however, the sighting distance to baited platforms was only 9 m. Moreover, by foraging in a wide front of some 40–50 meters, groups of foragers were almost twice as likely to encounter an unknown feeding site as a solitary forager. These data suggest that one advantage of group foraging and group spread is an increased "visual field," and therefore an increased likelihood of encountering new and widely scattered feeding sites. In areas of their range in which group members have limited spatial information about the location of productive feeding sites, increased group spread may serve an important function in resource detection. Patterns of group spread may also play a role in predator detection (see Boinski, Treves, and Chapman, chap. 3, this volume).

What Do Experimental Field Studies Tell Us about How Primates Use Spatial Information?

The results of these field experiments indicate that both moustached tamarins and white-faced capuchins relied primarily on spatial information (certainly in the form of a topological map and possibly in the form of a Euclidean map of small-scale space) in locating and selecting baited feeding sites. Despite the fact that tamarins are characterized by small brain size and limited cerebral complexity or fissurization (Hershkovitz 1977), while capuchins exhibit the "fullest expression of platyrrhine cerebral complexity" (Hershkovitz, 1977, 364) and the largest relative brain size of any New World primate, both species learned the spatial position of reward and sham feeding sites after only a single day's exposure to the test conditions. This is analogous to what has been described as "one-trial learning" (Menzel and Juno 1982). The ability to rapidly learn landmark cues and associate them with the position of a feeding site is probably widespread among primate and nonprimate species (S. A. Altmann 1974; Boesch and Boesch 1984; Robinson 1986; Garber 1988b, 1989; Balda and Kamil 1989; Biebach, Gordijn, and Krebs 1989; Chapman, Chapman, and McLaughlan, 1989; Jacobs et al. 1990; Vander Wall 1990; Henzi, Byrne, and Whitten 1992).

Capuchins and tamarins differed little in their ability to associate spatial information and temporal information to predict the availability of food rewards in small-scale space, although it did appear that capuchins attended to landmark cues directly associated with reward platforms more readily than did tamarins (Garber and Dolins 1996; Garber and Paciulli 1997). The ability of capuchins to associate a nearby landmark cue (visual information: a yellow block placed on platforms only when they contained a food reward; see Garber and Paciulli 1997 for details) with an increased likelihood of encountering a feeding site may relate to differences in the types of insect prey exploited by each species. White-faced capuchins typically forage in a wide front when searching for prey that are concealed and embedded within a substrate, such as dead wood, insect nests, tree holes, and bromeliads. These capuchins do not, however, explore every tree hole, insect nest, piece of dead wood, or bromeliad. Rather, nearby cues associated with these microhabitats probably serve as a foraging guide indicating the potential presence of prey. Foragers are often located 20–30 meters from their nearest conspecific, and thus areas of successful and unsuccessful foraging attempts are largely private information. In contrast, moustached tamarins travel and forage as a highly cohesive unit, and often several group members scan the same tree for exposed prey on the leaves or branches. In this case, the direct visual sighting of the prey itself is the guide or cue that influences foraging success.

The white-faced capuchins also were found to use public information in both selecting and avoiding feeding sites. Among the tamarins, social information appeared to be used principally as an attractant that permitted all or most group members to feed on the same platform at the same time. Similarly, Janson's research indicates that even in large-scale space, brown capuchins are able to integrate spatial, temporal, quantity, and social information to direct group movement and select feeding sites.

These experimental field studies provide empirical evidence that tamarins and capuchins retain and use landmarks and other forms of complex spatial and computational information during traveling and foraging. Whether such information is more likely to be represented as a route-based map, a coordinate-based map, or some set of multiple spatial strategies is discussed in the next section.

A Model of Spatial Mapping

Poucet (1993) has presented an innovative model of how animals might encode and represent information in small-scale and large-scale space that is consistent with recent ideas concerning how the mammalian brain organizes and stores information. In this model (fig. 10.5), the spatial positions of particular places or sites in the environment are represented using a combination of Euclidean (coordinate) and topological (route-based) networks. This model is described as a multiple-point reference system in which the coordinates of a set of "special" places are used to determine the relative distance and direction to other points in the environment. Poucet calls these special places "location-dependent representations." Each of these is an independent frame of reference used by the animal to navigate and orient within a local area (small-scale space), rather than part of "a global system of polar coordinates" (Poucet, 1993, 170). The distribution of location-dependent representations in real space results from the recurrent use of specific

Paul A. Garber

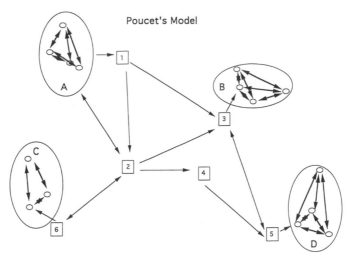

Figure 10.5 Poucet's model of spatial representation Areas A, B, C, and D represent location-dependent areas, in which animals may represent spatial information in terms of a coordinate system (Euclidean map). This is indicated by the darker lines with arrows at both ends. Small circles within each location-dependent area represent feeding sites. Boxes 1 through 6 represent landmarks or orientation cues used by the forager in large-scale space to travel between location-dependent areas. Arrows to and from these boxes indicate travel routes encoded by the forager and represented as a topological map. In moving between area A and area D, for example, the forager could take several routes, including 1→3→5, 1→2→4→5, 1→2→3→5, or 2→4→5. (Adapted from Poucet 1993.)

areas by the forager (e.g., nest site, feeding trees, water hole), its ability to integrate and remember different "views" of the same area, and its ability to associate salient features of the area, such as presence of feeding sites, topography, and landmarks, with these views. Over time, the forager might build up a mental representation of many such areas. Within each location, animals have relatively accurate knowledge in small-scale space from which to compute efficient routes of travel. Within large-scale space, however, the animal may not represent true distances and directions from one location-dependent area to another, because increasing distance and decreasing visibility, the forager is unable to obtain views of single and multiple landmarks from different points in the forest.

Based on this model, Poucet (1993) argued that many types of animals probably travel between location-dependent areas using topological information. Such information involves knowledge of the general direction from, and proximity of, one location-dependent area to another. This knowledge is acquired through processes of exploration and path integration that are built up over time, resulting in the construction of a learned network of fixed points, travel routes, landmarks, and interconnecting locomotor sequences (see Dyer, 1996 for a discussion of similar ideas in bees). These travel routes can link one location-dependent area to another. In such a network, there may exist multiple routes of travel between particular locations, and therefore animals possessing enhanced computational skills can select the shorter of two fixed routes to reach the target, or combine segments of two different routes to reach the target. According to Poucet, the number of routes an animal may utilize, and the processes by which route segments are selected or combined, relate directly to both neuroanatomy and the set of spatial rules developed by the forager to exploit its particular set of resources. By representing space in this way, a forager is expected to reuse particular path segments and remembered visual landmarks to travel between goals. However, the forager is unable to navigate using a global coordinate system, and therefore cannot encode the accurate geometric (actual distance and direction) information from one location-dependent area to another, nor can the forager compute novel straight-line travel routes between goals in large-scale space for which landmark, visual, auditory, or olfactory cues are presently unavailable (Poucet 1993; Benhamou 1996).

Using such a representation, the forager can rely on a variety of spatial strategies to travel effectively between sets of feeding trees and other goals in large-scale space (Poucet 1993). For example, starting at location-dependent area A in figure 10.5, the forager could reach goals B, C, or D using a number of different routes. Several of these would appear to be efficient in terms of limiting distance traveled and recrossing paths, but none would require global coordinate spatial knowledge. In the case of traveling from area A to area B, the forager would orient toward landmark 1 and then either landmark 2 or 3 in large-scale space in order to eventually encounter recognized local views of area B.

Travel routes similar to those described in Poucet's model are consistent with movement patterns that are associated with "multiple central place" foraging. Several species of primates (e.g., yellow baboons, pig-tailed macaques, yellow-handed titi monkeys, black spider monkeys, cotton-top tamarins) and nonprimates are reported to reduce the distance they travel to exploit feeding sites by selecting from a set of known sleeping trees the one that is nearest to their present feeding site (Chapman, Chapman, and McLaughlan 1989). Such an ability requires general knowledge of the location of many sleeping sites within a group's home range and of the overall spatial relationships between feeding and sleeping sites. These species use between ten and twenty sleeping trees, which may be represented by the forager as location-dependent areas. Movement between sleeping and feeding sites tends to be highly predictable and repetitive, which may suggest that centrally located sites or sets of landmarks are serving as orientation or connecting points between nonadjacent location-dependent areas.

In many instances, however, it is difficult to distinguish movement patterns consistent with Poucet's model from movement patterns predicted by other forms of spatial representation—in particular, Euclidean maps. Poucet's model predicts the use of Euclidean spatial information in small-scale space and topological information in large-scale space. Use of Euclidean spatial information in large-scale space would require an ability to integrate different local views of the same area with local views of other, out-of-sight areas. For arboreal rainforest primates, the structural complexity and denseness of the rainforest canopy causes even nearby local views of the same feeding site to differ markedly in the number, types, presence, or absence of distinguishing landmarks. This situation is very different from the set of local views a primate foraging in an open landscape would encounter. A terrestrial primate could enter a savanna grassland or woodland from the north, south, east, or west and potentially use different views of the same landmarks to determine its spatial location. An arboreal primate, however, would have to build up a much larger number of local views of each location-dependent area, many of which would have few or no landmarks in common. Poucet's model predicts that a forager capable of forming a Euclidean map in small-scale space could, over time, enter a location-dependent area from any number of different directions and travel directly to target feeding sites. In large-scale space, however, the model predicts route-based travel from one location-dependent area to another. It assumes that the computational difficulties of geometrically integrating local views of distant areas that have no or few landmarks in common preclude the use of a Euclidean map over large-scale space. In this case, the forager is expected to orient to a particular landmark or set of landmarks that serves to link areas of frequent use, and to repeatedly (but not exclusively) travel the same set of paths to reach the same goals. Although over time the forager may build a network of multiple routes to reach the same feeding site, these routes are expected to share common crossing points.

In the next section I evaluate the degree to which Poucet's model is consistent with the ranging patterns of two species of arboreal primates, moustached and saddleback tamarins. Tamarins are not central-place foragers (Garber 1993a, b). These primates rarely sleep in the same trees on consecutive nights, rarely visit the same set of feeding trees on consecutive days, infrequently backtrack or recross their paths, commonly take direct routes to distant feeding sites, and use many different routes of travel to reach the same target feeding tree (Garber 1988b, 1989). Previous studies have suggested that tamarins maintain detailed knowledge of the actual distance and direction between hundreds of feeding trees within a 40-hectare home range and use Euclidean spatial information to compute least-distance travel routes to reach feeding sites (Garber 1989; Garber and Hannon 1993).

Comparing Poucet's Model Directly with Data on Patterns of Group Ranging in Wild Tamarins

I analyzed ranging patterns in wild moustached and saddleback tamarins during twenty-seven full-day follows in an attempt to identify evidence of route-based travel, orientation to particular landmarks, and the use of location-dependent areas. A mixed-species troop (see Cords, chap. 4, this volume) of moustached and saddleback tamarins was tracked continuously from dawn till dusk, and its location on a 50 m by 50 m grid map was plotted every 2 minutes throughout the day. Thus, it was possible to examine patterns of quadrat use, distribution of feeding trees, travel sequences, and sequential changes in travel direction.

Tamarins traverse much of their home range each day. The troop moved through an area of approximately 8 hectares daily, visiting thirty-six of the 50 m by 50 m quadrats in their 40-hectare home range. Most of these quadrats (90%) were crossed only once each day. On average, only 28% of quadrats fed in or traveled in on one day were also visited on the next day. In order to identify movement patterns consistent with either Euclidean *(coordinate-based)* travel (e.g., ability to enter feeding sites from a large number of different directions and previous locations) or topological *(route-based)*

Paul A. Garber

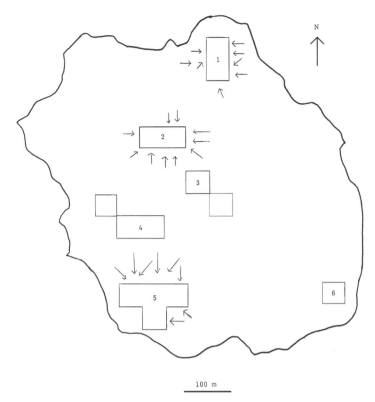

Figure 10.6 Diagram of the home range of a mixed-species troop of moustached and saddleback tamarins. Major feeding areas are numbered 1 through 6, and may be analogous to Poucet's location-dependent areas. These feeding areas exhibit a highly scattered distribution in the troop's home range. Arrows indicate the various directions from which tamarins traveled to enter these feeding areas.

travel (e.g., reuse of the same travel paths to visit a particular feeding site and evidence of intersection points from which foragers reorient their direction of travel), I examined the movement sequences used by the tamarins to reach quadrats containing feeding sites. During the 3-month period of this study, major feeding trees (whose fruits constituted the bulk of the tamarin diet) were located in six widely scattered localities within the troop's home range. The sizes of these localities ranged from one-fourth to one hectare (fig. 10.6). In total, these major feeding trees were visited on eighty-nine occasions, and the area represented by these six localities accounted for 3.5 ha, or 8% of the home range. Each of these six localities may be analogous to Poucet's location-dependent areas, or sites from which the tamarins could build up a multiple-point reference system and integrate different views of the same feeding trees.

An examination of tamarin ranging patterns indicated that once the troop was within 50–100 meters of a target quadrat (a quadrat containing feeding trees), the tamarins took direct routes and entered these trees from any number of different directions and quadrats (fig. 10.6). For example, there were eleven different quadrats that surrounded feeding area 1, and over the course of thirteen visits, seven of these were entered immediately prior to visiting the target trees. Similarly, there were sixteen different quadrats that surrounded feeding area 2, and over the course of twenty-three visits, twelve of these were entered immediately prior to visiting the target trees. Within small-scale space, the tamarins utilized a variety of direct, but different, travel routes to reach the same feeding sites. Given the evidence from the experimental field studies described earlier, indicating that olfactory cues do not play a role in tamarin navigation (Garber and Dolins 1996), these movement patterns suggest that the tamarins maintained and integrated a large number of local views of each of the six localities, and once in the general vicinity of the target trees, probably relied on Euclidean-based information to reach their goal. This conclusion is supported by experimental work on rodents (Collett, Cartwright, and Smith 1986) and primates (Dolins 1993) indicating the ability to encode the spatial relationships between cues or landmarks in small-scale space and to use this geometric information to locate additional points in the environment.

What about movement in large-scale space? How did the tamarins use spatial information to travel between the six feeding localities? Poucet's model would predict that travel routes would be internally represented as topological information and would involve the reuse of a limited number of pathways, or a series of straight-line segments between landmarks (nodes or intersection points). These landmarks would function as orientation cues.

Tamarin troop movement patterns in large-scale space did fit well with some of the expectations associated with a topological spatial representation. These primates did not employ a spatial strategy of reusing the same set of travel pathway or routes in going from one location-dependent area to another. What was observed, however, was a pattern of ranging in which the troop typically traveled in a relatively straight line over a distance of 150–350 m when moving

Paul A. Garber

from one location-dependent area to another. Approximately three such sequences were recorded every day, and over two-thirds of these were associated with turns of 90° or greater at the beginning or at the end of the travel sequence. Most of these changes in direction occurred and reoccurred in only 18 of the 144 quadrats traveled in by the tamarins. These were designated landmark quadrats because they may have contained salient landmark cues used by the troop for orientation in large-scale space. The ratio of 90° turns to straight-line travel in landmark quadrats was 1.2:1. The ratio of 90° turns to straight-line travel in all other quadrats was 0.3:1. Thus, the tamarins were four times more likely to reorient their path in a landmark quadrat than in a travel quadrat. The exact routes used by the tamarins during straight-line travel sequences were rarely repeated. Therefore, if these routes are represented as topological spatial information, the tamarins must also encode a large number of prominent landmarks throughout their range. These landmark cues serve as beacons, and each must be recognized from a number of different views.

Navigating by straight-line segments is not inconsistent with the use of Euclidean or global spatial information, and therefore it remains plausible that tamarins rely on coordinate-based maps in both small-scale and large-scale space. However, it is also possible that members of the tamarin troop used less exact and more generalized spatial information concerning the location of orientation cues and a direction of travel to move between major feeding localities. In this case, the tamarins may adopt a navigation strategy in which they have knowledge of the general direction of targeted feeding sites, and travel in a relatively straight line in that direction until encountering a known landmark or feeding locality. Such a pattern is consistent with Poucet's (1993) model of animal navigation in large-scale space. At present, it remains unclear (1) what specific landmarks tamarins and other primates might use to orient in large-scale space, (2) how many landmarks are involved, (3) whether Euclidean information is used in small-scale space, and if so, what sets the upper limits on the size of location-dependent areas, and (4) whether large-scale space is more commonly represented by topological maps, Euclidean maps, or some other form of spatial map. Future studies of group movement patterns in lemurs, monkeys, and apes must be designed to distinguish between the use of topological and Euclidean representations of spatial information in primates.

Conclusions

Ecological approaches to the study of learning provide important insights into the evolution of sensory capabilities and foraging behavior (Kamil 1994; Real 1994). In the wild, primates and other animals face the challenge of locating feeding sites distributed across broad spatial and temporal boundaries. *Where* the group travels, *when* it visits particular sites, and *how many* individuals travel together is determined, in part, by the ability of individual group members to store, categorize, and associate disparate types of social and resource information. By controlling the information available to the forager, experimental field studies, in conjunction with more traditional field studies, offer a highly productive approach to examining questions of primate cognition, information processing, and the role of the social environment in feeding ecology. Janson's (1996) and Janson and Di Bitetti's (1997) experimental field studies, for example, have identified a hierarchy of spatial and sensory cues used by capuchins to select feeding sites, as well as the benefits of group foraging in locating new feeding sites. Moreover, these and other experimental field studies (Garber and Dolins 1996; Garber and Paciulli 1997) indicate that many species of primates rely on social information regarding the location, identity, and individual behavior of nearby conspecifics to increase foraging success. In this regard, individual and group strategies for finding food must also include strategies for maintaining access to food (Janson 1990a).

An understanding of how primates use spatial and temporal information to generate foraging rules is also central to an understanding of the ecology of group movement. It is likely that, in species characterized by the ability to track and predict the future resource schedule or phenological pattern of particular trees, "naive" animals can substantially increase their foraging success by following "knowledgeable" conspecifics. Although this needs to be verified empirically, experimental field studies can be designed to create "knowledgeable" and "naive" foragers and use differences in individual information to examine its effect on patterns of group cohesion and individual movement. For example, a study by Menzel (1979a) on captive chimpanzees inhabiting a 1-acre enclosure indicated that naive foragers quickly learned the locations of hidden food resources by observing the movement patterns and postures of a knowledgeable forager.

There remain several challenges to future field studies on the ecology of group movement in primates. Researchers need to design

Paul A. Garber

studies that distinguish the ways in which spatial information is represented (e.g., Euclidean spatial map vs. topological spatial map) by foragers in both large-scale and small-scale space. In part, this could be accomplished by controlling the olfactory and visual cues available to the forager and distributing feeding sites in selected configurations across a group's home range. These configurations (e.g., a uniform distribution of platforms vs. a clumped distribution of platforms) might indicate whether foragers group feeding sites into larger spatial units or represent each site as a discrete entity. In addition, just as we found that white-faced capuchins attended to near-to-site landmarks (yellow blocks) to increase their ability to locate baited feeding sites (Garber and Paciulli 1997), artificially placed and manipulated distant landmarks also could be used to determine how primate foragers represent spatial information. Animals that represent spatial information as an internal Euclidean map would be expected to encode the actual metric relationships of landmarks to one another and to use this information to compute direct novel routes to feeding sites that lie outside their present field of view. Landmark manipulation experiments (see Dyer, chap. 6, this volume) would serve to identify species-specific differences in the ability to remember and associate different local views of the same object in large-scale space. Such an ability is a prerequisite for constructing Euclidean maps.

A second set of questions that needs to be addressed is the degree to which primates learn resource and dietary information by observing others rather than by trial-and-error processes (Visalberghi and De Lillo 1995). Recently, claims for the importance of observational learning as a factor in the spread of complex behaviors within a primate group have been challenged (Visalberghi and Fragaszy 1990; Visalberghi and De Lillo 1995). If observational learning by both immatures and matures is an important aspect of primate foraging strategies, then attention to the feeding activities of others (public information) may help explain patterns of group cohesion, ranging, and individual food choice. Experimental field studies on nonhuman primates provide a critically important research tool that can be used to examine these and related issues in the ecology of group movement.

Acknowledgments
I wish to thank Dr. Francine Dolins and Lisa Paciulli for their assistance and collaboration in collecting the experimental field data presented in this chapter. Insightful comments on earlier drafts of

this manuscript were provided by Sue Boinski, Fred Dyer, Jennifer A. Rehg, Anthea Yannopolous, and several anonymous reviewers. As always, I wish to thank Sara Garber and Jenni Garber for teaching me new rules and search strategies for locating their toys by randomly and systematically concealing them in both large-scale and small-scale space.

Paul A. Garber

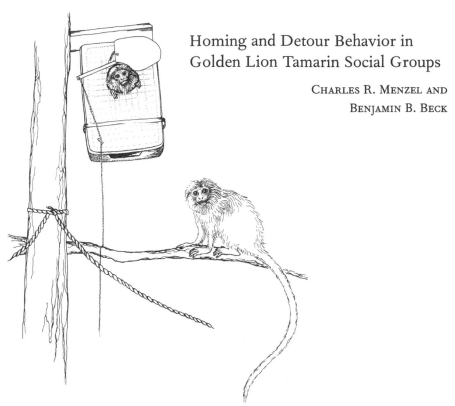

CHAPTER ELEVEN

Homing and Detour Behavior in
Golden Lion Tamarin Social Groups

CHARLES R. MENZEL AND
BENJAMIN B. BECK

Most primates live and travel as members of established social groups and must adjust their movements continually in relation to one another and the environment. For many species, group travel occurs in the forest canopy, and animals are confronted with thickets, gaps, and tree branches that do not simply point directly toward their desired goals (Cannon and Leighton 1994; McGraw 1996). To get to particular locations in their home range, animals must often make arboreal jumps or use indirect routes. Movement in the canopy thus presents group-living primates with frequent choices regarding where to go and whom to follow. Group members sometimes behave as if they were in disagreement over which direction to take (e.g., hamadryas baboons, *Papio cynocephalus hamadryas:* Kummer 1968a, 1995; Common chimpanzees, *Pan troglodytes:* Menzel 1974; golden lion tamarins, *Leontopithecus rosalia:* Boinski, chap. 15, this volume). Typically, however, conflicts over travel direction are mild, cohesiveness is maintained, and animals move together quickly over long distances to food, water, and shelter (Menzel 1973b, 194; Kummer 1968a, 1995; Garber 1989). Social cohesiveness during feeding and traveling is obvious in species of Neotropical primates that live in small, close-knit social groups, such as titi monkeys *(Callicebus)*, marmosets *(Callithrix)*, and tamarins (*Saguinus* and *Leontopithecus*).

Questions arising about coordinated movement include the following: How cohesive are the animals, and under what conditions does cohesiveness decrease or become greater? How do factors such as the spatial distribution of food or the presence of novel gaps in the forest canopy influence cohesiveness? Which animals control the group? What signaling systems and other variables bring about coordinated movement? What methods can be used to study social coordination and group movement simultaneously?

To study how the members of a social group select a travel direction, it is useful to have a basis for predicting where the group will go next. One method for manipulating animals' spatial priorities and for studying group movement and coordination is to examine where animals go after an experimentally imposed displacement of the group from its home range. Displacement of a social group can induce homing attempts by some or all group members. This method provides an opportunity to study how animals interact with one another and with environmental structures during travel. Experimental displacements have been used to simulate the round-trip excursions that animals make spontaneously outside their home range, and to examine how well animals can find their way home over distances that are much longer than a natural excursion. The difference between an imposed displacement and a natural excursion is that the experimenter controls "the features of the outward journey, e.g., its length, its sinuosity, or the type of sensory information available to the animals" (Bovet 1992, 334; see also R. R. Baker 1982).

Most experimental field studies of "homing" in mammals focus on single animals rather than social groups, and because of the methods used, little can be said about the exact paths of travel that the animals use to reach a goal, or about the details of their decision-making processes (Tattersall and Sussman 1985; Bovet 1992). Bovet states that, in comparison with work on invertebrates, fishes, amphibians, reptiles, and birds.

C. R. Menzel and B. B. Beck

The study of homing behaviour in mammals is marked by a steady reference to the concept of *home range* . . . i.e., 'the area over which an animal normally travels in pursuit of its routine activities' . . . during a stated period of time. . . . The evidence available for many mammals is that the smallest area in which they spend, say, 90% of their time is one or two orders of magnitude smaller than the surrounding area where they spend the remaining 10%. It is thus reasonable to postulate that, as a result of this difference in 'utilization density,' a mammal has much more familiarity with its small home range than with any part of the huge surrounding area where it travels occasionally; and that the strategies and mechanisms it uses to find its way within its home range are not necessarily the ones it can use outside. In the mammalian literature, homing usually refers to travelling back to one's home range. (Bovet 1992, 321).

When detailed records are available on the path of travel that an individual squirrel, fox, bear, or polecat *(Mustela)* uses during natural excursions outside the home range, the outbound and return trips are both typically fairly straight, the return trips are not simply backtracks of the outbound path, and the animal does not usually simply follow topographical features or corridors (Bovet 1992, fig. 8.4). General field observations on social groups of primates (e.g., Garber 1989; Sigg and Stolba 1981; Kummer 1995) suggest that they too can return in fairly straight lines to distant food, water, and sleeping sites without simply retracing their route. To date, however, there is little systematic information available on how entire social groups of primate or nonprimate mammals navigate in unfamiliar areas.

The goal of this chapter is to provide a description and locational analysis of homing and detour behavior by social groups of golden lion tamarins. A locational approach to group movement attempts to describe events in spatiotemporal terms. It takes as its analytical "unit" the simultaneous position of each animal and selected environmental features at particular moments in time. A goal of analysis is to control or predict in detail how the configuration of individuals within a social group will change next, and to determine the most appropriate spatial frame of reference for describing and explaining the animals' movements, including the pathways, landmarks, and other objects the animals use to reach goals (Carpenter 1964; Menzel 1971b, 1974, 1979b, 1987). Our interest, in part, was in travel competence in relation to the conservation task of reintroducing captive-born tamarins to their natural habitat (Kleiman et al. 1986; Price 1992; Beck 1995).

In this chapter, we describe how the members of an established family group of captive-born golden lion tamarins move with respect to one another and the environment following experimental displacements of the group from its home range. We address the following questions. (Menzel 1971a, 1974; Gibbons and Menzel 1980; Norton 1986; Lamprecht 1991; Boinski et al. 1994; Kummer 1995):

1. How will the translocated group move with respect to its home range?
2. How will the animals move when confronted with arboreal structures and gaps?
3. How far will the animals range from one another?
4. Do particular animals control the movements of the group?
5. How is this control achieved?

Naturalistic Background

In their natural habitat, golden lion tamarins typically live and travel in "extended family" groups. Groups consist of a reproductive pair and one or more sets of twin offspring; 40% of groups contain a second nonnatal adult male (Baker, Dietz, and Kleiman 1993), and approximately 10% of groups contain a second breeding female (Dietz and Baker 1993). Group size ranges from two to eleven animals, with a mean size of 5.4 tamarins (Dietz and Baker 1993). Groups travel and carry out their daily activities within an area of 41 ± 21 SD hectares (Dietz and Baker 1993). One social group for which ranging data were obtained traveled $1,480 \pm 322$ SD meters per day (Peres, chap. 5, this volume). Groups encounter one another more often than predicted on the assumption of chance movement. In the early morning, groups are beyond visual contact. They exchange long calls, and animals move rapidly to the range boundaries. Once neighboring groups approach to within about 25 meters of each other, animals engage in arch-walk displays, chases, and brief fights (Peres 1989). In the social group that Peres (chap. 5, this volume) studied, the locations of encounter zones along home range boundaries were stable across months, the number of encounter zones ($N = 7$) corresponded to the number of neighboring groups ($N = 6$ to 7), and the group spent 10% of its waking hours within 40 m of neighboring groups. The members of a golden lion tamarin group remain relatively close together during the day. A traveling group often has a maximum diameter of dispersion of less than 20 m (Boinski et al. 1994). As nightfall approaches,

C. R. Menzel and B. B. Beck

the animals terminate their foraging and make a sudden move to a distant sleeping site, usually a tree hole or some other shelter (Coimbra-Filho 1977) located toward the center of the home range (Peres, chap. 5, this volume).

Behavior of Reintroduced Golden Lion Tamarins

The immediate background for the present research was the observation by the second author and his colleagues (Beck et al. 1991) that groups of captive-born, recently reintroduced golden lion tamarins in the Poco das Antas Biological Reserve, Brazil, were deficient in their travel patterns compared with wild tamarins. Hungry animals sometimes failed to cross from one tree into an adjacent tree to obtain visible provisioned food. An animal might run repetitively up and down the tree trunk or back and forth on a branch for 15 minutes, then descend and run across the ground to the adjacent tree. The day range length of recently introduced captive-born tamarins was much shorter than that of wild-born tamarins living in similar forest. Finally, captive-born animals sometimes failed to return to their shelter (a hutch box) at dusk, after traveling away from the shelter during the day. In these cases, the animals moved about as if seeking a route and showed signs of agitation as nightfall approached, but they did not make an organized long-distance movement in the homeward direction. Frequently they moved quickly in one direction for up to 200 meters, stopped, and ran in another direction. Individuals often separated, and sometimes went in different directions. Failure to return to the shelter could result in failure to locate provisioned food and in exposure to rain, cold, and nocturnal predators, and was a source of mortality in reintroduced tamarins.

Were the reintroduced tamarins unable to discriminate the homeward direction after making an outward trip? This question is difficult to answer without a detailed record of how the animals moved in space. A detailed record of the travel paths of animals can clarify the kinds of information they possess about the position of resources (Garber 1988b, 1989) and their position in their home range. In a model experimental field study of homing in an arboreal mammal, Bovet (1984) displaced red squirrels *(Tamiasciurus hudsonicus)* individually up to 2,220 m from their nest or home range center. This distance was about four to sixteen times the diameter of a red squirrel home range, assuming the home range to be of mean size (1.44 ha) and a compact shape. The squirrels' initial straight movement away from the release site was oriented in the general direction of the home site. After moving a mean distance of 242 m, the squirrels returned to the release site area and made additional forays centered on the initial direction. Forays were approximately the length of the exploratory trips that red squirrels occasionally make away from their home range (100–400 m). Bovet suggested that the squirrels had some information about the direction of their home range, but not its distance; that during the first several hours the squirrels were reluctant to venture more than a critical distance in any one direction from their new starting point; and that this critical distance depended on previous ranging experience. Bovet suggested that the squirrels' behavior constitutes a conservative strategy that, under normal conditions, would guard against an animal becoming widely separated from its home range following a spontaneous outbound trip.

Experimental Methods
We used the direction in which experimentally translocated golden lion tamarins moved to assess whether they possessed information about the direction of their home range. We also presented the animals with novel canopy gaps along the homeward route to determine how they worked as a group to find roundabout routes (Köhler 1925; Menzel 1978; Fragaszy 1980; Menzel 1986; Chapuis 1987).

Subjects of the study were members of an established group of captive-born *Leontopithecus rosalia*. The social composition of the group was similar to that of wild groups. It included a male (4.52 yr) and female (6.26 yr) adult pair and their four offspring: a young adult female (2.39 yr), an adolescent/young adult male (1.51 yr), and juvenile male twins (0.58 yr). Prior to the study, the animals had lived in relatively small cages at the Riverbanks Zoo, Columbia, South Carolina. We released the group into a patch of mature oak-beech forest at the U.S. National Zoological Park 102 days before the present experiment began.

The experimental area is shown in figure 11.1. Manila ropes were hung between trees throughout the length of the forest, at the height at which wild golden lion tamarins normally travel (3–5 m). The ropes provided the animals with a physically continuous pathway that did not require them to make jumps. Branch networks and a modest understory also provided the animals with substrates for travel. All the monkeys could move quickly on the ropes and

C. R. Menzel and B. B. Beck

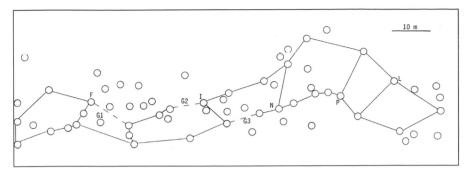

Figure 11.1 Experimental area. The circles represent trees. L, lodge tree. Release sites: P, perimeter; N, near; I, intermediate; F, far. Solid lines connecting trees represent ropes in phase 2. Dotted lines connecting trees represent ropes present in phase 1 but absent in phase 2; G1, G2, and G3 represent gaps 1, 2, and 3, respectively. An additional rope (not shown) extended beyond the left (east) side of the area shown.

seemed to be well habituated to the forest and to the test procedures. Prior to and during this study, the animals had 24-hour access to the forest, aside from a brief period of confinement prior to the start of experimental trials. In the 102 days of continuous, all-day observation prior to this experiment, the animals had traveled within an area that measured just 30 m in diameter. No tamarin had ventured outside of this area. The animals showed caution and gave vocalizations at the edge of the area. Thus, for the present experiment, we operationally designated the 30 m diameter area as the animals' home site. No other zoo animals were housed within this home site. At night the animals slept together in a hutch box, a modified ice cooler, which was located in a tree at the center of the home site ("lodge tree"). The animals had never been tested on homing tasks of the sort described next.

The test entailed confining the group to its overnight hutch box before sunrise, removing the box from the traditional lodge tree, moving the box to a new tree, and releasing the animals after sunrise. A second, physically identical hutch box, with which the tamarins had previously had an equal amount of physical contact, was installed at the original location, with the door open to preserve the exact appearance of the home site. The hutch box used for transport was left in place at the release site following the release, with the door open, to minimize the degree of disturbance to the animals and the risk that they would "panic," drop to the ground, and scatter into the underbrush at the start of the trial.

To present the group with different types of visual separations from its home site, the distance between the release site and the home site was varied. Four release sites were used (see fig. 11.1). The "perimeter" release site was located at the edge of the 30 m diameter home site, 14 m from the lodge tree. From this release site, at tamarin level, an animal could easily see the lodge tree and most of the home site. The animals had previously traveled across the rope segments that were suspended between the release site and the lodge tree. The "near" release site was 30 m from the lodge tree. From this release site, the lodge tree and other parts of the home site were easily visible to a human. However, unlike the situation at the perimeter release site, the monkeys had not previously traveled on any of the ropes that extended away from the release site, nor had they previously viewed their home area from this perspective. Thus, to reach the visible home site, the animals had to use new paths and travel through new surroundings. From the 48 m "intermediate" and 77 m "far" release sites, the animals almost certainly could not see any part of the home site directly, due to intervening foliage. From the far release site, animals also apparently could not see the area that lay within about 30 m of the home site, and they possibly could not see any of the other three release sites directly.

The experiment consisted of two phases, both of which addressed the five questions listed earlier. In phase 1, the displacement distance and compass orientation (N versus S) of the hutch box at the time of release were varied systematically. Compass orientation was varied to assess whether the animals' familiar, traditional visual perspective on the sun and surroundings (N) substantially aided homing, or whether the animals could also orient in the homeward direction following a 180° change in visual perspective(s). The animals were released together eight times from the perimeter location and six times each from the near, intermediate, and far locations, for a total of twenty-six trials. All segments of the ropeway were present in phase 1.

Phase 2 evaluated intertrial improvements in homing speed from a specific release site. The animals were released seven times from the far location. The hutch box always faced north. The animals' task of finding a route to the home site was made more difficult than in phase 1. A rope segment up to 11.3 m long was absent in each of three locations between the release site and the home site. At these three locations, the animals had to use an alternative, roundabout rope, find a route through tree branches, or descend

C. R. Menzel and B. B. Beck

from the trees and cross the ground. Precisely the same layout of ropes was presented on each of the seven trials in phase 2.

Precautions were taken to ensure that the animals could not see the surroundings during transport. We rotated the box gently about its vertical axis three full turns in one direction, at an average speed of about 90° per second, before we carried it away from the home area. On each trial, we carried the hutch box a standard distance of 80 m. The animals were released at about 8:00 A.M., somewhat more than 1 to 1.5 hours after their normal time of arising. The animals were hungry at the time of release.

The animals' locations were plotted at 1-minute intervals on 1:329 scale maps by the first author and three trained observers. The animals' paths of travel during each interval were also plotted. Animals were mapped at 1-minute intervals until at least 20 minutes after they had arrived in their familiar area. Thereafter, their locations were mapped once every 15 minutes until at least 4:00 P.M. No food or food pans were present in the home site during the displacement and homing portion of the trial. Morning rations of chow, fruit, and vegetables were delivered on a single tree about 20 minutes after the animals had arrived in the home site.

Long-Distance Travel and Cohesiveness
Figure 11.2 shows how the animals handled the problem on their first six releases from the farthest translocation site. Each graph in the figure indicates the location of the animals along the long (E–W) axis of the test area throughout one trial. Changes in location along the N–S axis are ignored here for simplicity. N–S deviations were usually less than 8 m from an E–W line through the center of the test area. The two lines show the spatial range of the group at the start of each 1-minute interval after the animals were released.

Figure 11.2 shows that the animals returned to the home site on each trial. The figure also shows that there was a high degree of spatiotemporal cohesiveness among the tamarins. In other words, the distances between the animals at any one instant in time were small compared with the variance of the group's location over one or more trials. The nonlinear correlation coefficient, Eta, between time interval number and the E–W location of the six animals provides a measure of "group cohesiveness" during a trial. A value of 1.0 would indicate perfect cohesiveness. Eta squared for each of the

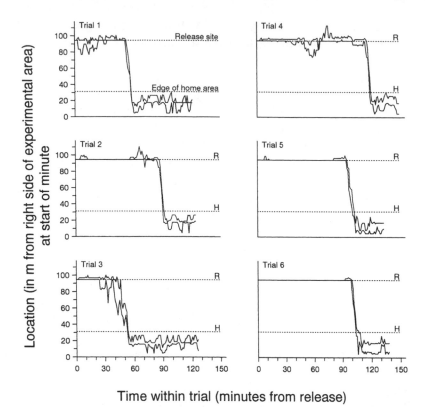

Figure 11.2 Social cohesiveness and site fidelity in a group of six golden lion tamarins during phase 1. The animals were transported and released together from the same location 77 m from their lodge tree. The lower and upper lines show the group's range during each 1-minute interval, that is, the locations of the nearest and farthest animals, respectively, relative to the right-hand side (west side) of the experimental area.

six trials in figure 11.2 was 0.987 or higher. Not shown in figure 11.2 were the twenty releases in phase 1 from the perimeter, near, and intermediate release sites. In all twenty of these releases, the group returned to the home site, and the spatiotemporal cohesiveness of the group was approximately the same as that shown in figure 11.2. After the group arrived in the home site, the animals invariably stayed within the home site during the remainder of the day and retired in the evening to whichever hutch box was present on the traditional lodge tree. On several trials, the animals showed exceptionally high levels of locomotor activity after they had arrived and fed in the home site, as if stimulated by some aspect of their homeward movement. These activities included arboreal jumps, play chases, and long calls.

C. R. Menzel and B. B. Beck

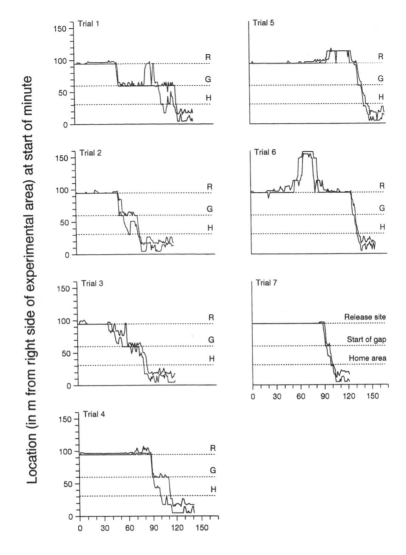

Figure 11.3 Social cohesiveness and homing during phase 2, when there was an experimentally imposed gap in the rope network.

Figure 11.3 shows how the animals handled the novel detour problem in phase 2. The first of the three novel gaps, which was located near the release site, posed virtually no difficulty for the animals. The second gap delayed the group for 20 minutes (minutes 39 through 58) during trial 3, but did not cause notable problems for the animals otherwise. The way in which the animals circumvented the second gap on trial 3 is described later. The third gap initially caused great difficulties for the group. Thus, on trial 1 of phase 2, the time elapsed from the first animal's arrival at the gap until the last animal crossed the gap was 72 minutes. However, the animals showed improvements across trials in their performance at the third gap. It may be seen by comparing figures 11.2 and 11.3 that group dispersion increased when the group encountered the third gap for the first time, then decreased across the six subsequent presentations of the same problem. Eta squared between time interval number and E–W location dropped to 0.954 on trial 1 of phase 2, but by trial 7 of phase 2 the animals remained highly cohesive throughout their homeward movement, Eta squared had returned to baseline levels (0.994), and less than 6 minutes elapsed from the time the first animal arrived at the gap until the last animal crossed the gap. All six animals showed a sharp increase from the last trial of phase 1 ("far condition") to the first trial of phase 2 in their time to travel through the region that contained the gap, followed by a sharp decrease in time across the six remaining trials of phase 2.

During delays at the third gap, animals made numerous forays out and back along ropes and tree branches in the vicinity of the gap. Most of these forays were visits to the adjacent tree along the ropeway, in the direction that led to the home site, and appeared to be attempts to circumvent or cross the gap. We defined a "vacillation" as a movement of at least 1 m away from the edge of the gap, followed by a return to within 1 m. (For phase 2, the edge of the gap was defined as the exact location at which the rope had previously been attached to a tree. For phase 1, the edge was defined as the same location, but with the rope attached.) During phase 1, the six animals together made a combined total of 1, 0, 11, 0, 1, and 0 vacillations on trials 1 through 6, respectively. During phase 2, the combined frequency of vacillations increased to 73 on trial 1, then decreased to 16, 23, 19, 0, 1, and 0 vacillations on trials 2 through 7, respectively. All six animals exhibited an increase (from phase 1 to phase 2) in the number of vacillations made prior to crossing the gap, followed by a decrease to phase 1 levels. During phase 2, the young adult female, young adult male, adult female, adult male, and two juveniles accounted for 38%, 19%, 19%, 10%, 9%, and 5% of the vacillations, respectively.

A young adult or juvenile that crossed the gap without the other animals crossing frequently returned toward the group and sometimes even re-crossed the gap, heading away from the home site to rejoin the group. Individual animals were fairly consistent across

C. R. Menzel and B. B. Beck

trials in which route they successfully used to cross the gap, but all six animals did not typically use the same route. Juveniles used thin branches as shortcuts more often than did adults. The adult female used a roundabout path along the rope. The adult male often went partway on the roundabout path, then descended and ran across the ground to the connecting rope.

The delays at the third gap and the vacillations did not appear to occur simply because the animals were interested in exploring the changed location per se. On all seven trials in phase 2, vacillations in the general vicinity of the gap stopped within about 2 minutes after the last animal in the group crossed the gap. At this time, all six animals moved to the home site quickly. To summarize, during delays at the gap, the tamarins searched for alternative routes and frequently approached one another. When the last animal crossed the gap, the group moved to the home site rather than making an additional extensive exploration of the gap region. After a few successful crossings, all six animals made fewer vacillations prior to crossing, all six animals crossed the gap more quickly, and the cohesiveness of the group increased.

How the Group Circumvented the Second Gap on Trial 3
The following sequence illustrates how the animals adjusted their movements in relation to one another and a novel gap in the ropeway.

> The group departed from the far release site and moved along the E–W ropeway in the homeward direction (W). The young adult female was in the leading position. The beginning of the group's movement onto a dead-end path occurred when the young adult female deviated 3.7 m from the E–W ropeway onto a dead-end rope, heading S. A juvenile followed her and extended her move 10.7 m toward the end of the dead-end rope, which after 2 m curved and pointed W. Five minutes later, the adult female extended the juvenile's W-oriented move on the dead-end rope by 3 m. The group began to gather near the W, i.e., homeward-oriented, end of the dead-end rope, and the animals were potentially stuck. The first animal to abandon the dead-end rope was the young adult female, who circled back to the E–W ropeway. She made a brief trip 15 m E to the far release site, then returned W to the dead-end rope, and then moved 26 m farther W toward the home site. From this position, she made a long call. This evoked a long call from the adult male but did not attract a following by the group. The young adult female returned E to the group and went to the end of the dead-end rope, where the five other animals were located. The first moment at which the "group as a whole" showed signs of getting off the dead-end rope occurred when the young adult female circled back (E then

N) to the E–W ropeway and all five group members followed her. However, the adult female and one juvenile (Pablo) failed to complete the detour and remained on the dead-end rope. The four animals that had completed the detour moved 26 m W, in the homeward direction, then stopped. The adult female and the juvenile, Pablo, remained behind on the dead-end rope. The adult female appeared to be stuck on the dead-end rope, rather than remaining there by preference, because she ran repetitively back and forth along the rope. After 3 minutes, the young adult female moved in a manner that produced a solution. Specifically, the young adult female moved 38 m E and stopped 12 m E of the adult female. The juvenile Pedro followed the young adult female and stopped 9 m E of the adult female. Within about 15 seconds, the other juvenile, Pablo, abandoned the dead-end rope and made a shortcut through the trees to the E–W ropeway. These changes in the positions of group members, especially including the fact that two animals were now located E rather than W of the dead-end rope, appeared to disengage the adult female from the homeward-oriented (W) end of the dead-end rope: the adult female moved N to the E–W rope and immediately moved 26 m W, where the adult male and the juvenile, Pablo, were located. The two other animals (the young adult female and Pablo) arrived at the same location moments later. Similar observations in other experiments suggested that a tree branch or a dead-end rope that pointed in the homeward direction, or in the direction of visible provisioned food, could trap animals. Furthermore, even if certain animals could solve the detour problem for themselves, they would return to the stuck animals, and the entire group might sit in contact without making forays for many minutes.

Onset of Travel Away from the Release Sites

As figures 11.2 and 11.3 suggest, the onset of the group's first travel progression was fairly abrupt. The onset of group travel from the four release sites was preceded by up to 211 minutes during which the animals sat in or near the release box, climbed in bushes, and made forays of a few meters away from the release site. Any individual, or even a subgroup containing five animals, could move away from the release site without the rest of the group following. That is, a movement by the adult male or adult female in the homeward direction did not necessarily ensure that a group move would follow. Animals that were not followed typically returned to the group within several minutes. Eventually, all six animals left the area of the release site within 1 or 2 minutes of each other and made a continuous travel progression to a distant location.

Which Animals Were the First to Leave the Release Site? Table 11.1 shows the number of times that each animal was the first in the group to move 10 m away from the release site and to arrive

C. R. Menzel and B. B. Beck

Table 11.1 Number of trials on which each animal was the first in the group to leave the release site or to arrive at various locations during the homeward movement

	Jenny AF	O'Reilly AM	Maria YAF	Carlos YAM	Pablo JM	Pedro JM	Total
Move 10 m away from release site after translocation	2	0	20	10	0	1	33
Arrive at edge of gap (phase 2)	3	0	4	0	0	0	7
Cross/circumvent gap (phase 2)	0	0	5	2	0	0	7
Arrive at lodge tree	3	0	30	0	0	0	33

at various locations during the homeward movement. The young adults accounted for most of the initial movements.

Orientation of the Group's Initial Travel Movement. The first movement of 2 m by the "group as a whole" was in the 180° semicircle facing the home site on thirty-one of thirty-three trials, and in the semicircle facing away from the home site on two of thirty-three trials. The two trials on which the group moved away from the home site occurred late in the experiment, on the thirty-first and thirty-second releases, and appeared to be exploratory forays. It may be seen in figure 11.3 that on these two trials (trials 5 and 6 of phase 2), the group traveled more than 15 m away from the release site. In sum, when all six animals were simultaneously located within the same semicircle on either the "toward" or the "away" side of the release site, travel subsequently occurred in the same general direction in which the six animals were located. This occurred on all thirty-three trials, and in the great majority of instances, this was in the direction of the home site.

External Factors Influencing the Time of Onset of Travel. As figures 11.2 and 11.3 show, the elapsed time from the start of the trial to the onset of group travel varied across trials. These variations in the time of onset of travel were not significantly correlated with variations in any physical factor that we could discern, including weather, temperature, wind level, time since sunrise at the start of the trial, or time of day at the start of the trial. Furthermore, neither the compass orientation of the translocated hutch box nor the replication number from a given location had a statistically significant influence on the latency of the group to leave the release site. However, the length of delay preceding the homeward movement did vary strongly with displacement distance. When the group was displaced just to the perimeter of its home site, the first animal took just 13.7 minutes to move 10 m from the release site. In contrast, when the group was displaced farther than the perimeter of its home site, the first animal took 72.6 minutes to move 10 m from the release site. Table 11.2 shows the mean latency of the six animals to move 10 m away from the release site on each of the twenty-six trials in phase 1. A two-way analysis of variance was conducted using the time scores in table 11.2. Trials served as subjects; data were $\log_{10}$ transformed. The increase in mean latency with increasing translocation distance was statistically significant (ANOVA: F $(3, 18) = 11.9$, $P < .001$). There were no significant or notable differences among the near, intermediate, and far release sites in the mean latency of the six animals to move 10 m away from the release site.

Responses of a Second Social Group to Novel Gaps
Were the intertrial improvements in crossing speed and in the number of vacillations shown at a novel canopy gap (see fig. 11.3) typical of other social groups and other gaps? We tested a second social group of captive-born golden lion tamarins in the same setting at

Table 11.2 Mean elapsed time (in minutes) from the start of the trial to move 10 m away from the release site in phase 1

Distance from lodge tree	Replication	Compass orientation of translocated hutch box	
		North	South
14 m	1	3.3	7.0
(Perimeter)	2	2.7	3.7
	3	3.8	74.2
	4	3.3	11.7
30 m	1	76.0	83.0
(Near)	2	5.3	189.0
	3	104.2	72.0
48 m	1	65.7	92.5
(Intermediate)	2	36.3	82.3
	3	114.2	81.8
77 m	1	76.0	13.8
(Far)	2	28.0	53.2
	3	99.5	94.8
	Overall median	36.3	74.2

C. R. Menzel and B. B. Beck

the National Zoological Park. The second group had lived in the forest for 4.5 months prior to testing. The first group was not present in the forest during the study. The second group traveled daily back and forth along a route of about 175 m. At one endpoint of this route was the group's hutch box. At the other endpoint was a group of golden lion tamarins in an exhibit cage. The study group showed a 42% or better decrease in average transit time from trial 1 to trial 2 on each of ten different, novel experimental gaps that we introduced along the group's familiar route. Each of the six animals in the group showed a decrease in transit time and a decrease in the number of vacillations at the gap from trial 1 to trial 2. These findings are in agreement with the data presented earlier and show that tamarin groups can, but do not always, show substantial intertrial improvements in the speed of crossing novel gaps.

Summary of Major Findings
Our findings suggest that the captive-born tamarins were highly philopatric, or had a strong attraction to established, familiar locations. The animals returned to their familiar home site after being trapped, transported, and released at relatively distant locations. Repeated displacements did not break down this pattern of movement. After arriving at the home site, the animals remained there throughout the day, despite the fact that the availability of trees and ropes in the home site was no greater than in the surrounding area. The animals retired together as a group to whichever hutch box was present in the traditional location, rather than to an identical, group-scented hutch box at the release site. Within the range of distances used in this study, the tamarins appeared to be acute in discriminating the direction of their home site. The group's initial movement away from the release site was oriented toward the home site, even on the first trial, from each of the four release sites, even when the animals almost certainly could not see the home site from the release site. Rotating the hutch box 180° relative to its traditional orientation changed the visual perspective that the animals had of the sun and surrounding terrain when they emerged, but this did not break down the group's ability to return to its home site.

Cues Available for Discriminating the Homeward Direction
An unsolved ethological problem is which internal and external cues the captive-born tamarins used to discriminate the homeward direction. In principle, internal cues of direction could have in-

cluded vestibular (inner ear) and somatosensory stimulation (signals from receptors located in skin, muscles, joints, tendons, etc.) caused by movements of the transport box just prior to and during the outbound trip. The tamarins were subjected to three full unidirectional rotations at the home site just before being translocated, and they experienced additional partial rotations while their hutch box was carried to the release site. To our knowledge, no studies have been conducted to determine whether tamarins can keep track of and compensate for as many as three full passive rotations; it would be surprising if they could, but the possibility remains open. Golden hamsters *(Mesocricetus auratus)* can be induced to leave a nest site and move in darkness to a food platform at the center of an experimental area. After the hamsters fill their cheek pouches, they return fairly directly to their nest at the periphery of the experimental area. Hamsters can orient in the general direction of the nest site if they are subjected to less than one full, unidirectional passive rotation while pouching food on the feeding platform. On average, hamsters can no longer orient in the direction of the nest site if they are subjected to more than two passive rotations (Etienne, Maurer, and Saucy 1988). Etienne and colleagues suggested that "path integration," the ability to compute a homing vector from self-generated route-based signals, is especially important for night-active central-place foragers, which have to return home even when external landmarks are not available or are not yet known. They state, however, that "without the help of external spatial cues, and in particular without an external compass, path integration by itself is not precise enough to be used beyond limited excursions" (Etienne, Maurer, and Séguinot 1996, 201). Tamarins are diurnal and relatively wide-ranging; they are not central-place foragers and do not start their excursions from just a single home base. If path integration does contribute to the optimization of long-distance navigation in tamarins, it might be predicted to do so in cooperation with the use of familiar visual or olfactory cues (Benhamou, Sauve, and Bovet 1990; Etienne, Maurer, and Séguinot 1996).

With regard to the use of visual landmarks and other external cues, several possibilities can be excluded or judged unlikely. The tamarins were displaced from the home site without having visual access to the surroundings. As previously stated, the home site itself was almost certainly not directly visible from the far and intermediate release sites. No acoustic stimuli, such as avian or mammalian

C. R. Menzel and B. B. Beck

vocalizations, emanated directly from the home site. The ropes beyond the perimeter of the group's home site had never previously been touched or scent-marked by any tamarin before the study began. Thus, these ropes did not provide a gradient of scent-mark odors for the animals to follow. (In any case, the tamarins did not merely approach group-scented objects: they moved away from the group-scented release box when they initiated their homeward movement.) It is possible that the tamarins used distant topographical features (e.g., a hilltop) that were visible from both the release site and the home site to orient in the homeward direction, but this is speculation. In summary, as in most field studies of mammalian homing (reviewed by Bovet 1992), the cues that the tamarins used to discriminate the homeward direction can be characterized mainly by exclusion. Discrimination of the homeward direction involved more than retracing the precise outbound route, and it almost certainly involved more than movement toward a familiar visible feature located within the home site.

In principle, displacing tamarins to gradually increasing distances in the same compass direction could train them to recognize and move in a particular direction and could result in an improvement in the speed and accuracy of their departure from release sites across replications. In the present study, the four release sites all lay in about the same compass direction from the home site. However, the accuracy of the group's departure direction and the elapsed time from the start of the trial until the group left the release site did not improve steadily across replications for any of the four release sites. The four displacement distances were presented in a balanced order, and the tamarins moved in the homeward direction even on their first release from each of the four release sites. In other words, our data on the speed and accuracy of the initial departure of the group from the release site do not suggest gradual learning of the homeward direction during successive experimental displacements. Nevertheless, it would be valuable in a future study to test the "directional learning" hypothesis more directly by determining whether tamarins show better initial orientation to the home site when they are released at increasing distances in the same compass direction than when they are released at comparable distances but in randomly selected directions. As stated earlier, the present study does provide clear evidence for intertrial improvements in crossing canopy gaps. Thus, if tamarins did return to the home site more efficiently when they were displaced in a consistent compass direction than when they were displaced in random directions, the question of whether their improved success was due to better homeward orientation, or to improved ability to cross the specific canopy gaps that they encountered repeatedly, or both, would remain.

Group Decision Making
Two fundamental goals of the analysis of group movement are to predict when the members of a social unit will initiate a cohesive, long-distance travel bout and to account for how the animals achieve a common direction (Kummer 1995). In the present study, the latency for captive-born tamarins to travel varied with translocation distance. The latency for the social group to initiate its homeward movement was low when the group was released on the perimeter of the home site and high when the group was released farther than the perimeter. There was no difference in the group's latency to leave the three farthest release sites. In other words, provided that the animals were displaced beyond the perimeter, they did not initiate their homeward movement any more quickly when they could see the home site (near release site) than when they apparently could not see it (intermediate and far release sites). Thus, the relationship between Euclidean distance from the lodge tree and latency to travel was not a simple linear one. Repeated releases from the near, intermediate, and far locations did not lead to a decrease in the group's latency to begin its homeward movement. It was as if the tamarins tended to remain near one another and the release box when they were faced with traveling as a group farther than some "reasonable" distance (e.g., 10 meters) to the home site. This occurred even when the home site was visible. That is, group latency apparently varied because its determinants were psychological and social, and only secondarily physical. Not all of the pretravel time was necessarily "decision-making" time; the group often rested inside the hutch box at the release site.

There was considerable variation in the time of the homeward movement from each release site. These variations were not correlated so far as we could discern with variations in weather or temperature. The tamarins appeared to establish and communicate the exact time of movement through their actions, including their travel intention movements and possibly vocalizations. The animals in this study had more than a single sensory channel available for communicating about when and where to move. Boinski et al. (1994; see also Boinski, chap. 15, this volume) have described how golden lion tamarins and other Neotropical monkeys use specific vocalizations in their natural habitat to initiate and lead group movements. In the present study, the captive-born animals were typically within 5 m of one another and in view of one another at the moment that a group travel movement began, and on many occasions we could not detect any vocalizations that appeared to prompt the onset of travel. Considering all the bouts of cohesive travel that we observed, it appeared that the tamarins, like titi monkeys (Menzel 1993), were capable of initiating travel without vocalizations or vacillations. On some occasions, one animal would move a few meters, a second animal would quickly follow, and within seconds a group move would be under way in the direction in which the first animal had moved. Boinski (pers. comm.) also notes that in situations of extreme danger, wild golden lion tamarins, squirrel monkeys, and capuchins can all "melt away" without travel calls. In these situations, the groups are tightly clustered, and the animals can rely on visual coordination.

These observations raise the point that most current studies of primate communication are signal-oriented. A study typically begins by identifying a set of specialized vocalizations, visual displays, or gestures that a species makes, then describes the contexts in which each type of signal occurs and tries to deduce the meaning and function of each type of signal. This signal-oriented approach is complementary to a sociologically oriented approach, which begins with an obvious case of coordinated action and works backward to identify the relevant social signals and other contributing factors (Menzel 1973b). In principle, both approaches should lead to the same conclusions regarding the patterns and processes of group movement. In practice, however, the question remains as to whether a list of specific signals and their contexts does account for the full set of observed group movements. E. Menzel (1971a) noted, in connection with his experimental studies of leadership and communication in young chimpanzees, that "No catalogue of communicative signals can be considered exhaustive until it accounts for how information is exchanged in the majority of instances of group travel, where animals achieve a highly coordinated, organized sequential pattern of action, with a fixed functional outcome, even *without* chattering and gesticulating at each other." The frequency with which the young chimpanzees he studied employed obvious signals (e.g., a vocalization, facial expression, or hand gesture) to initiate travel toward hidden goals decreased as they became more familiar with one another and their environment. Coordination of travel and communication about the environment by the young chimpanzees appeared to be achieved primarily by the animals' attention to the global features of one another's locomotion, including the direction and speed of partners' movements, and to the position and direction of movement of the group center relative to the environment (Menzel 1971a, 1973b). This does not mean that ethological displays are unimportant in primate group movement in general, or that specific vocalizations are not important for tamarin group movement in particular; the question is how to put such displays into a broader sociological context.

The determinants of the time of group travel in the captive-born golden lion tamarins we tested appeared to be multiple and varied. Young adults and adults seemed to establish the time of travel, in part, by making repeated movements in the homeward direction. Locomotion by several animals that coincided closely in time and in direction appeared to increase the probability that a move by the entire group would begin. The juveniles sometimes tipped the balance by joining in. The cohesive quality of travel was reflected in the tendency to follow other animals and to wait if not followed. The high degree of cohesiveness shown by captive-born golden lion tamarins during travel (see figures 11.2 and 11.3) is similar to that seen in semi-free-ranging, captive-born cotton-top tamarins (Price 1992). Additional experimental trials that we conducted suggested that the attraction among group members could override the attraction to the home site. In these trials, four of six animals were trapped and translocated, and two of six animals were left unconfined in the home site. On these trials, the unconfined animals left their home range and joined the translocated group within several minutes.

Wild versus Reintroduced versus Zoo Tamarins
Wild golden lion tamarins are relatively cohesive compared with capuchins or squirrel monkeys, and seldom have a group diameter of more than 20 m (Boinski et al. 1994). Nevertheless, during their foraging and travel, wild golden lion tamarins are often separated by 5 m or more and cannot always see one another due to foliage and tree trunks. As mentioned earlier, wild tamarins give frequent vocalizations during foraging, which appear to help the animals maintain contact ("whee calls") or to initiate and lead travel ("wah wah calls": Boinski et al. 1994). The ability of tamarins to forage

C. R. Menzel and B. B. Beck

separately might be important in reducing indirect food competition. Boinski (pers. comm.) reported that, in contrast to wild tamarins, the social groups of reintroduced tamarins that she observed in Brazil usually stayed in close visual contact and did not use travel vocalizations in what appeared to be appropriate contexts. Further data on reintroduced animals are needed to assess the hypothesis that wild-born golden lion tamarins use vocalizations to initiate and guide travel more frequently or effectively than do recently reintroduced tamarins living in similar forest, and to assess whether vocal behavior in reintroduced tamarins changes post-release. Data on whether reintroduced tamarins come, after months, to spread out more during foraging, to move farther in a day, and to obtain a larger proportion of their diet from naturally occurring foods are being collected and analyzed by the second author and his colleagues.

Unlike the tamarins at the National Zoo, the reintroduced tamarins in Brazil do not have the option of moving long distances along a continuous rope network. The fact that the animals have to use discontinuous, flexible natural substrates in Brazil might account for some of the dramatic instances of hesitation, confusion, and breakdowns in cohesiveness described earlier.

Spatiotemporal Routines
One way for group-living organisms to avoid interindividual conflicts over when and where to go is for individuals to form and adhere to shared spatiotemporal routines in their travel (Menzel 1993). The development of cohesive, regular travel sequences in an arboreal setting illustrates the interplay between social attractions, environmental structures, and experience in a given situation. When the captive-born tamarins we studied encountered a novel gap in the forest canopy, animals made exploratory forays along many different branches and ropes. Initial forays along a particular route were made by one or two animals rather than the group as a whole. A salient feature of the situation for the tamarins was that interindividual distances increased. Social attractions were evident in the tendency of animals to wait for one another, to return toward the group if not followed, to gather in close proximity to one another near the end of homeward-oriented branches and ropes, and to follow one another (at least partway) along new routes. Animals also extended other animals' direction of travel along a new route, taking the lead and exploring the new route even farther. With re-

peated releases, the animals came to cross the gap increasingly quickly and in an increasingly cohesive manner. All group members did not use the same pathways in crossing the gap; some made complete detours using the ropeway, whereas others circled partway but then cut straight across on the ground or using small branches. Nevertheless, animals came to arrive at the gap, and to cross it, at about the same time. The fact that each animal's crossing speed increased, and its number of vacillations decreased, with experience suggests that the animals picked up and retained information about the location of suitable paths. Presumably, the better each group member learned the available pathways and the social events that normally followed the group's arrival at the gap, and the better each animal became at ignoring irrelevant differences in the paths and the momentary compass bearings of other group members, the better equipped the group was to move quickly as a cohesive unit.

Adherence to established patterns of travel is just one example of the conventions that human and nonhuman primates use to avoid or resolve interindividual conflicts (Ullmann-Margalit 1977; Mason and Mendoza 1993; Menzel 1993; Cords and Killen 1998). An important question about primate group movement is how well individuals can learn behavioral and environmental signs that predict where group members will go next on a moment-to-moment basis in a familiar environment. Signs of where group members will go next can be somewhat indirect and arbitrary. A titi monkey appears to anticipate from its partner's travel path and from other features in the familiar environmental context that its partner is now moving toward a traditional sleeping site, even when its partner is "approaching" the sleeping site along an indirect route; if nightfall is approaching, the animal may jump over its partner's back, take the leading position, and move several meters ahead of the group to the sleeping site (Menzel 1993). Sigg and Stolba (1981) reported that the direction in which hamadryas baboons depart from the sleeping cliff in the morning, prior to splitting into subgroups for foraging, points toward the particular, distant water hole at which the animals will reassemble at midday. The animals' joint departure is preceded by a period in which different males propose and abandon alternative directions. In this example, the spatial relationship between the departure direction and the corresponding water hole is not simply arbitrary; the departure points toward the destination. In an example of a nonhuman primate's ability to learn and use totally arbitrary signs of group destinations in a familiar environment, Savage-Rumbaugh et al. (1986, 1993) report that a bonobo, or pygmy chimpanzee *(Pan paniscus)*, can touch arbitrary geometric forms (lexigrams) to indicate which one of seventeen destinations it will visit next in a 20-hectare forest, and can discriminate from lexigram "road signs" which location in the forest contains food or some other goal object (Menzel, Savage-Rumbaugh, and Menzel 1997). The bonobo appeared to learn the correspondence between particular places in the forest and particular lexigrams initially as an infant, in the context of daily travel with human caretakers to named locations in the forest. In a recent study (Menzel, in press), a female common chimpanzee that had previously learned a large number of lexigrams (Brakke and Savage-Rumbaugh 1995, 1996) used lexigrams to report the type of object she had seen hidden outdoors hours or days earlier and used gesture, and locomotion to recruit a person to the location. Her reporting was unprompted and suggested the use of recall or "episodic-like" memory (cf. Clayton and Dickinson 1998) in the context of recruitment. Thus, for tamarins, hamadryas, and apes, group travel is based on the ability of individuals to pick up, retain, and transmit information about potential travel directions or destinations, and the effectiveness of communication depends on the animals' prior knowledge of the common environment in which they are operating (Menzel and Johnson 1976). The behavior patterns and information processing capabilities involved in maintaining group cohesion, and the types of information that animals transmit during travel, can be expected to vary in interesting ways across primate species.

Conservation and Reintroduction
With regard to the reintroduction of captive-born golden lion tamarins into the wild, we identified at least three factors that could interfere with group travel. First, adults were often reluctant to use small, flexible supports for locomotion or to make long arboreal jumps. On some occasions, an adult would not step down 20 cm from one rope segment to another. Second, a dead-end rope pointing in the homeward direction could temporarily trap an animal. Third, the animals sometimes vacillated between the social group and an external goal, as if they were dependent upon the other animals and unable to resolve an approach-approach conflict in favor of moving decisively to the external goal. In other detour tasks that we presented, hungry tamarins sometimes quit trying to reach

a visible food source once they had made dozens of unsuccessful approaches along a spatially direct but incomplete ropeway. On these occasions, the animals sat in close proximity to one another, as if they had lost initiative. It is unlikely that wild-born tamarins would have made more than a few vacillations before all the members jumped to branches and moved to the food simultaneously. We emphasize that the captive-born animals did not always show a "loss of initiative" and were clearly capable of cohesive long-distance movements.

Comparable responses have been seen in reintroduced captive-born golden lion tamarins for up to 6 months after release, a period during which approximately 37% are lost (B. B. Beck et al. unpub.). Proximate causes of loss include predation, injury, starvation, exposure/hypothermia, bee stings, eating toxic fruit, snakebite, and theft/vandalism by humans (Beck et al. 1991; B. B. Beck et al. unpub.). But many of these losses can be traced ultimately to deficiencies in locomotion and spatial orientation such as those described here, and to small home ranges. For example, the two losses due to snakebite occurred while the tamarins were moving on the ground, and losses due to theft occurred when tamarins repeatedly returned to the same night shelter, allowing poachers to predict their whereabouts. The success and cost effectiveness of golden lion tamarin reintroduction may be improved by a better understanding of these deficiencies, and by remediation during pre-release preparation through trials such as those used in this research. In turn, these findings might be generalized to other taxa.

Conclusions
Captive-born golden lion tamarins show deficiencies in their travel patterns compared with wild tamarins. In the present study, factors that contributed to social groups getting stuck in the arboreal canopy included reluctance by adults to use flexible substrates or to make long arboreal jumps, reluctance by animals to deviate substantially from straight-line movement toward goals, and the tendency of animals to gather in contact and quit moving after making unsuccessful attempts to cross a canopy gap ("loss of initiative"). Factors shown by this study to contribute to successful homing included the ability to discriminate the direction of the home site, experience in a given situation (animals showed intertrial improvements in circumventing novel canopy gaps), and travel initiation movements by adolescents and young adults. The relationship be-

C. R. Menzel and B. B. Beck

tween Euclidean distance from the home site and latency to travel was not a simple linear one. The captive-born tamarins usually remained near one another and the release box when they were faced with traveling as a group farther than about 10 meters to the home site, even when the home site was visible. Group latency to travel apparently varied because its determinants were psychological and social, and only secondarily physical.

Research on primate spatial orientation is at a preliminary stage compared with research on birds and insects. Three directions for future research can be suggested. First, the immediate causes and ontogenetic development of the reluctance, confusion, and disorientation observed in some reintroduced golden lion tamarins merit additional study. Second, the movements of a primate can be expected to be very different when it is released with its group than when it is released individually. Social influences on homing could be studied in small-bodied Neotropical monkeys by removal of selected individuals or age classes from social groups prior to displacement, by comparison of an individual's performance when it is displaced alone to that when it is displaced as a member of a dyad or group, or by releasing individuals together that have had different types of homing experience. Relevant questions include the following: Are single animals more successful in homing than social groups? Does a social group make exploratory excursions that no animal makes alone? How does an animal's age and homing experience influence its ability to control the group? Can group members discriminate which animal has the best information about the location of connecting pathways and the home site? Third, comparisons of social groups of closely related species in the same test situation would clarify the distinctive aspects of group decision making in each species. For example, comparison of golden lion tamarin social groups with social groups of a more sedentary primate, such as dusky titi monkeys *(Callicebus moloch),* and with social groups of a wide-ranging primate, such as squirrel monkeys, *(Saimiri sciureus),* would allow predictions regarding group cohesiveness, the strength of animals' attachment to familiar places, and the distance and temporal organization of the homeward movement to be made in advance.

In conclusion, for most group-living animals, all behavior is in some sense social behavior, and homing behavior is no exception to this rule. While some students of homing behavior might feel that testing animals in groups rather than as isolated individuals introduces unnecessary complications to the analysis, in our opinion these "complications" are of the essence.

Acknowledgments

We thank E. Bronikowski for managing the tamarin exhibit and A. Baker, E. Bronikowski, J. Dietz, T. Grand, D. Kleiman, and E. Menzel for comments on the study. T. Allendorf, T. Hutson, and S. Roper assisted with the experiments. Friends of the National Zoo volunteers helped build the experimental ropeway. Research was supported by a Smithsonian postdoctoral fellowship to C. R. M. Manuscript preparation was supported by Swiss National Science Foundation grant 31.27721.89, U.S. National Science Foundation grant SBR-9729485 and U.S. Public Health Service grants HD06016 and MH58855.

C. R. Menzel and B. B. Beck

Comparative Movement Patterns of Two Semi-terrestrial Cercopithecine Primates: The Tana River Crested Mangabey and the Sulawesi Crested Black Macaque

MARGARET F. KINNAIRD AND
TIMOTHY G. O'BRIEN

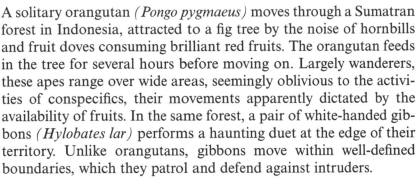

A solitary orangutan *(Pongo pygmaeus)* moves through a Sumatran forest in Indonesia, attracted to a fig tree by the noise of hornbills and fruit doves consuming brilliant red fruits. The orangutan feeds in the tree for several hours before moving on. Largely wanderers, these apes range over wide areas, seemingly oblivious to the activities of conspecifics, their movements apparently dictated by the availability of fruits. In the same forest, a pair of white-handed gibbons *(Hylobates lar)* performs a haunting duet at the edge of their territory. Unlike orangutans, gibbons move within well-defined boundaries, which they patrol and defend against intruders.

Halfway around the world in the llanos of Venezuela, two groups of wedge-capped capuchin monkeys *(Cebus olivaceus)* vigorously contest a fruiting guarataro tree *(Vitex orinocensis);* the smaller group, unable to defend the resource, alters its travel path and beats a hasty retreat. Although wedge-capped capuchin monkeys move within large home ranges that overlap extensively with those of their neighbors, resources are partitioned sequentially (they are "time-sharers," *sensu* Jolly 1972), and simultaneous arrivals inevitably result in a fight. In the Kanyawara section of the Kibale Forest in western Uganda, two groups of blue monkeys *(Cercopithecus mitis)* are disputing a highly prized fruiting ebony tree *(Diospyros abyssinica),* while only 10 km away at Ngogo, another group of blue monkeys dines alone. Movements of blue monkeys show extreme variability among subpopulations. Groups living in areas of low population density range widely, unconstrained by neighbors, who are rarely encountered, whereas groups living in areas of high population density are severely limited in their movements and display territorial behavior (Butynski 1990).

No matter where you look, primate movement patterns vary enormously, reflecting the effects of two broad classes of influences: characteristics of the species and characteristics of the environment. Species-specific features such as body size (Eisenberg, Muckenhirn, and Rudran 1972), group size (Waser 1977a), population density (Dunbar 1988; Butynski 1990), diet and digestive physiology (Bauchop and Martucci 1968; Struhsaker and Leland 1979; Milton 1980; Clutton-Brock and Harvey 1983), adaptations for arboreality (Rodman 1991), social structure, the degree of territoriality (Struhsaker 1975; Struhsaker and Leland 1979; Whitten 1982; Kinnaird 1992b; Jolly et al. 1993), and the ability to communicate information to neighbors (Waser 1976; J. M. Whitehead 1989; Kinnaird 1992b) all affect the tempo and distance of movement during a day. Likewise, the ecosystem in which a primate lives shapes movement patterns. Within forests, canopy structure and forest patch size may limit opportunities for movement or may channel movements along certain well-defined paths (Menzel 1986; Rodman 1991). A pair of siamangs *(Hylobates syndactylus)* brachiating through the forest canopy of peninsular Malaysia, for example, will be forced to take an unexpected detour when encountering a large forest gap resulting from heavy logging operations. Climatic conditions, especially temperature extremes and rainfall, affect the timing of movements (Marsh 1981). For some primates, movement patterns may be affected by the location of sleeping sites (DeVore and Hall 1965; Tenaza 1975; Rasmussen 1979) and by water availability (Altmann and Altmann 1970).

Movement and Food Resources
Perhaps the most important environmental determinant of primate movements is the abundance and distribution of food resources (Crook and Gartlan 1966; Milton and May 1976; Clutton-Brock and Harvey 1977b; Post 1981; Wrangham 1980; Gautier-Hion, Gautier, and Quris 1981; Isbell 1983; Bennett 1986; Boinski 1987).

M. F. Kinnaird and T. G. O'Brien

Primates feeding on invertebrates confront food items that are small, broadly scattered in the environment, and often camouflaged or concealed (Terborgh 1983), while fruit eaters exploit resources that occur in concentrated but widely scattered patches that may or may not be hard to find. When invertebrates dominate the diet, groups tend to move through their environment at a relative steady pace, sifting through leaves, peeling bark, and attempting to trap the unwary invertebrate. Because few invertebrate species offer enough of a meal to promote specialization, monkeys foraging for invertebrate prey tend to generalize (Terborgh 1983), picking from a myriad of species that together provide a relatively broadly scattered, uniformly distributed resource base (Terborgh 1983; Robinson 1986).

On the other hand, the tempo of groups exploiting fruits tends to be more "stop-and-go," with the focus of activity shifting from one spot to another as fruit trees go through periods of production and decline (Janson 1988b; Terborgh 1983). Because fruiting trees tend to create areas of high resource concentration separated by areas of low resource concentration, they are considered clumped, or "patchy," in distribution (Oates 1986), and the fruiting trees themselves are generally referred to as "patches" (Leighton and Leighton 1982; Chapman, Wrangham, and Chapman 1995). In this chapter, we also consider individual fruiting trees as resource patches. Fruit patches vary dramatically in size and quality, often in an unpredictable fashion (White and Wrangham 1988). "Stop-and-go" movement patterns may result from time spent searching for widely dispersed fruiting trees (between-patch foraging) and, once found, time spent in the canopy harvesting resources (within-patch foraging). Consider a group of spider monkeys *(Ateles geoffroyi)* sitting in a large fruiting *Spondias mombin* tree in the dry forests of Santa Rosa National Park, Costa Rica. Group members casually glance about, deciding on the next *Spondias* fruit to grab, while chewing mouthfuls of previously selected fruits. Movements of a few meters or less are sufficient for the spider monkeys to secure their next rewards, and little effort is spent foraging (within-patch foraging). Once the patch is sufficiently depleted, however, the group moves off with acrobatic speed and agility in search of another *Spondias* patch or other areas with high fruit density (between-patch foraging).

Studies worldwide have shown that when primates are foraging for fruits, groups tend to move shorter distances as the percentage of fruit consumed increases (spider monkeys, *Ateles paniscus:* Nor-conk and Kinzey 1994; brown capuchin monkeys, *Cebus apella:* Terborgh 1983; squirrel monkeys, *Saimiri oerstedii:* Boinski 1987b; olive baboons, *Papio anubis:* Barton et al. 1992). This is true, how-ever, only when food patches are big enough to accommodate the entire group and are not rapidly depleted. As relative patch size decreases, a group is less likely to satiate itself within a patch, and the only way to maintain or increase the amount of fruit in the diet is to visit more patches. Groups will spend less time in any one tree and travel farther in order to locate other fruit patches (Chapman, Wrangham, and Chapman 1995; Bennett 1986). Thus, patch size and distribution play a major role in shaping travel patterns (see Chapman and Chapman, chap. 2, this volume).

Movements and Neighbors
No matter how important the search for resources is, access is the key. Primate groups do not live in isolation, and access is often con-tested by neighboring groups and other species. There are many strategies for gaining access to food resources, and all of them in-fluence movement patterns. The most extreme strategy is to main-tain exclusive access to all resources within a well-defined territory, like the gibbons described above. Territoriality is likely to arise when resources (i.e., fruit trees) are in short supply and are econom-ically defensible (Brown 1964; Mitani and Rodman 1979; Waser and Wiley 1979; see also Peres, chap. 5, this volume)—in other words, when the energetic costs of defending boundaries do not outweigh the benefits accrued from resource ownership. Although such costs and benefits are often difficult to measure, the move-ments of territorial species are shaped by the need to patrol as well as the need to locate food, and are constrained by territorial bound-aries (Mitani and Rodman 1979; Lowen and Dunbar 1994).

Often, fruit resources are widely scattered and groups must use home ranges of several square kilometers, making the patrolling or defense of an area uneconomical (Waser and Homewood 1979). Several groups may overlap within these large areas, and avoidance of or confrontation with neighboring groups influences movement patterns. Overlapping groups of gray-cheeked mangabeys *(Lopho-cebus albigena)* ranging in the Kibale Forest, Uganda, use long-distance vocalizations to locate and avoid neighboring groups (Waser 1976). In the case of wedge-capped capuchin monkeys, neigh-bors do not avoid one another, and every encounter results in a confrontation (Srikosamatara 1986). Upon detection, groups ap-

M. F. Kinnaird and T. G. O'Brien

proach one another and interact agonistically. Wedge-capped capuchins employ a bully strategy in which group size reliably determines the winner: bigger groups predictably displace smaller ones.

Between the extremes of altering movements so as to avoid neighbors and the rush to confrontation, some species temper their response to intruding neighbors according to the continuously changing abundance and distribution of resources (Butynski 1990; Kinnaird 1992b). Subpopulations of some species occurring in dissimilar habitats show varying degrees of territoriality in response to differing resource abundance and distribution between habitats (banded langur, *Presbytis melalophus:* Bennett 1986; blue monkeys: Butynski 1990; Laws and Henzi 1995; vervet monkeys, *Chlorocebus aethiops:* Gartlan and Brain 1968; Kavanagh 1981; Chapman and Fedigan 1984). Along the Tana River, in eastern Kenya, the Tana River crested mangabey *(Cercocebus galeritus galeritus)* is unusual in that it shows variable resource defense within a single population (Kinnaird 1992b). Movements and responses to neighbors' long call vocalizations vary temporally, reflecting seasonal variation in food availability and distribution. When fruit resources are rare, neighboring groups tend to avoid one another, moving in separate domains. As fruit availability increases, resource defense becomes an issue. Groups collide with greater frequency, but resources are contested only if they occur in large, defensible patches.

In this chapter, we investigate the relative importance of several environmental factors, including food availability and distribution, in shaping the group movements of two primate species. We also examine the role of habitat constraints and social boundaries in modifying these movements. We chose to examine two Old World cercopithecines, the Tana River crested mangabey (fig. 12.1) and the Sulawesi crested black macaque (*Macaca nigra:* fig. 12.2), as independent tests of how socio-ecological factors affect ranging behavior in primates. We used data from two studies, each 2 years long, each focusing on three social groups, and each involving simultaneous observations of neighboring groups. By examining several groups in both species using the same observers and methods, we were able to distinguish differences in movement patterns within and between species.

Ecological and Behavioral Similarities

Tana River crested mangabeys and Sulawesi crested black macaques (hereafter referred to as Tana mangabeys and black macaques), have much in common. Both are medium-sized, semiter-

Figure 12.1 The Tana River crested mangabey foraging on *Phoenix reclinata* palm fruits.

Figure 12.2 The Sulawesi crested black macaque sitting.

Table 12.1 Group size, percentage of fruit in diet, and movement characteristics for the three study groups of Tana mangabeys and of crested black macaques.

Group	Size	% fruit	Home range (ha)	Daily path (m)	CV_{step}	Hours of observation
Tana River crested mangabey						
North	16	77	17	1,142	68	595
South	18	70	70	1,147	84	308
Nkano	28	77	35	1,252	63	374
Sulawesi crested black macaque						
Malonda	50	71	218	2,288	385	1,740
Dua	61	73	156	1,805	404	1,728
Rambo	97	63	406	3,013	322	1,716

restrial, forest-dwelling frugivore/omnivores (table 12.1). Tana mangabeys weigh about 10 kilograms, spend over half of their day on the ground, and consume a diet that is about 70% fruit (Kinnaird 1990). They live in multimale, female-bonded groups ranging in size from sixteen to twenty-eight individuals. Home ranges vary between 17 and 70 hectares, and neighboring groups have partially overlapping ranges. As described above, they exhibit facultative territoriality (Kinnaird 1992b) in which group spacing is mediated by spectacular loud calls that can be heard more than a kilometer away.

Black macaques are broadly similar to Tana mangabeys in body size and in many aspects of their behavior, including their semiterrestrial nature, multimale group structure, largely frugivorous diet, and facultative territoriality. Unlike Tana mangabeys, however, black macaques occur in large groups of up to ninety-seven individuals and cover home ranges between 150 and 400 hectares (O'Brien and Kinnaird 1997). Males give loud calls (Kinnaird and O'Brien, in press; Reed, O'Brien, and Kinnaird 1997), but their calls are much less impressive than the mangabey call, and we were able to hear them only at distances of 300 m or less. Although they may be used by neighboring groups to locate one another, black macaque loud calls appear to function primarily in within-group communication (Kinnaird and O'Brien, in press).

Whereas Tana mangabeys and black macaques share many characteristics, their environments are radically different (table 12.2). Tana mangabeys are endemic to a 65-kilometer stretch of riverine forest, mostly within the Tana River National Primate Reserve, Kenya (Homewood 1976; Kinnaird 1990). The region is semiarid, with a strongly seasonal rainfall of 400 mm per year. The forests

Table 12.2 Habitat characteristics of the Tana River Primate Reserve and the Tangkoko
Nature Reserve.

Tana	Tangkoko
Small	Large
Fragmented	Contiguous
Dry	Wet
Seasonal fruiting	Continuous fruiting
Depauperate vegetation	Diverse vegetation

depend on groundwater supply for their existence, which restricts them to the river's edge. They are naturally fragmented by river meanders, but fragmentation has increased over the last century due to human agricultural practices (Medley 1990). Forest patches range in size from as small as 10 ha to only 240 ha. Fruiting phenologies are highly variable over time and result in periods of resource abundance and scarcity (Kinnaird 1992a; fig. 12.3).

The situation is quite different for black macaques. They are endemic to the northern peninsula of the Indonesian island of Sulawesi, and the last viable population is found in the Tangkoko-DuaSudara Nature Reserve (Tangkoko) at the northernmost tip (Sugardjito et al. 1989; O'Brien and Kinnaird 1996). Tangkoko is an 8,867-hectare reserve of continuous forest ranging from sea level to 1,300 meters elevation. Rainfall is less seasonal than on the Tana River, averaging 2,000 mm per year, and fruiting phenologies showed less seasonality than those on the Tana River (fig. 12.3).

The Tangkoko forests have far greater species richness than the Tana River forests, and this is reflected in the diets of the two primates. Tana mangabeys fed on 58 fruit species, while black macaques consumed over 145 species. Additionally, the density of mid-sized and canopy-sized fruit trees (>10 cm dbh) was much higher in Tangkoko than on the Tana River. Densities of the top ten mangabey food species on the Tana ranged from 0.5 to 6.5 individuals/ha, although palm densities were higher because of their multistemmed growth form. *Phoenix reclinata,* an important but uniformly distributed palm species, was found at densities of 34 clusters/ha. In Tangkoko, the top ten macaque food species occurred in densities ranging from 4.6 trees/ha for the most frequently eaten fruit species, *Dracontomelum dao,* to 18.1 trees/ha for *Cananga odorata. Ficus* spp., the most commonly consumed genus, occurred at densities of 11.6 trees/ha in Tangkoko. In Tana River for-

M. F. Kinnaird and T. G. O'Brien

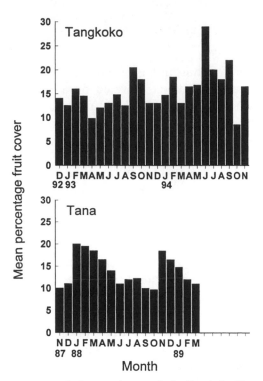

Figure 12.3 Mean percentage fruit cover by month for Tangkoko Nature Reserve (1,478 individuals in 111 species) and for the Mchelelo forest patches of the Tana River (126 individuals in 16 species). Mean percentage fruit cover represents the mean percentage of the tree canopy covered by fruit in a given month, and is calculated by first averaging the midpoint of four estimates of canopy cover (0–25%, 26–50%, 51–75%, >75%) for all individuals within a species, then taking a final average for all species (see Kinnaird 1992a).

ests, fig densities reached only 1.8 trees/ha. The average macaque group, therefore, has more than five times as many reproductive figs per square kilometer of home range as does the average Tana mangabey group.

In summary, black macaques live in a relatively large, continuous forest block compared with the small, fragmented forests inhabited by Tana mangabeys. Black macaque food resources occur at higher densities, are more common, are less temporally variable, and do not reach the low levels experienced by Tana mangabeys (table 12.2). In the following section, we explore how black macaque and Tana mangabey groups move through their environments, and how differences in group size, fruit resources, and habitat might influence similarities and differences in their movement patterns.

Patterns and Measures of Movement

Influence of Group Size

The number of individuals in a group has been shown in many primate species to affect distance traveled on a daily basis as well as time spent in various activities (Isbell and Young 1993; Chapman, Wrangham, and Chapman 1995; Chapman 1990b; Clutton-Brock and Harvey 1977b; Waser 1977a). Group size also affects a group's ability to secure resources (Srikosamatara 1986; Wrangham 1980) and deter predators (van Schaik 1983). It is widely accepted that a major cost of group living is reduced foraging efficiency (Terborgh and Janson 1986; Chapman, Wrangham, and Chapman 1995). Generally, larger groups require larger areas to overcome intra-group competition for food while meeting individual nutritional requirements; consequently, the distance that must be traversed to find adequate food will be greater (Terborgh 1983; van Schaik et al. 1983; Janson 1988b; Chapman 1990b; Chapman, Wrangham, and Chapman 1995).

Our data for black macaques and Tana mangabeys provide a good test of this relationship. Group sizes differ dramatically between mangabeys and macaques and are variable among groups of each species (see table 12.1). Given the larger average group size of black macaques, we predicted that their average daily path length would be greater than that of Tana mangabeys, and this was indeed the case. Black macaques traveled on average 2,343 m per day, while Tana mangabeys traveled significantly shorter distances, averaging 1,176 m per day ($F = 17.58$, d.f. $= 4.82$, $P = .0001$). Within each species, however, group size surprisingly had no effect on distance traveled, indicating that factors other than group size (e.g., micro-habitat differences) are modifying travel patterns among groups of Tana mangabeys and black macaques.

If group size does not explain within-species differences, it probably does not explain between-species differences, which weakens the support for group size as a primary factor determining species differences in day range. Isbell and Young (1993) found that group size overall had a positive influence on time spent moving in vervet monkeys, but the relationship was not supported by within-group analysis; time spent moving by individuals in larger groups and in smaller groups was not significantly different. They concluded that habitat was a confounding factor. Our results also point toward the possibility of habitat differences (e.g., physical constraints or

M. F. Kinnaird and T. G. O'Brien

differences in the resource base) playing important roles in differences in the daily path lengths traveled by mangabeys and macaques.

Influence of Habitat Geometry and Resource Abundance and Distribution

The small size of the forest blocks that Tana mangabeys inhabit physically limits the area over which a group may travel and may set bounds on daily travel patterns. Mangabeys might be expected to have shorter daily path lengths in smaller forest patches, or they might circle the forest patch repeatedly, which would be reflected in patterns of home range use and turning. Kinnaird (1990) showed that Tana mangabeys spend 50% of their time in less than 25% of the home range, which suggests that their movements are restricted.

Because long daily travel paths tend to be associated with forward movement rather than with backtracking along a path (Robinson 1986), we asked whether Tana mangabeys backtracked more than black macaques. Kinnaird (1990) used the number of times in a day that groups crossed their own paths, rather than turning angles, as a measure of backtracking because a series of small turning angles in the same direction may give the impression of a forward movement, but result in backtracking (e.g., three turning angles of 45° result in a 135° turn over 1.5 h). We found that Tana mangabeys cross their own daily paths three times more frequently than black macaques ($\bar{x}_{mangabeys} = 3.24$, $\bar{x}_{macaques} = 1.1$, $F = 14.25$, d.f. $= 5,242$, $P = .0001$). The difference may be due to repeated encounters with boundaries, or it may be because Tana mangabeys need to revisit fruit trees more often than black macaques. Backtracking to revisit food patches depends on resource abundance and the degree to which a group's passage through an area influences the rate of patch depletion and replenishment. Tana mangabeys may revisit fruiting trees more often during a day because they exploit a less productive and more seasonal resource base in which the abundance of fruiting trees at any given time is lower than in Tangkoko.

To distinguish between these alternatives, we tested for differences in the frequency of revisiting fruit trees during a day, and asked whether seasonality or the proportion of fruit eaten were important factors within and between the two species. Tana mangabeys revisit fruit trees slightly more often than black macaques (0.47/day mangabeys and 0.30/day macaques), but the differences are not significant between or within species. Seasonality of fruit resources might change the pattern of revisiting rates as resource abundance fluctuates, but we found that neither season nor the amount of fruit eaten affected revisitation. Tana mangabey turn-around rates, therefore, do not appear to be dictated by the need to revisit fruit resources. This does not, however, negate the importance of fruit resources in shaping the movement patterns of both species. As animals deplete fruit patches, the density, size, and distribution of patches may have a major effect on the time (and energy) spent searching for the next patch (Chapman, Wrangham, and Chapman 1995).

As fruit availability declines, patch size may or may not decline, but fruiting trees become more widely dispersed. For a primate group of a given size that is trying to maintain a given proportion of fruit in the diet, more trees must be visited and more distance must be covered as fruit availability declines (Chapman, Wrangham, and Chapman 1995); thus, declining fruit resources should result in longer daily travel paths. The higher density of large fruit trees in Tangkoko (256 trees $\geq$ 10 cm dbh/ha) than on the Tana (124 trees/ha) suggests that more fruit is available per hectare to black macaques than to Tana mangabeys. Monthly diversity of fruit in the diet was much lower on the Tana; the top four diet species accounted for 80% of the fruit consumed monthly by Tana mangabeys, but only 30% of the fruit consumed by black macaques, indicating a wider resource base for macaques. In addition, Tana mangabeys rely heavily on a uniformly distributed palm, *Phoenix reclinata,* throughout the year. While we would expect daily path length for both species to increase as the percentage of fruit in the diet declined, such an effect should be more apparent in the Tana mangabey in its highly seasonal environment, unless mangabeys shift from patchy to uniformly distributed fruit resources.

Black macaques, but not mangabeys, showed a significant decline in daily travel distance as the percent rage of fruit in the daily diet increased ($F = 4.33$, d.f. $= 1,426$, $P = .038$; fig. 12.4). When we examined the effects of season on the relationship between fruit in the diet and distance traveled, Tana mangabeys showed no seasonal differences in daily path length, due to their heavy reliance on *P. reclinata* during periods of general fruit scarcity (Kinnaird 1990, 1992b). Black macaque movements, however, were affected significantly by the seasonal availability of fruit ($F = 8.24$, d.f. $= 1,426$, $P = .0043$). Black macaques moved farther in a day during

M. F. Kinnaird and T. G. O'Brien

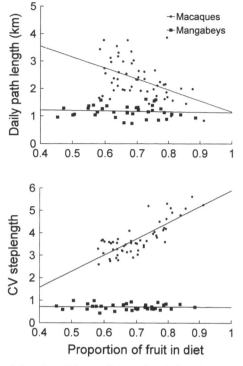

Figure 12.4 Daily path length and the coefficient of variation of step length as a function of proportion of fruit in the diet for black macaques and Tana mangabeys.

the wet season, but moved less per day as percentage of fruit in the diet increased. Figs and fruits of Rao *(Dracontomelum dao)*, account for approximately 30% and 15% of the total diet, respectively, for black macaques. The bright yellow fruits of Rao, the most commonly eaten species, occur in large trees (mean dbh = 86 cm) capable of accommodating the largest macaque groups. Rao fruits account for 28% of fruit feeding time during the dry season, but only 17% of fruit feeding time during the wet season. Figs were eaten year-round, but patch size changed between seasons; during the wet season, black macaques fed almost twice as long in large canopy figs than in the dry season, when fruits of smaller-crowned figs were eaten more often (O'Brien and Kinnaird, unpub.). Feeding less in Rao trees and more on widely dispersed, large-crowned figs accounts for some of the seasonal difference in movement patterns.

Diet and patch distribution may also affect the tempo of movement. Some primate groups move constantly in a continuous search for discrete prey packets (a steady tempo of movement), whereas others make long dashes between distant patches punctuated by periods of sitting (the stop-and-go tempo). One measure of tempo is the coefficient of variation in half-hour steps (CV). A Half-hour step is the distance moved during a 30-minute time interval and is measured by the shift in center of mass of a group during the time interval. Days in which movement is fairly constant throughout will have lower variance in half-hour steps (reflecting a constant rate of movement with few stops) than days of long journeys between patches followed by long rests. Uniform movement patterns result in low CVs, whereas long movements followed by long stops increase the range and variance of half-hour steps and increase the CV. The coefficient of variation scales the variation in half-hour steps to the mean half-hour step to allow direct comparisons of variation between groups that have different daily path lengths (and thus different average half-hour steps).

The CV of the average half-hour step length was five times greater for macaques than for mangabeys (see table 12.1) due to longer movements by macaques between patches of fruit coupled with long bouts of sitting and eating. Among black macaque groups, but not Tana mangabey groups, the coefficient of variation of step length increased as the percentage of fruit in the diet increased (fig. 12.4). This finding confirms that as macaques increase the percentage of fruit in the diet, the range of half-hour step lengths increases as more long-distance movements accompany more sitting. Thus, proportion of fruit in the diet affects the tempo of movement, by black macaques; as fruit increases, the distance moved declines, and the tempo becomes stop-and-go. Tana mangabeys, given their reliance on uniformily distributed *Phoenix reclinata* fruits and the well-defined boundaries of the forest edge, have fewer bouts of sitting and eating interspersed with long-distance movements, and their tempo of movement is more steady.

To recapitulate, Tana mangabeys, on average, are moving half as far as black macaques, but are turning around three times more. Macaque movements appear to be more closely tied to the fruit supply, whereas all measures of movement by Tana mangabeys seem to be relatively independent of fruit resources. These differences suggest that both species are constrained in their movements, macaques by fruit sources and mangabeys by both food and the configuration of their home range. Both species may also face social boundaries, created by neighboring groups, that impede forward progressions.

M. F. Kinnaird and T. G. O'Brien

Social Boundaries to Group Movements
The physical boundaries of the Tana River are obvious. Tana mangabeys encounter a crocodile-infested river on one side of their forest habitat and a dry, inhospitable savanna on the other. Although they may move between forest patches less than 500 meters apart, forest edges in general are hard boundaries for Tana mangabeys (Kinnaird 1990). Black macaque groups range in a forest far larger than the biggest home ranges of Tana mangabeys, and are not limited in this manner.

Influence of Neighboring Groups on Movement Patterns
Kinnaird (1992b) was able to take advantage of the fact that one of two neighboring mangabey groups in the 17 ha Mchelelo forest fragment frequently left Mchelelo and ranged in a nearby but separate forest. She was able to look at the influence of the presence or absence of the transient group (South group) on the movement patterns of the resident group (North group). North group shared the forest with the neighboring South group on 43% of the days monitored, indicating that the two groups tended to avoid each other by inhabiting adjacent forest patches (Kinnaird 1992b). After controlling for monthly variability in use of space, Kinnaird found that the presence of South group did not influence the distance traveled by North group, but did influence its travel patterns and use of space. The amount of backtracking by North group, as measured by mean turning angle and number of daily path crossings, was significantly higher when South group was present. North group also restricted its use of space when South group was present: the diversity of quarter hectare quadrats entered (measured by the Shannon-Wiener index of diversity, H') and the number of unique quadrats entered by North group were much lower when South group was present.

Similarly, black macaque groups tended to avoid one another most of the time. Pairs of macaque groups maintained intergroup distances greater than 500 m for 71% to 94% of the time. At these distances, we assume that neighboring groups have no effect on one another's movements. To look more closely at the effect of neighbors on group movements, we split our macaque data set into days when pairs of neighboring groups were close together (defined as passing within 100 m of each other during at least one 30-minute sampling period) and days when neighboring pairs were distant (at least 500 m apart all day, and the third group was at least 200 m

Table 12.3 Effect of neighboring groups on the turning angle distribution of the test group

		Chi-square		
Test group	Neighbor	IGD ≤ 100[a]	IGD > 100[b]	Interpretation
Rambo	Malonda	NS	NS	Does not modify turning
Rambo	Dua	NS	$P < .05$	More 45°–89°, more 135°–180°
Dua	Malonda	NS	NS	Does not modify turning
Dua	Rambo	$P < .05$	$P < .05$	Less 0°–44°, more 90°–180°
Malonda	Rambo	$P < .05$	$P < .05$	Less 0°–44°, more 135°–180°
Malonda	Dua	$P < .01$	$P < .05$	Less 0°–44°, more 90°–180°

[a] Days when neighbor was less than 100 m from test group.

[b] Days when neighbor was more than 500 m away and the third group was more than 200 m away from the test group.

away). We used the distribution of turning angles on days when groups were distant as the expected distribution of turning angles for movements unaffected by neighboring groups, and compared it with the distribution of angles when groups were close. We grouped turning angles into four categories: forward progression (0°–44°); veering from forward trajectory (45°–89°); turning back (90°–134°); and reversing (135°–180°), and tested for differences using chi-square tests (table 12.3). The identity of the groups influenced the distribution of turning angles at close intergroup distances. Rambo group, the largest group, did not alter its distribution of turning angles when neighboring groups were within 100 m. Dua group, the intermediate-sized group, increased its rate of turning back and reversing trajectory, and decreased its rate of forward progression, during close encounters with Rambo. Dua, however, did not alter its turning angle distribution in response to the proximity of Malonda group, the smallest group. Malonda reduced its rate of forward progression and increased its rate of turning back and reversing trajectory in response to close proximity of the larger neighboring groups.

Proximity of neighboring macaque groups also affected the half-hour step length (table 12.4). Rambo decreased its half-hour step length on days of close encounters with Malonda and Dua compared with days of no proximity. Dua displayed shorter half-hour step lengths when neighbors were within 100 m compared with more than 100 m on the same day. Malonda's response depended on the neighbor; Malonda increased its half-hour step length slightly ($P > .05$) in response to close proximity of Rambo, but reduced its half-hour step length in response to close proximity of

Table 12.4 Effect of proximity of neighboring groups on half-hour step lengths of the test group

Test group	Neighbor	IGD ≤ 100	IGD > 100	IGD ≥ 500 all day
Rambo	Malonda	86 ± 79[A]	117 ± 122[B]	133 ± 127[C]
Rambo	Dua	73 ± 54[A]	70 ± 73[A]	133 ± 130[B]
Dua	Malonda	56 ± 62[A]	68 ± 76[B]	60 ± 66[AB]
Dua	Rambo	87 ± 71[A]	109 ± 92[B]	71 ± 75[A]
Malonda	Rambo	102 ± 123[A]	92 ± 94[A]	90 ± 105[A]
Malonda	Dua	62 ± 71[A]	87 ± 102[B]	89 ± 107[B]

Note: The superscripts A, B, C denote results of Duncan's multiple range tests. Dissimilar letters indicate that intergroup distances (IGD) were significantly different ($P < .05$) from one another. Similar letters indicate no significant differences among the categories.

Dua. Shorter half-hour steps when neighbors were nearby were not due only to stationary confrontations in which group locations did not change. We recorded interactions that lasted less than a half-hour (fights in which groups collided from more than 100 m and separated to more than 100 m within a 30-minute period), as well as confrontations that lasted more than 7 hours. The average duration groups spent within 100 m was 2.3 intervals (70 minutes) for Rambo and Malonda, 2.7 intervals (80 minutes) for Rambo and Dua, and 3.0 intervals (90 minutes) for Malonda and Dua. During extended confrontations, groups would patrol common boundaries, sit across from each other, or move in parallel along the boundary. As in Tana mangabeys, these extended interactions between macaque groups often involved fighting.

Variability in Intergroup Encounters
Having shown that groups of Tana mangabeys and crested macaques tend to avoid one another, and that proximity alters their movement patterns, is it reasonable to assume that encountering neighbors is a chance occurrence? Tana mangabeys, moving within their highly constrained environment, would be expected to encounter their neighbors more frequently than macaques. And they do. Neighboring groups of Tana mangabeys encounter one another 3.5 times more often than neighboring black macaque groups (table 12.5). Although encounter rates for the two species are radically different, their rates of fighting between neighbors (number of fights per encounter) are similar, indicating similar temperaments (*sensu* Clarke and Boinski 1995) when encountering neighbors. The temperaments of both species can be described as bold and fearless during confrontations with neighbors.

Table 12.5 Number of backtracks/day, percentage of scan samples with neighboring groups less than 100 m apart, and number of fights/day for Tana mangabeys, black macaques, and the ratio of one species to the other.

Species	No. backtracks per day	% scan samples less than 100 m apart	No. fights per day
Tana mangabey	3.2	13.0	0.51
Black macaque	1.1	3.7	0.18
Mangabey: Macaque	2.9	3.5	2.8

We used the random gas model, first adapted for primates by Waser (1976), to ask whether or not groups encountered neighboring groups more or less often than one would expect by chance. Because Waser's basic model assumes that groups move at random within joint or completely overlapping home ranges, we adjusted the model to account for partially overlapping home ranges by incorporating the probability that neighboring groups would occur simultaneously in areas of home range overlap. The model predicts a value, Z, equal to the number of times in a half-hour interval that two neighboring groups will come within 100 m of each other:

$$Z = 2 \sqrt{v_i^2 + v_j^2} \, (d) \, (p_i) \, (o_{ij}),$$

where v_i is the average distance traveled by group i during one half-hour; v_j is the average distance traveled by group j during one half-hour; d is the distance between the center of mass of groups i and j; p_i is the density of group i; and o_{ij} is the probability that groups i and j are simultaneously present in the area of overlapping home ranges. We calculated the number of expected encounters per month by multiplying Z by the number of half-hour samples. Actual encounters were measured as the total number of times per month neighboring groups were observed within 100 meters of each other.

Both mangabeys and macaques showed extremely high variability in spatial proximity from month to month. The frequency of intergroup interactions for Tana mangabeys ranged from four times more than expected to as much as fourteen times less than expected due to chance alone, and that for macaques from three times less than expected to five times more than expected due to chance alone (fig. 12.5). These results indicate that although mangabey groups and macaque groups tend to avoid neighbors, there are circumstances that lead neighbors to spend more time in close proximity. Since groups are making context-sensitive decisions to alter their

M. F. Kinnaird and T. G. O'Brien

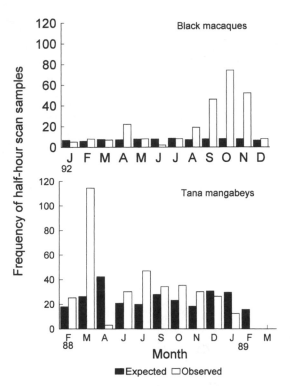

Figure 12.5 Observed numbers of 30-minute scan samples during which neighboring groups of Tana River crested mangabeys (North and South groups) and Sulawesi crested black macaques (Rambo and Malonda groups) were within 100 m of each other (open bars), and expected frequencies as calculated by the random gas model (solid bars).

movement patterns based on the proximity (in both species) and identity (in macaques) of neighboring groups, neighbors can be thought of as social boundaries for both species—boundaries that are just as real as the physical boundaries encountered by Tana mangabeys.

Social Boundaries and the Resource Base

The circumstances that mediate the social boundaries experienced by Tana mangabeys are related to the resource base, thereby creating a link between mangabey movement patterns and fruit resources. Kinnaird (1992b) showed that movements and responses to a neighboring group's long call vocalizations varied over time, and reflected seasonal variation in food availability and distribution. When fruit resources are scarce, neighbors tend to interact infrequently and use nonoverlapping areas (fig. 12.6); the most

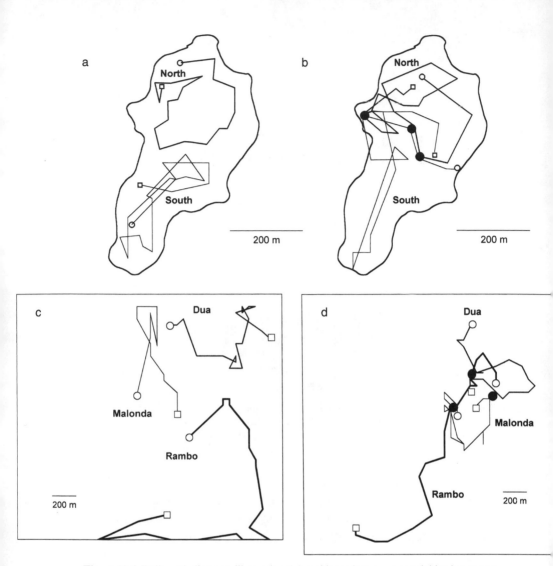

Figure 12.6 Daily path diagrams illustrating *(a)* avoidance between two neighboring groups of Tana mangabeys; *(b)* interaction between neighboring mangabeys; *(c)* avoidance among three groups of neighboring Sulawesi crested black macaques; and *(d)* interactions among neighboring macaques. Open circles and squares indicate beginning and end of daily travel paths, respectively. Solid circles indicate intergroup interactions.

common response to a long call is to turn and move away from a neighboring group. As fruiting increases, the distribution of diet species influences the type of interaction between groups. When mangabeys are eating abundant and uniformly distributed fruit species, such as *P. reclinata,* intergroup interactions are extended,

M. F. Kinnaird and T. G. O'Brien

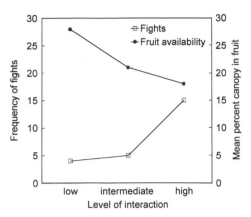

Figure 12.7 Number of fights among neighboring groups of Sulawesi crested black macaques, and mean percentage of the canopy in fruit for tree species constituting at least 2% of the macaque diet, by three levels of monthly interaction.

peaceful, and characterized by intermingling, side-by-side foraging, grooming, and sexual presentations to members of the neighboring group. When mangabeys focus on fruit species with patchy, defensible distributions, the most common response to a neighboring group's call is to approach and fight.

Similarly, black macaques engage in complex and highly variable intergroup interactions that appear to be tied to resource defense. As in mangabeys, interactions between neighboring macaque groups vary in intensity from extended, peaceful occupation of the same area with side-by-side foraging to aggressive confrontations characterized by screams, chases, grappling, and biting. Confrontations occasionally escalate to the point where deep, bloody wounds are inflicted, a level of intensity not witnessed in Tana mangabeys.

To examine the role of food resources in intergroup encounters among black macaques, we first classified months as low-, intermediate-, or high-encounter months based on observed and expected encounter rates according to the random gas model. During low-encounter months, groups met less often than or as often as expected by the random gas model. Intermediate-encounter months were characterized by one pair of groups meeting more often than expected by chance. If two or more pairs met more often than expected, the month was classified as a high-encounter month. Black macaque intergroup encounters differ from those of Tana mangabeys in that macaque groups approach each other more often, and are more willing to fight, as fruit availability declines (fig. 12.7).

In Tangkoko, as fruiting declines, fruit resources become more

Table 12.6 Analysis of variance results for the effects of non-fig fruit resource patchiness (measured by percentage of patchily distributed fruit in the diet), percentage of fig fruit in the diet, and crop size of figs on monthly frequency of intergroup interactions in black macaques

Variable	d.f.	Type III SS	F-value	$P > F$
Group pair	2	24.65	14.48	0.0001
Fig crop size	1	3.76	4.43	0.044
% figs in diet	1	6.86	8.06	0.008
% patchy fruit in diet	1	3.07	3.61	0.067

Note: Frequency of interaction is square root-transformed, and identity of group pair controls for variation due to macaque groups. All tests use type III sum of squares to control for effects of other factors in the model.

widely dispersed and more patchy in distribution. We found a trend toward increasing numbers of macaque intergroup encounters with an increasing percentage of patchy, or defensible, foods in the diet (table 12.6). Large, widely dispersed fruiting fig trees also appear to be a key to understanding the frequency of macaque intergroup encounters. The number of encounters increases as the crop size of fruiting fig trees in the habitat declines and the mean percentage of figs in the diet decreases (table 12.6). Fifty-four percent of all fights associated with food trees were over fruiting figs large enough to hold the largest macaque groups.

The relationship between resource availability and patchiness is quite the opposite for Tana mangabeys, who encounter increasing patchiness as fruit resources increase. Tana mangabeys experience months when very few fruit species are available other than undefensible thickets of *Phoenix reclinata* palm. Given the relatively richer, more diverse resource base in Tangkoko, black macaques probably never experience a situation in which fruit resources are so scarce that defensible fruit patches are nowhere to be found.

Given the higher variability in food resources on the Tana River, mangabey intergroup communication may be more important than that of macaques, and may result in better-defined social boundaries. Male Tana mangabeys begin each day by climbing high in the canopy and announcing their group's location with a booming loud call, and males may continue calling throughout the day. Male black macaques, in contrast, do not give such clear advertisements to neighboring groups. Macaque loud calls consist of a raspy, repetitive cackle, and are usually given during intragroup aggression; neighboring groups do not respond to these calls in any predictable manner (Kinnaird and O'Brien, in press).

M. F. Kinnaird and T. G. O'Brien

Conclusion

Our results demonstrate that the movement patterns of the Tana mangabey and black macaque are driven by the availability and distribution of resources, but are sculpted in very different ways by physical and social boundaries (table 12.7). Black macaque movements are directly related to measures of fruit availability and distribution, but they are unconstrained by habitat boundaries. Social boundaries are mediated by direct confrontation rather than by long-distance communication. The relationship between resources and movement patterns for Tana mangabeys is indirect; mangabeys modify movements in response to neighboring groups, and this strategy is driven by resource defense. The locations of social boundaries, although changeable, are always known through intergroup communication and further constrain patterns of travel. Ultimately, it is the physical boundaries imposed by small forest patches that limit opportunities for long-distance movements by Tana mangabeys.

We believe this comparison highlights the importance of intergroup interactions as a modifier of primate movements and underscores the importance of studying several groups simultan-

Table 12.7 Summary of responses of Tana mangabeys and black macaques to social and ecological factors affecting group movement

| | Response | |
Variable	Tana mangabeys	Black macaques
Range limitations		
Forest size	Small	Large
Boundaries	Restrict travel	No effect
Day range	1,150–1,250 m	1,800–3,000 m
Cross own path	3.24/day	1.1/day
Group size	No effect	No effect
Fruit		
Increase availability	No effect on travel	Reduce daily travel
Increase patchiness	No effect on travel	Increase daily travel
Revisiting patches	0.47/day	0.30/day
Intertroop interaction		
Avoidance	60% of follows	70–94% of follows
	Increase backtracking	Backtracking increases for smaller group
	No effect on day range	Travel rate lower during interaction
Antagonism		
Encounters	13.0% of samples	3.7% of samples
Fights	0.51/day	0.18/day
Loud call	Spacing mechanism	No apparent affect

eously. Primate groups do not live in isolation, and studies of primate group movements that focus on single groups are unlikely to capture the full range of factors influencing group movements. Despite the sometimes herculean efforts necessary to follow many groups simultaneously (see Jolly et al. 1993), the result is a better understanding of the social matrix in which primate groups live. Likewise, focusing only on the resource base may give misleading results. Had we looked at only one group of Tana mangabeys, we would have concluded that the resource base had no effect on group movements. Similarly, a study of a single macaque group might have given very different results depending on the group chosen. A large, dominant group like Rambo appears very responsive to the resource base; whereas a small group like Malonda may be responding to resource abundance and distribution, but its movements are constrained by its neighbors. Even among socioecologically similar species, generalizations of movements relative to the resource base do not always tell the whole story—we need to consider the neighborhood.

Acknowledgments
For permission to work in Kenya, we thank the Kenyan National Council for Science and Technology and the Office of the President (permit OP/13/001/16C282/12). We also thank the Institute of Primate Research, National Museums of Kenya, for local sponsorship. For permission to work in Indonesia, we thank the Indonesian Institute of Sciences, Puslitbang Biologi, and the Ministry of Forestry. Field research was supported by funds from the Wildlife Conservation Society, Sigma Xi, Wenner-Gren Foundation for Anthropological Research (grant 5543), the National Geographic Society (grant 4912–92), and the Conservation Food and Health Foundation. We thank all of our friends and colleagues who supported and helped us throughout our time in Kenya and Indonesia, and especially our technicians who helped collect the data. Finally, we warmly acknowledge Tom Struhsaker and Carel van Schaik for once again pushing us to think harder and write better.

CHAPTER THIRTEEN

Mountain Gorilla Habitat Use Strategies and Group Movements

DAVID P. WATTS

Variation in resource distribution should influence how primate social groups use their home ranges or territories. This prediction applies both to how much time groups will spend in different parts of
their home ranges over the long term ("area occupation densities":
S. A. Altmann 1974) and to short-term aspects of habitat use, such
as day journey length, patch choice and residency time, and distances traveled between patches (S. A. Altmann 1974; Pyke, Pulliam, and Charnov 1977; Stephens and Krebs 1986; Bell 1991). In
general, groups should allocate time among areas so that they gain
essential resources efficiently while minimizing exposure to hazards
(S. A. Altmann 1974; see Boinski, Treves, and Chapman, chap. 3,
this volume), and should take foraging paths that give the highest
nutrient yield per calorie expended (S. A. Altmann 1974).

Highly frugivorous and generalized primate species tend to eat
food that occurs in scattered, temporarily productive clumps or in
complicated mosaics. Often, they use most heavily those areas or

habitat types with abundant food (e.g., baboons, *Papio cynocephalus:* Altmann and Altmann 1970; Post 1978; Whiten, Byrne, and Henzi 1987; Barton et al. 1992; mangabeys, *Cercocebus* spp.: Waser 1977a; brown capuchins, *Cebus apella;* white-fronted capuchins, *Cebus albifrons:* Terborgh 1983; white-faced capuchins, *Cebus capucinus;* spider monkeys, *Ateles* spp.: Chapman 1988b). However, balancing foraging efficiency against risks may require trade-offs (S. A. Altmann 1974). For example, proximity to refuges from predation strongly influenced baboon area occupation densities at Laikipia (Barton et al. 1992) and at a site in Botswana (Cowlishaw 1997). The Botswana study group fed mostly in areas where food abundance was relatively low, but where predation risk was also low.

A few primate field studies have directly addressed questions about short-term foraging efficiency. For instance, Garber (1989) found that tamarins moving between food trees usually chose the closest available individual of the destination species, and that their movement was highly directional. A food intake rate maximization model explained much of the variance in patch residency time for baboons at Laikipia (Barton and Whiten 1994). The same baboon group made shorter day journeys (above a necessary minimum), and hence expended less energy on travel and food search, in areas where high-protein, relatively digestible food was abundant than in areas where such food was less abundant (Barton et al. 1992). However, food productivity has complex and variable effects on day journey length and foraging time within baboon populations (Bronikowski and Altmann 1996).

Day journey length often is positively related to group size for highly frugivorous species, but not for more folivorous ones (Isbell 1991; Janson and Goldsmith 1995; Chapman and Chapman, chap. 2, this volume). This frugivore-folivore difference presumably exists because the denser and more even distribution of nonreproductive plant parts (leaves, stems, etc.) than of fruit makes scramble competition less important for "folivores" than for "frugivores" (Wrangham 1980; van Schaik 1989; Chapman and Chapman, chap. 2, this volume).

Still, primates who feed heavily on nonreproductive plant parts do so highly selectively, thereby increasing dietary protein content and digestibility while limiting structural carbohydrate intake (e.g., mantled howler monkeys, *Alouotta palliata:* Milton 1979a; colobines: reviewed in Waterman and Kool 1995; baboons: Whiten et

David P. Watts

al. 1991; Barton and Whiten 1994). Primate folivores thus also face habitat variation and can benefit by locating food sources and harvesting nutrients efficiently, especially when highly digestible foods from which they can extract energy with relative ease are scarce and energy balance is difficult to maintain (Milton 1980; DaSilva 1992). Indeed, several studies of colobines and of howler monkeys show positive relationships between area occupation densities and food abundance (e.g., mantled howler monkeys: Milton 1980; Chapman 1988b; Thomas langurs, *Presbytis thomasi:* Sterck and Steenbeck 1997; banded langurs, *Presbytis melalophos:* Bennett 1986; black-and-white colobus, *Colobus guereza:* Oates 1977; red colobus, *Procolobus badius:* Clutton-Brock 1975; Marsh 1981a).

Mountain gorillas *(Gorilla gorilla beringei)* are gregarious primates. Females always live in social groups that also contain males, although many males are solitary for part or all of their adult lives (Stewart and Harcourt 1986; Watts 1990, 1996). Groups have overlapping, undefended home ranges. Mountain gorillas in the Virunga Volcanoes region of Rwanda, Zaire, and Uganda eat mostly leaves, pith, and stems from perennial herbs and vines and are among the most folivorous primates (Schaller 1963; Fossey and Harcourt 1977; Watts 1984; McNeilage 1995; Plumptre 1995). As expected for large generalist herbivores (Westoby 1974), they feed selectively with regard to plant nutritional quality (Waterman et al. 1983; Watts 1990; Plumptre 1995). Their food sources reach high densities in most of their habitat and are evenly distributed in time and space compared with those of most other primates. Still, fine-scale variation in food density and abundance is considerable (Vedder 1984; Watts 1984, 1998a; McNeilage 1995), and is influenced by the damage the gorillas do to plant stems as they forage (Watts 1987; Plumptre 1993).

In this chapter, I summarize the results of research on habitat use by mountain gorillas in the Virungas carried out at the Karisoke Research Centre, Parc National des Volcans, Rwanda. Detailed studies and a rich long-term data base provide a great deal of information on the influence of variation in food distribution on area occupation densities, on foraging effort within and across days, and on the duration of visits to foraging areas and of intervals between visits. They also address how male-male mating competition influences habitat use. Most analyses are correlational, and gorilla foraging energetics have not been modeled formally. I also consider how group members coordinate their movements and, briefly, whether they use spatial maps (see also Byrne, Dyer, chap. 17; chap. 6; and Garber, chap. 10, this volume).

Mountain Gorilla Habitat in the Virungas

Karisoke is at 30° E latitude and 1° S longitude, in the central part of the Virungas chain. Mean annual rainfall over 5 years was 1,811 mm (SD = 171 mm). March–May and September–December rainy seasons are typically separated by June–August dry seasons and shorter, less marked dry periods in January–February (Fossey and Harcourt 1977; Watts 1998a,b). Temperature variation within days exceeds that in mean daily minima and maxima across months (Schaller 1963; Fossey and Harcourt 1977; Vedder 1984; Watts 1984; McNeilage 1995). Temperature also varies inversely with altitude.

Gorilla habitat extends altitudinally from about 2,400 to 3,800 m. The vegetation, loosely categorized as tropical montane moist forest, varies with altitude, topography, and other factors. Researchers have identified various vegetation zones, most of which are valuable to the gorillas (e.g., Schaller 1963; Fossey and Harcourt 1977; Watts 1984; Vedder 1984; McNeilage 1995). No uniform classification exists, nor is there any based on statistical analysis of associations among plant species, although Vedder (1984), Watts (1984, 1988, 1991, 1998a), Plumptre (1993, 1995), and McNeilage (1995) sampled the frequency (percentage presence at sample points), species diversity, stem density, and biomass of gorilla food species and some nonfood species at points preassigned to observationally identified vegetation zones. Classifications generally include bamboo *(Arundinaria alpina)* forest; *Hagenia-Hypericum* woodland; dense herbaceous vegetation with sparse or no tree canopy, including areas dominated by *Mimulopsis solmsii;* subalpine and Afro-alpine vegetation; and several minor categories. Classifications differ mostly over subdivisions of *Hagenia-Hypericum* woodland and open herbaceous vegetation. Table 13.1 summarizes the most recent version (Watts 1998a). Important generalizations include:

1. Below about 3,400 m in the Karisoke study area, the frequency of gorilla food species at 1-square-meter sample points approaches 100% (Vedder 1984; Watts 1984; McNeilage 1995). A foraging gorilla thus finds some food almost anywhere it stops, except in open meadows.

Table 13.1 Characteristics of vegetation zones in the Karisoke study area

Zone	Altitude (m)	Relief[a]	Canopy[b]	Understory[c]	Herb[d]	Prot[e]	Food species[f]	Nonfood species[g]
High *Hagenia-Hypericum*	2,900–3,200	Slopes	Moderate/closed	Dense	523	111	Cn, Gr, La, Pl	Cs, Ib, Lg, St
Low *Hagenia-Hypericum*	2,700–2,900	Mild/slopes	Moderate/closed	Dense	594	104	La, Gr, Pl	Cs, Ib, Lg, St
High herbaceous	2,900–3,200	Mild/slopes	Open	Dense	506	127	Cn, Gr	Cd, Cs, Gs, Lg, St
Low herbaceous	2,700–2,900	Mild	Open	Dense	708	141	Gr, La, Pl	Ai, Dp, Lg, Sm, St
Secondary	2,700–2,850	Mild	Open	Dense	491	91	Gr, La, Pl	Cg, Pt, St, Vr
Nettles[h]	3,000–3,100	Mild	Open	Dense	503	123	Gr, La, Um	Ib, Lg, St
Giant *Lobelia*/thistle	3,200–3,300	Slopes/steep	Moderate	Dense	510	115	Cn, Gr, Ll, Rk	Lg, St
Lobelia	3,300–3,500	Slopes/steep	Moderate	Moderate	73	30	Cn, Ll, Rk, Sj	Ls, St
Giant *Senecio*	3,400–3,650	Slopes/steep	Open	Open	27	15	Hf, Ll, Sj	Aj, Ls
Brush ridge[i]	2,800–3,400	Steep	Open/moderate	Moderate/dense	199	50	Gr, Rk	Sr, St
Mimulopsis[j]	2,800–2,900	Mild	Open/moderate	Dense	279	91	Gr	Ms, Pg
Bamboo	2,700–2,800	Mild	Closed	Moderate	—[k]	—	Aa	

[a] Mild, level to undulating terrain, in saddle areas between volcano peaks and at bases of mountains; slopes, on mountain slopes, moderately to steeply inclined, but not dissected by ravines; steep, steeply inclined and more dissected slopes, and, for brush/ridge, ravine sides and narrow ridge crests between ravines.

[b] Canopy cover is mostly or entirely *Hagenia abyssinica* and *Hypericum revolutum* trees except in the *Lobelia* zone (a mix of stunted *Hypericum revolutum* trees and giant senecios) and the giant *Senecio* zone (stands of woody giant senecios that often form a dense, low woodland).

[c] Density of herbaceous vegetation and vines at ground level.

[d] Mean biomass (grams wet weight) of all herbaceous and vine foods per square meter.

[e] Mean biomass (grams wet weight) of all "high protein" foods per square meter (see text for definition).

[f] Important gorilla food species: Aa, *Arundinaria alpina*; Cn, *Carduus nyassanus*; Gr, *Galium ruwenzoriense*; Hf, *Helichryssum formossisimum*; La, *Laportea alatipes*; Ll, *Lobelia lanuriensis*; Pl, *Peucedanum linderi*; Rk, *Rubus kirungensis*; Sj, *Senecio johnstonii*; Um, *Urtica massaica*.

[g] Ai, *Agrocharis incognita*; Aj, *Alchemilla johnstonii*; Cg, *Cinerarea grandiflora*; Cs, *Coleus sylvatica*; Cd, *Crassocephalum ducis-aprutii*; Dp, *Discopodium penninervum*; Gs, *Gynura scandens*; Ib, *Impatiens burtonii*; Lg, *Lobelia gibberoa*; Ls, *Lobelia stuhlmanii*; Ms, *Mimulopsis solmsii*; Pt, *Pteris sp.*; Pg, *Pychnostachys goetzenii*; Sm, *Senecio mannii*; Sr, *Senecio mariettae*; St, *Senecio trichopterygius*; Sa, *Stephania abyssinica*; Vr, *Volkensia ruwenzoriense*.

[h] The Nettles zone is distinguished by local dominance of the nettle species *Urtica massaica* and *Laportea alatipes*.

[i] Dense thickets of *Senecio mariettae* cover much of the brush/ridge zone.

[j] Dense thickets of *Mimulopsis solmsii* cover much of the *Mimulopsis* zone.

[k] Food biomass in the bamboo zone varies greatly on a seasonal basis (see Vedder 1984, and McNeilage 1995).

2. Not all potential feeding sites are equally valuable, however. Food biomass per square meter varies over more than two orders of magnitude. The array of species, their stem densities and biomasses, and the availability of nutrients at potential feeding spots vary across areas of the same vegetation zone and among vegetation zones (Vedder 1984; McNeilage 1995; Watts 1984, 1998a; table 13.1). Nor does variation in food abundance correspond exactly to that in food quality. "High-quality" foods are defined as those relatively low in digestion inhibitors (acid detergent fiber plus condensed tannin content, by percent dry weight; Waterman et al. 1983) and high in protein (Vedder 1984; Watts 1984, 1998a; McNeilage 1995). For example, wild celery *(Peucedanum linderi)* and the thistle *Carduus nyassanus* contribute most food biomass in many areas with high herbaceous food biomass; gorillas eat the stems (low in protein and high in fiber) of both, but the leaves (high in protein) of thistles only, so protein availability is higher where thistles are more abundant.
3. Food species diversity and biomass decrease with altitude and are lowest for subalpine vegetation.
4. Bamboo shoots, important for some of the population, are seasonal (Fossey and Harcourt 1977; Vedder 1984; McNeilage 1995; Watts 1998b). Almost all other foods, and all other important ones, are perennial, although herb and vine productivity is lower during dry seasons (Plumptre 1993).
5. The proportion of woodland and open-canopied herbaceous vegetation, in which food is particularly abundant (e.g., the high herbaceous zone in table 13.1), is higher in the Karisoke study area than in much of the rest of the Virungas (Weber and Vedder 1983; McNeilage 1995). Only McNeilage has studied mountain gorilla feeding ecology outside of this area. His results mostly agree with those from Karisoke groups, but show that we need more work in other parts of the habitat (see below).

Habitat Use Data

The Karisoke data include detailed analyses of home range use by single groups or solitary males, usually for 1 to 2 years (Fossey 1974; Fossey and Harcourt 1977; Vedder 1984; Watts 1988, 1991; McNeilage 1995), and data from simultaneous monitoring of the movements of up to five groups during 1981 through 1987 (called the "long-term sample" below; Watts 1998b). Researchers have used two measures of area occupation density: the amount of time that groups spent in individual map quadrats (squares 250 m on a side), and the amount of time that they spent in particular vegetation zones. Some studies (e.g., Vedder 1984; Watts 1991) gave data on day journey length or on the number of quadrats entered per day (a good proxy for day journey length; Watts 1991). Watts (1991) gave data on the distance that individuals moved between consecu-

Table 13.2 Examples of significant ($p < .05$) relationships between aspects of habitat use by mountain gorillas in the Karisoke study area and characteristics of their food supply

Dependent variable	Independent variable[a]	Sources[b]
Total quadrat use	Herb and vine food biomass; high-protein food biomass	1, 2
Quadrat visit length	Herb and vine food biomass	1,3
	High-protein food biomass	1, 3, 4
	Protein per square meter; biomass of *Galium ruwenzoriense*	1
Intensity of vegetation zone use	Herb and vine food biomass; high-protein food biomass	1, 2, 5
Total vegetation zone use[c]	Herb and vine food biomass; high-protein food biomass; herbaceous food stem density; herbaceous food frequency; biomass and frequency of *Galium ruwenzoriense*	1
Day journey length	Herb and vine food biomass; high-protein food biomass; protein per square meter; biomass and frequency of *Galium ruwenzoriense;* food digestibility	1
		4
	Rain	
Distance between consecutive feeding spots	Herb and vine food biomass; high-protein food biomass; protein per square meter; food frequency; biomass and frequency of *Galium ruwenzoriense*	1

[a] "High-protein foods" were those with dry weight protein contents above the median for all foods; "frequency" is percentage of sample points at which food was present; "digestibility" is based on pepsin-cellulase assays (Waterman et al. 1983).

[b] 1, Watts 1991; 2, Watts 1998a; 3, Watts 1998c; 4, Vedder 1984; 5, McNeilage 1995.

[c] Corrected for zone area (Watts 1991).

tive feeding spots. Watts (1998b) used data from the long-term sample to measure intervals between visits that groups made to particular quadrats.

Mountain Gorilla Habitat Use Strategies

Quadrat and Vegetation Zone Use

All studies report that food distribution influenced long-term group movements, and food distribution also influenced movements within days and foraging periods (Watts 1991). For example, quadrat occupation densities for two different groups were strongly correlated with several measures of food abundance and quality, particularly the biomass of high-protein, highly digestible food per square meter (Vedder 1984; Watts 1991; table 13.2). All groups in the long-term sample also showed significant positive correlations between quadrat occupation densities and food biomass and quality per quadrat (Watts 1998a; table 13.2). These variables also were positively correlated with the intensity of vegetation zone use (the proportion of total time spent per zone divided by proportional zone area in the home range; fig. 13.1, table 13.2), as Watts (1991) had earlier found for a single group.

McNeilage (1995) found a strong relationship between vegetation zone use intensity and food abundance and quality for a Karisoke study group (table 13.2), but not for a group outside the study area whose home range contained mostly *Mimulopsis* vegetation and bamboo forest, in which food is less abundant than in most of the Karisoke study area (table 13.1). He did not sample food abundance and quality on the scale of 250 m × 250 m quadrats, and thus could not say whether this group used areas differentially in relation to food distribution on this scale.

The pervasive influence of food distribution on area occupation densities is counter to Fossey's (1974; cf. Fossey and Harcourt 1977) argument that, for one study group, interactions with other groups were a more important influence. However, Fossey had no quantitative data on food abundance. For another group, the mean day journey was longer on days of and after interactions with other groups or lone males than on other days, but this was a transient effect superimposed on the longer-term influence of food distribution (Watts 1991). Interactions among groups often involved attempts by males either to avoid or to follow other males, and thereby to retain or gain females, and increases in day journey length thus reflected mating competition among males.

Shorter-Term Movements
The mean length of day journeys that one group made in a particular vegetation zone decreased as herb and vine food biomass and

Figure 13.1 The relationship of the percentage of total occupancy time per vegetation zone (% use) to the percentage of total home range area that the zone contributed (% area) for groups in the Karisoke long-term sample. Vegetation zones (see table 13.1): GT, Giant *Lobelia*-thistle; LH, low herbaceous; HAG, high *Hagenia-Hypericum* woodland; HH, high herbaceous; NE, nettles; LHG, low *Hagenia-Hypericum* woodland; SEC, *secondary;* MIM, *Mimulopsis;* GS, giant *Senecio;* BR, brush/ridge; BAM, bamboo. Chi-square values, calculated from observed days of use and expected days of use based on zone area (1-tailed;* $p < .05;$** $p < .01$): group 5: 33.92** (d.f. = 9); group Nk: 19.72* (d.f. = 9); group P: 14.68* (d.f. = 7); group B: 15.78* (d.f. = 7); group Tg: 25.83** (d.f. = 11).

David P. Watts

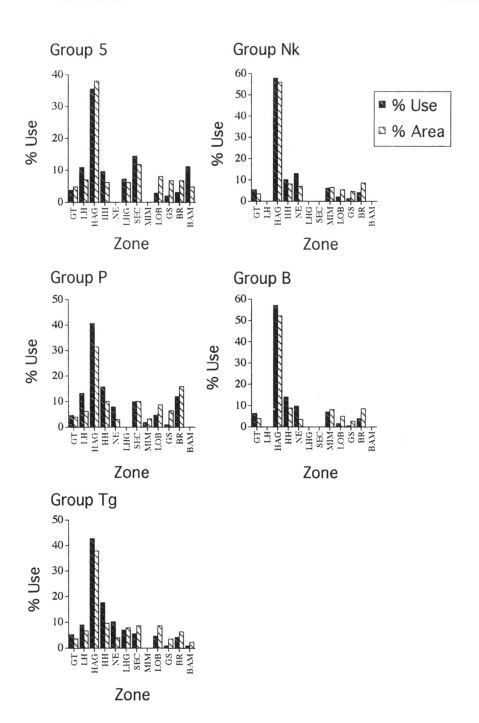

stem density, protein availability, and the mean digestibility of food in that zone increased (Watts 1991; table 13.2). Individuals moved shorter mean distances between consecutive feeding spots as food biomass and stem density, protein availability, and food digestibility increased, and as food frequency increased (Watts 1991 table 13.2). Group members spent less time feeding and searching for food per vegetation zone the more abundant food was in that zone and the higher its quality (Watts 1988). Because zone use intensity was positively related to food abundance and quality, the gorillas thus concentrated their activities in areas where they met foraging needs relatively quickly and with relatively low energy expenditure on food search (Watts 1988).

Indirect evidence shows that the groups studied by Vedder (1984) and McNeilage (1995) and those in the long-term sample (Watts 1998a) also spent most of their time in areas where foraging effort was relatively low. Their area occupation densities and mean quadrat visit lengths (table 13.2) were also positively related to food abundance and quality. Longer quadrat visits, and fewer quadrats entered per day, meant shorter day journeys.

Habitat Use by Solitary Males

Generalizations about movements that increase foraging efficiency fit solitary males poorly. They use areas far larger than needed to meet nutritional needs, and often search for and follow groups that contain females (Caro 1976; Yamagiwa 1986; Watts 1994). Unfortunately, we know the entire history of only one male who spent time both as a solitary and with his own breeding group. While solitary, he concentrated his activities in areas with abundant, high-quality food and usually made short day journeys, but he alternated this pattern with faster movement through larger areas and stealthy follows or open pursuits of groups (Yamagiwa 1986; Watts 1994, 1998a). His quadrat occupation time and intensity of vegetation zone use, and the distribution of his revisits to foraging areas, were unrelated to food abundance and quality. When he was with females, however, these variables showed the same relationships to food measures evident for other groups (Watts 1998a,b).

Solitary males are highly skilled at following trails left by groups (Watts 1989, 1994). This has profound implications for female reproductive strategies, because it effectively prevents females whose own males have died and who have vulnerable infants from evading infanticidal males for prolonged periods.

David P. Watts

Seasonality and Use of High-Altitude Vegetation

Karisoke study groups showed no consistent temporal variation in their use of vegetation zones with abundant herb and vine foods (Watts 1998c). Those with access to bamboo forest typically used it heavily during seasons when new shoots, which are high in protein and are a highly desired food, are growing (usually twice yearly: Fossey and Harcourt 1977; Vedder 1984; McNeilage 1995; Watts 1998c). The biomass of other foods in this zone is moderate, and gorillas in the study area used it much less when bamboo shoots were unavailable (Vedder 1984; Watts 1998c; McNeilage 1995). Bamboo forest is more extensive outside the study area, however, and some of the population used it heavily even when new shoots were unavailable (McNeilage 1995).

All foods in the high-altitude *Lobelia* and subalpine zones are perennial, but occur at low density and biomass (Vedder 1984; Watts 1984; McNeilage 1995; table 13.1). The pith and roots that form most of the diet there are nutrient-poor, based on available measures, compared with many foods at lower altitudes (Watts 1988, 1991; McNeilage 1995), although they may be major sources of some nutrient(s) for which we lack data. For example, *Senecio johnstonii* pith is low in protein, calcium, phosphorus, and vitamins A and C. Food intake rates are relatively low in these zones, particularly the subalpine, and the gorillas forage almost continuously when in them (Watts 1988). Use of these zones was inversely correlated with monthly rainfall for several groups in the long-term sample (Watts 1998c), although the gorillas used them little overall and, during months when they used them, still fed mostly in dense herbaceous vegetation at lower altitudes. Herb and vine foods grow during dry months, although relatively slowly (Plumptre 1993). Thus movements to high altitudes during dry months may have reduced resource depletion at lower altitudes at times of slow renewal, but were not responses to sharp declines in food abundance. Whether any rainfall-related changes in food nutritional value occur in the Virungas is unknown. The soluble carbohydrate content of gorilla herbaceous foods varies with rainfall at Lopé in Gabon, although it is actually higher during dry seasons (White et al. 1995).

Core Areas and Home Range Size and Stability

The three groups followed for at least 4 years in the long-term sample used areas of 20–25 square kilometers over that period of time (Watts 1998a). Annual home ranges for these and other groups varied from 4.4 to 11.5 square kilometers (Vedder 1984; McNeilage 1995, Watts, in press-a). However, groups spent most of their time within fairly small portions of their annual home ranges and their total home ranges. For example, groups in the long-term sample had annual and total core areas that were 30% to 40% of their annual and total home ranges, respectively (a "core area" included those quadrats that, in descending order of use, accounted for 75% of total quadrat occupancy time: Watts 1998a).

All core areas included several vegetation zones. Those of groups in the long-term sample had disproportionately large areas of zones in which food, particularly high-protein and highly digestible food, is abundant, compared with the percentage of the total home range that each zone contributed (fig. 13.1, table 13.1; Watts 1998a).

The areas that groups used varied considerably across years. Overlap between a group's home range for a given year and those for other years was usually less than 50%, and overlap between yearly core areas was usually less than 40% (Watts 1998a). Groups continued to enter new areas over periods as long as 7 years. One reason for low interannual overlap and limited site fidelity may have been a need for supplying areas larger than annual home ranges. Food resources other than bamboo shoots are renewed continuously, not in seasonal or unpredictable pulses, but renewal is slow (see below). An area just sufficient to meet annual needs may not sustain that level of use for longer periods.

Variation in habitat quality may have two somewhat opposing scale effects. The fine-scale spatial variation in food abundance and quality that influences movements between feeding spots, and the broader variation that influences quadrat and vegetation zone use, may recur regionally: large habitat swaths may offer multiple, comparably attractive alternatives for establishing a home range, which then floats somewhat over time. Also, the dense terrestrial vegetation in which the gorillas mostly feed is a temporally shifting mosaic. Slow changes in the vegetational composition of particular areas (including those that the gorillas induce; see below) may increase or reduce their attractiveness as foraging areas, so that groups add them to, or drop them from, their movement circuits.

Extreme male-male mating competition can override even the long-term influence of food on group movements and lead to home range shifts (Watts 1994). Group Nk abruptly and completely shifted its home range in 1984 after an encounter with another group and a series of prolonged encounters with a solitary male,

David P. Watts

during one of which he chased them for a week. Two females emigrated from the group and one infant was killed during these encounters, and some group members lost contact with others for as long as 5 weeks (Watts 1994, 1998a). No noticeable change in home range quality accompanied the shift, which seemed to be a response by the group's single male to mating competition. The impression that he was responsible for the shift was reinforced one day when he made several trips to escort frightened group members across the largest stream in the study area (J. Rafert, pers. comm.), which they had almost certainly never before crossed. When he died in 1985, part of his group merged with part of an all-male group. By 1992, this group had used about 20 square kilometers that overlapped with the home ranges of both parent groups, but also included areas that neither had entered (Watts 1998a).

Gorilla "Trampling" and Vegetation Dynamics
Bamboo shoots emerge and grow at high rates (Vedder 1984), but heavy use of these feeding areas during foraging makes them unattractive for at least several months while damaged vegetation regenerates (complete biomass regeneration of herbaceous vegetation takes about 8 months; Plumptre 1993). However, the slow but continuous growth of mountain gorilla foods raises the possibility that they "crop" areas (Bell 1991) or practice some form of "rotational" browsing, like the rotational grazing of Serengeti ungulates (Fryxell, Greever, and Sinclair 1988; Fryxell 1995). They might adjust intervals between visits to trampled areas to the extent of trampling and the rate of regeneration, so as to maintain high nutrient acquisition rates on successive visits and maintain long-term habitat productivity (fig. 13.2). By stimulating tissue production and vegetative reproduction by food species, gorilla trampling can even enhance food species productivity and stem density over the time needed for biomass regeneration (fig. 13.2; Watts 1987; Plumptre 1993). The stem density of food plants in a trampled area, and the biomass of young growth, which may have relatively high nutritional quality, can be higher 6 months after trampling than it was before (Watts 1987). How long this effect persists, and how it affects heterogeneity in vegetation composition and forage quality, is unknown.

When gorillas feed in an area, they do not deplete food and trample the vegetation uniformly. Individuals take separate, sometimes overlapping, foraging paths, feed at some spots and bypass others, and leave much vegetation standing. The extent of tram-

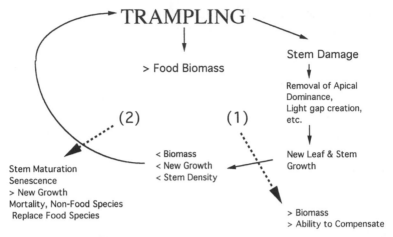

Figure 13.2 Hypothetical cycle of the effects of mountain gorilla feeding and movements on the dynamics of vegetation in the herb layer in the Virungas. Gorillas remove plant biomass by feeding and damage stems by folding, tearing, walking, sitting, and lying on them. Damaged plants then produce new growth and send up new side shoots. Trampling can increase food species stem densities within the time needed for biomass regeneration (240–270 days; Plumptre 1993). Revisits and retrampling at suitable intervals would allow the gorillas to harvest abundant new growth and to maintain high food species densities (cycle that starts and ends with trampling). Repeated trampling at intervals too short compared with regeneration time (dashed arrow at 1) could possibly impair the compensatory ability of damaged plant stems, lead to replacement of food species by undamaged (or less damaged) nonfood species, and reduce biomass over the long term. Conversely, prolonged absence of trampling (dashed arrow at 2) could lower the productivity and nutritional quality of food plants as stems mature and senesce, and perhaps lead to replacement of food by nonfood species, if trampling maintains food species productivity.

pling damage should depend on how long they stay in the area, and should in turn influence how efficiently they could harvest nutrients if they returned there before much regeneration had occurred. If they do crop areas, the length of intervals between visits should be positively related to the duration of previous visits, at least up to the point at which intervals are long enough for complete regeneration. However, the long-term sample, plus a 17-month sample on a single group, provided preliminary data on the duration of visits to quadrats and on revisit intervals that do not support this prediction (Watts 1998b). Intervals between visits to a quadrat increased in proportion to the duration of previous visits in only a few cases, and groups often revisited quadrats sooner than the time needed for complete regeneration in Plumptre's (1993) trampled study plots. Variation in food abundance and quality tended to counteract any effect of trampling on revisits. The gorillas visited quadrats with

David P. Watts

abundant high-quality food often, stayed relatively long per visit, and revisited them at relatively short intervals. Mean revisit intervals were actually inversely related to food biomass and quality (Watts 1998b).

The relatively short intervals between visits could have allowed the gorillas to harvest abundant new growth of food. Repeated trampling at appropriate intervals could maintain high productivity in an area (fig. 13.2). "Appropriate intervals" could be less than what would allow for peak vegetation biomass if productivity were highest at some intermediate biomass (Noy-Meir 1978; Fryxell, Greever, and Sinclair 1988; Fryxell 1995; Wilmhurst, Fryxell, and Hudson 1995). This seems to be true for regeneration from gorilla trampling in the Virungas (Plumptre 1993). Plant nutritional quality may also be higher at intermediate biomass; such areas may then be most attractive to herbivores because energy gain rates are highest, as Wilmhurst, Fryxell, and Hudson (1995) showed for wapiti. Conversely, prolonged absence of trampling may make foraging areas less valuable because food nutritional value declines, and perhaps because food species become less productive and nonfood species replace them (fig. 13.2). My impression was that over 14 years, the vegetation changed little in some areas that gorillas used regularly (e.g., some of the nettles zone at the western base of Mt. Visoke). Other areas changed more, perhaps because the gorillas used them less. For example, one group often used a small side crater on Visoke, where *Carduus nyassanus,* an important food species, was especially common, in 1978–79, but gorillas rarely entered this area during my observations in 1984–87 and 1991–92, by which times *Senecio trichopterygius* (a nonfood species) had replaced most of the *Carduus.*

We need more precise data on visits to specific, smaller trampled areas to analyze the effects of the gorillas on the vegetation more fully (Watts 1998b). Plumptre (1993) showed that typical levels of trampling by all large mammals (including gorillas) combined probably did not degrade the habitat over 2 years, but stressed the need for longer-term monitoring of plant species composition, stem density, and biomass in trampled and untrampled areas. Establishment of exclosures, some subjected to simulated gorilla damage, would allow controlled measurement of the effects of different levels of trampling on plant productivity and a test of the prediction that habitat quality declines in the absence of trampling.

Spatial and Temporal Knowledge

Humphrey's (1976) observation of the ease with which mountain gorillas acquired food helped inspire his elaboration of the hypothesis that social factors have been the main impetus for the evolution of primate cognitive abilities. Conversely, the ecological intelligence hypothesis (Milton 1981a) holds that the challenges of foraging efficiently in complex and highly variable habitats have driven cognitive evolution. Folivores can presumably remember food source locations more easily than frugivores, and track their productivity better, for several reasons: they need relatively smaller supplying areas, use more evenly distributed and less ephemeral food sources, and need to travel shorter daily distances to meet nutritional needs (Milton 1981a; cf. Dunbar 1992; Byrne 1995b; Barton 1996b; Barton, chap. 8, this volume). They also can more easily switch to foods that are low in quality (high in structural carbohydrates), but abundant and easy to locate, when fruit is scarce (Byrne 1995b).

These proposed ecological differences often occur. Highly folivorous species generally have relatively small home ranges (Clutton-Brock and Harvey 1977a), and some studies show folivores to use more evenly distributed foods than sympatric, more frugivorous species (e.g., red colobus and redtail monkeys, *Cercopithecus ascanius:* Struhsaker 1980; howler monkeys and spider monkeys: Chapman 1988b; Thomas langurs and long-tailed macaques, *Macaca fascicularis:* Sterck and Steenbeck 1997). Few quantitative data on spatial knowledge exist, however (Byrne, chap. 17; Garber, chap. 10, this volume, for a review), and available data (e.g., Garber 1989) do not address possible frugivore/folivore differences. Comparative analyses of relative neocortex size (considered by some a proxy for spatial knowledge or for "intelligence" more generally), with taxa broadly categorized as "frugivores" or "folivores," actually support both hypotheses, which are not mutually exclusive (Barton, 1996; Barton, chap. 8, this volume).

Mountain gorillas seem to know the locations of good foraging areas, but many animals adjust patch choice, residence time, and revisitation rates to patch quality, as optimal foraging models predict (reviewed in Bell 1991). Circumstantial evidence suggests that gorillas have sophisticated spatial knowledge, using "network maps" based on orientation to topography and landmarks and perhaps "vector maps" that allow them to plan novel pathways based on topological knowledge (Byrne, chap. 17; Garber, chap. 10, this volume). For example, groups often crossed ravines and streams at particular safe points, often after quick movements from locations that were out of sight. Group 5 often crossed the two main streams in the study area at several points where fallen trees or large rocks allowed them to do so without contacting water even when the streams were at peak volume. Many approaches to these crossings were along well-worn trails the gorillas had reached through dense vegetation. Retrospective analysis of movement directionality could test whether groups typically used whichever crossing was closest to their morning starting point. Even if they did, however, demonstrating use of novel pathways would be extremely difficult (Byrne, chap. 17, this volume).

Groups sometimes rapidly traveled long distances, sometimes over a kilometer, from feeding areas in herbaceous vegetation to others in subalpine vegetation or bamboo forest. They did not feed during this travel, and were usually excited; an experienced observer could predict their general destination. On clear days, the gorillas could orient to the subalpine zone by climbing toward the summit of Visoke or another volcano, but fog or clouds sometimes obscured visual landmarks. Some of their rapid movements to bamboo happened when bamboo shoot production should just have been starting, and when the gorillas had not visited the bamboo zone for up to many weeks. They behaved as if operating on both spatial knowledge (where to find bamboo shoots), using at least "richly connected network maps" (Byrne, chap. 17, this volume), and temporal knowledge (when shoots should be available).

The importance of bamboo shoots in the diet indicates that mountain gorillas could benefit from sophisticated spatial and temporal knowledge, even though most of their foods are evenly distributed in space and time. Knowing when to look for bamboo shoots, how to find those still mostly underground, and where to find even small, isolated bamboo patches is presumably harder than finding celery stems. In some respects, bamboo shoots are to mountain gorillas what seasonal fruit crops are to other, more frugivorous gorilla subspecies. The diets of other subspecies overlap highly with those of sympatric chimpanzees, especially in their fruit components (Rogers et al. 1988; Williamson et al. 1990; Mwanza et al. 1992; Tutin and Fernandez 1993; Yamagiwa and Mwanza 1994; Remis 1997b, Yamagiwa et al. 1996; Kuroda et al. 1996; Doran and McNeilage 1998). Western *(Gorilla gorilla gorilla)* and eastern lowland gorillas *(Gorilla gorilla graueri)* may face less temporal variation in the number, size, and dispersion of food patches than chimpanzees, for several reasons: they eat fruit less selectively than chimpanzees, feed more on herbaceous vegetation as they move between fruit trees (Yamagiwa et al. 1996; Kuroda et al. 1996), and can better exploit perennial herbs and bark when fruit is scarce (Yamagiwa et al. 1996; Kuroda et al. 1996; Rogers et al. 1988; Doran and McNeilage, 1998). Also, variation in gorilla frugivory can be high even within populations (e.g., eastern lowland gorillas at Kahuzi Biega: Yamagiwa and Mwanza 1994; Yamagiwa et al. 1996). Still, gorillas may have experienced a history of selection for the ability to locate fruit sources efficiently, although relationships between spatial and temporal memory, navigation skills, and food type and distribution in primates and other animals remain unclear. At the least, the Karisoke study population should not serve as the only source for inferences about the importance of spatial and temporal memory and navigation skills in gorilla evolutionary ecology.

Coordination of Group Movement
Mountain gorilla groups use vocalizations to help coordinate and control their movements. Especially when group members are not all in visual contact, vocalizations may reduce exposure to risks associated with being too far from adult males. For example, the rate of "double-syllabled close calls" increases toward the end of siestas (more individuals call, and the call rate per individual increases), and the gorillas often exchange these calls. Among other possible functions, these calls may signal the caller's own intention to move, stimulate responses, allow individuals to assess others' tendencies to resume feeding, and generate a consensus about departure. This behavior facilitates coordinated departures and helps individuals to avoid wasting time and energy on false starts (Harcourt and Stewart 1994; cf. Harcourt, Stewart, and Hauser 1993; Boinski 1991, 1993). Other, "non-syllabled close calls" may promote group cohesion during foraging by conveying information about the attractiveness of feeding sites and mediating feeding competition (Harcourt, Stewart, and Hauser 1993).

Males are the main decision makers and almost always the ultimate arbiters of movement direction (see Byrne, chap. 17, this volume). Male displays, common at the end of siestas and during fast movement (Schaller 1963; D. P. Watts, pers. obs.), give auditory and visual information about male location, intention to move, and movement direction. Males follow females at the females' pace,

David P. Watts

rather than the reverse, only in unusual situations (e.g., when the top-ranking of four males in group 5 was mate-guarding a female who continually avoided him: D. P. Watts, pers. obs.). Males also respond to females' movements during interactions with other groups or solitary males, and sometimes herd females (Sicotte 1993).

Virunga Mountain Gorillas and Other Gorillas
Dietary variation, especially variation in frugivory, contributes to variation in home range size and use across gorilla populations and subspecies (reviewed in Doran and McNeilage 1998). For example, western lowland gorillas at Lopé (Tutin 1996), Dzanga Sangha (Remis 1997a), and Mondika (Doran and McNeilage 1998) make longer average day journeys than do mountain gorillas in the Karisoke study area, at least partly because they often move far between fruit sources. A similar contrast occurs within the Kahuzi-Biega population of eastern lowland gorillas: gorillas at low altitudes have abundant seasonal fruit resources, while those at high altitudes do not; those at low altitudes travel much farther per day than those at high altitudes during these seasons (Yamagiwa and Mwanza 1994). Data from habitats such as Lopé (Tutin 1996) and Dzanga Sangha (Remis 1997a,b) indicate that where gorillas eat a lot of fruit, fruit patch size, dispersion, and temporal availability strongly influence area occupation densities when fruit is abundant.

Gorillas outside the Virungas use leaves, stems, and pith from terrestrial vegetation as fallbacks when fruit is scarce, but also eat these perennial foods throughout the year (e.g., Lopé: Williamson et al. 1990; Ndoki: Kuroda et al. 1996). Their habitats are heterogeneous, and the spatial distribution, abundance, and nutritional value of herbaceous foods can vary markedly across vegetation types (White et al. 1995). Associated variation in area occupation strategies should occur, with the expectation that the gorillas should concentrate their activities in the more attractive parts of the "fine mosaic" of vegetation (White et al. 1995), at least when fruit abundance is low. Comparative quantitative data on home range use in relation to the distribution of herbaceous foods are limited, but provide some support for this prediction. For example, encounter rates with gorillas at Lopé were highest in areas dominated by Marantaceae forest, in which herbaceous food biomass is highest (White 1994). Also, eastern lowland gorillas at high altitudes in Kahuzi-Biega use bamboo forest heavily when new shoots are available (Casimir 1975; Casimir and Butenandt 1973; Goodall 1977). The open swamps where some lowland gorillas regularly eat aquatic plants provide an intriguing case in point. Groups visit these regularly, and multiple groups often feed simultaneously (Olejniczak 1994). Along with the relatively sparse distribution of fruit resources, greater variation in the distribution of herbaceous foods than in the Virungas and lower overall food abundance (White et al. 1995) could contribute to longer day journeys and larger home ranges per individual at sites such as Lopé (Tutin 1996) and Mondika (Doran and McNeilage 1998). These differences might also contribute to less even home range use than in the Virungas.

Solitary male western lowland gorillas at Lopé sometimes pursue groups for up to several days (Tutin 1996). This implies that they influence group movements, and that mating strategies influence their own habitat use more than considerations of foraging efficiency, as in mountain gorillas.

Mountain Gorillas and Other Large Herbivores

Use of broad vegetation swards that contain relatively small but densely packed food sources, like the herbaceous vegetation of the Virungas, is unusual among primates, but characterizes many ungulates. Trampling effects are also important for many ungulates. Buffalo, for example, also cause trampling damage in the Virungas, and regeneration time for herbaceous vegetation is the same as after gorilla trampling (Plumptre 1993). The effects of buffalo trampling on vegetation composition and on gorilla habitat use have not been measured. Dietary overlap between gorillas and buffalo is low (Plumptre 1995), but the gorillas may avoid areas that buffalo have recently used heavily.

Plumptre (1995) found higher dietary overlap between mountain gorillas and elephants. Overlap in habitat use and diet between western lowland gorillas and forest elephants may typically be high, or might have been high in the past (White, Tutin, and Fernandez 1993; White et al. 1995). Elephant trampling in the Virungas kills many plant stems, unlike trampling by gorillas and buffalo (which mostly bend stems and thereby promote regeneration from side shoots), and vegetation takes longer to regenerate from elephant damage than from gorilla trampling (Plumptre 1993).

Few, if any, elephants survive in the Virungas, but they could previously have had major direct and indirect effects on gorilla

David P. Watts

habitat use: gorillas might have avoided recently trampled areas, and elephant feeding and trampling probably strongly influenced vegetation composition. Long-term data from Kibale in Uganda (Struhsaker, Lwanga, and Kasenene 1996) confirm the hypothesis (Eggeling 1947; Laws 1970) that elephant browsing and trampling can prevent forest regeneration and maintain areas with dense herbaceous ground cover. Elephants in Kibale use habitat types and foods selectively. They particularly use areas of herbaceous tangle—many created by treefalls or logging—where they eat herbs, trample seedlings, and damage saplings and poles (either incidentally or by browsing them). This activity maintains their favored habitat, and elephants may perpetuate secondary growth and herbaceous tangle vegetation in forests across Africa in this way (Eggeling 1947; Jones 1955; Langdale-Brown, Osmaston, and Wilson 1964; Struhsaker, Lwanga, and Kasenene 1996); perhaps this formerly included the Virungas.

Virunga gorillas may create local concentrations of food plants by stimulating vegetative reproduction and then maintain these through subsequent use, as described above. This strategy would be similar to the proposed maintenance of nutritional "hot spots" in the Serengeti by ungulate grazing (MacNaughton and Banyikwa 1995). Serengeti hot spots can persist for at least 20 years, but occur in a context of greater temporal heterogeneity in food abundance and quality than for herbaceous vegetation in the Virungas.

Female Transfer and Habitat Use in Mountain Gorillas
High home range overlap and low site fidelity also characterize some other primates that, like mountain gorillas, are highly folivorous and in which female transfer is common (e.g., Thomas langurs: Sterck 1997; capped langurs: Stanford 1991). Like the females of group Nk, females in these species may shift their home ranges without leaving their groups ("locational dispersal": Isbell and Van Vuren 1996); they may also change groups, but continue to use large parts of their previous home ranges (reviewed in Isbell and Van Vuren 1996). These strategies should reduce the ecological costs of dispersal. Low within-group contest competition for food among females, and the limited advantages of long-term alliances between related females, correspondingly limit the social costs of dispersal and of acceptance of female immigrants. Along with the low ecological costs of dispersal, these factors facilitate female transfer in mountain gorillas and in some other primates (Watts 1990, 1996;

Isbell and Van Vuren 1996; Sterck 1997; Sterck, Watts, and van Schaik 1997). Age at first reproduction is the same for female mountain gorillas regardless of whether they have transferred or are still in their natal groups; this is strong evidence that transfer is not costly (Watts 1996).

Summary

Thirty years of research at Karisoke have provided a wealth of data on mountain gorilla habitat use and group movements and an excellent basis for comparison across gorilla subspecies and populations. They have also left many questions unanswered, and we can only hope that resolution of the ongoing political, economic, and military crises in the region will allow conservation efforts to continue and research that addresses these and other questions to resume. Some of the major findings and remaining questions include the following:

1. Mountain gorilla groups in the Karisoke study area concentrated their activities over the long term in areas, and in habitat types, where food abundance was relatively high. The availability of high-protein, relatively digestible food was particularly important. Limited evidence points to similar effects outside the study area, but vegetation heterogeneity in the Virungas raises questions about how consistent these are.
2. For one gorilla group, day journey length, distance between consecutive feeding spots, and feeding time varied inversely with food biomass and quality across habitat types. The group spent the most time in areas with abundant high-quality food, and used vegetation types with these characteristics heavily in relation to their availability in the habitat; thus they concentrated their activities in areas where total foraging effort was low. Other groups apparently behaved similarly, but replication of these analyses would be valuable.
3. Seasonal effects were limited except for the heavy use of bamboo forest at times when new bamboo shoots were appearing. Bamboo shoots are an important food for much of the population, but not all groups have bamboo forest within their home ranges.
4. Male mating strategies also influenced movements. Groups sometimes moved far to avoid other groups or lone males, and lone males sometimes followed or pursued groups. The effect of these encounters on groups was usually short-lived, but in one case it led to a complete home range shift. The one solitary male for whom data were available put sufficient effort into mate search that his long-term habitat use was independent of variation in food distribution. Once he acquired females, food abundance and quality strongly influenced his movements.

David P. Watts

5. The home ranges of Karisoke study groups continued to expand over 5 years or more, and overlap between annual home ranges and core areas was often low. Several factors may limit home range stability; in turn, long-term home range expansions and shifts, plus high home range overlap between groups, should reduce the ecological costs of dispersal for females.
6. Circumstantial evidence suggests that the gorillas have good spatial and temporal knowledge of their habitat; good spatial mapping ability should be important for other, more frugivorous gorilla populations.
7. Mountain gorillas use several short-range vocalizations as well as male displays to coordinate movements. Males are the leaders of group movements.
8. Like other large African herbivores that browse on tall herbaceous vegetation, and like grazing ungulates, mountain gorillas deplete resources while they forage, but also can stimulate food production. Study groups showed some adjustment of rates of revisits to areas to the extent of resource depletion, but revisit rates were more strongly tied to variation in total food abundance. The gorillas may be agents in a cycle of vegetation trampling/regeneration/reuse that maintains, or even improves, habitat quality. Long-term monitoring of trampled areas and experiments using vegetation exclosures could address this issue more completely, and are an important research priority.
9. Existing data on other gorilla subspecies and populations show that variation in fruit availability influences their habitat use predictably. How similar their responses to variation in the distribution of perennial herb and vine foods are to those of Karisoke mountain gorillas remains to be determined.

Acknowledgments

I thank L'Office Rwandaise du Tourisme et des Parcs Nationaux for permission to work in the Park National des Volcans, and L'Institut Zairois pour la Conservation de la Nature for permission to work in the Parc National des Virungas, in 1978–79, 1984–85, 1986–87, 1989, 1991, and 1992. I am indebted to the late Dian Fossey for permission to work at Karisoke in 1978–79 and 1984–85. My research was funded by NIMH grant NIMH 5T32-15181-03 and by the L. S. B. Leakey Foundation, the Dian Fossey Gorilla Fund, the Eppley Foundation for Research, the Chicago Zoological Society, the World Wildlife Fund U.S., and the Wildlife Preservation Trust International. Sue Boinski, Paul Garber, and three anonymous reviewers made helpful critical comments on earlier versions of this chapter, and Dick Byrne provided an illuminating exchange about cognitive mapping. Many people contributed to the Karisoke long-term habitat use records. Sandy Harcourt, Kelly Stewart, Juichi Yamagiwa, Richard Barnes, and Jan Rafert deserve special thanks. Those records exist, and my research at Karisoke was possible, only because of the dedicated and skilled assistance of A. Banyangandora, K. Munyanganga, C. Nkeramugaba, S. Kwiha, L. Munyanshoza, the late F. Nshogoza, E. Rukera, F. Barabgiriza, L. Kananira, and, especially, A. Nemeye and the late E. Rwelekana. I dedicate this chapter to them, their Rwandan colleagues at Karisoke, and their families in gratitude for their efforts to keep Karisoke alive, and in the hope that the future is brighter than the present for them and for the gorillas.

CHAPTER FOURTEEN

Quo Vadis? Tactics of Food Search and Group Movement in Primates and Other Animals

KATHARINE MILTON

> When comparative psychologists want to study a particular cognitive skill . . . they would do well to search the animal kingdom for taxa that solve similar problems.
>
> (P. Tyack, 138) 1993

The question posed in this section of this volume is whether taxonomic differences in the group movements of primates reflect differences in the ways in which species process information. In primates, group movement (travel) almost invariably occurs in the context of foraging. Thus, to explore this question, I begin with a detailed analysis of food search behavior in mantled howler monkeys *(Alouatta palliata),* then use information generated by this analysis to examine features of group movement in four other primate taxa: spider monkeys, chimpanzees, bonobos, and human foragers. I then extend this examination beyond primates, seeking common attributes of food-associated group movement and cognitive processes across ordinal boundaries by looking at the foraging and travel behavior of members of two other taxa; namely, dolphins and parrots.

The different sections of this chapter are somewhat disjunct because in each case, the type and amount of available information varied. For howler monkeys, I had detailed field data that could be used to examine quantitatively some features of foraging behavior, whereas for other taxa, both primate and nonprimate, I relied largely on information from the literature, generally extensive. When I turned to foraging behavior and its associated cognitive processes in my own species, the amount of potentially relevant data increased exponentially. For this reason, comment on group movement and food search behavior in human foragers is limited to a small subset of case studies.

Background

In 1974, when I began my study of the dietary ecology of wild howler monkeys *(Alouatta palliata)* in Panama, primatologists working with more "lively" species often remarked that they couldn't understand how I could spend so much time with such an inactive monkey. Didn't I get bored? they asked. Couldn't I just predict most events in a howler monkey's day? Suffice it to say that I did not find howlers predictable (in my view, the predictable animal is extinct since predators and parasites would have little trouble tracking a predictable organism), and probably for this reason, did not find them boring. Particularly in the early months of my study, the monkeys were continually eating food plants that were new to me, and it was extremely rare for a sample period to pass in which I was not led to one or more huge fruiting or flushing trees whose existence I had never suspected. Indeed, I often found myself hard pressed to keep up with all of the information my "inactive" study subjects were continually supplying in terms of their dietary activities.

By the conclusion of my one-year study, I'd learned the general extent of the home range area of each of my two study troops and where clusters of productive food trees could be found such that, when a howler troop began to travel, I could at times predict where they were going and what they might eat when they got there (Milton 1977, 1980). But compared with the monkeys, I was still largely a novice at finding rich food sources. Furthermore, it was still not uncommon for my study subjects to travel in a directed manner to huge fruiting trees whose existence I had never suspected.

The lowland tropical forest on Barro Colorado Island (BCI) in Panama, where my study took place contains about 200 large ($\geq$ 60 cm in circumference breast height, CBH) trees per hectare (100 ×

100 m) as well as numerous smaller trees, saplings, shrubs, bushes, vines, and other vegetation. Visibility, both on the ground and in the canopy, is limited, and thus my lack of information as to the locations of some of the thousands of trees in the 32-hectare home range of my two study troops isn't that surprising. After all, I spent only 5 days per month with each of my two study troops. If I had spent an entire year with only one troop, following it about the forest every day for 365 days, I'm sure I would have learned considerably more about the locations of most preferred fruit trees in its home range. But even based on only 10 days of fieldwork per month, it was clear that howler monkeys had a food search strategy that often appeared goal directed.

Goal-Directed Travel
When howlers set out to travel (actively moving from tree to tree for a distance of 100 meters or more), they did not move about the forest in a hesitant or random manner, but rather traveled slowly but steadily through the canopy in single file, typically finishing their journey at a large tree that contained edible fruits or a large crop of new leaves—a tree that could not be seen from our starting point and whose location and phenological condition appeared to be anticipated by the monkeys before they began to move (Milton 1980; see also Smith 1977). As this happened with new leaf as well as fruit crops, it seemed doubtful that olfactory cues were being utilized, though this remains to be demonstrated (Able 1996; Garber, chap. 10, this volume).

Howler monkeys are not the only primates that have been observed to travel to food trees in this manner. Many primatologists report that their study subjects travel confidently and directly to food sources located some distance away from the initial starting point (Kummer 1968a; Sigg and Stolba 1981; Garber 1989; Garber, chap. 10, this volume; Byrne, Whiten, and Henzi 1990; Byrne, chap. 17, this volume; Boinski 1996; Boinski, chap. 3, this volume). In particular, Garber's (1989) study of foraging patterns of two *Saguinus* species in Peru showed that a high percentage of travel to food trees was goal directed and that animals could move directly to the nearest conspecific neighbor of a particular fruiting tree over long distances; the monkeys could also reach these particular trees by various arboreal pathways, depending on where they were in their home range when travel was initiated.

In the laboratory, controlled experiments have shown that some monkey and ape species have well-developed and highly flexible memory strategies for remembering food locations (Menzel 1973a; MacDonald and Wilkie 1990; Gallistel and Cramer 1996). Experimental subjects of several species also minimized the travel distance to food locations even though there were no obvious experimental constraints on either time or distance traveled (MacDonald and Wilkie 1990; Gallistel and Cramer 1996). As goal-directed travel behavior and computation of the optimal (most efficient) travel route to one or more foraging goals has been demonstrated for primates both in the wild and under captive conditions, the problem thus becomes one of determining how this might be accomplished and what benefits may accrue to its practitioners as a result. The theory of foraging strategy offers testable predictions that aid in such investigation.

Foraging Strategy Theory
The theory of foraging strategy (or optimal diet theory) predicts that the efficiency of foraging is maximized by natural selection (Schoener 1971). In early models associated with this theory, the optimal diet was viewed as the set of food choices that maximized net energy yield/foraging time, or some other units believed to reflect fitness. The usual variables were the caloric value of foods and the time spent in search and pursuit of edible prey. However, researchers interested in the foraging behavior of plant-eating animals soon pointed out that the foods of primary consumers (plant eaters) differed in certain important respects from those of secondary consumers (flesh eaters) and that such differences often called for somewhat different foraging solutions (Westoby 1974; Pulliam 1975; Pyke, Pulliam, and Charnov 1977; Milton 1980, 1981a,b; Sih and Milton 1985).

One important difference is that for secondary consumers, potential foods (animal prey) tend to be mobile and evasive, while the plants and plant parts eaten by primary consumers are sessile—tree locations tend to be predictable over the lifetime of a given primate. Thus, for primary consumers (and most anthropoids and some prosimians can be viewed in this light: Schultz 1969; Milton 1980, 1987), the pursuit of mobile prey is generally not the most critical variable in foraging efficiency (Westoby 1974), and indeed, in many cases (e.g., gorillas, howler monkeys, spider monkeys), is not important at all.

A second important difference between the foods of secondary and primary consumers is that the foods of primary consumers vary far more in their nutritional (and toxic) content than foods from the second trophic level (Freeland and Janzen 1974; Westoby 1974; Milton 1980). Leaves, for example, particularly young leaves, tend to be high in protein though low in available energy; in contrast, ripe fruits tend to be high in digestible energy but low in protein (Milton 1980, 1981b). Yet primates require both adequate protein and energy (as well as other nutrients and water) to remain in good health.

Because of the characteristics of their foods, the foraging objective of many primary consumers should be to optimize the nutrient mix within a given fixed bulk of food, rather than simply maximizing net energy yield/foraging time (Westoby 1974). Being selective in feeding, however (selectivity being called for to obtain the best possible dietary mix of high-quality plant parts), would increase the time spent searching for food (since high-quality plant parts are generally less abundant than low-quality plant parts). This would increase foraging costs, both directly in terms of the energy expenditure required to find high-quality foods and indirectly in the sense of opportunity lost—time spent seeking food could be devoted to some other beneficial activity, such as seeking out mates. Prolonged food search could also increase the risk of exposure to predators (e.g., van Schaik 1983; Terborgh and Janson 1986) or accidents (Milton 1980).

For all of these reasons, we would predict that primary consumers might exhibit features (both behavioral and morphological) that function to reduce the costs of their selectivity (Schoener 1971; see Milton 1980, chap. 6 for discussion of behavioral and morphological features that reduce the costs of selectivity in *A. palliata*). In particular, when preferences for certain dietary items are marked and such items are patchily distributed in space and time—which is precisely the case for most higher primates—selection should favor behaviors that improve the forager's probabilities of encountering desirable high-quality foods with the lowest expenditure of time or energy. Examination of the food search strategy of howler monkeys provides insight into how certain travel behaviors might function to enhance their foraging efficiency.

Study Subjects
Howler monkeys on BCI live in relatively closed social units (troops) averaging nineteen individuals and composed of some

three to four adult males, five to eight adult females, and five to seven immature animals. Adult males weigh between 7 and 9 kg, about 20% more than adult females. During most of the year, howler monkeys eat a mixed diet composed of about 48% leaves, 42% fruits, and 10% flowers. The number of food species used per day is about 8: 5.1 leaf species, 1.7 fruit species, and 0.8 flower species (Milton 1977, 1980). However, during the transition period between the wet and dry seasons on BCI, when ripe fruits are in short supply, howler monkeys rely heavily (85–100% of feeding time per day) on leaves as a dietary staple; the daily number of food species during this period is also about 8, but the diet is now composed of 7 leaf species, 0.6 fruit species, and 0.9 flower species (Milton 1977, 1980). Howler monkeys can live on diets high in leaves for weeks at a time (Glander 1978; Milton 1980). Feeding data show that, generally, howlers orient their feeding around one or two primary food trees (a primary food tree being a food source that is used for >20% of feeding time per day) and a variety of secondary food sources fed in for shorter, often extremely brief, periods of time (Milton 1977, 1980).

A howler monkey troop tends to perform all of its daily activities as a unit. When one animal is eating, there is a strong probability that most or all other animals in its troop are eating too; when one monkey is traveling through the canopy, there is a high probability that other members of the troop are likewise traveling. On BCI, howler troops generally do not fission into subgroups during the day to forage except in rare instances, as on peninsulas, whose narrow configuration makes foraging as a cohesive unit difficult, or when feeding on scattered leaf sources. For most of the year, a typical howler troop on BCI can be thought of as a tight-knit social unit whose constituent parts move through the forest together, feed and rest together, howl together, do everything together—howler monkey troops on BCI are the essence of togetherness.

Travel Initiation
The decision as to which direction a howler troop will travel in seems to be made by the alpha male (Carpenter 1934). This male generally makes a few low clucking or rumbling vocalizations to alert other troop members; he may walk a few steps in a particular direction, then sit down again. These behaviors, in some manner, appear to indicate to other troop members the direction in which the troop will travel (Carpenter 1934). Other animals will then

Katharine Milton

slowly begin to file out of the tree, passing by the alpha male, who may sit until most troop members are already moving along the travel route. My observations indicate that the alpha male does not then run to the head of the procession, but may be found traveling at almost any position. This travel pattern indicates that though one animal may suggest or choose the travel destination, once this is somehow "decided," many other troops members "know" the arboreal pathway to take to move directly to that new locale and feeding tree, as animals can and do travel to their destination in almost any processional order.

However, the subject of how group movement is initiated in howler monkeys warrants further study. Prins (1996), for example, observed buffalo cows reaching travel direction consensus through subtle stretching and gazing behavior during periods of resting; this behavior (termed "voting behavior") served to coordinate later group movement (see also Kummer 1968a; Sigg and Stolba 1981; Byrne, chap. 17, this volume; Wilson, chap. 9, this volume). In howler monkeys, it is possible that travel path decisions result from subtle troop consensus rather than the decision of a single dominant member. As the knowledge possessed by the group as a whole should add up to more than the knowledge of any one member, animals might do better (be more efficient foragers) by using some type of consensus mechanism to direct their major episodes of food travel, rather than just playing follow the leader (e.g., Wilson 1997a).

Foraging Efficiency on Patchily Distributed Foods

Examination of the distribution pattern of potential howler diet items on BCI showed that the seasonal items howler monkeys prefer to eat (tender young leaves, soft ripe fruits, flowers) are far more patchily distributed in space and time in the BCI forest than largely ignored perennial foods such as mature leaves. Study of the relative availability of foods in particular dietary categories (mature leaves, new leaves, flowers, green fruits, ripe fruits) showed notable differences, with ripe fruits showing by far the most patchy distribution, both by species and by individual tree (tables 14.1 and 14.2; see Milton 1977, 1980 for additional details).

Given the uneven distribution of preferred howler foods, procurement costs for these foods, as measured in the time or energy spent seeking them out, must be considerably higher than if the monkeys focused more of their feeding time on more uniformly dis-

Table 14.1 Number of months in a year when seasonal foods are available from twelve tree species drawn at random from the Barro Colorado Island forest

Species	Young leaves	Number of months Fruits	Flowers
Hirtella triandra	10	4	4
Jacaranda copaia	7	5	2
Tabebuia rosea	10	2	6
Ceiba pentandra	12	2	2
Apeiba membranacea	9	11	7
Triplaris cumingiana	9	2	4
Palicourea guianensis	11	6	3
Ormosia coccinea	8	0	2
Ocotea cernua	6	3	3
Paullinia turbacensis	4	7	3
Zuelania guidonia	4	3	3
Cecropia insignis	8	8	6

Table 14.2 Mean number of months in a year during which seasonal foods were observed on twelve tree species drawn at random from the Barro Colorado Island forest

Food category	Species	Individual trees
Young leaves	6.81 ± 2.53	5.26 ± 2.53
Mature leaves	11.75 ± 0.46	11.03 ± 1.22
Green and ripe fruits	3.67 ± 2.92	2.08 ± 1.80
Ripe fruits	1.13 ± 1.27	0.78 ± 1.00
Flower buds and flowers	2.73 ± 2.00	1.84 ± 1.12

tributed and abundant foods such as mature leaves. These additional procurement costs must compensated for, at least in part, by the higher nutritional quality of these preferred seasonal foods (Milton 1980). Even so, the potential return in nutrients is limited, and the howler digestive tract can hold only a limited amount of food at any one time (Milton 1980; Milton 1981b). In particular, during seasonal low points in the availability of energy-rich fruits, howler monkeys should be under particularly strong pressure to minimize the energy they expend in food search. How are howler monkeys able to afford the costs of their selectivity?

One important way to lower the costs associated with food procurement would be to minimize the time and energy spent searching for food. This could be achieved by a food search strategy that maximized the probabilities of encountering preferred high-quality foods. The optimal search strategy would depend on the distribution patterns of preferred foods. If, for example, preferred foods were uniformly distributed in time and space, the most efficient

Katharine Milton

search strategy would be to cover the entire supplying area in a "lawn mower" pattern, moving back and forth in even, nonoverlapping swaths until the entire area had been covered. But if the distribution of preferred foods was extremely patchy in space and time, as is the case for howler monkey foods on BCI (Milton 1977, 1980), the most efficient strategy would be to use a pattern of goal-directed travel—that is, to move directly to important sources of preferred foods (primary food species) when and where they were available without wasting valuable time and energy in random search.

As noted above, considerable field data suggest that howler monkeys often show a pattern suggesting goal-directed travel to primary food sources. To travel in this fashion, however, monkeys would somehow have to know, prior to travel initiation, *when* a particular highly desirable food source was available as well as *where* the particular tree was in their home range *and* the most direct route to it. Thus, to be effective, the food search strategy of howler monkeys would have to have at least two essential components: (1) the monitoring over a fairly substantial (32 hectares) area of a large number of specific trees such their phenological state could be tracked, and (2) the ability to determine each day which primary food trees were to be used such that monkeys could travel by the most economical route (straight line) to them.

Use of Two Ficus *Species*
To investigate the pattern of food search used by howler monkeys, I focused on the use of two important and highly preferred food species (*Ficus yoponensis* and *F. insipida,* Moraceae) by one howler monkey troop (14–18 individuals) living in the Lutz Ravine area on BCI (Milton 1980). This area of the forest consists of various large areas of old second growth interspersed with patches of mature primary forest, undisturbed perhaps for centuries (Hubbell and Foster 1990). Monkeys in the Lutz Ravine study troop spent 25.3% of their annual feeding time eating fruits and leaves from these two fig species (Milton 1980). Out of a total of 40 sample days, howlers ate foods from one or both *Ficus* species on 26 days (65% of total sample). Yet the relative density of these two fig species together was only 0.00371% (relative density was calculated from an examination of the number of trees ≥ 60 cm CBH in 30,000 m^2 of the 32 ha home range of the monkeys and a direct count of the total number of adult fig trees of these two species within this total supplying area).

Phenological data (Milton 1991) showed that, unlike many tree species on BCI, these two *Ficus* species were intraspecifically asynchronous in fruit production. Further, individual trees of each species produced fruit crops at different times each year. Trees of both species often produced small quantities of new leaves and, once or twice per year, large crops of new leaves (Milton 1991). Thus there appeared to be a strong incentive to howlers for visiting ("keeping an eye on") individuals of these *Ficus* species, since they tended to offer nutrient "rewards" for vigilance on a species-wide basis more frequently than most other tree species in the habitat. Most other tree species in the BCI forest are intraspecifically synchronous in their production of ripe fruits or new leaves, producing these seasonal items at best only once per year and producing them at approximately the same time each year (Milton 1980).

If howler monkeys have features that increase food search efficiency, field data should show that they are encountering individuals of *Ficus yoponensis* and *F. insipida* more frequently than if they were traveling around their home range at random. To test this assumption, I used data from my study in a search path model, as indicated in table 14.3. The supplying area of this study troop was determined by measuring all travel routes used on each of the 40 sample days (8 months of 5-day sampling designed to cover all portions of an annual cycle) (Milton 1977, 1980). These travel routes were referenced to a baseline point and plotted by computer onto a map of the study area. The area of the least convex polygon that enclosed all movements of the troop in the 40 sample days was re-

Table 14.3 Search path model parameters

Number of days in sample	40
Mean daily travel time (t)	1.22 hrs $\pm$ 0.46
Mean daily travel distance (d)	392 m $\pm$ 127
Mean radius of tree crowns (r_1)	7.00 m
Mean radius of fig tree crowns (r_2)	11.70 m
Width of effective search path (s)	37.40
Area covered by troop per day (A)	14,660.80 m^2
Mean density of trees per m^2 (q)	.0185
Number of trees "sampled" in 40 days (n)	10,848.99
Relative density of fig trees (p)	.0037
Number of encounters with fig trees (X)	75

Note: r_1 is from a random sample of 20 trees in the Barro Colorado forest with a circumference at breast height (CBH) of 60 cm or more. r_2 is from a random sample of 20 mature fig trees (10 *Ficus yoponensis* and 10 *F. insipida*) in the Barro Colorado Island forest. $s = 2 (r_1 + r_2)$. $A = ds$. q is for trees with a CBH of 60 cm or more from three 1 ha sample plots. $n = 40 Aq$. p is from 3 ha of the troop's 32 ha supplying area in which all *Ficus* were tagged and mapped. X was counted from trees used as food sources or encountered on travel routes by the study troop.

Katharine Milton

garded as the supplying (home range) area (Waser and Floody 1974; Milton 1980). All data on tree distributions were calculated within this 32 ha area.

Since the monkeys almost invariably traveled in single file through the forest canopy, rather than spreading out and covering a wide swath, I used the mean crown diameter of canopy trees in their supplying area as the width of the troop's search path. The measurements of crown diameter were taken in a random sample that measured 20 trees (≥60 cm CBH) at the widest points of branch extension, so the results should be an overestimate. The "effective search path" should include the width of the food patches (i.e., the crown diameters of the two species of fig trees), as pointed out by Holling (1966) and Pulliam (1975). The effective search path for this howler troop was estimated at 37.4 meters (table 14.3).

It is possible that the actual search path was wider than my estimate, but a circle with the assumed radius of effective search would include a mean of 20 trees with a CBH of >60 cm. It would also include smaller trees, vines, and lianas with a heavy overlap of crowns, so that most of the time the visual range of the monkeys must have been severely restricted by the dense foliage. The area covered by this troop in an average day's travel ($\bar{X}$ daily travel distance = 392 ± 127 m) should also be an overestimate, since the calculation assumes that the troop never covered the same area twice in the same day. However, in reality, whenever the troop changed direction, there would have been an overlap of the area covered, as demonstrated by Morrison (1975).

Results showed that howlers were encountering individuals of *Ficus yoponensis* and *F. inspida* significantly more frequently than if they had been "sampling" the forest at random ($Z = 5.51$, $p < .0001$), where

$$Z = \frac{X - np}{\sqrt{np(1 - p)}},$$

with X, n, and p as defined in table 14.3. This result supports the hypothesis that howlers have features that increase their efficiency in food search. The question thus posed is, what might such features be?

Features of Food Search

In observing their daily travel patterns, I noted that howlers appeared to concentrate travel in areas of their home range with relatively high densities of fig trees. An analysis of the ranging data showed that three of the four heaviest concentrations of travel routes also contained relatively high numbers of *F. yoponensis* and *F. inspida* (Milton 1980). There were an estimated twenty-one individuals of these two fig species in the 32 ha home range of the study troop. The 3 ha with maximal travel concentration contained ten of these twenty-one individuals (48%), six individuals in one hectare, four in another, and one in the third. Thus howler monkeys were concentrating travel in areas of their home range where chances of encountering these preferred foods were relatively high. As Tullock (1971) has phrased it, they were "shopping in the cheapest market."

Howler monkeys elsewhere have likewise been reported to forage in areas where fig tree densities are high (Sekulik 1981). Indeed, the home range size of one red howler monkey *(Alouatta seniculus)* troop studied by Sekulik (1981) in the llanos of Venezuela was so small (3.9 ha), particularly in terms of the sections containing trees (about 49% of the home range was open grassland), that rather than shopping in the cheapest market, these howlers might well be said to have been living in the market itself.

Another key feature of howler travel both on BCI and elsewhere is the repeated use of particular arboreal pathways that tend to connect areas of the home range where the densities of preferred food trees are relatively high (Milton 1977, 1980; Sekulik 1981). Similar arboreal pathways have also been noted for some other primate species and are likely to be characteristic of all primate species. For example, MacKinnon (1974, cited in Bell 1991) remarked on the use of distinct arboreal "highways" by orangutans and noted that such highways often followed natural environmental features such as ridges or streams and were used by various individuals.

As noted, howlers in the Lutz Ravine area of BCI traveled an average of 392 m per day (table 14.3). During travel, they typically visited one or more primary ($\geq$20% feeding time per day) food sources and various secondary (<20% feeding time per day) food sources, often moving in a loop away from a primary source and then returning to it later that same day, or on subsequent days within a 5-day sample period (Milton 1977, 1980). Primary food sources are almost invariably fruits, whereas secondary food sources generally are leaves. Further, as discussed above, howlers tended to concentrate foraging activity in areas of their home range where fig tree density was relatively high.

The combination of these foraging behaviors could be very effi-

Katharine Milton

cient. By using a known food source as a primary reference point and traveling a relatively predictable distance away from it each day, howlers would never be more than a normal day's travel from any productive food source in their entire home range (Milton 1977, 1980). Thus they avoid unpleasant "surprises" in terms of unexpectedly long day ranges and are able to keep their daily energetic expenditures (i.e., travel costs) both predictable and relatively constant. At the same time, they can also take advantage of any other high-quality food sources encountered while traveling as well as monitoring the phenological condition of other important food trees in their home range. Wrangham (1977), for example, has noted that chimpanzees at Gombe show daily travel patterns that at times suggest they are monitoring the phenological condition of particular fruit trees and can return to them at the optimal moment for harvest (see also van Roosmalen 1985).

Operations Research and Search Theory
Operations research is a field of study which, by assessing the overall implications of various alternative courses of action in a given management system, tries to predict the optimal policy or suggest an improved basis for decisions (see, for example, Pocock 1956). It has been termed "the science of generalized strategies and tactics" (Camp 1956, 102). Operations research seeks to discover regularities in apparently unrelated or random activities, drawing on techniques such as linear programming, game theory, search theory, and dynamic programming, among others (Pocock 1956; Hillier and Lieberman 1967). Barnett (1976) used search theory to construct a model designed to predict the optimal search pattern for preventive activities of police officers in cars who were interested in visiting potential crime sites in such a way as to intercept crimes in progress. Visiting every potential crime site was not feasible; further, some locales were far more likely to be (had a higher probability of being) the scene of a crime than others. In his model, Barnett showed that for events of limited duration (i.e., crimes) arising randomly in time at several discrete points, if an optimal search policy exists, then there is a cyclic optimal policy; that is, a regular pattern of visiting points that generate events less frequently every k units of time and busier points the rest of the time (see Barnett 1976 for a full description of the parameters of this model). In essence, Barnett's model appears applicable to the problem of howler foraging behavior in terms of finding fruiting fig trees.

Fig trees are not totally random in fruit production, but individual trees do alter the time(s) of year at which they produce fruit crops and, to a lesser extent, leaf crops as well (Milton 1991). This variation makes it difficult for howlers to predict precisely when an individual fig tree in their supplying area will initiate a fruit crop or produce new leaves, but it makes fig trees relatively worth monitoring (fig trees, in this sense, are like areas with higher probabilities of violent crime). From the point of view of howlers, it would appear that fig trees (and perhaps other tree species with similar traits, such as other members of the Moraceae; Milton 1991) can be regarded as "random busier points" to be visited on a fairly frequent basis and other, either less productive or more predictably productive food species, points to be visited every k units of time. Therefore, if there is an optimal food search strategy for howlers, it could well resemble the one shown by this study troop and predicted by Barnett's search theory model.

Estimates of Efficiency
Relative efficiency in howler food search can be measured in terms of time spent searching. Based on the data in table 14.3, the expected time needed for howlers to encounter one fig tree in random search would be $t/dspq = 1.22$ hours. The actual mean search time per fig tree was $40t/X = 0.65$ hours. Thus for fig trees, these howler monkeys were 47% more efficient in food search than if they had been traveling at random. This is a quite substantial improvement over a policy of random search.

It is easy to see the effect of this food search behavior on the predictions of some models for optimal diet. In the graphic models of MacArthur and Pianka (1966), a reduction in search time would have the same effect as increasing food density—that is, it would reduce the predicted number of different items in the diet. Thus the animal could be more *selective* in its choice of food. In the models of Emlen (1966), Schoener (1971, 1974) and Pulliam (1975), a reduction in search time would also result in a prediction of greater selectivity on the part of the forager. Thus in all four models, the animal can be more selective in its choice of diet at the same level of food density.

A corollary effect of a reduction in search time would be to lower the level of food density at which an item would be excluded from the diet. This is in accord with Westoby's (1974) prediction that the

Katharine Milton

choice of diet by large herbivores will be constrained by availability of foods only at rather low levels. As he points out, for such animals, pursuit time is not a variable, being small and similar for all food items, and the probability of capture is essentially 1 (a plant part cannot escape from an herbivore). This should be true for all terrestrial primary consumers, though the probability of capture may be somewhat less than 1 for arboreal primary consumers due to the possible inaccessibility of some food items (e.g., at the tips of small, flexible branches). Search time, however, is a variable, and since the "prey" items of primary consumers are stationary and also relatively predictable in time, there is clearly much room for improvement over a strategy of random search.

Given the potential payoff, it is reasonable to expect that primary consumers such as howler monkeys should have behavioral adaptations that enable them to locate preferred foods more efficiently than if they were searching at random. Application of Barnett's (1976) model to my data on howler monkey foraging behavior suggests that they do. These features may have been selected for early in primate evolution and should be characteristic of many other primate species eating plant-based diets (Milton 1979b, 1981a, 1988).

Examination of food search strategy in howler monkeys leads naturally to consideration of the cognitive processes involved in such behavior. When a given troop sets out in a purposeful manner and travels for 300 m or more through the canopy to a fruiting tree by a direct route, how is the travel route being determined?

Memory Strategy in Howler Monkeys
As discussed above, laboratory experiments and actual field data indicate that primates use memory strategies to locate preferred foods (Menzel 1971a; MacDonald and Wilkie 1990; Gallistel and Cramer 1996). Currently, there seems fairly general agreement that many mammals, including primates, navigate to particular travel goals by means of path integration (or dead reckoning) in combination with landmark information (e.g., Sherry 1996, 163; Etienne, Maurer, and Séguinot 1996; Gallistel and Cramer 1996). Recently, Gallistel and Cramer (1996) proposed a model suggesting that cognitive maps (mental representations of space possessed by the animal and used in navigation: Sherry 1996) consist of landmarks placed on a geocentric frame of reference by vector addition. Ac-

cording to this model, animals determine their geocentric position by path integration and take positional fixes on known landmarks to deal with the error accumulation of path integration.

For many researchers, one test for a cognitive map has been whether an animal can use its map to derive a novel shortcut—a direct route between two places that are familiar but have never before been visited successively (Tolman 1948; O'Keefe and Nadel 1978; Sherry 1996). Bennett (1996) recently reviewed evidence for novel shortcuts (hence proof of cognitive maps) for a variety of different organisms, including nonhuman primates and humans, and concluded that until other simpler explanations for such behavior have been tested and rejected, all claims for cognitive maps based on the premise of a novel shortcut are premature.

In the case of howler monkeys, the question of a cognitive map *sensu* Tolman and O'Keefe and Nadel cannot even be addressed, as I have no data on the use of novel shortcuts. Nor can I address the utility of the model proposed by Gallistel and Cramer (1996), though it has much to recommend it (Bennett 1996). Even the use of long-term spatial memory by howlers is uncertain, as they may actually not have to remember a great deal about the locations of potential foods in their supplying area. Rather, in line with the systems model discussed above, howler monkeys could simply keep moving steadily for a fairly set distance day after day, over traditional arboreal pathways that lead directly from one cluster of important food trees ("random busier points") to another, checking less busy points as they pass by them. For howlers and other primates with similar travel patterns, such daily group movement would actually kill two birds with one stone. Howler monkeys not only have to know *where* particular trees are and how to travel efficiently to them, but they also have to monitor particular key food trees in their home range such that they know *when* to visit them. Both of these demands could be satisfied in large part by the system of travel shown by howler monkeys, as discussed above.

Data show that the Lutz Ravine study troop covered between 32% and 64% of its total home range (32 ha) in six 5-day sample periods (April, July, Aug., Sept., Oct., Nov.), averaging 47% home range coverage per sample. In the dry season, average home range coverage per 5-day sample was 64%, while in the rainy season it was 43%. In January of 1975 (dry season), the travel movements of one howler troop were observed for 20 continuous days; at the end of this period, the troop had covered almost exactly two-thirds of its total home range area. Average *daily* coverage of the home range during 3 months of the rainy season (Aug., Sept., Oct.) was 19.3%, with a low of 8% and a high of 32% (Milton 1977, 1980).

These estimates indicate that a howler troop routinely covers about 50% of its home range every 5 days or less. Ripe fruit crops do not appear overnight, but rather can be seen on individual trees for a period of weeks before immature fruits are sufficiently ripe to harvest. The travel pattern shown by howlers should permit a troop to keep a fairly close eye on phenological activity within its total home range without the need for strong dependence on long-term memory. However, though howlers could be relying on more or less continuous monitoring of "potential hot spots" in their home range to detect most important food crops, their skill in moving across their home range directly to distant food sources at the opportune moment for harvest suggests that they do make some use of longer-term memory to enhance foraging efficiency (Milton 1979b, 1981a, 1988). Howler monkeys show a smaller brain size for their body mass than similar-sized frugivorous anthropoids (Milton 1979b; 1981a, 1988). Since they have a demonstrably efficient system for locating high-quality dietary resources within their home range (at least in terms of the two *Ficus* species considered above), larger brain size might not net them a higher return in foraging efficiency, but could prove costly due to the energetic demands of brain tissue (Milton 1988).

Long-Term Occupancy of Home Ranges
The main travel routes (arboreal pathways) that each howler troop follows within its home range may have been worked out through trial and error learning over many successive generations and then passed on through social facilitation and other means to new generations of descendants (Milton 1981a). Field data from BCI covering more than 25 years show that particular howler monkey troops and their descendants use the same general home range area over many consecutive generations; particular troops in particular locales also tend to remain at approximately the same size over time (K. Milton, unpub.). As howler monkeys live in cohesive troops composed of animals of different ages, and as some animals in each troop are philopatric, there are always some monkeys familiar with the principal arboreal pathways and primary food sources in that home range. Foraging efficiency can be greatly enhanced by pooled information about features of diet as well as detailed knowledge of one's

foraging range and resource locations (Bell 1991). Furthermore, as has been suggested for social carnivores (e.g., lions: Packer and Ruttan 1988), group living and group size in primates may relate, at least in part, to the need to have a social group of a particular size and composition such that key dietary information can be passed on smoothly and efficiently from generation to generation.

Group Movement and Food Search in Spider Monkeys (*Ateles* spp.)

> Brains are metabolically expensive and don't get bigger (phylogeneti-
> cally) unless in some fashion they are more than paying for their up-
> keep. (Hockett 1978, cited in Ridgway 1986)

What happens when we turn to consideration of a primate species that shows a different pattern of group movement than howler monkeys while living in precisely the same forest and even using many of the same fruit trees; namely, spider monkeys *(Ateles geoffroyi)* on BCI? Spider monkeys, which as adults are approximately the same size as howler monkeys, eat a diet composed primarily of ripe fruits, the most patchily distributed high-quality plant food in the BCI canopy see (tables 14.1 and 14.2). Ninety percent or more of the daily feeding time of spider monkeys may be devoted exclusively to ripe fruits, and the monkeys generally manage to devote 65% or more of their feeding time to ripe fruits at all times of the year (Milton 1981a,b, 1993b; Symington 1988a,b).

The daily travel path of spider monkeys is considerably longer than that of howlers. On BCI, daily travel by spider monkeys averages 900 (as contrasted with 392 m for howlers), and on occasion, they may travel more than 3 km in a single day (Milton 1991; K. Milton, unpub.). At other sites, spider monkeys have been observed to travel as much as 5 km per day in search of food (van Roosmalen 1985). Thus, for spider monkeys, the use of visual or olfactory cues to locate ripe fruit crops at a distance seems even less probable than for howler monkeys.

In contrast to the fig trees preferred by howlers, many tree species used as primary fruit sources by spider monkeys ripen only a small portion of the total fruit crop each day (Milton 1982, 1991). In effect, this pattern forces these specialist fruit eaters to visit many different trees of a given species each day to secure sufficient fruits to meet their nutritional demands (van Roosmalen 1985). Probably also because of this fruiting pattern, members of a spider monkey troop tend to forage in small subgroups of only a few animals, rather than in a single cohesive troop like howler monkeys (Cant

Katharine Milton

1977; van Roosmalen 1985; Symington 1988a,b, 1990). If all members of a given spider monkey troop were to remain together over the course of the day, many individuals probably would not be able to obtain sufficient ripe fruit, or the entire group would have to travel farther. Thus, in essence, the diet of spider monkeys can be said to underlie their fission-fusion pattern of social organization (e.g., Cant 1977; van Roosmalen 1985; Symington 1990).

Traplining

On 7 consecutive days in October 1978, I followed spider monkey subgroups as they foraged in the BCI forest seeking fruits of *Spondias mombin,* a tree species that produces a sugar-rich, soft-pulped fruit with a single large, hard seed. On BCI, *S. mombin* shows a clumped spatial distribution (also characteristic of many other important fruit species utilized by spider monkeys), and generally one encounters several large trees of this species within a small area (Milton 1977, 1980). The spider monkey subgroups I observed were, in essence, "traplining" different patches of *S. mombin.* The term *traplining* describes a pattern of foraging behavior analogous to the movements of fur trappers. The trapper knows where his traps are, and follows a set schedule designed to minimize his travel path such that all traps are visited, and yet the trapper never wastes effort by doubling back over his path or revisiting traps already checked that day (see Thomson, Slatkin, and Thomson 1997 for a discussion of the difficulties both of defining and demonstrating traplining with free-ranging animals).

In seeking fruiting trees, spider monkey subgroups moved directly from one patch of *S. mombin* to another, harvesting fruits from one or more trees in each patch. They were not observed to double back on their trail or revisit patches already visited that day, though on occasion, other spider monkeys from their community were already present in a patch or entered a patch after the subgroup I was following arrived there (see also van Roosmalen 1985). Because of the tendency of many tree species on BCI to show a clumped spatial distribution, the presence of another subgroup in a patch would not necessarily imply that all fruit had been eaten. On successive days, animals generally visited at least some *S. mombin* patches utilized the previous day, but one, two, or more new patches were visited as well. This foraging pattern would permit spider monkeys to visit individual trees or species patches that they knew had fruit available (from their visit the day before, or in the very recent past) and, at the same time, monitor or feed from new patches of *S. mombin*. But, of course, they would have to know in advance where such patches were—otherwise they could not have moved directly to "new" patches by the shortest route—which, like Garber's (1989) *Saguinus* species, they did. Often, too, travel to new patches was extremely rapid, a feature likewise suggesting prior knowledge of the route and destination. Janson and Di Bitetti (1997) have noted that increased travel speed appears to decrease the probability of encountering food patches by chance unless animals pass within 10 m of a large patch.

Sex Differences in Travel Patterns of Spider Monkeys

Because of its dependence on ripe fruit, a spider monkey troop has a huge home range relative to a howler troop of similar size. On BCI, adult male spider monkeys may forage over most portions of the entire island. In my view, a 300 ha supplying area is a conservative estimate for an adult male spider monkey on BCI (> 500 ha might be more often the case), and 200 ha is conservative for an adult female. In Manu National Park, Symington (1988b) estimated home range size for troops of some 18–24 spider monkeys at 200 ha, while in Suriname, van Roosmalen (1985) estimated home range size (usable supplying area) for a troop of 15–20 spider monkeys at 220 ha.

Field data from BCI, Suriname, and Manu show that spider monkey females, particularly females with infants, have a shorter average day range than males (van Roosmalen 1985; Symington 1988a,b, 1990; Milton 1993b); furthermore, each adult female (and her dependent offspring) has a core area in which she tends to forage, particularly when ripe fruits are in short supply (van Roosmalen 1985; Symington 1990; Milton 1993b). Individual females have an extremely detailed knowledge of food sources in their particular core area (van Roosmalen 1985). Individual male spider monkeys also have core areas; these are larger than those of females and may overlap the core areas of two females (van Roosmalen 1985; Symington 1990).

Though males and females may forage together when fruit is abundant, generally on BCI and at some other sites, male spider monkeys tend to be found frequently in all-male associations; on BCI, on average, males spend about two-thirds of their daylight hours apart from females and young (Eisenberg and Kuehn 1976; Milton 1991). Adult males may forage for a short time each day with females and young, but typically at some point, generally sooner rather than later, males leave females and travel away at great speed to feed in other areas of the forest, where they may remain for most or all of the day (Symington 1988a,b; K. Milton, pers. obs.). Male spider monkeys are at times observed foraging alone, though on BCI this is rare. In Suriname, spider monkeys show a pattern somewhat similar to that reported for bonobos *(Pan paniscus)* (Hohmann and Fruth 1994): particular adult males often appear to associate with particular leading females, and all-male bands or male-male associations are not common (van Roosmalen 1985).

"Leading Females"

In his anecdotal descriptions, van Roosmalen (1985) differentiated between "leading females," generally older females who tended to lead foraging subgroups, and "other females" (see also Rowell 1972a). Leading females generally left sleeping trees first, traveled in front of a subgroup most of the time, always fed on the food sources that determined the route that was taken, and initiated most travel activity. Leading females took the shortest routes between important food sources and usually did not hesitate while traveling. They also appeared to check particular trees, often over a long period of time, in order to incorporate them into the foraging itinerary at the proper moment in the days or weeks to come.

Moreover, van Roosmalen suggested that a leading female probably roughly mapped out a foraging route in her mind for a particular day early that day, or even the day before. If, however, a leading female temporarily joined another subgroup that was following a different foraging route, later that same day she might have difficulty picking up her former route and could be observed backtracking to a particular food source already visited or moving back and forth between trees already visited earlier that same day—apparently trying to pick up the proper environmental cues to continue with her initial planned travel itinerary. Non-leading females (probably younger and less experienced) were described as foraging with less confidence than leading females, hesitating and traveling back and forth between the same food sources, rather than striking out rapidly and directly to new distant food patches (van Roosmalen 1985).

Because of the variable and uneven characteristics of their principal items of diet and the fact that they often forage in small subgroups (which may contain only a single adult) or alone, to be efficient, spider monkeys must formulate individual foraging paths each day such that they maximize encounters with abundant, high-quality fruit resources and, at the same time, monitor potential "hot spots" over a large area. As discussed above, the speed and skill with which some individuals move from patch to patch of the same fruiting tree species shows that they clearly recognize that when one individual of that species is in fruit, other members of that species are likewise in fruit, and suggests that they possess knowledge of specific tree locations that enables them to map out a travel route and then move between a certain number of these food sources by the most direct means. But, as noted above, female spider monkeys with dependent young do not, on average, forage as far each day as male spider monkeys, nor do they have as large a supplying area. How, then, can females and young benefit from the wider pool of dietary information available to males in their social unit?

Food Long Calls in Spider Monkeys
It is worth the effort to travel to a rich food source if no similar food patches are nearby and you know that fruits will be there when you arrive. On the other hand, if you do not pass through a given area sufficiently frequently to monitor its trees, traveling to that area could be a waste of energy if you do not find ample food there when you arrive. Perhaps to counteract this potential problem, individual spider monkeys have been observed to give loud calls at fruiting trees (van Roosmalen 1985; Chapman and Lefebvre 1990; Chapman and Weary 1990; K. Milton, pers. obs.)—calls that ostensibly signal to other group members the location of the food source. Individuals or subgroups interested in such information can then travel to that tree if they choose.

Chapman and Lefebvre (1990) investigated food calling behavior in free-ranging Costa Rican spider monkeys *(Ateles geoffroyi)*. Though both males and females gave food calls (a "whinny," described as a "long-range vocalization"), females were reported to call more frequently than males (Chapman, Wrangham, and Chapman 1995). On occasion, other subgroups responded to the calls and traveled to the food tree. There was a positive relationship between subgroups with dominant individuals and frequency of calling, and the frequency of calling positively affected the number of conspecifics that arrived (Chapman and Lefebvre 1990). There was some suggestion that particular females formed coalitions to monopolize rich food resources (Chapman, Wrangham, and Chapman 1995).

Observing another loud, long-distance vocalization, the "whoop," Van Roosmalen (1985) noted a clear sex-based difference in the food calls of *Ateles paniscus* in Suriname. At this site, only male spider monkeys gave "food long calls," generally just before or during feeding on a primary food source. However, in contrast to the situation in Costa Rica, other subgroups usually did not join the caller. In van Roosmalen's view, the food long call signaled to other group members that they should *not* travel to the location as the food source was being depleted. The different calling patterns and subgroup reactions to calls that have been noted between Costa Rican and Suriname spider monkeys are puzzling and warrant further study (that fact that two different calls are being discussed may also relate to the different reactions described). But it does not really matter whether the calls are meant to attract or repel other community members—the end result is still the passing of key dietary information over long distances, information that can either attract group members to rich food sources (provide energy) or save them the trouble of traveling to foods being depleted (conserve energy), in either case enhancing their foraging efficiency. This energetic savings could be particularly critical for females, as it could permit them to invest more energy in reproductive efforts than otherwise would be the case as well as improving their own nutritional status.

Food Long Calls in Chimpanzees and Bonobos

Spider monkeys are not the only primates observed to use long calls (apparently) to communicate dietary information to conspecific group members. Both bonobos *(Pan paniscus)* and common chimpanzees *(P. troglodytes)* are reported to give loud food calls (Goodall 1986; Hohmann and Fruth 1994; VanKrunkelsven et al. 1996). Like spider monkeys, both chimpanzee species show a fission-fusion pattern of social organization, tend to forage primarily in small subgroups, and focus the majority of their feeding on ripe fruits (Goodall 1986; Symington 1990; VanKrunkelsven et al. 1996).

Captive male bonobos are hypothesized to utter food calls to attract potential mates, and apparently are willing to give up the discovered food resources in return for sex; food call frequency increases when finding larger amounts of food (VanKrunkelsven et al. 1996). Free-ranging common chimpanzee males likewise are reported to give loud calls—pant-hoots—to signal the location of rich food sources, with more intense calling at richer, divisible food sources (Hauser and Wrangham 1987; Mitani et al. 1992; Clark and Wrangham 1993; Hauser 1996), and females and young are reported to respond to these calls (Goodall 1986). It is adult male chimpanzees who call most often and most loudly at desirable food sources (Goodall 1986; Mitani 1996); females tend to utter only soft food grunts (Goodall 1986). The pant-hoot call of adult males has a third phase, the climax phase, that adult females typically do not give (Marler and Hobbett 1975; Mitani et al. 1992).

The pant-hoot vocalizations of two geographically distinct chimpanzee communities (Gombe and Mahale) differed statistically, suggesting that different communities might possess different "dialects" (Mitani et al. 1992), a feature that could help identify community members (Green 1975). Chimpanzees apparently can also identify distant long calls as being those of particular individuals (Marler and Hobbett 1975, Mitani et al. 1992), and community members react differently to calls of known versus strange conspecifics (Mitani et al. 1992; Mitani 1996; Baker and Aureli 1996). Loud calls of individual chimpanzees are said to vary depending on social context (Clark and Wrangham 1993). Similarly, human observers can identify particular spider monkeys or chimpanzees from their long calls (Marler and Hobbett 1975; van Roosmalen 1985). Masataka (1986) suggested that spider monkeys might both recognize one another as individuals by their calls and direct calls to particular individuals.

In effect, there appears to be a continuum of call complexity in anthropoids (see, for example, Marler 1970, 1976b; Cheney and Seyfarth 1982; Boinski and Mitchell 1997, Boinski, chap. 15, this volume). Most calls are "intragroup calls"—that is, they are intended to be heard and reacted to by one's own group members (Marler 1970; Boinski, chap. 15, this volume), who generally are in fairly close proximity to the caller. Calls directed over a long distance, on the other hand, such as the roars of howler monkeys, tend to be aggressive intergroup calls, broadcast to repulse strange conspecifics at a distance (Harrington 1987; Whitehead 1987). However, in fusion-fission species such as spider monkeys, chimpanzees, and bonobos, it would appear that various loud long-distance calls have been elaborated into an intragroup call system—a system of calls directed at one's own community members.

Katharine Milton

Long-distance intragroup calls might require more skill to "decode" than most other intragroup calls, which are often simple calls exchanged between individuals in fairly close proximity (proximity permitting the use of other cues simultaneously—facial gestures, hand or body gestures, environmental context, and so on; e.g., Marler 1976b). In support of this view, Marler and Hobbett (1975) have commented that in common chimpanzees, individuality is especially marked in those sounds used for long-range communication, while Hohmann and Fruth (1994) describe the high-hoots of bonobos as the major device used to regulate and maintain the social network of the bonobo community. Mitani et al. (1992) speculate that the social system of chimpanzees (and, by analogy, other fusion-fission species) may have created an appropriate selective milieu particularly favoring the evolution of vocal learning.

Avoiding Cognitive Overload in Foraging

One striking characteristic of most extant primates, which appears to set them apart from many other plant-eating arboreal mammals, is the large number of food species they exploit, both on a daily and an annual basis (Milton 1987). It is not unusual to find that a given group of monkeys or apes has taken foods from more than 150 different plant species over an annual cycle. As tropical forests have a high diversity of species, and most species occur at very low densities, and as each plant species has its own phenological pattern, the amount of information a primate might theoretically have to deal with is immense.

But primates have ways of getting around potential information overload. Field studies show that though primates eat a large number of plant species, typically only a small percentage of these make up the bulk of the daily or annual diet. For example, for howler monkeys on BCI, the overall pattern in the differential use of food species, as measured in percentage of feeding time per species, is that a few tree species are used rather heavily but most are used hardly at all. The same pattern is seen if differential use of food species is measured by the number of days a species is eaten. A few species are eaten on many days, but many are eaten on only a single day of observation (Milton 1977, 1980). Though on BCI, howlers eat an average of 8 plant species per day, only 1.5 of these are primary food sources (i.e., contribute $\geq 20\%$ of feeding time). A species used as a primary food source in more than one sample month of my study was considered a *staple* resource. Very few species could be regarded as staples because of the high turnover rate of most howler food sources, particularly leaf species. Combined data from both study troops showed that, overall, only 9 plant species were used as staple foods by howler monkeys over the total study (Milton 1977, 1980). Similar feeding patterns are found in a large number of other primate species.

Howler monkeys and other primates thus appear to counteract the great diversity of their food species by concentrating heavily on a select number of staple species and using many others largely opportunistically. This strategy greatly reduces the amount of information an animal needs to retain, both in terms of species worth monitoring and in knowing where particular individuals of those species are located. Howlers, for example, use an average of only 1.5 primary food sources per day; a large amount of spatial memory does not seem necessary to travel to 1.5 trees per day, particularly when well-traveled arboreal pathways connect many important resource clusters (Milton 1977, 1980).

In spider monkeys, a very similar pattern is found. As spider monkeys trapline particular fruit species, they typically deal with only a single or perhaps two primary fruit species per day, and one (or two) species may serve as their primary source of ripe fruits for weeks at a time (K. Milton, pers. obs.). However, in contrast to howler monkeys, each spider monkey or spider monkey subgroup has to visit a number of different trees of a primary species each day to get sufficient fruit, an activity that could prove costly. But since many such tree species show a clumped spatial distribution, travel to one fruiting individual generally results in finding a number of other individuals of that species in close proximity (Milton 1980).

Feeding data suggest that, on average, a spider monkey subgroup travels to three or four patches of primary fruit trees per day. Like howler monkeys, spider monkeys have arboreal pathways through the forest such that an observer can often predict where a subgroup is going and the route it will take as it begins travel. To supplement their primary foods, as noted, spider monkeys and howler monkeys (and probably most primate species) feed opportunistically on any desirable dietary items encountered as they travel between primary sources (Milton 1977, 1980).

As chimpanzees are considerably larger than spider monkeys, an individual chimp needs absolutely more ripe fruit per day than a spider monkey. All else being equal, this could necessitate travel to more fruit trees per day, perhaps the use of more primary fruit spe-

Katharine Milton

cies or trees per day, and a larger home range area. Furthermore, chimpanzees at some sites—apparently more so than spider monkeys—may often have to forage alone or only in the company of dependent young. These various travel and foraging pressures would seem to call for enhanced navigational abilities and the need for more memory, both short and long term, in these considerably larger fruit-eaters.

Cognitive Abilities with Respect to Multi-Destinational Routes

To examine some features of spatial memory in primates with respect to such foraging challenges, Cramer (Gallistel and Cramer 1996) carried out experiments with vervet monkeys (*Chlorocebus aethiops*) to test their ability to choose a multidestinational foraging route. Each subject watched in a holding area as grapes were hidden in a number of holes within its outdoor enclosure. The monkey was then released into the enclosure. Because the grapes were hidden, the monkeys had to rely on memory to compile a multidestinational route and collect the grapes. The vervets never remembered more than six locations (Gallistel and Cramer 1996).

In contrast, similar experiments by Menzel (1973a) using young chimpanzees as subjects showed that these apes were able to retrieve pieces of hidden fruit from as many as eighteen different hiding sites in a single rapid foraging expedition (described in Gallistel and Cramer 1996). The route the ape took bore no relationship to the one the experimenter had taken while hiding the fruits, and it appeared to minimize the distance traveled. Earlier work by Tinkelpaugh (1932), in which the memory capacities of rhesus monkeys (*Macaca mulatta*), common chimpanzees, and humans were tested gave the same result (six locations for the monkeys; sixteen or more for both chimps and humans).

The ability to map out a mental multidestinational route in vervets seems to be predicated on what has been termed "a three-step look-ahead"—that is, when planning its route, the monkey is able to consider at least two further destinations beyond the next destination (Gallistel and Cramer 1996). Most arboreal primates are not particularly large in size and, like vervet monkeys, may therefore require no more than a three-step look-ahead to plan their daily foraging activities. The food requirements of larger-bodied anthropoids such as great apes, however, would appear to be considerably higher, and this might account for their enhanced ability to plan more complex, multidestinational foraging routes and their larger brain-to-body ratio.

Group Movement and Food Search in Humans

> At the simplest level, the significance of material culture lies neither in the establishment of chronology nor as a measure of relationships, but as an indicator of efficiency in obtaining food. (Bartholomew and Birdsell 1953, 492–93)

With humans, much of the speculation that marks the above discussion is avoided. Humans can describe how they will travel to or locate particular dietary resources, and they can explain why they seek certain foods in preference to others, or why they choose to hunt alone or with others. Though Bennett (1996) has called into question evidence supporting the use of cognitive maps, *sensu* Tolman and O'Keefe and Nadel, even for humans, we know that humans can mentally visualize complex geographic areas and describe the most efficient travel route from point to point (mentally calculate direction and distance). But generally in such examples we are dealing with human knowledge of a familiar landscape, or an experiment in which the human subject has previously studied some scaled layout of landmarks or been led blindfolded over a simple route and asked to return to the starting point (Fujita et al. 1993). What happens when blindfolded human subjects are taken some distance into an unfamiliar area—an area in which they have not been before? Can they, for example, point to the direction of home?

Formerly, it was speculated that humans, like some other vertebrates, might be able to orient themselves toward home or familiar landmarks (goal orientation) without instruments or celestial cues through the use of a magnetic sense organ (Baker 1980). Repeated experiments have consistently failed to provide any convincing evidence that humans utilize a magnetic sense organ (Gould and Able 1981). Humans, therefore, must learn certain landmarks, celestial cues, or navigational "rules" in order to move about in a goal-directed (directional) rather than a haphazard manner (see, for example, Etienne, Maurer, and Séguinot 1996; Gallistel and Cramer 1996) when carrying out daily activities over more than a short distance, including movement associated with food procurement. How, then, do human foragers (as, for example, indigenous hunters) behave when traveling through an unfamiliar landscape in search of food?

Orientation When Foraging

The indigenous hunters I work with in Brazil (lowland tropical forest dwellers in the Amazon Basin) generally select their hunting area the night before and depart in a group well before daybreak.

Katharine Milton

If they leave from the village, they follow a trail, and even distant areas may contain trails. If hunters leave in a boat, they may travel by river for several hours, then enter the forest where, often, there are no trails. If three or fewer men are hunting, they may all walk along together in single file while looking and listening for game. However, if more men are hunting, they typically fan out once they enter the hunting area, with each man taking a somewhat different route. The hunters may hunt all day (8 hours or more), and each one is somehow able to return to the boat or trail by dusk. Hunters often call out loudly on their return, both to signal their location and to aid others in walking by the most direct route to the reunion site—calls from hunters already near or at such sites are particularly useful here.

Indigenous hunters tell me they use the angle of the sun to help orient themselves while in the forest. Apparently, use of the sun is very important, for one Mayoruna hunter (Pacha) told me that if it were a cloudy or rainy day, hunters might not be able to return to the boat or trail and would have to sleep in the forest. Amazonian hunters may also mentally keep track of landscape details—hills, streams, the direction of the main river, and so on—to help orient themselves when moving through the forest.

Yost and Kelley (1983), in working with the Waorani in Ecuador, noted that unless tracking a large terrestrial animal, hunters almost always hunt by trail, and trails almost always follow the tops of ridges, paralleling the streams and rivers. Though cutting across ridges and rivers would take the hunter through a richer variety of biotopes, hunters prefer to hunt along the ridge tops (where trails have been created through such use). Apparently the extra physical effort required to cross rivers, climb hills, and so on is viewed as too costly energetically, especially when one is burdened with game (Yost and Kelley 1983; K. Milton, pers. obs.). Thus, hunters are more efficient (i.e., net a higher return in exchange for their hunting efforts) if they follow the ridges (Yost and Kelley 1983), and for this reason, over time, trails have been created in these areas rather than others.

Food-Related Calls and Other Acoustical Cues

To aid in prey procurement, groups of Matis hunters use calls in order to keep themselves spaced well apart as they move through the forest (e.g., Hill and Hawkes 1983). A hunter also calls to "tell" other hunters if he encounters game and what type of game it is; these calls are cryptic and not detected as human calls by game (monkeys, wild pigs, etc.), but all of the hunters understand their meaning (just as spider monkeys and chimpanzees "understand" the meaning of their long calls). Hunters will then close in to try to make a kill.

The use of calls to space hunters at some regular (presumably optimal) distance over the landscape should improve the probability of encounters with game. If a hunter encounters a singleton game animal, he can pursue and hopefully kill it himself, but if he encounters more game than he can kill by himself, or game more likely to be killed in large numbers if surrounded by hunters, he can summon others to the area with calls before he begins to shoot.

Amazonian (and other) hunters use subtle features of the environment (overturned leaves, hollow logs, animal prints or scats, plant fragments on the forest floor, odors) to determine the presence or passing of game species (Winterhalder 1981; Henley 1982; O'Dea 1992; K. Milton, pers. obs.). They also know the calls of literally every species in the forest as well as the significance of particular noises such as falling fruits or moving branches. Little escapes their attention, and their reflexes seem amazingly quick relative to mine. As one nonindigenous man who had done considerable hunting with the Mayoruna remarked to me at the end of a successful hunt, "I pity the animal who crosses the path of a Mayoruna."

In reflecting on the sensory modality most important for hunting success in Amazonia, Yost and Kelley (1983) made the following observation:

> Undoubtedly, the Wao(rani) hunter depends on hearing more than upon any other sense to locate potential game. He learns to distinguish among the animals in the canopy by the sound they make as they move; the frequency of the movements, the loudness of the rustling leaves, the distance between movements, the kinds of trees the animals are in are all clues to the species. It is not unusual for hunters to know what kind of animal is present long before they see it or hear it call. The dense growth of the canopy often obscures animals from view, but knowledge of what species is present, combined with an understanding of that species' behavior, can make it possible for a hunter to predict an animal's next move without ever seeing the animal. Obviously, the better the hunter understands the species' behavior, the greater his chances of success in the hunt.

Vision is generally viewed as the most important sensory modality for primates, including humans, but reliance on auditory skills may be underrated in terms of its influence on the trajectory of human evolution. Anthropoids have a long evolutionary history of relying strongly on acoustical signals to aid in carrying out routine activities each day (Mitani et al. 1992; Boinski, chap. 15, this volume), and human language continues this trajectory.

Learning as the Key to Foraging Success
In his book *Hunters of the Northern Forest,* Nelson (1973) provides a detailed examination of the hunting behaviors of the Kutchin, indigenous inhabitants of the Chalkyitsik region, Alaska. This book helps emphasize for the urbanized Westerner the amazing number of different skills a human hunter (and, for that matter, with certain modifications, a human gatherer) has to master to be successful (see also Gladwin 1970; O'Dea 1992). The required knowledge includes the characteristics of the hunting landscape, the behaviors of the many different prey animals, the different hunting tactics and the manufacture and use of the weapons or tools needed to capture particular animal species, highly honed navigation skills of many different types (e.g., Gladwin 1970), the logistics of hunting, survival tactics under difficult or critical environmental conditions, and hundreds of other things. The life of "natural man" is most emphatically not a "remarkably easy one" (i.e., Humphrey 1976, 307). And all of this information has to be learned—none is encoded genetically in the human forager (e.g., Milton 1992).

A Division of Labor in Foraging
With human foragers, we also find some unusual behavioral adaptations related to group movement and food procurement that have antecedents, but no parallels, among other extant mammals, including other primates; namely, a division of labor between the sexes and food sharing (Lancaster 1978; Isaac 1978). Some other mammals show a division of labor in terms of food acquisition— for example, a pride of lions, a pack of wolves—and there are examples of food sharing in some bats and particularly among the social carnivores, which may also bring food back to a den (home base) to feed pups. But in no case do we find a sex-based division of labor in terms of food acquisition like that which apparently characterized ancestral hunter-gatherers *(Homo)* and still characterizes most hunter-gatherer groups today, a division of labor typically directed at foods from two trophic levels and in which foods obtained are shared with most or all members of the social unit. Human young are also provisioned entirely or partially for a very

protracted time period, which should greatly facilitate their survivorship to reproductive adulthood (Lancaster 1978). Though the particularly human type of labor division was undoubtedly initiated in response to foraging pressures, it can be extended into almost all spheres of activity. As Hutchins (1995) points out:

> In anthropology there is scarcely a more important concept than the division of labor. In terms of the energy budget of a human group and the efficiency with which a group exploits its physical environment, social organizational factors often produce group properties that differ considerably from the properties of individuals . . . a particular kind of social organization permits individuals to combine their efforts in ways that produce results . . . that could not be produced by any individual . . . working alone. (Hutchins 1995, 175)

In short, the human division of labor is ergonomically efficient. One interesting aspect of group foraging in humans is that, typically, men hunt and women gather. Animal prey, unlike plant foods, is not sessile but mobile. Various aspects of food search strategy optimal for sessile plant foods would not be equally efficient for mobile animal prey, perhaps necessitating some fairly radical modifications in behavior and perhaps morphology to deal efficiently with these new foraging challenges (e.g., Milton 1981b).

A Foraging Complexity Continuum
In effect, howler monkeys, spider monkeys, chimpanzees, and human foragers can be viewed as successive points along a foraging complexity continuum. In general, spider monkeys and chimpanzees do not appear to have elements in their foraging behavior that could not be found in less complex form in howler monkeys or most other primates. Even the long call given at rich food sources by spider monkeys and chimpanzees appears to have its antecedents in calls some other primates give on encountering unusually rich food sources (e.g., toque macaques, *Macaca sinica:* Dittus 1984). Differences noted in group movement among anthropoids appear to originate from a common pool of cognitive potential shared by all anthropoids, elements of which are expressed or developed to a greater or lesser degree in a particular taxon in response to the challenges posed by its particular dietary niche. In the case of spider monkeys and chimpanzees, such challenges appear to have resulted in a set of very similar behavioral solutions (e.g., Symington 1990) in spite of phyletic distance, a considerable difference in body size,

Katharine Milton

and presumably, differences in the two genera's respective forest environments.

Are Primates "Special"?

As an order, Primates are noted for their large brains and enhanced capacity for learning and retention, as well as their remarkable degree of behavioral plasticity (Eisenberg 1973; Milton 1981a). However, though monkeys and apes may seem particularly clever and special to us, their closest living relatives, it is important to realize that there are a number of other animal taxa that show many similar characteristics. We have seen that distantly related primate lineages can exhibit very similar behaviors on a number of levels, apparently in response to similar dietary challenges. To test the broader validity of this observation, it is useful to examine group movement and its attendant behaviors in two nonprimate taxa that, like spider monkeys, chimpanzees, and human foragers, eat very high quality foods, live in social groups, and show a fission-fusion pattern of group movement; namely, dolphins and parrots. For brevity, my discussion is focused primarily on a single species in each taxon, but many observations are likely to be equally valid for other members of that taxon as well.

Group Movement and Food Search in Bottlenose Dolphins

Bottlenose dolphins *(Tursiops truncatus),* like all Cetaceae, are secondarily aquatic, their ancestors (and ancestral neocortex) having evolved on land (whales are thought to have returned to the water some 70–90 mya) (Morgane, Jacobs, and Galaburda 1986). Dolphins (unless otherwise specified, the term "dolphin" here will refer only to *T. truncatus*) feed on mobile prey—principally fish. Some prey species are solitary or form small schools, while others form large schools. Thus, depending on the particular prey type being exploited, dolphins sometimes forage alone or in small parties, while at other times, frequently alerted by conspecifics, numerous dolphins converge on large schools of fish (large food patches) and pursue them, capturing individual fish, apparently aided by their echolocation system.

Dolphins are regarded as fission-fusion foragers (Smolker, chap. 19, this volume). This fluid grouping pattern appears to relate to their fluctuating and unevenly distributed food supply, which calls for constantly fluctuating subgroup size and composition. Data suggest that, as in spider monkeys and chimpanzees, mother-dependent offspring may be the only constant social association in bottlenose dolphins. However, it has been repeatedly observed that within specific locales, there are long-term association patterns between particular dolphin females and particular dolphin males, and that male-female localized associations or communities appear to exist—communities, however, that may also be constantly entered and left by noncommunity (or less frequently seen community) members (Smolker, chap. 19, this volume).

Individual Recognition

As dolphins appear to live in fairly discrete, localized communities but do not forage as a cohesive social unit, they are faced with the problem of maintaining and efficiently coordinating the activities of their group members at a distance. Like the above-discussed fission-fusion primate foragers, they appear to rely heavily on acoustic signals (and attendant cognitive processes) to accomplish this task.

Observational data suggest that dolphins can recognize conspecifics as individuals. Until recently, it was also believed that each dolphin produced a distinctive "signature whistle" developed during ontogeny (Tyack 1993). This observation has been challenged as recent research on whistle production by individuals from three different dolphin communities failed to detect evidence for individual signature whistles, though some data did support the possibility of community-based dialects (McCowan and Reiss 1995a). The dolphin whistle system is regarded as an open-ended system of communication (Tyack 1993) analogous to the open-ended system of verbal communication in humans (and similar in this respect to the "graded" calls of some nonhuman primates).

The Echolocation System

In addition to whistles, dolphins possess an echolocation system—a mental system of amazing precision that enables a given dolphin not only to locate an individual fish, even in a large school, but also to follow its rapid flight and (apparently) "stun" it by emitting a burst of signals (Wood and Evans 1980). However, blindfolded dolphins can capture live prey without producing any detectable echolocation sounds, apparently using their keen sense of hearing (Wood and Evans 1980). Within the dolphin brain, the echolocation system appears distinct from the signature call system. The call system appears to be localized primarily in the thin cerebral cortex, particularly the temporal cortex, while the echolocation system occupies a large area in the midbrain (Tyack 1993). Ridgway (1986) suggests that much of the great hypertrophy of the dolphins auditory system—and perhaps the entire cerebrum—results from the animal's need for great precision and speed in processing sound—a key attribute of human speech (and the human brain).

These and other data suggest that the large brain size of dolphins is functionally linked to the rapid processing of sound. I can appreciate the need for such abilities when foraging for rapidly moving prey, particularly in turbid water. But rapid sound processing seems less urgent in terms of *social* signaling, unless acoustic signals facilitate hunting efficiency, predator avoidance, or some other critical behavior(s) that requires a rapid response. Until recently, conspecific food calls had been reported only for animals foraging on plants. However, Janik (1997) has presented data describing a unique low-frequency two-part call in wild bottlenose dolphins that is associated with feeding on large fish. Dolphins produced these low-frequency calls in 94% of all observed feeding events, and in each case, the feeding dolphins were rapidly joined by conspecifics. It was suggested that such calls could function to alter prey behavior, to increase food intake of close kin, or to recruit conspecifics to approach and thus chase fish backward toward the callers (improving prey catch efficiency). To feed efficiently, dolphins must also possess navigational skills such that they can orient themselves within their foraging range, travel to seasonal foraging sites, travel rapidly and directly to fast-moving schools of prey from a distance, and so on.

As discussed, dolphins appear to know one another as individuals, and Janik's (1997) data suggest that they communicate information about prey size and location to one over long distances. As noted for primates, there appear to be long-term, multigenerational communities of dolphins loyal to particular foraging areas. Dolphin young take some 10 years to mature. This suggests that, like the young of some other large-brained mammals, they require a long period of practice and learning to become successful adults.

As yet, it is difficult to speculate on what dolphins are doing with the various components of their brains because so little is known about the specifics of communication feedback among dolphins when foraging or pursuing schools of fish (or doing anything else). The dolphin case is particularly interesting because dolphins clearly can do so many things besides echolocate and whistle (Tyack 1993).

Captive dolphins easily learn elaborate "trick" routines, often cued by signals from their trainers; they appear able to observe an action and replicate it; they are reported to imitate sounds, mimic the behavior of other animals, and so on (Herman 1980, discussed in Griffin 1992). Does signaling about and honing in on mobile prey require such a rich repertorie of behavioral capabilities? Perhaps enhanced cognitive capabilities with respect to a particular set of foraging challenges can be generalized to many other situations.

Group Movement and Food Search in Parrots

Parrots are highly social birds, noted for their complex vocalizations and their ability to acquire new sounds from the environment (Nottebohm 1970; Wright 1996, 1997). They are not the only birds possessing this ability, however, and more than one factor almost certainly can select for it. Data on wild parrots are not abundant, as they are extremely hard to study in the natural environment. Here comments are limited largely to information on the foraging and social behavior of the yellow-naped amazon *(Amazonas auropalliata)* in Costa Rica, supplemented with more general information on some other psittaciformes.

In the wild, many parrots (e.g., *Amazonas* spp., *Ara* spp.), like primates, have diets made up primarily of patchily distributed plant foods, principally fruits and seeds—foods that vary over an annual cycle in abundance and quality. Many seeds parrots feed on are protected by a thick, hard husk or shell. Parrots use their powerful bills to pry or bite open these hard-shelled fruits. Unlike those of most other birds, the tongues, feet, and toes of parrots are highly manipulative and serve as important foraging aids. They are used to position fruits for opening, remove food from the shell, and turn and position fruits and seeds for ingestion. This manipulative behavior is quite precise and enables parrots, like many primates, to discard the lower-quality, less digestible portions of their food items, ingesting only the most select, high-quality material (K. Milton, pers. obs.).

Grouping and Foraging Patterns of the Yellow-naped Amazon and Some Other Species

Parrots are long-lived birds that form long-lasting pair bonds. Available data suggest a three-tiered social structure for some *Amazonas* species. At the simplest level there is a mated pair; various pairs appear to associate in a type of loose "flock" (some thirty to

Katharine Milton

fifty individuals who associate with one another frequently); and finally there is a large communal roost at which dozens to hundreds of conspecific parrots that appear to be members of the same community congregate at night (Ridgely 1982; Wright 1996, 1997; K. Milton, pers. obs.). On leaving the night roost each morning, parrots deploy themselves in various groupings, which can change over the course of a day, such that only a single pair, a flock, or a large congregation may be seen at a given feeding site. As members of a larger social unit (night roost) that appear to forage in subgroups of changing size and composition (other than the bonded pair), at least some parrot species would appear to be fission-fusion foragers. As such, they are faced with the challenge of maintaining social bonds, coordinating foraging activities, defending themselves from predators, and sharing important information with fellow group members, though often apart. Much of this social coordination and maintenance may be achieved through vocalizations (Saunders 1983).

Regional and Night Roost Dialects
The "contact" call is the most frequently uttered component of the vocal repertorie of many parrots. Contact calls appear to initiate group activities and maintain contact between flock members and mates (Saunders 1983; Wright 1996). In yellow-naped amazons, the contact call is used by individuals of both sexes and all ages in a number of social contexts, and is particularly common at night roosts and nest sites (Wright 1996, 1997). As discussed for primates and dolphins, data suggest that various parrots (*Amazonas* spp., some cockatoos) can recognize individual family members or flockmates by their calls (Saunders 1983; Wright 1996). It is possible for a human observer to sex and identify individual black cockatoos *(Calyptorhynchus funereus latirostris)* from their calls (Saunders 1983).

Study of the contact calls of free-ranging yellow-naped amazons in Costa Rica indicates that this species exhibits call variation at two geographic scales: (1) at the regional level (termed "dialects" and shared by members of various night roosts in a given geographic area) and (2) at the level of the individual night roost within a dialect ("within-dialect variation," Wright 1996, 1997). Regional dialects are defined by large-scale shifts in the structure of the contact call, while within-dialect variation exists in the fine-scale structure of the contact calls at each particular night roost (Wright

1997). Some birds at roosts bordering two regional dialects use the calls of both neighboring dialects interchangeably. Dialect borders thus appear to act as barriers to the spread of "foreign" calls (e.g., Hardy 1966; Nottebohm 1970; Wright 1996).

Night roost size in yellow-naped amazons ranges from twenty to three-hundred birds, and night roosts occur in highly traditional sites that are used throughout the year (Wright 1997). In this species, birds respond strongly to, and may interact only with, birds from their own night roost, ignoring other conspecifics regardless of dialect (Wright 1996, 1997). This observation strengthens the contention that features of each particular night roost subdialect define its members and help identify them as such to one another. This recognition could facilitate the transmission of survival information (e.g., food sites, nest sites, etc.) to roostmates. As yet, our understanding of avian dialects is hampered by a lack of information as well as by the possibility that they may serve different functions in different lineages (see Hardy 1966; Nottebohm 1970; Wright 1996, 1997 for discussion of hypotheses related to avian dialects).

Night Roosts as Information Centers?
Why many parrot species form communal roosts is not known. Ward and Zahavi (1973) suggest that night roosts serve as "information centers" for the social group, functioning primarily in the sharing of information about food and only secondarily as any type of antipredator defense. It has been noted that avian species that feed in flocks upon an unevenly distributed food supply tend to roost communally, while those that feed solitarily on more evenly distributed foods roost alone (Ward and Zahavi 1973). Other environmental characteristics also appear to influence flock size, as J. D. Gilardi and C. A. Munn (unpub.) report that *Amazonas* species in their study area (Manu, Peru) foraged in small family groups composed of three to five birds—apparently a mated pair and their immature offspring—and, at least in the dry season, did not form large communal roosts. However, communal roosts have been noted for *Amazonas* species in the Bahamas, Puerto Rico, Guatemala, Panama, and as noted, Costa Rica (J. D. Gilardi and C. A. Munn, unpub.; Wright 1996, 1997; K. Milton, pers. obs.). Information sharing at communal roosts is hypothesized to confer "massive" foraging enhancement (Ward and Zahavi 1973). Leaving each day from a communal roost could also help to deploy groups of

parrots over the landscape in the most favorable configuration for maximizing individual foraging returns.

The high-quality foods many parrots depend on may be distributed over huge home ranges, estimated to cover many square kilometers (Ridgely 1982). Some flocks (ten to fifty individuals) of the pink cockatoo *(Cacatua leadbeater)* in Australia are estimated to forage over more than 300 square kilometers (Rowley and Chapman 1991). It has been noted that the largest associations occur at night roosts when food is scant—the time when parrots need the largest possible pool of information (Ward and Zahavi 1973). At the night roost, parrots are speculated to gain information on food sources over a wide area through association with roostmates who have fed in productive areas that day (Ward and Zahavi 1973). It is not known how birds distinguish successful foragers, but well-fed parrots may have behaviors, special vocalizations, or even olfactory cues that indicate this to their conspecifics.

Because particular parrot pairs come to a common roost site at night (or to a particular tree hole when nesting), they can be considered central-place foragers. As pointed out by Galef (1993), social birds or mammals that forage from a central site can benefit from exchange of information with conspecifics about the availability and distribution of foods. Such information pools may be particularly helpful for younger individuals by aiding them in finding their widely scattered dietary resources. Because parrots form long-term pair bonds and live for many decades in the same area, it would seem that, over time, a particular area might become populated by many parrots sharing strong kinship ties (first a pair, then a flock, and ultimately, a community), who would sleep at a communal roost, share important information with one another, and often associate at the same feeding sites during the day. Rowley and Chapman (1991) provide considerable information on the complex flock structure and feeding patterns of the pink cockatoo in western Australia, but in general, data on this topic are scant.

The widely appreciated longevity of parrots suggests that the accumulation and transfer of information from generation to generation, presumably at least in part about types, locations, and production patterns of particular food species, is a critical feature in the eventual foraging success of offspring. Rowley and Chapman (1991) described the behavior of pink cockatoos who took their dependent progeny to places where food and water were conveniently available. Young parrots appear to require a long period of maturation, learning, and practice to develop the skill and strength to find, manipulate, and open the variety of hard seeds and other foods that provide them with much of their diet, as well as to fly long distances between food sites. Like spider monkeys, chimpanzees, bottle-nosed dolphins, and humans, parrots show a large brain-to-body ratio (Pearson 1972; MacPhail 1982).

Overview

With the exception of howler monkeys, all of the vertebrates discussed above are fission-fusion foragers. This pattern of social organization appears to be directly related to the uneven patterning of their high-quality items of diet, a trait that also appears to necessitate a large suppling area. In all of these taxa, individuals known to one another (and many probably closely related) are part of distinct, often closed, social units—troops, groups, communities, pods, flocks, roosts—whose members come together and drift apart in subgroups whose size and composition (other than mother-dependent offspring or, in parrots, the mated pair) varies. These taxa all have a large brain in relation to their body mass, they are all highly social (for a discussion of the difficulty of defining sociality and social complexity, see Blumstein and Armitage 1997), and they are all strongly dependent on a highly elaborated system of acoustic communication aid in foraging coordination and efficiency and the maintenance of their social network. All of these species have a lifestyle that calls for a relatively long period of maturation during which a considerable amount of learned information and skill must be acquired for successful adulthood.

Within its particular lineage, each of these fission-fusion taxa appears to represent a somewhat extreme position in terms of body size *and* dietary quality—two traits that in many mammals, particularly herbivores, tend to be antagonistic and mutually exclusive rather than complementary, as they are in the fission-fusion species discussed above (e.g., Milton 1987). To counteract this seeming paradox, the niche each species occupies appears to require unusually well-developed cognitive abilities of various types in order for its holders to be successful—such behaviors bound up in the large brain size of the species.

For such species, it seems maladaptive for most information to be coded genetically, since environmental conditions (including those of the social environment) are constantly changing (Milton 1981a, 1988; Provenza and Cincotta 1993). Rather, what seems to

Katharine Milton

be called for is an increasing reliance on cognitive skills, particularly learning and memory, as well as sufficient behavioral plasticity to respond rapidly to changes in the environment (Milton 1981a, 1988). Provenza and Cincotta (1993) have stressed that learning is vital to securing rapid adaptation, and point out that foraging models have not yet incorporated learning as a within-or between-generational adaptive process. And, if you are going to have to learn a great deal before you can function successfully as an adult, it also seems of utility to have a long life span to give you the chance to put some of this knowledge to good use.

As these fission-fusion taxa are all highly social, I conclude that their basic requirements (i.e., obtaining food, avoiding danger, re-producing successfully) or some other requirements (e.g., infor-mation transfer between generations, provisioning of offspring, passing on a supplying area to descendants) are best realized in association with known conspecifics. However, because these spe-cies forage in subgroups rather than cohesively, benefits postulated to accrue to cohesively foraging groups (e.g., van Schaik 1983; van Schaik and von Hooff 1983; Bell 1991) might have to be obtained by somewhat different methods.

The calls discussed above (loud calls, whistles, food calls, contact and other calls) seem particularly important in this respect—they are speculated to be communication devices that permit scattered community members to "e-mail" one another during the day, touching bases and apparently sharing information (often about dietary resources but also about reunion sites, strange conspecifics, predators, and other features) that may be critical to group mainte-nance and coordination and foraging efficiency. In some species, total group size is not particularly large, and average subgroup size, where known, tends to be small (e.g., Wrangham 1977; van Roos-malen 1985). This suggests that it may not be the quantity of indi-viduals in the social unit that is important in the development of complex cognitive abilities, but rather the quality of the interactions between them.

These behaviors seem familiar because they are also representa-tive of our own species, *Homo sapiens*. Particularly in our hunter-gatherer past, it would appear that we too lived in relatively closed social units with a high degree of relatedness and showed a fission-fusion pattern associated with food procurement, a well-developed acoustic repertorie, and highly elaborated affiliative behaviors. Symbolic language, the hallmark characteristic of the human species, is most parsimoniously viewed as a labor-saving device. Human language is a low-cost behavior relative to its potential for enhancing ergonomic efficiency. Medawar (1976) views human language as essential for "the human system of heredity," that is, cultural evolution: "this characteristic human system of heredity calls for and depends upon the existence of language and other forms of conceptual communication." (Medawar 1976, 502)

The Other Side of the Coin
The identification calls, food-related calls, communication repertoires, and "dialects" of parrots, spider monkeys, dolphins, and chimpanzees have another aspect that needs to be addressed. On the one hand, these calls identify and provide a shared communication matrix for group members, but on the other hand, these acoustic traits also serve to identify nongroup members, whose detection, at least in nonhuman primates, typically elicits highly negative, defensive behaviors (Mori 1983; Goodall 1986; Baker and Aureli 1996). Similarly, most hunter-gatherer groups I have worked with in Amazonia show only the most intense animosity toward neighboring groups, and any suspicion that members of a neighboring group have been seen within the supplying area of its residents engenders an immediate hostile reaction (Milton 1992). Moore (1981) provides a detailed discussion of the particular benefits potentially accruing to centrally based human foragers through exclusive use of their supplying area.

Just as dialects may emerge between distinct social groups of nonhuman conspecifics and aid them in distinguishing kin or affiliated group members from nonkin/outsiders, language differences between human societies may emerge because a distinctive dialect or language that cannot be easily understood by one's neighbors could help a given human population safeguard particular facets of its unique behavioral (cultural) heritage, thereby securing or maintaining a competitive advantage over its neighbors. Nottebohm (1970), Barbujani (1991), Lieberman (1992), and others have pointed out that dialects and languages may also act as barriers to gene flow between human populations, perhaps contributing to microevolutionary processes.

In terms of our own species, the enormous number of different languages and dialects representative of human societies living at the same time in the same geographic region (e.g., Papua New Guinea, pre-contact South America) may well have developed not

Katharine Milton

only through isolation or drift, but rather to some extent though an active interest in impeding intergroup communication and, in this way, discouraging potential "information parasites" ("free-riders" *sensu* Enquist and Leiman 1993; see also Wright 1997; Wilson 1997a). If this notion seems far-fetched, let me point out that it is not uncommon to find such behavior even *within* subgroups of particular human populations. Cockney, pig latin, and the languages invented by twin siblings are good examples of this. Among the Kutchin, hunters who live in small, economically self-sufficient, kin-based groups, the traplines of different families often pass through portions of the same geographic area. In such cases, trappers have been noted to hide or bury their trapping devices so that members of the neighboring group cannot see them and learn their unique design (Nelson 1973). Western fly fishermen exhibit the same behavior, developing "secret" lures that they do not wish to show to anyone for fear the lure will be copied and their individual fishing prowess eroded. The old saying "monkey see, monkey do," as we now realize is most correctly applied to our own species— and with good reason. As has been suggested, true imitation (*sensu* Galef 1988) may well be our most uniquely "human" trait (Meltzoff 1988).

Acknowledgments

I wish to thank Timothy Wright and James Gilardi for providing me with unpublished information from their respective studies of wild parrots.

PART FOUR

Social Processes

The social processes inherent in coordinated travel by troops of monkeys and apes were featured in the reports of many of the classic field studies published before 1970. It was clear to earlier generations of field-workers that differences in group size and composition, habitat and resource characteristics, and age, sex, and individual motivation, as well as individual displays, could predict spatial outcomes of interactions between and within species. Yet because many of these reports were anecdotal, later researchers seemed to view them with incredulity—the social and cognitive machinations hinted at verged on anthropomorphism. Because of their exemplary documentation, however, the studies by Sigg and Stolba and their mentor, Kummer, of travel decisions made by hamadryas baboons effectively taunted laboratory and field-workers with questions that could not be dismissed. The chapters in this section detail both the vast amount of knowledge that is now well substantiated and that which remains uncertain regarding the social processes that contribute to travel decisions by primate troops. In turn, deductions are offered on the cognitive abilities that these social manipulations might reflect.

CHAPTER FIFTEEN

Social Manipulation Within and Between Troops Mediates Primate Group Movement

SUE BOINSKI

Waltzing couples negotiate the space of a dance floor with different stratagems and abilities. Some selfishly monopolize prime areas. Others range widely over the floor, politely and deftly avoiding approaching couples. Invariably, at least one couple lingers in the seldom-visited corners, thereby evading observation and potential interactions. Of course, the occasional collision between couples occurs. Variation is also apparent in the interplay between partners. The gentleman smoothly leading his pliant and responsive dance partner may be followed by another couple in which the leading partner is difficult to identify.

The interactions observed within the microcosm of a dance floor are directly comparable to the social processes underlying group movement decisions in primates. Individuals within a troop must reach a group-level consensus as to *when* and *where* the troop will travel. Social mechanisms somehow operate to coordinate cohesive travel among dispersed group members. Moreover, social interactions may extend beyond the periphery of a troop to affect the movement patterns of nearby troops. Troops may avoid or approach each other, often based on vocal signals that propagate over long distances.

As with many aspects of the behavior of primates in the wild, Carpenter (1934) was the first to describe explicitly signals that mediate travel by and spacing between troops. Based on field observations of mantled howler monkeys *(Alouatta palliata)* during 1932 and 1933 on Barro Colorado Island, Panama, he concluded that a "deep, hoarse, cluck" acts to initiate and lead travel within howler troops.

> The sound is almost invariably given just prior to the movement of a clan from a lodge or food tree, and it is repeated at intervals while the clan is moving. . . . The probable direction in which the clan would move could be estimated from the position of the male giving the vocal signal and the reference of his behavior. The animals of the clan respond specifically to the vocal clucking by preparing to follow the animal giving the signal. If he begins to move, the rest of the clan follows him, and it seems that by his movements and vocalizations he controls the initiation, direction, and rate of progression. (1934, 110)

Howler monkeys employed a distinctly different vocal signal, however, when communicating with other howler troops.

> One of the most impressive periods of the day in the forest of Barro Colorado island is that of dawn, and one of the most conspicuous aspects of this period is the roaring howls of the howling monkeys. Invariably, as light comes some clan or clans will break forth into howls, and these will be followed by similar roars from groups in adjacent areas. The vocalizations then spread from clan to clan until the majority of the groups on the island have howled. . . . These vocalizations function, apparently, as stimuli of inter-group stimulation at a distance and signalize the location of various groups. I believe the roars are important in partially determining the direction of group progression and in regulating the territorial ranges. (1934, 115)

Carpenter's seminal observations raise a fundamental issue: the intent of the communication between an individual and others that constitutes a social interaction. Implicit in his descriptions is a sense that howler monkeys cooperate in travel and spacing. This view is consistent with a long-prevalent paradigm in animal behavior studies: communication is honest and mutualistic in that it increases the predictability of the individual signaler's current and future behavior, and the self-interest of individuals is enhanced

Sue Boinski

through accommodation and cooperation with others (Smith 1969). This perspective emphasizes that coordination of activities provides far-reaching benefits to the participants. Unity of purpose and consensus is predicted in the movement patterns among, and especially within, groups.

On the other hand, more recent theory argues that signalers attempt to manipulate the behavior of signal recipients to their own benefit (Maynard Smith 1974; Krebs and Dawkins 1984). In other words, communication exists to extract benefits from one's social companions. A divisive scenario of extensive between-individual and between-group conflict is expected to prevail under this hypothesis. Evidence mounts that communication in a wide range of contexts, including contests over resources (Parker 1974) and signals to potential mates (Ryan 1994), is often best interpreted as social manipulation. Yet the extent of manipulation by signalers is constrained. In many instances, signals may be "truly honest" because they are constantly evaluated for their validity by recipients. If cheaters were prevalent, the communication system would become unstable (Zahavi 1975), although if sufficiently uncommon, a low level of cheating may be tolerated (Johnstone and Grafen 1993).

This apparent discord between paradigms used to generate functional interpretations of social communication is superficial. The multiple motivations predicted for communication are merely another refrain of the theme of conflict and cooperation pervading the behavior of individuals in groups. Primates, for example, are thought to live in groups in response to predation risk. Cooperation among individuals enhances their ability to detect and deter potential predators (Terborgh and Janson 1986; Boinski and Chapman 1995). However, as group size increases, these benefits may be outweighed by the cost of increasing competition for food (van Schaik 1989; see Chapman and Chapman, chap. 2, this volume). Similarly, the social relationship of any two individuals over time can be characterized by varying proportions of social aggression over resources such as foraging sites and mating opportunities, and affiliative social interactions such as grooming and alliances (Walters and Seyfarth 1986).

Within the context of the social processes mediating group movement, a major focus of this chapter is the observation that within a cohesive primate troop, individual differences in motivation, ability, and participation are rampant, as is individual conflict. My operational definition of a cohesive group is an aggregation of individuals that in most circumstances remain in visual or vocal contact with most other group members and travel together as a concerted unit. On reflection, the apparent unanimity of the outcomes of troop movement decisions is surprising. Fission-fusion social organizations (Chapman, Wrangham, and Chapman 1995) would seem to be a more likely response to the potential divisiveness of group movement decisions. Individuals within primate troops may have widely disparate mating and foraging strategies, nutritional needs, locomotor and communication abilities, spatial knowledge of the ranging area, and susceptibilities to predation. The gamut of social mechanisms used to determine access to mates or other resources, including dominance, reciprocation, acquiescence, concurrence, and deception, are evident among group members conferring on the selection of a travel route. In like fashion, individuals within the same social group may have different objectives in interactions with other troops. Female behavior, in particular, is usually consistent with the idea that females seek to enhance foraging success (Emlen and Oring 1977). Male behavior is less strongly predicted by foraging contexts. Instead, male contributions to determination of troop movement, if overt, are often linked to enhancing or protecting mating opportunities. Finally, the distances over which troop members and troops are dispersed can be considerable (see Garber, chap. 10, this volume; J. M. Whitehead 1989). This in itself can present obstacles to communication regarding individual and troop intent and the maintenance of troop cohesion; as interindividual distances increase, opportunities for signal failure or misinterpretation are common (Wiley 1994).

Many data pertinent to the social processes underlying group movement decisions are either anecdotal or collected at the level of the group. Aside from individuals that produce long-distance vocal signals, quantified data on the contributions of recognized individuals to travel decisions are also scanty. The absence of systematic information does not mean, however, that a comparative framework cannot be assembled. The scope of this chapter is broad in an effort to collate for the first time many issues related to group movement that are usually considered separately, if at all. The proximate and ultimate functions of loud calls are first reviewed. Next, the use of visual and acoustic signals by individuals to coordinate group movement is documented. Third, five field studies of New World primates that explicitly address the mechanisms by which cohesive troop travel is achieved are considered in detail. Fourth, I discuss known and probable sources of within-species individual and troop variation in the mechanisms of travel coordination. Fifth, some of the many cognitive implications of the observed manipulations of group travel by individuals are introduced, including learning, memory, intent and deceit, representational communication, and motivation. Finally, the relative influences of phylogeny and ecology on variation in travel coordination mechanisms among monkey and ape taxa are evaluated. The reader should also be forewarned that this chapter does not explicitly consider prosimians. Little detailed information pertinent to the individual behavioral strategies of prosimians in the wild was available until recently. Marked disparities between the social behavior and ecology of prosimians and that of monkeys and apes (Anthropoidea) are now recognized (Kappeler 1997; van Schaik and Kappeler 1996). Therefore, social aspects of group movement in prosimians are addressed elsewhere in this volume (see Kappeler, chap. 16).

"Loud Calls" Are the Usual Mode of Between-Troop Communication

Description

Loud calls, also known as long or intertroop calls, represent an eponymous category of vocalizations in the vocal repertoires of many primates (table 15.1, fig. 15.1a). Although most commonly studied in primates, loud calls are found as well in nonprimate social groups, including wolves (*Canis lupus:* Harrington and Mech 1983), coyotes (*C. latrans:* Lehner 1978), lions (*Panthera leo:* Schaller 1972), and spotted hyenas (*Crocuta crocuta:* East and Hofer 1991a). Loud calls, when present in a species' repertoire, are most frequently produced by mature, dominant group members, usually only adult males or both adult males and females (Gautier and Gautier-Hion 1977).

These intense, highly stereotyped, and species-specific acoustic signals (Snowdon et al. 1986; Zimmerman 1995b) may be transmitted over distances exceeding a kilometer, sometimes sufficiently far to be audible to at least several other conspecific groups (Brown 1989; Whitehead 1995). In fact, the range of transmission of a species' loud calls is positively correlated with the size of its home range (Brown 1989). Multiple categories of information are broadcast in a loud call. First, accurate information on the location of the caller, including both the angle and distance from recipients, is

Table 15.1 Use of long calls in manipulation of group movement patterns

Species	Emitter	Responses/ascribed functions	Data[a]	References
Indri *Indri indri*	Both sexes	(1) Advertises location to other groups (2) Allows cohesion of dispersed group members	Q	Pollock 1986
Agile managbey *Cercocebus galeritus*	Adult males	Promotes intergroup spacing and maintains territorial boundaries	S	Kinnaird 1992b
Gray-cheeked mangabey *Lophocebus albigena*	Adult males	(1) Promotes intergroup spacing (2) Maintains intragroup cohesion (3) Directs troop movements	S Q A	Waser 1975, 1976, 1977b
Diana monkey *Cercopithecus diana*	Adult males	Promotes intergroup spacing and maintains territorial boundaries	S	Hill 1994
Black-and-white colobus *Colobus guereza*	Adult males	Defense of feeding sites	S	von Hippel 1996
White-handed gibbon *Hylobates lar*	Both sexes	Promotes intergroup spacing and maintains territorial boundaries	A	Ellefson 1968
Agile gibbon *H. agilis*	Adult males	Territorial defense	Q	Mitani 1988
Mueller's gibbon *H. muelleri*	Both sexes	Promotes intergroup spacing	S	Mitani 1985a

Species	Age/sex class	Function	Data[a]	Reference
Orangutan *Pongo pygmaeus*	Adult males	Represents male-male competition and promotes male spacing	Q	Mitani 1985b
Mantled howler *Alouatta palliata*	Adult males	Promotes intergroup spacing	A	Carpenter 1934; Altmann 1959; Chivers 1969; Baldwin and Baldwin 1976
	Adult males	(1) Stimulates avoidance or approach by recipient troops	S	J. M. Whitehead 1987, 1989
		(2) Promotes intragroup cohesion	A	
		(3) Broadcasts information on location of food patches	S	
		(4) Attracts females as mates	A	
		(5) Initiates and indicates trajectory of troop travel	Q	
Dusky titi *Callicebus moloch*	Adult males	(1) Promotes group spacing and maintains territorial boundaries	S	Robinson 1979
		(2) Increases callers' group cohesion	Q	
Yellow-handed titi *Callicebus torquatus*	Both sexes	Promotes group spacing and maintains territorial boundaries	S	Kinzey and Robinson 1983
Golden lion tamarin *Leontopithecus rosalia*	Both sexes	Promotes group spacing and maintains territorial boundaries	Q	Halloy and Kleiman 1994

[a] Indicates the nature of the data on which descriptions are based: A, anecdotal; Q, quantified but without extensive statistical analyses; S, quantified data with conclusions supported by robust statistical analyses.

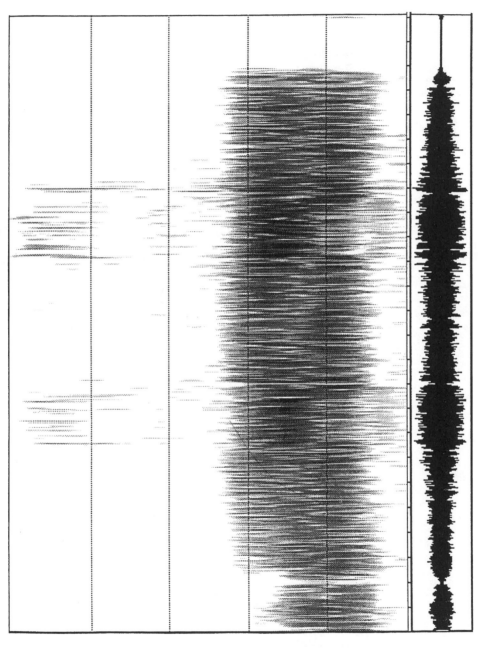

Figure 15.1 (A, this page) Spectrogram (*top*) and waveform (*bottom*) of a representative loud call of a male red-handed howler (*Alouatta belzebel*). Vertical divisions of the spectrogram are in 0.5 kHz intervals, and the duration of the loud call is 5 sec. (Spectrogram courtesy of J. M. Whitehead.) (B, opposite page) Spectrogram of representative travel calls (left to right) of the Costa Rican squirrel monkey (twitter), white-handed capuchin (trill), and golden lion tamarin (wah-wah).

A

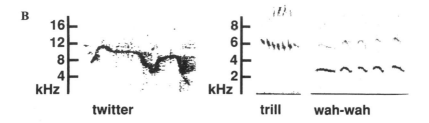

twitter **trill** **wah-wah**

transmitted (Brown 1982; J. M. Whitehead 1989). In forested environments, visual contact between individuals of different troops typically is possible only at close distances (<100 m). In most circumstances, loud calls provide the only cues to the location of neighboring troops. Second, fine-scale features of the acoustic structure of loud calls also contain reliable information on the individual identity of the caller (Marler and Hobbett 1975; Waser 1977b; Snowdon, Cleveland, and French 1983; East and Hofer 1991b; Butynski et al. 1992). Members of neighboring groups probably recognize the voices and track the long-term behavior patterns of the individuals emitting loud calls. Third, Heinsohn (1997) found that lionesses *(Panthera leo)* are able to assess the number of individuals producing roars in opposing prides, and approach aggressively only if the number of opponents is smaller than their own pride. Although no nonhuman primate has yet been shown to assess the number of prospective opponents based on the number of individuals producing loud calls, there is no obvious reason why lions, but not primates, should have this ability.

Proximate Functions of Loud Calls and Predictors of
Troop Behavior
The relative energetic cost of producing loud calls compared with intratroop calls remains uncertain (Horn, Leonard, and Weary 1995), as does the possibility of exacerbated predation risk associated with the highly detectable loud calls. Yet the direct costs associated with producing loud calls are probably low compared with the potential benefits. Field data strongly support the interpretation that loud calls announce the occupancy of a resource, and often the willingness to defend that resource aggressively. *Food,* both in the form of territories (Mitani and Rodman 1979) and in the form of specific food patches, such as a fruit-laden tree crown (Kinnaird 1992b), and *mates* (Dunbar 1988) are the two types of resources commonly advertised. Reliable access to these commodities, espe-

cially when they are in short supply, can have a critical effect on reproductive fitness. In other words, the proximate function of loud calls in many situations is to deter the close approach of other troops and minimize competition for resources. Visual displays are usually accorded a secondary, transitory role in between-troop communication among those species that use loud calls. Loud calls permit, but do not necessitate, direct intertroop encounters by providing clear locational cues to a troop's position within a large area. If visual confrontation occurs, agonism between the troops often quickly intensifies to direct aggression (Mitani 1985a; Hill 1994). When intertroop encounters do occur, at least one troop is often thought to have "intentionally" engaged the other (Robinson 1979; Hill 1994).

Maintenance of a nearly exclusive ranging or buffer area into which conspecific troops seldom venture is characteristic of species wielding loud calls (Waser 1976; Robinson, Wright, and Kinzey 1986). Moreover, in many of these species, the rate of encounter between adjacent troops is significantly lower than expected (Waser and Wiley 1979; Whitehead 1987). Based on detailed contextual analyses (i.e., Kinnaird 1992b; Hill 1994) and playback experiments (i.e., Waser 1975; Robinson 1979; Whitehead 1987), production of and response to loud calls contribute to three basic mechanisms of stable between-troop spacing: territory defense; site-dependent defense; and avoidance. These mechanisms are phylogenetically labile; different mechanisms often predominate among congeners. First, species that actively monitor and patrol defended territories (Mitani and Rodman 1979) often employ loud calls as a component of behaviors demarcating their territory, including the white-handed gibbon (*Hylobates lar:* Ellefson 1968; Chivers 1974), dusky titi (*Callicebus moloch:* Robinson 1979), golden lion tamarin (*Leontopithecus rosalia:* Halloy and Kleiman 1994), and diana monkey (*Cercopithecus diana:* Hill 1994). However, a lack of territoriality does not mean that specific resources are not defended. Aggressive defense of resources depends on the position of the calling troop. A troop hearing a loud call from another troop near a dense food patch (agile managbey, *Cercocebus galeritus:* Kinnaird 1992b; black-and-white colobus, *Colobus guereza:* von Hippel 1996) or deep within their buffer area (wolves: Harrington and Mech 1983) readily approaches the calling troop. Finally, avoidance of other troops when loud calls are detected, in combination with fidelity to a general ranging area, is typical of mantled howler monkeys

(Carpenter 1934; Altmann 1959), yellow-handed titis (*Callicebus torquatus:* Kinzey and Robinson 1983), gray-cheeked managbeys (*Lophocebus albigena:* Waser 1976), and Mueller's gibbons (*Hylobates muelleri:* Mitani 1985a).

The primary audience of loud calls is individuals in other social groups, and their apparent primary intent is affecting those groups' movement (and exploitation) patterns. Some species, however, produce loud, distinctive vocalizations with a more ambiguous target audience and function, including barks by baboons in Guinea, West Africa (Byrne 1982), spider monkey *(Ateles geoffroyi)* whinnies (Chapman and Lefebvre 1990); chimpanzee *(Pan troglodytes)* pant-hoots (Mitani et al. 1992; Clark and Wrangham 1993; Mitani and Nishida 1993), and numerous vocalizations by sperm whales (*Physeter macrocephalus:* Weilgart, Whitehead, and Payne 1996) and elephants (*Elephas maximus* and *Loxodonta africana:* Payne, Langbauer, and Thomas 1986; Poole et al. 1988). In other words, these calls are the bane of review writers: instances that do not tidily fit into heuristic categories. The boundaries distinguishing within-group from between-group communication and manipulation are particularly difficult to perceive in these species because of (1) their fission-fusion social organizations, (2) the often great dispersion of even subgroup members, (3) long-distance transmission of these acoustic signals, and (4) numerous challenges in documenting changes in the behavior of both callers and distant recipients subsequent to call production. Many of these calls are thought to affect movement patterns, but whether they increase, decrease, or maintain spacing between individuals, subgroups, or groups, and in what contexts, remains open for discussion. That baboons, spider monkeys, and chimpanzees are among the largest primate taxa, and elephants and whales represent the largest terrestrial and aquatic social taxa, probably underlies the functional ambiguity of these vocalizations. Large species are thought to benefit less from maintaining cohesive group formations due to their reduced susceptibility to predation (Boinski and Chapman 1995). Therefore the relation of specific vocal signals to the behavior of groups and individuals in these large animals is often difficult to determine.

Based on playback experiments and detailed field observations of diana monkeys, Zuberbühler, Noë, and Seyfarth (1997) provide strong evidence that further expands the intended audience of loud calls in this species from conspecifics to encompass potential predators. Male diana monkeys produce loud calls upon detecting those sit-and-wait predators, such as leopards *(Panthera pardus)* and crowned hawk eagles *(Stephanoaetus coronatus),* that rely on surprise attacks to successfully capture prey. In addition to producing loud calls, troops approach the predator, apparently to thwart surprise attacks by notifying the predator that it has been detected.

Exploitation of Loud Calls by Individuals

Individual strategies and conflicts underlying loud call production and response are receiving more attention in recent years. In early studies of loud calls, the behavior of group members other than those individuals emitting the vocalizations was often poorly documented. Close scrutiny does not substantiate the often presumed unanimity of motivation among members of a group in contexts associated with loud calls. Behavior during intergroup encounters is particularly revealing of underlying motivational differences among troop members (Cheney 1986). Given that female reproductive success is usually limited by the distribution of food resources in the environment, whereas male reproductive success is more closely limited by mating opportunities (Schoener 1971; Emlen and Oring 1977), strong within-species sex differences in behavior during intergroup encounters are not surprising. Males are usually the only sex to act in a manner consistent with mate defense hypotheses, while one or both sexes may exhibit behaviors consistent with food resource defense, such as vigorously displaying at, fighting, and chasing animals from other troops away from food patches. Individuals that are seemingly oblivious to the close presence of individuals from other troops, or that approach them and engage them in play or other affiliative interactions, are interpreted as exhibiting nondefensive behavior in intertroop encounters. For example, females in both the capped leaf monkey (*Trachypithecus pileata:* Stanford 1990) and olive baboon (*Papio anubis:* C. Packer in Waser 1982a) are seldom active participants in aggressive or defensive interactions between troops; males herd females away from calling males during an encounter. Agile mangabey males herd females with estrous swellings *away from* and anestrous females *toward* the front line of troop members engaged in an aggressive intertroop encounter (Kinnaird 1992b; M. F. Kinnaird, pers. comm.). Among gibbons, only in *Hylobates muelleri* do females actively participate in agonistic interactions during boundary encounters, although females in both *H. muelleri* and *H. lar* engage in duets with males that appear to be functional equivalents of loud calls (Mitani

Sue Boinski

1985a). Only males produce loud calls in *Colobus guereza,* but both sexes are aggressive during intertroop encounters that defend food patches (von Hippel 1996).

Females can exploit males as megaphones to manipulate intergroup interactions. Hill (1994) describes male diana monkeys as being incited to produce loud calls by the chatter-scream vocalizations of females. If and when intertroop encounters are subsequently engaged, only the female and juvenile diana monkeys participate in the scrimmages. Similar "cheerleading" by females that instigate and reinforce loud call production by males is described for blue monkeys (*Cercopithecus mitis:* Butynski 1982a), Kloss's gibbons (*Hylobates klossii:* Tenaza 1976), and red howler monkeys (*Alouatta seniculus:* Sekulic 1982; J. M. Whitehead 1989). Mate and food resource defense, however, are not mutually exclusive functions of male loud calls (van Schaik, Assink, and Salafsky 1992). In twelve populations of Southeast Asian langurs (representing *Presbytis aygula, P. cristata, P. melalophos, P. obscura, P. rubicunda,* and *P. thomasi*), only males produce loud calls. van Schaik, Assink, and Salafsky (1992) suggest that female langurs exploit males and their loud calls and aggressive mate defense in a manner such that these displays also effectively function to defend food resources. Male aggression in intertroop encounters is directed at males, and the extent of aggression increases as the number of males in an opposing troop increases. Females are merely bystanders in agonistic intertroop encounters. Yet female langurs lead and appear to dictate group travel decisions, and thus determine the timing and location of intertroop encounters (van Schaik, Assink, and Salafsky 1992).

Do Loud Calls Function in Sexual Selection?
Loud calls plausibly influence individual behavior in arenas other than resource defense. Remarkable parallels exist between loud calls and the flashy signals and extreme morphological traits associated with sexual selection (Kirkpatrick and Ryan 1991). For example, the loud calls of male red deer *(Cervus elephas)* signal fighting ability, attract females, and induce early ovulation in females (Clutton-Brock and Albon 1979; McComb 1987, 1991). Individual strategies of loud call production and travel coordination, and the resultant movement patterns of troops might be affected by the varying quality of potential mates and their loud calls in nearby troops. Potential mates and competitors for mates in neighboring

troops plausibly employ loud calls to evaluate the quality of callers in terms of natural selection and sexual selection benefits (Darwin 1871; Fisher 1958). Loud calls could provide information on the resource-holding abilities and thus the potential parental investment of the callers in terms of direct material benefits such as food and protection from predators (Boake 1986). The acoustic structure, intensity, and rate of emission of loud calls could also allow potential mates to select "good genes" for their offspring (Zahavi 1975). Similarly, these acoustic traits could be selected under runaway sexual selection (Fisher 1958), and runaway models based on sensory exploitation (Ryan 1990) may be particularly important.

Among primates, however, the potential role of loud calls in sexual selection has thus far been tested only in the orangutan *(Pongo pygmaeus)* and gibbons (*Hylobates* spp.). Rodman (1973a) first suggested that the loud calls of solitary adult male orangutans may increase spacing between males, while also serving to attract females over long distances. Mitani (1985b) tested these hypotheses with field playback experiments and observations. Although loud calls do appear to have a role in male-male competition in orangutans (dominant males approach and subordinate males avoid a male orangutan emitting loud calls), no evidence indicated that male calls elicited the approach of females that were prospective mates. A comparable field study of the complex loud calls or "songs" produced by male agile gibbons *(H. agilis)* reached similar conclusions. Mitani (1988) found that these elaborate vocalizations were probably important in mediating male-male competition for territories, a crucial prerequisite for male reproductive success in gibbons, but were not employed by females to evaluate potential mates. More recently, Mitani's (1988) findings were reinforced in a reevaluation of previously published field data on loud call production among nine gibbon species (Cowlishaw 1996). Cowlishaw (1996) argued that territorial male, but not female, gibbons use loud call displays to advertise their current ability to defend territories from nonterritorial "floater" male gibbons.

Drum Majors Use Whistles and Batons to Lead Marching Bands

Organizing the movement of numerous people at a particular time and in a specific direction is not an easy task. Teachers leading their classes on field trips, school crossing guards, scout leaders, and tour guides can all attest to the many challenges in keeping their charges in a cohesive group traveling in an efficient, prompt fashion. These human leaders rely heavily on various combinations of visual and vocal signals to manage movement patterns. Hand gesticulations and shouting are so ubiquitous that their crucial role in communication is underappreciated. Perhaps the clearest example is a drum major leading a marching band. Despite endless hours of drill (I write with the experience of the lowly fourth French horn in my high school marching band), drum majors must continually provide information to band members regarding direction, pace, and musical endeavors. Not coincidentally, the visual and acoustic signals employed by drum majors are highly ritualized and exaggerated: elongated batons raised high so that even the last ranks of marchers have a clear view, and ear-splitting shrieks from vigorously blown whistles. Moreover, the herds of humans mentioned here all have predesignated leaders. The problems in achieving coordinated movement are exacerbated among groups that must reach a consensus as to their objective before travel proceeds (see Wilson, chap. 9, this volume; Dyer, chap. 6, this volume).

Visual and acoustic cues were first recognized as pivotal to the process of coordination of travel by troops of nonhuman primates in Carpenter's (1934) original anecdotal description of mantled howlers. Mechanisms underlying the coordination of troop movement have received sporadic attention since then. The next mention of travel coordination phenomena in the primate literature occurred in 1963. Itani (1963) reported that one specific coo variant produced by Japanese macaques *(Macaca fuscata)* was closely linked with the initiation of travel by a troop, and that another coo variant was emitted specifically by individuals at the leading edge of a traveling troop. The initiation of travel in mountain gorilla *(Gorilla gorilla)* troops was described by Schaller (1963, 1965). Silverback males indicate the trajectory of travel by assuming a stiff posture while staring fixedly in a specific direction, and soft grunts produced by all group members sometimes accompany this visual display (see Watts, chap. 13, this volume). Numerous primate field studies during the 1960s and the 1970s included matter-of-fact descriptions of the social interactions with which the study species in question achieved coordinated troop movement (table 15.2).

The most influential and most often cited accounts from this early period are three anecdotal studies describing how individual members or subsets of baboon troops employ intention movements to initiate troop movement and to lead a traveling troop (*Papio anubis:* Hall and DeVore 1965; Rowell 1972b; *P. hamadryas:* Kummer

Table 15.2 Use of within-troop signals in the coordination of troop movement

Species	Displays/intention movements	Call	Emitters	Initiate travel	Indicate trajectory	Lead travel	Data[a]	Reference
Gorilla		X	All group members call, only adult males display				S	Stewart and Harcourt 1994
Gorilla gorilla	X			X			A	Schaller 1963
Bonobo			Adult males		X		Q	Ingmanson 1996
Pan paniscus	X		Adult males (?)	X	X		A	Savage-Rumbaugh et al. 1996
Chimpanzee								
Pan troglodytes	X		Adult males	X	X		Q	Boesch 1991b
Olive baboon	X		Adult males indicate trajectory and lead, but females decide	X	X		A	Rowell 1972b
Papio anubis	X		Group monitors dominant males	X	X	X	A	Hall and DeVore 1965
Hamadryas baboon								Kummer 1968a,b, 1995
Papio hamadryas	X		Males initiate, females influence, but dominant male decides	X	X		A	Solba 1979 (as recounted in Kummer 1995)
Chacma baboon							S	Stolz and Saayman 1970
P. ursinus	X	X	Adult males	X	X	X	A	Buskirk, Buskirk, and Hamilton 1974
Yellow baboons			Adult males indicate trajectory and lead, but females decide	X	X	X	A	Norton 1986
P. cynocephalus	X			X	X	X		
Hanuman langur	X		Adult males	X			A	Sugiyama 1976
Semnopithecus entellus	X	X	Adult males	X		X	A	Vogel 1973
Capped leaf monkey			Usually adult females, rarely adult males					
Trachypithecus pileata	X			X	X	X	Q	Stanford 1990, 1991
Red colobus								
Procolobus badius		X	Both sexes	X	X	X	S, A	Struhsaker 1975
Black-and-white colobus								
Colobus guereza	X		Adult males	X	X		A	Marler 1968

Common name / Species			Caller				Reference
Proboscis monkey *Nasalis larvatus*		X	Adult male	X	X	A	C. Yeager, pers. comm.
Patas monkey *Erythrocebus patas*	X	X	Usually females, occasionally males	X		A	J. Chism, pers. comm.
Drill *Mandrillus leucophaeus*		X	Adult males	X		A	Jouventin 1975
Mandrill *M. sphinx*		X	Adult males	X		A	Kudo 1987
Grey-cheeked mangabey *Lophocebus albigena*	X	X	Both sexes, often adult females	X	X	A	P. Waser, pers. comm.
Crowned guenon and moustached guenon *Cercopithecus pogonias* and *C. cephus* mixed-species groups		X	Adult male *C. pogonias* for both species	X		A	Gautier and Gautier-Hion 1983
Vervet monkey *Chlorocebus aethiops*		X	Both sexes	X		A	Struhsaker 1967a
Japanese macaque *Macaca fuscata*		X	Both sexes	X	X	A	Itani 1963
Barbary macaque *M. sylvanus*	X	X	Adult males	X		A	Mehlman 1984
			Adult males		X	Q	Mehlman 1996
Mantled howler monkey *Alouatta palliata*	X	X	Adult males	X	X	A	Carpenter 1934; Milton 1980; K. Glander, pers. comm.
Dusky titi *Callicebus moloch*	X	X	Both sexes	X	X	A	Mason 1968
		X	Both sexes	X	X	S	Menzel 1993
Black uakari *Cacajao melanocephalus*		X	Both sexes	X	X	A	Boubli, in press; J. Boubli, unpub.
White-faced saki *Pithecia pithecia*		X	Both sexes	X	X	A	E. Cunningham, pers. comm.
Golden lion tamarin *Leontopithecus rosalia*	X	X	Both sexes	X	X	S	Boinski et al. 1994
Red-handed tamarin *Saguinis midas*		X	Both sexes	X	X	A	Boinski, unpub. data

continues

Table 15.2 (*continued*)

Species	Displays/intention movements	Call	Emitters	Initiate travel	Indicate trajectory	Lead travel	Data[a]	Reference
Saddleback tamarin *S. fuscicollis*		X	Both sexes	X	X		A	C. Peres, pers. comm.
Moustached tamarin *S. mystax*		X	Both sexes	X	X	X	A	C. Peres, pers. comm.
Pygmy marmoset *Callithrix pygmaea*	X	X	Both sexes	X	X		A	Soini 1981
Costa Rican squirrel monkey *Saimiri oerstedii*	X	X	Adult females	X	X	X	S	Boinski 1991
Peruvian squirrel monkey *S. boliviensis*	X	X	Troop follows movements of sympatric *Cebus* spp.					Boinski and Mitchell 1992
Surinamese squirrel monkey *S. sciureus*	X	X	Mixture of calls and displays by both sexes; troop also follows *Cebus* spp. on occasion	X	X	X	A	S. Boinski, unpub. data
White-faced capuchin *Cebus capucinus*	X		Adult males	X	X	X	A	Freese 1978
	X	X	Both sexes	X	X	X	S	Boinski 1993; Boinski and Campbell 1995
Brown capuchin *C. apella*	X		Troop monitors dominant male			X	A	Boinski and Janson, unpub. data

[a] Indicates the nature of the data on which descriptions are based: A, anecdotal; Q, quantified but without extensive statistical analyses; S, quantified data with conclusions supported by robust statistical analyses.

1968a; see Byrne, chap. 17, this volume). *Intention movements,* as defined by these authors for baboons and by subsequent workers among other primate species, encompass those displays, postures, orientations, and changes of position exhibited by one or a few troop members that in specific contexts appear to human observers to indicate the animals' preferred trajectory of troop travel. These baboon accounts were particularly important because they suggested that social dominants made the final travel decisions. Troop members effectively filled "initiator" versus "decider" roles in the decision process. An initiator was described as suggesting a trajectory and a probable goal for troop travel, but the decider made the eventual decision on which of the alternative trajectories, if any, would be followed. This was a somewhat surprisingly subtle social manipulation given the then prevalent emphasis on male dominance and aggression as explanatory variables in the primate literature. Vocal signals to initiate and lead group travel, which I term *travel calls* for simplicity, are much less commonly used among baboons compared with many forest primates; visual contact among group members is usually a sufficiently effective communication mode in the terrestrial baboons (Kummer 1995; but see Byrne 1981).

Loud calls were accorded a major role in coordinating within-troop travel in the 1970s. Loud calls were described as "rallying" troop members before travel, conjuring an image of a commander on horseback mustering his troops before an onslaught on the common enemy. Two observations contributed to this interpretation. First, group cohesion often increases subsequent to loud call production (Gautier and Gautier-Hion 1969; Struhsaker 1970; Waser 1977b; Robinson 1979). Second, in many situations, loud calls strongly predict not only that the troop will start traveling, but also give information on the direction of travel. The inference is that the individual(s) emitting loud calls also determines the subsequent troop travel trajectory. Yet, as discussed above, loud call production neither determines group movement decisions and patterns nor necessarily reflects which group members instigated the loud calling. Discussions of the motivations of signals exchanged between troops are best disengaged from subsequent troop movements, as they may well represent different decision processes. At the minimum, only one troop member needs to decide to produce a long call, but as more recent investigations document, group movement decisions are often much more complex and may require the concurrence of most troop members (see below).

Since 1990, abundant new data on travel coordination by troops have appeared. Three of these new quantitative studies confirm findings from earlier anecdotal observations. Stanford (1990, 1991) determined that in capped leaf monkeys, of the more than 446 travel initiatives in which a leader or initiator could be identified on the basis of intention movements, adult females were the leaders in more than 90% of the instances. Sugiyama's (1976) description of the comparable process in the hanuman langur *(Semnopitbecus entellus)* differed only in that adult males were the group members predominating in the use of intention movements. Menzel's (1993) captive study showing how dusky titis achieve cohesive group movement using intention movements in combination with traditional, intensely familiar travel routes closely corroborates Mason's (1968) descriptive study of this species in the wild. Similarly, Stewart and Harcourt's (1994) field study of the use of "close calls" by gorillas to initiate group movement when an adult male silverback does not exhibit intention movements also substantiates Schaller's (1963, 1965) original descriptions.

Travel coordination processes have been documented only in the past few years for a number of primate species. In all of these reports, distinct acoustic signals or flamboyant visual displays are the predominant communication channels used by individuals to signal their travel decisions and preferences to other group members. Mehlman (1996) elaborates earlier anecdotal descriptions (Mehlman 1984) of the use of branch-shaking displays and dead-log thumping by Barbary macaques *(Macaca sylvanus)* to coordinate travel. A series of field studies detailing the use of species-specific travel calls among four species of New World monkeys has recently been published (Costa Rican squirrel monkeys, *Saimiri oerstedii:* Boinski 1991; Peruvian squirrel monkeys, *S. boliviensis:* Boinski and Cropp 1999; white-faced capuchins, *Cebus capucinus:* Boinski 1993; Boinski and Campbell 1995; golden lion tamarins, *Leontopithecus rosalia:* Boinski et al. 1994). Boesch (1991b) suggests that within one community of chimpanzees in the Taï forest, information on changes in direction and starts and stops of group progressions is transmitted to dispersed group members by loud drumming on tree buttresses. At Gombe, however, Goodall (1986) could not confidently identify the means by which chimpanzees coordinated group travel. Bonobos *(Pan paniscus)* are well documented by Ingmanson (Ingmanson and Kano 1993; Ingmanson 1996) to exploit noisy, highly apparent branch-dragging displays to greatly emphasize intention movements associated with the mediation of group travel.

I suspect that these new data became available and, importantly, publishable for these species because their discrete signals are easy to define and quantify objectively. They stand in contrast to those species whose signals are more subtle and therefore more challenging to quantify. For example, in brown capuchins *(Cebus apella)*, the alpha male, the preeminently dominant troop member, appears responsible for nearly all travel decisions, but makes negligible efforts to signal his intent overtly to other troop members (S. Boinski and C. Janson, unpub.). Instead, troop members closely monitor his activities. Not surprisingly, Boinski and Janson have not published statistically substantiated conclusions because we were stumped at how to efficiently and convincingly quantify the subtle social interactions involved. Baboons are a better-known example of a situation in which human observers "perceive" specific social interactions as signal exchanges to coordinate troop travel, but documentation relies heavily on descriptive anecdotes (Kummer 1995; see Byrne, chap. 17, this volume). Clearly, the more elusive travel coordination signals in primates and other taxa merit greater attention from biologists.

Within-Troop Travel Coordination: Field Studies of New World Monkeys

Recent comparative field studies of the intragroup vocal behavior of Costa Rican and Peruvian squirrel monkeys, white-faced capuchins, and golden lion tamarins detail the diversity of individual contributions to the coordination of travel within a troop, as well as several ecological constraints underlying species differences in the process of achieving coordinated troop travel (Boinski 1991, 1993; Boinski and Mitchell 1992; Boinski et al. 1994; Boinski and Campbell 1995). Field methods and data analyses were nearly identical across the five studies. Long-term research on the foraging behavior and social organization of the relevant population preceded each vocal study. Every member of each study troop was individually recognized. All four species maintain stable troops, are highly arboreal, prefer small-branched, often heavily foliated, traveling and foraging substrates, and are broadly similar in regard to diet (predominantly fruits and arthropods and the occasional small vertebrate). Troops are cohesive in that all group members remain in constant auditory contact and never fission into separate subgroups. But the dispersion of individuals within the troop perimeter can be considerable, often exceeding an area of 1 ha for squirrel monkeys and capuchins. Females allocate more time to foraging than do males in the squirrel monkey and capuchin (Boinski 1988a; Mitchell 1990; Rose 1994b; Fedigan 1993). Comparable foraging data have not yet been published for the golden lion tamarin, but within age classes, sex differences in foraging are probably minimal (C. Peres, pers. comm.).

Field Protocols
Qualitative observations made during my first field study of the vocal behavior of Costa Rican squirrel monkeys (Boinski and Newman 1988) suggested that vocalizations were associated with group movement. I returned to my primary study site, Parque Nacional Corcovado, in 1988 for another vocal study of this Costa Rican squirrel monkey population, the first of the series. This time, however, I was armed with a hypothesis: that the usage of a specific vocalization, the twitter, was restricted to the initiation and leading of troop movement. I sought to learn how twitters were used, which troop members overtly participated in travel decisions, and, if within-troop conflict over travel route was detected, how consensus for the selected route was achieved. A major concern in this and subsequent studies was to quantify individual participation in travel coordination and to devise objective definitions of individual efforts to initiate and lead troop travel.

In focal animal samples of individual behavior (J. Altmann 1974), a continuous sample of a monkey's vocal behavior was recorded with a tape recorder along with descriptions of the monkey's position within the troop, its activity (such as looking for food or resting), and nearby troop members. A monkey was described as making a "start attempt" if it produced at least one of the designated twitter travel calls while the troop was stationary. If the troop began traveling within 10 minutes after the first travel call, the azimuth of troop travel relative to the approximate center of the stationary troop was estimated with the aid of a compass and patient field assistants. When a troop was stationary, it was noted whether the focal monkey was at or beyond the troop periphery in the "edge" or in the "core," the area of troop dispersion interior to the edge. In like fashion for all the studies, two positions were defined within a traveling troop: the "vanguard," a 5 m deep zone starting at the leading edge, and the "rearguard," the area of troop dispersion

Sue Boinski

following the vanguard. Vanguard zones normally occupied far less than 10% of the area of troop dispersion. A monkey was considered to be "leading" a traveling troop if it was positioned within the vanguard.

How Travel Calls Are Used: Spatial and Contextual Patterns
The initial qualitative observations of how Costa Rican squirrel monkeys use twitters to coordinate troop travel were corroborated by the follow-up field study (Boinski 1991). Furthermore, the vocal repertoires of both golden lion tamarins (Boinski et al. 1994) in Brazil and white-faced capuchins at Parque Nacional Santa Rosa (Boinski 1993) and La Selva (Boinski and Campbell 1995) in Costa Rica also had distinct travel call categories, the wah-wah and trill, respectively (fig. 15.1B; table 15.3). Travel calls are wielded in an identical fashion across the three species to initiate travel in a specific direction and to lead a traveling troop. The acoustic structure of all three travel calls—either rapid frequency modulation (Costa Rican squirrel monkeys) or numerous starts and stops (white-faced capuchins and golden lion tamarins)—provides much information to the standard mammalian ear as to the location of the caller in three-dimensional space (Boinski et al. 1994; Boinski 1996). Locatability would seem to be a requisite feature of a vocalization coordinating travel among dispersed animals in a forest environment. Foliage and wood can impede reliable transmission of vocalizations due to reverberation and reflection of acoustic signals by these plant materials (Waser and Brown 1986). Although intention movements, such as staring or pacing outward, sometimes accompany the production of travel calls, the calls alone are probably the only signal that can be reliably transmitted more than about 10 m from the caller in the densely foliated branch-end microhabitats exploited by these small monkeys.

Travel is initiated when a monkey (occasionally two, and rarely three) moves to the edge of the troop and produces the species-typical travel call. Even one travel call produced from the troop edge predicts that the troop will initiate travel within 10 minutes. In each species, the designated travel call was the only vocalization in the repertoire associated with the initiation of travel within a period of 10 minutes or less. A 10-minute criterion was selected based on my judgment that this interval represented a pragmatic balance between the minimum time necessary for group members to concur with or to reject the start attempt, but not so long that troop travel

Table 15.3 Summary of troop size and dispersion and features of travel call usage in three species of New World monkeys

Species	Costa Rican squirrel monkey Saimiri oerstedii	White-faced capuchin Cebus capucinus [Santa Rosa]	White-faced capuchin Cebus capucinus [La Selva]	Golden lion tamarin Leontopithecus rosalia
Travel call name	Twitter	Trill	Trill	Wah-wah
Number of troop members	65	14	19	Mean = 6.6 for five troops
Typical dispersion of troop members				
Traveling	1.5 ha	0.5 ha	0.7 ha	20 × 20 m
Stationary	1 ha	0.2 ha	0.4 ha	10 × 10 m
% "successful" travel initiatives	86.7%	92.3%	78.3%	69.0%
% "contested" travel initiatives (multiple simultaneous trajectories indicated)	4.3%	2.6%	8.7%	7.0%
Mean (SE) minutes to travel after first travel call in a "successful" travel initiative	3.3 (1.6)	2.5 (1.1)	5.3 (0.8)	4.7 (2.0)

Source: Boinski 1991, 1993; Boinski et al. 1994; Boinski and Campbell 1995; Boinski, unpub. data.

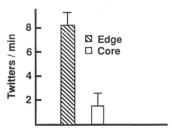

Figure 15.2 The mean individual rates at which adult Costa Rican squirrel monkeys produced twitter travel calls on the edge versus in the core of a stationary troop.

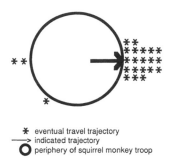

✻ eventual travel trajectory
⟶ indicated trajectory
O periphery of squirrel monkey troop

Figure 15.3 Summary of the indicated and subsequent troop movement trajectories of the twenty-three travel initiatives documented in the Costa Rican squirrel monkey study troop. All indicated trajectories were centered on the large arrow. The two asterisks opposite the arrow (at the 9 o'clock position) indicate the two travel initiatives in which the troop did not travel within a 10 min period after the first travel call was emitted.

would occur independently. This decision was substantiated (table 15.3).

In the Costa Rican squirrel monkey troop, for example, the rate of travel call production by focal subjects at the edge of a stationary troop was significantly higher than in the core (fig. 15.2). Nearly all the travel calls by individuals on the edge were produced in the course of successful travel initiatives (fig. 15.3). Twenty of twenty-three start attempts made by Costa Rican squirrel monkeys were successful, and the trajectories were significantly clustered about the twittering squirrel monkey's position on the edge. On hearing travel calls (but not other species-typical vocalizations), individual squirrel monkeys, capuchins, and tamarins in stationary troops visually isolated from all other troop members were commonly observed to orient and travel almost immediately in the appropriate direction.

The major travel routes used by the study troop of Costa Rican squirrel monkeys at Corcovado throughout their 2 km^2 range had been extremely stable for the previous decade (Janson and Boinski 1992), and the range and its major food resources and sleeping trees had been mapped in detail. Daily movement patterns consisted largely of straight travel segments punctuated by locations where troops stopped and engaged in foraging activities or resting. Abrupt changes of trajectory when the troop was traveling were rare when there was no impediment to continuous arboreal locomotion (i.e., canopy gaps). Instead, major alterations in trajectory were achieved when a troop ceased traveling and travel was reinitiated in a new direction. The trajectory of every travel initiative made by Costa Rican squirrel monkeys during the vocal study intersected with a specific, well-established foraging, resting, or sleeping site. Based only on the position of a squirrel monkey twittering at the edge of a stationary troop, I was able to predict with complete accuracy where the troop would go if it did start traveling (S. Boinski, unpub.). I was less intimately acquainted with the ranging areas and habits of the study troops of the other species, but both capuchins and golden lion tamarins also strongly tended to travel in straight paths between stops. After several weeks of close association with these latter study troops and thus greater knowledge of their ranging patterns, I could predict with great reliability (>80%) where a troop was going to stop next even before travel was initiated. If I was able to accurately deduce the apparent objective indicated by travel initiatives, then presumably other troop members could too.

Travel calls were produced by individuals within the leading edge, or vanguard, of a traveling troop of Costa Rican squirrel monkeys and capuchins at rates approximately four times higher than the same focal animals when in following, or rearguard, positions (fig. 15.4). Within the vanguard, travel calls appear to function as acoustic beacons indicating travel trajectory to troop members following in their distant wake. Among golden lion tamarin troops, however, only twenty-one of the fifty-eight (36%) observed travel call (wah-wah) bouts in a traveling troop were emitted by focal subjects in the vanguard. This appears to be due to the much smaller dispersion and number of members in a golden lion tamarin troop. As a result, spatial position categories are probably less clear-cut, both to the observers and to the tamarins, especially during travel. In a traveling tamarin troop with a maximum diameter of dispersion often less than 20 m, travel calls emitted from any position may be better interpreted as "continue traveling" rather than the more specific "travel in this particular direction."

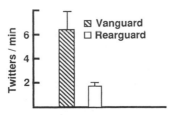

Figure 15.4 The mean individual rates at which adult Costa Rican squirrel monkeys produced twitter travel calls in the vanguard versus rearguard positions of a traveling troop.

Overt conflict over travel decisions and travel route selection is also a feature of coordination of travel among Costa Rican squirrel monkeys, white-faced capuchins, and golden lion tamarins (table 15.3). Troop members may produce travel calls simultaneously from multiple positions on the periphery of a stationary troop. Costa Rican squirrel monkey troops on occasion "stall" for periods exceeding an hour because monkeys persist in twittering simultaneously from different positions. On several occasions I observed a troop evenly partitioned in supporting three different trajectories, and every adult was emitting travel calls loudly and at high rates. Eventually the troop headed off in the trajectory indicated by the squirrel monkeys calling last and loudest. Instances in which troop members are unresponsive to any travel initiatives are more subtle (and quieter) examples of conflict. A small proportion (<10%) of the travel initiatives documented in each study failed because the troop did not begin traveling within the 10 min criterion, despite vigorous and persistent production of travel calls by the initiator(s). The possibility was nil that the signals simply were not successfully transmitted to other troop members.

Why Peruvian Squirrel Monkeys Are Different
Peruvian squirrel monkeys do not have an identifiable travel call. Despite ardent efforts by Carol Mitchell and myself, we were unable to identify any aspect of the vocal behavior of Peruvian squirrel monkeys in Manu that hinted at coordination of travel in a manner similar to that evident in Costa Rican squirrel monkeys, white-faced capuchins, and golden lion tamarins. Yet the Peruvian study troops swiftly and deftly traveled throughout large portions of their 6 km² home ranges with a facility that was truly staggering to the field-workers trailing in their wake. They accomplished this by "catching capuchin troops" (the sympatric brown capuchin and, less often, white-fronted capuchin, *Cebus albifrons*) much as surfers

"catch waves." Parasitism of the capuchins' more detailed knowledge of the location of local fruit resources is the apparent motive for the squirrel monkeys' persistent association with capuchins (Terborgh 1983; Boinski and Mitchell 1992). Their reliance upon capuchins to direct the movement of the mixed-species troop in Peru is apparently so complete (>95% of diurnal activity) that no specialized call to coordinate travel independently of capuchins has evolved or been maintained in the repertoire of the study population (Boinski and Mitchell 1992). In the limited time squirrel monkeys travel apart from capuchins in Peru, they appear to determine travel paths and maintain a loose troop cohesion by producing contact calls at exceptionally high rates and volumes, albeit with apparent confusion and many false starts; travel trajectories are best described as bumbling, erratic walks, not the straight, direct maneuvers of the other study species.

Costa Rican squirrel monkeys, in contrast, avoid troops of the sympatric white-faced capuchin (Boinski 1989), and always determine travel routes by a within-troop decision process (Boinski 1991). The basis for this avoidance appears to be the lack of foraging benefits squirrel monkeys procure from association with capuchins under the food availability regime present in the Costa Rican habitat (Boinski 1989). Yet a third distinct, more complex, and barely studied system of travel coordination is evident among squirrel monkeys in Suriname *(S. sciureus)*. Surinamese squirrel monkeys associate with sympatric brown capuchin and wedge-capped capuchin *(Cebus olivaceus)* troops only 50% of the time (Mittermeier and van Roosmalen 1981). These squirrel monkeys have a call used in travel coordination similar to the twitters of Costa Rican squirrel monkeys (S. Boinski, unpub.). Squirrel monkeys in Suriname, however, are able to coordinate mixed-species associations with their travel calls; in some situations the capuchins respond to the travel leadership of the squirrel monkeys in a manner indistinguishable from that of the squirrel monkeys in the association.

Sources of Individual and Troop Variation in the Process of Travel Coordination

In this section I review major sources of within-species variation in the coordination of troop movement. Participation in travel decisions by group members is neither random nor egalitarian. Leadership, or the ability to influence group movement decisions, is just another reflection of competitive ability and differences in individ-

Sue Boinski

ual motivation. The same individual differences in needs, acquisition abilities, and competitive regimes that are detectable in social interactions mediated by loud calls are also reflected in the coordination of travel within a troop. The outcome of a day's group travel decisions potentially has far a greater effect on an individual's success in food acquisition than the results of that same day's squabbles over access to specific food items. Variation should also be expected among troops of the same species in the configuration of troop travel. The characteristics and overall quality of a group's ranging area, as well as each group's composition and history, offer further opportunities for individual strategies and behaviors to diverge.

Age

Many reports specifically state that only adults initiate and lead travel or decide whether to acquiesce to travel initiatives by other troop members. Subadults, juveniles, and infants appear to passively follow the lead of older group members or their immediate kin and other associates (i.e., Itani 1963; Struhsaker 1967; Mehlman 1984, 1996; Boinski 1991, 1993; Kummer 1995). Although subadult and younger group members overtly participate in travel coordination among a diverse set of species, full adults are most active, and their decisions predominate (Struhsaker 1967; Menzel 1993; Boinski et al. 1994; Boinski and Campbell 1995).

Sex

Strong, often complete, within-species sex biases have been documented in the extent of active coordination of troop movement. Female biases, when present, are attributed to the greater female energetic investment in reproduction (Stanford 1990; Boinski 1991, 1993). Adult female primates in many primate species typically spend more time foraging than do adult males (Rose 1994b; Fragaszy and Boinski 1995). All else being equal, females should be most motivated to influence the many travel decisions that affect foraging efficiency. For example, Sugiyama (1976) describes instances among hanuman langurs when males appeared to want a stationary troop to initiate travel but the females' and immatures' apparent preference was to remain and continue foraging. In at least one species in which females predominate in the expression of travel preferences and males only occasionally exert an influence, the white-faced capuchin, males usually direct the troop to locations with few foraging opportunities. The endpoints of male-directed group movement usually feature water holes or cool, shady horizontal branches ideal for napping (Freese 1978; Boinski 1993). Consistent with this energetic argument, both sexes assert travel preferences in species with high levels of direct parental investment by males (Mason 1968; Soini 1981; Menzel 1993; Boinski et al. 1994).

Curiously, the two subspecies of Costa Rican squirrel monkeys differ markedly in the extent of female bias. Among *Saimiri oerstedii oerstedii* male production of travel calls and overt participation in travel decisions are rare, observed less than once a month (Boinski 1991, 1996). For the most part, adult males in this subspecies are reminiscent of human fathers following their families around a shopping mall: not expressing any particular preference as to which stores his family enters, but lingering about if problems arise. While female prerogative usually prevails among *S. o. citronellus,* adult males commonly produce travel calls and herd troop members to mobilize travel (S. Boinski, unpub.). An ecological basis for this difference has yet to be identified, but in those situations in which male *S. o. citronellus* herded a troop, other troops of squirrel monkeys were in the near vicinity (<0.5 km).

Among populations and species in which male preferences and involvement consistently supersede those of females in travel decisions, male travel leadership is generally a corollary of male social dominance (i.e., Schaller 1963; Hall and DeVore 1965; Boesch 1991b; Buskirk, Buskirk, and Hamilton 1974; Vogel 1973; Jouventin 1975). Perhaps males in these species exert their dominance in the domain of travel in part to reduce competition for mates. Ron's (1996) documentation of male-initiated and maintained troop fission in chacma baboons *(P. cynocephalus ursinus)* is the best evidence that males influence troop travel in a manner consistent with reducing the exposure of troop females to mating opportunities with nonresident males. A subset of baboon studies depicts a more complex dialogue between the sexes on travel decisions. Individuals of one sex typically make travel initiatives, but the other sex ultimately selects the initiative followed (Kummer 1968a; Rowell 1972b; Norton 1986). This lability might be linked to the unusual energy budget of adult male baboons. Males are twice the size of females and require 60–80% more calories each day than do adult females, and mate guarding of estrous females limits male foraging success (Alberts, Altmann, and Wilson 1996). Future studies should investigate the possibility that male baboons exercise tighter

Sue Boinski

control of group travel in seasons and at sites in which males experience more stringent energy constraints. I also predict that estrous females in baboon troops wield more influence in travel decisions than other classes of adult females.

Wild horses *(Equus caballus)* (Rubenstein 1986, 1994), like gelada baboons (Kummer 1995), have a social organization based upon male-dominated harems. As is also true of gelada baboons, these horses illustrate the potentially complex interaction between sex, dominance, and male mate competition in travel leadership. Although all harem-tending male horses are able to control the movement patterns of their harems, females in the harems of dominant males in the community have much greater freedom to determine travel routes and foraging activities than do females in the harems of low-ranking males; dominant males often simply follow the movements of their harem. High-ranking males are able to deflect rival males away from their females, leaving the females effectively unimpeded in their ranging. On the other hand, subordinate males are about twice as likely as dominant males to commandeer control of harem travel so as to lead females away from potential encounters with male rivals. One significant consequence is that females from the harems of dominant males obtain about 10% more foraging time than females allied with subordinate males (Rubinstein 1994).

Individual Preferences
Each day's travel path represents many travel decisions. Conflicts between the preferences of individual troop members as to the trajectory of troop travel and whether the troop should cease or initiate traveling are rampant (e.g., Kummer 1968a,b, 1995; Rowell 1972b; Struhsaker 1975; Sugiyama 1976; Norton 1986; Boesch 1991b; Boinski 1991, 1993; Boinski et al. 1994; Boinski and Campbell 1995; Ingmanson 1996; Mehlman 1996). I am thus surprised that the agonism apparent in the resolution of leadership conflicts is so muted as to be negligible. Typically, individuals vying for their directives to be followed simply persist in producing travel signals, oftentimes increasing signal intensity. Signaling continues until either at least one party abandons the effort or the troop accedes to one of the competing signals. No mention exists of injuries incurred as a result of physical aggression used to determine the outcome of leadership conflicts.

To the best of my knowledge, I witnessed the sole instance yet published of an aggressive threat in the course of a leadership conflict (Boinski 1993). Blanche, an adult female, was in the vanguard leading a white-faced capuchin troop on a trajectory that intersected with a favored fruit-foraging area, emitting occasional trill-travel calls. Junior and Winston, the alpha and beta males, rapidly ran forward from the center of the troop to a position about 1 m ahead of and directly facing Blanche in the vanguard. Several seconds later when Blanche emitted another trill, both Junior and Winston uttered loud threat barks and directed threat faces to Blanche. Junior lunged at Blanche with bared canines. Blanche ran back into the rearguard of the troop and produced no additional trills. Junior and Winston, now alone in the vanguard, immediately turned 120°, trilled, and led the troop directly to a water hole.

Dominance

Dominance relations within a troop may be sufficiently hierarchical that dissension among troop members in travel decisions is seldom detected. In these cases, the most dominant troop member makes all travel decisions (Schaller 1963; Hall and DeVore 1965; Gautier and Gautier-Hion 1983; Buskirk, Buskirk, and Hamilton 1974; Jouventin 1975; brown capuchin: S. Boinski and C. Janson, pers. obs.). Resolution of individual preferences in less steeply hierarchical social groups is also influenced by relative rank. Within a white-faced capuchin troop at Santa Rosa, Costa Rica, for example, the proportions and absolute numbers of successful travel initiatives were perfectly concordant with relative rank within each sex (Boinski 1993). Researchers should be aware of possible masking by dominants of the leadership abilities of subordinates. At La Selva, Costa Rica, a subordinate young adult male capuchin that had emigrated into the troop as an adult and had never been observed to overtly participate in travel coordination events immediately started emitting travel calls appropriately and being the most active travel leader when the dominant male wandered away from the group for several days (Boinski and Campbell 1995). When the dominant male returned, the subordinate male resumed his former mute and docile ways.

It is tempting to predict that travel decisions will be reached and implemented more rapidly within those groups with rigid social hierarchies than among more egalitarian groups. However, the evidence suggests that the time lag between the exhibition of a travel signal, be it vocal or visual, and the subsequent initiation of travel is usually 5 minutes or less in many species (i.e., Struhsaker 1967a, 1975; Carpenter 1934; Menzel 1993; Boinski et al. 1994), even in the most egalitarian primate species with typically large troops, the Costa Rican squirrel monkey (Boinski 1991; Boinski and Mitchell 1994). In fact, some of the most hierarchically ranked species appear to have typically long "deliberation" periods among group members prior to travel onset, especially baboons (Hall and DeVore 1965; Kummer 1968a, 1995; Rowell 1972b), and chimpanzees (Boesch 1991b).

Susceptibility to Predation

A peripheral position in a primate troop, even if transitory, is a prequisite to leading and initiating travel in many species. Numerous studies indicate that these same peripheral positions incur an enhanced exposure to predation (van Schaik and van Noordwijk 1989; Janson 1990a; Ron, Henzi, and Motro 1996; see Boinski, Treves, and Chapman, chap. 3, this volume). Immatures may participate less extensively than adults in travel coordination not because of lesser skills or knowledge, but because the overall benefits following from participation are reduced by their increased susceptibility to predation (Boinski and Campbell 1995). Mothers with newborn or otherwise vulnerable infants are another category of troop members predicted to at least temporarily withdraw from travel interactions in many species. My study of travel call usage among Costa Rican squirrel monkeys (Boinski 1991) encompassed the synchronized 2-week-long annual birth period of this population (Boinski 1987a). Pregnant and nonreproductive females did not differ in the extent of their involvement, and were extremely active in coordinating travel. As soon as a female gave birth and was carrying an infant, however, she ceased exhibiting travel leadership and joined spatial clusters of other mothers with infants abiding in the center of the troop (S. Boinski, unpub.). Infants are highly susceptible to predation by raptors in this population (50% mortality by 6 months of age), and squirrel monkey mothers join in antipredator vigilance and aggressive predator deterrence with other mothers in these spatial clusters (Boinski 1987a).

Habitat Quality and Structure

The two studies of white-faced capuchins (Boinski 1993; Boinski and Mitchell 1995) offer the first detailed between-site comparison of travel call usage in primates (see also discussions of effects of habitat differences in Weilgart, Whitehead, and Payne 1996 and Smolker, chap. 19, this volume). Capuchins at both sites are assigned to the same subspecies, and the acoustic structure of the travel calls produced at the two sites is indistinguishable (Boinski and Mitchell 1995). Santa Rosa is a tropical dry forest, and my study was conducted at the peak of the dry season, when most of the deciduous trees had lost their foliage; during the study period foods available to the capuchins were sparse and of low quality. La Selva, in contrast, is a relatively aseasonal, densely foliated tropical wet forest with abundant fruiting tree crowns. The capuchins at the two sites employed structurally indistinguishable travel calls in a manner that differed in two major ways (Boinski and Campbell 1995).

First, at La Selva, all troop members older than 6 months of age used trills in the coordination of troop travel, whereas at Santa Rosa only full adults, primarily adult females, employed trills as travel calls. The strongest hypotheses to explain this disparity lie in the abundance and quality of fruit at La Selva compared with the scant rations at Santa Rosa. The primary and highly preferred fruit source for La Selva capuchins (*Welfia georgii,* Palmae) is the second most common tree at La Selva, and this fact contributes to fruit distribution at La Selva being much less patchy than at Santa Rosa. Under this food availability regime, La Selva dominants are plausibly more likely to acquiesce to foraging initiatives by subordinates or other troop members with different foraging strategies or knowledge of resource distribution. In other words, wherever the troop goes, acceptable foraging will be encountered. A second, related hypothesis is that the numerous large, dense fruit patches available during the La Selva study, especially from the synchronously fruiting *Dipteryx panamensis* (Papilionoideae) trees (with >10 m diameter crowns laden with fruit), made the travel route preferences of troop members concordant. Of course, these arguments also apply to other potentially limiting commodities. Areas of "reduced exposure to predators," "easy locomotor substrates," and "enhanced likelihood of encountering mates" can readily be substituted for "tasty food patches."

Second, the two sites differed markedly in the apparent function of trill travel calls produced in the rearguard of a traveling troop. Capuchins at Santa Rosa altered the trajectory of traveling troops with trills produced at the periphery of the rearguard, even successfully reversing the trajectory 180°. At La Selva, trills were commonly produced in the rearguard (still at significantly lower rates than in the vanguard), but these trill emissions had no discernible effect on travel path. Numerous lines of evidence suggest that these patterns are a facultative response to the different forest structures at La Selva and at Santa Rosa. Vocalizations propagating through a reflective medium, such as a forest, are degraded by reverberation in the form of echoes of the acoustic signals (Waser and Brown 1986). Reverberation obscures acoustic signals to a greater extent in a dense forest (La Selva) than in a forest with sparse foliage and branches (Santa Rosa). At Santa Rosa, the sparsity of plant material that could reduce transmission distances of trills is likely to have facilitated extreme responsivity to trill production, thus allowing the alteration of trajectory while a troop was traveling. Not only were La Selva capuchins incapable of such flexible maneuvers, but trills produced by La Selva capuchins in the rearguard seemed necessary and effective in relaying the travel signal to distant troop members that were unlikely to hear trills produced within the vanguard. Further observations at Santa Rosa strongly bolster this interpretation emphasizing acoustic propagation qualities. Lisa Rose recently completed extensive observations of the former study troop during the wet season, when the density and abundance of foliage most closely approaches that of La Selva. Usage of the trill travel call in the rearguard of troop progressions by Santa Rosa capuchins during the wet season became indistinguishable from that at La Selva; trills increased in the rearguard, apparently to relay the signal, and the troop changed trajectories only after stopping and then reinitiating travel (L. Rose, pers. comm.).

Thus, in summary, these differences between the two capuchin studies can be attributed to site variation in (1) quality and distribution of food patches and (2) habitat acoustics—the effects of forest structure on the distances over which acoustic signals can be reliably transmitted (Waser and Brown 1986). As more between-site and between-season comparisons of the usage of travel calls and other primate vocalizations become available, habitat characteristics will probably be given increasing importance in explaining within-species variation in primate vocal behavior (see also Boinski and Campbell 1996).

Group Composition
Individual knowledge, experience, and temperament plausibly affect the outcomes of group travel decisions and the social processes by

which those decisions are reached. Thus far, only two reports suggest that such group differences in individual membership may have significant repercussions on the form of travel coordination. The earliest was Buskirk, Buskirk, and Hamilton's (1974) study of five troops of savanna baboons. Adult males in only one of the five troops observed regularly used calls and herding behavior to mobilize and direct troop travel. Another example comes from Cristophe and Hedwige Boesch's long-term study of chimpanzees in the Taï forest. In 1982 Boesch qualitatively recognized that drumming on tree buttresses by the most dominant male, Brutus, closely predicted alterations in travel by the chimpanzee community, including alterations in trajectory and the initiation and cessation of travel (Boesch 1991b). Four of the ten then recognized adult male chimpanzees (which were only a subset of a much larger number of adult male members of the community) disappeared in early 1984. Shortly thereafter, Brutus ceased the unique drumming technique that apparently dictated community movement patterns. Boesch (1991b) argues that the only way that Brutus could effectively coordinate travel among a community containing ten rambunctiously independent adult males was to employ an extremely intense, unambiguous signal such as drumming. However, this post hoc explanation is dependent on many unexamined premises, such as the recognized male chimpanzees having extremely different personality profiles compared with the more numerous undistinguished adult males.

The vast majority of researchers who have observed travel coordination among multiple troops of the same population simply make no mention that any between-troop differences were evident (i.e., Carpenter 1934; Struhsaker 1975; Kudo 1987). The single statistically substantiated demonstration of homogeneity of travel call usage among multiple troops in a population comes from a study of five wild troops of golden lion tamarins. These groups exhibited at least moderate variation in sex and age composition (Boinski et al. 1994). Three other studies of Neotropical primates, however, also expressly sought, but failed to find, indications of between-troop differences that might be accounted for by group composition (white-faced capuchins; three troops at La Selva: Boinski 1993; 2 troops at La Selva: S. Boinski, unpub.; Costa Rican squirrel monkeys, five troops: S. Boinski, unpub.). Future investigations might well document different styles of wielding dominance in travel decisions, as Sapolsky (Sapolsky and Ray 1989) described among ba-

Sue Boinski

boons for other forms of social interactions. For example, males with more relaxed styles of dominance generally in domains of social interactions are predicted to be more accommodating to female foraging needs and permit females more leadership in travel decisions.

Some Cognitive Implications of Loud Calls and Travel Signals

Learning

Circumstantial evidence indicates that competent leadership of troop movement requires familiarity with the ranging area. In addition to older, and presumably more experienced, troop members predominating in leadership roles (see above), observations from one study site suggest that recent immigrants into a troop are handicapped by inexperience with their new ranging area (L. Fedigan, cited in Boinski 1993). Presumably, as an individual's navigation skills within the ranging area increase, the effectiveness of its travel initiatives and loud calls increases. Ungulates provide stronger evidence than primates that leaders are the oldest group members and also the most complete repositories of knowledge of the ranging area. Upon the death of their leader, family units of African elephants become disoriented, seemingly rudderless, and lose their normal ranging patterns (Laws, Parker, and Johnstone 1975). Thinhorn mountain sheep *(Ovis dalli)* acquire their adult ranging patterns as a direct consequence of which adult leader they elect to follow as juveniles (Geist 1971). In fact, Geist argues that the reliance of mountain sheep upon between-generation transmission of traditional, highly conservative ranges and migratory routes is a central, easily disrupted feature of their behavioral biology; areas avoided due to transient hunting pressure are permanently lost because sheep are never introduced as juveniles to those ranging areas.

Still fewer quantitative data are available on the acquisition of the displays and other performance and social skills needed to implement travel leadership and produce effective loud calls. Among nonhuman primates, the acoustic structure of vocalizations is genetically determined, although trial and error learning is necessary for their appropriate expression (Symmes and Biben 1992). Youngsters are commonly reported to complement the efforts of older troop members producing loud calls with vocal and nonvocal behavior (Hill 1994; Halloy and Kleiman 1994). Similarly, juveniles are involved, but in a reduced role, in the coordination of travel among at least some species employing travel calls (Struhsaker 1975; Stewart and Harcourt 1994; Boinski et al. 1994; Boinski and Campbell 1995) and intention movements (Menzel 1993). Further studies might profit from close attention to potential associations between apparent early efforts of young primates to mediate group movement by means of vocalizations and displays and the active cooperation and reinforcement of those efforts by older, and especially related, troop members. Perhaps the participation of young primates reflects "practice" and skill acquisition (Gouzoules and Gouzoules 1989). The concurrent expression of such visual and vocal signals by immature and mature group members may provide us with the first examples of social experience affecting the acquisition of a vocal behavior by means of *social influence*—another individual stimulating or facilitating the expression of a vocal behavior—and the direct acquisition of knowledge from another via *social learning* (Whiten and Ham 1992).

Memory

The mediation of group movement is a social interaction on a grandiose scale. When an individual transmits a signal, be it a travel signal or a loud call, the desired response may hinge on negotiation, domination, concurrence, reciprocation, and acquiescence among a large proportion, if not the entire complement, of members in the callers' or recipient groups. In terms of the parallel processing and integration areas of the brain constituting memory (see Barton, chap. 8, this volume), the cognitive demands of group movement interactions are expanded still further by the incorporation of alternative travel goals and their respective motivations, distances, and sundry travel costs. Quite plausibly, the memory needed to process a complete travel decision algorithm, incorporating both the social and foraging components, could be enormous. Rules of thumb may, however, permit adequate shortcut approximations, thus simplifying complex multivariate cost-benefit decisions.

In any case, the memory demands of foraging decisions coupled with the social skills needed to implement them are certainly greater than the memory demands of foraging decisions alone. This conclusion blunts Milton's (1981a) influential argument that frugivory created a selective regime favoring enhanced intelligence in primates (see Milton, chap. 14, this volume). Only the cognitive challenges of travel decisions in terms of foraging are considered in her model, not memory costs for social interactions and other travel motivations. The latter costs may be considerable (Dunbar 1993; Barton and Dunbar 1997; Barton, chap. 8, this volume). A more realistic evolutionary scenario for enhanced memory capacity in primates (incorporating comparative information on mating patterns, group size, dominance relationships, energy budgets, susceptibility to predation, and social communication for each taxon) could predict a ranking of memory capacity among primates markedly different from that of a model restricted to consideration of seasonal and spatial patterns in the distribution of food resources.

Are Loud Calls and Travel Signals Simply Long-Range Spacing Signals?

Loud calls and travel signals do not necessarily require new spatial and cognitive abilities. They can be viewed as an extension of a communication system that serves to regulate interindividual distance among conspecifics. Such communication systems appear to be ubiquitous among primates and many other mammals. Even solitary animals, according to Eisenberg (1977, 42), "in order to remain apart from one another . . . must know the position of their neighbors and thus some form of communication or monitoring of conspecifics must take place." Many types of discrete signals emitted by individuals in a broad range of contexts directly influence the spacing and movement patterns of others, including those that (1) mobilize group members to food patches (Wrangham 1977; Chapman and Lefebvre 1990; Galef and Buckley 1996; Judd and Sherman 1996); (2) facilitate separated group members rejoining either their social group or particular individuals in that social group (Boinski 1991; Cheney, Seyfarth, and Palombit 1996); (3) prompt nearby group members to increase their distance from a caller (Robinson 1982; Boinski and Campbell 1996); (4) function to summon group members, often infants, to the near vicinity of the caller (Boinski and Mitchell 1995; Maestripieri 1996); and (5) recruit group members to assist in agonistic alliances (Gouzoules and Gouzoules 1995) and predator mobbing (Curio 1978). These categories of spacing signals are distinguished on the basis of function. From a cognitive standpoint, however, it is important to ask whether the *intent* of the signaler producing loud calls or travel or other spacing signals corresponds to the signal's function. In the other words, does the howler monkey producing loud calls from a fruiting fig tree in the morning intend to deter the close approach of other troops to this valuable food patch? Or are the loud calls it produces better described as reflecting its excitement that the prospects are good for that day's foraging?

Intent and Deceit

The use of specialized signals in the initiation and leading of travel, as well as in broadcasts to other troops, suggests a recognition of their effect on recipients, what Dennett (1983) and Cheney and Seyfarth (1990) describe as first-order intentionality. Zero-order intentionality is the rubric for a purely emotive response system, in which communication and mental processes are motivated by fear, hunger, and other basic emotions. It is the simplest, most conservative null hypothesis on the proximate motivations underlying the behavior of birds, reptiles, and mammals. One such (inadequate) hypothesis to explain travel calls, for example, is that these vocalizations merely reflect the caller's peripheral position in a troop, and its increased alarm as a consequence of increased susceptibility to predation. A first-order intentionality, in contrast, implies that an individual has "beliefs and desires, but no beliefs about beliefs" (Cheney and Seyfarth 1990, 143). In other words, the animal producing the travel signal or loud call wants other troop members or another troop to move toward or away from a specific location. The critical distinction between first and higher orders of intent is a "theory of mind" (Premack and Woodruff 1978), an individual's ability to recognize that others also possess minds with beliefs, knowledge, and intents. Deceit, the ability to provide false information and recognize the false beliefs of others, is possible only once second- and higher-order intentionality is attained. Efforts to document deceit in nonhuman primates began in earnest with collations of anecdotes of apparent deception by primates observed in the wild (Whiten and Byrne 1986, 1988).

One way to assess whether full first-order intentionality can be invoked for travel signals is to determine whether the signaler subsequently monitors the behavior of its audience, and reinforces it with additional signals as necessary. Monitoring is a general feature of travel arbitration within and between troops, and monkeys and apes clearly indicate specific intents and goals. Three examples will illustrate the point to be made here. First, individual troop members acting in the role of scouts, as well as entire troops, attend to the effect of loud calls on the movements of other troops, often increasing surveillance for a period of days (Chivers 1974; Waser 1977b, Mitani 1985a; Whitehead 1987). Second, in my field studies, Costa Rican squirrel monkeys, white-faced capuchins, and golden lion tamarins emitting travel calls as part of a travel initiative repeatedly looked toward the main body of the troop and scanned for evidence that the troop had begun to travel. No monkey gave travel calls in the periphery of a stationary troop and then moved forward more than a short distance by itself. Among the former two species (no distinct vanguard position could be identified in the tamarin), monkeys vocally leading a troop in the vanguard commonly stopped and looked backward, especially if troop members traveling in the rearguard had slowed or slightly deviated in trajectory. Similar accounts of an individual closely tracking the effectiveness of its travel signals come from baboons (Rowell 1972b; Kummer 1968a,b, 1995) and bonobos (Ingmanson 1996). Herding of other troop members to achieve troop mobilization, an elaborated form of intention movement, also implies monitoring and intent on the part on the herder (i.e., hanuman langur: Sugiyama 1976; chacma baboon: Buskirk, Buskirk, and Hamilton 1974; Ron 1996). Third, seasonal fluctuation in the use of travel calls by white-faced capuchins in traveling troops at Santa Rosa is linked to changes in the distance travel calls are able to propagate and be perceived due to seasonal foliage density variation at this site (see above). The flexibility of travel call usage presumably hinges on the perceived effectiveness of the travel calls, both by leaders in the vanguard producing the "real" travel call and by followers in the rearguard making decisions as to whether the travel calls should be relayed to group members still more distant from the vanguard.

Although arguments can be made that monkeys and apes exhibit first-order intentionality in the mediation of group movement, evidence for second-order intentionality, at least in the form of deception, in this context remains weak. Nothing even hinting at sneakiness has yet been reported in Old World monkeys and apes, squirrel monkeys, or golden lion tamarins in the use of travel signals or loud calls. Yet, in rare instances, white-faced capuchins at both of my study sites employed their travel call in tactical maneuvers suggestive not only of intentionality, but also of deception and the anticipation of behavioral effects. This genus's apparently exceptional ability is consistent with the abundant evidence that capuchins are by far the most clever New World monkeys (Torigoe 1985; Boinski

1988b; Fragaszy and Boinski 1995). In part, however, the lack of plausible instances of deception in other taxa in regard to travel signals, although not loud calls, can be attributed to sampling bias. Few investigations of the coordination of group travel in other species have so closely scrutinized each instance of travel signal use as have these capuchin studies.

The three most suggestive examples of capuchin deception follow. First, an adult female capuchin "hijacked" the leadership of a traveling troop at Santa Rosa, taking the troop on a very different trajectory by emitting travel calls at an exceptional high rate and intensity, effectively masking the travel calls of the original leader (Boinski 1993). Second, La Selva capuchins twice employed trills in a possibly deceptive manner, in efforts to gain access to desirable fruit patches (Boinski and Campbell 1995). In neither case was the effort successful. In the first such instance, two adult females, Mom#2 and Chinga, were foraging in a *Welfia georgii* palm with a dense crop of ripe fruits. Mom#2 was pacing several meters outside the tree and staring at the clumps of ripe fruit, and appeared eager, but extremely hesitant, to enter the palm crown. After several minutes, Mom#2 started producing trill travel calls and giving intention movements from a position 3 meters outside the palm; she acted as if she was trying to start a stationary group, but her attention and efforts remained focused on the fruiting palm and its two occupants. After 24 trills over a 7-minute period, Mom#2 abandoned this activity and moved rapidly forward in a direction more than 90° away from the trajectory indicated by her position and intention movements while trilling. A second anecdote from La Selva of travel calls being used in a highly manipulative, if not deceitful, manner also involves an adult female. Mom#3 was the only capuchin in a fruit-laden *Hampea appendiculata* (Malvaceae) tree. She suddenly looked into the distance, and moved out of the *Hampea*. About 30 seconds later, Mom#4 walked into the vicinity from the direction that Mom#3 had looked, but remained more than 10 meters distant from the *Hampea*. Mom#4 appeared to be scanning desultorily for foraging opportunities and never glanced toward the nearby fruiting tree. Mom#3, while looking directly at Mom#4, made intention movements away from the *Hampea* and emitted two trills. Mom#4 approached and displaced Mom#3 from the general area. Only after more than 2 minutes of closely visually scanning vegetation for foraging opportunities did Mom#4 discover the *Hampea* fruit patch.

Sue Boinski

Representational Content of Travel Calls

A major issue in primate vocal communication is whether or not certain vocalizations can be termed "representational," meaning that the vocalization refers to an event or object external to the caller (Gouzoules, Gouzoules, and Ashley 1995). External reference is a theoretical Rubicon. Until recently one fundamental distinction between human language and the vocal communication of nonhuman primates was that only the former were thought to be able to communicate about events and objects in the real world. Nonhuman primates were limited to expressing emotional or motivational internal states (Lancaster 1968; Premack 1976). Several primate studies in particular prompted serious reconsideration of the formerly secure separation of the communication (and therefore thought) processes of human and nonhuman primates. Acoustic variants of the alarm calls of vervet monkeys *(Chlorocebus aethiops)* were shown to distinguish three different classes of predators (Cheney and Seyfarth 1990). Similarly, the agonistic screams of rhesus (*Macaca mulatta:* Gouzoules, Gouzoules, and Marler 1984) and pig-tailed macaques (*M. nemestrina:* Gouzoules and Gouzoules 1989, 1995) encompass, respectively, four and five subcategories of acoustically distinct recruitment screams. Within both these macaque species, the scream variants are reliably emitted with different categories of opponents (depending on the opponents' relative dominance rank and kin relationship) to elicit assistance from other group members in aggressive encounters.

Travel calls expand the evidence for representational vocal signals to yet another arena of a primate's external environment, the social interactions coordinating group travel. Classification of travel calls as representational is consistent with two criteria set forth by Macedonia and Evans (1992): that the signal (1) be closely associated with specific contexts in the environment external to the caller, and (2) be sufficient in itself to provoke the appropriate response. First, all reports of travel call usage among primates tightly link the designated travel call with coordination of travel. This generalization must be somewhat qualified because in at least one species, the white-faced capuchin, developmental changes in the use of the travel call are reported (Boinski 1993). Yet developmental changes are also found in the proficiency with which vervet monkeys produce alarm calls (Seyfarth and Cheney 1980) and pig-tailed macaques produce recruitment screams (Gouzoules and Gouzoules 1995). The second criterion is met because in at least three species,

Costa Rican squirrel monkeys, white-faced capuchins, and golden lion tamarins, troop members were able to respond appropriately to travel calls within brief time periods, even though the typical dispersion of troop members often provided only limited opportunities for visual contact with the signalers (Boinski 1991; Boinski et al. 1994; Boinski and Campbell 1995). Among all three species, individuals judged to be visually isolated from all other troop members commonly oriented and traveled in the appropriate direction after hearing travel calls, but not other species-typical vocalizations. Descriptions of travel call usage among other primate taxa do not specifically state that visually isolated troop members respond appropriately to travel calls, but nothing in these descriptions suggests otherwise (i.e., Itani 1963; Struhsaker 1975; Stewart and Harcourt 1994).

The definition of a representational call, in contrast, does not comfortably accommodate the relative sloppiness of the contextual associations of, and responses to, loud calls. The loud calls of numerous species are produced in a wider array of contexts than those readily assigned to the category of social manipulation. Instead, a subset of loud calls in these species are better described as startle or alarm responses, or simply cannot be associated with any obvious eliciting stimuli (i.e., Bornean gibbon: Mitani 1985a; indri, *Indri indri:* Pollock 1986; guenons, *Cercopithecus* spp.: Struhsaker 1970). Notably, male crested black macaques *(Macaca nigra)* produce loud calls most predictably during contexts of female-female aggression among troop members, and the calls appear to announce the male's intention to intervene in the aggression (M. F. Kinnaird and T. G. O'Brien, pers. comm.). Moreover, even when loud calls are emitted in contexts consistent with arbitration of spacing between troops, the responses of recipient groups, although predictable (see Kinnaird 1992b), are not as tightly linked to the loud call stimuli as is true of alarm calls, recruitment screams, or travel calls.

Motivation to Coordinate Troop Travel
Any vocalization produced by a nonhuman primate is likely to be neither purely representational nor purely motivational (Marler, Evans, and Hauser 1992). More realistically, the information content of a vocalization is best considered on a continuum between these two endpoints. Some calls exhibit a highly specific relationship to an external referent, while others reflect a more emotional, nonsymbolic information content. In each of the four field studies

(and three species) in which I investigated travel call use, it was much easier to quantify the referential component of this signal than what I perceived as the motivational component. In other words, apparent attempts to initiate travel, successful and failed attempts to initiate travel, and leading of troop travel were easy to define objectively. My frustration was and continues to be how to quantify "how much" the monkeys wanted the troop to move: their level of motivation.

At the level of my qualitative and intuitive understanding of travel call usage, bouts of travel call production can be reliably ranked by the callers' urgency. Sometimes individuals producing travel calls appear to be abjectly beseeching other troop members to follow their leadership. Each of the three species with which I am most familiar evinced the same constellation of traits as their apparent motivation increased: (1) calls are emitted at a higher rate; (2) calls become more intense (loud); (3) duration of travel signal increases; (4) fluctuation and modulation of the acoustic signal increases; and (5) more exaggerated intention movements and monitoring of recipients' response occur. Typically, the longer the duration of a travel initiative, the more likely these qualities will intensify over time. On occasion, highly motivated individuals producing travel calls seemed to succeed only because other troop members finally acceded to their dogged persistence. Variation documented in the acoustic structure of human speech across different emotional contexts (Scherer 1992) is reminiscent of this qualitative description for three New World monkey species. Thus far I have failed in devising a system with which to simply and intelligibly quantify and statistically analyze these behavior patterns. This is not equivalent to stating that motivation is irrelevant to understanding the behavioral ecology of group movement. I wish future researchers more success than mine in this endeavor.

Phylogenetic and Ecological Variation among Primates in Travel Signals

Phylogenetic constraints on the mechanisms by which a taxon coordinates group movement are weak. Congeners commonly vary across the two alternative communication modes: travel calls versus visual displays and intention movements (table 15.2: *Saimiri* and *Cebus*). Even in genera that comprise species employing the same communication mode, individual strategies, particularly sex differences, can differ markedly between species (table 15.2: *Papio, Cer-*

copithecus, and *Macaca*). Yet monophyletic taxa using the same communication mode and exhibiting similar age, sex, and rank patterns certainly exist, and include *Alouatta* (Whitehead 1995), *Pan,* and *Mandrillus* (table 15.2). Usage of a wah-wah travel call in golden lion tamarins can be tentatively extrapolated to other species of lion tamarins, the closely related *Saguinus* tamarins, and possibly all callitrichines. Other species of lion tamarins have vocalizations with acoustic structures clearly assignable to the wah-wah (Snowdon et al. 1986). Observations of wild red-handed tamarins *(S. midas)* in Suriname (S. Boinski, unpub.) and moustached *(S. mystax)* and saddleback *(S. fuscicollis)* tamarins in Brazil (C. Peres, pers. comm.) indicates that calls acoustically similar to the wah-wah are used to coordinate troop travel in a manner indistinguishable from that of golden lion tamarins. Moody and Menzel (1976), working with captive saddleback tamarins, and Snowdon, Cleveland, and French (1983), in studies of captive cotton-top tamarins *(S. oedipus),* also concluded that a vocalization structurally similar to that of the golden lion tamarin wah-wah functions in intragroup cohesion. Based on his field observations, Soini's (1981) description of the travel calls of another callitrichine, the pygmy marmoset *(Callithrix pygmaea),* is also remarkably consistent with those of golden lion tamarins.

Consideration of higher taxonomic levels, such as the superfamily Ceboidea, which includes all extant New World primates, emphasizes the apparent within-taxon diversity of mechanisms employed to coordinate travel. Based only on communication mode and the extent to which social dominants monopolize travel leadership, taxa within this superfamily can be assigned to at least five different categories (table 15.2): (1) visual displays and weak dominance: dusky titis; (2) travel calls and weak dominance; Costa Rican squirrel monkeys, white-faced capuchins, golden lion tamarins, pygmy marmosets, red-handed tamarins; (3) visual displays and male dominance: brown capuchins; (4) travel calls and male dominance: mantled howler monkeys; and (5) intention movements of another species: Peruvian squirrel monkeys. Moreover, our learning curve on travel coordination in primate social groups remains steep. Given that less than 15% of the approximately 220 species and 37% of the 52 genera of primates have had travel coordination described even superficially, the range of travel coordination mechanisms exploited by primates undoubtedly will enlarge further.

Vocalizations and visual displays are exploited by both the great apes and the monkeys as travel signals (table 15.2). Great apes, despite claims to the contrary by Boesch (1991b), Ingmanson (1996), and Savage-Rumbaugh et al. (1996), do not appear to evince a more complex manner of coordinating travel signals than do monkeys. Mehlman (1996) describes branch-shaking displays and dead log thumping by Barbary macaques to initiate and lead group travel that are indistinguishable in substance from reports on tree-trunk drumming by chimpanzees (Boesch 1991b) and branch-shaking displays by bonobos (Ingmanson 1996) to coordinate travel. Savage-Rumbaugh's conclusions that bonobos leave broken branches as signs for group members to the rear (the "pointed," broken-off end is interpreted to designate trajectory) are not convincing. First, her observations come from a brief, 6-week-long study of bonobos; branches were found on the ground as she and her assistants were tracking usually unseen bonobos. Second, the broken branches encountered in the wake of bonobo groups could have been employed in branch-shaking displays consistent with Ingmanson's (Ingmanson and Kano 1993; Ingmanson 1996) observations on this species. The consistent orientation of the dropped branches could be merely incidental to their original purpose. More simply still, the locomotion of bonobos through the forest undergrowth could plausibly result in broken branches with a predictable orientation.

Ecological and Social Sources of Variation
Because phylogeny and crude measures of relative intelligence certainly leave much variation unexplained, emphasis is also profitably placed on identifying the ecological and social contexts that affect the form of travel coordination. Several factors seem to predict the presence of a specialized travel call within a species' vocal repertoire. The first and preeminent is exploitation of an arboreal habitat or other situation in which visual contact among troop members is impeded. Second, among species with an egalitarian social organization or a weakly ranked dominance hierarchy, such as Costa Rican squirrel monkeys and white-faced capuchins, more advantage is likely to accrue to those individuals able to widely broadcast their travel preference. In contrast, among social organizations characterized by one or a few clear dominants, such as brown capuchins (Janson 1990a) and baboons (Kummer 1995), the dominants that make and impose travel decisions are undoubtedly already being closely monitored by other troop members. Third, all else being equal, large group size should bias a species toward travel calls and

away from dependence on intention movements. Larger groups plausibly engender more confusion and impede visual communication. Small groups of dusky titi monkeys do not employ travel calls, and generally coordinate travel by closely following one another and restricting travel to customary routes (Mason 1968; Menzel 1993). But the group size of golden lion tamarins is also small, averaging five and seldom exceeding eight (Dietz and Baker 1993). Perhaps a better predictor of the presence of travel calls in a species' repertoire is the extent of typical dispersion and nearest-neighbor distances. The larger the typical between-individual separation, and thus impediments to visual contact among troop members, the greater the likelihood of travel calls. Finally, foraging behavior, through its effect on the dispersion of troop members, is also expected to influence the expression of travel calls. Species in which troops typically move as cohesive groups directly between specific foraging sites of sufficient size to hold all or most troop members, such as fruiting trees or crowns with fresh-flushed foliage, would probably receive less benefit from travel calls than would species that often form highly dispersed troops composed of nearly independent foragers avoiding one another's close proximity because of indirect food competition. In the former case, intention movements and visual displays would probably suffice to negotiate or direct travel, whereas travel calls would probably be useful to ensure communication among dispersed group members in the latter.

Conclusions

The amazing nuances and intrigues now recognized in the communication of social animals promise continuing occupation for researchers in the coming decade. Manipulation by individuals of the movement patterns of their own and other troops is an arena of social interaction much larger in scale and complexity than the usual behavioral paradigm. Travel signals and loud calls are both manifestations of individual strategies for access to food, mates, and other commodities that constrain reproductive success. If theoretical biologists using game theory models are finding the solutions to two-person contests nearly dizzying in complexity (Heinsohn and Packer 1995), the diversity and complexity of individual strategies in more realistic situations, such as manipulation of travel patterns, will be truly awe inspiring.

Concomitantly, communication is being increasingly exploited as a window into the thinking of primates and other social animals.

The mechanisms of travel manipulation by means of loud calls and travel signals described in this chapter belie the tidy dichotomy of selective regimes commonly used to consider the adaptiveness and evolution of primate intelligence: spatial foraging decisions (Milton 1981a) versus social manipulations (Byrne and Whiten 1988). In the coordination of troop travel, complex social manipulations and complex travel and foraging decisions are utterly confounded. Yet if socially more complex processes are eventually found to characterize the mediation of group movement among primates compared with other taxa, social hypotheses on the evolution of primate intelligence are strengthened.

Mechanisms of travel mediation need to be documented in a broader range of primate and nonprimate species. Travel signals are not necessarily obvious until researcher discrimination is honed and observations are carefully quantified. In particular, playback studies of travel calls should be initiated; cleverness, perseverance, and multiple assistants will be needed. The behavioral ecology of loud calls, although much more often studied than travel signals, largely remains a black box in the sense that group behavior too often overshadows that of the individual callers and noncallers constituting the group. Long-term studies of loud call and travel signal usage would also permit new insights on the development of effective signal production and the long-term social relationships and strategies that influence signal production.

Acknowledgments
Much of my fieldwork on the coordination of troop movement among New World monkeys was made possible by the Laboratory of Comparative Ethology, NICHD, National Institutes of Health, and a National Research Service award. More recent support for this field research comes from the National Science Foundation (SBR 9722840) and the National Geographic Society. The many colleagues who shared comments and illustrations for this chapter include Colin Chapman, Paul Garber, Kay Holekamp, Margaret Kinnaird, Bill Leonard, Gary Steck, Peter Waser, and Jay Whitehead.

CHAPTER SIXTEEN

Grouping and Movement Patterns in
Malagasy Primates

PETER M. KAPPELER

According to current socioecological theory, the social system of a
species reflects the outcome of optimal individual strategies of survival and reproduction. Depending on body size, activity, life history, diet, various social factors, and a species' evolutionary history,
a particular spatiotemporal distribution of adult males and females
will be adaptive under a given set of circumstances (Clutton-Brock
1989; Davies 1991). The most fundamental distinction between
possible outcomes concerns grouping pattern: individuals may either lead a solitary life or associate at least temporarily with other
adult conspecifics (Krebs and Davies 1992). In the latter case,
trade-offs involving, for example, reproductive strategies or predation risk or feeding competition will favor groups of a particular
size and composition (van Schaik 1989; Wrangham, Gittelman, and
Chapman 1993). If group formation is favored, trade-offs will also
determine the spatiotemporal distribution of group members, and
individuals will have to communicate and coordinate their interests
to maintain the level of social cohesion needed to respond in an
adaptive manner to current local conditions. While the first two
questions, whether animals should live in groups and what their
optimal size should be, have been investigated in detail in diverse
taxa (e.g., Bertram 1978; van Schaik 1983; Zemel and Lubin 1995),
the problems of mechanisms and strategies of coordinated group

movement have received comparatively little attention until recently (see Boinski, chap. 15, this volume).

The general costs of group living become obvious to anyone studying patterns of group movement. First, and most importantly, individual foraging strategies and schedules are expected to be heterogeneous and, therefore, a source of conflict. Growing juveniles, pregnant or lactating females, and adult males should have divergent overall activity budgets and different diets—that is, they should differ in the types of food items eaten and the time devoted to foraging on each item (see, e.g., Altmann 1980; Dunbar and Dunbar 1988). Second, depending on the type and distribution of resources used, within-group feeding competition may threaten group cohesion and influence individual and subgroup movements (van Schaik 1989; van Noordwijk et al. 1993); for example, when low-ranking individuals would have higher intake rates in other feeding patches (Pulliam and Caraco 1984). Finally, a conflict of interest may also arise between the sexes; for example, when between-group encounters have different costs or benefits for males and females (Cheney 1986), or when mating competition interferes with foraging efforts (Alberts, Altmann, and Wilson 1996). Ultimately, the extent of group cohesion and coordinated activity manifested in the face of conflicting individual interests may largely depend on the risks of predation and infanticide, creating a trade-off between safety and other fitness components (van Schaik 1983; van Noordwijk and van Schaik 1986; Wrangham 1986, Dunbar 1988; van Schaik 1996).

Primates are an excellent group in which to investigate behavioral mechanisms relevant to the resolution of these trade-offs because they stand out among mammals for the diversity and complexity of their social systems (Eisenberg 1981; Smuts et al. 1986). The content of other contributions to this volume, as well as most previous research in this field, pays tribute to this fact. Among other things, this work illustrates various behavioral mechanisms, such as specific vocalizations, that exist to coordinate individual movements of group members at the proximate level (e.g., Boinski, 1991 and chap. 15, this volume).

In this chapter, I will focus on the lemurs of Madagascar (Lemuriformes), which provide a particularly interesting group for comparison in this context because they represent the endpoints of an adaptive radiation following a single colonization event in the Eocene (Yoder et al. 1996), providing a natural experiment in social evolution that can help to identify instances of convergent evolution (Martin 1972). In this context, it is significant that group living evolved (at least) twice independently among lemurs, compared with only once among all other primates, as suggested by parsimonious phylogenetic reconstructions (Kappeler 1998). Thus, lemurs can provide important comparative information on several questions related to coordinated group movement that have already been addressed in studies of anthropoids. In this chapter I will review the available information on lemur grouping and movement patterns and their underlying mechanisms with the twofold aim of providing preliminary answers to some of these questions and of stimulating additional focused research on this topic.

Patterns and Dynamics of Lemur Social Systems

The social organization of a primate society can be defined by its size, composition, and genetic structure (Struhsaker 1969). At the coarsest level, lemur social organization parallels that of other primates in that solitary, pair-living, and group-living taxa are found among the thirty-two living species (Richard 1986). The diversity of lemur societies is impressive, especially when considering the possibility that some of the at least fifteen recently extinct lemur species (Tattersall 1993) may have exhibited one of the social organizations not represented among extant species. However, characterization of lemur social organization is still hampered by a scarcity of long-term studies of known individuals, so that attempts to classify individual populations or taxa have to be largely based on census data (e.g., Kappeler 1997a). Moreover, it has been difficult both to define social boundaries of groups in some gregarious lemurs (e.g., Tattersall 1977; Richard 1985b) and to classify their social organization because pairs, single-male, multimale, and polyandrous groups are common in some taxa (summarized in Kappeler 1997a).

Despite these difficulties, some generalizations are possible. For example, the majority of lemurs are nocturnal and solitary: adults typically forage alone but may interact with conspecifics during the night and form sleeping groups with them during the day. This type of social organization characterizes all members of two families (sportive lemurs, Lepilemuridae, and aye-ayes, Daubentoniidae) and the majority of mouse and dwarf lemurs (i.e., the genera *Microcebus, Cheirogaleus, Allocebus,* and *Mirza*). Compared with other primates, a relatively large proportion of lemurs live in pair-bonded

Peter M. Kappeler

family groups (Jolly 1998) consisting of only one adult male and female. They include the genera *Indri, Avahi,* and *Phaner,* as well as most grey gentle lemurs *(Hapalemur griseus),* golden bamboo lemurs *(H. aureus),* red-bellied lemurs *(Eulemur rubriventer),* mongoose lemurs *(E. mongoz),* and some ruffed lemurs *(Varecia variegata).* A relatively small proportion of lemur species live in multimale, multifemale groups, which are on average smaller than those of anthropoids of similar body size (Kappeler and Heymann 1996). They include sifakas *(Propithecus* ssp.), ringtailed lemurs *(Lemur catta),* greater bamboo lemurs *(H. simus),* and several eulemurs *(E. coronatus, E. fulvus,* and *E. macaco).* Of the pair- and group-living species, *Avahi* and *Phaner* are nocturnal, *Propithecus, Lemur,* and possibly *Varecia* predominantly diurnal, and *Hapalemur* and *Eulemur* cathemeral (i.e., regularly active both day and night) (Tattersall 1987), even though all lemurs appear to possess morphological adaptations to nocturnal activity (Pereira 1995; van Schaik and Kappeler 1996).

In the following section, I will review some of the relevant observations on individual and group movements and their underlying behavioral processes that characterize lemurs with different social organizations (see also Pollock 1979a for a similar review of earlier studies). In discussing gregarious species, I will make a distinction between pair- and group-living species because the former need to coordinate their movements with only one other adult, apart from possibly present offspring, whereas the latter have to consider several adult group co-residents.

Movements of Solitary Species
In solitary species, individual male and female home ranges overlap to various degrees, and most individuals form daytime sleeping groups (Bearder 1986). Both of these aspects of social life require interindividual coordination and communication. Moreover, solitary foragers should also try to reduce the costs of travel between foraging patches and remember the location of their multiple nests or tree holes. In addition, females face the task of remembering and finding their infants, which they park while foraging and may move several times per night (see Russell 1977). Thus, while solitary lemurs may not have to compromise and coordinate individual foraging activities, they nevertheless have to coordinate home range defense with foraging while remembering the location of shelters and young and communicating acoustically with neighbors or pro-

cessing deposited olfactory signals. They may, therefore, face cognitive challenges similar to those of group-living primates (see also Sterling and Richard 1995).

Some preliminary information on how solitary lemurs deal with these multiple tasks can be gleaned from previous studies, but very little is known about their actual movements and the factors determining them, presumably because all-night follows of these animals are difficult to accomplish. Some summary statistics on ranging patterns and movements are presented in table 16.1, but relevant specific issues have not yet been studied explicitly. For example, all solitary lemur females park their dependent young in nests or tree holes for up to several months, and later in parking places in the vegetation, to which they return periodically to nurse or relocate the infants. Immense cognitive insights could be gained from careful examination of the movements of mothers with dependent young, but only one anecdotal description is available from the wild (Russell 1977). A study of the same problem might be logistically easier in the diurnal *Varecia,* the only nonsolitary lemur that parks its young (Klopfer and Dugard 1976; Pereira, Klepper, and Simons 1987; Morland 1990). Solitary species also offer an opportunity to study how infants and juveniles coordinate their movements with their mothers—a fundamental aspect of coordinated movement in all mammals.

Solitary lemurs rely on communication signals that have the potential to affect the movements of conspecifics and to facilitate the signaler's orientation within its own home ranges. The use of these signals in relation to individual movements has not yet been studied in the wild, however. For example, olfactory signals from urine and a number of scent glands are used for territorial marking, and may hence mediate individual spacing (Schilling 1980) and help individuals to orient within their home ranges (Schilling 1979), but we do not know how these signals are distributed within home ranges, how long they persist, or whether marking sites are (re-) visited by the sender and/or which neighbors.

Results of such studies in other nocturnal primates (e.g., Charles-Dominique 1977a, Clark 1978, Harcourt 1981) indicated that olfactory signals are not used for individual orientation within home ranges. Charles-Dominique (1977b), for example, followed the movements of several lorisid species and concluded that "an animal rarely retraces the same path when returning to an area previously visited. The maps of itineraries show that *Galago alleni* moves in a

Peter M. Kappeler

Table 16.1 Summary of lemur ranging patterns and group movements

Species	GS	HRS	DPL	NPL	Reference
Daubentonia madagascariensis	1	35.6 (2 F)		1378 (12)	Sterling 1993
		170.3 (2 M)		1904 (34)	
Lepilemur edwardsi	1	1.1 (4)		343 (4)	Warren and Crompton 1997
Lepilemur leucopus	1	0.2		270	Charles-Dominique and Hladik 1971
Avahi laniger	3	1.4 (1)		457 (7)	Harcourt 1991
Avahi occidentalis	3.5	1.6 (4)		1175 (4)	Warren and Crompton 1997
Indri indri	3.1	17.9 (2)	300–700		Pollock 1979b
Eulemur rubriventer	3.4	19 (2)	440 (120)	197 (46)	Overdorff 1993a
Eulemur fulvus rufus	9.5	97.5 (2)	961 (126)	230 (56)	Overdorff 1993a
		0.8 (6)	138 (6)		Sussman 1974
Eulemur mongoz	3.1	2.9 (2)	1185 (2)		Curtis 1997
Lemur catta	16.9	18.9 (8)	1377 (8)		Jolly et al. 1993
		8.8 (1)	920		Sussman 1974
		25 (6)			Sussman 1991a
Varecia variegata rubra	5	24.6 (2)	436 (47)		Rigamonti 1993
Propithecus v. verreauxi	6.3	6.9 (2)	775		Richard 1978
coquereli	5.5	7.5 (2)	925		Richard 1978

Note: Group size (GS) refers to the mean foraging party size (Kappeler 1997a). Home range size (HRS) provides an estimate (ha) of the average area used by an individual or group. Daily (DPL) and nightly path length (NPL) report the average distance (m) traveled during one daily activity period. Only taxa for which data on HRS and DPL or NPL are available were included. Sample sizes for the number of individuals or groups studied to determine HRS and the number of days or nights to determine DPL and NPL, respectively, are presented in parentheses, when available.

series of loops even when travelling in an area which has not been visited for a week. . . . The dependence of an animal on an odorant 'thread of Ariadne' would considerably reduce its possibilities for exploitation of the milieu. In addition it would require enormous quantities of urine to mark all of the possible itineraries in a domaine which may exceed 100,000 m², and which is frequently washed by tropical rains" (134). Students of lorisid behavior concur in the belief that visual cues are more important for individual orientation and that scent marks are deposited for the information of conspecifics (e.g., Clark 1978).

Most species of solitary lemurs have elaborate repertoires of acoustic signals (Zimmermann 1995a), some of which may function in the context of influencing individual movements. Estrous females of grey mouse lemurs *(Microcebus murinus)*, fat-tailed dwarf lemurs *(Cheirogaleus medius)*, and Coquerel's dwarf lemurs *(Mirza coquereli)*, for example, advertise their brief periods of receptivity with specific vocalizations (Stanger 1993), which should influence the movements of neighboring males. Similar behavior by estrous female aye-ayes *(Daubentonia madagascariensis)* in the wild may indeed attract up to six males (Sterling 1993). Male mouse and dwarf lemurs also produce advertisement calls during the breeding season that may encode individual characteristics and result in approach or withdrawal of particular males and females (Stanger 1993; Zimmermann and Lerch 1993). Because these signals have been studied only in captivity and/or because only one animal has been followed at a time, we do not yet know how far these calls propagate and how strongly they affect recipients' movements. In one species with such estrus advertisement calls *(M. coquereli)*, males move quickly and range several hundred meters beyond their usual home ranges during the brief mating season (fig. 16.1). This roaming behavior, which has also been described for aye-ayes (Sterling 1993), may maximize their encounter rate with females, and males may use these calls to home in on estrous females during these forays.

In *Lepilemur edwardsi,* a particular advertisement call, which was not restricted to the breeding season, commonly resulted in the approach of the calling animal by a conspecific and their joint subsequent movement (Warren 1994). The calls can be heard over several hundred meters. Moreover, some species, such as *M. coquereli,* utter a soft contact call, which may indicate individual position and identity, and hence mediate interindividual spacing more or less

Peter M. Kappeler

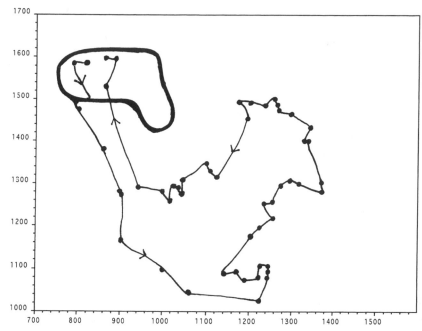

Figure 16.1 Movement of an adult male *Mirza coquereli* during the mating season. The heavy solid line indicates the normal home range; connected dots represent a single excursion between 6:00 and 11:00 P.M. on October 18, 1993. The *x* and *y*-axes indicate coordinates of the study area in meters. (For additional details see Kappeler 1997b).

constantly during their activity (Stanger 1993). Similarly, up to three *Daubentonia* may occasionally coordinate their movements and forage as a unit, also using accompanying calls as they move from resource to resource (Sterling 1993). It is not known whether visual signals are involved in coordinating movements over such short ranges in these species as well.

Thus, there are empirical and theoretical indications that the patterning and coordination of individual movements in solitary species occur within a social network that also requires complex decision rules and effective communication. Simultaneous all-night follows of neighbors (see, e.g., Jolly et al. 1993) and removal experiments (see Charles-Dominique and Petter 1980 for an example), along with detailed observations, should reveal more details about how solitary lemurs organize their movements.

Movements of Male-Female Pairs

Compared with solitary animals, pair-living males and females face the additional challenge of coordinating their activities and movements with a partner, in addition to their possible offspring. Existing studies indicate that vocal signals are primarily responsible for mediating the coordinated movements of pair-living lemurs, which are largely controlled by females.

This pattern is illustrated by nocturnal fork-marked lemurs *(Phaner furcifer)*, who remain in constant vocal contact by exchanging a particular call (Charles-Dominique and Petter 1980). As long as the members of a pair are in close proximity, their joint movement is coordinated by a simple rule: the male follows the female at a distance of 1–30 m. When she enters a feeding tree, the male waits behind until she has left the tree. If members of a pair become separated farther, they exchange another call, sometimes antiphonally, which may facilitate their reunion (Charles-Dominique and Petter 1980). These animals appear to possess information about the location of the (gum) trees they feed in because they have been reported to go directly to certain trees after leaving their sleeping places (Petter, Schilling, and Pariente 1975). This assertion is nicely illustrated by the following anecdote: "On the next to last day of our observations, we cut down a favorite feeding tree to obtain samples for analysis. As soon as night had fallen, we witnessed the arrival of two animals at this spot at an interval of 5 min. Running and jumping rapidly through the branches, they appeared completely disoriented at failing to find the familiar tree. They stayed immobile for at least a minute, looking at the place where the tree had stood; the first of them returned twice to search again before finally moving off towards other feeding places" (Petter, Schilling, and Pariente 1975, 214).

Woolly lemurs (*Avahi laniger* and *A. occidentalis*) received their onomatopoeic genus name from an advertisement call that is commonly exchanged during intergroup encounters (Harcourt 1991). Another specific vocalization (whistle) is used by these lemurs to locate each other after being separated and to meet at sleeping trees (Harcourt 1991). The male of a pair was observed repeatedly to leap off immediately and to join the female after she uttered such a whistle. Other calls (purr) are used by *A. occidentalis* to maintain cohesion during traveling and foraging (Warren and Crompton 1997), but they are given intermittently without any evidence of reciprocation or coordination.

Although pairs of *A. laniger* jointly defend their exclusive range, they may spend most of the night (i.e., up to 60%) far enough apart that an observer cannot locate them simultaneously (Harcourt

Peter M. Kappeler

1991). Similarly, an adult female *A. occidentalis* with a dependent infant was regularly observed traveling and feeding separately from her group and joined them only after the infant was no longer being carried (Warren and Crompton 1997). These observation could indicate that the pursuit of divergent individual foraging needs can override the assumed benefits of close cohesion, but perhaps more easily at night when vigilance benefits of cohesion are low.

Cathemeral red-bellied lemurs *(Eulemur rubriventer)* commonly travel and feed as cohesive units; all two or three group members are within 5 m of one and other most of the time. Females also initiate and lead the vast majority (75%; $n = 40$) of group progressions, with the other group members following in single file (Overdorff 1988), but it is not known which behavioral mechanisms they use to coordinate movements. Red-bellied lemurs are territorial, use their home range evenly, and frequently travel linearly from border to border (Overdorff 1993a). Some of the seasonal and daily variation in their ranging behavior (measured as daily path length, DPL) could be explained by variation in patch characteristics (Overdorff 1993a; Overdorff and Rasmussen 1995). At night, red-bellied lemurs travel less than during the day (table 16.1), but behavioral details about their nocturnal movements have not yet been reported.

Group movement in the largest extant lemur, the indri *(Indri indri),* is also largely controlled by females. Despite some differences among groups, adult females, especially those carrying young infants, lead most group progressions and initiate feeding at the next site (Pollock 1979a). Imminent movement is signaled by a soft "hum" vocalization (Pollock 1975). Males commonly wait until the female leaves a feeding tree and then feed briefly in the exact same location before following her. However, group members also frequently overtake each other, making it difficult for observers to predict the direction of movement. As a result, the adult pair spends a higher proportion of their time far away from each other than other dyads, sometimes more than 100 m, which is a considerable proportion of their average daily path length of about 350 m (see table 16.1). In these situations, any individual may initiate the famed *Indri* song or another (lost) call, resulting in rapid aggregation of all group members (Pollock 1986).

Pollock (1979b) measured several directional aspects of indri ranging behavior. Food availability is one important factor affecting the movements of groups (fig. 16.2). When available, important

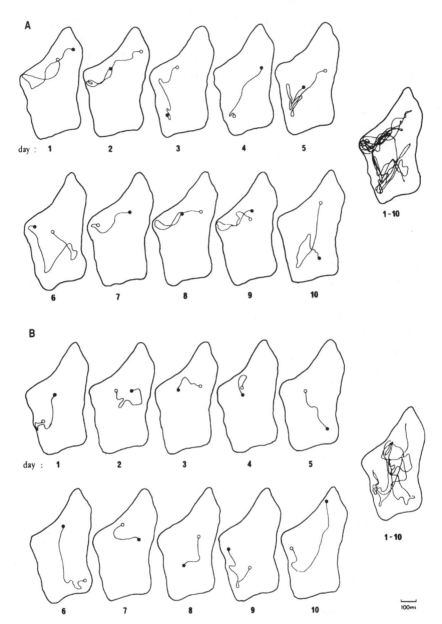

Figure 16.2 Ranging patterns of an *Indri indri* group on 10 successive days during two times of the year. (A) Repeated visits to a few fruiting trees (solid circles) resulted in predictable daily travel routes from central sleeping sites (open circles). (B) When no food plant species was in fruit or new leaf, a more random pattern of movements among widely distributed sleeping sites occurred (solid circles). (From Pollock 1979a, with permission of Academic Press.)

food trees are visited predictably, whereas movements are more undirected when no concentrated source of food is present. In neither situation are daily ranges organized according to any territory circumscription principle, even though a large portion of the ranging area is used exclusively and defended against intruding groups.

Pollock (1979b) also found that two neighboring groups differed greatly in the directedness of their movements, measured as the frequency with which a group moved into an adjacent 50 m quadrat by changing its previous direction of movement while entering from the previous quadrat. He attributed the fact that one group moved on average three times as much as the other before entering the next 50 m quadrat of the study plot, measured as the number of tree changes, to their poorer knowledge of their range. This group had apparently procured their territory only recently and subsequently spent much time visiting new trees in the search for food. Thus, site familiarity may be an important determinant of group ranging patterns in this and other primate species.

Very little is known about the movements of other lemurs that live and move in small family units. Groups of *Hapalemur griseus* are spatially tight, the animals being found within a 5–10 m radius, moving in single file (Pollock 1979b). During the movement of a *Eulemur mongoz* group, Tattersall and Sussman (1975) recorded the adult male and female about equally often as group leaders. In addition, animals frequently overtook each other during movements, so that no obvious leader emerged.

Movements of Group-Living Lemurs
Among group-living lemurs, cathemeral and diurnal taxa exhibit strikingly different grouping patterns and dynamics (e.g., Kappeler 1997a; Pereira and Kappeler 1997). Cathemeral lemurs may oscillate between living in pairs and living in groups, or form groups characterized by fission and fusion of varying subgroups on different temporal scales. The predominantly diurnal lemurs, on the other hand, form comparatively stable groups, which may, however, occasionally dissolve in a larger neighborhood, as, for example, during the mating season (Jolly 1966; Richard 1985b). How either one of these types of societies is actually structured and organized is still unresolved, but long-term field studies of marked individuals and genetic studies are beginning to reveal association patterns and their underlying rules (e.g., Sussman 1991a; Richard, Rakotomanga, and Schwartz 1993; Wright 1995; Overdorff 1996).

Groups of cathemeral *Eulemur* spp. with more than two adults are often difficult to identify because of their low cohesion and the fluidity of subgroups. In crowned lemurs *(E. coronatus)*, this pattern was described succinctly: "A group of individuals (behaving like a troop) would often move closer together, apparently communicate or even rest together, but then would split up and leave individuals behind. In some instances the remaining individuals would continue resting for an hour or more before following the individuals which had departed" (Wilson et al. 1989, 16). How these subunits find each other again is not known, but contact grunts and scent marks may facilitate their aggregation. It has also been observed, however, that some groups followed the same routes day after day while visiting *Ficus* trees with ripening fruit (Wilson et al. 1989), so that subgroups would congregate independently at certain meeting points. As in pair-living lemurs, traveling groups or subgroups of crowned lemurs were usually led by adult females (Wilson et al. 1989).

Brown lemurs *(E. fulvus)* exhibit a very similar social organization, as exemplified by Tattersall's (1977, 446) description of *E. f. mayottensis:* "During the course of a single day, the number of individuals in the association under observation varied considerably, as animals left or joined those being followed . . . usually a number of individuals would arrive or depart together, or at least in succession. Such departing associations appeared generally to remain as separate units for some time, although occasionally they would break off from the association under observation and almost immediately join another; this was usually in response to vocalization by the latter." Some concept of group identity must exist, however, because on occasion encounters between certain associations resulted in "frenzied chasing and fighting." Vick and Pereira (1989) proposed that groups consist of communities composed of subgroups of formerly associated matrilines that dissociated to forestall or resolve periods of targeted aggression.

Use of space by brown lemur groups is highly flexible. Up to hundredfold differences in home range size have been reported among populations (table 16.1), which may be partly explained by variation among habitats and seasons. In addition, groups may abandon their normal ranges for extended periods in response to seasonal ecological challenges. For example, one population of *E. f. rufus* groups had large ranges (100 ha) that completely overlapped between groups. Nevertheless, these animals moved up to 5 km

Peter M. Kappeler

away from their normal ranges for over a month to exploit extraordinary food abundance (a guava plantation) elsewhere (Overdorff 1993a).

Similarly, I observed numerous *E. f. rufus* groups congregating at the last remaining water holes toward the end of the dry season (Sept.–Oct.) at Kirindy forest over the past 2 years. The area surrounding one of these water holes is used by three or four (marked) groups during the rest of the year, but regular simultaneous counts of more than thirty unmarked animals at this water hole indicated that at least five other groups were present. Some of these animals were later radio collared and regularly located near the water hole for several weeks before completely disappearing within a few days. None of them could be relocated within a few kilometers, despite intensive searching and radio tracking from a vehicle. Moreover, one of our study groups had its center of activity about 500 m from the same water hole. In September and October, this group made daily straight excursions to the water hole, drank, and returned directly to their point of departure, following the same foot trail (S. Tombomiadana, pers. comm.). Both observations indicate that these lemurs possess knowledge about the location of this important resource. The movements of individual groups or subgroups of *E. f. fulvus* were also rapid and direct between distant and "obviously well-known feeding sites" (Tattersall 1977).

Groups of *E. f. fulvus* are almost always led by one of the adult females, who also control the time of departure (Andriatsarafara 1988). Groups of *E. f. sanfordi*, which almost always traveled one behind the other, were also always led by a female (Wilson et al. 1989). Seasonally increased traveling activity coincided with the presence of young infants in *E. f. rufus* (Overdorff 1993a), suggesting that female energetic needs determine ranging behavior of brown lemurs at least part of the year. Reports of consistent female leadership in the little-known black lemur (*E. macaco*, Petter 1962) indicate that female control of group movements is widespread among eulemurs.

Although the functions of most vocalizations of *E. fulvus* (Pereira and Kappeler 1997) remain unstudied, some appear to coordinate movements of individuals and groups. For example, their frequently emitted grunts occur disproportionately frequently just before and after individuals start and terminate stationary positions (M. Pereira, pers. comm.). Moreover, their spectacular croak call is used antiphonally between groups and may also function as a lost call.

The predominantly diurnal ruffed lemurs *(Varecia variegata)* exhibit overall variability in grouping patterns similar to that of eulemurs, but also present some unique features. In this species, some populations are pair-living, whereas others live in groups or communities of five to thirty individuals (Morland 1991; White 1991; Rigamonti 1993). The latter may form subgroups of varying size and composition that frequently consist of or include pairs of adult males and females. All members of a community were never found together in one location at the same time (Morland 1991), not even when the community consisted of only five animals (Rigamonti 1993). There was both daily and seasonal variation in the association patterns of *Varecia.* During the dry season, foraging subgroups of two or three animals met for a midday rest and again in the evening in the same or adjacent trees (Rigamonti 1993). During the wet season, however, relatively stable subgroups, mostly consisting of pairs, formed for several weeks (Morland 1991; Rigamonti 1993). Each subgroup restricted its ranging to a particular small portion of the communal home range. The shared home range, which overlapped relatively little with those of neighboring groups, was exclusively defended by females. Ruffed lemur groups did not use all parts of their home ranges equally; at a given time, only about one-third of the total area was intensively used. Every 2 or 3 weeks the range shifted partially to new areas, so that other areas became completely neglected (Rigamonti 1993).

Several vocalizations appear to be important in communicating and coordinating the movements of individuals and subgroups (Pereira, Seeligson, and Macedonia 1989), but their precise function remains to be studied. For example, ruffed lemurs produce an impressive group chorus that is promptly answered by neighboring groups (Macedonia and Taylor 1985). In addition to this apparent territorial function, this call could also signal the location of individuals and subgroups, and thus facilitate their coordinated movements.

Ring-tailed lemurs *(Lemur catta)* form the largest groups among lemurs, comprising on average about fifteen animals, including several adult males and females, of which only the former regularly transfer between groups (Sussman 1992). Movements may take the form of linear progressions in a single file, but amoeboid progressions, in which the group is dispersed as a broad front lateral to the direction of travel, have also been described (Klopfer and Jolly 1970). Jolly (1966) describes preprogression vocalizations (clicks), and other vocalizations are frequently heard during group progressions (Macedonia 1994). So-called meows are emitted by separated group members and inevitably answered by others, thus facilitating group aggregation (Jolly 1966). This function was recently confirmed in a study that demonstrated that this call was specifically exchanged between and within two subgroups of females whenever they were overdispersed (Oda 1996).

Adult female ring-tailed lemurs initiate and lead group movements (Sauther and Sussman 1993). They are joined by juveniles and the highest-ranking male, and move clearly ahead of the subordinate males, the "drones club" (Jolly 1966; 1972; Sussman 1977b). This spatial separation is maintained during foraging, during which the group can be dispersed over several trees at several heights. Adult females also take the lead in initiating intergroup encounters (Jolly et al. 1993) and during excursions into neighboring group ranges (Budnitz and Dainis 1975). Group movements before, during, and after such encounters have been analyzed in detail in one population by following several groups simultaneously (Jolly et al. 1993). Interestingly, groups were found to encounter each other less often than expected by chance, but remained in each other's vicinity after a confrontation for some time. This finding shows that group movements in this and possibly other species are to some extent determined by social factors emanating from other groups.

The primarily diurnal sifakas (*Propithecus* ssp.) live in cohesive groups of about five members (Richard 1978; Wright 1995). Group size and composition are stable over most of the year (Richard, Rakotomanga, and Schwartz 1991). Male group transfer and visits by roaming males during the mating season are responsible for virtually all social changes (Richard, Rakotomanga, and Schwartz 1993). Members of a group can be widely dispersed during foraging, but usually join one another again for extended resting periods. Movements are led by adult females in *P. verreauxi* (Kubzdela, Richard, and Pereira 1992), and preliminary analyses revealed that females initiate and lead over 90% of all successful group movements in *P. diadema* as well (D. J. Overdorff and E. M. Erhart, unpub.). Infants and juveniles travel in central positions and males lag 25–50 m behind (Wright 1998). Interestingly, in a group with three adult females, it was always the same female that initiated these movements (J. D. Overdorff and E. M. Erhart, unpub.). However, males in groups of both species sometimes moved ahead while traveling, once the direction had been picked, perhaps because females were slowed down by dependent offspring, and only isolated males gave contact calls (Kubzdela, Richard, and Pereira 1992).

Because phylogenetic reconstructions indicated that permanent groups of three or more adults evolved independently among the Lemuridae and these members of the Indridae (Kappeler 1998), these preliminary observations on sifakas suggest that parallels in basic aspects of group movements and their control may provide an example of convergent evolution within the Malagasy primate radiation.

Conclusions

This review of lemur grouping and movement patterns reveals that female interests consistently take precedence in determining the movements of lemur pairs and larger groups. Despite the anecdotal nature of most observations and the small number of groups observed, I found no reports of consistent male leadership. Given widespread female dominance and sexual monomorphism, such female leadership is not unexpected. Because females of most pair- and group-living lemurs have dependent offspring during the austral winter and spring, when food, and in some areas water, availability is reduced, their energetic needs are probably the driving force behind this pattern. Moreover, female philopatry is the rule among group-living species, so that females can develop a more profound knowledge of their home range and the distribution of important resources therein.

Second, vocal signals are the primary behavioral mechanisms used to coordinate individual and group movements, but possible roles of other modalities have not been explicitly examined in this context. There are anecdotal reports of both preprogression vocalizations, which may facilitate coordinated movements, and vocalizations during progressions, which may facilitate group cohesion. Olfactory signals are deposited during group movements, but it appears that lemurs mark where they go, rather than going where they marked before. Moreover, studies of the spatial distribution of scent marks revealed that they are concentrated in areas of range overlap (Mertl-Millhollen 1988), and may thus primarily influence movements of neighboring groups. Finally, some visual signals, such as the tails of ring-tailed lemurs or the leaps of sifakas, may aid in maintaining group cohesion. However, given their limited visual acuity (Pereira 1995) and the rarity of visual signals in their communicative repertoires in general (Pereira and Kappeler 1997),

Peter M. Kappeler

group-living lemurs are unlikely to rely heavily on this modality in the context of group movements.

Causes and Mechanisms
What factors shape the grouping and ranging patterns of lemurs described above? To address this question, I will briefly discuss some possible effects of ecological, social, and life history variables on group cohesion and movements.

Resource Distribution
The spatiotemporal distribution of food resources affects the group size, cohesion, and movement patterns of primates (van Schaik 1983; Dunbar 1988; Wrangham, Gittleman, and Chapman 1993; Chapman, Wrangham, and Chapman 1995). Small food patches that are rapidly depleted or patch characteristics, such as low density and scattered distribution, that incur high travel costs should favor small groups and reduced group cohesion (Chapman 1990b; Barton, Byrne, and Whiten 1996; Chapman and Chapman, chap. 2, this volume). The formation of temporary subgroups by some lemurs, especially during the season of low resource abundance, may thus be a mechanism to reduce feeding competition, because members of small subgroups incur lower traveling costs (if they are stable for sufficient periods of time) and experience less competition over the existing patches (see, e.g., Overdorff 1996). It remains to be determined whether the relatively small size of lemur groups also reflects an adaptive response to these resource characteristics.

Resource conditions also influence group movements. Long excursions by brown lemurs away from the usual home range to exploit superabundant or extremely limited resources elsewhere are one example. Movements of neighboring groups relative to one another as a function of resource condition provide another example (Kinnaird 1992b). Among lemurs—for example, in *Lemur catta, Eulemur fulvus,* and *Propithecus verreauxi* (Sussmann 1974; Richard 1978; Jolly et al. 1993; Sauther and Sussman 1993; Overdorff 1996)—differences in the size and degree of exclusive use of home ranges at different sites, and even among microhabitats within sites, predict corresponding variation in movement patterns. Quantifying ecological differences among and within sites and relating them to the ranging patterns of the respective animals could provide insights into proximate ecological determinants of ranging behavior in these taxa.

Predation

There is strong evidence that grouping pattern, group size, cohesion, and movement are influenced by predation pressure (Boinski, Treves, and Chapman, chap. 3, this volume). Extant lemurs are under a constant threat of predation from raptors and viverrids (Goodman, O'Connor, and Langrand 1993; Rasoloarison et al. 1995; Wright 1995), although predation pressure may have recently been reduced with the disappearance of several eagles (Goodman 1994; Goodman and Rakotozafy 1995). The comparatively small sizes of most lemur groups and the frequent fissioning into even smaller subgroups in some taxa are therefore surprising, unless they reflect a strategy that aims at reducing detectability (Wright 1998).

Some observations are in agreement with predicted effects of predation pressure on lemur behavior, however. For example, ring-tailed lemurs, which may be exposed to the widest range of predators because of their semiterrestrial lifestyle (see also Goodman and Langrand 1996), form the largest groups among lemurs. Similarly, those lemurs that were introduced to the predator-free Comoro Islands show the most pronounced formation of temporary subgroups (Tattersall 1977). Furthermore, experimental predator encounters induced closer aggregation of ring-tailed lemurs (Pereira and Macedonia 1991). Finally, diademed sifakas choose different resting sites during the day and at night, and form small, spatially dispersed sleeping parties at night, in response to the predominant threat of predation from aerial and terrestrial predators, respectively. Thus, the spatial behavior of lemurs appears to be affected by predation risk, but additional observations, and especially field experiments (see, e.g., van Schaik and Mitrasetia 1990), are needed to better describe its effects on group cohesion, travel order, and ranging behavior.

Social Influences

Social factors such as sex, age, and dominance mainly affect group cohesion and movement patterns of individuals or classes of individuals. For example, the formation of temporary subgroups and ranging patterns are often not random with respect to sex (e.g., Chapman, Wrangham, and Chapman 1995). In most lemur groups, sex is indeed the primary variable structuring patterns of association among individuals (Kappeler 1993b). In contrast to most anthropoids, however, associations are between male-female pairs and not between same-sex individuals. Males and females also differ in

Peter M. Kappeler

the frequency with which they initiate and lead group travel. Across primates, sex appears to have a relatively strong, but not consistent, effect on group travel (Boinski, chap. 15, this volume). It is striking that in all lemurs in which leadership has been recorded, females are primarily responsible for the timing and direction of group movement. It is tempting to attribute this bias to the general social dominance of females, which is widespread among lemurs (Richard 1986; Kappeler 1993a), but variation in the degree of sexual size dimorphism may provide a more general explanation across all primates.

The age and social status of individual group members may also affect their role in group movements. Infants and juveniles, which are most vulnerable to predation, are often found near the group's center, where safety from predators may be highest (Rhine and Westlund 1981; van Noordwijk et al. 1993). In other species, high-ranking individuals forage near the center of the group because they are avoided by their lower-ranking groupmates (Robinson 1981a; Janson 1990b). Corresponding data for lemurs are generally lacking, however, except for qualitative observations on *Lemur catta* (see above), so that they cannot contribute comparative information on these effects.

Differences in dominance styles, post-conflict behavior, and bonding patterns may also contribute to variation in group cohesion. These traits usually do not vary independently of one another across taxa, and they summarize many of the crucial behavioral mechanisms that counteract the dispersive effects of intragroup competition (de Waal 1989; de Waal and Luttrell 1989). For example, in groups with despotic dominance relations, limited tolerance, and rare post-conflict reconciliation, it may be more difficult to maintain group cohesion, compared with groups that display the opposite combination of social characteristics, because members of many dyads may avoid each other. Future studies of group cohesion in lemurs and anthropoids could test these predictions with interspecific comparisons.

Life History
Both grouping and ranging patterns are potentially influenced by various life history traits, such as body size, activity, mode of infant care, and brain size, but more detailed data, especially on individual and coordinated movements, are needed for rigorous evaluations of these effects. Because of their diversity in these life history

traits, lemurs can provide the necessary data for these comparisons. Cathemeral lemurs, for example, provide unique opportunities to document the effects of diurnal and nocturnal activity on group cohesion, communication, and ranging patterns (see, e.g., Curtis 1997). Moreover, a comparison between *Varecia* and sympatric *Eulemur* could illuminate the effects of different infant care patterns on female ranging behavior. Females that carry their infants with them are not constrained to return to them periodically, and thus are predicted to range much farther than females with parked infants.

Conclusions
This review reveals that behavioral aspects of individual and group movements of lemurs are poorly documented, and that most available information is based on anecdotal observations. The few focused empirical studies cannot provide a basis for firm conclusions either about lemur-specific patterns or about consistent differences or parallels with other primates. On the other hand, this review also indicates that lemurs can contribute interesting, if not unique, comparative information on central issues in this context. Most remaining lemur habitats contain both solitary and gregarious, as well as diurnal and nocturnal, species with different feeding strategies, so that studies of different sympatric species should allow detailed examination of relevant variables.

Acknowledgments
I want to thank Sue Boinski and Paul Garber for the invitation to contribute to this volume, Carel van Schaik for stimulating discussions, and Sue Boinski, Jörg Ganzhorn, Paul Garber, Alison Jolly, Eckhard Heymann, Deborah Overdorff, Dietmar Zinner, and an anonymous reviewer for their comments on this chapter. Michael Pereira deserves special thanks for his friendship, collegiality, and 25-page review of an earlier 24-page version of this chapter.

Peter M. Kappeler

CHAPTER SEVENTEEN

How Monkeys Find Their
Way: Leadership,
Coordination, and Cognitive
Maps of African Baboons

RICHARD W. BYRNE

The African baboons (*Papio* spp.) are in many ways an ideal group of primates for asking questions about travel and its coordination. Living in many different ecologies, their patterns of ranging are highly variable—from cohesive travel in large bands to continual fragmentation and reaggregation. Their home ranges are often large, and their habitats vary from open grassland and desert to dense and featureless bush and forest. Baboons are also unusually well-studied primates, so there is a wealth of information available on their ranging behavior. And finally, as the largest monkey, their absolute brain size (Stephan, Frahm, and Baron 1981) and relative neocortex size (Dunbar 1992) are the greatest of any non-ape. If any monkey species has sophisticated cognitive mechanisms that may be employed in coordination and navigation, baboons should be it.

This chapter will examine some of the evidence concerning how baboons move around in the environment and what this evidence might mean for baboon cognition. The standpoint will deliberately be rather theoretical. I will begin by asking "What, in principle, are the options for a primate that needs to travel with others?" Starting with the most minimal assumptions, I will try to justify each elaboration of cognitive complexity from this minimal position by using specific evidence.

Perhaps the simplest rule to follow, in terms of the cognition demanded of the individual in order to remain with others when traveling, would be "keep within sight of at least one other group member." (Vision, as well as being a dominant modality for primates, requires no deliberate "emission of stimuli" to all allow orientation, unlike audition and often olfaction.) All that is required to follow this rule is the ability to recognize a number of associates as "familiar" individuals. If such an evolved algorithm were present in every member of a social network, they would perforce travel as a group, as baboons usually do—without any need for leadership, spatial knowledge, or any form of specialized ethological display. This, then, is an appropriate "null hypothesis" against which to assess evidence for other cognitive abilities; for reference, let us call it the *visual contact model*. The characteristics to be expected from this most basic system of group coordination include the following:

1. group spread should vary predictably with visibility
2. no one individual should have more influence than any other on the route taken
3. use of specialized vocal or visual displays should not correlate with any aspects of ranging
4. the travel path should show no signs of directedness toward particular locations; formally, it should be a random walk
5. individuals should always remain in sight of others; or, if one accidentally loses sight of all others, it should range independently until (by chance, if at all) they meet again
6. particular locations should not recur more often than expected by chance on successive days' marches, nor vary systematically in frequency of visit with season

In fact, of course, baboon group movements differ from this picture in a number of ways. Some features are so obvious and well known that they will be described with no attempt to reference their "discovery"; others will be briefly reviewed. Evidence for group cohesion (point 1), leadership (2), communication associated with group movements (3), and baboons' knowledge of large-scale space (4, 5, and 6) will be examined separately, as far as this can be done. Where deviations from the visual contact model *are* found, we can then go on to examine whether or not these constitute evidence for more complex cognitive mechanisms involved in group movement.

Richard W. Byrne

Group Cohesion

If individuals did simply behave according to the null hypothesis, attempting when traveling to stay in visual contact with members of their own group, we might expect group spread to be a simple function of (1) group size and (2) the visibility at baboon height in any particular habitat (i.e., the average distance that can be seen by an animal with eyes only 0.5 m from the ground). Unfortunately, rather few studies have systematically recorded group spread, but one that did, a study of mountain baboons, will serve to show how far reality can lie from these predictions.

"Mountain baboons," a population of chacma baboons *(Papio ursinus)* that inhabit the upper montane and subalpine zones of the Drakensberg Mountains in South Africa, allow us to examine the effect of variation in group size with visibility held constant. In this very open grassland habitat, visibility at baboon height is limited— year-round—only by folds of the hillside. The dispersion of individuals is unusually large; sometimes a baboon group as small as nine individuals was found to be spread over 1 km of hillside (Byrne, Whiten, and Henzi 1990). This is just as the visual contact model would predict. However, "total group spread" is a poor measure of individuals' dispersion, since it is overly influenced by any outlying individuals, even when the majority of the group remains clumped. "Mean nearest-neighbor distance" is worse; in a group in which pairs of animals tend to range together, far larger dispersions *between* pairs would go unrecorded. It is therefore preferable to measure group dispersion by the mean distance between all pairs of adult individuals, where this can be reliably seen. When this was done for two mountain baboon groups of nine and fourteen animals, sharp *seasonal* differences in spread emerged, from little more than 20 m in the fertile spring to over 60 m in the dry winter (Byrne, Whiten, and Henzi 1990). This effect could not be a consequence of visibility, which was constant. Food supply is a more likely explanation: the available nutrition did vary seasonally, enabling the seasonal difference in group spread to be explained as a consequence of variation in food competition (Byrne et al. 1993). This population is known to be nutritionally stressed in the dry winters, spending longer in travel-feeding than any other baboon population for which published data are available (Whiten, Byrne, and Henzi 1987) (Note that travel time and feeding time need to be treated together in time budget analyses because baboons may rapidly alternate between feeding and walking, or pick up foods while hardly breaking step.) When food-stressed, baboons would be expected to benefit from decreased competition, and aggression over food was virtually absent at the high interindividual dispersion found in winter.

However, group spread was seldom anywhere near the maximum that would be possible while remaining in sight of one other individual. In a habitat where food is sparse, far greater dispersion would be predicted if staying in visual contact were the only force for cohesion. Evidently, there is an additional tendency to remain close to other baboons that partially offsets the need to avoid foraging competition. This tendency might be explained as an antipredator adaptation, although predation risk was essentially nil at that time (leopard had been removed many years earlier, leaving only jackal, serval, and caracal). In the light of baboons' known flexibility toward predator risk (Cowlishaw 1997), an explanation relying on such conservatism of behavioral traits is rather unsatisfying. Nor does slightly altering the minimal assumption—to "ranging in sight of *several* familiar conspecifics"—help salvage the null hypothesis for long. In the same habitat as the very small groups on which the study concentrated, larger groups of baboons are also found. These groups were noted to range *more* cohesively than the small groups; thus, group spread, even in this unusually simple environment, does not necessarily increase with group size. If, however, contrary to our null hypothesis, one or a few individuals have particular influence in determining group movement (that this is the case becomes abundantly clear below), the cohesive tendency is explained. Such individuals would act as pivots for the group, reducing its spread and imparting a slight cohesive tendency, even in the absence of an overriding need for cohesion to minimize predation.

Visibility at baboon height fares no better in predicting group spread. Indeed, in some sites with poor visibility (e.g., Mikumi, Tanzania; Mont Asserik, Senegal; Awash, Ethiopia), where the null hypothesis would predict highly cohesive ranging in the absence of any means to regain contact once lost, regular fission-fusion of groups is quite common.

The many variations in group cohesion that are found among baboons simply do not match those predicted by the visual contact model; however, neither do these differences give any obvious clue to baboon travel mechanisms, and they are better explained in other ways. Group spread is a complex function of social organization, including male constraints on females (Byrne, Whiten, and Henzi

Richard W. Byrne

1990), mate guarding (Alberts, Altmann, and Wilson 1996), and the maintenance of relationships among females (Dunbar 1988), and ecological constraints—in particular, the actual predation risk (Cowlishaw 1997) and pattern of food availability (S. A. Altmann 1974; Barton, Byrne, and Whiten 1996).

Leadership

Most observers of primates that range as cohesive groups have always been convinced that some individuals have more influence on group movements than others (e.g., DeVore and Washburn 1963; Rowell 1972b) and that group members pay attention to the behavior of pivotal animals (Chance 1967; Chance and Jolly 1970). There is now considerable evidence to support these beliefs. Much of this evidence relies on descriptive accounts, rather than quantitative analyses. The reason for this may be—paradoxically—the methodological sophistication of modern studies. Most modern studies rely on focal animal sampling (J. Altmann 1974) to avoid several kinds of bias in data collection; though a great improvement on *ad lib* sampling for most purposes, this method directs the observer's attention firmly toward single individuals. Since few studies can afford the luxury of multiple observers collecting data synchronously on different individuals, understanding of group coordination has been the casualty. As a result, I must apologize at the outset for the level of descriptive detail that will be necessary in order to present and evaluate the evidence for leadership in baboons.

If there is leadership of some sort, the simplest algorithm would be to follow a single "leader" individual. This would only require a baboon to be able to discriminate among some familiar individuals and remember which one is the leader, abilities well within the range of known achievements of many species of monkeys. The movements of one animal, alone, would then predict group movements. All other individuals would follow the leader, and the attention of the group should focus on this individual. Considering this straightforward picture, it seems at first sight strange that leadership has proved difficult to study in baboons.

A major problem, first realized by Kummer, is that to *lead,* in the sense of "determine group travel direction," does not require going out in front; one can "lead from behind." Kummer introduced a useful distinction between the *decision-making* individual of a group, the true leader, and *initiators,* individuals who "suggest" possible changes of direction, for instance, by setting off in front of the others with apparent determination (Kummer 1968a). This pattern is evident in the travel coordination of the species that Kummer himself studied, hamadryas baboons *(Papio hamadryas)* in the deserts of Ethiopia (Kummer 1968a, 124–28). Although the basic social unit of hamadryas is a single male with his females, hamadryas baboons often range as a semipermanent alliance of two single-male units: a "team." The male baboon striding out in front, usually the younger of the two, superficially appears to be controlling where the team goes. However, careful examination of the routes of the two males shows that very often the decision-making baboon is not the male at the front, but the one in the rear. (In hamadryas baboons, the decision-making and initiator individuals are always adult males, although females are said to sometimes influence the decision somewhat by immediately following the initiator male [127]; often the older of the two males is the decision-making individual.) When the decision maker takes the route "suggested" by the initiator, no difference in ranging paths is seen; however, every now and then, he does not. Then, evidently realizing he is not being followed, the initiator has to detour onto the path chosen by the decision-making male. He therefore takes a more sharply angled and longer route, and there is a lag before his path matches that of the decision maker. The reverse does not occur: if the decision maker moves first on any occasion, his lead is always followed. The distinctive result is that the decision maker's path is always slightly shorter and more economical of effort than the initiator's. Also, initiators, when observed carefully, can be seen occasionally to check the movements of the decision maker, rather than walking confidently ahead without looking back.

Similar kinds of evidence for the leadership of group travel were found in mountain baboons (Byrne, Whiten, and Henzi 1990). Decision makers could be identified by their smoother and more economical travel paths during ranging, and by the fact that when they began to move after a rest period—whatever their spatial position in the group—the whole group would begin to move off simultaneously, in the same direction. In this population, also, the decision-making individual was often found not to travel in the front of the group. And in almost all cases, the decision maker was an adult male, but not necessarily the dominant individual. This pattern matches the hamadryas system, except that in mountain baboons, initiators could be male or female. Byrne et al. studied the relations within two small groups in detail. In one, whose pattern of social affiliation closely resembled that of a hamadryas unit of a prime male, the sole fully adult male—the dominant baboon—determined group travel. However, in the other group, two adult males were present, one (DV) visibly older than the other (HL). Again, a social parallel with hamadryas was noted, this time with the unit of an old male with a young follower. The young, dominant male HL was able to monopolize mating with adult females. However, HL did not determine the group's travel direction, which remained the prerogative of the older male DV despite his lack of power.

A subtly different pattern of leadership has been described in geladas *(Theropithecus gelada)* (Dunbar 1983). In the unusually good observation conditions permitted by this species' behavior and habitat, Dunbar was able to record visual monitoring among individuals, and this was by no means random. Females glanced at their harem male and faced toward him more than toward other females, whereas males tended merely to face whichever female happened to be nearest to them at the time, although they glanced most often at the dominant female. Progressions were generally initiated by lactating females, but whether an initiation was followed by the rest of the group was largely a function of the behavior of the dominant female: if she followed the lead, then the male and the rest of the group would also. Dunbar argued that the male, by monitoring the dominant female's behavior and coordinating his behavior with hers, is able to capitalize on the strength of the female-female bonds within gelada society and so prevent his harem being dispersed. In contrast, in both hamadryas and mountain baboons, the male keeps his group together by actively herding females (Byrne, Whiten, and Henzi 1987; Kummer 1968a).

A division of role between initiator and decision-making individuals is also seen in mountain gorilla groups, as can be shown using unpublished data of my own. In 1989, Group 5 at Karisoke, Rwanda, contained two silverbacks, Ziz and his younger brother Pablo. Although the two males cooperated effectively in intergroup encounters, Ziz was dominant and Pablo was able to mate only secretly; and Ziz was clearly the decision-making individual when it came to group movements. Contrasting this pattern with earlier observations of the group shows that, as in baboons, leadership in gorillas is not simply a function of dominance, and the extensive experience of old males may give them a key role long after they have lost rank and power. In the same group 5 years earlier, in 1984, the very old male Beethoven was subordinate to his son Ziz, then a large blackback (adolescent), and Ziz entirely monopolized mating. Nevertheless, when the group came to a place where some sort of decision about travel needed to be made—such as whether to cross a river—the whole group, including Ziz, waited, and only when Beethoven caught up and crossed the river did the rest follow and continue. David Watts's long-term observations of the group (pers. comm.) show that after 1984, Ziz also began to acquire a role in leadership, but at first this applied to only a part of the group, comprising the younger or recently immigrated females. In early 1985, Ziz had just attracted four females to transfer into Group 5 within a 6-week period, resulting in a sudden enlargement to ten females. The group began to split into two subgroups that were sometimes completely out of sight of each other, and a number of times over the course of about a month, the subgroups nested in different places. The immigrant females, a few immatures, and occasionally one or more of the resident females stayed with Ziz. The others, including the oldest and dominant female Effie, stayed with the old, now subordinate, but experienced male Beethoven. This may be a common pattern of change in leadership in mountain gorillas, since in Group 4 during 1989–1991 I found a similar situation. The oldest and dominant female, Papoose, always remained with the old, newly subordinate male Beetsme, while the youngest females followed the much younger and more powerful male Titus; other females varied in which male's lead they followed. In this case, the subgroups did not sleep separately, but could be as much as a kilometer apart during the day. Again, Watts's long-term records show what happened next. By 1991, Papoose was part of the main group around Titus. She still interacted with Beetsme more than the other females did, but he was often quite peripheral, and usually alone in that position; Titus had become the leader. (See Watts, chap. 13, this volume, for further aspects of gorilla group coordination and habitat use.)

At times, the situation may be more complicated than the rule of following the actions of a single individual. Hamadryas baboons in Ethiopia sleep on cliffs in large groups. Each morning, the baboons leave the safety of the cliffs for a period of quiet grooming and resting on open areas by the cliffs. Then a mass exodus by the band from the area of the sleeping cliffs occurs. A hamadryas "band" is a grouping roughly equivalent to a "troop" of savanna baboons, (e.g., *P. anubis* or *P. cynocephalus,* comprising many single-male units; several bands may share a sleeping cliff, but are socially quite

Richard W. Byrne

independent of one another.) After this exodus, the band splits during ranging into two-male teams and clans, composed of a number of frequently associating units whose males are thought to be closely related. It is males who organize the band's movement during the exodus. Kummer has described how one hamadryas male will often set off in a particular direction, and some other males will follow him, with females following them in turn (Kummer 1968a). Units from the center of the group often follow and overtake the original leaders, so the male at the front often changes. Kummer vividly likens this process to watching an amoeba under a microscope, with a pseudopod developing. Frequently the main mass of the band does not follow, and the initiative dies; gradually those who started to move off drift back into the mass, and another "pseudopod" of baboons begins to form in another direction. Then, one particular initiation causes the band to set off: suddenly, every individual seems to be moving at once. Kummer likens this to the moment when the whole amoeba "flows" down one particular pseudopod. A similar "amoeboid" process occurs with the morning departure of Guinea baboons *(Papio papio)* in Senegal (Byrne 1981), a species whose social organization is thought to be similar to that of hamadryas (Boese 1975).

The number of animals in a pseudopod is not the decisive factor, as sometimes Kummer found that a large movement would abort in favor of a rather small-scale move in another direction. One interpretation of this pattern is that the failed initiations are ones that some single decision-maker individual for the whole band (located somewhere in the mass of the group) does not follow, and that the crucial difference when the group does set off is that this key individual also moves, and this is noticed by the rest. This interpretation is supported by Kummer's observation of a few cases in which a red-faced old male in the band's center began to move with a peculiar, rapid, swinging gait toward the periphery, even if there was no pseudopod in that direction. This male did not look back or hesitate, and within a few seconds the entire band moved off in the direction he initiated. However, in more typical departures, there was much hesitation; scratching, a sign of conflict (Maestripiaeri et al. 1992), was found to peak in the males at this time.

Stolba examined the initiation process in more detail. He found that before the exodus, the males paid close attention to one another, particularly to those in their own clan; when a female and a male moved together, it was the male that other males watched

(Stolba 1979). When movements that might initiate travel were made, varying numbers of other males began to follow, or indicated that intention by their stances. Initiatives by younger clan males prevailed less often than those of older males, and usually prime males managed to enlist many supporters and so apparently determined the direction of travel. However, as Kummer had found, if old males did intervene, they were particularly influential. Thus it seems that there is more to the process of pseudopod formation than attempts by initiator males to "second-guess" a single decision-maker male who has not yet indicated the direction in which he alone will take the troop. At the very least, the decision maker must be somewhat influenced by the "pull" of many individuals setting off in one direction. It may be that there is not always a single final arbiter, but rather sometimes several influential individuals may be involved.

The coordination of hamadryas baboon movements has been described in the literature as a process in which majority agreement is sought by individuals "voting with their feet" (Kummer 1995; Stammbach 1986), a sort of negotiation process. However, we need to be wary of importing baggage from the normal human usage of *negotiation:* "to confer with a view to finding terms of agreement." We would not normally describe as negotiation a conflict between two visceral urges, perhaps a young man's deciding whether to join his friends to watch an important football match or to join his bride to celebrate his wedding night (the example is taken from a recent newspaper, and in that case the former urge won). Nor would we call it negotiation if a single powerful figure always determined the next move of a group, however much nervous milling about took place before the decisive signal were given, as in the effect a famous scientist might have on his young postgraduates. Kummer's descriptions match these two cases, and he did not use the term *negotiation,* rather describing the influences as a matter of "pull." To claim that baboons, or other primates, "negotiate" about travel direction requires that some form of relevant information, or some commodity, be exchanged in return for a modified decision. So far, there seems to be no clear evidence of either in baboon behavior.

In hamadryas baboons and geladas in Ethiopia, and in mountain-living chacma baboons in South Africa, the small size of groups helped in identification of the decision-making individuals. Savanna baboons live in relatively large groups and there is good reason to expect their leadership pattern to be different, at least

Richard W. Byrne

with respect to the sex that leads. In hamadryas, and probably also Guinea baboons in Senegal, social organization is based around males, with female transfer between units. In mountain baboons, there is a partial convergence toward this system, with small, male-focused groups created by fission (Byrne, Whiten, and Henzi 1990; Henzi and Lycett 1995). All savanna baboon populations, by contrast, show male transfer, such that groups generally have a stable core of matrilineally related females. Thus, while under good conditions for studying leadership, baboons have proved to be male-led, we cannot assume that this finding will generalize to other populations. Unfortunately, the troops typical of savanna baboon populations are often thirty to sixty individuals in size, making identification of decision-making individuals very difficult. It is an observational task of daunting dimensions to attempt to record the actions of many potential decision-maker animals at once.

Generally, researchers have avoided any mention of this tricky problem, but Rowell (1972b) has argued that typical baboon movements are determined by females. She describes a pattern of failed initiations of the day's travel in olive baboons *(Papio anubis)* in Uganda that is strikingly reminiscent of hamadryas baboons—until it comes to the final determination of the group's travel. While males led the initiative, it was only when one of the old females of the group set off after one of them that the group moved as a whole. The lack of subsequent discussion of leadership in savanna baboons presumably implies a willingness to follow Rowell's conclusions, accepting female decision making for all savanna baboons. However, evidence to back this acceptance is sorely needed, especially as no other researchers seem to have described "failed initiations" in the morning departures of savanna baboons; the possibility that the Ugandan woodland population Rowell described was in some way exceptional in social structure cannot be discounted.

In searching for strong evidence of leadership in savanna baboons, a distinction made when discussing spatial order in baboon progressions (Collins 1984) may be useful. A distinct progression order, with the different age-sex classes occupying different places in the order of march, was originally described as invariant in savanna baboons (DeVore and Washburn 1963; Washburn and DeVore 1961), but later both the precise pattern and the existence of any order at all were disputed (Altmann 1979; Harding 1977; Rhine and Westlund 1981). Collins points out that most progressions are slow, with baboons dispersed over a broad front, whereas "full progressions" are infrequent, but in these the troop travels as a compact group. In slow progressions, baboons are feeding as well as moving, and the influence of the spatial distribution of food resources may well outweigh the influence of particular decision-making individuals most of the time. Thus future research on leadership in savanna baboons may profit by concentrating on the relatively rare progressions in which the group travels without feeding.

We may conclude that in at least some and probably in all baboon populations, some individuals have more influence than others on group movement. Particular "decision-maker" individuals exist; these are adult males in hamadryas and mountain baboons, but may be females in the matrilocal troops of savanna baboons.

Communication about Movement and Location
An individual may effectively lead a group without using directed communication, merely by setting an example that is followed. However, if leadership is a regular feature of primate behavior, we might expect specialized communicative displays to have evolved, facilitating the process to the benefit of its individual participants (Haven Wiley 1983; Krebs and Dawkins 1984). There is evidence of such displays in baboons, but the question of whether these monkeys intend to communicate remains a vexed one.

Mechanisms of Communication
In small, two-male teams of hamadryas on the march, Kummer (1968a) identified a specific visual communication pattern, "notifying," somewhat resembling a sexual presentation but performed only by one adult male to another. In its clearest form, one male slowly approaches a seated neighbor, with each looking at the other's face; when very close, the approacher wheels round and presents his highly colored anal area, often while showing an erection. The seated baboon may bend forward or touch the other's scrotum, and continues to gaze at the anal area of the other—now rapidly departing—baboon. Other, weaker variants of this display are possible; in the mildest, the males simply look steadily at each other from a few meters away for several seconds. In many cases, this behavior corresponds to a change in travel direction, although notifying is not done solely by initiator to decision-maker individuals, and may be performed on arrival as well as departure, or without any group movement (Kummer 1968a, 130–31). The distinctive, exaggerated swagger of an old male able to precipitate and determine

Richard W. Byrne

a whole band's morning departure has already been mentioned; Kummer considered this, too, a form of "notifying" behavior, to everyone (139). Stolba (1979) was able to show that notifying displays among males peak at the moment that the band leaves the sleeping cliffs, and also identified other signals used at this time. When a sitting male lowers his head to his chest abruptly, it predicts that he will not move for at least 2 minutes, and his neighbors "also change position less, as though his refusing to see were a No vote" (Kummer 1995, 268). Conversely, a male standing on all fours in a rigid posture is likely soon to set off in the direction shown by the axis of his body; this display is reminiscent of Schaller's description of the silverback gorilla's means of announcing time and direction of intended departure (Schaller 1963). No visual signal used in such within-troop coordination has been described in other baboon populations, but the possibility exists that subtle communicative acts remain to be described.

It is considerably easier to discover the means of movement coordination of individuals in a group scattered in dense vegetation: loud vocalizations. Of course it is not the case, as our original null hypothesis would predict, that individuals out of visual contact cease in general to be group members. Separated individuals are routinely able to rejoin the group within hours. It has often been noted that vocalizations are used in doing so, but the details have been studied less often (see Boinski, chap. 15, this volume). Guinea baboons in Senegal present particularly good opportunities for observing this process of re-location. Large groups of Guinea baboons fission daily in the dry season into very small parties, much as in hamadryas (Byrne 1981). However, in Senegal, the visibility is low in most habitat zones because of tall grass and scrub. Predator density is also high, to judge from the many sightings and calls of lion, leopard, hyena, and wild dog, and the baboons certainly behave as if regrouping into the large group is a matter of some priority. Each morning, I followed 150–200 baboons setting off as a group across the open laterite plateau, but once they reached feeding areas they began to split up. When I used factor analysis to relate vocalizations to the circumstances of the baboons' daily lives, a clear association emerged between barks and three specific events (Byrne 1981): (1) when a group fissioned; (2) when one group passed another; and (3) when two groups fused. Increased rates of calling in these circumstances were found with both the double-phase "type 2 loud call" of adult males—the *wahoo* bark—and the single-phase "sharp barks" given by females and younger individuals (Waser 1982a). These associations were confirmed by comparing the calls given during these events with those given at other times during travel. For both single-phase barks and double-phase *wahoo* barks, significantly more barks were given with fission, fusion, or passing another group than when group composition was stable (pooling all barks given by the focal party, there were 4.2, 3.8, and 4.0 times as many barks in the three cases, respectively).

This is what was happening. When exploiting fruit trees, the baboons' party size remained at about thirty to fifty, and the several parties remained in auditory contact through the use of loud barks. Typically, the double-phase *wahoo* call was given at a rate of three or four per minute by each subgroup. These calls evidently enabled regrouping, since they acted as an auditory "beacon" to indicate the others' whereabouts. Often a large group would fission and fuse several times in the course of a morning's travel. However, when the baboons were feeding on smaller, herb-layer material, the party size was much smaller, typically only five to eight animals, and few calls were made. When following one party, it was in fact usually impossible to know where the other parties were. Then, after some period of travel-feeding, events such as the following happened (Byrne 1981):

> 9:45 A.M. (subgroup of five, including two males) One male climbs to 4 m in a leafless bush and begins a deliberately paced series of double-phase *wahoo* barks, delivered loudly at a rate of about two per minute, scanning round and not feeding.
>
> 9:50 A.M. Well-spaced, double-phase *wahoo* barks continue, looking round after each, until a distant bark to the W is heard. Then the male descends and all set off toward the calls.
>
> 9:55 A.M. Group heading toward stand of tall trees, still barking at intervals, with another small group now behind me and another to the S adding their calls.
>
> 10:00 A.M. Group pauses and gives six to eight single-phased barks, then two double-phase *wahoo* barks are heard from different locations to the NW, and the group sets off again, toward them.
>
> 10:05 A.M. A number of groups are now funneling into the area of shade trees, joining up as they move.
>
> 10:15 A.M. The same size group as originally set off is now assembled, resting and feeding.

I observed a similar pattern, with variations, many times during the study, although I was more often with a party that did not initiate the whole process of regrouping. It was quite evident that exchanges of long-range vocalizations functioned to facilitate re-unions of scattered parties of baboons in these low-visibility habi-tats with high predation risks. Precisely how was less clear. At the time, my puzzlement was with how the baboons knew which way to head when they heard a steady series of *wahoo* barks: if they simply approached the caller, then it seemed a lucky coincidence that they always tended to aggregate at a suitable shady resting place, as they did. A likely answer to this question has since been provided by Stolba and Sigg's work with hamadryas, described be-low. Writing in the early 1980s, I took it as an obvious matter that the initiator of the calling *intended* to provoke other baboons to call in order to join with them, and that the hearers *understood* that the caller wanted to reaggregate, as they most often did themselves. However, the assumption of such intentional behavior in a monkey would not be taken so lightly today.

Intentions of Communication

In a range of spheres other than movement coordination, monkeys have shown no definite signs of any capability to communicate with intent to change others' beliefs (Byrne 1995b; Tomasello and Call 1997). An understanding of the intentions of others would presum-ably allow baboons increased efficiency, but does any evidence from movement coordination warrant revision of our current under-standing of baboon cognition?

Working with chacma baboons in Botswana, Cheney, Seyfarth, and Palombit (1996) looked specifically at the question of whether separated female baboons make vocalizations with the intention of informing others of their location, in order that other group mem-bers will understand their plight and help them. They found no evidence of such an intention. Barks were most often given during troop movement through woodland or by separated individuals, and hardly at all when the group was tightly coalesced or feeding on open plains. Also, experimental playbacks of barks (but not other types of calls) evoked barks in response. Thus, as in the Guinea baboons of Senegal, there was an association between the emission of barks and circumstances in which there would seem to be a need for maintaining contact; similarly, barks were in practice "exchanged" by separated parts of a troop. However, other features were at variance with the intentional hypothesis. If baboons some-times intended to help others rejoin the group, they would be ex-pected to specifically after hearing calls from separated individuals, especially if these were their relatives. No statistical support could be found for any tendency to give calls shortly after hearing another female call; but instead, females were more likely to call soon after they themselves had just called. To simulate the case of hearing a separated relative, Cheney et al. played barks to target females who were close kin to the individual on the audiotape. Although in most cases the response was simply to look in the direction of the call, females sometimes did respond to their relative's call by barking themselves, whereas other, unrelated females never did so. However, once again, the tendency to give these answering calls was strongly influenced by whether the target female was alone or trailing the main group—circumstances in which she herself was at risk of separation. Cheney et al. concluded that these contact barks "seemed more often to reflect the signaller's own state and position than the state and position of her audience," and that "baboons give contact barks when they are at the group's periphery and at risk of becoming separated from others. Through experience, listeners learn that they can maintain contact with at least a subset of the group simply by listening to other individuals' calls" (Cheney and Seyfarth 1996). According to this interpretation, the barks do no more than reflect the internal state of the caller, but baboons learn to use the calling of other individuals (albeit perhaps lost ones) to keep in contact with others.

There are difficulties with this analysis. Recall the description of Guinea baboons (Byrne 1981). During most of a morning in the dry season, small, separated subgroups foraged out of sight or sound of one another in dense habitat; very few calls were given. When an individual did finally begin a series of barks, it continued for several minutes, but stopped once it heard one or a few distant barks. In such a system, it would indeed be the case that most barks would be preceded by a bark made by the same individual, rather than by a distant individual. Using a null hypothesis of "random calling from any individual" would be inappropriate. The crucial test is whether barking continues if it is unanswered, but declines in frequency (or ceases) once an "answer" has been heard and converging travel begun. In principle, one could perhaps define a string of barks from one individual as a putative "request," and predict that barks from other baboons out of sight would be more likely toward the end of, just after, this string than at other times. And Cheney et al. *did* find that females were more likely to call in response to close kin. That this particularly occurred when they themselves were sep-

(needing food). This finding was quite unexpected, as Kummer himself once specifically noted that "being limited to the here and now, primates cannot designate a time and place for meeting again after they have separated" (Kummer 1971, 31).

Stolba and Sigg's work has profound implications, since referential communication appears to be highly limited in primates (Cheney and Seyfarth 1990), and no other evidence suggests that they can communicate about remote locations, in the way that bees can by means of their "dance language" (von Frisch 1967). Furthermore, this seems to be the only evidence at present that a monkey can anticipate a future event and engage in anticipatory planning toward it (Byrne 1995b). Given the importance of displaced reference in theories of the origin of human language, it is surprising that this single baboon study has not been followed by more detailed quantitative work.

Knowledge of Large-Scale Space
It has long been noticed that baboon daily marches are not entirely random walks. Sites where baboons have been successful in feeding on one day tend to be visited again on succeeding day's marches, whereas at another season the same areas may not be visited for weeks. Evidently baboons have some knowledge of the large-scale space in which they range ("large-scale space" refers to areas that cannot be seen entirely from a single vantage point). Also, their travel sometimes shows clear signs of directedness toward specific, distant objectives. Most obviously, marches end up at sleeping sites, even in habitats where sleeping sites are relatively rare, and baboons sometimes move directly to semipermanent rain pools even when they could not possibly see the pools until they were close by. These facts led Altmann and Altmann (1970) to suggest that baboons have a "cognitive map," a mental representation of the spatial layout of their home range. Sigg and Stolba's study of hamadryas (1981), examined in the previous section, not only confirmed that baboons do indeed know the layout of their range, but also showed that they can rely on *other* individuals' knowledge of large-scale space when they communicate about reaggregation places.

What form might a baboon's cognitive map take? One possibility, unlikely to be realistic, is that a complete knowledge of the distribution of stationary resources is accurately encoded and used to guide ranging. For baboons, S. A. Altmann (1974) noted that "the foraging pathway with the shortest average distance per useable calorie would be optimal," and discussed algorithms a baboon might in principle use to approximate optimality. He considered the problem of choosing an optimal route within a "perceptual field." A minimal algorithm would simply compare all visible items and head for the nearest, but a more sophisticated forager might "look ahead" to each food item and estimate which was closest to other items, thus beginning to detect clumps. Still more sophisticated algorithms would look several steps ahead and incorporate information about food item quality and quantity (see Menzel 1997 for experimental evidence bearing on this issue). Oddly, Altmann's analysis of the problem did not explicitly consider the possibility that baboons could compute potential routes beyond their immediate perceptual field. However, if baboons do possess a cognitive map, then in principle, they could compute an optimal foraging route for the whole day's ranging, not just for what was in view at a given time (although in species that range cohesively, what is optimal for one individual may not be for another). Attempts to apply optimal foraging theory to animal behavior typically do not assume any limits on mental competence, in which case baboons should choose foraging routes that are also optimal at the large scale. However, there is as yet no definite evidence to support this conviction, and studies of optimal foraging have failed to address the possibility, let alone discovering what level of detail might be represented on the baboon's cognitive map.

Even when we return to the gross level of large-scale spatial information that is certainly known to baboons, the relative locations of sleeping sites and water sources, we must distinguish between different ways in which they might be known. Unfortunately, the term "cognitive map" (equivalently, "mental map") is ambiguous. In animal behavior, "cognitive map" has often been used specifically for a representation that is *isomorphic* to the two-dimensional environment: a view from above, like the maps we unfold to look at, but constructed mentally after experience at ground level only (Gallistel 1990; Poucet 1993). Thus the claim that "bees have a cognitive map" was a strong one, implying that bees acquire richly organized mental representations of space, allowing simultaneous access to information about distances and directions between all known locations. (This strong claim is in fact now rather fully disproved; see Dyer 1991; Dyer, Berry, and Richard 1993; Kirchner and Braun 1994; Wehner, Michel, and Antonsen 1996.) However, in cognitive psychology, the same term has generally been used for any mental substrate that exists to allow navigation (and the various primatologists that have talked about a "cognitive map" seem likewise to have meant it in this sense). Cognitive psychologists would take it for granted that any creature that can navigate in space has some sort of cognitive map, and then ask, "what sort?" In particular, they would ask, what *form* does any particular cognitive map take, what *freedoms* does it allow, what *limits* does it place on spatial knowledge?

Vector Maps

Let's start with the possibility that would probably occur first to the nonspecialist: a mental representation rather like a real map, in that it would routinely and obligatorily include Euclidian spatial relations between all remembered places. In cognitive psychology, to distinguish it from other, computationally simpler kinds of representations, this representation has been called a *vector map* (Byrne 1979). The need to distinguish among different kinds of representations has also been realized independently in insect ethology: Dyer (chap. 6, this volume) has suggested the term *metric map* for the same concept. Whatever it is called, the important point is that this kind of cognitive map provides very rich spatial information. If several places are known, and are out of sight of one another on the ground, then the distances and bearings of one place from the others are automatically available, just as if the places could be viewed from above. Thus, novel shortcuts can be taken: efficient new routes to known locations can be computed. This ability should be diagnostic of the presence of vector map representations. However, the fact that we ourselves often take shortcuts should not lead us to overestimate human reliance on vector map representation of large-scale space. In nontraditional environments, experience with physical maps is so widespread that our apparent vector map abilities may often be a consequence of map experience. But everyday experience shows that we can at least compute approximate directions of out-of-sight places on the basis of locomotor experience alone (an ability often called a "sense of direction"), reassuring us that humans should be capable of developing vector maps.

Equally, it has always been believed that other mammals (such as rats) can develop this sort of global orientation ability after gaining on-the-ground experience of a large environment. Provocatively, Benhamou (1996) has recently pointed out that in all cases, the evidence can be explained otherwise. For instance, it has been found that a rat in a water maze can learn to swim directly toward an invisible platform (the water is made opaque and milky, so a platform just below the surface is hidden until stepped on). This is so even when the rat is placed in the water at a novel point, not near the places where it was released during the learning trials (Morris 1981). The rat appears to possess vector map knowledge. However, the views from the new starting place and the platform do have many features in common. An alternative possibility is that the rat compares the (new) panorama it can see with the (remembered) panorama from the platform; then, by swimming around so that the mismatch is always reduced, it homes in on the platform. Benhamou redesigned the water maze to test this possibility. He found that when no single feature was visible from both points, rats failed to show any ability to navigate to the platform (Benhamou 1996). So it seems that even rats, for which Tolman originally introduced the term "cognitive map" (Tolman 1948), may lack vector map ability.[1] Evidently rats—and baboons—do have some knowledge of large-scale space, since their everyday lives depend on exploration and return to home bases. If not vector maps, what sort of representations might these be?

Network Maps
One possibility is a model developed to explain some anomalies in human behavior. When long-term residents of a city were asked specifically for the layout of road junctions they knew very well, they universally drew them as right angles, despite the fact that in reality the roads made either 60° or 120° angles (Byrne 1979). Not surprisingly, when asked to draw maps of the city, they frequently came to (and recognized) "impossibilities" in what they had drawn, since they drew almost all road junctions as 90°! A similar lack spatial knowledge was found when residents were asked to estimate distances between locations in town. Their estimates were almost entirely a function of the number of changes of direction along the route between locations and the complexity of visual information along the way: straight routes in the suburbs were estimated to be short; winding routes in the center were estimated to be long (Byrne 1979). Perhaps human vector map knowledge, like that of rats and bees, has been overestimated. Yet the people who acted as subjects in these experiments did not get lost—far from it. I proposed, therefore, that the underlying mental representation used for most everyday navigation is not isomorphic to a two-dimensional map, but

Richard W. Byrne

is based instead on propositional instructions for getting around (Byrne 1979, 1982). Imagine this representation as a network of many strings of *commands* that govern navigation, interconnecting at some common nodes. One such string might look like this:

AT (post office)	→	START & GO (right)
AT (library)	→	GO (left)
AT (four-way junction)	→	GO (straight on)
AT (police station)	→	GO (half right)
AT (John's home)	→	STOP

The commands contain only the minimal directional information needed to choose between paths (e.g., left, half right), explaining the dramatic tendency to recall junctions as right angles. They also lack distance information, explaining the tendency to estimate distance by the amount of remembered route information. Once several intersecting routes are learned, the "string" of commands becomes more like a net, with several routes leaving certain nodes by following different commands: hence the term *network map*. A network map preserves topological relations among locations, but lacks Euclidean or vector information. An organism navigating with this system would not get lost, but might sometimes fail to choose the route of least effort and would not be good at taking novel shortcuts to known locations. In the built environments of most readers of this chapter, these are unlikely to be serious disadvantages.

Although visualized here as if they were verbal instructions, the commands should be understood as propositions in a nonverbal code, with each set forming a **program**, which if *implemented* would lead to movement, but which can also be *scanned* for the implicit information contained about recognizable places and their approximate relative positions. That is, not only can we reliably find our way to work without a road map (running the program), but if we are asked for novel information—"How many traffic lights are there on your route to work?"—we can work it out (scanning the program). This typically gives us a phenomenal impression of a "movie" in the head, pictures of travel along the route. Each step in the program contains a *pattern* for the location at which a change of movement is required and a *procedure* to carry out there. In the jargon of artificial intelligence, each conditional command is a "production" and the whole string is a "production system" (Newell, Shaw, and Simon 1958; Newell and Simon 1972). Note that this representation has been used to describe primate competences in deception and food processing (Byrne 1995a), and so would require little modification to represent primate navigation in space.

Converging evidence comes from the study of animal orientation, suggesting a similar need to recognize a kind of mental representation that allows navigation but generally lacks most of the distances and angles that are an essential part of a vector map (Poucet 1993; Thinus-Blanc 1989). For instance, dogs take novel shortcuts between locations they have visited only from a single starting point, showing that they form and use vector maps; but their travel turns out to be nonoptimal, systematically biased (Chapuis and Varlet 1987). As in humans, the directions of their biases suggest that they are additionally using a representation that encodes only topological information, as in a network map. From functional studies of brain structures comes a theory that cognitive maps may consist of stored descriptions of places, together with encodings of the movements required to get from one to the other (McNaughton, Chen, and Markus 1991; McNaughton, Leonard, and Chen 1989)—a formalism similar to that of network maps.

Independent derivation of similar theories to explain widely differing data is strong evidence of the reality of the basic concept, but tends to make for a confusing plethora of names. For instance, in the social insect literature, the term "route map" has been used for a representation of locations and the bearings to succeeding locations along a route, attributed originally to Baerends (1941; cited in Dyer, chap. 6, this volume). Unfortunately, this term has been used slightly differently in cognitive psychology, which has proposed a mental representation of large-scale space falling short of full Euclidean knowledge and based on a set of linear strings (Shemyakin 1962; Siegel and White 1975). Lacking orientation information or any other way of interrelating separate strings, this representation would appear of little use for real navigation in a complex world, and indeed, was devised to explain children's failures of drawing rather than competent adult behavior. In child psychology, the acknowledged originator of the idea of two mental formalisms for encoding space is Piaget, who proposed that children begin with topological representations of space and develop Euclidean representations only at the age of 6–7 years (Piaget and Inhelder 1956). However, this distinction—between a representation that obligatorily encodes all spatial relationships among places, and one in which strings of locations are represented along with directions for reaching the next landmark along a route—has a much longer history. Medieval navigators did not in general possess complete maps of faraway seas, but used *rutters:* annotated lists of locations, described so that they could be recognized, with ship's bearings and other sailing instructions that enabled a course to be set for the next intermediate destination along a long route. While rutters lacked the general vector information that makes a real map isomorphic to the physical world, they were nevertheless highly prized and useful—and closely similar in form to network maps.

The Cognitive Maps of Primates

Although nonhuman primates live in environments without barriers or roads, and thus could benefit from vector map representations of knowledge, there is little clear evidence in published reports to differentiate such representations from the use of the computationally simpler network maps. Moreover, for a few species of primates, there are distinct suggestions that they treat their worlds as networks of paths. Hamadryas baboons tend to use familiar paths rather than novel shortcuts, although the multiply interconnecting nature of their pathways allows parties to wander independently of one another (Sigg and Stolba 1981). Milton (1981a, 1988) described howler monkeys moving "over what appear to be traditional arboreal pathways connecting important clusters of food trees," and added that "like howlers, spider monkeys often travel over the same arboreal pathways through the forest." A consequence of using regular pathways between landmarks is that the landmarks are always met from the same direction; most people have had disconcerting experience of reaching a familiar landmark from the "wrong way" and failing to recognize it. Similarly, Visalberghi (1986) described an infant gorilla that failed to recognize a scene from different orientations, evidently unable to compensate for any rotations of its body apart from a full 360°. Confronted with this rotational task, the young gorilla developed a regular route of searching to find the right position, following the rule "turn right and check each place in turn."

Given the lack of evidence for vector map knowledge of space in other mammals, and the finding even in humans of a dominance of network map (propositional) knowledge, the safest assumption is perhaps that all primates navigate predominantly by means of network maps. Would this mean that primates have no vector map ability? The difficulty in answering this question without experiments is that the predictions the two models make for everyday ranging are similar. For exploring a new environment, the two systems are radically different in what they would permit. With a vector map, travel from A to B, then B to C, then C to D (each out of sight of the others) allows travel directly from A to D on the next occasion—a novel shortcut; with a network map, subsequent travel to D could be done reliably only along the original path used, and direct return travel back to A would be largely guesswork. However, most primates range over large portions of their range every day, and are long-lived animals. Also, most nonhuman primates travel as cohesive social groups, and young individuals could thus safely learn a myriad of paths that the group traditionally uses. By adulthood, their network map knowledge would be richly interconnected—perhaps the famous route knowledge of London taxi drivers is the best analogy. Identification of examples of genuinely novel shortcuts is thus problematic. Thus, it may be difficult, if not impossible, to conclusively detect vector map knowledge of space in free-ranging primates. Choice of least-effort routes, while showing the possession of good cognitive maps of some kind, does not differentiate the two models either. The evidence for least-effort routes in primates come from orangutans' use of sparse fruit trees (MacKinnon 1978), chimpanzees' transport of hammer stones (Boesch and Boesch 1984), spider monkeys' fruit foraging (Milton 1981a), baboons' orientation to water and sleeping sites (above), and fruit and exudate foraging by tamarins (Garber 1989); all of these findings fall into this ambiguous class. Similarly, the experimentally induced travel of Japanese macaques to out-of-season fruit trees (Menzel 1997) does not demonstrate vector map ability. Garber's analysis of the highly seasonal nectar feeding of tamarins (*Saguinus mystax* and *S. fuscicollis*) provides the closest to definite evidence of vector map knowledge in nonhuman primates. The nectar sources are widely spaced, and feeding on them is intensive, bypassing food trees with lesser rewards; the routes taken by the tamarins are direct and straight, from one nectar source to the next (Garber 1988b). However, even here we cannot rule out the possibility that the tamarins are following *familiar* paths, used each year for this purpose alone, rather than computing novel routes on the basis of Euclidean spatial knowledge. Only for humans, to whom specific questions about entirely novel (and indeed impossible) routes can be posed, such as a "shortcut" through a blank wall, can we be sure that vector map knowledge is present. And even for humans, its use is probably a great deal less common or valuable than popularly imagined.

Conclusions

The null hypothesis of group movement without direction or coordination can be decisively rejected for baboons. Groups are relatively cohesive when they range, or if they split, they are able to rejoin each other later. Attention is paid to particular individuals who have a disproportionate share in deciding travel direction. In hamadryas and mountain baboons, these "decision-making" individuals are males, whereas in savanna baboons they are probably females. Specialized visual and vocal displays are used in influencing group movements. Ranging is not random, but shows signs of orientation toward out-of-sight goals.

The mental mechanisms that underwrite these behaviors in baboons are less clear. In most cases, the decision making is apparently in the hands of a single individual. In the few cases in which this individual has been identified reliably, it is an old—rather than a high-ranking—animal in hamadryas and mountain baboons, but the dominant female in geladas. Other individuals regularly initiate moves that may be accepted or rejected by the decision maker, like a process of question and answer, and these initiations seem likely to have influence on the final decision, but how much cannot be specified. There is also evidence that "weight of numbers" favoring a particular direction of movement has some influence on the final decision. However, there is no evidence for real negotiation, in which relevant information is offered to change others' minds, or other currencies traded for "agreeing" to one particular initiation.

When baboon troops fragment during ranging and become separated in dense vegetation, exchanges of loud barks facilitate their reunion. The calls are given in a manner that in humans would suggest that hearers understand that callers are lost, and reply in order to help them. This interpretation is unnecessarily complex, and a simpler picture is more likely, in which individuals understand only that giving calls tends to elicit calling from groups they wish to rejoin, and that if they reply to calls they hear, they will be joined. In a region of high predation pressure, when foraging necessarily has to be in small groups because food is scarce, giving calls allows efficient reunion—to the benefit of all. Suggesting instead that calls are emitted only on the basis of callers' motivational states does not account for the frequent silence of small, separated foraging groups, nor the tendency of females to "answer" barks of their close kin.

There is no evidence that the cognitive map of baboons is isomorphic to the world as viewed from above; more likely most travel is based on the use of interconnected route knowledge, a propositional representation called a network map. In some populations, baboons communicate about out-of-sight locations that are not immediate goals by mass initiation of movements in the direction of these locations, usually shady water holes in desert habitat. The direction of movement eventually chosen predicts which water hole will be visited some hours later, but not the paths of baboon groups in between. This seems to be the strongest evidence in wild nonhuman primates for an ability to conceive and communicate about events displaced in time and space, and surely deserves further investigation.

Acknowledgments

I would like to thank Sue Boinski and Paul Garber for inviting me to be part of the seminar on this topic at the IPS Congress, Madison 1996, and for constructive comments on an earlier manuscript. In addition, I appreciate the helpful criticisms of this chapter by Susan Alberts, Sue Boinski, Paul Garber, Charlie Menzel, David Watts, and two anonymous reviewers, but the opinions and errors within it remain my own responsibility.

Note

1. In rats, individual hippocampal neurons fire in response to stimulation at specific points in vector space, and lesions of this area destroy orientation ability selectively (O'Keefe and Nadel 1978). These "place cells" have generally been seen as part of a rat's cognitive (vector) map, and considerable reformulation would be needed to interpret them as components of a propositional network map. However, Poucet has recently argued that the hippocampus handles topological information about space, whereas metric information is dealt with parietally.

Richard W. Byrne

PART FIVE

Group Movement from a Wider Taxonomic Perspective

Nonhuman primates are shown in previous chapters to share many similarities in their coordination of group travel. We admit that our initial motivation for soliciting chapters for this section was that we had failed in our personal efforts to critically review the literature and, especially, the most current research on group movement in other animals. Although we had expected to be enlightened by the nonprimatology contributions here and elsewhere in the volume, these chapters amaze and delight us in their presentation of rich literatures and conceptual issues virtually untapped by primatologists. Moreover, these chapters incontrovertibly demonstrate that group movement is a selective regime of critical importance in taxa other than nonhuman primates. In terms of cost of locomotion, modes of communication, life history patterns, food competition, dependence on learned culture, and the apparent minimally required cognitive skills, however, fundamental differences are evident in the processes by which nonprimates, nonhuman primates, and humans achieve coordinated group travel. Comparative studies examining these differences in detail are somewhat hindered because data on group movement for nonprimates are not yet as abundant as for primates. Yet we alert readers to the fact that even now sufficient information is available to justify more ambitious analyses across major taxonomic groups.

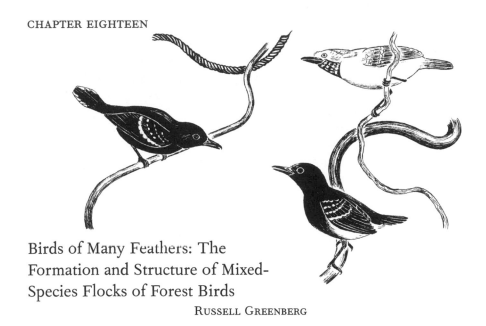

CHAPTER EIGHTEEN

Birds of Many Feathers: The Formation and Structure of Mixed-Species Flocks of Forest Birds

RUSSELL GREENBERG

Considering the differences in the social and ecological requirements of the different participants, the cohesive movement of animal groups is an amazing phenomenon. Associations composed of many species, each with its own intraspecific social organization and species-specific requirements, seem even more improbable. Yet mixed-species fish schools (Ehrlich and Ehrlich 1973; Alevizon 1976), ungulate herds (Fitzgibbon 1990), primate troops (Gartland and Struhsaker 1972; Terborgh 1983; Peres 1992a; Cords, chap. 4, this volume), and bird flocks are commonplace.

Nowhere else in the animal kingdom is the phenomenon of mixed-species associations more widespread than it is in birds. Mixed-species associations of birds are particularly well developed in forests, occurring from the most diverse tropical forest to the species-poor taiga. During the nonbreeding season in particular, a majority of bird species participate in mixed-species flocks. Mixed-species forest flocks contrast with flocks found in open habitats (blackbirds, shorebirds, finches, etc.) in their smaller and more consistent sizes, more regular membership, the presence of a consistent flock home range, and participation by a small number of individuals of each species (Terborgh 1990). Because of these attributes, forest flocks make a logical focus for a chapter on the regulation of group formation and movement.

I approach this topic as a researcher who published a small amount on mixed-species flocking a number of years ago; the phenomenon—which I encounter at all my study sites—still never fails to fascinate me. However, as exciting as it is to see a species-rich bird flock, it is surprising to note that, a spattering of good studies aside, the field of mixed-species flock studies has stagnated. The composition of many flock systems has been inventoried, the behavior and ecology of different species has been described for a few flock systems, and the costs and benefits accrued to individuals have been examined for a few focal species. However, the integration of these different levels of flock organization has rarely been achieved. Therefore, the goal of this chapter is in some small way to renew interest in flocking research by first reviewing our knowledge and assumptions about how mixed-species flocks work and then suggesting ways to build new integrated approaches to mixed flock behavior.

I begin by critiquing traditional approaches to defining species-specific roles in flock formation. I then discuss the rich literature on the potential costs and benefits of flock participation and how these relate to the proximate mechanisms of flock formation and movement. Finally, I sketch out a vision of future work on the mixed-species flock phenomenon that integrates considerations of proximate mechanisms of formation with those of ultimate causation.

An Encounter with a Mixed-Species Flock
Mixed-species flocks are incredibly varied in their composition and structure. However, given that there is no "typical" mixed flock system, I will give the reader a more tangible feel for what it is like to encounter and follow mixed flocks by describing an encounter with an understory flock on Barro Colorado Island, Panama—the site where I have conducted most of my mixed-species flock work.

The first impression most people have when walking though a lowland tropical forest is how few birds there seem to be. However, as you continue through the forest, the drone of cicadas is occasionally interrupted by the soft contact notes and the rustling of foliage made by a flock of birds in the dense canopy. At first it seems that

Russell Greenberg

only a few birds are darting between trees, but as you watch patiently, more and more slip silently ahead of you.

Although we usually notice bird flocks at their most conspicuous—moving rapidly through trees—antwren flocks spend much of the day slowly milling about areas of dense viny vegetation around old treefall gaps. At the core of the flock are three species of small insect-eating birds known as antwrens—so named because someone thought they looked like wrens and they are related to birds that follow army ants. In fact, they are ecological equivalents of temperate zone warblers. Checker-throated antwrens (*Myrmotherula fulviventris*) usually occur in pairs and quietly chip as they spend their lives probing into dead, curled leaves that hang in the understory vegetation. In family groups, white-flanked antwrens (*Myrmotherula axillaris*) move rapidly through the low- to mid-forest strata, giving an oddly musical "whit" note while acrobatically gleaning arthropods from live foliage. Dot-winged antwren (*Microrhopias quixensis*) pairs or families forage through the densest vegetation, keeping up a constant patter of loud contact notes. For hours, the pairs and family groups of the three antwren species and a few other birds—perhaps a hulking slaty antshrike (*Thamnophilus punctatus*) or a diminutive olive flycatcher, the southern bentbill (*Onocostoma olivaceum*)—perch quietly at the edge of a black hole of tropical vegetation.

Then, seemingly inexplicably, the antwrens increase their chipping and pewing and begin to move rapidly toward the sound of a neighboring flock. During the rapid advance to the territorial interaction, more species join the flock—a tiny, ochraceous ruddy-tailed flycatcher (*Terenotriccus erythrurus*), a green and yellow black-throated trogon (*Trogon rufus*) in the canopy, perhaps a few spotted antbirds (*Hylophylax naivioides*) near the forest floor. Occasionally you catch a glimpse of a robin-sized woodcreeper as it flies from trunk to trunk, disappearing around the back of a large tree. Struggling to keep up, you find the antwrens in an intense interaction with another antwren flock. Two checker-throated antwren males are perched side by side chipping, bodies waving back and forth, with throat feathers expanded so that the checkering looks more like a large black patch. Occasionally the two chase each other while producing loud "keek" notes. Dot-winged antwrens puff up and fan their wings and tails while singing (sweet notes ending in a musical trill), and chase through the dense foliage—males facing off with males, and females with females, while the young birds watch. Out of corner of your eye you can see the brown flash of woodcreepers and furnariids chasing as well.

The flock drifts away from the border interaction as the last dot-winged antwren continues its raspy agonistic vocalizations. Moving back to the dense vine tangle at the core of their territory, the ant-wrens enter the thickest vegetation, where they and a few other species fluff up to spend the night. If you return to the roost shortly before dawn, you will hear the antwrens and some of the other species singing—for many species, it is the only time of the day (apart from border interactions) they utter a song. As light begins to filter into the somber understory, the antwrens and their attendant species start to mill slowly about the vine tangle, searching for their arthropod prey.

Feeding Aggregations versus Cohesive Flocks
Mixed-species flocks consist of many species moving through the forest together in search of dispersed food. However, although multispecies associations are common among forest birds, many are actually feeding aggregations, brought together by a specific resource, rather than the presence of other birds (flocks). Two examples of feeding aggregations that have been studied in detail are birds visiting fruiting and flowering trees (Leck 1971) and those following swarms of army ants in tropical forests (Willis 1972; Willis and Oniki 1978). In the latter case, birds jockey for position in front of the swarm, where arthropods are flushed by waves of ants scouring the leaf litter and lower stems of understory saplings. That the ant-following aggregation is not a cohesive social grouping of birds does not signify an absence of social interactions. The long-term research of Willis (1972) has exposed a complex interspecific dominance hierarchy, in which the more specialized "professional" ant-following birds (usually members of the true antbird family, Formicariidae) dominate the foraging zone directly in front of the swarm and the many subordinate species are found farther above or to the sides of the swarm. In addition, some species—particularly antbirds—are quite noisy, providing a cue for a variety of bird species to locate swarms (Willis 1972; Willis and Oniki 1978).

The hallmarks of a true mixed-species flock are that the birds travel together and that the benefits accruing to an individual joining the association are derived primarily from the behavior of other

Russell Greenberg

birds rather than from the presence of a food source. Furthermore, individuals of different species in mixed flocks usually use different food resources (Diamond and Terborgh 1967). However, the distinction between aggregation and flock is often not a clean one: birds may readily switch between moving in flocks and feeding in aggregations as the distribution of food changes from scarce and dispersed to abundant and patchy. The frugivorous birds that occur in aggregations at fruiting trees (Leck 1971) may also occur in highly organized canopy foraging flocks that move between fruiting trees (Moynihan 1962; Munn 1985). Experiments with food supplementation show that in temperate zone forests, mixed flocks of chickadees and other species break up and form local aggregations with more aggressive interactions (Bock 1969; Berner and Grubb 1985). Although aggregations brought together by a particular resource do not have complex mechanisms for maintaining group cohesion, aggregations still comprise complex dominance interactions, and animals potentially communicate information regarding the approach or presence of predators.

Flock Composition
The species richness and composition of mixed flocks can be characterized in two ways: by the total number of species regularly participating in mixed-species flocks in a given locality, or by the mean number of species participating in a particular flock at one time (table 18.1). Tropical localities, particularly in humid and lowland areas, may boast as many as 50–100 participating species versus 10–15 for temperate flocks (table 18.1; Odum 1942; Davis 1946; Gibb 1954; McClure 1967; Morse 1970; Buskirk et al. 1972; Munn and Terborgh 1979; Gradwohl and Greenberg 1980). However, individual flocks in the Tropics may be on average only marginally more diverse than those found at temperate zone sites; tropical flocks vary and change often as less common species join and drop out. In addition, most tropical forests have two or more basic flock types; for example, Willis (1972), Greenberg (1984), and Munn (1985) describe the existence of canopy and understory associations in lowland Neotropical forests. A third association—one of honeycreepers and small tanagers—moves between flowering and fruiting trees of the outer canopy. Often the first two flock types come together, and occasionally these are joined by the third, forming "mega-flocks" of remarkable diversity (Greenberg 1984; Munn 1985). A similar array

Table 18.1 Characteristics of some mixed-species flocks

Flock type and location	Flock size	Total species	Flock core[a]	Some core species	Reference
European tit	15	9	N	Tits, goldcrest	Morse 1978
North American tit	13–23	15	2N	Titmouse, chickadee	Morse 1970
Panamanian antwren flock	7	34	MST	2–3 antwrens	Gradwohl and Greenberg 1980
Amazonian understory	30–35	48	MST	Antshrikes, antwrens, furnuriids	Munn and Terborgh 1979
Amazonian canopy	?	53	MST	Lanio, Tachyphonus tanagers, greenlet	Munn 1985
Highland Costa Rica understory	8	43	N	Basileuterus warbler	Bell 1983
New Guinea understory					Bell 1983
Peninsular Malaysia	35	109	?		McClure 1967
Borneo	12	62	N	Babblers, drongos, flycatchers	Laman 1992
Australia	10	59	N	Thornbill, fairy wren	Bell 1980
African savanna	10–20	56	N	Tit, Eremomela warbler	Greig-Smith, 1978
Kashmir	?	15	N	Phylloscopus warblers, tits	McDonald and Henderson 1972

[a]N = nuclear species; 2N = two or more nuclear species; MST = several species with same territory.

of different flock types has been described for the forests of New Guinea (Bell 1983; Diamond 1987). On the other hand, McClure (1967) found that mixed flocks in Malaysia consist of understory, mid-level, and canopy species that move together consistently.

The average number of individuals participating in mixed flocks is fairly consistent throughout the world, usually ranging from ten to thirty (Davis 1946; Morse 1970; Buskirk et al. 1972; Munn and Terborgh 1979; Powell 1979). Although larger flocks have been reported (McClure 1967; Bell 1983), most forest flocks are moderate in size, suggesting an upper limit determined by the ability to maintain group cohesion and consistent movement (Buskirk 1976). In addition, some observers (Morse 1970) have found that flock speed greatly increases with flock size, further suggesting that resource disturbance or depletion may be operating.

A high proportion of forest species join mixed-species flocks at least occasionally. This proportion varies: Hutto (1994) found that canopy birds participated in mixed-species flocks in proportion to their overall abundances in a tropical dry forest habitat. Greig-Smith (1978) found that a majority of species did not join the savanna flocks he studied. In moist tropical forests, species that are not found in flocks are often stationary sentinel foragers, have exceedingly small home ranges, or consume large fruits or other resources not used by mixed flock species (Buskirk 1976).

Behavioral Roles in Flock Formation

Although considerable effort has gone into classifying species according to the "role" they play in mixed flock formation, most of this work has been conducted without marked individuals or attempts to follow the activities of flocks for long periods of time (but see Buskirk et al. 1972; Munn and Terborgh 1979; Gradwohl and Greenberg 1980; Munn 1985; Powell 1985). Five distinct concepts have been incorporated into different classification schemes: (1) intraspecific gregariousness of the species; (2) "attractiveness" to other species (Moynihan 1962; McClure 1967); (3) whether the species follows or leads other species (Moynihan 1962; Morse 1970; Buskirk et al. 1972; Powell 1979; Diamond 1987); (4) the frequency of the species in flocks (regular versus accidental: Davis 1946; Buskirk et al. 1972; Bell 1983); and (5) the percentage of time a species occurs in mixed-species flocks (Morse 1970; Willis 1972; Powell 1979). The terminology commonly used to describe mixed flock behavior can be found in table 18.2.

Table 18.2 Common terminology used in mixed-species flock research

Term	Definition
Aggregation	A group of birds at a resource patch that does not travel together between patches
Alarm call	A conspicuous call given in the presence of a predator (usually a flying predator)
Attendant species	A species that joins and follows flocks
Beat	A predictable daily movement pattern
Contact calls	Soft call notes frequently given without agonistic or defense context
Core species	The set of species found in most individual flocks in a particular habitat or habitat stratum
Flock	A group of birds that move together while foraging for dispersed food
Flock territory	Territory of the nuclear or core species
Leader species	A species that coalesces the flock or is followed by other species
Multispecific territory	The territory of a flock made up of more than one species sharing the same boundaries; generally does not involve cooperative defense
Nonflocking species	Species that seldom join flocks
Nuclear species	Intraspecifically gregarious species joined by solitary or pair-forming species
Sentinel species	Species that produce conspicuous and early alarm vocalizations upon detecting a predator
Social mimicry	Hypothesized interspecific convergence of social signals to promote gregariousness

Nuclear Species

Because of the often superficial nature of mixed-species flock research (i.e., no marked birds or long-term following of flocks), the most commonly used term to describe the species around which flocks form—nuclear species—has permeated the literature without a rigorous definition. The nuclear label is usually applied to species that are in most of the flocks at a particular location and forest stratum, show evidence of leading the flock, and are intraspecifically gregarious and noisy. Most researchers believe there to be considerable intercorrelation between these attributes. For example, intraspecifically gregarious species, particularly those whose own flocks comprise family groups, often are more vocal than species that occur solitarily, with well-developed mobbing behavior and predator alarm notes (Mönkkönen, Forsman, and Hella 1996). This vocal behavior may make them attractive to other species.

Examples of nuclear species include chickadees and tits, various species of babblers and thornbills, bulbuls, *Chlorospingus, Tachyphonus,* and *Tangara* tanagers, *Campylorhynchus* wrens, antwrens,

Basileuterus warblers, and *Hylophilus* greenlets. All of these form single-species flocks that attract a following of other, more solitary species. The demographic composition of these single-species flocks is quite diverse. Some (antwrens, wrens, greenlets, African tits, and American titmice) occur in nuclear or extended family groups. Temperate zone tits (Ekman 1979; S. M. Smith 1991) and *Chlorospingus* tanagers (R. Greenberg, pers. obs.) often form non-breeding groups of unrelated individuals or pairs that separate during the breeding season. In some areas, flocks of presumably unrelated migratory wood warblers appear to form a nucleus that attracts other species (Greenberg 1984; Hutto 1994). The diversity of social organization within the flock nucleus reminds us that perhaps, under many circumstances, mixed flocks form opportunistically around any gathering of birds that reaches critical mass and momentum. Discussions of possible adaptations that promote interspecific gregariousness should always entertain this, more mundande, hypothesis.

Moynihan (1962) defined nuclear species as those whose members lead and seldom follow other birds. Morse (1970) developed matrices based on the relative proportion of individuals and species led or followed. The relative frequency of leading and following has not been quantified for birds in most flock systems, but is accepted on an intuitive level as the fundamental attribute for defining flock roles. One senses the frustration of determining flock leadership— particularly in large, diverse tropical associations—in a comment by McClure (1967, 135): "at times babblers would lead or they might be overrun by the flycatchers, cuckoo-shrikes and minivets above them, and at other times pseudopodes of species stream ahead at different levels." Observations of flocks crossing habitat gaps may facilitate quantification of the leadership role of flock members (T. Grubb, pers. comm.). Such observations would lead to more representative conclusions in areas where dominant flock members freely cross gaps than in tropical flocks, where only a select subset of species do so (Poulson 1994).

Most workers have restricted the classification of behavioral roles to nuclear and attendant species—that is, leaders and followers. However, Moynihan's original system was a bit more complex, including a second tier of (active) nuclear species that followed the primary (passive) nuclear species but were sufficiently gregarious and noisy to attract other, more solitary species. Regardless of whether this more complicated classification system is warranted, it is common to find mixed flock systems with two or more species that are found in most of the flocks and are themselves intraspecifically gregarious. Examples include *Basileuterus*-led flocks in Central America (Buskirk et al. 1972; Powell 1985), multispecies tit flocks in temperate Eurasia and North America (Morse 1970; Ekman 1979; Pravosudov 1987), antwren flocks of the Central American lowland tropical forest understory (Willis 1972; Gradwohl and Greenberg 1980), and *Hylophilus decurtatus/Tachyphonus luctuosus* flocks of the lowland forest canopy (Willis 1972). Finally, some tropical flocks have a nucleus that consists of individuals of a number of species that are not themselves intraspecifically gregarious (Munn and Terborgh 1979; Powell 1989).

Home Ranges of Nuclear and Attendant Species
A few studies of Neotropical flocks (Buskirk et al. 1972; Munn and Terborgh 1979; Powell 1979; Gradwohl and Greenberg 1980) have examined flock participation based on long-term following of color-marked individuals. These studies have, by and large, concluded that intraspecific social interactions are paramount in determining the participation of individuals in mixed flocks. The movement pattern and range of a particular flock system is defined by those of the nuclear species, whose space use is ultimately determined by intraspecific interactions. The joining by attendant species is highly constrained by their intraspecific spacing behavior as well. Working in the understory of the Central American highland forests, Buskirk et al. (1972) and Powell (1979) found that most attendants had smaller territories than the nuclear species and were found in flocks when the nucleus spent a significant amount of time within the smaller attendant territories. Gradwohl and Greenberg (1980) reported that in the lowlands of Panama, attendant species joining a core of two to three antwren species had both smaller territories (e.g., slaty antshrike) and larger home ranges (wedge-billed woodcreeper, *Glyphorhynchus spirurus*) or territories (buff-throated woodcreeper, *Xiphorhynchus guttatus*) than the antwrens. Birds with small territories joined flocks when the nuclear species passed through their territory. Those with larger ranges were seen in mixed-species flocks more regularly, since they moved through the forest by hopping from one flock to the next when the nuclear species interacted at territorial boundaries. Territoriality often regulates the numbers of a particular species in a mixed-species flock: either one, a pair, or a family group depending upon whether indi-

Russell Greenberg

viduals or pairs maintain territories in the species. In addition to the above observations of territoriality in tropical attendant species, Sullivan (1984) reported that downy woodpeckers *(Picoides pubescens)*, a common associate of North American tit flocks, defended individual territories against conspecifics. Even in nonterritorial species, such as the wedge-billed woodcreeper, intraspecific intolerance restricts the number of individuals in a flock to one (Gradwohl and Greenberg 1980).

The difference between the highland Central American *Basileuterus* and lowland antwren flocks may relate to the specific relationship between the home range size of the nuclear species and that of other common forest species. Apparently most highland forest species have small territories, whereas lowland species often have large ranges, and the antwren territories are relatively small themselves. We will not be able to generalize any patterns until detailed mapping of territories is completed for many more flock systems.

Seasonality of Flock Formation
Because mixed flocks form out of a complex interaction of the home ranges of attendant and nuclear species, the size and range of flocks varies considerably with seasonality of movement of the nuclear species. Temperate zone flocks are primarily a winter phenomenon. Flocks break up entirely during the breeding season and late summer, and fall flocks are unstable in composition, subjected to postbreeding influxes of migratory birds (Morse 1970). In contrast to winter flocks, where overt aggression is rare and dominance interactions take the form of subtle supplantations, interspecific chasing is common in postbreeding flocks. Tropical flocks persist throughout the year (Davis 1946; Moynihan 1962; Willis 1972; Gradwohl and Greenberg 1980). Powell (1979) found that highland flocks almost completely disappeared during the breeding season of the nuclear species *(Basileuterus)*. However, Munn and Terborgh (1979) and Gradwohl and Greenberg (1980) found that even when breeding, members of the nuclear species often remained with the flock. Furthermore, the long and asynchronous breeding season of many tropical birds means that, particularly in flock systems with more than one nuclear species, there will usually be species between nesting efforts that can form the core of a flock. However, the flocks are generally smaller during the breeding season. For example, the home ranges of attendant species are often smaller, so opportunities to join flocks are reduced (R. Greenberg, pers. obs.). More criti-

cally, the velocity and range of movement of the nuclear species can change profoundly between seasons. J. Gradwohl and R. Greenberg (unpub.) followed antwren flocks during the nonbreeding and breeding seasons in 1985 and found that the nuclear species (antwrens) moved considerably more slowly and covered a much smaller area during the breeding season—even when the pair was between nesting efforts. In the nonbreeding (dry) season, antwrens covered more area at a much greater velocity, increasing their range by trespassing on neighboring territories. The reduced movement and home range size during the breeding season probably can be ascribed to the increase in food abundance, which allows nutritional requirements to be met in a smaller area. The behavioral adjustments birds make when they join flocks are summarized in table 18.3.

Vocalizations Promoting Flock Formation and Cohesion
Although McClure (1967) and Gradwohl and Greenberg (1980) found that the core species of flocks roost together, some flocks apparently spend the night apart and come together during the morning. T. Grubb (pers. comm.) and Munn (1985) both reported that a particular species (titmice and antshrikes, respectively) produces loud calls in the morning that attract other species.

Once the flock has coalesced, the activity of the nuclear species appears to promote further flock cohesiveness and momentum. Although this has yet to be rigorously tested, vocalizations appear to play a central role. Many of the gregarious nuclear species are "noisy" as they move through the forest; a familiar example is the complex vocal repertoire of flocking tits *(Parus)* (S. M. Smith 1991). In a tropical example, the dot-winged antwren—one of the nuclear species of Central American antwren flocks—gives short "pew" notes almost continuously, which are punctuated by longer "pachuka" notes when it moves between trees. It is these notes that allow ornithologists, and presumably birds, to locate flocks as they move slowly through dense vegetation. Loud notes of the *Thamnomanes* antshrikes promote flock coalescence and movement in understory flocks in Amazonian forests (Munn and Terborgh 1979; Wiley 1980), as do those of tufted titmice *(Parus bicolor)* in temperate tit flocks (T. Grubb, pers. comm.). Bell (1983) and Diamond (1987) made reference to a leader call in babblers of New Guinea, and McDonald and Henderson (1972) reported that the continuous notes of the western crown-warbler *(Phylloscopus occipitalis)* of

Table 18.3 Possible behavioral adjustments associated with mixed-species flock formation

Behavior	Example	Intraspecific function	Interspecific function
Frequent contact calls	Antwrens, tits	Facilitates group movement or spacing	Used by attendants to track flock
Loud and early alarm call	Tits, *Thamnomanes* antshrike	Reduces risk of predator attack	Used by attendants; perhaps allows reduced vigilance and greater time for foraging
Territorial adjustment	Antwrens, core of Amazon flocks, some tit flocks	Territory determined by flocking behavior and may adjust territory size from optimal with respect to foraging or density-dependent nest predation	Ensures consistent association with other flock species
Heterospecific following	Many attendant species		Facilitates the maintenance of association with mixed flocks
Adjustment of forgaging velocity	Unknown	Speed of movement adjusted, perhaps away from optimal as determined by prey depletion or required search time	"

Table 18.3 (continued)

Behavior	Example	Intraspecific function	Interspecific function
Low interspecific aggression	Tropical understory flocks		"
Reliance on territorial boundary interactions	Tropical forest flocks	Allows individuals to assess status of consexual conspecifics in neighboring territory (flock switching is the dominant form of dispersal)	Temporally restricts the use of song; reduces conspicuousness
Irregular flock movement	Tit flocks, antwren flocks, highland tropical flocks	Reduces the predictability of flock location in time and space	
Regular flock beats	Possibly Malaysian flocks	Minimizes temporal overexploitation of patches	
Foraging site divergence	Temperate zone flocks	?	Reduces probability of being attacked by dominant; may increase predation risk for subordinates; reduces probability of foraging at already visited site; allows dominant to monopolize best site
Foraging site convergence	Some temperate flocks Highland Costa Rica	?	May result from copying or to facilitate following

Kashmir attracted attendant species. In my experience, the contact notes given by individuals of gregarious nuclear species are similar whether they are in a mixed or a single-species flock (see also Greig-Smith 1978). However, I am unaware of a detailed and quantified study of the call repertoire of a species in and out of a mixed-species association. In contrast to calls that promote gregariousness, Powell (1979) suggested that certain vocalizations (song and sibilant flight call) of the three-striped warbler *(Basileuterus tristriatus)* discourage attendants from joining because they indicate likely movement out of the attendant's small territory.

If all this discussion of the role of vocalizations in the formation of mixed-species flocks seems anecdotal, it is because it *is* largely anecdotal. The function of particular vocalizations, or of the vocalizations of particular species, could be tested with playback experiments or vocal muting (McDonald 1989) experiments. The one experimental study of the effect of a particular vocalization was on the formation of loose mixed feeding groups during the breeding season in Finland, using the territorial song of the willow tit (*Parus montanus:* Mönkkönen, Forsman, and Hella 1996). The researchers studied foraging aggregations among different species of boreal forest birds during the breeding season. Species were differentially attracted to the song of the willow tit—a species that they demonstrated was joined disproportionately by other boreal forest insectivores. The degree of distinction made by the joining birds is unclear, since the experimental comparisons were made with the song of a North American tyrant flycatcher and a piece of classical music. However, the results of the experiment do not support the idea that special vocalizations are required to promote the formation of tit flocks.

Do Plumage Patterns Promote Interspecific Gregariousness?
A number of authors have pointed to plumage characteristics that might promote interspecific gregariousness. Moynihan (1962) argued that dull coloration may reduce the visual conspicuousness of vocally conspicuous species while promoting interspecific gregariousness. His idea, which has received no rigorous experimental testing, is that dull coloration promotes the approach of other species by reducing hostile reactions that might be released in the presence of bright colors. Moynihan illustrated the potential importance of dull coloration with the example of the red-legged honeycreeper *(Cyanerpes cyaneus),* the only species in the honeycreeper complex

in which males molt into a dull nonbreeding plumage. (This is, in fact, quite unusual for nonmigratory tropical birds.) This bird is quite gregarious intraspecifically and is an important member of canopy tanager-honeycreeper flocks. Although his findings were not quantified, Moynihan argued that this species is considerably more vocal than other honeycreepers. The argument that dull coloration promotes gregariousness was developed in an analysis of plumage changes in Neotropical migratory songbirds by Hamilton and Barth (1962) as well. These hypotheses are intriguing, but are very difficult to test. Looking over the sample list of nuclear species provided above (see table 18.1), which range in color from black and white to olive and gray or brown to yellow, it is hard to see a strong common thread. It would be interesting to see whether the proportion of brighter colors (for example, bright reds) among these species is statistically different from the relative abundance of other dominant colors in forest avifaunas.

Gregarious species also often have flash patterns in their plumage that are conspicuous when the bird flies. This can be easily seen in the white outer tail feathers of a number of flocking pipits and wagtails (Motacillids) and emberizid finches, mainly open-country species; such flashes are uncommon among leaders of forest bird flocks. Often these plumage features are inconspicuous until piloerection or flight exposes the feature. Examples of such plumage characteristics are the white flanks of the white-flanked antwren, the white shoulder patches of the otherwise black white-shouldered tanager *(Tachyphonus luctuosus)*, the blue shoulder patches (greater wing coverts) of the plain-colored tanager *(Tangara inornata)*, and the wingbars of the white-winged tit *(Parus leucomelas)* found in African savanna flocks (Greig-Smith 1978). Although some authors (Moynihan 1962; Wiley 1971) have argued that flash patterns facilitate mixed flock cohesion, the association between such patterns and mixed flocking is weak. These patterns are particularly important in maintaining flock formation in synchronized flights, but such displays are not characteristic of forest flocks. Many of the nuclear species of forest flocks lack any such pattern, and for those that have them, these patterns may function in a number of contexts, such as foraging and intraspecific agonistic interactions (common in antbirds, for example).

Signal Convergence in Mixed-Species Flocks
Members of mixed-species flocks might show convergence of signals to promote group cohesion (Moynihan 1962, 1968). Based on

what he posited is the economy of signals, Moynihan argued that species should converge on similar plumage patterns to promote positive associations. Although it is reasonable that a bird would have an easier time following a limited number of color patterns, the mechanism for individual selection of a convergent color pattern is more problematic. One could posit that a bird that sports a coloration more similar to that of a species it regularly follows is allowed to approach more closely than one that has a strikingly divergent pattern. Although no evidence from forest bird flocks is available to evaluate this idea, Caldwell (1981) presented some intriguing experiments on mixed flocks of foraging herons. When she presented lifelike models, she found that snowy egrets *(Egretta thula)* attracted the most individuals and species of foraging herons. Herons foraging near snowy egrets improve their foraging success, perhaps because the snowys stir up the substrate and flush fish. Immature little blue herons *(Hydranassa caerulea)* are white and are allowed to approach snowy egrets, resulting in increased foraging success. However, adult little blues are dark-colored, are attacked by snowy egrets, and as a result, are more successful when solitary than when foraging in association with egrets. All this occurs without a measurable decrease in the success of the snowy egrets.

Moynihan (1968) cited several examples of flocks that consist of apparently conspicuously and similarly colored birds. In the canopy of lowland forests in Panama, flock members are generally blue, green, or blue-gray; in the second-growth forests of the highlands, flocks consist of black and yellow species. The social mimicry hypothesis rests on the assumption that shared plumage patterns are disproportionately more similar than one would expect of random assemblages of species from the same habitat—an analysis that has not been conducted for any flock system. With the exception of plumage patches that are covered by erectable feathers, avian coloration is not readily changeable and must function in many ways simultaneously. The social mimicry hypothesis would be most convincing if color patterns could not be demonstrated to have some other function, such as crypticity, with more obvious adaptive value. For example, the blue and green colors of canopy honeycreepers might appear to constitute convergence on a bright color. But these colors may actually be difficult to detect in a tropical forest canopy (Endler 1985). For this reason, it has proved difficult to rigorously distinguish between the social mimicry hypothesis and other reasonable explanations.

In another example of an alternative explanation for convergent plumages, Greig-Smith (1978) argued that the core species in African savanna flocks often share predominantly black and white patterns, possibly because these species are unpalatable. Hypotheses can also be advanced for the convergence on black and brown patterns described for understory flocks in New Guinea (Diamond 1987). The nuclear species (babblers) in these flocks are predominantly black or brown, or both, and the followers (for example, drongos) are black. This, Diamond argued, was a striking example of apparent social mimicry, and he reiterated the arguments of Moynihan (1968). Of course, brown and black could be cryptic coloration for tropical understory vegetation. Furthermore, at least two leader species of the genus *Piohui* were more recently found to be poisonous (Dumbacher et al. 1992) and involved in an apparent Müllerian mimicry complex. This finding leaves open the possibility that the shared brown and black patterns that Diamond described are actually part of a Batesian mimicry complex.

The likelihood of convergence in vocal signals is also unclear. The economy of signals idea would best apply to situations in which a mistake that causes a slow or inappropriate response results in a large cost to the responding individuals. Therefore, calls associated with predator detection and avoidance are an obvious place to look. Alarm notes to flying predators and mobbing notes to stationary predators often share acoustic properties among mixed flock members. However, as Marler (1957) argued long ago, there are functional explanations for similarity in vocal responses to predators—an argument that, with some modifications, stands today. Alarm notes to aerial predators are often short in duration and given at a high and narrow frequency band. These sharp notes usually cause other flock members to freeze. Presumably the bird giving the alarm gains some advantage by reducing the conspicuousness of the entire flock, and for some species, by warning relatives. Mobbing notes are generally longer in duration, with a broader and lower frequency band. Alarm and mobbing calls are selected to provide the minimum and maximum amount of locational information, respectively, to a potential predator, which accounts for much of the interspecific similarity in their respective structures.

It is less clear why birds cannot easily respond to a variety of "contact" notes—calls given by foraging birds not necessarily in the presence of a predator or in an overtly aggressive context. One could argue that giving a convergent call note might facilitate closer following by heterospecifics. However, it is likely that such contact notes function primarily as intraspecific signals within a flock—to promote either following or avoidance. Senders of so-called contact notes may also be advertising to conspecifics that a flock is already occupied. If intraspecific interactions dominate, then one would expect the contact notes of different species to diverge to minimize confusion or misidentification on the part of the intended conspecific receivers. As far as I know, the divergence or convergence of contact notes within the context of mixed-species flocks has not been systematically studied. However, the likely reason for the lack of such studies is that the contact notes of flock members are readily distinguished by human observers. Begging notes of juveniles also show great species-typical variation, at least among members of antwren flocks (R. Greenberg, pers. obs.). It could be a costly waste of time (and in the presence of kleptoparasites, time may be of the essence) for adults carrying prey to young in a mixed flock not to be able to readily and rapidly determine the location of their own young.

With regard to intraspecific convergence, there is evidence (Nowicki 1983, 1989) that the "chickadee-dee" calls used for mobbing and territorial interactions by members of chickadee flocks tend to converge rapidly among flock members and diverge from those of other flocks. Similar copying has been found in intraspecifically gregarious finches (Mundinger 1970) and budgies (Farrabaugh, Linzenbold, and Dooling 1994). These convergent calls in budgies and finches are thought to synchronize group movement—and an individual may benefit by not looking or sounding different from other flock members as the group flies in unison. However, in the case of chickadees, convergence in an aggressive vocalization may reduce mistaken intragroup attacks in intergroup agonistic interactions.

Do Species Evolve in Response to Flock Role?
Does the mixed-species social environment shape specific adaptations, such as visual or vocal signals? Or do mixed-species flocks develop idiosyncratically around any group of birds that provides critical mass and momentum, so that no long-term alteration of communication systems is necessary or advantageous? Moynihan (1968), Cody (1971), and Diamond (1987) have all proposed that some species have evolved signals and behavior patterns specifically adapted to promote interspecific gregariousness. Moynihan's argument is somewhat contradictory on this point. On the one hand, he

posits that certain features of nuclear species are general adaptations, shared by many nuclear species, to facilitate other species joining them and hence to promote the formation of mixed-species flocks. On the other hand, he suggests that all species commonly occurring together in a particular flock type converge upon special patterns that distinguish that flock type. The most parsimonious mechanism of flock formation is that the various attendant species evolve plumage, vocalization, or behavior patterns that allow them to approach and follow the nuclear species. Both of these suggestions place paramount importance on interspecific interactions in shaping signals. Alternatively, the adaptations of nuclear species that promote gregariousness may have primarily been shaped by intraspecific interactions.

The assumptions of these two views are different and rest strongly upon the degree of symmetry in the advantages gained by nuclear and attendant species through mixed-species flock formation. The evolution of features in individuals of nuclear species that somehow encourage other species to follow them presupposes that the individuals of nuclear species enjoy some net benefits from the presence of the attendant species. It is easier to assume that the species that actively join or follow the nuclear species derive benefits from this association. The social mimicry hypothesis is reasonable if the signals of the attendant species converge to facilitate their following response, even if—or particularly if—the nuclear species does not benefit from the association.

Dominance and Flock Role

Although it is probably safe to assume that the attendant species gain a net advantage from associating with the nuclear species, the converse assumption is more problematic. The relative dominance of flock participants and the importance of dominance-mediated access to resources are critical to understanding the costs and benefits accruing to the nuclear species. There has been at least one experimental confirmation that members of a small and subordinate nuclear species (Carolina chickadee, *Parus carolinensis*) do in fact experience a cost in association with a larger, dominant species (tufted titmouse: Crimprich and Grubb 1994). The cost was measured by removing titmice from isolated woodlots and measuring the regrowth of chickadees' plucked outer tail feathers. Although the experiment was conducted on only one associated species, the results at least suggest the possibility that sometimes nuclear species must merely tolerate the presence of other species, rather than

Russell Greenberg

actively attracting them. Although the experiment raises this possibility, the results were more narrow. The experiment tested the effect on chickadees of the presence of titmice in the same habitat patch, not the cost of actually flocking together. In addition, the results of an analysis of possible benefits (vigilance) were ambiguous.

Mixed flocks are often characterized by few overt agonistic interactions (Morse 1970; Munn and Terborgh 1979; Powell 1979) except intraspecific territorial skirmishes. The few studies that have examined the dominance of nuclear versus attendant species paint a diverse picture for different flocks. For example, the *Thamnomanes* antshrikes and *Lanio* shrike-tanagers are thought to be critical in the coalescence of the complex Amazonian flocks and are strongly dominant to other species. In North American winter flocks, the two tit species are subordinate to some common attendants (woodpeckers and nuthatches: Waite and Grubb 1988; but see Sullivan 1984 for a report of black-capped chickadees [*Parus atricapillus*] kleptoparasitizing woodpeckers) and dominant to kinglets. Suhonen (1993) and Alatalo (1981) found tits to dominate goldcrests and treecreepers, causing the latter to forage farther out on limbs, thereby increasing their predation risk. Since nuclear species vary greatly in size, and size is known to be an important determinant of dominance, I see no reason to believe that they are consistently dominant or subordinate to their common attendants.

Possible Foraging Benefits of Participation in Mixed-Species Flocks
When considering hypotheses about the benefits and costs of mixed flocking, it is critical to remember that the presence of species-typical foraging specialization is one of the main ways in which these flocks differ from single-species flocks. Mixed-species flocks, particularly in the Tropics, often contain resource specialists that hunt insects in particular strata or microsites, such as dead leaf curls, bark, dead or live twigs, or tree boles, or attack prey in specialized ways (long-distance strikes, hanging, reaching, etc.).

Foraging niche generally correlates with flock role. Nuclear species are often foraging microhabitat generalists. Tits provide a good example of this; they use a diversity of foraging maneuvers to extract prey from a wide variety of sites. Attendant species use specialized foraging techniques (such as long-distance aerial strikes) or specific foraging sites. In tropical flocks, the intraspecifically gregarious nuclear species are most often simple gleaners of live foliage—an abundant microhabitat, but one that supports relatively low prey abundances. Species that either capture arthropods with long-distance attacks or search relatively rare microhabitats, such as dead twigs, epiphytes, dead curled leaves, or bark, are generally solitary or occur in pairs and fit the attendant species behavioral syndrome (Buskirk et al. 1972; Willis 1972; Greenberg and Gradwohl 1985). These authors argued that their smaller intraspecific group size is a response to greater sensitivity to intraspecific competition (rarer foraging sites, or longer foraging attack radii).

Location of Specific Foraging Sites
One commonly suggested potential benefit of flocking is learning where food is located from other flock members. There is little field evidence that birds in mixed flocks learn about the distribution of food from other flock members; however, experiments with captive tits showed that members of two generalist species (black-capped and chestnut-backed chickadees [*Parus rufescens*]) might gain information on productive foraging sites from each other (Krebs 1973). The possibility that tit species learn from each other in the wild is particularly interesting in light of the specialization that has been reported among conspecifics in a flock (Vanbuskirk and Smith 1989). Convergence of foraging niches of birds in flocks has been put forward as evidence that birds join flocks to learn about the specific locations of food (local enhancement) or, perhaps more commonly, to learn what types of foraging sites are productive at a particular time "site enhancement" (Morse 1978). Waite and Grubb (1988) present evidence from aviary experiments that both types of copying can occur in mixed-species flocks consisting of tits, woodpeckers, and nuthatches. Their experiment showed that attendant species (woodpeckers and nuthatches) pay attention to the foraging success of the more generalized tit species, at least in a confined aviary situation. However, more data from the field are required to determine the importance of copying behavior.

Limits to Social Facilitation: Stereotypy and Neophobia
Perhaps rather than focusing on the small body of evidence for copying—which at this point is largely restricted to tit species—a more interesting question to consider is how individual birds retain their species-specific specializations in the context of a mixed flock. What prevents birds, particularly young birds, from attending to the foraging success of species with highly divergent foraging strategies? Responses of birds to the foraging behavior of other birds should be a fruitful area for research. One could hypothesize that

the specialized attendant species of tropical flocks are highly resistant to social facilitation. Although learning mechanisms between birds with different degrees of foraging specialization have not been well studied, I have conducted a few experiments that are relevant.

The Carolina chickadee is a nuclear species that shows a high degree of foraging plasticity. The worm-eating Warbler *(Helmitheros vermivoros)* is a highly specialized attendant species during the nonbreeding season, occurring solitarily in mixed flocks and feeding almost exclusively from dead curled leaves. I presented a large number of objects (including dead leaves, live leaves, boxes, tubes, chenille stems, etc.) to naive fledgling warblers and tits housed in flight cages and recorded their preferences for exploring these objects (Greenberg 1987a) (these were dependent young and no food reward was involved). The worm-eating warblers all showed highly similar preference rankings, whether they had been reared with conspecifics or with chickadees. The chickadees showed highly divergent preferences between rearing groups, but converged upon a preference ranking within groups. The mechanism for the convergence was unclear—only a small portion of visits to objects were made simultaneously by more than one chickadee. The experiment suggests that chickadees have relatively unstereotypic preferences that are easily shaped by their social environment, as opposed to the highly stereotypic preferences of the warbler.

It has been hypothesized (Greenberg 1984) that neophobia (fear of novelty) plays a central proximate role in controlling foraging specializations: individuals of more specialized species have an aversion to approaching unfamiliar foraging sites or feeding on unfamiliar foods. I tested the hypothesis that the more specialized chestnut-sided warbler *(Dendroica pensylvanica)* would approach a productive but novel feeding site more readily in the presence of the intraspecifically gregarious and generalized bay-breasted warbler *(Dendroica castanea),* a species that it commonly associates with on the wintering grounds (Greenberg 1987b). The chestnut-sided warblers did not decrease their aversion to novel stimuli in the presence of the non-neophobic species.

Avoiding Exploited Sites
Foraging competition is discussed below as a probable cost of mixed flock participation. However, since interspecific competition may occur anyway, participation in flocks may be a way to monitor the resource use of potential competitors (Morse 1970; Austin and

Smith 1972). By avoiding substrates or sites already being used by another species, birds in mixed flocks may actually enhance their foraging efficiency. Divergence in foraging sites has been reported for temperate zone flocks, but this is a difficult hypothesis to distinguish from a behavioral competition hypothesis (see below under foraging costs) or one regarding dominance and predation (Suhonen, Halone, and Mappes 1992; Suhonen 1993).

Flushing of Prey and Kleptoparasitism

Kleptoparasitism—stealing food from subordinates—can be an important source of food, and hence a benefit, for a few behaviorally dominant flock species, at a substantial cost to subordinate flock members. J. Gradwohl and R. Greenberg (unpub.) observed that the slaty antshrike frequently stole large food items from the smaller antwrens. The association between antshrikes and antwrens is particularly strong during the breeding season, when adults handle many large prey items. Antshrikes tended to station themselves between male and female antwrens during periods of nuptial feeding, and the antwrens moved into dense vine tangles and performed evasive maneuvers to avoid attacks by antshrikes. Munn and Terborgh (1979) and Munn (1985) found that the leader species of Amazonian flocks, *Thamnomanes* antshrikes in the understory and white-winged shrike-tanagers *(Lanio versicolor)* in the canopy, depend heavily upon prey flushed by or stolen from other flock members. Munn (1986) further suggested that the antshrikes use a deceptive alarm call to induce subordinate birds to drop prey, which are then kleptoparasitized. Similar "deceptive" use of alarm calls has been reported for tits (Moutsouka 1980; Møller 1988). Møller found that during periods of inclement weather, great tits *(Parus major)* gave alarm calls when no predators were seen by the human observer. These calls were given at feeders in the presence of finches and allowed the tits access to food or facilitated food theft. The published studies on deceptive alarm calls are suggestive, but leave a number of unanswered questions. As both Munn (1985) and Møller (1988) point out, the advantages of using deceptive alarm calls are highly frequency dependent. If foraging is common and predation is rare, the caller risks habituating the target species, which would reduce the efficacy of the "false alarm" as well as the efficacy of the real alarm call.

Kleptoparasitism of live prey has not been documented in temperate flocks, where winter prey is usually small. However, Waite and Grubb (1988) suggested that kleptoparasitism by large atten-

Russell Greenberg

dant species against subordinate nuclear species (chickadees and titmice) could be common. In addition, it seems possible that birds pilfer food from caching species. One captive study looking for interspecific pilfering failed to find interspecific interactions, however (Suhonen and Iniki 1992).

Optimal Movement Patterns

In one of the few attempts to apply optimization models to bird flocks, Cody (1971) proposed that mixed finch flocks move through a range in such a way as to optimize the harvesting of resources with respect to their renewal. When resources were abundant, he found that flocks moved more rapidly, turned more frequently, and that the angles of flock turns were larger, which brought the flocks back to previously visited areas more frequently. The desert flocking system was studied further by Eichinger and Moriarty (1985), who correlated movement patterns with topography, specifically the presence of desert washes. Even in a seemingly simple system, the construction of movement models that disentangle resource use strategies from patchiness of the environment is difficult.

The movement of mixed-species flocks has so rarely been charted over a number of days that we cannot assess the possibility that there exist long-term movement "strategies." A few observers, such as McClure (1967) in Malaysia, have found that flocks follow regular "beats." In general, mixed flock movements are irregular and seemingly haphazard, with no tendency to avoid previously visited areas at regular intervals. Based on his own experience and earlier accounts, Gaddis (1980) concluded that chickadee/titmouse-led flocks are irregular in their movement, with no clear pattern of speed or direction. He proposed that this irregularity was in itself a strategy to make the location of the flock unpredictable to predators. In a particularly careful study of flock movement, Powell (1979) found that flock movement approximates a random walk with a bias toward forward motion, and that a dominant feature of flock movement is movement to territorial boundary interactions. The nuclear species of mixed-species flocks maintain territories, to a large degree, by face-to-face interactions with neighboring groups that reestablish traditional territory boundaries. Gradwohl and Greenberg found that in the dry season, antwren flocks move slowly through a few core areas (dense vine tangles) and intermittently make forays to territorial boundaries (or into neighboring territories if neighboring flocks are temporarily absent from the boundary area). In many core flock species, such as the antwrens, song is restricted to a brief predawn period and to boundary interactions. It is tempting to speculate that song is restricted to these times to reduce the possibility that the location and movement of the flock will be monitored from a distance by predators.

The overall patterns of territory use by Powell's highland flocks and by Gradwohl and Greenberg's lowland flocks are quite different. The former flocks use the territory evenly, suggesting some mechanism for regulating space use, whereas the latter flocks' space use is very patchy, with individuals concentrating their activity on a few core areas. Without more detailed following in conjunction with habitat sampling, it is impossible to generalize about strategies of space use by mixed flocks.

Spatial Memory of Flocking Species
Perhaps certain species are joined by others to take advantage of their special spatial cognitive abilities. The development of spatial memory has been an important area of recent research in avian behavioral ecology. However, this work has not been conducted in the context of group movement per se, but rather in the area of food storage and caching. Food caching is not known for participants in tropical flocks, but is widespread in common flock species in temperate forests (Sherry 1989; Krebs et al. 1996). Spatial memory over a few days has been demonstrated, and longer-term memory is possible as well. Species that store food have a well-developed hippocampus compared with related nonstoring species. It seems possible that if mixed flocks have an emergent strategy for optimizing the time between visits to particular parts of the flock territory, this strategy might be based on the spatial memory of individuals of the nuclear species.

Learning from the Locals
Species may join a mixed flock because the locally territorial species are familiar with favorable or safe areas in which to forage. This hypothesis works best for attendant species that have large home ranges compared with the nuclear or other attendant species. Where the nuclear species has a territory larger than its attendants', it seems unlikely that the attendants would join to "learn" about features of their territory.

Possible Antipredation Advantages of Participating in Flocks
Although it has been proposed that flocking can reduce the risk of predation to participants in a number of ways, for forest flocks, the

primary mechanism is increased vigilance and the ability to take advantage of the alarm calls of certain species (see Powell 1985 for review). The nuclear species, with their well-developed and conspicuous vocalizations, may be particularly attractive sources of information on the presence of predators. Both Morse (1970) and Gaddis (1980) showed that individuals of two tit species consistently alarmed first and most vociferously, with a loud whistle, when accipiters attacked mixed flocks. The whistled alarm causes all flock members to freeze, thereby reducing the conspicuousness of both the individual and the overall flock. Attendants in flocks are often species that forage intensively by examining surfaces or probing into specific microhabitats (Willis 1972; Buskirk 1976), whereas species with less intensive foraging behavior join flocks less commonly, and presumably are more vigilant.

A number of studies have examined the role of group vigilance within single-species flocks (Powell 1974; Siegfried and Underhill 1975), and interspecific changes in vigilance have been documented for simple mixed flock systems (Barnard and Thompson 1985). Sullivan (1984) first carefully tested the vigilance hypothesis in forest flocks by examining video footage of foraging downy woodpeckers in and out of mixed flocks, and found that vigilance time was much reduced in this bark-searching insectivore when it associated with other species. The decreased need for vigilance by attendants can also be seen as a feeding advantage, as the downy woodpeckers in flocks spent more time foraging. Areval and Gosler's (1994) finding that the solitary attendant treecreeper *(Certhia familiaris)* restricted its foraging zone, but increased its feeding rates, as mixed flock size increased suggested that the treecreeper depends upon the greater surveillance for predators found in larger flocks (see also Henderson 1989). However, it has never been demonstrated that intraspecifically gregarious nuclear species change their vigilance behavior when joined by attendant species. In fact, Carolina chickadees showed no significant change when tufted titmice were removed from flocks (Crimprich and Grubb 1994). However, Hogsted (1988) reported reduced vigilance in willow tits *(Parus montanus)* when they participated in mixed flocks.

As the number of birds increases in a flock, not only does the probability of detection of a predator increase, but so does the possibility of "false" alarms—calls given mistakenly when no predator is actually present. There is a cost in lost foraging time to responding to inappropriate alarms. Recent work by Lima (1995) suggested that assessing whether an alarm is valid is a complex process in sparrows in mixed flocks, and that alarms gain salience if several birds alarm and flush. The ambiguity of alarm information is therefore an issue—one that has not been addressed in forest flocks.

Costs of Flock Participation

Foraging Disturbance and Competition
Several studies of temperate zone flocks (Morse 1970; Alatalo 1981; Crimprich and Grubb 1994) have shown that birds adjust their foraging sites when they join flocks. These adjustments appear to be a result of aggression or the threat of aggression, or of exploitative competition—the superior ability of some species to harvest arthropods from particular foraging sites. Crimprich and Grubb (1994) demonstrated through a removal experiment that Carolina chickadees expand their foraging sites in the absence of titmice. However, the choice of foraging sites may be based on relative safety from predation as well as on food resource availability (Suhonen 1993). Buskirk (1976) indicated that birds actively converge upon a single foraging height stratum to join flocks. But there have been no reports of changes in the use of particular foraging sites.

Competition from other species should be viewed as a cost of mixed flock participation. However, because different species tend to forage in different sites, mixed-species flocking allows birds to gain the benefits of flocking with less competition than they would experience in single-species flocks. The actual resource depletion by insectivorous flocking species has only rarely been measured, but may be great (Gradwohl and Greenberg 1982).

The effect of temporary disturbance may be profound as well. Based on observations of captive checker-throated antwrens (J. Gradwohl and R. Greenberg, unpub.) we found that, upon disturbance, arthropods may retreat into a dead curled leaf after an initial visit and become more difficult to find. Although long-term prey abundance may not be seriously depressed, birds may have to let areas "rest" before arthropods are easily located again. The effect of this type of competition should be most pronounced between members of the same species, hunting in the same microhabitats. Valburg (1992) found that the common bush tanager *(Chlorospingus ophthamlicus)* formed large, single-species flocks when feeding on fruit (where such disturbance-based competition is minimal), but participated in smaller single-species groups within mixed-species flocks when searching for arthropod prey. A number of authors (e.g., Ekman 1979) have argued that in temperate zone tit flocks, within-species group size declines as more congeners occur together, suggesting that species gain flocking advantage with reduced intraspecific competition. Aggression is generally much greater within tit species than between, and between tit species than between tits and attendants (Morse 1970). Waite and Grubb (1988) suggested that tits gain an advantage by flocking with dominant attendants because interactions with them are fewer than with conspecifics or congeners.

Nonoptimal Movement Patterns
I have discussed the possibility that movement across flock territories is optimized when birds join flocks. However, movement with flocks may impose a cost on birds that are required to move through areas at speeds or with return times that are inappropriate for their particular foraging strategy. Munn and Terborgh (1979) argued that flocks consist of species that have generally compatible movement patterns when foraging. Incompatible movement patterns may explain why some gregarious species (such as dusky-faced tanagers, *Mitrospingus cassinii,* in Panama) attract few followers. However, as pointed out by Hutto (1988), even for species that move through the habitat at similar speeds, the adjustment of movement patterns away from the optimal may be a cost of flocking. Buskirk (1976) reported that slate-throated redstarts *(Myioborus miniatus)* use simple gleaning rather than aerial maneuvers when they join flocks; he proposed that the aerial maneuvers cannot be accomplished in a rapidly moving flock.

The issue becomes more complicated because the speed of flocks is not constant. In fact, Morse (1970) showed that tit flocks with more individuals move faster. Although Powell (1980) found no such relationship in his highland tropical flocks, Greenberg and Gradwohl (1980) found that antwren flocks with more forward momentum contained more species. The reason for this flock size/velocity relationship is unclear. One explanation might be that more birds cause more disturbance and force birds to change foraging sites more quickly. In this case, the upper limit of flock size may be reached when birds can no longer sustain their foraging at high flock speeds. Another explanation is that more species require higher flock speeds to sustain their foraging, and as the nuclear species moves more rapidly, more attendants are able to join. Obviously both the generality and the mechanism of a flock size/velocity relationship require further study.

Flock Conspicuousness

Despite the well-developed alarm behavior of some flocking species, flocks themselves provide a tempting target for predators. A predator locating a flock has a number of targets, and it is difficult for a flock to melt away into the forest the way a solitary bird can. It is interesting that observers have generally not found raptors that track mixed-species flocks (I am assuming that nonavian predators would be unable to track flocks over great distances). Predators are able to station themselves at food resources or track feeding aggregations (such as at fruiting or flowering plants, bird feeders, ant swarms). For example, ant-following aggregations are often joined for long periods by barred forest falcons (*Micraster ruficollis:* Willis and Oniki 1978; R. Greenberg, pers. obs.). Foraging flocks may be a less tempting target for a following raptor because participants have more options for different patterns of movement than birds whose movements are determined by the vagaries of an ant swarm. Furthermore, ant-following birds produce noisy agonistic vocalizations as they jostle for position.

Movement and behavioral strategies for discouraging the following of flocks by avian predators may include spending time in dense and perhaps safer patches (e.g., the vine tangles of antwren flocks), moving rapidly and irregularly away from these areas, and reducing the use of long-distance vocalizations in favor of short-distance contact notes and border interactions. The strategies flocks employ to discourage following by avian predators would be a fascinating line of research.

Costs and Benefits and the Nuclear/Attendant Paradigm

There are many varied potential costs and benefits to joining mixed flocks, centering on detection of predators and enhancement of foraging efficiency on the one hand and increased competition on the other. It is not surprising that different costs and benefits accrue to birds in different flock systems and to birds of different species within a flock. One can probably safely assume that the attendant species that actively join flocks receive a net benefit from the association. It is more challenging, however, to understand the costs and benefits experienced by the individuals that are joined. It is conceivable that nuclear species benefit from attracting individuals of other species through an increase in the diversity of vigilant individuals and the ability to monitor a wider range of foraging sites. It is equally conceivable that nuclear species gain minimally from the presence of attendant birds and actually experience a net cost due to the increase in competition or the attraction of predators. Carefully exploring the costs and benefits for species with different flock roles will be a formidable task for future research.

Decision Making Within Nuclear Species Groups
The strong species-specific role in apparent flock leadership begs the question of which individual birds determine the pattern of flock movement and how this is conveyed to other flock members. Little research has focused on the decision-making process within the social group of the nuclear species. However, Mostrom (1993) and S. M. Smith (1991) examined the relationship between dominance status and tendency to initiate movements in groups of Carolina and black-capped chickadees, respectively, and found no relationship. Their work, as well as other studies (notably, Odum 1942), failed to detect a strong tendency for particular individuals to initiate group movements. Without such a tendency, the possibility of correlating demographic attributes with group leadership is meaningless. In addition, although chickadees are followed by most attendant species and are generally considered to be nuclear, it appears that they, in turn, usually follow titmice (Morse 1970). T. Grubb (pers. comm.) suggested that the dominant male titmouse may play the greatest role in flock movement and coalescence.

Beyond the Nuclear/Attendant Paradigm: Flock Leaders versus Sentinel Species
Wiley (1971) suggested that the nuclear/attendant dichotomy may be inadequate to describe the potential complementary roles species play in the functioning of mixed-species flocks. His work focused on flocks with a nucleus of two species of *Myrmotherula* antwrens. He argued that one congener, the white-flanked antwren, by virtue of its sociability and conspicuous plumage pattern (black with white flank flashes, as described above), promotes the cohesion of flocks. A second species, the checker-throated antwren, is cryptically colored and follows the white-flankeds, but acts as a sentinel species, giving a sharp "peeeet" call in the presence of predators (a call that is often used in high-intensity territorial interactions as well). It is not clear how selection would act on individuals of the two species to shape these complementary roles. Perhaps the association of the two species did not involve any evolutionary adjustments, and it is just fortuitous that two species with these character-

istics are found commonly in the same forest. However, the details of the particular relationship outlined by Wiley are not well established. Further work has shown that the strongest association is between the checker-throated and dot-winged antwrens. Attendant species probably follow the continuous contact notes of the dot-winged antwren. Checker-throated antwrens are highly specialized dead leaf foragers and are probably less effective at detecting predators than are the other antwrens (J. Gradwohl and R. Greenberg unpub.).

The concept of sentinel species, however, has been supported by research on mixed-species flocks in Amazonian forests. Munn and Terborgh (1979), Wiley (1980), and Powell (1989) found that antshrikes in the genus *Thamnomanes* were sentinel foragers with the best-developed alarm call in the flocks. They provided some anecdotal evidence that *Thamnomanes* give alarm calls to which other flock members respond. However, as I stated previously, they also suggested that these antshrike species function as flock leaders: their calls coalesce dispersed flocks, and other flock members follow when these species give calls away from the flock. Therefore, since the call notes of one species serve both as a warning system and a mechanism of group coalescence, the concept of complementary roles as suggested by Wiley for antwren flocks does not apply to the Amazonian flocks.

Bell (1983) and Diamond (1987) provide another possible example of complementary roles in New Guinea bird flocks. The leader species (often babblers) gain from the sentinel behavior and loud alarm calls of drongos. Croxall (1976) found that the loud calls of drongos are important in flock formation, but did not discuss alarm calls in particular.

It should be remembered, however, as we discuss stable and perhaps coevolved relationships between species in the core of a mixed flock system, that it is likely that many attendant species join mixed-flock systems opportunistically—that they are attracted to any group of birds moving through their forest.

Beyond the Nuclear/Attendant Paradigm: Multispecific Territories
The most surprising discovery concerning the formation of complex mixed-species flocks is that the core often consists of a number of species that defend the same territorial boundaries. This phenomenon was found in understory flocks in Panama (Gradwohl and Greenberg 1980) and Peru (Munn and Terborgh 1979), where three

Russell Greenberg

and sixteen species respectively occupied multispecific territories. Munn (1985) further demonstrated that canopy flocks in the same Peruvian forest consisted of a core of species that codefended territories, although the territories were larger and less aggressively defended than those of understory flocks. Multispecific territories are not restricted to tropical forests: temperate tit flocks sometimes have two or three species in such a system (Ludescher 1973; Ekman 1979; Bardin 1984; Pravosudov 1987). Studies of marked members of the core of these multispecific associations have been carried out for from a few to up to 14 years (Greenberg and Gradwohl 1997). All studies suggest that the boundaries of these associations are extremely stable, maintaining their salience far longer than the tenure of particular birds.

Multispecific Territories and Population Ecology

We can assume that species that join multispecific territorial systems do so at some energetic cost. If optimal territory size is dependent upon the abundance and distribution of resources, and resource abundance and distribution vary considerably with the body size, foraging strategy, nesting behavior, and so forth of individual species, then it is unlikely that the optimal territory size will be identical for all participating species. Added to this assumption is the observation that the sizes and locations of multispecific territories are extremely stable, despite annual fluctuations in rainfall and food supply. Greenberg and Gradwohl (1986) reported that annual adult survival showed considerable yearly variation in two species (checker-throated antwren and slaty antshrike), and that this variation was correlated with the amount of rainfall during the late rainy season (the more rain, the higher the disappearance rate). If territory did not precisely regulate the breeding population, one would not expect stable populations and territories unless the number of young produced each year exactly matched the mortality rate. However, at least in the checker-throated antwren, the number of recently fledged young recruited into flocks showed very little year-to-year variation and thus could not balance the fluctuating turnover in adults. These observations suggest that territory size in these species is not finely adjusted to resources, either between species or through time. Greenberg and Gradwohl (1986) argued that this system is a prime example of populations that are socially regulated well below the carrying capacity for the participating species. Powell (1989) extended this argument to suggest that the high diversity of flock-participating species may be a result of large territory sizes with respect to resource abundance, which would reduce intraspecific competition and allow a number of specialist species to coexist on the same territories.

The core species of multispecific territorial associations (Munn and Terborgh 1979; Gradwohl and Greenberg 1980) appear to be highly consistent and ecologically regulated in composition. Simulation studies by Graves and Gotelli (1993) based on flock censuses by Munn (unpub.) uncovered some strongly nonrandom patterns, in which ecologically similar congeners (*Xiphorhynchus, Myrmotherula, Hylophilus,* and *Automolus*) co-occur significantly less often than expected in the same flocks. This analysis suggests that interspecific competition may regulate flock membership. However, no habitat data were presented in this study, and the species could instead be responding to subtle variations in forest structure.

Decision Making in Multispecific Territorial Associations
The existence of multispecific territories greatly complicates the question of how group movement is coordinated between a number of species. Up to this point I have argued that the movement pattern results from behavioral asymmetries between nuclear and attendant species. However, when a number of species occupy precisely the same home range, then movement coordination becomes the equivalent of a single-species coordination problem. The question is, which bird decides where in this complex territory to go next? The question is more complex than that for a single-species system because the participants have a much greater diversity of resource needs. In our (Greenberg and Gradwohl) studies of antwrens in Panama, we were unable to answer this question quantitatively. However, after many thousands of hours spent observing these groups, I am left with the impression that each antwren species is capable of initiating movement. Movement is often pseudopodic, with one or another of the antwren species moving out from the core. When the different species of antwrens are separated, each species calls antiphonally with the other to facilitate flock coalescence.

In contrast, as I stated earlier, the Amazonian understory flocks are formed and led by the activity and vocalizations of one or two antshrike species. The details of flock roles in these multispecific territorial associations may provide some insight into the evolution of these systems. The coalescing of a number of core species around one or two sentinel species turns out to be quite similar to the nuclear/attendant model. It differs only in that some of the attendant species have become sufficiently dependent upon the information provided by the sentinels to adjust their territories so as to assure constant association. The roles in antwren flocks, which appear to have a core of several species with more egalitarian or even complementary role relationships, suggest that these flocks develop from the mutual adjustment of the territorial boundaries of the core species. Detailed study of the bi- or multispecific territorial tits (Ludescher 1973; Ekman 1979; Bardin 1984; Pravosudov 1987) would be instructive. V. V. Pravosudov (pers. comm.) believed that the Siberian tit *(Parus cinctus)* was more numerous, dominant, and tended to lead in its association with willow tits.

Mixed flock territorial systems may represent a near obligatory mutualism (Munn and Terborgh 1979), which would set the stage for the evolution of complementary roles. On the other hand, it is equally plausible that the associations are the result of learned facultative adjustments of individuals. For example, the migratory chestnut-sided warbler spends up to 7 months of the year in the antwren flocks of Barro Colorado Island, where it defends the same territorial boundaries as the core species. Yet in nearby second-growth woods, warbler territories are one-tenth the size and unrelated to the territories of any other species. It has been argued that by adjusting their territory location and boundaries to those of ecologically similar resident birds, migrants can use the resident birds as cues for locating stable resources (Morton 1980; Greenberg 1984).

Proximate Factors in Multispecific Territory Establishment and Maintenance
What determines the locations of the stable and shared boundaries of multispecific territories? In the short run, replacement of individuals is slow, and the territories are maintained by the behavior of the remaining flock members and their neighbors. In the long term, what maintains these territories has yet to be determined. Are their boundaries traditional ones carried over from when the participating species first colonized an area of mature forest, or are they somehow determined by structures within the forest? Over a much shorter time frame, settlement pattern has been shown to have a profound effect on the size and configuration of territories. Krebs (1971) conducted a removal experiment on great tits *(Parus major)* and found that if a series of territorial occupants were removed simultaneously, new birds established a new array composed of

smaller territories. However, one-by-one removal of birds resulted in an array similar to the original. In stable multispecific territorial systems, we could be observing territorial arrays constrained by settlement patterns over very long periods of time. The alternative is the possible presence of stable habitat features. This possibility seems unlikely considering the continued occurrence of treefall gap succession. However, the habitat features that form the core of the antwren territories are areas of vine tangles, which may have a much longer existence than other parts of a tropical forest. In our studies of antwrens, there was one period when three sets of territorial individuals of both core species of antwrens disappeared. The territorial array then consisted of fused territories within the already existing external boundaries. When the population recovered and the fused territories split, the same internal boundary was re-formed. That pairs of both of the core species disappeared and then were replaced to re-form the exact original territories suggests that boundaries are at least in part formed around some intrinsic structure in the forest. Perhaps, then, the different species in these flock systems defend the same territories because of individual responses to similar cues. This idea needs to be tested, but seems unlikely, given the different ecological requirements and body sizes of all the species involved. Furthermore, convergent territory boundaries do not seem to occur in bird communities where no mixed-species flocking is involved.

Future Research Directions
In my mind, the fundamental issue facing mixed-species flock research is: to what degree does participation in mixed-species flocks select for behaviors or attributes that do not evolve in response to intraspecific interactions? A research program that addresses this fundamental question would integrate the analysis of flock behavior at the level of the individual, species, and whole flock: a daunting task—particularly for a Ph.D. candidate looking for a "doable" topic. Therefore, in table 18.4, I lay out some key questions that, if answered, should improve our understanding of mixed flock formation.

Conclusions
For forest birds, involvement in multispecific associations is the rule during the nonbreeding season in the temperate zone and throughout the year in the tropics. For many intraspecifically solitary spe-

Russell Greenberg

Table 18.4 Some hypotheses or questions for future research on mixed-species bird flocks

A.	Do flocks use their territories in an even, random, or patchy manner? If the latter, do the peaks of usage correspond to particular habitat features, territory centers, or territory boundaries?
B.	Do flocks move about the territory in predictable beats or variable and irregular paths? If the beat is regular, does this result in regularly spaced return times to specific areas? If the path is erratic, is the movement relatable to interactions with neighboring groups, or is it a tactic to evade tracking by predators?
C.	Is short-distance flock progression generally led by particular individuals and species, or is leadership shared evenly among a number of species? Do the leadership roles of species change with context (i.e., moving toward boundary, in core area, etc.)?
D.	Does flock formation depend upon the presence of particular species, as determined by removal experiments or their equivalent, and if so, when numerous flock systems are compared, do these species share behavioral traits? Alternatively, are there a number of species whose presence is sufficient to generate flock formation?
E.	Are certain vocalizations particularly efficacious for flock formation and movement?
F.	Does net benefit correlate with importance in flock formation as determined in C–E?
G.	Are roles in food finding, predator detection, and flock coalescence complementary between core flock members?
H.	In flocks in which core members have complementary roles, are net benefits more similar than in flocks in which one species performs most of the key functions?
I.	Are there vocalizations or behaviors of core flock members that function exclusively or primarily in an interspecific context, or are most behaviors shaped primarily by intraspecific interactions?
J.	Do mechanisms underlying social learning of resource distribution vary with flock role?

cies, such associations form the dominant social unit. These associations vary in their cohesion and stability. At their simplest, associations form around an attractive food resource, and interspecific interactions center primarily on conflict resolution and dominance relationships. Mixed-species flocks range from those that form when a number of species join a gregarious leader species, to those that have a leadership core of several species, to those that have a core of species that share and defend the same territorial boundaries. The last appears to reflect the greatest behavioral commitment to a mixed-species association because, unlike other mixed flock systems, it requires a complete adjustment of movement patterns and home range use to participate.

Mixed-species associations demand many of the same behavioral adjustments that are at the core of single-species group formation. However, the nature of these adjustments is often qualitatively different between single- and mixed-species groups. Participants in single-species groups of vertebrates are morphologically similar and share the same basic ecological requirements. Differences in resource and home range use are based largely on quantitative differences or on variation in social status. Participating species in mixed-species groups show large qualitative differences in morphology and ecological requirements. These differences, essentially, make coordination of movement potentially more expensive, but reduce the effect of exploitative resource competition. They also open the possibility for individuals to play markedly different roles in detecting and mobbing predators and locating food sources. Furthermore, they underlie potentially large asymmetries in the costs and benefits of group formation.

Asymmetries of advantages and disadvantages between species have been postulated in numerous studies of mixed-species flocks, but rarely analyzed experimentally. This is an area of mixed flock formation research that is ripe for the picking. Asymmetry in roles in the maintenance of group cohesion is another area in which exciting work could be accomplished. Most researchers have viewed flocks as consisting of leader species, which play a central role in group movement and decision making, surrounded by a constellation of species that follow these species. However, within the stable core of mixed-species flocks we find the possibility of co-leadership—a phenomenon that has been rarely studied.

The most exciting possibility is that the multispecific social unit has exerted an important selective force on the communication systems and color patterns of the participating species. So far, evidence for the possibility that flock members have undergone significant coevolutionary changes is scant. It is my impression that mixed-species flocks form where the selective pressure to flock is great and under conditions consistent with the demands of intraspecific interactions. However, the natural historical observations suggest over and over again that these flock systems have overriding importance in the survival of forest birds, so coevolution of traits in flocking species remains a compelling possibility.

Acknowledgments
I would like to thank Ann B. Clark, Dick Byrne, Mandy Marvin, Sue Boinski, and Paul Garber for valuable comments on an early draft of this chapter. Long-term research on mixed-species flocks was supported by the Environmental Sciences Program of the Smithsonian Institution.

CHAPTER NINETEEN

Keeping in Touch at Sea: Group Movement in
Dolphins and Whales

RACHEL SMOLKER

A Dolphin's Day

We first come across Snubnose and Bibi, two male bottlenose dolphins as they are traveling rapidly out to the north, diving, surfacing, and breathing in perfect synchrony. Just out of synchrony,
slightly farther apart, but keeping pace with them, is Sicklefin, another male and the third member of the alliance. Eighty meters to
the west, and moving in parallel, are two more males, Wave and
Shave. They are Snubnose, Bibi, and Sicklefin's "second-order alliance" partners. These two alliances will shadow each other's movements from a distance for much of the day, occasionally joining
forces. We know that Snubnose, Bibi, and Sicklefin are "up to

something" just by the way they are moving—in perfect synchrony, directly and purposefully toward some distant dolphins.

We follow along. They speed up as they approach a group of females, their arrival creating a stir of excitement in the otherwise restful group. They single out Poindexter, an older female who is probably cycling this year. She has no calf, and the males have been interested in her off and on over the past few months. They chase her from the group at top speed, and when we catch up, there is a lot of splashing. The tension in their movements and the explosiveness of their breathing reveals the males' excitement as they swim just behind and to either side of Poindexter, "flanking" her. Now and again they rush up to her, rub along her sides or point their rostra toward her genitals as though inspecting her with their echolocation.

From their flanking position, Snubnose and Bibi suddenly and synchronously lurch forward and dive, then come leaping high out of the water on either side of Poindexter. Both of the males resurface and dive again at exactly the same moment, slapping their tails synchronously on the water surface as they go down. A moment later they pop to the surface in flanking position again, performing in perfect unison, like Olympic synchronized swimmers.

After a time, they calm down, seem to lose interest in Poindexter, and begin to drift apart. First Sicklefin, then Bibi, and then Snubnose head in different directions, with their tails breaking the surface as they dive, marking the beginning of a bout of foraging. Over the next 2 hours they scatter outward, separated by as much as 200 meters or so, occasionally snapping at small fish near the surface. Snubnose and Sicklefin end up on intersecting paths, join up, and begin to travel together, no longer tail-out diving. A few minutes later, Bibi, still 100 meters away, whistles—a whistle uniquely his own, which Sicklefin and Snubnose apparently recognize. They stop and wait while Bibi catches up to them, and all three then travel very slowly to the west, in a close rank abreast formation. They are resting.

About an hour later they stop and hang motionless at the surface. Snubnose is pointed west in the direction they had been going, but Bibi and Sicklefin are pointing toward the north now. They seem to have different ideas about which way to go. Perhaps they hear distant sounds of other dolphins? Eventually Sicklefin and Bibi turn back to the west with Snubnose. This happens twice more,

Rachel Smolker

but in the end, it is Snubnose who ends up turning to the north with Sicklefin and Bibi.

A few minutes later they are all flying through the air, racing at top speed out to the north, making long arcing leaps clear over the water to catch a breath as they go. They race like this over a distance of a kilometer or so, and when we catch up, they have joined at least twenty other dolphins scattered around in a huge feeding aggregation. Dolphins are tearing around in all directions, slicing through the water and leaping, occasionally snapping up fish in their jaws at the water surface. We catch a few glimpses of the dark shadow of a large fish school down below. Cormorants and pelicans have also converged on the spot and join in the feeding.

Eventually the activity dies down, and Sicklefin, Bibi, and Snubnose, who are scattered once again among many other dolphins, relocate each other, as do other small groups of dolphins, perhaps by listening for each other's signature whistles, and head off together. Traveling on an intersecting path with some other dolphins, they soon join Wave and Shave and two other males, Lucky and Pointer, who are with them now. All of these males will socialize for the next hour before splitting up into smaller groups again to rest.

Snubnose, Bibi, and Sicklefin are members of a population of bottlenose dolphins (*Tursiops truncatus*[1]) resident in Shark Bay, Western Australia, subjects of long-term research by me and my colleagues (Connor and Smolker 1985; Smolker et al. 1992; Connor, Smolker, and Richards 1992a,b; Smolker, Mann, and Smuts 1993; Richards 1996; Mann and Smuts 1998). This account of a day in their lives brings to light the many ways in which they must coordinate their movements, and some of the ecological and social forces that favor their doing so.

But before analyzing these in more detail, let me first provide readers with some context by briefly reviewing the behavior and ecology of cetaceans in general, and the Shark Bay bottlenose population more specifically.

Marine Mammals and Their Habitat

The order Cetacea, like the order Primates, includes a wide range of species that are morphologically, ecologically, and behaviorally diverse (see Leatherwood, Reeves, and Foster 1983; Klinowska 1991). The cetaceans are broadly divided into two major groups, the odontocetes and the mysticetes. The six families of odontocetes include delphinids (e.g., bottlenose dolphins, orcas, spinner dolphins), phocenids (porpoises), platanistids (river dolphins), monodontids (beluga and narwhal), physeterids (sperm whales), and ziphiids (beaked and bottlenosed whales). The three families of mysticetes include balaenids (right and bowhead whales), balaenopterids (blue, fin, sei, humpback, minke, and Brydes whales), and eschrichtiids (grey whales).

The mysticetes are generally the very large "great whales." Instead of teeth, they have fringed plates of a horny material, baleen, that hang from the upper jaw. These largest of mammals use their baleen to filter the smallest of cetacean prey: various marine invertebrates and tiny fish. The odontocetes are also referred to as "toothed whales," though the teeth are greatly reduced and modified in some, such as the various beaked whales (e.g., the "strap tooth" whale, *Mesoplodon layardii*).

The marine habitat presents some interesting challenges that differ from those of terrestrial environments. Except where the bottom topography can be exploited, the ocean is a vast and relatively featureless habitat. Light is much more rapidly attenuated in water than in air, limiting vision to short distances, particularly in murky waters. Through this often relatively dark and featureless environment, many cetaceans must navigate relative to one another and to their habitat, meanwhile avoiding predators such as sharks and securing the resources they require.

Since light is so quickly scattered and attentuated under water, visual signals are useful primarily at close range. In some species that inhabit very murky waters, such as the Ganges river dolphin *(Platanista gangetica)*, the eye is greatly reduced. In others, vision is well developed and functions both under water and in air (reviews in Dawson 1980; Madsen and Herman 1980; Nachtigall 1986). Some species, such as orcas *(Orcinus orca)*, Dall's porpoise *(Phocoenoides dalli)*, and humpback whales *(Megaptera novaeangliae)*, have boldly contrasting patterns in black and white that are visible at some distance and are undoubtedly useful in signaling individual, sexual, or age-related features. Contrasting patterns may also be used to communicate changes in position and posture that facilitate coordinated group movement at close range (Madsen and Herman 1980; Norris and Dohl 1980b; Mobley and Helweg 1990; Würsig, Keikhefer, and Jefferson 1990).

Sound, on the other hand, travels fast and far in the dense me-

Rachel Smolker

dium of salt water. Cetaceans in general are highly vocal (Herman and Tavolga 1980). Mysticetes, where they have been studied, produce a range of call types (reviews in Thomson, Winn, and Perkins 1979; Clark 1990). Almost all are known to produce some sort of low-frequency simple call, or "moan," used in long-distance communication. Some also produce more complex calls, including the "songs" of humpback (Payne and McVay 1971), bowhead (*Balaena mysticetus:* Ljungblad, Thompson, and Moore 1982) and fin (*Balaenoptera physalus:* Watkins et al. 1987) whales.

Odontocetes use sound to echolocate, and also produce a tremendous range of social sounds. The extremely sophisticated hearing and sound processing abilities of odontocetes (reviews in Popper 1980; Johnson 1986), as well as their remarkable morphological adaptations for producing and hearing underwater sound (Norris 1968; Norris and Harvey 1974; Brill et al. 1988; Cranford 1988; Amundin and Cranford 1990; chapters in Thomas and Kastelein 1990) testify to the importance of this sense. At distances greater than a few meters, acoustic cues probably are involved in most coordinated group movement.

In addition to using sounds produced "deliberately" (for their communicative benefit), cetaceans undoubtedly make considerable use of passive hearing. For example, dolphins are probably aided in locating prey by making use of the sounds produced by some fish species. The splashing and slapping sounds caused by breaching, leaping and hitting conspecifics are likely to be useful indicators of activity. Finally, echolocating species may listen to the clicks of other individuals to gain information.

Although cetaceans lack an olfactory sense per se (Jacobs, Morgane, and McFarland 1971), they are capable of detecting chemicals in water (Friedl et al. 1990; Kuznetzov 1990; Caldwell and Caldwell 1977). Some dolphins (including bottlenose dolphins) possess anal glands and urinate frequently, raising the possibility that they can make use of chemical trails to locate or avoid one another or to communicate about dominance or reproductive state.

The energetics of traveling through water differ markedly from the costs of locomotion for terrestrial mammals. While the effects of gravity are reduced, the high density of water presents resistance. The cetacean body form is designed to reduce hydrodynamic drag, allowing the animal to glide through the water with apparent effortlessness. Travel costs are probably considerably less than for terrestrial mammals (Williams et al. 1992). This assumption is further supported by the fact that some species range over tremendous distances.

For some cetaceans, this dark, slippery world of sound has favored lifestyles that are difficult for terrestrial and visual humans to comprehend. The sperm whale *(Physeter macrocephalus),* for example, roams from the Arctic to tropical latitudes, diving to depths of 2,000 meters or more (Watkins 1993), where, under tremendous pressure, it hunts for squid in the dark abyss, often remaining below for periods of an hour or more before resurfacing to breathe.

Because cetaceans move almost constantly, and most are at least somewhat social, some form of coordinated group movement is probably characteristic of most species. However, the difficulties associated with finding, identifying, observing, and keeping track of individual cetaceans at sea have limited our understanding of their behavior. For some of the more accessible populations of cetaceans, such as those that frequent coastal waters, we have reasonable information concerning the outcomes of group movement patterns; that is, we know at least something about what size groups they live in, how animals of different ages and sexes are distributed, and how these groups are deployed in time and space. But we still have very little understanding of the mechanisms by which group movement patterns are achieved or the functions that they serve.

In particular, we are far from understanding the costs and benefits to individuals of observed group movement patterns. Some forms of group movement may result from mutual, active coordination, in which case participants are likely to receive mutual benefits (cooperation). Examples of such mutually actively coordinated group movements in cetaceans involve some types of cooperative foraging, particularly those in which individuals adopt different roles and "take turns," incurring costs in the short term (e.g., forgoing opportunities to feed) but benefiting overall in the long run (e.g., successfully containing a fish school so that predation by all participants is possible for a longer time).

Other forms of group movement may appear to be coordinated, but actually result from individuals pursuing their own independent strategies without mutual coordination. Examples are individuals converging on a food resource, or forming a "selfish herd" (Hamilton 1971) in response to predation pressures. Finally, one individual may benefit by taking advantage of the movements of another individual, who thereby suffers some cost or is unaffected. An example

Rachel Smolker

might be one animal following another to take advantage of food it locates. In the case of active, mutually beneficial coordination, we expect to see social signaling designed to facilitate coordination, while in the latter cases we do not.

Mysticetes and odontocetes differ fundamentally in the degree to which they live in social groups. For the most part, mysticetes tend to travel singly or in small groups with temporary membership. Long-term social relationships (other than mother-calf bonds) are not characteristic of the great whales. Many odontocete species, on the other hand, are group living and highly social. Some oceanic delphinids, such as the common dolphin *(Delphinus delphis)*, dusky dolphin *(Lagenorhyncus obscurus)*, and spotted dolphin *(Stenella plagiodon)*, aggregate in groups numbering hundreds of individuals. Yet within these very large groups, distinct subgroupings are usually evident and somewhat stable over time (Pryor and Kang 1991; Würsig, Cipriano, and Würsig 1991). Long-term social relationships are common among odontocetes, and it is from this group that most accounts of complex social behavior (alliance formation, epimeletic behavior, cooperative hunting) are derived. Thus, in odontocetes, more so than in mysticetes, the costs and benefits of coordinating movements with conspecifics must be accounted for in the social as well as the ecological domain.

Odontocetes

Odontocete social systems vary among species, and even within species from location to location. A few species, particularly the phocoenids and platanistids, tend to be relatively solitary, occasionally forming small groups with temporary membership. Other species, such as orcas and sperm whales, live in stable groups comprising females and their kin. Most others live in fission-fusion societies, with individuals joining and leaving temporary groupings frequently during the course of a day.

A second factor that differs among odontocete species and populations within species lies in their habitat use. Some are resident in particular locations, while others make large-scale movements on a seasonal or occasional basis.

Our knowledge of the odontocete cetaceans is directly correlated with species accessibility. Those that inhabit coastal waters are better known than those that inhabit the open ocean. Some open ocean species, such as the beaked whales, are virtually unknown except from rare carcasses that wash ashore. Others, such as the bottlenose dolphin and orca, both of which frequent nearshore waters, have been subjects of long-term research.

Dolphin Social Systems and Behavior

Our research on bottlenose dolphins in Shark Bay was initiated in 1982. Since then, my colleagues and I have learned to recognize about 400 dolphins within our study area (approximately 150 km²). Some of these dolphins are seen on a regular basis, while others are seen more occasionally.

Range sizes for female dolphins in Shark Bay, calculated by A. Richards (1996), show considerable individual as well as seasonal variation, ranging from 5 to 58 km²· Male ranges may tend to be somewhat larger, but have yet to be analyzed. These ranges are based on sightings of animals we see regularly. There are many dolphins that appear to wander into our study range only occasionally, raising the possibility that we can measure ranges only for individuals with relatively small ranges that overlap our own. These range sizes are comparable to those reported for bottlenose dolphins in Florida (Wells, Scott, and Irvine 1987).

Bottlenose dolphin populations living along open ocean coasts range more extensively. In Argentina, for example, some of the bottlenose dolphins studied by Würsig (1978) traveled 300 km up the coast and later returned. Similarly, off southern California, bottlenose dolphins range along a several hundred kilometer stretch of coastal waters (Hansen 1990; Weller 1991), and show little site fidelity. On a few occasions, researchers have documented very long distance (600–1,200 km) round-trip movements by individual bottlenose dolphins (Würsig and Harris 1990; Wells et al. 1990).

Basically, wherever they have been studied, bottlenose dolphins live in fission-fusion societies, forming small groups whose membership changes frequently as individuals join and leave. This pattern differs markedly from that in many primates, in which discrete groups or troops basically all travel together or nearby most of the time.

The dolphin fission-fusion grouping pattern is analogous in many ways to that described for chimpanzees *(Pan troglodytes)* and spider monkeys *(Ateles geoffroyi:* Symington 1990; Chapman, Wrangham, and Chapman 1995) and some carnivores, including coatis, hyenas, and social canids (Holekamp, Boydston, and Smale, chap. 20, this volume). In chimps, spider monkeys, and social carnivores, individuals that join and leave one another are usually mem-

Rachel Smolker

bers of a relatively closed community. Individuals who are not members of these communities are often met with aggression. Male chimps, for example, engage in "border patrolling" and are hostile to individuals, with the exception of estrous females, from other communities (Goodall et al. 1979; Goodall 1986).

The evidence for discrete communities in dolphins is weak. Although the dolphins in Shark Bay are basically resident within largely enclosed embayments, they do not appear to form discrete communities. Rather, they seem to be part of a very large "social network" involving hundreds of individuals, without any obvious boundaries. Wells, Scott, and Irvine (1987), found some evidence for resident "communities" in the Florida population, but regular associations between individuals from different "communities" occur.

In the Shark Bay bottlenose population, at any given time, dolphins may be scattered around as solitary individuals, (usually foraging) or feeding in aggregations. When not feeding, individuals form small groups (typically four to five, ranging to a maximum of about twenty individuals), either resting, traveling, or socializing.

Spatial cohesion within these groups is highly variable. Sometimes groups are clearly discrete; interindividual distances are small within groups and large between groups. At other times, within- and between-group distances result in less discrete groupings. It is necessary, therefore, to use specific spatial criteria for defining group membership. We use a "10-meter chain rule": a dolphin is considered to be a member of a group if he or she is within 10 meters of another group member.

Group membership changes frequently over time as individuals join and leave, and it is therefore necessary to use specific temporal criteria for group membership. We assess membership during the first 5 minutes of our encounter with the dolphins (long enough for us to sort out who is present, but short enough to approximate a point sample). These temporal and spatial criteria are necessary precisely because associations are so fluid.

We have used the frequency of co-occurrence in these groups as a measure of association and as an indicator of social "preferences" (Smolker et al. 1992). Since animals may forage together for nonsocial reasons, we restrict our analysis of associations to nonforaging activities.

Even within the context of their fluid fission-fusion social system, Shark Bay dolphins exhibit very marked association preferences.

Not surprisingly, for example, mothers and their infants (up to at least 4 years of age) are consistently found together (although they do separate frequently, as discussed below). More surprising are the very consistent associations between some males. Typically two or three individuals form stable pairs and triplets. We refer to these associations between males as alliances (Connor, Smolker, and Richards 1992a,b).

Alliances are usually stable over long periods of time. Some that were evident in the beginning of our research are still together more than a decade and a half later. Associations between allied males are not circumscribed by particular joint activities, but rather the males do most everything together: resting, traveling, socializing, foraging, and joining and leaving groups. They associate with each other as consistently as do mother-infant pairs; in other words, they are virtually always together.

Males within alliances cooperate to aggressively herd females, presumably a strategy for monopolizing mating opportunities (see Connor, Smolker, and Richards 1992a,b). Many male alliances consist of three males, yet when they are herding a female, two of the three will be primarily involved, while the third ("odd man out") remains nearby but participates less in herding activities. The roles taken by males during herding change from time to time.

Male alliances also associate with particular other male alliances on a regular basis, forming "second-order alliances" (Connor, Smolker, and Richards 1992a,b). Second-order alliances often shadow each other's movements at some distance, occasionally merging to travel together for a time. Male alliances attempt to take females from other alliances. On such occasions, second-order alliance partners sometimes join forces to aid their partners in "stealing" females, or in defending against such attacks. Relationships within and between male alliances are complex, sometimes cooperative and affiliative, sometimes competitive and aggressive, and apparently highly dependent on social context (Connor, Smolker, and Richards 1992a,b).

Other than with their calves, females tend to associate most consistently with other females (Smolker et al. 1992; Richards 1996). However, even the most consistent female-female associations are less consistent than those among allied males. Females vary in their tendency to associate with others. While some are relatively solitary, others are usually found in groups and may associate with a wide variety of other dolphins. In a few cases, consistent female-female associations (perhaps kin) remain stable over many years.

Rachel Smolker

Females exhibit a wide range of responses to the attentions of male alliances. In some cases, a female may appear extremely eager to get away from the males, in which case there may be considerable aggression (chasing, hitting, aggressive vocalizations, etc.), while in other cases she seems unconcerned and may go about her business, usually foraging, as the males follow her around closely.

To summarize, in Shark Bay, as in other places where they have been studied, bottlenose dolphins (and this is probably true for other odontocete species) maintain a diverse array of social relationships, both affiliative and antagonistic, with many different conspecifics. These relationships are maintained within the context of a fluid fission-fusion social system. While dolphins join and leave temporary groups many times in the course of a day, some individuals nonetheless exhibit very strong association preferences and engage in complex forms of cooperation and competition.

The dolphin pattern of fission-fusion grouping may represent a compromise between forces that favor group living (e.g., protection from predators, cooperative hunting, etc.), and those that favor a more solitary lifestyle (e.g., feeding competition). Fission-fusion grouping may have the effect of relieving some of the conflicts associated with making travel decisions as a group (or troop, in the case of primates: see Boinski, chap. 15, this volume). Fission-fusion grouping also makes the task of relocating and rejoining ones' associates a pervasive aspect of day-to-day life, and results in a continually shifting social context as group memberships change frequently.

Coordination of Movements in the Day-to-Day Lives of Dolphins
With this background in mind, let me now return to the account of a "day in the life" of Snubnose, Bibi, and Sicklefin with which this chapter opened and discuss in greater detail the specific ways in which movements are coordinated in the day-to-day lives of Shark Bay bottlenose dolphins, and odontocetes in general.

Synchrony
When we first came upon them, Snubnose and Bibi were performing "synchronous surfacings," rising to the surface to breathe precisely at the same moment, maintaining an ordered "rank abreast" formation, and diving together (see chapter opening illustration). Surfacing synchrony is a striking aspect of dolphin behavior, particularly among allied males.

The changing social dynamics among these males are manifested in their surfacing synchrony. For example, when a triplet alliance (like Snubnose, Bibi, and Sicklefin) is herding, the primary pair (in this case, Snubnose and Bibi) will usually surface in perfect synchrony, while the "odd man out" (in this case, Sicklefin) will be slightly apart and asynchronous (see chapter opening illustration).

In addition to synchronous surfacing during travel, males precisely synchronize their behavior during elaborate displays, as Snubnose, Bibi, and Sicklefin did when they first came upon Poindexter. These displays are reminiscent of some of the "tricks" dolphins are trained to perform in marine parks. Exactly how the males determine what action to take next and coordinate these diverse displays remains an intriguing mystery. The displays appear so perfectly synchronized that it is not clear whether there are "leaders" and "followers."

A recent study by Braslau-Shneck and Herman (1993) attempted to address this question by training a pair of captive dolphins to perform novel behaviors in tandem. She concluded that the dolphins achieved synchrony by means of one pair member precisely and rapidly imitating the other.

Reunions

Even dolphins that associate consistently separate from each other frequently to forage, travel, or socialize with other dolphins. They often end up several hundred meters or even kilometers apart, as Snubnose, Bibi, and Sicklefin did as they split up to forage. They must then find and reunite with each other in a vast expanse of water, often with many other dolphins around to potentially add confusion. Dolphins use individually distinctive signature whistles to facilitate such reunions. Mothers and infants provide a useful model to study this phenomenon, discussed further below, but the same general task is faced by all dolphins that associate with each other on a consistent basis.

Direction Changes

After their bout of rest, Sicklefin and Bibi began surfacing oriented toward the north, while Snubnose continued orienting toward the west. After several such surfacings, Snubnose apparently "gave in" and also turned to the north. Such behavior often precedes direction changes or group fissioning, and appears to result from conflicting motivations concerning the direction of travel.

A related phenomenon described by Norris et al. (1985) is "zig-

Rachel Smolker

zag" swimming by Hawaiian spinner dolphins *(Stenella longirostris)*. After a day of resting in coastal bays around the islands, the dolphins begin to rouse, as indicated by increases in travel speed, aerial behavior (leaps and spins), and vocalizations. The group may start out with some very active members in the lead while others, still apparently resting, trail behind. The group moves out toward the mouth of the bay, but then loses momentum, mills around for a bit, and then turns back into the bay, where the process is repeated, often several times. Ultimately, when the entire group reaches the same level of arousal, all head very rapidly and directly out to sea, to the feeding grounds where they will spend the night. Norris et al. suggest that this zig-zag swimming is a period of social facilitation during which all members become alert and ready to move offshore: "The school as a whole tests its own integrity as an alert unit capable of functioning in relative safety in deep water" (1985, 73). Zig-zag swimming by spinner dolphins may be analogous to the "social rallies" of many social carnivores, as discussed by Holekamp, Boydston, and Smale, chap. 20, this volume.

Shadowing
Often individuals and groups of dolphins in Shark Bay appear to coordinate their movements while traveling at some distance from each other (see chapter opening illustration), as was the case throughout the day as Snubnose, Bibi, and Sicklefin shadowed (and/or were shadowed by) Wave and Shave. This sort of shadowing is particularly evident between two alliances of males (second-order alliance partners), who may travel separated by a distance of 200 meters or so, but move in the same direction, making the same changes in direction at about the same time, and joining up occasionally. Similarly, it is not uncommon for a single dolphin to break away from a group to forage, shadow the movements of the group he or she left, and later rejoin it. Again, it is unclear who is responsible for coordinating position.

Group Geometry and Activity
The activity a group is engaged in is often indicated by its geometry (the relative positioning of individuals within the group) and the manner in which animals approach the surface or dive below. For example, when resting, Snubnose, Bibi, and Sicklefin formed a tight-knit group, traveled slowly, and spent considerable time below the surface between breaths. While traveling, they maintained a

rank abreast formation, all moving parallel to one another in the same direction, and remained close to the surface between breaths. When feeding, they spread out and swam in multiple directions, with their tails breaking the surface when they dove. When racing to a feeding aggregation (or social event), they spread out and leapt repeatedly.

A number of researchers (Norris and Dohl 1980b; Würsig and Würsig 1980; Whitehead 1989) have reported "rank formation swimming"—individuals traveling parallel to one another but separated by some distance—by various odontocete species. Rank formation foraging has usually been hypothesized to function by providing the group as a whole with a larger search area, enhancing the likelihood that some individual(s) will locate a school of prey (on which others can then converge). The best studied example involves sperm whales off the Galápagos Islands (H. Whitehead 1989). Female sperm whales live in fairly stable groups of about twenty adults (often kin) with their offspring (Whitehead, Waters, and Lyrholm 1991; Richard et al. 1996), generally in tropical waters. Females within these groups may share nursing and "babysitting" duties (Whitehead 1996). Males disperse as juveniles, travel to high-latitude feeding grounds, and spend a number of years separate from the female groups. As mature bulls (much larger than mature females), they travel singly in search of females, roving from group to group (Whitehead 1993). This social system has been aptly compared to that of the African elephant (*Loxodonta africana:* Weilgart, Whitehead, and Payne 1996).

H. Whitehead (1989) reports that when foraging, female groups of sperm whales spread out in a rank formation perpendicular to the direction of travel. Whales or small clusters of whales may be about 40 m apart in these ranks, spread over about half a kilometer. The whales dive to depths of about 400 m to feed (on squid). They surface periodically, about every 40 minutes, to breathe for about 10 minutes before descending again. Whales in rank tend to dive at the same time. Occasionally the whales stop foraging, come together, and remain near the surface, where they socialize and rest before spreading out into a rank formation again for another bout of foraging. Whitehead considers three possible functions of rank formation foraging: that it increases prey search area, that it enables individuals to take advantage of prey fleeing from the pursuits of neighbors, or that it serves to minimize feeding competition. He concludes that the latter is the most likely explanation.

Rachel Smolker

Aggregations and Cooperative Hunting

During some types of feeding activities in Shark Bay, many individuals, scattered over a large area, may aggregate and travel in what appears to be a coordinated fashion. Dolphins may suddenly break off what they are doing and race off, periodically leaping out of the water as they travel swiftly to a distant location (up to a kilometer or two distance). Upon arrival, they may spread out in an area where there are already many other dolphins "leap and porpoise" feeding. For a while, all of the dolphins converge within an area, zooming around in different directions, leaping and porpoising as they chase prey. After some time, the activity slows, and scattered individuals begin to reunite, form small groups, and travel. At this point, there may be quite a few small groups and some scattered individuals spread over a large area, all traveling roughly in concert, perhaps in the direction of the fish school. Then, at some location within this broad phalanx of dolphins, a few individuals will begin racing around, leaping and porpoising, and the remaining dolphins will converge on that area.

This phenomenon appears similar to the hunting behavior of dusky dolphins off Argentina described by Würsig and Würsig (1980) and Würsig (1986). Early in the day, these oceanic dolphins are found in small groups (6–15 individuals). The groups are widely scattered (20–30 groups over a 100 km² area) and tend to avoid one another. At some point one group begins diving, and shortly thereafter a school of southern anchovies (*Engraulis anchoita*) becomes visible as it is driven up against the surface by the dolphins. The dolphins swim around and under the school and leap around the edges, forcing the school into a tight ball. Occasionally an individual dolphin passes into the fish ball, coming out the other side with several anchovies in its mouth. (A similar strategy has been used against dolphins by orca: Brown and Norris 1956.) This activity attracts both birds and other dolphin groups.

Other groups of dusky dolphins may perceive the feeding activities of distant conspecifics using acoustic cues such as the echolocation clicks of the feeding dolphins or the loud cracks produced when a leaping dolphin slaps against the water surface on reentry (these leaps are frequent during feeding). The Würsigs hypothesized that a "food call" of some sort might be produced by members of the group that have discovered fish. Visual cues also may be useful; in particular, large aggregations of birds and the aerial behavior of the feeding dolphins may be visible from some distance. Whatever the mechanism(s) of attraction, many other dolphin groups converge on the area, leaping in from all directions and from distances as great as 8 kilometers away, resulting in aggregations of as many as 100–300 dolphins.

Small groups of dusky dolphins apparently cannot contain the fish school for long, and once it begins to break up, feeding ends. With more dolphins present, the fish school can be contained for longer. Hence the group that discovers the fish school may benefit from attracting others. After feeding, the dolphins engage in socializing and sexual activity, and gradually disperse back into their small, separate groups.

Observations of apparently coordinated hunting by bottlenose dolphins are described by Würsig (1986) and Tayler and Saayman (1972). These accounts include apparent "division of labor," with different members of the group attacking or herding a school of fish from different directions, and "turn-taking," with dolphins alternating time spent actually eating and herding the fish school while others eat. How the participants in such hunting efforts coordinate their activities is unclear.

Hoese (1971) and Rigley (1983) described bottlenose dolphins hunting in the salt marsh channels of the southeastern United States. Several dolphins form a rank abreast oriented toward the shore, with the prey fish (mullet) between themselves and the shore. They then rush simultaneously toward the fish, forcing them up onto the banks. The dolphins strand themselves partly out of water on the banks, snapping up the flopping fish off the beach, before they roll back into the water. Similar "strand-feeding" by lone individuals occurs as well.

Many accounts of odontocete cooperative hunting involve orcas. Orcas are widespread, occurring throughout the world's oceans. Their social systems seem to vary considerably from place to place, and even within the same locality. The best-studied groups occur off the coast of the Pacific Northwest of the United States. Here there appear to be two distinct lifestyles (Bigg et al. 1987; Jacobsen 1986; Morton 1990). "Resident" orcas live in very stable "pods" (the term used here to distinguish them from the more ephemeral "groups" of many other cetaceans). Pods consist of related females and their offspring, including adult sons (Bigg et al. 1990; Olesiuk, Bigg, and Ellis 1990). Although different pods do meet and interact at times, they remain discrete much of the time and have distinctive

Rachel Smolker

vocal dialects (Ford 1989, 1991). These resident pods feed primarily on salmon and other fish species (Heimlich-Boran 1986).

In addition to the residents, there are "transient" whales. Transients are usually found as single individuals or in small groups. They appear to range much farther than residents, occasionally passing through the ranges of residents in the area. Rather than a primarily piscivorous diet, these orcas feed on other marine mammals and birds (Baird and Stacey 1988; Baird and Dill 1995; Felleman, Heimlich-Boran, and Osborne 1991).

Orcas have frequently been observed to cooperate while hunting large baleen whales, pinnipeds, and fish (Martinez and Klinghammer 1970; Condy, van Aarde, and Bester 1978; Smith et al. 1981; Steltner, Steltner, and Sargent 1984; Ljungblad and Moore 1983; Lopez and Lopez 1985; Hall 1986; Bigg et al. 1987; Guinet 1991; Silber, Newcomer, and Perez-Cortez 1990; Tarpy 1979; Steiner et al. 1979; Simila and Ugarte 1993; Hoelzel 1991; Jefferson, Stacey, and Baird 1991). For example, Tarpy (1979) reported that, when attacking a 60-foot blue whale in the Gulf of California, orcas "exhibited a distinct division of labor. Some flanked the blue on either side, as if herding it. Two others went ahead, and two stayed behind to foil any escape attempts. One group seemed intent on keeping the blue underwater to hinder its breathing. Another phalanx swam beneath its belly to make sure it didn't dive out of reach. The blue whale's dorsal fin had been chewed off and its tail flukes shredded, impairing its movement."

Steltner, Steltner, and Sargent (1984) reported seeing several whales swimming rapidly in a coordinated rank abreast formation toward a seal *(Lobodon carcinophagus)* hauled out on a small ice floe. Their rapid approach created a wave that washed the seal into the water. Hoelzel (1991) and Lopez and Lopez (1985) observed orcas stranding themselves on the beach in pursuit of pinnipeds *(Otaria flavescens, Mirounga leonina)*, which were sometimes passed back and forth between whales after capture.

Simila and Ugarte (1993) described small groups of orcas in Norwegian waters encircling and herding schools of herring *(Clupea harengus)*, causing the fish to "ball up." The whales then hit at the edge of the fish ball with their tail flukes, stunning fish, which were consumed not just by the whale that hit them, but by any member of the feeding group. Similar behavior has been observed in Shark Bay bottlenose dolphins (Smolker and Richards 1988).

Mechanisms by Which Group Movements Are Coordinated

Sensory Integration

Norris and Dohl (1980b) and Norris and Schilt (1988) proposed that dolphin groups "serve to integrate the sensory inputs of its many members to provide important environmental information to part or the entire school" (Norris and Schilt 1988). This information, they suggest, is transmitted through the school as "waves of influence," signaled by means of body movements (e.g., a slight shift in the plane of the pectoral fin), by shifts in visibility of the color patterns of neighboring animals, and by acoustic cues. These schooling mechanisms are not unlike those occurring in fish.

Although "waves of influence" per se have not been studied, some sort of mechanism along these lines must be involved in the remarkable split-second coordination of movements during such activities as synchronous surfacings and displays and some directional changes that occur at close range.

Coordination of movements occurring at distances greater than a few meters most likely involves acoustic cues. All odontocetes echolocate, and it is possible that at medium ranges, they may "coordinate" movements passively simply by listening to one another's clicks. Certainly information about the presence of another individual, what she is doing, and perhaps her direction of travel is available simply by listening. Listening to echolocation clicks may play a role in the maintenance of position in rank formation foraging by sperm whales, for example.

Territorial "Long" Calls

There is little evidence for territoriality in any cetacean. However, two examples of cetacean behaviors appear to serve the function (at least in part) of maintaining spacing between individuals. The first comes from sperm whales. Mature bulls emit extremely intense, slow-repetition-rate clicks while roving in the vicinity of female groups. Weilgart and Whitehead (1988) suggest that these might function to maintain spacing between bulls and, if females can assess the "quality" of these clicks, perhaps also in female choice. Similarly, the songs of humpback whales are thought to be a male display used in male-male competition or perhaps serving to maintain spacing between males in addition to attracting females (Payne and McVay 1971; Tyack 1981, 1983; Frankel et al. 1991; Helweg et al. 1992).

Rachel Smolker

Group-Specific Calls

Group-specific call dialects have been reported for two cetaceans: orcas and sperm whales, both of which live in stable "pods." Presumably these dialects function in part to maintain cohesion among the individuals sharing a dialect. Orcas produce strident, often frequency-modulated calls. Call dialects in orcas are so marked that even human listeners can identify different whale pods aurally, and it is probable that the whales do so as well. Pods with different call dialects interact, yet still maintain their distinctive call repertoires (Ford 1991).

Sperm whales produce loud clicks, audible several kilometers away, some of which probably function in echolocation. In addition, some clicks are produced in "codas" (Watkins and Schevill 1977): several clicks emitted in series with distinctive timing patterns (e.g., two clicks at equal long intervals followed by three at shorter intervals). Different groups of sperm whales have different coda repertoires (Weilgart and Whitehead 1997).

Individual-Specific Calls

Probably the best-known mechanism used to coordinate movements—in this case, separations and reunions—among dolphins involves whistles. Melba and David Caldwell recorded whistles from captive dolphins at various aquaria and marine parks, and discovered that each dolphin produced a unique whistle type. They hypothesized that dolphins use these whistles to announce their identity to other group members, and referred to them as "signature" whistles (Caldwell and Caldwell 1965, 1968; review in Caldwell, Caldwell, and Tyack 1990).

The use of signature whistles makes considerable sense considering the task dolphins are faced with. They have very strong ties with particular other dolphins and therefore must differentiate among individuals, yet they are perpetually moving about in a visually opaque medium. An acoustic signal that unambiguously identifies the sender to others would be useful, if not essential. The receiver must recognize the signal—that is, discriminate between it and other audible signals—and also associate it with a particular individual. This capacity has been demonstrated by Caldwell, Caldwell, and Hall (1973) and Sayigh (1992).

The ability to produce a signature whistle to announce one's identity and location (and perhaps other information) could be beneficial if it permitted a preferred associate to relocate and rejoin the whistler. But coordinating travel could be a lot more effective if whistles contained additional information—information that seems necessary given the degree of coordination that dolphins appear capable of. For example, dolphins might benefit from being able not only to say "I am here," but also to ask "are you there?" and to reply "I heard you (and, by the way, I am over here)." I will refer to these different types of signals as (1) announcement, (2) query, and (3) acknowledgment.

Thus far, we have much evidence for announcements. Most dolphins that have been recorded have been found to have a signature; that is, one whistle type that predominates in the repertoire, usually accounting for most whistles produced by that individual. This has been true for dolphins both in captivity and in the wild (Caldwell, Caldwell, and Tyack 1990; Sayigh et al. 1990; Smolker, Mann, and Smuts 1993; but see also McCowan and Reiss 1995a and R. Smolker and J. W. Pepper, in press). In the contexts in which they were recorded, most dolphins emitted their signatures repetitiously, sometimes with occasional "variants" (whistle types that differed from the predominant "signature" type). In sum, it seems that most dolphins can and do announce their identity, often repetitiously.

Queries and acknowledgments necessarily involve an exchange of information. Whistle exchanges have been reported in a few cases (Sayigh 1992; Gish 1979; McBride and Kritzler 1951; Tyack 1993), but these have generally involved captive animals, held forcibly apart. We remain quite ignorant about how whistles are normally used in the day-to-day lives of dolphins in the wild.

This ignorance is partly due to the difficulties inherent in studying dolphin communication. Since dolphins produce their sounds internally (in a system of air sacs and other structures located in the forehead), there are no externally moving mouthparts that can be used to attribute specific sounds to individuals. Therefore, when many dolphins are present, it is difficult to determine who is responsible for any given whistle. Studies of dolphin whistles have thus primarily made use of captive animals held out of water (Caldwell, Caldwell, and Tyack 1990) or free-swimming (Tyack 1986; McCowan and Reiss 1995a,b), or of temporarily captured wild animals (Sayigh et al. 1990; Sayigh 1992).

In the wild, dolphins move almost constantly. To observe and record them, it is therefore necessary to move with them. This generally requires an outboard motor, which is noisy and interferes with the recording (Sayigh, Tyack, and Wells 1994). Thus, it is difficult to simultaneously hear and observe wild dolphins.

Rachel Smolker

Mother-infant Reunions

We studied signature whistle use in wild, freely-swimming mother-infant pairs in Shark Bay (Smolker, Mann, and Smuts 1993) by following and recording infants from a small boat powered by an electric motor, which is largely quiet underwater, using a hydrophone towed alongside. This method allowed us to hear and record simultaneously much of the time.

Mother-infant pairs provide a tractable system in which to study whistle use because they often travel as a lone pair, rather than in the company of many other dolphins. Thus, we could be quite certain that the whistles we recorded were produced by either the mother or the infant. They also separate from each other frequently, travel to distances up to 300 meters apart, and then later reunite, just as adult associates do. When they were separated, we could stay close to one or the other, allowing us to further narrow down the source of whistles. We almost always chose to stay with infants when they separated from their mothers.

Infant mortality is very high (Richards 1996). Presumably infants are safer when in close proximity to their mothers, yet they spend considerable time apart from them. That mothers and infants, even very young infants, engage in such separations is curious, and we assumed that they must, at least, have very effective ways of keeping in contact.

We predicted that mothers and infants would exchange whistles to facilitate reunions; for example, an infant might whistle to "query" its mother, and she might respond with an "acknowledgment" to provide the infant with information about her whereabouts. We suspected that they would rely on these whistle exchanges to relocate each other.

In line with our predictions, we found that infants rarely whistled when with their mothers. Whistles were emitted primarily during separations and reunions. They were, in fact, concentrated around the time when infants were *in the process* of reuniting with their mothers (traveling directly and rapidly toward them). The longer the infants were away from their mothers, and the farther apart they were from their mothers, the more whistles they produced. These findings provided support from the wild for the hypothesis that signature whistles are used to mediate reunions.

Surprisingly, though, we rarely heard whistles that were likely to be from mothers. This could simply be an artifact of the fact that we stayed closer to infants during separations. Mothers may respond to their infants, but produce only one or a small number of whistles, making them easy for us to miss. In some cases mother and infant did whistle back and forth in an ongoing "query and response" exchange. Often, however, it seemed that infants were "announcing" themselves, and apparently relying on other cues to navigate back to their mothers' sides.

Whistle Mimicry

If dolphins do "query" and "acknowledge," and if the communications of a large number of conspecifics are audible at any given time (as is undoubtedly the case much of the time), it might make sense that they direct their queries and responses to a specific desired audience. Tyack proposed one possible mechanism by which this could be achieved (1986). Recording pairs of captive dolphins, he found that each had a primary and a secondary whistle (the most frequently and second most frequently emitted whistles). Each dolphin's secondary whistle was the other's primary. This, Tyack suggested, might result from each dolphin's emitting its own signatures (primary whistles) and also imitating the partner's signature (secondary whistles), which could serve to establish contact.

This suggestion is exciting not only because of parallels with human use of names, but also because it offers an explanation for the unusual vocal imitation capabilities of dolphins based on their natural behavior. Dolphins are exceptional among mammals, other than humans, in having the capacity to mimic vocalizations (in this case whistles) they are exposed to. This capacity was first discovered when dolphins imitated the training whistles used by handlers (Caldwell and Caldwell 1972), and was later investigated systematically by Richards, Wolz, and Herman (1984) and Reiss and McCowan (1993), who trained dolphins to imitate computer-generated whistles.

If signature imitation proves to be real, a sender emitting another dolphin's signature could query a very specific receiver for a response. Alternatively, a receiver could respond by emitting the sender's signature, thereby acknowledging that he/she heard that specific query. Overall, such a system fits well with the facts that (1) dolphins have stable long-term bonds with specific others (2) set within a fission-fusion social system (3) in a habitat in which sound is the most useful medium for signaling and (4) where the acoustic signals of many individuals are audible, resulting in a high potential for confusion.

What are the potential cognitive implications of the ways in which dolphins use signature whistles to coordinate their movements? First, because many animals are audible at any given time, the acoustic environment is likely to be confusing. Dolphins must somehow filter through many irrelevant signals in order to pinpoint and attend to those whistles that are relevant to their own purposes. Because of the "wide availability" of acoustic signals, individual dolphins may be apprised of the whereabouts of many of their conspecifics at any point in time. They may in essence maintain a continually updated cognitive map of the distribution of many of their associates.

The "downside" of being able to hear sounds at great distances is that it probably makes it difficult to conceal information. A signature whistle intended to provide information regarding one's whereabouts to a particular conspecific could easily be "parasitized." For example, it may be in a female's interest to broadcast her whereabouts to her calf, but she may also wish to conceal that information from a male alliance. Similarly, an alliance of males herding a female may well lose the female to another alliance if they come into contact. It may therefore be quite costly for a male to signal his whereabouts to his alliance partners if he thereby makes this information available to a wider audience, including males from competing alliances.

Smolker et al. (1992) hypothesized that the apparently higher degree of male-male competition evident in Shark Bay as compared with Sarasota Bay, Florida, may result in part from differences in the habitat that affect acoustics. Sarasota Bay is subdivided by several bodies of land and barrier islands, resulting in a more fragmented body of water, with greater acoustic isolation of its parts. Shark Bay is basically one large body of water. In Shark Bay, therefore, it may be that males can easily hear sounds produced by other males and females during herding. Thus, they can monitor and interfere with each other's behavior more readily.

The capacity for whistle mimicry could present further complications. Unless dolphins are capable of discriminating between authentic and imitation renditions of a whistle, the potential for confusion would seem tremendous. Given that sounds are distorted over distance (e.g., by irregularities in the transmitting properties of seawater caused by temperature and salinity gradients), such discriminations would seem quite difficult (though dolphin hearing is extraordinary). This raises the possibility, though purely hypothetical at this point, that dolphins might be able to manipulate and perhaps deceive one another by imitating whistles. In short, whistle mimicry raises new possibilities that bring into sharp relief our ignorance of dolphin behavior.

Mysticetes

As mentioned earlier, the most fundamental difference in the behavior of odontocetes and mysticetes of relevance to this discussion is that mysticetes are generally far less social. Mysticete whales are typically found singly or in temporary aggregations, and long-term associations between particular individuals (other than mother-calf pairs) appear to be rare.

Individuals of most species of mysticetes travel huge distances as a matter of course. For example, one radio-tagged bowhead whale traveled 2,670 miles in just 34 days (B. R. Mate, pers. comm.). Blue whales have been followed traveling 250 miles between inshore and offshore feeding grounds in a matter of a few days (B. R. Mate, pers. comm.), and one individual was tracked acoustically as it traveled 1,700 miles in 43 days (Gagnon and Clark 1993). A right whale *(Eubalaena glacialis)* cow-calf pair was tracked as they traveled a circuit from the Bay of Fundy in Nova Scotia to New Jersey and back again, a distance of 2,500 miles in 6 weeks (Mate, Nieukirk, and Kraus 1997).

Some mysticete species undertake annual long distance migrations between cold water feeding grounds and warm water breeding grounds. Grey whales *(Eschrichtius robustus),* for example, travel up and down the west coast of North America between breeding lagoons in Baja California to feeding grounds in the Bering, Chukchi, and Beaufort Seas, a journey of approximately 6,000 miles. Humpback whales also migrate long distances. For example, in the Pacific Ocean, they move between feeding grounds off Alaska and breeding areas in Hawaii and Mexico. Probably the best-studied migratory species, humpbacks show little evidence of consistent associations between individuals (Mobley and Herman 1985; Darling, Gibson, and Silber 1983; Tyack and Whitehead 1983; but see D'Vincent, Nilson, and Hanna 1985; Weinrich 1991). Thus, it appears unlikely that migrations are "coordinated."

Some evidence for coordination of migratory movements comes from bowhead whales. Bowheads migrate through ice-laden Arctic waters, apparently making use of the reverberation of their calls off ice bodies to navigate (Ellison, Clark, and Bishop 1987; George et al. 1989). While migrating, groups of three to five whales, spread out over a large area, may exchange calls periodically (Clark 1989, 1990).

Little evidence for cooperative foraging exists for mysticetes, though in some species groups of whales do swim in roughly coordinated fashion while feeding on the same prey school (e.g., humpbacks: Whitehead 1983). One example involving more complex coordination involves "bubblenet" foraging by humpback whales (Jurasz and Jurasz 1979; Hain et al. 1982). In this technique, eight to ten whales dive together underneath a school of prey. The whales then rise slowly toward the surface, releasing a steady stream of bubbles to surround part of the school. Once the prey are entrapped, the whales lunge into the center of the enclosure, gulping great mouthfuls of the fish. Interestingly, whales that bubblenet forage together sometimes associate with one another over periods of at least several days (D'Vincent, Nilson, and Hanna 1985). How the whales coordinate this activity remains unknown. They do produce vocalizations, but it is not clear whether these function to influence the behavior of prey or that of other whales (Sharpe et al. 1998).

Given the large scale of whale movements, it makes sense that the mysticete whales might communicate over very long distances, an idea pioneered by Payne and Webb (1971). Recent evidence derived from recordings made by the U.S. Navy's "integrated undersea surveillance system" (IUSS), recently opened to civilian use, have supported this hypothesis (Clark, Gagnon, and Mellinger 1993). Whale sounds (primarily blue, finback, minke, and humpback) were detectable to the Navy's hydrophones (and potentially, therefore, by other whales) at distances of hundreds of miles—in the case of blue whales, over a thousand miles. Individual whales were tracked acoustically as they traversed large areas of the Atlantic Ocean basin.

We do not yet know that whales make use of distant sounds. Singing humpback whales have been observed to stop singing and travel directly toward active groups at distances up to 9 km (Tyack and Whitehead 1983). The possibility remains that whales are in contact and in effect members of a "group" even when spatially separated by very large distances. We may ultimately be forced to reconsider how we define groups for mysticetes and whether they engage in "coordinated group movement" over such distances.

Conclusions

Mysticetes and odontocetes differ greatly in the degree to which they form stable and complex social relationships. Mysticetes spend much time on their own, or in the company of those other whales that happen to be nearby. Stable associations are apparently lacking. There are some observations of whales traveling in an apparently coordinated manner while feeding, particularly during bubble-net foraging. It is not yet clear whether these actions are actively coordinated or the whales are simply acting independently. It is quite feasible that whales might, in some situations at least, enjoy the benefits of increased foraging success by coordinating their actions regardless of whether they have anything to do with one another outside of that context.

Many odontocetes, on the other hand, form stable associations: for example, matrilineal "pods" in the case of orcas and sperm whales and "preferred" partners (male allies or female kin) in the case of bottlenose dolphins. Complex social relationships involving both competition and cooperation are the norm.

Reflecting the importance of these relationships in determining who goes where with whom and how, most dramatic examples of coordinated movement involve individuals with a history of social interaction. Such examples include the spectacularly coordinated displays of allied male dolphins, the precisely synchronized surfacing of social partners during travel, and the use of signature whistles to mediate the comings and goings of a fission-fusion grouping system.

Given these examples, it seems likely that evolutionary pressures favoring coordination of movements among odontocentes have been strong, and that coordination mechanisms have evolved hand-in-hand with sociality. We would expect individuals to actively coordinate their movements when doing so results in mutual benefit. Thus, in Shark Bay, for example, mothers and infants are likely to benefit from successfully reuniting when they have been apart. Females are likely to benefit from coordinating their travel with one another when remaining nearby and in contact, possibly through allomaternal care or decreased vigilance against threatening predators such as sharks. Males may benefit from coordinating their movements with both first- and second-order allies, since their continued cooperation seems to be an essential factor in obtaining access to females and protecting themselves against thefts by other alliances.

Rachel Smolker

R. Smolker and J. W. Pepper (in press) have hypothesized that synchronization may be subject to sexual selection. Females may prefer to mate with males who are skillful at performing behaviors that display alliance solidarity, since alliance formation and performance are apparently critical to male mating success (and thus the future success of their sons). It is in fact difficult to interpret the elaborate displays of allied males as anything other than a sexually selected characteristic.

The flip side of the situation in which cooperative social relationships dictate who coordinates movements and how is that individuals with competing interests may actively seek to avoid one another. In Shark Bay, competing male alliances may attempt to steer clear of one another to avoid conflict. An alliance in the process of herding a female may attempt to conceal that fact from another alliance that could take the female from them. It may be in the interest of a female to conceal her whereabouts from males who might harass her.

The highly individualized nature of signature whistles and their use during separations and reunions appear to be the result of strong selection pressures favoring accurate recognition and relocation of social partners. Yet, keeping in mind that it may at times be advantageous to conceal one's identity, or even to deceive another individual with regard to one's identity, the signature system has the potential for far greater complexity. Though it is purely conjecture at this point, such cognitively sophisticated social manipulation might be expected from dolphins given what we know of their behavior.

We are far from understanding much about the behavior of most cetaceans, but one thing is clear: all move almost constantly, and, particularly for the odontocetes, keeping in touch with other individuals is an ongoing, pervasive part of daily life. At distances of more than a few meters, acoustic signals are probably the primary, if not the only, means for coordinating travel. Perhaps with improved technologies for recording sounds and attributing them to specific individuals, we may get a better handle on how, with whom, and why dolphins and whales coordinate their movements. In the process, we may discover new dimensions to the social intelligence of dolphins.

Acknowledgments
Thanks are due to my colleagues on the Shark Bay dolphin research project, particularly Andrew Richards, Richard Connor, and Janet

Mann. Research support from the National Geographic Society, the Dolphins of Shark Bay Research Foundation, the National Science Foundation, the University of Michigan, and the Seebie Trust is gratefully acknowledged. Logistic support from the rangers at Monkey Mia, the Western Australia Museum, the University of Western Australia, the Department of Conservation and Land Management, and the Monkey Mia Resort has been invaluable. Thanks also to Peter Tyack, Hal Whitehead, John Pepper, and Andrew Richards as well as anonymous reviewers for commenting on the manuscript.

Note

1. The genus *Tursiops* is currently considered to comprise one widely distributed species *(T. truncatus)* with variable geographic races, including the Shark Bay dolphins, which were formerly classified as a separate species, *T. aduncus* (Ross and Cockroft 1990). Recent molecular work (e.g., Curry et al. 1995) is pointing to a revision that could ally the Shark Bay dolphins more closely with the stenellids than with other members of the genus *Tursiops.*

CHAPTER TWENTY

Group Travel in Social Carnivores

KAY E. HOLEKAMP, ERIN E. BOYDSTON,
AND LAURA SMALE

Endowed with massive jaw muscles, specialized bone-crushing teeth, and the ability to run at over 60 km per hour (Kruuk 1972), spotted hyenas hunting and feeding together can bring down an adult zebra and reduce it to a few scattered bone fragments in as little as 20 minutes. Cooperative hunting permits hyenas and other gregarious carnivores to capture prey animals many times larger than the body mass of any individual hunter. How are the movements of individual hunters coordinated during these group hunts? How does the distribution and abundance of prey influence group movements in carnivores? Which individuals make decisions about the distance and direction traveled? In this chapter we will address these and related questions about group movements in mammalian carnivores.

The problems confronting carnivores during foraging and reproduction differ from those confronting primates in several ways that might be expected to promote dissimilarities in movement patterns between the two taxa. Therefore, we will first review the basic differences between carnivores and primates with respect to feeding ecology and reproduction. Second, we will summarize current knowledge about group travel and its initiation and coordination in five types of gregarious carnivores for which substantial amounts of field data have been collected. Third, we will identify the common themes that emerge from our appraisal of group travel in these five carnivore taxa. Fourth, we will consider two special types of group movements in carnivores, den moves and hunts, that will help to elucidate the mechanisms by which group travel may be coordinated. We will also compare group hunts in carnivores and nonhuman primates. Fifth, we will summarize the insights gained about the cognitive abilities of social carnivores from the study of their group travel. We define cognition broadly, as those mental processes that are presumed to be occurring within an animal but which cannot be directly observed. Finally, we will inquire whether carnivores differ qualitatively from nonhuman primates in the coordination of group-level activities.

The Contrasting Natural Histories of Carnivores and Primates

The sensory world of any mammalian carnivore is usually dominated by olfactory stimuli. Specialized scent glands abound in the skin of all carnivores, and glands located on the soles of the feet leave odor trails on the substrate (Ewer 1973). The sense of smell is acute in most carnivores (e.g., Bradshaw 1992). Scents remain present in the environment far longer than do visual or acoustic stimuli, and they offer an effective means of information transfer under conditions in which visual communication is impaired. In contrast to primates, most carnivores exhibit nocturnal or crepuscular patterns of activity (Gittleman 1989), so even when foraging in open habitat, carnivores may not be able to see their companions. Thus, compared with most primates (e.g., Garber, chap. 10, this volume), carnivores rely far more heavily on olfactory cues to recognize conspecifics, maintain contact with other group members during travel, navigate to feeding sites, and maintain territorial boundaries.

Most carnivores feed on prey animals that can detect the hunter's presence and take evasive action in response, so groups of terrestrial carnivores generally forage very quietly. Fruits and leaves don't run away when herbivores approach. Therefore, except when hunting vertebrate prey (e.g., Boesch and Boesch 1989; Rose 1997) or when noisy travel increases their own vulnerability to predators (e.g., Terborgh 1983; Wright 1994; Boinski, Treves, and Chapman, chap. 3, this volume), monkeys can make a tremendous racket during group travel without reducing their foraging success.

The costs of duplicating foraging paths may be similar for primates and insectivorous carnivores such as mongooses, since days or weeks may be required for invertebrate prey to restock an area. However, these costs are probably low for large carnivores due to the mobility and foraging habits of their vertebrate prey.

Prey animals represent large, ephemeral packets of energy-rich food occurring unpredictably in space and time. Compared with the small, stationary, relatively abundant packets of vegetable material consumed by most primates, prey animals are rare and widely dispersed, forcing carnivores to travel relatively long distances while foraging. Thus home range size in carnivores usually exceeds that in primates of comparable body size (McNab 1963). Moreover, to minimize intraspecific competition for food, most carnivores defend individual territories containing resources adequate only for their own life support (Gittleman 1984). Thus, in contrast to primates, most carnivores are solitary creatures that interact exclusively with their mates, their offspring, and alien conspecifics at territorial boundaries (Ewer 1973; Eisenberg 1966, 1981). In fact, only 10–15% of extant carnivore species live in stable social groups (Gittleman 1984; Bekoff, Daniels, and Gittleman 1984). Group living has apparently been favored in some of these species by improved ability to detect and evade predators, and in others by improved ability to acquire or defend territories, young, mates, or food (Kruuk 1975; Pulliam and Caraco 1984; Clark 1987; Houston et al. 1988; Packer, Scheel, and Pusey 1990; Rasa 1986b; Rood 1986, 1990; Burger and Gochfeld 1992; Gompper 1996).

Various authorities have suggested that gregariousness has been favored in some large carnivores by improved energy intake due to feeding on large-bodied prey or by reduced energetic costs of prey capture (e.g., Kruuk 1972, 1975; Tilson and Hamilton 1984; Creel and Creel 1995). Others have presented data that appear to contradict the predictions of these hypotheses, particularly among carnivores reliant on stealth while stalking prey (e.g., Packer, Scheel, and Pusey 1990; Caro 1994; Packer and Caro 1997). This controversy will undoubtedly persist until appropriate energetic data are collected from a wider variety of carnivore species. At present the possibility cannot be ruled out that increased per capita energy yield in larger hunting groups has favored gregariousness in long-distance coursers such as canids (Creel 1997) and spotted hyenas (Holekamp, Smale 1997) (see appendix 20.1 for Latin names of species mentioned in this chapter). However, other selection pressures must clearly also have promoted sociality in carnivores, even in those species that do not rely on stealth while stalking prey (Creel 1997; Packer and Caro 1997).

The individual units of food exploited by most primates tend to weigh a few grams, whereas those exploited by large carnivores may weigh hundreds of kilograms, and several hunters may be required to secure a single prey animal. However, because ungulate carcasses represent such incredibly rich but ephemeral energy sources, contest competition (also called direct or interference competition) is often intense among the carnivores feeding on them. Even within the most stable carnivore groups, this intense competition functions as a disruptive force, often prompting individuals to leave their companions temporarily in order to eat without being disturbed. In gregarious carnivores, per capita rates of food intake may increase when individuals separate from their social groups to forage alone (e.g., lions: Packer, Scheel, and Pusey 1990; spotted hyenas: Holekamp, Smale et al. 1997; coatis: Gompper 1996; European badgers: Kruuk 1989).

Another disruptive force in carnivore society is introduced when young are born, because the special needs of neonates affect adult foraging behavior. As a result, the gregariousness of female carnivores tends to vary with their reproductive state (e.g., Holekamp et al. 1993; Holekamp, Cooper et al. 1997). Most carnivores are blind at birth and far more helpless than are newborn monkeys. Furthermore, because their limbs are specialized for running, infant carnivores cannot cling to their mothers as young primates do. Therefore carnivore litters are secreted in dens or creches until young are capable of traveling with older conspecifics. Adult female carnivores often go off alone to bear their young and tend them while they are immobile and most vulnerable. When young carnivores are sedentary in dens, adult group members are often forced temporarily to become central-place (radial) foragers, which means that they must return regularly to a fixed nest or den instead of simply using whatever handy refugia or rest sites they encounter each day along their foraging routes. Central-place foraging usually ceases once pups are mobile.

The upshot of these taxonomic differences in foraging and reproduction is that most gregarious carnivores live in societies that can be characterized to some extent as fission-fusion in nature, and in general, carnivore groups are less cohesive than are troops of pri-

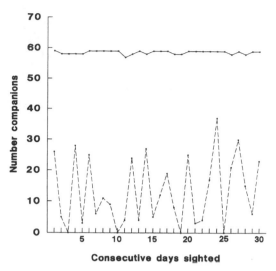

Figure 20.1 Mean number of other group members likely to be found during a month-long series of daily observations in close physical proximity to an adult female olive baboon (solid line) and a sympatric adult female spotted hyena (dashed line) in the Serengeti ecosystem. This number varies little from day to day for the female baboon, but it varies enormously for the female hyena, due to demands on the latter of foraging and reproduction. Furthermore, although these two females are both members of groups containing sixty conspecifics, they regularly associate with very different numbers of groupmates. The hyena data were collected systematically from a middle-ranking adult female member of the Talek study population; the baboon data were derived from our casual observations of a troop of *Papio anubis* living sympatrically with the Talek hyena clan.

mates. For example, consider a typical female olive baboon inhabiting the Serengeti ecosystem of eastern Africa. This female is likely to belong to a troop containing sixty or more members, with whom she sleeps, forages, and travels hour after hour, every single day (fig. 20.1). In her entire life there is unlikely to be even so much as a 10-minute period during which she is more than a few hundred meters from everyone else in her troop. Now consider a female spotted hyena living in a clan with sixty members in the exact same habitat as that baboon troop (fig. 20.1). Although she may join thirty clanmates to win a dawn war with a neighboring clan, the female hyena then wanders off to rest with a few companions, and by noon she may be found hunting gazelles several kilometers from the nearest hyena. That evening she is once again likely to meet up with other clan members, but not necessarily with any of the same animals with whom she interacted that morning. Thus, in contrast to that of the female baboon, her life is characterized by an endless series of relatively brief meetings with other clan members, all of whom she recognizes, and with whom she defends a group territory.

There are, of course, exceptions to the general patterns described here for carnivores and primates. For example, chimpanzees and spider monkeys live in fission-fusion societies (e.g., Symington 1990), while dwarf mongooses live in tight-knit packs (Rood 1983). Furthermore, group living among predators does not necessarily entail group travel, and many gregarious carnivores that defend group territories often forage alone (e.g., European badger: Kruuk 1989). Our discussion here will be restricted to those terrestrial carnivore species in which individuals regularly engage in group movements (table 20.1). We shall define group movements as situations in which two or more animals travel simultaneously in the same direction for extended periods while remaining within approximately one hundred body lengths of each other. We will focus on coatis, dwarf mongooses, social canids, lions, and spotted hyenas because the field data documenting group movements in these species are most abundant. For each different type of carnivore, we will describe the circumstances under which group movements occur, how group movements are initiated, how group cohesion is maintained during travel, and which group members appear to make decisions about the timing and direction of travel. We will define group leaders as those individuals who depart first from a rest site and/or who travel at the front of the group (see Boinski, chap. 15, this volume).

Case Studies of Five Gregarious Carnivores

Coatis
Social Life and Group Travel.   Coatis are raccoonlike creatures inhabiting Neotropical forests, where they eat invertebrates, small vertebrates, and fruits (table 20.1). Coatis form bands containing four to thirty adult females and their offspring of the previous 2 years, which sleep together in a common roost tree at night and travel together during the day (Kaufmann 1962; Lanning 1976; Russell 1983; Ratnayeke, Bixler, and Gittleman 1994). Such groups, which may contain up to fifty-six coatis (Kaufmann, Lanning, and Poole 1976; Timm et al. 1984), represent extended families, but often also contain some unrelated individuals (Gompper, Gittleman, and Wayne 1997). Adult males are usually solitary except during the breeding season, when they may temporarily join a band

K. E. Holekamp, E. E. Boydson, and L. Smale

Table 20.1 Space use and group travel among gregarious carnivores

Species	Habitat[a]	Diet[b]	Mean body mass (kg)	Population group size[c]	Mean travel group size	Daily linear distance (km)	Home range size (km²)	References
Coati	F	I, Fr	5.0	8	7.3	1.7	0.4	Kaufmann 1962; Russell 1983
Dwarf mongoose	O	I, SV	0.3	10	10.0	0.5	0.8	Rood 1983; Rasa 1986b, 1987
Banded mongoose	O	I, SV	1.3	14	14.0	?	?	Rood 1983
Meerkat	O	I	0.7	12	12.0	?	?	Rood 1983
Common cusimanse	F	I, SV	0.8	13	13.0	?	?	Ewer 1973; Goldman 1987
Giant river otter	FS	F	26.5	5	4.0	?	35	Duplaix 1980
Wolf	O, F	SV, LV	33.1	7	5.7w, 3.7s[d]	14.4	255–390	Mech 1970; Messier 1985
Coyote	O, F	SV, LV	10.6	5	2.3w, 1.4s[d]	8.0	5–14	Bowen 1982; Bekoff and Wells 1986; Andeldt 1985; Gese, Rongstad, and Mytton 1988; Camenzind 1978
Golden jackal	O	SV	8.8	3	2.0	?	8	Kleiman and Eisenberg 1973; Macdonald 1979a
Dingo	O	SV, LV	14.0	11	2.3	3.3	37–80	Corbett 1995; Thomson 1992
Dhole	F	LV	15.8	6	6.0	?	69	Davidar 1975; Johnsingh 1982; Venkataraman, Arumugan, and Sukumar 1995

continues

Table 20.1 (*continued*)

Species	Habitat[a]	Diet[b]	Mean body mass (kg)	Population group size[c]	Mean travel group size	Daily linear distance (km)	Home range size (km²)	References
Cape hunting dog	O	LV	22.0	10	10.0	10.0	250–2,500	Estes and Goddard 1967; Schaller 1972; Fuller and Kat 1990; Kuhme 1965a,b; Frame et al. 1979; Creel and Creel 1995
Lion	O	LV	166.0	9	4.0	11.0	29–400	Schaller 1972; Rudnai 1973
Spotted hyena	O, F	LV	51.9	55	3.0	19.0	30–1,500	Kruuk 1972; Mills 1985, 1990; Gasaway, Mossestad, and Stander 1991; Henschel and Skinner 1991; Holekamp, Smale et al. 1997; Frank 1986

[a]F, forest; O, open woodlands and grasslands; FS, forest streams.

[b]I, invertebrates; Fr, fruit; SV, small vertebrates; LV, large vertebrates; F, fish.

[c]Data from Gittleman 1989.

[d]W, winter; S, summer.

(Smythe 1970; Russell 1981). Females leave their bands during pregnancy and make tree nests, where they bear their litters alone. During this period bands are composed of juveniles and any adult females deferring reproduction that year (Ratnayeke, Bixler, and Gittleman 1994). Breeding females rejoin the band only when their young can run well. Bands routinely split and rejoin during the course of a day, and individual females may forage alone for up to 24 hours before rejoining the group. Bands move over their range without using any consistent travel routes (Kaufmann 1962). Although coatis feed almost constantly as they travel, band members do not cooperate in obtaining food. However, females sometimes join forces to displace larger solitary males from food (Gompper 1996). There is considerable home range overlap between bands, and there is little or no friction when bands meet (Kaufmann 1962). Instead, bands simply tend to avoid one another.

Initiation of Group Movements and Maintenance of Group Cohesion during Travel.   Group movements in coatis are preceded by much social grooming among band members (Kaufmann 1962). Coatis emit quiet gruntlike contact vocalizations more or less continuously during group travel. They typically spread out while moving and feeding on the forest floor, with some members up to 100 m from any others (Kaufmann 1962). Thus, when coatis forage in groups, they are often out of sight of one another, and keep in touch by contact vocalizations. They also tend to move with their ringed tails elevated, which presumably helps them to maintain visual contact. When a band member becomes separated from its group, it typically utters a chittering "lost call," but may also simply follow a scent trail to relocate the group. Foraging coatis bunch up tightly and increase their rate of contact calling during travel at times when visibility is particularly poor (Kaufmann 1962).

Leadership.   After reviewing several different lines of observational evidence, Kaufmann (1962) concluded that there is no consistent leadership within coati bands. While traveling, juveniles are seldom in the vanguard, but any adults or subadults may be. When any part of the band changes direction, the rest may follow or not. No specific individuals or subgroups habitually exert even passive leadership by choosing their own route and leaving it up to other band members to maintain contact. Similarly, choice of roost trees is not consistently made by any one band member.

Dwarf Mongooses
Social Life and Group Travel. Dwarf mongooses travel and den together in cohesive packs containing up to twenty-seven members (Rood 1983). Only one dominant pair of adults breeds in each pack, but subordinate animals participate in care of offspring (Rood 1978, 1983; Creel 1996; Creel and Creel 1991; Creel and Waser 1994). The home ranges of these small carnivores are studded with termite mounds, which they use as den sites and refuges from predators. Dwarf mongooses follow familiar travel routes between termite mounds, and take approximately 3 weeks to complete a circuit of their home range (Rasa 1987). Home range overlap between dwarf mongoose packs may be slight (Rasa 1987) or considerable (Rood 1983), but territorial boundaries are marked with scent deposits and remain stable over many years (Rasa 1987). Alien mongooses encountered near boundaries are often mobbed by all adult pack members (Rasa 1986a, 1987).

Initiation of Group Movements and Maintenance of Group Cohesion During Travel. Among dwarf mongooses, group movements are initiated after much social grooming and communal scent marking. Eventually the alpha female sets out first, and emits a distinct "moving out" call, which Rasa and colleagues interpret as meaning "I am leaving. Come with me" (Rasa 1977b, 1986b; Meier, Rasa, and Scheich 1983). A mongoose pack appears to move as a single organism, undulating across the plain like a huge amoeba. Indeed, in contrast to groups of larger carnivores, the cohesiveness of mongoose packs rivals or exceeds that of any primate troop. Dwarf mongooses, banded mongooses, and meerkats all communicate with frequent contact calls while foraging (Ewer 1973; Meier, Rasa, and Scheich 1983; Messeri et al. 1987). Group travel is coordinated in dwarf mongooses by a variety of distinct vocalizations, each of which can be used in several different ways (Rasa 1986b, 1987; Meier, Rasa, and Scheich 1983; Beyon and Rasa 1989). Rasa (1986b) suggests that the vocalizations emitted by a foraging dwarf mongoose transmit information not only about the whereabouts of the caller, but also about its mood, what it has found, and where it is headed. Pack members typically forage individually within several meters of one another, but often lose visual contact in tall grass or scrub. Foragers periodically rise up onto their hind legs ("posting" behavior) to peer over low vegetation. More importantly, however, individual pack members continually emit quiet, beeping contact

K. E. Holekamp, E. E. Boydson, and L. Smale

calls with which group cohesion is maintained. Each individual's contact call is unique in its frequency, duration, and modulation (Rasa 1986a; Marquardt 1976; Peters and Wozencraft 1989) so that pack members can keep track of everyone's location at all times. When an individual becomes separated or lost, call volume increases, but animals may also use scent trails to relocate the main group (Rasa 1987).

Leadership. A dwarf mongoose pack is led by the alpha female, who usually appears to make all decisions about foraging routes, distances traveled, and refuge sites (Rasa 1987). Unlike other female pack members, the alpha female produces up to four litters in 6 months, each averaging 22% of her body mass (Creel and Creel 1991; Creel et al. 1992). Because the energy demands on the alpha female are usually greater than those on other pack members, she may consistently be the hungriest individual in the group. This notion is supported by the observation that the alpha female spends more time foraging than do other pack members (Creel and Creel 1991; Rood 1978). Perhaps because she is hungriest, she is typically the first to emerge and depart from the den each morning (Rood 1983). The alpha female emits a "moving out" call as she sets forth after a rest period. Although other individuals occasionally emit the moving out call as they move away from a refuge site, only high-ranking callers are followed by other pack members. Dwarf mongooses appear to attend selectively to the behavior and vocalizations of the alpha pair, particularly those of the alpha female. Experimental removal of the alpha pair causes a dwarf mongoose pack to become immobile for up to 48 hours, during which time contact calling occurs at unusually high rates until a new pair assumes the alpha positions (Rasa 1987).

Social Canids
Social Life and Group Travel. The basic social unit in most canid species is a monogamous pair with a long-term bond. However, packs may form when siblings or offspring of the mated pair are retained within the group, and when immigrants join it. Most large-bodied canids, including wolves, dholes, and Cape hunting dogs, usually live in packs. Many small canids, including foxes, jackals, and feral dogs, usually live in pairs but may form packs where food abundance and distribution permit (Macdonald 1979a,b, 1981, 1983; Bekoff 1975; Wyman 1967; Eaton 1969; Golani and Keller

1975; Moehlman 1989; Daniels and Bekoff 1989b; Geffen and Macdonald 1992). Pack members cooperate in defense of food and territories, and when food resources permit, they also travel, sleep, and forage together (Mech 1970; Bowen 1981; Bekoff and Wells 1986). Large packs tend to dominate smaller ones during intergroup encounters (Harrington and Mech 1979). Within most canid packs, there are distinct dominance hierarchies for males and females (Schenkel 1947; Mech 1970; Zimen 1976). Subordinate pack members often help with the care of pups produced by the alpha pair.

Pack-living canids may forage alone, in pairs, or with all other pack members, depending on the type and distribution of available prey. Such intraspecific variability in foraging group size occurs in bush dogs, jackals, dingoes, dholes, coyotes, and wolves (Davidar 1975; Bekoff and Wells 1986; Bekoff, Daniels, and Gittleman 1984; Bowyer 1987; Bowen 1981; Gese, Rongstad, and Mytton 1988; Messier 1985; Lamprecht 1978). Foraging group size in coyotes and wolves varies with season, prey abundance, and prey type hunted. Mean foraging group size in both species is smaller in summer, when these canids prey mainly on rodents and hares, than in winter, when deer, elk, and moose are the most common prey (see table 20.1: Camenzind 1978; Bekoff and Wells 1982, 1986; Bowen 1981; Gese, Rongstad, and Mytton 1988; Messier 1985). Indeed, wolves and coyotes often forage alone during the summer, and wolves exhibit amicable behaviors less frequently in summer than in winter (Zimen 1975; Gese, Rongstad, and Mytton 1988; Jordan, Shelton, and Allen 1967; Mech 1970). Within seasons, traveling groups of wolves are larger in habitats containing abundant prey than in other areas (Messier 1985). Wolves typically travel single file along regular travel routes that are roughly circular in shape. These routes may cover more than 200 km (Banfield 1951) and may be used over several decades by multiple generations (Mech 1970).

The only canids that invariably occur in packs are Cape hunting dogs (Creel and Creel 1995; Kuhme 1965a,b; Estes and Goddard 1967; Schaller 1972; Malcolm and Marten 1982; Fuller and Kat 1990, 1993; Fanshawe and Fitzgibbon 1993; Kruuk and Turner 1967). These are the most gregarious of all canids (Bekoff 1975), and the only canids that specialize on large-bodied prey throughout the year (Moehlman 1989). In this species, the pack functions as a cooperative unit during rearing of young, mutual defense, and hunting. Cape hunting dogs generally occupy vast home ranges that often partially overlap those of neighboring packs (Frame et al.

K. E. Holekamp, E. E. Boydson, and L. Smale

1979; McNutt 1996). On average, a pack travels 10–12 km per day (Creel and Creel 1995; Fuller and Kat 1990). During group travel, pack members trot or lope along more or less abreast at about 10 km per hour, usually spread over 10–100 m (Creel and Creel 1995).

Initiation of Group Movements and Maintenance of Group Cohesion during Travel. In most episodes of the classic television series *Lassie,* the canine heroine actively manipulates the movements of her owners, prompting them to follow her to the place where their little boy requires rescue from a life-threatening situation. Are canids truly capable of such feats, or are these scenarios merely Hollywood fiction? Observations made in the wild suggest that Lassie's televised attempts to initiate movement by her owners resemble the actions of individual wolves attempting to initiate movements by their packmates. Often, after a pack has been sleeping for a long period, the alpha male gets up, walks to each dozing wolf, and awakens it (Mech 1970). Pack members then engage in a "group ceremony," in which they assemble closely, wag their tails, touch noses, and groom the alpha male (Mech 1970). Once the entire pack has been aroused, the alpha male trots off with his tail elevated, and this apparently indicates to his subordinates that they should follow. Upon scenting prey during group travel, all pack members usually huddle again to engage in another brief group ceremony before initiating a chase (Mech 1970).

Similarly, group movements are preceded by a ritual of muzzle sniffing, tail wagging, and greeting behavior in golden jackals (Macdonald 1979a). Among Cape hunting dogs, group movements are regularly preceded by a period during which pack members greet and actively submit to one another while emitting whines and high-pitched twittering vocalizations (Estes and Goddard 1967; Schaller 1972; Creel and Creel 1995). If one pack member continues to rest while the others greet, they nudge, nip, tug, and jab at it as if trying to induce the recalcitrant animal to join them (Schaller 1972). Because these activities arouse all pack members simultaneously to a state of high excitement, these sessions are called "pep rallies" (Estes and Goddard 1967) or "social rallies" (Creel and Creel 1995). The pack usually sets out immediately after the rally, and typically makes a kill within 20–30 minutes (Kuhme 1965a; Estes and Goddard 1967; Schaller 1972; Malcolm 1979; Fuller and Kat 1993).

Some canids emit quiet contact vocalizations more or less continuously during group travel, whereas this does not occur in other canid species. Cape hunting dogs and dholes keep in touch while foraging when visibility is poor with soft whimpering or twittering vocalizations (Davidar 1975; Kuhme 1965a,b; Estes and Goddard 1967). Similarly, golden jackals keep in touch while foraging with frequent vocalizations (Macdonald 1979a; Moehlman 1983). However, black-backed jackals, dingoes, coyotes, and wolves do not continuously communicate with conspecifics while foraging. Instead individual foragers may occasionally hoot, yap, or howl to announce their whereabouts (Lehner 1978, 1982; Schaller 1972; Peters and Wozencraft 1989).

Leadership. There is considerable controversy in the literature regarding leadership in canid packs. Most authorities assert that the alpha male directs and controls the behavior of subordinate animals, and that he initiates coordinated group activities, including travel (wolves: Rabb, Woolpy, and Ginsburg 1967; Mech 1970; Fox 1972; Cape hunting dogs: Estes and Goddard 1967; golden jackals: Macdonald 1979a). However Cape hunting dogs are also described as having an adult leadership core within each pack, but no consistent leadership by any single individual (Kuhme 1965a,b; Malcolm and Marten 1982; Schaller 1972). Mech (1970) and Fox (1972) describe the wolf leadership pattern as a combination of autocratic and democratic systems, in which the alpha male sometimes acts independently of his packmates, who are dependent upon him for direction, whereas at other times he is influenced by the behavior of subordinate animals. In contrast to this view, Scott (1965) argues that there is no strong system of leadership in packs of wolves or domestic dogs, but rather that group activities in pack-living canids are coordinated by what he calls "allelomimetic behavior." Scott defines this as "a tendency to do what other animals in a group are doing via some degree of mutual stimulation."

Lions
Social Life and Group Travel. Lions live in prides consisting of two to eighteen related females, their dependent offspring, and one to seven adult immigrant males (Schaller 1972; Bertram 1975, 1978, 1979; Packer and Pusey 1982; Bygott, Bertram, and Hanby 1979; Packer et al. 1988). There is no apparent dominance hierarchy among pride females (Schaller 1972; Bertram 1978, 1979). All adult female pridemates typically breed at similar rates, and they may sometimes even nurse one another's cubs (Packer and Pusey 1983,

K. E. Holekamp, E. E. Boydson, and L. Smale

1984; Pusey and Packer 1994). Lion prides are fission-fusion societies, in which individuals travel and forage in subgroups ranging in size from one to twenty-five animals. The size and composition of these subgroups change from day to day (Schaller 1972; Packer 1986). Lions typically spend about 2 hours per day traveling, during which time they cover distances of 1–22 km (Schaller 1972). Most lion prides defend group territories: males defend the pride against intrusions by other males (Packer et al. 1988; Grinnell, Packer, and Pusey 1995), and females defend the territory against incursions by other females (McComb, Packer, and Pusey 1994; Heinsohn and Packer 1995). Large prides dominate smaller ones in interpride encounters, which on average occur once every 5 days (Schaller 1972; Packer, Scheel, and Pusey 1990). Intergroup competition occurs primarily over land rather than over specific carcasses (Packer 1986). The sizes of lion territories vary widely, depending on pride size and prey abundance (see table 20.1; Schaller 1972).

Initiation of Group Movements and Maintenance of Group Cohesion During Travel. The most common cohesive behaviors seen among members of lion prides are social licking and greeting behavior in which one individual rubs its head against the head or body of another. At dusk, before setting out on a hunt, group members usually greet one another with unusually high frequency and intensity, and engage in a great deal of social licking (Schaller 1972; Rudnai 1973). Although lions normally carry their tails fairly limply, a lioness attempting to initiate a group hunt tends to elevate her tail (Schaller 1972). Rudnai (1973) describes the activities of lions before a hunt as "contagious behavior" functioning to promote group cohesion and to synchronize the mood of all the lions present. Lions can distinguish the roars of their pridemates from those of alien conspecifics (McComb et al. 1993; McComb, Packer, and Pusey 1994), and when traveling under conditions of low visibility, may occasionally emit a soft roar, which apparently functions as a contact call (Schaller 1972). Otherwise, lions forage silently.

Leadership. There is usually no consistent leadership in lion prides, but adult females appear to make most decisions about group travel (Schaller 1972). During encounters with neighboring prides, certain female lions typically lead the way while others consistently lag behind (Heinsohn and Packer 1995). Although lion travel processions in other contexts are usually led by an adult female (Schaller 1972; Rudnai 1973; Stander 1992a), several different adult females may lead a procession, even during travel over distances as short as 1 km (Schaller 1972). Cubs and subadults typically walk in the middle of a travel procession, and if any adult males are present, they usually bring up the rear (Schaller 1972; Rudnai 1973). Lionesses and cubs seldom respond to the movements of adult males, which therefore often come and go without followers. Instead, cubs, subadults, and adult males orient much of their activity to the movements of adult females (Schaller 1972; Rudnai 1973).

Spotted Hyenas
Social Life and Group Travel. Spotted hyenas live in clans containing four to ninety members (Kruuk 1972; Whately and Brooks 1978; Tilson and Henschel 1986; Frank 1986; Henschel and Skinner 1987; Gasaway, Mossestad, and Stander 1989; Mills 1990; Hofer and East 1993). Clan members use a common territory and defend it against intrusions by alien hyenas (Kruuk 1972; Henschel and Skinner 1991). Although lactating females all maintain their young cubs at a communal den, clan members do not cooperate in rearing young (Mills 1985). Clans contain multiple adult immigrant males and one to several matrilines of natal females and their offspring (Frank 1986; Mills 1990). Spotted hyena clans are structured by linear rank relationships that determine priority of access to food during competition at kills with other clan members (Tilson and Hamilton 1984; Frank 1986). Adult females and their young outrank all adult immigrant males (Kruuk 1972; Smale, Frank, and Holekamp 1993). Although all adult female clan members breed, they do so at rates that increase with social rank (Frank, Holekamp, and Smale 1995; Holekamp, Smale, and Szykman 1996). Hyena clans are fission-fusion societies in which individuals travel, rest, and forage alone or in subgroups (fig. 20.1, 20.2). Subgroup composition typically changes several times during the course of a single day, and all clan members seldom, if ever, aggregate simultaneously (Holekamp, Cooper et al. 1997).

Since June 1988, we have been studying a single hyena clan in Kenya that usually contains about seventy members. Although we have observed hyenas traveling in groups of up to twenty-five individuals, these animals usually travel alone, and the average travel group size is 3.2 ± 0.3 hyenas (fig. 20.2). Hyena travel group size varies significantly with context (fig. 20.3). The smallest travel

K. E. Holekamp, E. E. Boydson, and L. Smale

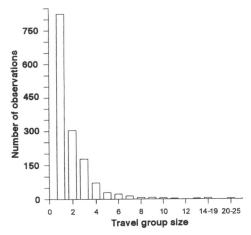

Figure 20.2 Travel group sizes in the spotted hyena, based on 1,486 observations made between 1992 and 1996 in the Masai Mara National Reserve, Kenya.

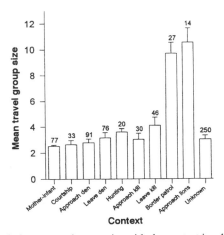

Figure 20.3 Variation in hyena travel group size with the context in which group movement occurs. Data represent 664 cases in which two or more hyenas were observed traveling together, excluding all observations in which one or more individuals carried food.

groups contain only a mother and her cubs. Cubs traveling between dens and kills without their mothers typically tag along behind an adult setting off in the desired direction, much like a hobo catching a ride on a boxcar. Courtship groups contain an adult female and an entourage of adult males. When hyenas approach dens or ungulate carcasses in groups, they typically do so with two or three companions. Travel groups that cannot be assigned to any obvious context (i.e., we simply find a group of hyenas moving along together) typically contain three hyenas. The largest travel groups are observed when hyenas approach lions or patrol the borders of the clan's territory, both situations in which the presence of allies is particularly important. During a border patrol, several hyenas maintain postures reflecting high levels of excitement as they move to or along a territorial boundary, where they engage in high rates of scent-marking behavior and defecation in "latrine" areas (Kruuk 1972). Border patrols are most frequently observed in populations characterized by intense intrusion pressure from alien hyenas. Thus border patrols occur far more frequently in habitats where clans are tightly packed and occupy small home ranges (Kruuk 1972; Henschel and Skinner 1991) than in areas where hyena density is low and home ranges vast (Mills 1990). Under the latter circumstances, border patrols may be pointless since a huge territory cannot effectively be defended by a small group of hyenas.

In order to compare the defensibility of primate group territories across populations and species, Mitani and Rodman (1979) calculated a "defensibility index" for each territory, roughly defined as the ratio of the group's daily path length to the area of the group's home range. They found that all territorial primate groups have defensibility indices of 1.0 or greater, but that most nonterritorial species have an index of less than 1. 0. Data available from three studies suggest that defensibility indices for hyena group territories range from 0.75 to 1.99 (Kruuk 1972; Mills 1990; Henschel and Skinner 1991). Whereas territorial behavior is commonly observed in areas where defensibility indices are greater than 1.0 (Kruuk 1972; Henschel and Skinner 1991), it is relatively rare in less defensible areas where hyena home ranges are enormous and neighboring clans seldom encounter one another (Mills 1990).

Initiation of Group Movements and Maintenance of Group Cohesion during Travel. Initiation of group travel in spotted hyenas usually takes one of two general forms. First, one individual simply gets up from a feeding or rest site and walks off, and others then get up and follow the departing animal. Such movements appear to involve no active effort on the part of the leader to induce other group members to join it, and the leader rarely even glances back to see who is following. This is by far the most common pattern. However, movement initiation appears far better coordinated during border patrols, group hunts, and approaches to lions. Under these circumstances, one individual gets up, moves a short distance away, and then looks back as if to determine whether it is being followed. If so, it may continue on or turn to greet its followers. But if that

K. E. Holekamp, E. E. Boydson, and L. Smale

individual is not immediately followed, it may lie down to wait until other hyenas start moving away from the den or resting spot. Ultimately a group sallies forth to a hunt or border patrol after a prolonged period of vigorous social activity. The hyenas engage in extensive greeting behavior (Kruuk 1972; East, Hofer, and Wickler 1993), scent marking, and "social sniffing," a peculiar activity in which several hyenas lean against each other as they excitedly sniff a spot on the ground together with their manes erect and their tails elevated and bristled like bottlebrushes. Since olfactory information could be acquired just as well by hyenas sniffing the ground alone, it appears that the primary function of social sniffing may be to rally hyenas in preparation for an activity requiring several individuals (Kruuk 1972).

Like dwarf mongooses, spotted hyenas emit about a dozen distinct vocalizations, each of which can be modulated in a variety of ways to communicate many different types of information. For example, the long-distance call of the spotted hyena, the whoop, is extremely variable in form, and appears to have a number of different functions (East and Hofer 1991b; Holekamp et al. 1999). Although each individual hyena has a unique whoop (East and Hofer 1991a), these animals emit no regular contact calls during group travel, even when foraging in thick bush at night. Indeed, hyenas seldom vocalize while foraging or on border patrols (Henschel and Skinner 1991), but at such times they typically engage in much bristling, scent marking, and social sniffing. Members of groups approaching lions or alien hyenas tend to whoop and "low" frequently. Whoops in this context appear to function as recruitment calls, resulting in the prompt arrival of additional clan members (East and Hofer 1991b), but lowing appears to synchronize individuals to attack as a group via concurrent excitation of all participants (K. E. Holekamp and L. Smale, pers. obs.). Once the attack has occurred and the conflict has been settled, whooping and lowing cease. Adult males traveling with an estrous female during courtship often bristle and social sniff, and occasionally emit slow whoops (East and Hofer 1991b). Otherwise, however, group movements by spotted hyenas appear to involve little acoustic communication among travelers.

Leadership. Although dominance relationships are extremely stable in hyena societies, leadership during hyena group travel is highly variable. High-ranking females tend to appear in the vanguard more frequently than do lower-ranking females, but all adult

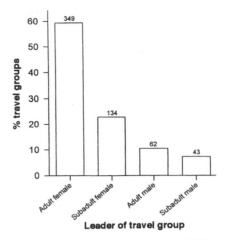

Figure 20.4 Leadership during hyena group travel, based on 588 progressions in which we recorded the identity of the leading animal. During the sampling period, adult females constituted approximately 29% of the clan's total membership, subadult females 32%, adult males 17%, and subadult males 22%.

females lead some processions, and no single female leads all processions in which she is observed. As in lions, adult female hyenas appear to make most decisions about group travel, and they lead more group movements than do members of any other age/sex class (fig. 20.4). Travel groups led by adult females tend to be larger than groups led by adult males ($t = 1.72$; $p = .09$), immature females ($t = 2.92$; $p = .004$), or immature males ($t = 2.52$; $p = .016$). We have found no significant differences between the sizes of travel groups led by adult and immature males ($t = .544$; $p = .589$), nor between those led by immature males and immature females ($t = .113$; $p = .911$). Adult females lead all border patrols and 80% of groups seen approaching lions.

Common Features of Group Travel in Gregarious Carnivores
Despite the great variation in body size, diet, and social organization among the various carnivores considered above, group travel in these animals shares certain common features. First, "social rallies" precede group travel in all the species considered (table 20.2). That is, group movements are initiated in gregarious carnivores after a period at a den or rest site during which individuals engage in extensive mutual sniffing, communal scent marking, social grooming, or greeting behavior. Because these activities typically begin after group members have been in close spatial proximity at the rest site

K. E. Holekamp, E. E. Boydson, and L. Smale

for several hours, they apparently function not to reintroduce group members, but rather to elevate and synchronize individuals' arousal levels and to promote group cohesion (see also Glickman et al. 1997).

The social rallies and "allelomimetic behaviors" observed among gregarious carnivores appear to reflect the operation of mechanisms more specific than Zajonc's (1965) "social facilitation," yet more general than Thorpe's (1956) "contagion." Instead, these behavior patterns most closely resemble Byrne's (1994) "response facilitation," in which the presence of a conspecific performing an act increases the probability of an observing animal emitting the same behavior. Not all group members are equally effective at facilitating the responses of others, nor can the responses of any one individual be facilitated equally well at all times. Rather, the probability that response facilitation will occur is likely to depend on the observing animal's own physiological state, its relative social rank, and its current knowledge about prey availability within the group's home range.

In most gregarious carnivores, specialized vocalizations (e.g., wolf howls or lion roars) and olfactory signals (e.g., scent-marking posts and "latrines" at territorial boundaries) occur frequently in the context of, and appear to mediate, intergroup spacing (Harrington and Mech 1979; Peters and Wozencraft 1989; Lehner 1978, 1982; East and Hofer 1991b; McComb, Packer, and Pusey 1994). However, vocalizations are also used regularly in some species to maintain cohesion within groups during travel (table 20.3). Coatis, social mongooses, giant river otters, and some canids emit quiet contact vocalizations more or less continuously during group travel,

Table 20.2 Events immediately preceding group movements in various social carnivores

Species	Events preceding group movements
Coati	Social grooming session
Dwarf mongoose	Social grooming session; group scent marking; "moving out" vocalization
Wolf	"Group ceremony": greeting and other affiliative behavior; social grooming
Golden jackal	Group sniffing and greeting behavior
Cape hunting dog	"Pep rally": greeting and appeasement behavior, quiet vocalizations
Lion	Social licking; greeting; group scent marking
Spotted hyena	Social sniffing; group scent marking

Table 20.3 Signals used to maintain group cohesion during group travel in gregarious carnivores

Species	Acoustic	Olfactory	Visual[a]
Coati	Continuous quiet vocalizations Louder "lost calls"	Odor trails	Tail elevation
Dwarf mongoose	Continuous quiet vocalizations Louder "lost calls"	Odor trails	Posting behavior
Meerkat	Continuous quiet vocalizations	Odor trails	Posting behavior
Cape hunting dog	Intermittent quiet vocalizations	Odor trails	
Dhole	Intermittent quiet vocalizations	Odor trails	
Giant river otter	Frequent vocalizations		
Golden jackal	Frequent vocalizations	Odor trails	
Black-backed jackal		Odor trails	
Dingo		Odor trails	
Wolf		Odor trails	Tail elevation
Coyote		Odor trails	
Lion			Tail elevation
Spotted hyena		Odor trails	Tail elevation

[a] Other than the mere sight of other individuals.

whereas other canids, lions, and hyenas do not. Coatis and social mongooses appear to rely on odor trails to relocate other group members when acoustic and visual cues are unavailable. In most of the larger carnivores, however, both visual and olfactory signals function more importantly than do acoustic cues in maintaining group cohesion during travel. Domestic dogs can detect the scent deposited by human fingerprints on a glass microscope slide left to weather outdoors for over a week (King, Becker, and Markee 1964). Thus it should be a relatively simple matter for a canid to follow recent odor trails deposited via the digital scent glands located on the feet of its packmates.

Decision making about the timing, distance, and direction of group travel in carnivores varies greatly among species (table 20.4). In some there is consistent leadership by a single individual, and in others there is no apparent leadership whatsoever. In some carnivore species it appears that decisions about when and where to travel are often made by those individuals currently experiencing the greatest energetic challenges. The most cohesive carnivore groups (those, like dwarf mongooses, in which all members continuously remain in close physical proximity) exhibit the clearest evidence of group leadership. A mongoose pack is led by the alpha female, and canid packs are often led by the alpha male. In these species, there is some indication that subordinate pack members may sometimes

K. E. Holekamp, E. E. Boydson, and L. Smale

Table 20.4 Group leadership in gregarious carnivores

Species	Group leadership
Coati	None apparent
Dwarf mongoose	Alpha female
Meerkat	Alpha female
Common cusimanse	Alpha female
Giant river otter	Alpha male/female
Wolf	Alpha male
Coyote	Alpha male/female
Golden jackal	Alpha male
Cape hunting dog	Adult leadership core or alpha male
Dingo	Alpha male
Lion	Adult females
Spotted hyena	Adult females

influence decisions the leaders' about where to go and when to rest, as occurs in some primates (e.g., Boinski 1996). By contrast, in the fission-fusion societies of coatis, lions, and spotted hyenas, group movements are not regularly initiated or coordinated by any single individual, and decisions about timing, distance, and direction of travel usually involve no negotiation, concurrence, or reciprocation among group members. Rather, one individual simply moves off, and other group members either follow or go their own way, depending on which course of action is in their immediate best interest.

In carnivore societies structured by hierarchical dominance relationships, relative success in initiating group travel is fairly well predicted by an individual's social rank (table 20.4). Successful leadership may be mediated by the selective attention of subordinates to the behavior of high-ranking animals. Among canids and social mongooses, alpha animals may actively manipulate subordinates to follow their movement decisions, whereas leadership in other gregarious carnivores usually appears to be more passive.

Den Moves

After parturition, many gregarious carnivores maintain their young in dens or creches for periods ranging from 6 weeks (dwarf mongooses: Rood 1983) to 8 months (spotted hyenas: Kruuk 1972; Holekamp and Smale 1993), after which the youngsters travel and forage with adults. In social mongooses, social canids, lions, and spotted hyenas, den sites are frequently shifted before the young become fully mobile. Den moves may occur in response to some disturbance at the den site (e.g., visitation by predators, flooding, etc.), to accommodate the foraging needs of older group members (e.g., Peterson 1977), or for no apparent reason (Ryon 1986). Den moves present members of carnivore groups with the challenge of coordinating the timing of departure, maintaining cohesion during the move, and making a collective choice of a new den site. Thus den moves represent a special type of group travel in carnivores, and their study sheds light on the mediation of group movements in these species.

Among wolves and coyotes, transfer of pups between den sites may be initiated by adults of either sex (Ryden 1975; Camenzind 1978; Hilton 1978; Andeldt, Althoff, and Gipson 1979; Joslin 1967; Harrington and Mech 1979; Ryon 1986). However, lactating females appear to make the final decision about where the pups end up, even when adult males initiate pup transfer (Mech 1970; Ryon 1986). Both wolves and coyotes return two to four times to dens from which they have just moved pups, looking inside as if checking to ensure that no pups were left behind (Ryden 1975; Ryon 1986). Disturbances at the den, and den moves themselves, generate much social excitement among pack members. Subordinate animals shift their focus of activity to the new den once the site has been selected by the alpha pair, to whose behavior they attend closely.

Dwarf mongooses frequently transfer their immobile young between den sites in different termite mounds (Rood 1983; Rasa 1986b, 1987). Den moves may be initiated either by the alpha female or by immigrant or natal adult males (Rood 1983). Rood (1983) observed that den moves in this species appear haphazard and confused. For example, one animal may carry an infant to a new den only to have another carry it back to the old den. Sometimes young are carried in relays in which one pack member drops an infant en route, leaving another to pick it up and complete the journey. Den moves in this species always involve high levels of general excitement within the pack and increased frequencies of scent-marking behavior (Rood 1983).

Female spotted hyenas bear litters of one or two cubs in isolated natal dens, then transfer their cubs to the clan's communal den when the cubs are 1–4 weeks old (Kruuk 1972; East, Hofer, and Turk 1989). Cubs are then kept at the communal den until they are 8–9 months old. On average, communal den sites are moved every 32 to 45 days (Mills 1990; E. E. Boydston, unpub.). Some den sites are used repeatedly over periods of several years (Mills 1990; Kruuk

K. E. Holekamp, E. E. Boydson, and L. Smale

1972). Den moves are typically accompanied by a great deal of greeting, social sniffing, and scent marking. They usually take place at night, with different mothers moving their cubs on each of two to four consecutive nights. Den moves may be initiated by any adult female hyena with cubs residing at the communal den. However, we have found that the highest-ranking female with cubs at the den is among the first to move her cubs in 73% of den moves. Small cubs are carried delicately in the mother's jaws, one at a time. After moving her cubs, the mother always returns once to the old den as if checking to ensure that none of her cubs remain there.

To summarize, den moves are accompanied by high levels of social excitement among group members, and are most likely to be initiated by dominant animals. Decisions during den moves appear to be made by the individuals with the most to lose from inappropriate site selection. However, den moves often appear uncoordinated and dangerously haphazard. Finally, when gregarious carnivores move N infants to a new den site, they always make at least $(N + 1)$ trips to the old den. This behavior suggests that, instead of being able to count, gregarious carnivores operate during den moves according to a rule of thumb dictating "Revisit the old den until you find no more of your infants there, then restrict activity to the new den site."

At this point it is worthwhile to distinguish between rules of thumb, like the one suggested here, and more complex mental calculations based on the organization of sensory data into relatively detailed internal representations of events and relationships in the external environment (Dyer 1994). Postulating a rule of thumb implies that the animal registers and responds exclusively to raw sensory impressions of events and relationships. The animal's decision making is thus based strictly on local information organized into a simple algorithm as, for example, a single "if, then" contingency. By contrast, hypotheses invoking more complex mental events postulate that the animal's internal representation of the relevant parts of its external world may supplement or supersede local information in determining what decision the animal will make next. To complete our computer metaphor, a simple "if, then" contingency is probably replaced here by a more complex algorithm involving multiple "if, then" statements processed in serial or parallel fashion. Constructing a complex internal model of the outside world should theoretically enhance the animal's ability to respond appropriately to familiar stimuli even when experiencing them in novel combinations or contexts (e.g., Cheney and Seyfarth 1990; Gallistel 1990), and thus should increase behavioral flexibility.

Hunting in Groups

Group Hunting in Gregarious Carnivores

Among canids, lions, and spotted hyenas, two to several individuals often join together to hunt large mammals. Coordination of individuals' movements during stalking or pursuit of prey may pose unusual cognitive problems for members of these species. Among canids, the tendency to engage in group hunts increases with body size, and cooperative hunting is observed most commonly in wolves, dholes, and Cape hunting dogs (Kleiman and Eisenberg 1973; Macdonald 1983; Moelhman 1989). Except in dholes and Cape hunting dogs, carnivore hunting groups often do not include all members of the social unit. Hunting groups of canids inhabiting seasonal environments often contain all pack members in winter, but seldom in summer (see table 20.1). Among both lions and spotted hyenas, the average hunting group contains only two to four individuals (Packer 1986; Stander 1992b; Mills 1990; Cooper 1990; Gasaway, Mossestad, and Stander 1991; Holekamp, Smale et al. 1997). In many carnivore species, hunting group size varies with the prey species stalked. Dingoes, for example, hunt birds and rabbits solitarily, but all pack members join hunts for large kangaroos (Corbett 1995). Similarly, among other social carnivores, larger prey are generally hunted by larger foraging groups (Lamprecht 1981).

In most large carnivores examined to date, hunting group size varies with the difficulty involved in prey capture. When individual hunters can straightforwardly bring down a particular prey species, group hunts of that prey are not observed (e.g., Fanshawe and Fitzgibbon 1993; Scheel and Packer 1991). Thus, Serengeti lions, for example, hunt difficult prey such as Cape buffalo in groups, but hunt individually for easy prey such as warthogs (Scheel and Packer 1991). By contrast, in Namibia, where lone hunters are rarely able to capture any antelope species, most hunts are undertaken by groups of lions (Stander 1992a,b; Stander and Albon 1993).

When stalking large or difficult prey, hunting success among carnivores generally increases asymptotically with increasing group size (Fanshawe and Fitzgibbon 1993; Packer and Ruttan 1988; Stander and Albon 1993). For example, among spotted hyenas, the addition of second and third hunters each increases the probability

K. E. Holekamp, E. E. Boydson, and L. Smale

of a hunt being successful by approximately 20% (Holekamp, Smale et al. 1997). In addition to improving foraging success, group hunting behavior permits carnivores to add larger prey species to their diet (Fanshawe and Fitzgibbon 1993). Thus group hunting facilitates capture of gazelles by jackals (Wyman 1967; Lamprecht 1978), moose by wolves (Burkholder 1959; Mech 1966; Zimen 1976), deer by coyotes (Bowyer 1987), and zebra by Cape hunting dogs (Malcolm and van Lawick 1975; Creel and Creel 1995).

The behavior of many gregarious carnivores indicates that one or more members of a hunting group often select a prey species far in advance of when the group actually starts pursuing a particular target animal. For example, preselection of prey type is suggested in coyotes, wolves, and spotted hyenas by the common observation that larger numbers of hunters leave rest sites together when large mammals are abundant in the group's home range than when the home range contains mainly small game (Kruuk 1972; Bowyer 1987; Bekoff and Wells 1986; Gese, Rongstad, and Mytton 1988; Messier 1985). Among Cape hunting dogs and spotted hyenas, preselection of prey is also indicated by the fact that hunting groups often leave a rest site and walk for long distances through herds of other antelope species until they come upon a group of individuals of the particular prey species in which they are interested (Schaller 1972; Kruuk 1972). Like Hans Kruuk (1972) before us, we have often been surprised to see hyenas walk through vast herds of wildebeest on their way to hunt zebra, without showing interest even in any of the limping individuals that cross their path. Their selective attention to particular ungulates and variation in the size of hunting groups leaving rest sites allow observers to predict with a high degree of certainty not only that these carnivores are hunting, but also what prey species they seek. Based on anecdotal observations, we suspect that behavior exhibited by a single hyena as it leaves the rest site (e.g., looking back, waiting for others to join it, returning to the rest site to greet other animals, etc.) might function to invite other hyenas to join a hunt. If this is correct, then the animal initiating such a group movement might, in fact, be the only hunter in the group with an immediate prey preference. Thus the followers may not be deciding to hunt a particular type of antelope, but rather simply deciding whether or not to do what the leader is doing.

While pursuing large mammalian prey, members of a canid pack may exhibit some division of labor in that specific individuals assume more or less regular roles in the hunt. For example, among wolves and Cape hunting dogs, the alpha or beta male usually leads the pack in a fast chase after a fleeing antelope, and he also typically makes the first grab at it (Mech 1970; Malcolm and Marten 1982). Throughout the chase, the leader remains as close as possible on the heels of the prey, while other pack members stay slightly behind, and flank the leader to grab the prey if it changes direction. Among coyotes, wolves, and spotted hyenas, several hunters often rush in to join a chase initiated by one individual (e.g., Bowyer 1987; Mech 1970; Kruuk 1972). When group members all focus on a single target animal, flanking movements are not uncommon. Hypothetically, the running speed and location of each pack member might be readjusted continuously, based on updated assessments of the behavior or position of both the leading hunter and the prey animal. However, an alternative hypothesis is that all hunters run as fast as they can, and the usual leader is just the fastest or hungriest member of the group. All other participants in the hunt might simply adjust their behavior according to the rule of thumb dictating "Take your own best line of approach to the target prey animal, unless another hunter already occupies that position."

For decades a controversy has raged about whether lion movements during group hunts reflect intelligent coordination and division of labor among hunters (Stevenson-Hamilton 1954; Guggisberg 1962; Bridge 1951; Stander 1992a,b), or represent individually selected movement strategies in relation to a common prey object that merely *appear* to involve sophisticated coordination (Schaller 1972; Kruuk and Turner 1967). Most lion watchers agree that encirclement of prey frequently occurs during group hunts (Rudnai 1973; Bridge 1951; Schaller 1972; Guggisberg 1962; Stander 1992a) and that hunters commonly divide themselves into "driver" and "catcher" categories (Guggisberg 1962; Schaller 1972; Griffin 1984). Drivers rush the prey, which flee toward concealed catchers, who perform the actual capture.

Stander (1992b) found that 40% of hunts by Namibian lions exhibited a surprisingly high level of coordination and what appeared to be intentional cooperation among hunters. He found that all participating lionesses watched both the prey and other pride members during the stalk, as if to obtain information for guidance of their own movements. Stander's data show that individual lionesses consistently assumed roles as either "wings" or "centers" during group hunts, with "wings" encircling prey and "centers" moving shorter distances into positions from which they could ambush prey fleeing

K. E. Holekamp, E. E. Boydson, and L. Smale

from one or more encircling lions. Stander (1992a) found that heavier, stockier lionesses tended to prefer "center" positions. It is possible that such morphological variables as mass and girth affect individual hunting skills, including running speed or concealment ability, and thus cause particular individuals to perform one role more effectively than another. However Stander (1992a) also found that individual hunting behavior was not inflexible, but rather varied with different hunting group compositions and with variations in the behavior of other stalking lions. Thus, for example, during 145 hunts in which the same two individuals foraged together, the stalking role occupied by one was dependent on the role occupied by the other. Stander observed that any two individual hunters coordinated their relative positions in what appeared to be consistent attempts to keep the selected prey animal on an imaginary straight line between them. Thus the behavior of lions participating in group hunts truly is coordinated, but this coordination could be effected by the operation of a few simple rules of thumb, such as "Move wherever you need to in order to keep the selected prey animal between you and another lion." That the ability to modify hunting behavior according to this sort of rule of thumb is not unique to lions is suggested by occasional observations of wolves using drivers and catchers to ambush deer (Mech 1970).

To summarize, it appears that social carnivores often select a particular prey species before leaving the rest site to hunt. Hunting group size is influenced by several different factors, the most important of which appears to be how difficult it is to capture the particular prey species being stalked. Prey that can be brought down by lone hunters are rarely hunted by groups. The success with which difficult prey can be hunted often increases asymptotically with predator group size. Thus it appears that cooperative hunting by gregarious carnivores usually occurs facultatively; group hunts occur only when individual hunters experience success rates too low to satisfy their own energy requirements. Division of labor is commonly observed among hunting carnivores, but this can be explained more parsimoniously than by postulating the occurrence of complex mental calculations in these animals. Specifically, individual differences in motivation level, running speed, and morphology may account for individual occupation of particular positions during hunts, and a few simple rules of thumb may direct the movements of participants in group hunts such that individual hunters merely appear to be engaged in complexly organized activity with others. With respect to social cognition, the key question is the extent to which individual hunters understand the roles being played by others in the cooperative endeavor (Tomasello and Call 1997). Falsification of the simple "rules of thumb" hypothesis will require experimental evidence that individual carnivores not only monitor both their prey and their fellow hunters (e.g., Stander 1992a), but also accurately anticipate the behavior of the latter based on knowledge of their goals. This would be demonstrated in the case of hunting lions, for example, if individual lions were able to make the appropriate movements for prey capture even when their fellow hunters were not visible or audible to them. Specifically, under these conditions of limited sensory input, an individual hunter should be able to place itself in line with the prey animal and the place where it *expects* another hunter to be.

Group Hunting Compared in Carnivores and Primates

Since the Upper Paleolithic, humans have successfully joined forces to hunt wild animals larger than themselves (Byrne 1995b; Stanford 1995a). Lions, spotted hyenas, and pack-living canids accomplish this feat on a daily basis. Both nonhuman primates and gregarious carnivores have been proposed as useful models for understanding the behavior of early hominids (Schaller and Lowther 1969; Washburn and Lancaster 1968; Hill 1982; Tooby and DeVore 1987). Here we compare group hunting in these taxa and inquire whether coordination among hunters requires special cognitive abilities in either group. Byrne (1995b) suggests that the uniquely "intelligent" aspects of group hunts by humans are the advance planning of cooperative behavior, the organization of the hunt, and the sharing of the spoils. Are these features also typical of group hunts by nonhuman primates or gregarious carnivores?

Advance Planning.   In many species of nonhuman primates, free-living individuals have been observed to catch and eat mammalian prey (reviewed in Boesch and Boesch 1989; Rose 1997). However, evidence is scant that hunts by nonhuman primates are planned in advance of encountering prey. Indeed, hunts by all nonhuman primates except chimpanzees and capuchins are most parsimoniously explained as mere opportunistic grabs at animals flushed accidentally while traveling or foraging for other foods (e.g., Strum 1981; Hausfater 1976; Stanford 1995a; Rose, 1997). Even chimpanzees usually initiate hunts only when they detect the presence of

K. E. Holekamp, E. E. Boydson, and L. Smale

nearby prey (Boesch and Boesch 1989; Stanford et al. 1994). Chimpanzees and capuchins in some areas engage in hunting "binges" during which they hunt every day for one to several weeks (Stanford et al. 1994; Rose 1997). However, capuchins often hunt for nesting prey, which vary seasonally in availability (Rose 1997), and daily hunting rates among chimpanzees are strongly influenced by group composition (Stanford et al. 1994; Stanford 1995). When sizeable groups of male chimpanzees are present simultaneously due to the availability of ripe fruit or estrous females, the probability of hunting increases. Otherwise hunting occurs fortuitously, and chimpanzees make no active efforts to search for a particular prey species before they hear or see it (Busse 1978; Boesch and Boesch 1989).

Similarly, little advance planning of hunts occurs among gregarious carnivores. The hunting "binges" observed among chimpanzees resemble the hunting "streaks" for a particular prey species that we observe among spotted hyenas, in which the hyenas target the same prey species every day for several days (often even at the same time of day) in the absence of a concurrent increase in the relative abundance of that prey species (Kruuk 1972). As described above, the particular prey type to be hunted by large carnivores often appears to be decided upon by at least one group member long before hunters sense prey, and carnivores' hunting behavior often enables human observers to predict which prey species is being sought (e.g., Kruuk 1972). However, this advance prey selection by carnivores no more resembles advance planning by human hunters than does the decision by a baboon troop leaving its sleeping cliff to head toward a distant fruiting fig tree. The extraordinary flexibility apparent in group hunting behavior by humans (e.g., Byrne 1995b), which is based largely on advance contingency planning and an accurate mental representation of the "game plan" of the hunt, appears to be entirely absent among both nonhuman primates and large gregarious carnivores.

Organization of the Hunt.   Like carnivores stalking prey, chimpanzees participating in a hunt are usually silent until a kill is made (Stanford et al. 1994; Boesch and Boesch 1989; Teleki 1973). In both situations, hunting success increases with hunting group size, but the mean size of hunting parties varies among habitats (Stanford et al. 1994; Boesch and Boesch, 1989). Due to different habitat structures, colobus monkeys at Gombe are more vulnerable to chimpanzee predation than are colobus in the Tai forest, and hunting success on colobus is higher for solitary hunters at Gombe than at Tai (Wrangham 1977; Busse 1978; Goodall 1986; Boesch and Boesch 1989; Boesch 1994; Wrangham, Gittleman, and Chapman 1993; Stanford et al. 1994). Similarly, hunting success for solitary lions is far higher in the Serengeti than at Etosha, and Etosha lions hunt alone far less frequently than do Serengeti lions (Packer and Ruttan 1988; Stander 1992b; Stander and Albon 1993). Mean hunting group size is larger among chimpanzees at Tai than among those at Gombe, just as mean hunting group size is larger among Etosha lions than among Serengeti lions (Boesch 1994; Stander 1992b; Stander and Albon 1993; Packer and Ruttan 1988). Thus prevailing ecological conditions appear to shape hunting behavior similarly in both chimpanzees and carnivores. Neither chimpanzees nor carnivores use tools to facilitate capture of vertebrate prey, but members of both taxa have been observed to use particular features of the environment to their advantage during hunts. For example, hunt initiation by spotted hyenas often occurs at the mouth of a cul-de-sac formed by watercourses or vegetation, such that the fleeing antelope must double back toward its hunters to escape (K. E. Holekamp et al., unpub.).

During group hunts by chimpanzees, Boesch and Boesch (1989) observed that particular adult males occupied regular roles in what appeared to be organized "team play." Group members encircled prey, and some males acted as drivers while others blocked the prey animal's escape routes and played the role of "catchers." Similarly, among lions and some canids, participants in group hunts regularly assume particular hunting positions relative to the prey and to their companions, and hunts tend to be most successful when individuals occupy their preferred positions. These aspects of hunting behavior by large carnivores and chimpanzees suggest that their group hunts genuinely represent more complexly organized phenomena than do the opportunistic grabs at vertebrate prey made, for example, by baboons. Positional preferences during hunts in both taxa may be based upon morphological differences among individual hunters that result in differential hunting success, and thus different reinforcement schedules, in various hunting positions. Both chimpanzees and lions participating in such coordinated group hunts as those observed by Stander (1992a,b) appear to utilize information about the locations of other hunters, as well as about the target animal, to guide their own behavior while stalking prey. However, no evidence currently exists that either lions or chimpanzees utilize mental algorithms more complex than simple rules of thumb to surround and capture prey. If this is the case, then these hunts pose less complex intellectual problems than does cooperative hunting among humans. To date it appears that human hunters are unique among mammals in their ability to anticipate the future movements of their fellow hunters while stalking prey. Language presumably facilitates the advance contingency planning that allows humans to anticipate responses of their fellow hunters to rapid changes in prey behavior (Aiello 1996).

Sharing of the Spoils. Many carnivores share the fruits of their combined labor, but usually only with mates and kin, except when they are simply unable to monopolize a carcass (e.g., Frank 1986; Holekamp and Smale 1990). Among both large carnivores and nonhuman primates (e.g., Busse 1978; Rose 1997), sharing of meat is influenced by the quantity of it available. A small carcass is usually monopolized in both taxa by a single hunter, but larger carcasses are often consumed by several individuals. Contest competition over carcasses can be as intense among chimpanzees as it is among large carnivores (e.g., Busse 1978). However, some voluntary sharing of meat apparently occurs among chimpanzees (Teleki 1973) in that conspecifics who beg for food are often allowed to take a bite or two (Boesch and Boesch 1989). If the current owner of the meat is becoming satiated, he may also actually offer the remains to another animal (Goodall 1968; Teleki 1973; Boesch and Boesch 1989). Boesch and Boesch (1989) noted that male chimpanzees shared meat most often with other adult males. Unfortunately, however, these observers knew very little about the kin relations among their subjects, so it is not clear whether food sharing occurred mainly among kin, as occurs in gregarious carnivores.

To date, observations of meat sharing in both carnivores and chimpanzees have virtually all been consistent with expectations based on theories of kin selection (Hamilton 1964), sexual selection (Darwin 1871), or reciprocal altruism (Trivers 1971). Voluntary sharing of meat with unrelated animals may occur more frequently among chimpanzees than among social carnivores. However, even the most magnanimous chimpanzee looks very selfish indeed when his sharing behavior is compared with that observed in human hunter-gatherer societies after a successful hunt. For example, among the !Kung or the Ache people of Paraguay, most killing of game is accomplished by only a very few male hunters, who then share the spoils remarkably equitably with mates, relatives, and nonrelatives alike within their band (e.g., Hawkes, Hill, and O'Connell 1982; Hill 1982; Hill and Hurtado 1996). The generosity of these human hunters is usually rewarded in ways that appear to improve their reproductive success, but the rewards often take the form of currencies other than food, and often appear only after a considerable delay. Thus it seems that the ability to engage in delayed reciprocal altruism via alternative currencies after food sharing may be far better developed in humans than it is in gregarious carnivores. Whereas chimpanzees appear to fall on this continuum between humans and carnivores (e.g., de Waal 1992), their food sharing behavior more closely resembles that exhibited by carnivores (Boehm 1992).

Mechanisms Mediating Group Movement in Carnivores and Primates

If social complexity favors the evolution of intelligence (reviewed by Byrne and Whiten 1988), then we might expect gregarious carnivores to exhibit some remarkable cognitive abilities. Although big game hunters have often claimed that large carnivores exhibit considerable intellectual capabilities, little is actually known about cognition in these species (Byrne 1995b; Holekamp et al. 1999). Similarly, studies of the relationship between brain size, group size, and social complexity in carnivores have lagged far behind comparable work on primates. For example, recent research has revealed that neocortex size in primates increases with group size (Sawaguchi and Kudo 1990; Dunbar 1992) and social complexity (Byrne 1993), but not with home range size (Dunbar 1992). The ratio of cortical volume to total brain volume is approximately equal in carnivores and primates (Macphail 1982). Until recently, however, relative neocortex size among various carnivore species had not been specifically evaluated. Using the cruder measure of overall brain size, some workers have examined its relationship to social and ecological variables in carnivores. When corrected for body size, relative brain size in carnivores is apparently unrelated to home range size or mating system (Gittleman 1986). Although Hemmer (1978) found that gregarious carnivores have larger brains than do solitary species, Gittleman (1986) found no such pattern. Gittleman found, in fact, that bears have the largest ratio of brain size to body size of any carnivore, despite the fact that they are mostly solitary. Interestingly, however, carnivores living in single-male groups have significantly

K. E. Holekamp, E. E. Boydson, and L. Smale

smaller brains than do those living in multimale groups (Gittleman 1986). Finally, using estimated neocortex volumes, Dunbar and Bever (1998) have recently shown that the ratio of neocortex volume to total brain volume is positively correlated with mean group size among twenty-seven carnivore species, as is also true among primates (Sawaguchi and Kudo 1990; Dunbar 1992; 1995; Barton 1996a). Future study of the size or complexity of specific brain areas (e.g., frontal cortex) in relation to social variables may reveal other trends in carnivores similar to those found among primates.

Our overview of group movements in gregarious carnivores suggests some interesting and paradoxical preliminary conclusions about the mental abilities of these animals. Carnivore group travel during group foraging, and particularly during group hunts of large-bodied prey, often appears beautifully coordinated, and lures us toward the conclusion that these animals must be utilizing sophisticated cognitive mechanisms to effect this coordination. By contrast, carnivore den moves generally appear so chaotic and disorganized that we are amazed that any carnivore infants ever survive transfer between dens. Nevertheless, both group hunts and den moves appear to function effectively in carnivores due to their common mediation by social facilitation and specific rules of thumb. How do the factors shaping patterns of group travel differ between carnivores and primates?

Basic ecological variables such as predation pressure and food availability profoundly influence ranging patterns, territory size, and foraging group size in carnivores, as they do in primates (Grant, Chapman, and Richardson 1992; Chapman, Wrangham, and Chapman 1995; Lowen and Dunbar 1994; Wrangham, Gittleman, and Chapman 1993). However, except for small-bodied species such as the dwarf mongoose, predation pressure on carnivores tends to be less intense than it is on primates, due to that fact that mammalian carnivores occupy trophic niches at the top of most terrestrial food pyramids. The ranging patterns and foraging group sizes observed in most larger carnivores are shaped by the type, distribution, and abundance of prey. When abundant food is clumped and defensible, carnivores show increased group size and decreased territory size compared with conspecifics for whom resources are less abundant and more widely dispersed (Macdonald 1983; Bekoff and Wells 1986). Home range size is generally greater for carnivores than it is for primates of comparable body size (McNab 1963; Mace, Harvey, and Clutton-Brock 1983). Home range size tends to increase with body size in both primates (Milton and May 1976; Clutton-Brock and Harvey 1977a) and carnivores (Gittleman and Harvey 1982). Undefended home ranges are larger than defended ones for carnivores, but not for primates (Grant, Chapman, and Richardson 1992). Although data from carnivores are more limited than are those available from primates, it appears that the same relationships hold in both taxa between ranging patterns, home range size, and the monitoring or defense of territorial boundaries (Mitani and Rodman 1979). Day range and population size are consistent predictors of travel group size in both carnivores and primates, and it appears that group size is constrained in both taxa by the metabolic costs of travel (Wrangham, Gittleman, and Chapman 1993).

The sensory modalities used to maintain group cohesion differ between primates and carnivores, and this appears to be attributable to constraints imposed by both phylogenetic and ecological variables. The need for group cohesion varies with time and circumstance in most carnivores, whereas it remains relatively constant in most primates. Due to reproductive demands and intense feeding competition among gregarious carnivores, the best interests of individual group members are not always served by remaining with conspecifics. Therefore social groups of large carnivores routinely splinter into smaller subunits during foraging or early care of young. On the other hand, when togetherness benefits most or all group members concurrently, such as during conflicts with other carnivores, den moves, or hunts of large prey, group cohesion is promoted with excitatory behaviors that synchronize all individuals present at a high level of arousal (e.g., Glickman et al. 1997).

Comparable social facilitation of specific responses certainly occurs in human societies when the desired result is to have large numbers of people concurrently exhibiting identical behavior. We have pep rallies before athletic events, political rallies before elections, and military rallies before attacks during wartime. Similarly impressive social facilitation occurs among nonhuman primates when, for example, howler monkeys engage in an infectious bout of dawn "singing" (Carpenter 1934), or when a chorus of nervous grunts sweeps through a baboon troop in the middle of the night (Kummer 1971). Although social rallies may, in fact, precede group movements in anthropoids, these generally seem more subtle than are those observed in gregarious carnivores. For example, the hour or so after dawn is a good time to observe a "social session" in

K. E. Holekamp, E. E. Boydson, and L. Smale

hamadryas baboons, in which troop members engage in much grooming, presenting, and sniffing of other animals (Kummer 1971). The social session ends only when the troop leaves the vicinity of its sleeping cliff to forage. These primate "rallies" might more appropriately be called "conferences," since they tend to last longer, and are lower-key, than those in carnivores (e.g., Kummer 1971). Nevertheless, such sessions may function in both taxa to elevate and synchronize the arousal levels of all individuals present in preparation for group travel. Anyone who has ever attempted to get a group of human campers moving after a night spent in the woods can attest to the fact that substantial variation exists among individual humans in the time they require for complete arousal and readiness to travel. Similarly, differences among individuals in the rates at which they achieve some threshold readiness to travel may account for observations of a 30–60 minute time lag in nonhuman primates between waking up and departing from the sleeping site (e.g., Smuts 1985).

Carnivore leadership patterns vary interspecifically, as they do in primates. In both taxa, social groups with only one or a few clearly dominant individuals are generally led by those high-ranking animals (e.g., Kummer 1971; Boinski 1993, 1996). Subordinate animals in both taxa attend carefully to the behavior and whereabouts of alpha animals, and follow them when they move. In more loosely structured social groups of carnivores and primates, it is often impossible to identify a group leader, although adult females appear to make most decisions about when and where to travel (e.g., lions: Schaller 1972; Stander 1992a; savanna baboons: Rowell 1972b; squirrel monkeys: Boinski 1996). In both taxa we should examine the hypothesis that group leadership is usually assumed by the individuals having the greatest energy requirements and thus perhaps also the greatest motivation to get moving.

The mental processes required to initiate and maintain group travel need not be complex in either carnivores or nonhuman primates, and do not appear to differ fundamentally between these two taxa. Even chimpanzees stalking monkeys appear to be exercising cognitive abilities no more sophisticated than those used by wolves joining forces to disembowel a moose. Thus, although group movements in social carnivores and primates alike often appear to be beautifully coordinated, these seemingly complex movements can often be explained by the operation of a few simple decision rules.

Conclusions

Despite important fundamental differences between gregarious carnivores and nonhuman primates with respect to their sensory worlds, feeding ecology, and requirements for reproduction, the group movements observed in these two taxa reveal many remarkable similarities. Basic ecological parameters such as the abundance, quality, and distribution of food influence the sizes and ranging patterns of travel groups in both carnivores and nonhuman primates. Some carnivores utilize consistent travel routes, whereas other species do not, as is also true in primates. The movements of groups of social mongooses or Cape hunting dogs are no less cohesive than are those of most primate troops, nor is group leadership any less apparent. In carnivore societies structured by hierarchical dominance relationships, relative success in initiating group travel is strongly influenced by an individual's social rank, as is also true in many primates. In both taxa, successful leadership may be mediated by the selective attention paid by subordinates to the whereabouts and activity of high-ranking animals.

Group movements in carnivores are generally preceded by "social rallies," which apparently function to elevate and synchronize arousal levels in all group members and to promote group cohesion. These behavior patterns seem most appropriately described as "response facilitation" (Byrne 1994). It remains to be determined to what extent this type of social facilitation functions in the initiation of primate group movements. Perhaps less extensive facilitation activity is required by primates than by carnivores because primates tend to watch one another more carefully, or because primates are more sensitive to the travel initiation movements of others (e.g., Menzel 1993). This latter possibility seems likely in light of the fact that it is rarely in the best interest of an individual primate to separate from its group, so natural selection should have favored individuals with low sensitivity thresholds to travel initiation movements by others. Social rallies may be needed in carnivore groups to override individual inclinations to wander off alone. However, before concluding that primates and carnivores differ with respect to the importance of social facilitation in the mediation of group movements, primatologists should rule out the response facilitation hypothesis by focusing their attention on the behavior of their subject animals during the periods prior to departure of the troop from night sleeping sites and midday rest sites. Only if no social interactions occur at these times that might function to stimulate and syn-

K. E. Holekamp, E. E. Boydson, and L. Smale

chronize activity by multiple animals will we be able to conclude that the mechanisms mediating initiation of group travel are fundamentally different in carnivores and primates.

Although we know of no group movements in primates that are strictly comparable to the collective den moves observed in carnivores, group hunts of vertebrate prey occur in both taxa, and the behavior of individual hunters sheds light on their cognitive abilities. The available data suggest that neither gregarious carnivores nor nonhuman primates exhibit the advance contingency planning, the complex mental representation of hunt organization, or the unique pattern of sharing the spoils that characterize group hunts by humans. Furthermore, we conclude that the data currently available do not support the hypothesis that chimpanzees or capuchins utilize more complex mental calculations during group hunts of vertebrate prey than do gregarious carnivores. It is clear from the available data that lions and chimpanzees simultaneously monitor both the prey animal and their fellow hunters, and that they somehow integrate information from both of these sources in deciding what to do next. However, in neither taxon has it been demonstrated that individual hunters can adjust their own position or behavior based, in the absence of immediate sensory input, on where they anticipate their colleagues to be or what they expect them to do. Thus, although the group hunting behavior of chimpanzees and lions certainly *looks* like organized team play, we believe that these behavior patterns demand no more complicated mental activity than the use of one or more simple rules of thumb.

How might it be shown that in fact the "rules of thumb" hypothesis is inadequate to account for the group hunting behavior of carnivores or nonhuman primates? Cooperative hunting in both taxa has often been likened in the literature to the organization of a football play, so we shall pursue this analogy to suggest a test of the "rules of thumb" hypothesis. Prior to a football game, the coach (leader) sketches each play on a blackboard and shows his players where they should go and what they should do to advance the ball toward the end zone, even when confronting a variety of different strategies that might be adopted by the opposing defense. Despite the fact that football players are not known for their intellectual prowess, each successful player internalizes a mental image of the play drawn on the blackboard, and he behaves on the field in accordance with that master plan. Thus the quarterback may throw the ball to a particular spot on the field, even in the absence of information regarding where his receiver is located at the time the ball leaves his hand. If the play works as anticipated, the receiver appears at the spot where the football drops to earth. The quarterback throws the ball to that spot based on his expectation of where his receiver will end up, and the receiver runs to that spot based on his expectation that the ball will appear there. Thus we can see how critical it is to the success of this particular group movement that each player bases his own behavior on his internalized representation of the play.

The "rules of thumb" hypothesis assumes that neither carnivores nor nonhuman primates have any such mental representation of a "game plan" during group hunts, and therefore predicts that individual hunters will fail to end up in the right place at the right time for prey capture if there is an interruption in the stream of sensory data informing them of the immediate whereabouts of their colleagues and the prey. Although detailed observations of hunts by free-living animals will be extremely difficult to collect, it should nevertheless be possible to identify the relevant sensory information each hunter is receiving at each instant during the hunt. If individual hunters deprived of immediate sensory data nevertheless end up in the right place at the appropriate time, we will be able to reject the "rules of thumb" hypothesis. At present, however, we are forced to conclude that the global patterns of behavior occurring during a hunt result strictly from local interactions among hunters, and between hunters and their prey. Only if the "rules of thumb" hypothesis can be ruled out will it be reasonable to conclude that group movement patterns in carnivores or nonhuman primates result from direction by a specific leader, an internalized "game plan," or any other complex mental representation.

Acknowledgments

We thank the following individuals for their excellent assistance in the field and in the laboratory: N. E. Berry, S. M. Cooper, M. Durham, A. Engh, J. Friedman, P. Garrett, T. H. Harty, C. I. Katona, G. Ording, L. Sams, M. Szykman, and K. Weibel. This work was supported by NSF grants BNS9021461, IBN9309805, and IBN9630667 to K. E. H. and L. S. and by fellowships to K. E. H. from the David and Lucille Packard Foundation and the Searle Scholars Program/Chicago Community Trust.

Appendix 20.1 Species mentioned in this chapter

Family	Common name	Latin name
Procyonidae	Coati	*Nasua nasua, Nasua narica*
Viverridae	Dwarf mongoose	*Helogale parvula*
	Banded mongoose	*Mungos mungo*
	Meerkat	*Suricata suricata*
	Common cusimanse	*Crossarchus obscurus*
Mustelidae	Giant river otter	*Pteronura brasiliensis*
	European badger	*Meles meles*
Canidae	Wolf	*Canis lupus*
	Coyote	*Canis latrans*
	Domestic dog	*Canis familiaris*
	Dingo	*Canis familiaris*
	Golden jackal	*Canis aureus*
	Black-backed jackal	*Canis mesomelas*
	Dhole	*Cuon alpinus*
	Cape hunting dog	*Lycaon pictus*
	Foxes	*Vulpes* spp.
Felidae	Lion	*Panthera leo*
Hyaenidae	Spotted hyena	*Crocuta crocuta*
Cebidae	Spider monkey	*Ateles geoffroyi*
	Capuchins	*Cebus* spp.
	Squirrel monkeys	*Saimiri* spp.
	Howler monkey	*Alouatta* spp.
Cercopithecidae	Red colobus	*Piliocolobus* spp.
	Hamadryas baboon	*Papio hamadryas*
	Olive baboon	*Papio anubis*
Pongidae	Chimpanzee	*Pan troglodytes*
Cervidae	Deer	*Odocoileus* spp.
	Moose	*Alces alces*
	Elk	*Cervus elaphus*
Bovidae	African Cape buffalo	*Syncerus caffer*
	Wildebeest	*Connechaetes taurinus*
	Gazelle	*Gazella* spp.
Suidae	Warthog	*Phacochoerus aethiopicus*
Equidae	Zebra	*Equus burchelli*
Macropodidae	Kangaroos	*Macropus* spp.
Leporidae	Rabbits	*Sylvilagus* spp.
	Hares	*Lepus* spp.

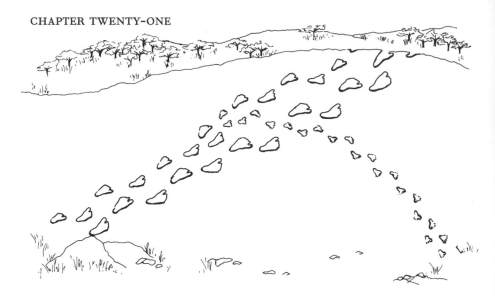

CHAPTER TWENTY-ONE

Ecological Correlates of Home Range Variation in Primates: Implications for Hominid Evolution

WILLIAM R. LEONARD AND MARCIA L. ROBERTSON

Given the limited and fragmentary nature of the human fossil record, insights into the patterns and processes of hominid evolution must be gained from the study of extant variation. Research from primate and mammalian ecology provides a particularly useful framework for interpreting trends in human evolution. Indeed, by looking at how diverse living species adapt to different ecosystems, we can make inferences about how environmental and climatic change may have shaped the biology and behavior of our ancestors. The use of modern analogues has been particularly useful in drawing inferences about changes in diet, foraging behavior, and movement pattern in hominids. Leonard and Robertson (1992, 1994), for example, demonstrated that among extant primates (including humans), those species with relatively large brains for their body size have higher-quality, more nutrient-dense diets (see also Aiello and Wheeler 1995). Such a pattern of variation suggests that the rapid expansion of brain size in the hominid lineage between 2 and 1 million years ago (mya) was likely associated with important shifts in diet and foraging behavior. Wheeler (1991) has used data on ecological variation in heat dissipation among mammals to suggest that the evolution of bipedalism may partly reflect an adaptation for reducing thermal stress among hominids as they expanded to an increasingly open (savanna) environment.

These studies underscore the utility of examining aspects of hominid evolution from a comparative, ecological perspective. However, to date, most such analyses have focused on the biology and behavior of individuals, rather than of groups or populations. Given the importance of group and population parameters for understanding long-term adaptive patterns of species (e.g., Roughgarden 1997), we believe that models of hominid ecology can benefit from a greater emphasis on group dynamics. This chapter will therefore examine the ecological and energetic correlates of variation in home range size, a key correlate of group movement patterns. Observed patterns of variation in home range size among living primates (including humans) will be used to explore how evolutionary changes in body size and foraging behavior may have influenced home range size and population density among early hominids.

A home range is the area typically traversed by an individual animal or group of animals during activities associated with feeding, reproduction, rest, and shelter seeking (Burt 1943). As such, a species' place or role within its ecosystem strongly shapes the size of its home range (Peters 1983). Comparative data on diverse mammalian species have shown that body size and dietary patterns, in particular, are strong predictors of variation in home range size (e.g., McNab 1963; Harestad and Bunnell 1979; Milton and May 1976; Peters 1983). Evidence from the human paleontological record indicates that hominid evolution has been characterized by changes in both body size and diet. Indeed, estimates from McHenry (1994) indicate that between 4 and 1.5 mya, average adult body weights for early hominids increased by about 56%, from about 37 kg in "Lucy" and her contemporaries *(Australopithecus afarensis)* to almost 58 kg in early African *H. erectus.* Archaeological and skeletal (e.g., craniofacial and dental) evidence suggests that the diet of these early hominids shifted from a fibrous, largely vegetarian diet in the australopithecines to a higher-quality, more omnivorous diet in early *Homo* (Potts 1988; Wolpoff 1980). Consequently, in light of these apparent changes in body size and dietary patterns, it is likely that there were important shifts in patterns of group size and range as well.

Previous attempts to estimate early hominid home ranges and population densities have met with limited success. Boaz (1979) calculated early hominid population densities based on their proportional representation in mammalian fossil assemblages from Pliocene levels from the lower Omo Basin in Ethiopia. This approach provided low resolution, yielding density estimates of between 0.001 and 2.48 individuals/km^2. The low level of resolution was in part due to the limited number of hominid fossils represented.

R. A. Martin (1981) and McHenry (1994) took a different approach to the problem, estimating hominid home range sizes and population densities based on the strong empirical relationships between home range and body weight presented by Harestad and Bunnell (1979) for mammalian herbivores, omnivores, and carnivores. This approach appears to be a more effective way of estimating population parameters in early hominids since it relies on highly predictive relationships (i.e., $r^2 > .75$) observed among extant species. However, the use of Harestad and Bunnell's equations is problematic, since their data set was restricted to North American mammals and thus contained no primate species. Indeed, work by Milton and May (1976) has shown that the relationship between home range and body weight among nonhuman primates is different from that seen among other mammals. How home range size in human foragers compares to that in other primates, however, remains unexplored. Hence, there is a need to examine hominid population parameters using more appropriate modern analogues (i.e., humans and nonhuman primates).

In this chapter we will examine the influence of body size and dietary strategy on variation in home range size among extant primate species, including human hunter-gatherers. We will then use these empirical relationships to estimate changes in home range size and population density over the course of hominid evolution. These estimates are combined with earlier estimates of hominid energy demands (from Leonard and Robertson 1997) to examine changes in trophic relationships during hominid evolution. Such an approach allows us to evaluate how the relationship between hominids and their ecosystem changed between 4 and 1.5 mya. These changes can provide insights into the factors that promoted the evolutionary trends seen with the emergence of early *Homo* in Africa and the later expansion of hominids from Africa to other parts of the Old World.

W. R. Leonard and M. L. Robertson

Primate and Human Ecological Data

Nonhuman primate species and traditional human foraging populations show a great deal of variation in home range size. Data on body weights (g), home range sizes (ha), and diets of forty-seven nonhuman primate species and six modern human tropical foraging groups were compiled from the literature and are presented in table 21.1. The nonhuman primate body weight and home range data were derived from Wrangham, Gittleman, and Chapman (1993), Sailer et al. (1985), and Milton and May (1976). Data on human hunter-gatherer home ranges and weights were derived from Lee (1979) for the !Kung, Hill and Hawkes (1983) for the Ache, Blurton Jones et al. (1992) for the Hadza, Yost and Kelley (1983) for the Waorani, Clastres (1972) for the Guayaki, and Kelly (1995) for the Mbuti. Body weight estimates for the human groups are mid-sex averages derived from Leonard and Robertson (1994) and Katzmarzyk and Leonard (1998).

Quantifying differences in dietary patterns among primates is difficult, since most species tend to be eclectic feeders, relying on a variety of different food sources. Most previous attempts to measure interspecific dietary differences have relied on simple indices such as percentage of feeding time allocated to one or more resource types (e.g., foliage, fruit, insects) (see Richard 1985a; Clutton-Brock and Harvey 1977b). In response to the difficulties in effectively assessing differences in primate diets, Sailer et al. (1985) developed a dietary quality (DQ) index based on a weighting of the percentages (by weight or feeding time) of three key resource types in the diet: (1) structural plant material (e.g., leaves, bark), (2) reproductive plant parts (e.g., fruit) and (3) animal material (including insects). In this index, each of the three dietary components is weighted by its relative energy and nutrient density as follows:

$$DQ = 3.5(a) + 2(r) + s$$

where

a = percentage of diet derived from animal material
r = percentage of diet derived from reproductive plant parts
s = percentage of diet derived from structural plant parts

Hence, this index provides a measure of the nutrient availability per unit weight of the diet. The DQ index is a continuous variable ranging from a minimum of 100 (i.e., a diet of 100% foliage) to a maxi-

Table 21.1 Body weight, home range size, and diet quality of selected primates

Species	Weight (g)	HR$_i$ (ha)	Diet quality
Prosimians			
Galago demidovii	60	0.80	305.0
Lemur catta	2,300	0.37	166.0
L. fulvus	2,370	0.10	129.0
L. mongoz	1,800	0.28	198.0
Lepilemur mustelinus	650	0.24	149.0
Propithecus verreauxi	3,800	1.14	159.0
Ceboidea			
Alouatta palliata	6,875	1.78	146.0
A. seniculus	7,250	0.38	177.5
Ateles belzebuth	5,800	14.00	181.5
Callicebus moloch	600	0.15	175.0
C. torquatus	1,100	5.00	208.5
Cebus albifrons	2,600	4.17	295.0
C. apella	2,100	2.50	310.0
C. capucinus	3,100	6.10	215.0
Saguinus geoffroyi	550	2.40	246.0
Saimiri oerstedii	600	0.76	245.0
S. sciureus	660	0.57	323.0
Cercopithecoidea			
Lophocebus albigena	7,900	5.10	226.5
Cercocebus galeritus	5,500	1.67	190.3
Chlorocebus aethiops	3,800	1.40	213.5
Cercopithecus			
ascanius	2,900	0.47	156.1
C. cephus	2,900	4.00	215.9
C. mitis	6,000	2.80	201.5
C. talapoin	1,130	4.00	222.5
Colobus badius	8,000	2.50	121.5
C. guereza	8,000	2.25	118.0
C. satanas	9,500	3.85	163.0
Macaca fascicularis	5,000	1.83	200.0
M. nemestrina	8,300	5.00	184.0
M. sinica	5,130	3.00	203.0
Nasalis larvatus	15,100	6.50	105.0
Papio anubis	21,400	10.00	207.2
P. hamadryas	13,850	33.33	199.0
P. ursinus	20,600	24.00	189.5
Presbytis cristatus	6,300	0.63	163.0
P. entellus	17,200	11.86	153.5
P. johnii	8,170	18.00	122.0
P. melaphos	6,600	1.65	163.0
P. obscura	6,500	3.22	152.0
P. senex	5,980	3.00	140.0
Theropithecus gelada	13,600	2.17	159.0
Hominoidea			
Gorilla gorilla	145,000	150.00	114.0
Hylobates agilis	5,700	5.29	162.5
H. lar	5,540	11.00	181.0

Pan troglodytes	40,700	23.00	178.0
Pongo pygmaeus	36,500	65.00	183.5
Symphalangus			
syndactylus	10,500	10.00	167.0
Homo			
Ache	55,700	714.00	263.0
Guayaki	55,700	387.00	263.0
Hadza	50,000	333.00	260.0
!Kung	43,500	1,896.00	255.5
Mbuti	40,700	588.00	252.5
Waorani	55,000	2,583.00	255.0

mum of 350 (100% animal material). This index is superior to such commonly used measures as "percent foliage" because it captures more of the variation in primate diets. Data on DQ for the nonhuman primate species were derived from Sailer et al. (1985). Diet quality indices for the human foragers were calculated based on data presented by Lee (1968, 1979) for the !Kung, Hill et al. (1984) for the Ache, Hill (1982) for the Mbuti, Blurton Jones et al. (1992) for the Hadza, Yost and Kelley (1983) for the Waorani, and Clastres (1972) for the Guayaki.

Fossil Hominid Data
In order to draw inferences about changes in group parameters during hominid evolution, estimates of body size are necessary. Obtaining such estimates is problematic given the small and fragmentary skeletal samples that are available for many fossil hominid groups. It is also clear that these estimates vary considerably depending on one's choice of which skeletal element and what comparative sample (e.g., all hominoids, humans only) are used (cf. McHenry 1988, 1992; Jungers 1988). Hence, as McHenry (1994) points out, these estimates should be viewed as best approximations that are subject to revision with the addition of new fossils or other analytic techniques. We should therefore be wary of ascribing too much precision to body weight estimates or the parameters they are used to predict (see Smith 1996; Barton, chap. 8, this volume). Nevertheless, it is clear from the fossil record that major changes in skeletal size (and thus presumably mass) did occur during the first 2–3 my of hominid evolution. Thus, by using consensus body weight estimates, we can begin to explore how changes in size may have interacted with other parameters (e.g., environmental change, dietary change, brain expansion) to shape the course of human evolution.

For the analyses in this chapter, we use the sex-specific body weight estimates presented by McHenry (1992, 1994) and derived from measurements of hindlimb joint size, based on regression equations developed on a sample of modern humans. These estimates incorporate a large and diverse fossil sample, and are the most widely utilized in the paleontological literature.

Determinants of Variation in Home Range Size among Extant Primates

Like many physiological and ecological parameters, interspecific variation in home range size among mammals is strongly related to body weight. Interspecific relationships between home range and body weight are expressed with the following general "scaling" equation:

$$HR = aWt^b$$

where a and b are constants (b = "scaling exponent").

Early work by McNab (1963) had initially found that home range size among mammals scaled to body weight with an exponent of approximately 0.75. Such a relationship was remarkable since the exponent was the same as that demonstrated between basal metabolic rate (BMR) and weight among mammals by Kleiber (1961). Thus, McNab's analyses suggested that variation in home range size varied directly as a function of differences in metabolic (food energy) requirements. Subsequent work by Harestad and Bunnell (1979), however, found mammalian home range versus weight scaling exponents of about 1.0, significantly greater than the so-called "Kleiber exponent" of 0.75. These authors argued that a scaling coefficient greater than 0.75 probably reflected the fact that larger mammals tend to live in less productive (generally more "open") habitats. Hence, larger mammals need to have disproportionately larger home ranges for their body size because they must spend disproportionately more energy above BMR to meet their daily food requirements.

Our results for primate species alone appear to be consistent with Harestad and Bunnell's model. Figure 21.1 presents the plot of home range versus body weight for the sample of forty-seven non-human primate species and six groups of human foragers living in tropical ecosystems presented in table 21.1. We have followed the approach advocated by Milton and May (1976) and divided home range size for each species by group size to determine the home

W. R. Leonard and M. L. Robertson

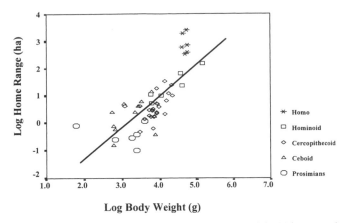

Figure 21.1 Log-log plot of home range size (ha) versus body weight (g) in a sample of forty-seven nonhuman primate species and six tropical human foraging groups. The observed scaling exponent of this relationship (1.21 ± 0.14) does not significantly differ from 1.

range per individual (HR_i). This approach is warranted given the great disparity in group size seen among primate species. Additionally, the data were log-transformed to allow us to use linear regression analysis to quantify the scaling relationship between home range and body weight (see Peters 1983).

Our analyses indicate that the relationship between weight and home range size is $HR_i = 0.0001Wt^{1.21}$ ($r = .763$; $P < .001$). The observed scaling exponent for this relationship (1.21 ± 0.14; 95% CI = 0.92–1.51) is significantly greater than 0.75, but does not statistically differ from 1.0. Previous work by Leonard and Robertson (1994) and Kurland and Pearson (1986) has shown that the scaling relationship between BMR and weight is the same among primates as it is among other mammals (exponent of about 0.75). Consequently, the explanation offered by Harestad and Bunnell (1979) for variation in home range size among mammals would appear to apply for primates, in particular, as well. In effect, it appears that larger primates must "work harder" in the pursuit of food, thus expending relatively more energy above their minimal "basal" costs than smaller primates.

Despite the strong relationship between home range and body weight, there are notable deviations from the regression. Human foragers, for example, have substantially larger home ranges than predicted for their size, whereas colobus (*Colobus* spp.) and howler (*Alouatta* spp.) monkeys tend to have relatively small home ranges. Some of these deviations are likely due to differences in dietary pat-

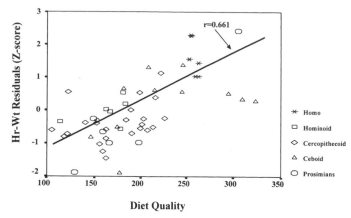

Figure 21.2 Plot of standardized residuals (z-scores) from the home range versus body weight regression (from figure 21.1) versus diet quality. The strong correlation between these measures ($r = .661$) indicates that species with higher-quality, more nutrient-rich diets have disproportionately larger home range sizes for their body weights than species with poorer-quality diets.

terns. In fact, both the Harestad and Bunnell (1979) and Milton and May (1976) studies demonstrated marked differences in the home range versus weight relationships across species with different dietary regimes, with more omnivorous and carnivorous species having substantially greater home range sizes than largely herbivorous species. These earlier studies, however, used only broad, relatively crude categories to assess diet. Thus, to more effectively assess the influence of diet on home range size, we are using the index of Sailer et al. (1985) to quantify DQ.

Figure 21.2 examines the relationship between deviations from predicted home range size and DQ among the fifty-three human and nonhuman primate groups from figure 21.1. As expected, deviations from the home range versus weight relationship in figure 21.1 are strongly associated with DQ ($r = .661$; $P < .001$). In other words, much of the variation in home range size that is not explained by weight alone can be explained by differences in diet. Species with higher-quality, more energy- and nutrient-dense diets (i.e., more fruit and animal material) have disproportionately larger home ranges for their size than do species with poorer-quality (largely folivorous) diets.

Table 21.2 presents the results of a multiple regression analysis that explores the joint influence of weight and DQ on home range size. Note that together, DQ and weight explain almost 80% of the variance in home range size. These findings are remarkable in show-

W. R. Leonard and M. L. Robertson

Table 21.2 Multiple regression analyses of the predictors of variation in Log-HR size among extant primates

Independent variable	Regression coefficient ($b \pm$ SE)	Beta weight	P - value
Log - weight	1.36 ± 0.11	0.854	< 0.001
Diet quality	0.009 ± 0.001	0.446	< 0.001
Constant	-6.08 ± 0.54	—	< 0.001

$R = 0.88$; $R^2 = 0.77$; $P < .0001$.

ing that the home ranges of human foragers are comparable to what would be expected for other primate species of similar size and dietary habits. Hence this relationship is particularly useful in providing insight into our evolutionary past, since it applies equally well to human and nonhuman primates species. In the next section we will explore likely changes in home range size and population density over the last 4 million years of human evolution.

Estimating Home Range Size in Fossil Hominids
Based on the patterns of variation we have outlined above, we can now explore alternative models for estimating HR_i in fossil hominid species. We used the regression equation shown in table 21.2 to predict home range size for hominids based on weight and DQ. For each species, table 21.3 presents two estimates of home range size based on different assumptions about DQ. The "ape model" assumes a diet quality comparable to that of modern large-bodied apes (i.e., DQ = 170), whereas the "human model" assumes a diet quality at the low end of the range for modern human foragers of the tropics (DQ = 250). We find that the differences in assumptions about diet quality produce a five- to sixfold difference in estimated home range size for each species. Estimated home range sizes using the ape model range from 40–45 ha for the early australopithecines to 80–90 ha for *H. erectus* and early *H. sapiens*. In contrast, the human model suggests a home range size ranging from about 200–250 ha for the gracile australopithecines to about 450 ha for early *Homo*. Consequently, changes in diet and foraging regimes would have had a dramatic effect on home range size in our hominid ancestors.

These home range estimates are markedly different from those presented by R. A. Martin (1981) and McHenry (1994) based on the equations of Harestad and Bunnell (1979) for mammals. These

Table 21.3 Estimated body weight and home range size for fossil hominid species

Species	Male wt (kg)	Female wt (kg)	Av wt (kg)	HR_i-Ape[a] (ha)	HR_i-Human[b] (ha)
A. afarensis	44.6	29.1	37.0	46	240
A. africanus	40.8	30.2	35.5	43	227
A. robustus	40.2	31.9	36.1	44	232
A. boisei	48.6	34.0	44.3	58	307
H. habilis	51.6	31.5	41.6	54	281
H. erectus	63.0	52.3	57.7	84	440
H. sapiens	65.0	54.0	59.5	86	451

Note: Body weight estimates are from McHenry (1992, 1994).

[a]Home range estimates assuming a diet quality similar to that of a modern great ape.

[b]Home range estimates assuming a diet quality similar to that of a modern tropical human forager.

differences result, in part, from the fact that the Harestrad and Bunnell equations, unlike the present analyses, make no adjustments for group size. The typical group size of modern chimpanzees appears to be about 50 individuals (Goodall 1986), whereas that for modern hunter-gatherers averages about 25 (Kelly 1995). Thus, estimates of group home ranges for early hominid species range from 2,000 to 4,000 ha when the ape model is used and 6,000 to 11,000 ha with the human model. When adjusted for group size, the estimates from both models are systematically higher than those obtained by McHenry (1994). These differences reflect the fact that, as noted by Milton and May (1976), home ranges for groups of primates tend to be relatively larger than those of comparably sized mammals consuming similar diets, probably because primate groups display greater social and behavioral complexity, thus allowing them to more efficiently exploit relatively larger home ranges.

Clearly, evolutionary changes in hominid body size and diet would have had a strong interactive effect on home range sizes. If we assume that hominid diets shifted from an apelike pattern in the early australopithecines to a more humanlike omnivorous diet with early *Homo* (especially *H. erectus*), this implies a tenfold increase in home range size (from about 45 ha to about 450 ha). The influence that these changes in home range size would have had on the interplay between hominid populations and their changing ecosystems is examined in the next section.

W. R. Leonard and M. L. Robertson

Energy Dynamics and Ecosystem Change

Many of the key evolutionary changes seen with the emergence of the genus *Homo,* such as the increases in brain and body size and the changes in body proportions, are thought to be the result of large-scale ecosystem change in Africa (Vrba 1993; Leonard and Robertson 1997; Isbell et al. 1998). Between 2.5 and 1.5 mya, there was a marked decline in forested areas throughout eastern and southern Africa (Behrensmeyer and Cooke 1985; Vrba 1985, 1988, 1995). This transition from woodland to open savanna environments resulted in changes in both the abundance and distribution of food resources. That is, the energetic structure of these ecosystems radically changed, with plant productivity declining and animal foods becoming a much more attractive resource. These increasingly open habitats likely allowed for changes in both diet and ranging behavior among early members of the genus *Homo.*

To get a better sense of how ecosystem change may have influenced resource availability for early hominids, we can begin by looking at variation in the yearly energy productivity of plants (net primary productivity; NPP) in modern ecosystems (from Leith 1975; Begon, Harper, and Townsend 1990; Odum 1971). Figure 21.3 shows mean values of NPP in megajoules per square meter per year (MJ/m[2]/yr) for selected tropical ecosystems.[1] The annual productivity of tropical rainforests averages more than 40 MJ/m[2,] as compared with 12–17 MJ/m[2] in shrubland and savanna environ-

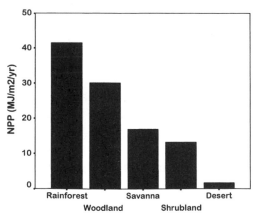

Figure 21.3 Net primary productivity (MJ/m[2]/yr) of selected tropical ecosystems. Yearly energy production ranges from less than 5 MJ/m[2] in desert environments to more than 40 MJ/m[2] in tropical rainforests.

ments and less than 5 MJ/m^2 in deserts. Recent paleoenvironmental reconstructions suggest that the earliest australopithecines lived in relatively wooded ecosystems (see White, Suwa, and Asfaw 1994; WoldeGabriel et al. 1994). Conversely, by 1.5–2 mya, it appears that open savanna had replaced much of the woodland in eastern and southern Africa (e.g., Behrensmeyer and Cook 1985; Vrba 1988, 1993). Therefore, to evaluate how resource availability changed for hominid groups between 4 and 1.5 mya, we will assume that a moderately productive woodland and an average savanna bracket the types of ecosystems inhabited by the earliest australopithecines through the emergence of African *H. erectus*.

Table 21.4 shows the yearly productivity of energy at different trophic levels for tropical woodland and savanna ecosystems. Secondary productivity measures the annual production of new energy by herbivores (plant eaters) of the ecosystem, whereas tertiary productivity measures the energy produced by carnivores. We see that while the primary productivity of the savanna is little more than half that of the forest (16,950 vs. 30,140 kJ/m^2), the level of herbivore (secondary) productivity is almost three times greater on the savanna (42.4 vs. 15.1 kJ/m^2). This difference is due to differences in usable forage for herbivores in forest versus grassland environments. Since much of the plant production in the forest (e.g., roots, woody vegetation) cannot be utilized by herbivores, consumption rates are typically low, about 5% of net primary productivity. In contrast, consumption rates by herbivores in grassland environments are much higher, in some cases as high as 50% of primary productivity (Begon, Harper, and Townsend 1990; Leith 1975; Heal and McLean 1975). Thus, a greater proportion of available energy is transferred to higher trophic levels in savanna ecosystems, allowing for a proportionally larger component of herbivores, such as antelopes. This increased relative abundance of herbivores underscores why ecosystem change in Africa between 2.5 and 1.5 mya would have made animal foods increasingly attractive resources for early hominids.

With our estimates of home range size, we can further examine how the interplay between early hominids and their environments changed over time. The energy demands of early hominid populations can be determined based on estimates of population density (individual/km^2) and individual-level daily energy requirements (kJ/individual/day). In our previous work we have estimated daily energy needs (i.e., energy expended over the course of a day) for

W. R. Leonard and M. L. Robertson

Table 21.4 Ecosystem productivity: forest/woodland vs. savanna

Ecosystem	Primary productivity (kJ/m²/yr)	Secondary (herbivore) productivity (kJ/m²/yr)	Tertiary (carnivore) productivity (kJ/m²/yr)
Forest/woodland	30,140	15.1	0.12
Savanna	16,950	42.4	0.34

hominids using comparative data on living primates and human foragers (Leonard and Robertson 1997). Population density, on the other hand, can be calculated simply as the inverse of HR_i (Peters 1983):

$$D = 100*(1/HR_i)$$

where

D = population density (individuals/km^2)
HR_i = individual home range (ha)

Estimates of daily individual energy demands (kJ/day), population densities, and annual population energy demands for early hominids assuming ape versus human HR_i sizes are presented in table 21.5. Using the ape model, population densities range from 1.2 to 2.5 individuals/km^2 resulting in annual energy demands of between 5.3 and 7.3 kJ/m^2 In contrast, the larger estimated home ranges of the human model imply population densities between 0.20 and 0.45 individuals/km^2 and annual energy demands of 1.00–1.40 kJ/m^2

Table 21.6 shows estimates of annual energy demands for *A. afarensis* and *H. erectus* compared with the estimates of primary and secondary productivity in wooded and savanna environments, respectively. Energy demands for *A. afarensis* are those derived from the ape model, whereas those for *H. erectus* were obtained using the human model. Note that for *A. afarensis,* the annual energy needs, assuming an ape-sized home range, are almost half the *total* herbivore productivity (7.14 vs. 15.1 kJ/m^2). In contrast, for *H. erectus,* with a considerably lower population density, predicted energy demands represent only about 2% of herbivore productivity.

These estimates highlight two important issues. First, they suggest that for the earliest australopithecines, living in a more closed environment, the degree of carnivory was probably low, being limited by the level of secondary productivity. That is, because the consumable vegetation in more wooded environments can support only a relatively small proportion of herbivores, hominids would likely have exploited plant resources to a much greater extent than animal foods. Second, these results suggest that with the expansion of grasslands in Africa, the energetic costs of exploiting large animal resources were probably outweighed by increases in energy availability and overall dietary quality. Indeed, human foraging patterns necessitate large home ranges and, as we have demonstrated previously, relatively larger day ranges, higher activity levels, and

W. R. Leonard and M. L. Robertson

Table 21.5 Estimated population densities and energy demands of selected fossil hominid species

Species	Individual energy needs (kJ/indiv/day)	Ape model		Human model	
		Population density (indiv/km²)	Energy demand (kJ/m²/yr)	Population density (indiv/km²)	Energy demand (kJ/m²/yr)
A. afarensis	8,960	2.18	7.14	0.42	1.36
A. africanus	8,680	2.31	7.32	0.44	1.39
A. robustus	8,790	2.26	7.25	0.43	1.38
A. boisei	10,266	1.71	6.41	0.33	1.22
H. habilis	9,788	1.86	6.65	0.35	1.27
H. erectus	12,543	1.19	5.46	0.23	1.04
H. sapiens	12,555	1.16	5.32	0.22	1.01

Table 21.6 Hominid population density, energy demands, and ecosystem productivity

Species	Population density (indiv/km²)	Energy demand (kJ/m²/yr)	Secondary productivity (kJ/m²/yr)	Primary productivity (kJ/m²/yr)
A. afarensis	2.18	7.14	15.1	30,140
H. erectus	0.23	1.04	42.4	16,950

greater energy expenditure than seen in other primate species (Leonard and Robertson 1997). Thus, large-bodied foragers, especially those with higher-quality diets, must "work harder" to procure food; however, these increased metabolic requirements are offset by the reliance on a higher-quality, more energy-rich diet containing greater amounts of animal material.

Primate and Early Hominid Ranging Patterns in Comparative Perspective

The results presented here underscore key similarities and differences in the home range versus body weight patterns of primates relative to other mammalian species. In both primates and other mammals, the scaling coefficient of home range to weight is significantly greater than the Kleiber exponent of 0.75 for basal metabolism versus body weight. This implies that home range size does not increase as a simple (linear) function of BMR. Rather, it seems that larger animals must spend relatively more energy on movement to traverse these larger areas. Such an interpretation is supported by recent analyses of total energy expenditure in primates and other mammals. Nagy (1987) found that among twenty-three species of mammals, the scaling coefficient of total energy expenditure to weight was significantly greater than 0.75. Similar results were obtained for eighteen primate species by Leonard and Robertson (1997). Consequently, it appears that McNab's (1963) initial insight about variation in home range size being linked to differences in metabolic requirements was correct. However, subsequent analyses, including the ones presented here, show that it is an animal's (or group's) *total* energy budget, rather than its minimal (basal) needs, that determines home range size. Thus, on a day-to-day basis, larger individuals must work harder (expend more energy above their basal needs) than smaller animals to meet their subsistence needs. This translates into relatively larger day and home ranges for larger

W. R. Leonard and M. L. Robertson

animals (see Steudel, chap. 1, this volume for a discussion of the link between locomotor energy costs and daily ranging patterns).

Home range size in primates and other mammals is also influenced by dietary regimes. Among mammals in general, carnivores have much larger home ranges than either omnivores or herbivores (Harestad and Bunnell 1979). The results presented here for primate species also show a positive correlation between dietary quality and home range size, suggesting that more nutrient-dense diets are associated with larger ranges. This association between home range size and diet is likely a consequence of differences in energy availability at different trophic levels. For animals consuming more omnivorous or carnivorous diets, the amount of usable energy per unit area (e.g., kJ/m^2) is greatly reduced, necessitating much larger home ranges to support their energy requirements. Among primates, for example, howler monkeys (*Alouatta* spp.) represent one end of the spectrum, living in relatively closed habitats, relying on a highly folivorous diet, and having small home ranges for their body size. In contrast, many baboon species (*Papio* spp.), living in more open environments, have relatively high-quality diets and disproportionately large home ranges for their size.

Patterns of variation among extant species provide a useful framework for evaluating hominid evolutionary trends, particularly between 2.5 and 1.5 mya. During this time, changes in three key ecological variables—body size, ecosystem productivity, and diet—all would have contributed to substantial increases in hominid home range size. According to the most recent estimates, the mean body weight of early *Homo erectus* was some 40–50% greater than that of the late australopithecines (e.g., McHenry 1992). Additionally, paleoecological reconstructions indicate increasing aridification of African habitats during this time period. Using modern tropical ecosystems as analogues, we can estimate that the transition from a mixed woodland to an open savanna environment was associated with declines in net primary productivity of about 40%. Harestad and Bunnell (1979) demonstrate that even within trophic classes (i.e., herbivore, omnivore, carnivore), habitat productivity has a profound effect on mammalian home range size. For hominids, this influence would have been amplified by the change in diet that also appears to have occurred based on the morphological and archeological evidence. As grassland environments expanded in Africa, the density of bovids and other herbivorous fauna actually increased (see table 21.4), making animal foods a more attractive resource and likely contributing further to the expansion of home ranges in early *Homo*.

Together, these changes suggest a ten-fold increase in home range size with the emergence and evolution of early *Homo* in Africa. Our estimates suggest that later australopithecines had home range sizes on the order of 40–50 ha/individual, while those for early *Homo* ranged from 400–500 ha. This home range expansion would likely have been associated with greater daily activity and energy budgets. Such increases in energy demands likely would have been offset by the increased quality and caloric density of a diet richer in animal foods (see Leonard and Robertson 1992, 1997).

These large-scale changes in home range size with early *Homo* may well have contributed to the dispersal of hominids out of Africa. Recent evidence from Southeast Asia indicates that *H. erectus* appeared in Java at about 1.8 mya, approximately the same time as its first appearance in Africa (Swisher et al. 1994; Huang et al. 1995; Antón, 1997). Thus the emergence and dispersal of *H. erectus* occurred at a time of a major shift in foraging behavior, to greater reliance on animal foods. The higher quality and greater stability of the diet, coupled with the larger ranges necessitated by a more carnivorous diet, therefore seem to have provided conditions favorable for the initial expansion of hominids from Africa.

Further, it appears that the increased behavioral complexity necessary for exploiting larger home ranges may have promoted the rapid increase in brain size observed with the emergence and early evolution of *Homo* (see Milton 1993a). Previously, we have demonstrated that among extant primates, dietary quality is positively associated with relative brain size, suggesting that the shift to a more omnivorous/carnivorous diet was likely important for fueling brain evolution in early *Homo* (Leonard and Robertson 1994). Analyses of the twenty-five primate species for which we have data on brain size as well as home range size show that relative home range size (i.e., residuals from the log-HR vs. log-wt regression) is significantly correlated with relative brain size ($r = .721$; $P < .001$). This association remains even after the influence of variation in diet quality is adjusted for (partial $r = .496$; $P < .01$) (but see Barton, chap. 8, this volume). These results suggest that relatively larger brain sizes are seen among primates with large home ranges, even after differences in diet are adjusted for. Hence, they imply that changes in home range size (along with the associated changes in day range

W. R. Leonard and M. L. Robertson

and total energy expenditure) and diet may have played a role in promoting the increases in brain size with *H. erectus.*

Conclusions

Home range size reflects many key aspects of a species' adaptive regime. Previous research on diverse mammalian species has shown that home range size is strongly shaped by variation in body weight and dietary patterns. The analyses presented in this chapter have shown similar associations among primate species (including humans). Indeed, variation in body weight and dietary quality explained 75–80% of the variation in home range size in our primate sample.

The scaling coefficient of the home range versus body weight relationship in primates is approximately 1.0, and is similar to that observed among other mammals. This coefficient is also comparable to that found for the relationship between total daily energy expenditure and weight in both primates and other mammals. As such, it appears that interspecific variation in home range size is largely influenced by differences in total food energy needs in primates and other mammals.

In addition, animals with higher-quality, more nutrient-rich diets tend to have larger home range sizes than expected for their body weights. Among mammals, carnivorous species have, by far, the largest home range sizes. Our analyses for primates show similar trends, as human foragers, who consume larger quantities of animal foods than other large-bodied primates, display disproportionately large home ranges for their size. In contrast, primate species consuming highly folivorous, low-quality diets (e.g., *Alouatta* spp. and *Colobus* spp.) have very small home ranges for their body size.

The relationships demonstrated between body weight, diet quality, and home range size in our primate sample have important implications for interpreting trends in hominid evolution. In particular, the apparent increase in body weight among hominids seen with the evolution of early *Homo,* coupled with the apparent environmental drying and resulting changes in resource use between 2.5 and 1.5 mya, suggest a tenfold increase in home range size in early *Homo* relative to that of the later australopithecines. The subsistence strategy of *H. erectus* was one that probably required substantially higher levels of energy expenditure associated with movement over much larger day and home ranges. These elevated energy demands, however, were likely offset by the utilization of a higher-quality diet, containing greater amounts of animal material. Such a dietary shift appears to have been promoted by changes in ecosystem structure that increased the availability of bovids and other herbivores, as well as by social and technological advances that allowed for more efficient exploitation of those herbivores. Further, we believe that these interrelated changes in body size, diet, foraging patterns, and home range size were critical for promoting both the rapid expansion of brain size in early *Homo* and the spread of *H. erectus* from Africa to other parts of the Old World.

Acknowledgments

We are especially grateful to Sue Boinski for her patience, encouragement, and insightful comments during the preparation of this chapter. The chapter also benefited greatly from discussions with Susan Antón.

Note

1. Joules, rather than calories, are the preferred scientific units for measuring energy: 1 MJ = 1,000 kilojoules [kJ]; 1 kilocalorie [kcal] = 4.184 kJ.

W. R. Leonard and M. L. Robertson

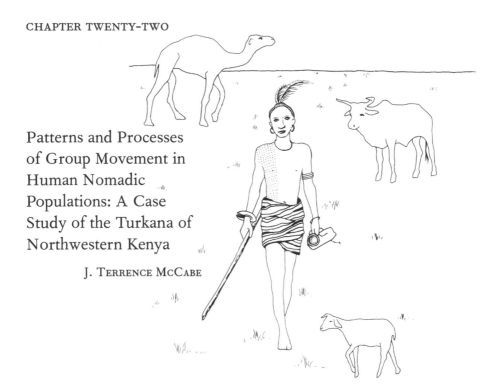

CHAPTER TWENTY-TWO

Patterns and Processes
of Group Movement in
Human Nomadic
Populations: A Case
Study of the Turkana of
Northwestern Kenya

J. TERRENCE McCABE

Throughout human history, the movement of groups has been an important adaptive strategy, and remains so today for millions of people. Foragers, hunters, fishers, pastoralists, gypsies, and tinkers all employ mobility as a means of obtaining spatially dispersed resources; human groups also move to get away from pests, disease, or danger. The most extensively studied of these groups are hunter-gatherers and nomadic pastoralists. In this chapter I consider patterns and processes of group movement among humans. In order to make this daunting task a bit more manageable, I will focus much of my discussion on a group of nomadic peoples with whom I have been working for over 15 years: the Turkana of northern Kenya.

Understanding mobility among modern humans presents many conceptual problems similar to those encountered by researchers studying mobility among nonhuman primates and other migratory species. Readers may find similarities between my treatment of the Turkana and the analyses of mobility in nonhuman primates contained in other chapters of this volume. However, the vital importance of cultural, social, economic, and political concerns adds further complexity to the study of humans. This complexity also allows for multiple explanations and interpretations, some of which I discuss in my review of the nomadic literature.

The literature concerning mobility among human populations is extensive and in many ways reflects the multiple perspectives found in anthropology. Researchers in this field engage one of the most contested domains in the study of human behavior: the extent to which the underlying determinants of human behavior are biological or cultural (for various opinions see Barth 1961; Rappaport 1968; Lee 1979; Wilmsen 1989; Descola 1994; Descola and Palsson 1996). Much of this debate concerns the extent to which Darwinian theory can be applied to human decision making, and is not restricted to understanding the pattern and process of group mobility (e.g., E. A. Smith 1991; Smith and Winterhalder 1992; Richerson and Boyd 1992; Vayda 1995; Ingold 1996).

With respect to mobility per se, there are those who view human mobility principally in ecological terms, and those who view this perspective as overly functionalistic and reductionistic. There are those who feel that real understanding of human mobility can come only from comparative studies based on typologies, and those who feel that this approach masks variability and detracts from our understanding of the causes, processes, and consequences of human mobility.

Some scholars feel that mobility among hunters and foragers is so fundamentally different from that of pastoralists that comparison is not possible (see Khazanov 1994); others, while recognizing the differences between foragers/hunters and pastoralists, feel that both groups utilize mobility in a similar enough manner to allow comparison (see Ingold 1987). Aside from the recent edited volume by Casimir and Rao (1992), few social scientists have tried to bring gypsies, tinkers, and other peripatetics (people who move from place to place) into this debate.

In this chapter I will first discuss some of the most important issues addressed by researchers involved in the study of human mobility; then I will turn to an examination of the pattern of mobility among one of the world's most mobile peoples, the Turkana pastoralists of northwestern Kenya. I have worked among the Turkana for over 15 years, focusing on mobility and the use of natural resources. Much of my previous work has focused on individual-level analysis, but here I will concentrate on group movements and deci-

J. Terrence McCabe

sions relating to mobility. I argue for an ecological explanation for mobility among the Turkana, one that understands movement as a means of both increasing livestock numbers and reducing risk. However, I also hope to point out that decisions related to mobility and land use can be understood only as being embedded within a cultural and social system. This system, of course, concerns relationships among individuals, but also extends to the cultural incorporation or "appropriation" (Ingold 1987) of natural resources into the system of social relations.

Mobility: Causes and Patterns

Anthropologists have long been fascinated by nomadic peoples, but until Evans-Pritchard's work among the Nuer (1940) and Lattimore's historical analysis of pastoralism in China (1940), nomadic peoples were viewed as "wanderers" whose movements were "irrational," at least from a Western perspective. During the 1950s and 1960s, many pastoral and hunter-forager peoples in Africa, Asia, and the Middle East were the subject of anthropological research, and for the first time a comparative approach to different pastoral and nomadic peoples was possible. Detailed case studies revealed nomadic peoples who were rational, intimately engaged with the environment, and thoughtful decision makers. Across regions and ethnic groups, the influence of the environment (in particular, the spatial distribution of rainfall) on mobility and decision making was striking, especially so among African pastoralists (Barth 1961; Dyson-Hudson 1966; Dyson-Hudson and Dyson-Hudson 1969; Gulliver 1951, 1955; Stenning 1957, 1969; Salzman 1969; Swidler 1969).

This emphasis on ecological factors lying at the heart of the decision-making process was not restricted to the study of pastoral peoples. Ecological anthropology was one of the principal theoretical paradigms within the field at this time, and human involvement in ecosystems was a major concern for anthropological research. One theoretical focus of this new ecological anthropology was the examination of ecological efficiencies among indigenous peoples. Lee's (1979) work among the !Kung San foragers and hunters and Thomas's (1976) analysis of adaptation in the high Andes are among the best known of these studies.

Anthropologists working with pastoral peoples in the Middle East and West Africa had to incorporate the relationship of pastoral peoples with settled agricultural communities, with whom they were engaged in trading relationships and upon whom they depended for access to forage following the harvest of crops. In these circumstances, movement patterns were more structured, and often followed a preset route along a "tribal road." Relationships between pastoralists and farmers were seen as the critical variable in understanding the pattern of pastoral mobility. The dominant influence of the environment was questioned, in many ways reflecting a growing disaffection in the anthropological community with ecologically based explanations for human behavior. Studies positing that many aspects of human behavior could best be understood as attempts to maximize energetic efficiency were the subject of particularly strong criticism (see Ellen 1982 for details).

Many anthropologists working with both pastoral and hunting and gathering peoples proposed nonecological factors as being the underlying causes for their mobility. This view is expressed for pastoral peoples by Bates in his study of the Yoruk:

> . . . nomadic pastoralism is often best intelligible as a political response to other communities and the state. The migratory cycle, residential pattern and even aspects of the internal organization often become clearer when approached from this perspective. (Bates 1971, 127)

The role of economics and exchange relationships was viewed as strongly influencing the mobility pattern followed by Bedouin camel herders (Marx 1967). Other nonecological factors that strongly influenced pastoral mobility patterns were also proposed, such as mobility as a response to political oppression (Elam 1979), or to facilitate social relations (Gulliver 1975). The opinion that mobility was primarily a means by which human populations realign themselves based on social relationships was shared by some anthropologists working with hunting and gathering peoples. For example, Woodburn (1972) argued that among the Hadza of Tanzania, movement provided no advantage in terms of improved access to resources, but by moving, Hadza families could change bands or restructure the organization of a camp, and thus diffuse social tensions that might lead to social strife. In some ways this debate reflected a division in anthropology between those who felt that humans should be considered part of ecosystems and studied as such, and those who felt that humans are unique and that our behavior cannot be understood in ecological terms.

Although all of the studies conducted during this time were based

J. Terrence McCabe

on direct field observations, there is surprisingly little detailed information concerning the exact movement patterns utilized by nomadic peoples or the processes by which decisions concerning mobility were made. One exception is the work carried out by the Dyson-Hudsons among the Karimojong of northeastern Uganda. In an important article published in *Scientific American,* the Dyson-Hudsons examined the mobility patterns of individual herd owners (Dyson-Hudson and Dyson-Hudson 1969). In contrast to the rather regular patterns common to much of the literature, the Dyson-Hudsons found that among individual Karimojong, many factors needed to be taken into account in order to understand individual decision making, and that there was marked variability in movement patterns among individual herd owners. They also found that the Karimojong used all available resources immediately and did not set aside grazing resources for use at later times.

Many anthropologists, especially those working with pastoral peoples living in arid and semiarid rangelands, supported the Dyson-Hudsons' position that pastoral movement is highly unpredictable and is characterized by significant variation in mobility patterns among households. However, anthropologists working in the wetter pastoral areas of Africa and those working with nomadic peoples in the Middle East found movement patterns to be fairly regular and predictable. For example, both Jacobs (1965) and Arhem (1985) found that the Maasai regularly utilized highland pastures in the dry season and moved down to the lowlands following the onset of the rains.

Thus, by the mid-1980s, there were primarily three schools of thought concerning mobility among pastoralists and hunters and foragers. One school felt that mobility was a means by which human populations adapted to spatial and temporal fluctuations in the resource base, and that mobility patterns were highly variable and unpredictable. Another school argued that for pastoralists and hunters-foragers, movement was a response to environmental factors, but that it was fairly regular and predictable. The third school of thought proposed that mobility among humans was best understood as a response to social, economic, or political conditions.

For many anthropologists studying pastoral and hunting and foraging peoples, the 1980s marked a significant shift in research emphasis. The droughts in the Sahel and the resulting famine conditions, which captured much of the world's attention, influenced anthropologists working with pastoral peoples to address issues concerning food security, and the theoretical issues that dominated the debates of the previous two decades were left unresolved. Many of the accounts concerning pastoral peoples written during the 1980s focused on the failure of pastoral systems and the degree to which these systems could be understood as being embedded in larger cultural, economic, and political structures.

Even though the focus of anthropological research had moved away from a concern with mobility, some geographers (Bassett 1986) and ecologists continued to examine the underlying causes of pastoral mobility (Western 1982; Sinclair and Fryxell 1985). Within anthropology, the view that mobility and decisions related to the use of natural resources are primarily responses to ecological variability has been taken up by those working within the theoretical framework referred to as human behavioral ecology or evolutionary ecology. Much of this body of work examines decision making among hunting and foraging peoples, and its advocates make use of optimization or maximization models (Winterhalder 1981; Smith 1991; Smith and Winterhalder 1992; Kelly 1995). Individuals are used as the basic unit of analysis, and aggregate behavior is modeled based on the analysis of individual decision making. The use of Darwinian theory in behavioral analysis has reinvigorated the debate among ecological anthropologists concerning the biological or cultural basis for behavior.

Although analysis of the behavior of hunting and gathering peoples has been predominant among anthropologists working within a behavioral ecology paradigm, some anthropologists have also applied this perspective to the analysis of decision making among pastoral peoples. Probably the best-known studies are those that have used behavioral ecology to examine the relationship of wealth and reproductive success in pastoral societies (Irons 1979; Borgerhoff-Mulder 1988, 1992; Cronk 1989). Others have used this paradigm to examine the rationale for pastoralists manipulating the species structure and reproductive pattern of their herds (Mace 1990; Mace and Houston 1989). Mace (1993) has also applied a "dynamic optimality model" to the process of transition from a pastoral subsistence strategy to an agropastoral or purely agricultural subsistence strategy. The use of a dynamic optimality model allows a sequence of decisions to be included in the analysis, and thus incorporates change. To my knowledge, only one attempt has been made to apply a behavioral ecology perspective to livestock movements, and that has involved herd management strategies of sedentary livestock keepers (De Boer and Prins 1989).

The idea that individuals optimize their use of natural resources in order to gain reproductive advantage over other individuals has a growing number of adherents, but also a growing number of critics. To many anthropologists, the dissolution of the cultural in attempts to understand why people do what they do is as an egregious an error as ignoring the biological (see Rappaport 1990; Vayda 1995). One strategy to bring culture into behavioral ecology has been proposed by Richerson and Boyd (1992) for foragers. A more recent proposal has been made by Cronk, who argues that "there is no escaping the fact culture is a crucial element in any full understanding of human behavior, including one with evolutionary biology as its starting point." He suggests that its inclusion may be accomplished by the use of cultural transmission models and by recognizing that culture is the "context of human social interaction and to examine its role as the raw material out of which signals— the medium of social interaction—are formed" (Cronk 1995, 200).

Before turning to the case study of the Turkana, I want to briefly discuss the categorizing of nomadic patterns. The idea that mobility patterns among nomadic peoples can or should be categorized has been debated since the late 1960s. Terms such as nomadic, semi-nomadic, or semi-sedentary were often used to describe the degree to which a particular people changed locations. Probably the best of the early attempts at categorizing nomadic movements was that proposed by Derrick Stenning based on his work among the Fulani of West Africa (Stenning 1959).

Some anthropologists have objected to the very idea of categorizing a set of behaviors that could be understood as existing on a continuum[1] and that often change through time (Dyson-Hudson 1972). From this perspective, any attempt at categorizing movement patterns masks variability and obscures critical behaviors by which human populations adapt to their environments. Others have felt that cross-cultural comparisons are essential for understanding nomadic peoples, and that categorizing mobility patterns is a necessary first step in this process. I feel that the most thoughtful of these categorizations have been proposed by Johnson (1969), Ingold (1987), and Khazanov (1994). I refer the reader to these works for more detail. In table 22.1 I summarize some of the major movement types and terms used by different authors as a guide to readers not familiar with the literature.

Table 22.1 Categories of pastoral movements

Type of movement	Subtype	Other terms and references
Movement within one ecological zone		Horizontal nomadism (Johnson 1969)
		Ingold 1987
	Unconstrained movement	Tethered nomadism (Ingold 1987)
	Mobility tied down to a central place	Semi-nomadism (Khazanov 1994)
Movement between ecological zones		Transhumance (Stenning 1957)
		Vertical nomadism (Johnson 1969)
	Periodic shifts to seasonal pastures	Nomadism (Krader 1959)
		Transhumance (many authors)
	Shifts between different ecological zones	Transhumance (Krader 1959)
	to exploit different resources	Multiple resource nomadism (Salzman 1969)
		Yaylag pastoralism (Khazanov 1994)
		Displacement (Johnson 1969)
Major shifts of nomadic ranges or migratory orbits		
	Gradual shift	Migratory drift (Stenning 1959)
	Permanent/abrupt shift	Migration (Stenning 1959)

Note: The major categories used in the table above are taken from Ingold 1987. For a classificatory scheme based on the degree to which agriculture is incorporated into the pastoral system, see Khazanov 1994.

The Turkana: A Case Study

Turkana District: Topography, Climate, and Vegetation

The case study that follows concerns group movements among Turkana pastoralists. Before presenting any data on Turkana group movements, I will briefly describe the Turkana environment and the means by which the Turkana manage livestock. Two bodies of data will be drawn upon in the analysis of group movements: detailed studies based on households that I have conducted among one section of the Turkana, and coarser data concerning large-scale movements collected on three other sections.

The Turkana number approximately 200,000 and inhabit the arid and semiarid rangelands of northwestern Kenya (figure 22.1). They raise camels, cattle, sheep, goats, and donkeys, and rely on the products of their herds for the bulk of their subsistence, although they do sell livestock and purchase grains.

Turkana rangelands are hot and dry, with an average annual rainfall of 180 mm in Lodwar, the district capital (Ellis and Swift 1988). Precipitation most often occurs during March through June, and sometimes during November, but rainfall is unpredictable in timing, duration, and intensity. Drought is a regular feature of the Turkana climatic regime, and analysis of precipitation records shows that there have been thirteen significant drought episodes in Turkana district during the last 50 years (Ellis and Swift 1988). A "significant drought episode" is defined here as a drop in precipitation of 33% or more from the long-term average.

The topography of Turkana district consists of low-lying plains juxtaposed to mountain massifs. The plains form the floor of the Rift Valley. Mountains form the western wall of the Valley and also rise abruptly from the plains throughout the district. The highlands are both cooler and wetter than the plains, and the vegetation there consists of perennial and annual grasses, large shrubs, bushes, and trees. The vegetation of the plains consists of annual grasses, small shrubs, and bushes; trees, primarily acacias, grow along the watercourses that flow out of the numerous mountain ranges.

The northern part of the district is topographically more homogeneous and wetter than southern Turkana. Vast plains dominate northern Turkana, while the south is more dissected and the vegetation more varied. The variations in topography, climate, and vegetation influence the species mix of livestock herded by the Turkana: the people living in the northern part of the district are primarily

Figure 22.1 Turkana District and major towns.

dependent upon cattle, while those living in the south keep mixed herds but rely principally on goats and camels.

Social Organization and Livestock Management

The Turkana are divided politically into eighteen sections, each of which is associated with a defined area with particular rights to grazing and water resources. In general, as in most pastoral communities in Africa, grazing is free to all members of the local population (in this case, the members of the section), while water resources may be owned depending upon the reliability of the water source and the difficulty of obtaining water. Deep wells are owned by individuals, who may permit close kin and friends to use the well. Open sources of water may be used by members of the section as long as they are available.

Herd management and decisions related to mobility and natural resource use usually occur at the smallest level of social organization, the household or *awi.* The awi consists of a herd owner, his wives and their children, and often a number of dependent women. This unit, together with the associated herds of livestock, is also the basic unit of production among the Turkana. The principal decision maker is the herd owner, and although he may discuss options concerning where or when to move with other male members of the household, the ultimate responsibility for the welfare of the family and its herds rests with him.

Although each awi acts as an independent unit, herd owners typically join together for periods of time that may last from a few weeks to many years. The basis of these associations may be kinship (either agnatic or affinal) or simply friendship. Within this "large awi" some food may be shared and labor pooled. Each time the awi moves provides an opportunity to change the human and livestock composition of this association of herd owners. As the dry season progresses, herd owners often segregate the nonmilking stock of each species, which are managed by young adult men who follow migratory patterns independent of the principal awi.

During the wet season all of the livestock and people rejoin the awi, and most awis cluster together in large, loose neighborhood associations called *adakars.* These adakars remain together as long as forage and water resources permit. People join together to form an adakar because they enjoy the social interaction that close proximity offers and for increased security. When people are living within an adakar, groups of men discuss management and movement options on a daily basis. Members of an adakar tend to move together, although decisions concerning movement still rest with each herd owner. Rarely is it the case that a herd owner will act completely independently and isolate his awi from the larger group.

Although there is no "leader" of an adakar with authority to tell individual herd owners what to do, adakars do tend to form around a few wealthy, influential men. Every Turkana man has a home area or *ere,* where he owns wells and which he usually returns to in the wet season. Thus, adakars form within specific geographic areas and move from there as forage resources diminish. There is no official coordination of movements among adakars, but herd owners are aware of the movements of other adakars in the area. Men and women frequently visit friends or relatives, many of whom may be living in different adakars. Information concerning the intended moves of an adakar is openly discussed, and thus this information spreads quickly. When men come together to discuss movement options, the current and possible future locations of other adakars are considered along with the availability of forage and water.

Adakars begin to break up as forage resources diminish and open water sources begin to dry up. During the dry season, forage resources are more efficiently exploited by small herds, and herd owners draw water from individually owned wells. Under these conditions it is difficult to maintain large aggregations of people and livestock.

There are circumstances, however, in which large aggregations remain together despite the difficulties imposed by the environment. These social formations are referred to as an *arum rum,* and are specifically designed to act as a military or defensive organization. Many of the Ngisonyoka Turkana with whom I work are currently living in an arum rum due to extensive raiding by the neighboring Pokot. Within an arum rum there is a clear line of command, and movements are organized and coordinated. Those individuals who act independently are punished. This is clearly a very different type of social organization than is normally found among the Turkana, and rarely occurs. However, the formation of an arum rum underscores the ability of human groups to quickly and radically alter their social organization to respond to particular circumstances.

The Process of Decision Making and Land Use
I have previously mentioned that most decisions relating to mobility are made by individual herd owners. Men frequently discuss movement options with one another, but only in rare circumstances do households move together as a group. Men make decisions based on the available information concerning the condition of forage and water in the areas that they are considering moving to, the predictions of soothsayers or *emerons,* the number and species mix of the livestock in their herds, the availability of labor, and their personal inclination concerning the amount of risk that they are willing to take. This leads to considerable variability at the level of individual households. The extent to which individual household mobility patterns may vary from one year to the next, and among households in any one year, is discussed in detail in McCabe 1994 and in Dyson-Hudson and McCabe 1985. In the sections below I consider the process by which information is gathered as well as the importance of ecological and social factors in decisions relating to mobility and land use.

J. Terrence McCabe

Information.   Information concerning the condition of forage and water and the presence of disease or danger from enemies is critical in the decision-making process. At times information is considered a public good that should be shared; at other times information is private, and a herd owner will share information only with close relatives or friends. When people are living within an adakar, a few young men will be selected from different families and sent to investigate areas that are being considered as possible places to move to. This information is considered public, and herd owners can make individual decisions based on it. The same holds true when people are living in dangerous locations, even if households are not organized into an adakar. This typically occurs for the Ngisonyoka when households are located in the far south of their territory and the rains have begun. It is generally accepted that livestock cannot move back to the wet season grazing areas until the grass is 3–5 inches tall and there is new growth on the shrubs and trees. Scouts will investigate the route to be taken back to the wet season range and report on forage conditions for the entire route.

During much of the dry season, when households are living on their own, information concerning the condition of forage is guarded. Typically a herd owner will send a young adult male to investigate the condition of forage in an area that he is considering moving to. If others ask about forage conditions, a herd owner will often say nothing, or even lie. People take the opportunity to collect information while visiting friends or relatives or while looking for lost livestock, which happens far more frequently than one would expect. Young men who are traveling through an area often inquire at water holes about the condition of forage in the surrounding areas. Rarely will they be given good information. I have observed strangers to an area asking children about forage and water, knowing that they may not be adept at assessing forage conditions, but that "they have not yet learned to lie."

Herd owners without access to labor or those who are not inclined to gather information themselves will watch the movements of other herd owners, especially those who are known as good decision makers. Thus the movement of a single herd owner often triggers the movements of many others.

In addition to the condition of the resource base, herd owners also consider how many people are using a particular area. Thus many decisions are density dependent. Areas that have access to common water sources can be densely populated in the dry season, and a herd owner may choose to move to an area where the forage

Table 22.2 Summary of responses concerning determinants of movement

Category	No. of responses	%
Forage	151	63
Water	9	4
Security	45	19
Prophetic prediction	2	1
Social	8	3
Transitional move	24	10
Miscellaneous	2	1
Total	241	

conditions are not quite so good, but access is restricted to those who own wells there.

Another issue relating to the process of decision making is the size of groups within which information is shared and the size of groups moving together. During the wet season, when herd owners are living in adakars, there may be 100–150 families and their livestock (numbering in the tens of thousands) moving together and exchanging information. (It should be noted that awis will be dispersed across the landscape even though herd owners consider themselves to be living in an adakar.) During much of the dry season awis are separated from one another, and thus the size of a group may be 10–50 and a few hundred livestock. Although households may be separated, herd owners will usually converge under a shade tree, called the tree of the men. Most herd owners within a 1- to 2-hour walk will gather there several times each week to talk to one another and exchange information.

Ecological Factors Affecting Decision Making. There is a real paucity of data available concerning the importance of various factors in decisions relating to pastoral movements. From 1979 to 1981, I collected data on the stated reasons for moving to and away from specific locations among four herd owners. The data set consists of 241 responses, and the results of this study are presented in table 22.2. Forage was mentioned in 63% of the responses as the principal reason for movement, and was thus the most important variable in deciding where to move. About 15% of the responses that fell into this category mentioned specific forage resources, such as moving to an area where leaves of a particular tree were available for the baby goats. Water, of course, is an important consideration, but in southern Turkana, where this research was conducted, water

J. Terrence McCabe

sources are numerous. Therefore, moving specifically for water was not an important consideration.

I have previously mentioned that Turkana herd owners avoid areas where disease is present, but responses related to disease appear only in the "miscellaneous" category (constituting less than 1% of the total responses). This is due to the fact that these responses related to locations people moved to rather than areas they avoided. (It should be mentioned that I am referring to livestock disease here; I have never heard of an area being avoided due to human disease.) The incidence of livestock disease is greatest in areas where the vegetation is the most abundant. Ticks thrive in these areas and are the vectors for numerous livestock diseases. Tsetse flies are the vectors for livestock trypanosomiasis (not for human trypanosomiasis in this area) and live in dense bush and thickets. Thus, areas with the most abundant forage resources are also those with the greatest risk of livestock disease. These areas are generally used only when all other forage resources have been exhausted, and herd owners know that they will suffer significant livestock losses when moving into these areas.

Social Factors Affecting Decision Making
Only 3% of the responses were directly related to social factors. The principal reasons given here were to move close to friends or relatives or to be back in one's home area (again, close to one's friends and family). I feel that this percentage actually misrepresents the importance of social factors in decision making. Clearly the needs of the livestock come first, but the location of one's friends and relatives is always considered in deciding exactly where to move. Also, it is important to realize that one of the most important reasons that herd owners move into an adakar is for the social context provided in daily interactions with other herd owners and their families. Thus most wet season moves should be understood as having a strong, but often unstated, social component.

Other Factors Affecting Decision Making
The predictions of soothsayers, called emerons in Turkana, also affect decisions regarding movement. An emeron is someone who is known to be able to predict the future, and emerons have played a very important role in Turkana history (see Lamphere 1992). Although many Turkana are skeptical about the ability of many emerons to accurately predict future events, there are a few emerons to

whom everyone pays attention. Looking back at my field notes for the early 1980s, I am struck by the extent to which discussions concerning moving took into account the predictions of an emeron called Kaling. If Kaling dreamt that there was likely to be a raid or an outbreak of disease in a particular area, most people would avoid that place. Again, the ultimate decision was left to individual herd owners, but Kaling's predictions definitely had an effect on the movement pattern as a whole.

The quest for security exerts a major effect on decisions relating to movement. Security concerns were mentioned in 19% of the responses. In these cases, herd owners chose to move away from a particular location because of the presence of bandits or intertribal enemies (in this instance, the Pokot). Areas considered especially dangerous are avoided as long as forage is available elsewhere. Once a raid occurs, herd owners may reverse the direction of their migration, or may leapfrog in back of other families so that their awi is not on the front line between the Turkana and the Pokot.

Short-Term and Long-Term Strategies
A Turkana herd owner must balance the needs of the livestock (for forage, water, and protection) with the needs of the people (for food). He must also balance short-term management strategies with long-term strategies. Short-term strategies may involve maximizing the production of food available for humans and the amount of forage available for livestock. Long-term strategies may involve maximizing the long-term growth potential of his herds and his own reproductive capacity. Often these strategies are in conflict with each other.

During the dry season, the most efficient way for a herd owner to utilize the varied forage resources available to Turkana livestock is to divide his livestock into species-specific herds. Herds composed of one species can concentrate on resource patches that are made up of one type of vegetation. For example, by separating the cattle from the other livestock, a herd manager can take the cattle to an area where there is good herbaceous vegetation, but which may be deficient in the shrubs and trees needed for camels and goats. However, this strategy may leave the family without an adequate supply of milk and other livestock foods. The herd owner must decide the extent to which each population will compromise its needs. The immediate need for human food may also compromise the long-term viability of the herd. The viability of the herd

J. Terrence McCabe

depends upon the survivorship of the calves, kids, and lambs, and both humans and immature livestock compete for milk.

In my description of long-term strategies I linked the survival and growth of the herds with the herd owner's reproductive capacity. I shall expand upon this point as it is important in understanding livestock management and mobility. Among the Turkana, children are considered as belonging to a man only within a marriage contract and after all the bridewealth (a payment made by the groom's family to the bride's family to compensate them for the loss of their daughter) has been paid. Turkana bridewealth payments are among the highest recorded for any people. In a typical bridewealth payment 40–60 large animals and 100–200 small stock are transferred. Until this payment is complete, all children are considered part of the woman's father's or brothers' family. A fine of up to 10 large animals is charged for impregnating an unmarried girl.

Successful Turkana men often marry three, four, or more wives. A friend of mine said that he hoped to be able to bring a young new wife into his household each time one of his current wives could no longer conceive. Thus his reproductive capacity depends upon his having enough livestock to distribute in bridewealth payments, and if he is to continue to reproduce, his livestock holdings must increase.

It is frequently the case that in order to improve access to forage, herds must be moved to areas where the danger of raiding is high. A man can lose all of his livestock in a single raid; thus a herd owner must decide whether a short-term improvement in access to forage resources, and thus energy intake, is worth the long-term risk of losing everything.

I bring up these points to emphasize the importance of the role of culture in understanding human reproduction and the importance of incorporating risk into the analytic framework. Some studies that examine the migratory behavior of animal species posit that maximizing energy intake is linked to reproductive success (Fryxell 1995). In many studies of human hunters and foragers, energy intake is used as the "currency" to be measured in optimization models (Smith 1991; Kaplan and Hill 1992). The necessity of incorporating risk into the behavioral ecology of humans has been explored by Mace and Houston (1989) and by all the authors in the volume *Risk and Uncertainty in Tribal and Peasant Economies,* edited by Cashdan (1990). In the Turkana case, we see that reproductive success is embedded within a cultural context (marriage and bridewealth), and that maximizing energy intake is only one of a number of variables that must be taken into account in deciding where and when to move. It also needs to be remembered that the risk factor is not constant, but varies seasonally and annually. In my analysis of individual herd movements, it is clear that some individuals have adopted a strategy that emphasizes energy intake, while others have adopted a strategy that minimizes risk.

Group Movement Study
The data presented in the previous sections resulted from a long-term detailed study of herd owners in one section of the Turkana, the Ngisonyoka. In 1984, two colleagues and I had the opportunity to conduct research among three other sections of the Turkana. In addition to myself, the team consisted of a human nutritionist and a range ecologist. This approach allowed for a comparison of movement patters on a much broader scale than would be possible if data were collected for the Ngisonyoka section alone.

Group movement patterns were constructed based on interviews with twenty or more individual herd owners from each section. Data were collected for household movements from 1980 through 1984, thus providing 5 years of data. Although there was marked variation among individual households' mobility patterns (it is important to remember that households may consist of 10–30 people and hundreds of livestock), once the data for individual households were plotted on a map, it was clear that households were moving roughly in concert with one another.

One component of the study was an examination of the productivity of the vegetation in the area occupied by each of the Turkana sections. I was able to use this information in my analysis of mobility patterns. The general patterns of mobility for the three sections included in this analysis are presented below.

Ngisonyoka Movement Patterns
The Ngisonyoka sectional territory encompasses approximately 8,600 km² in the southwestern portion of Turkana district. It is topographically diverse, containing mountains, lava flows, dry sandy plains, and wetter bush-covered plains. There is a strong rainfall gradient running from southwest to northeast across the territory. The southernmost area receives a mean annual rainfall of 601 mm per year, while the northernmost recording station averaged 240 mm per year.

The Ngisonyoka conceptualize their territory as containing a wet

J. Terrence McCabe

season home area, a dry season grazing area, and a drought reserve. The wet season range includes the sandy plains located between the central mountains and the eastern plateau area. During the 2–3-month wet season the majority of the Ngisonyoka and their animals congregate there. As the dry season progresses, they move southward along the central mountains, and in very dry years, or if drought conditions prevail, they move into the mesic southwestern plains (figure 22.2).

The cattle may follow a slightly different migratory path than the other livestock due to the fact that they are dependent upon grasses for forage. They therefore leave the plains area as soon as the vegetation begins to dry out and move into the higher elevations. In normal years nonmilking small stock and nonmilking camels are also separated from the major homestead, allowing for a more efficient use of the resources available and also saving forage for livestock in the vicinity of the awi. Although separate, these satellite herds follow the same general movement pattern of north to south migration until the rains begin and then move northward until all herds rejoin during the wet season.

The analysis of primary productivity revealed that the Ngisonyoka wet season range had an estimated herbaceous production of 1,800 kg/ha and an estimated woody foliage and shrub production of 100 kg/ha. Herbaceous production in the dry season and drought reserve range was estimated at 2,800 kg/ha, while woody foliage and shrub production for this area was estimated at 200 kg/ha. The Ngisonyoka thus utilize one of the least productive areas during the peak of its productivity and move into the more productive areas as the dry season progresses and the nutritive value of the vegetation diminishes. During dry years the Ngisonyoka move all the way down to their southern border, while in good years they may move only half that distance.

Although access to good forage and water strongly influences Ngisonyoka movement, so does the relationship of the Ngisonyoka to their southern neighbors, the Pokot. Since the early part of the twentieth century, the Turkana and the Pokot have engaged in raids and counter-raids; there have also been periods of peace. However, during most of the time that I have been working with the Turkana (1980–1996), Turkana-Pokot relations have been quite hostile. During raids, many people may be killed and thousands of livestock stolen. Thus moving close to the intertribal border engenders considerable risk.

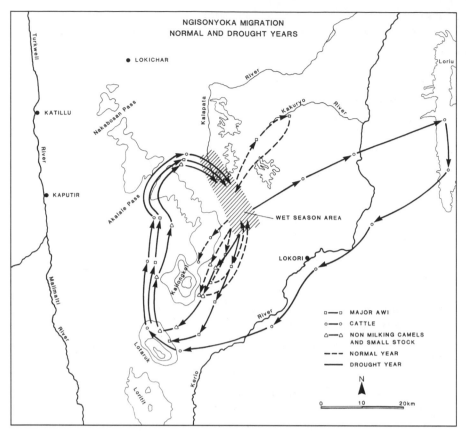

Figure 22.2 Generalized migration cycle for Ngisonyoka in normal and drought years.

In addition to the risk of raiding, livestock using the most southern and most productive rangelands are exposed to far more disease than they are in other areas. The denser vegetation in the southern part of the Ngisonyoka range supports populations of ticks and tsetse flies, both of which are vectors for livestock disease.

The pattern described above is accurate for most years. However, departures from this pattern do occur. In 1982 the Ngisonyoka did not move south as the dry season progressed, but moved north into a rather unproductive area that was unfamiliar to many Ngisonyoka. I accompanied a number of Ngisonyoka households during these movements and thus understood the reasoning behind this departure from the usual pattern. Raiding between Turkana and Pokot had been especially bad during the previous dry season; a

J. Terrence McCabe

number of Ngisonyoka men had been killed, and many families had lost large herds of livestock. There were good rains in 1982, and a consensus was reached among most herd owners that would they move away from areas prone to raiding and hope that the forage would be sufficient to sustain the livestock throughout the dry season. There were dissenting voices, and those individuals who wished to move south could do so, but they would be isolated and largely unprotected from attack. No households moved south that year.

Although I do not have detailed mobility data for the years 1990–1996, I have recently returned from Ngisonyoka, where I was told that in the years 1993 and 1994, people altered their movement patterns because of the distribution of famine relief from a few centers in southern Turkana. People located their awis within a half day's walk to these centers and moved very little. This was a pattern that I observed in northern Turkana in the mid-1980s, also associated with the distribution of famine relief food (McCabe 1990).

Ngisonyoka movements can be viewed as a strategy to even out the spatial and temporal availability of resources, but one constrained by the risks entailed in moving into the most productive part of the range available to them. The examples from 1982 and 1993–1994 illustrate the danger of basing results on short periods of fieldwork, and the extent to which a mobility pattern can vary according to special circumstances.

Ngikamatak Movement Patterns
The rangelands occupied by the Ngikamatak contain some of the highest mountains in Turkana district and are bordered on the west by the Rift Valley escarpment. An extensive area of dry lowlands constitutes the eastern part of Ngikamatak territory. Precipitation varies with elevation; the lowlands receive as little as 200 mm of rainfall per year, while the higher elevations may receive as much as 600 mm per year.

The Ngikamatak utilize the lowlands during the wet season and move into the higher elevations as the dry season progresses. During drought years, the Ngikamatak move above the escarpment into Uganda, sharing the rangelands of the Karimojong, with whom they have relatively friendly relations (figure 22.3).

Our production estimate for herbaceous vegetation in the wet season range is approximately 1,300 kg/ha, and woody foliage and dwarf shrub production is estimated at 120 kg/ha. For the dry sea-

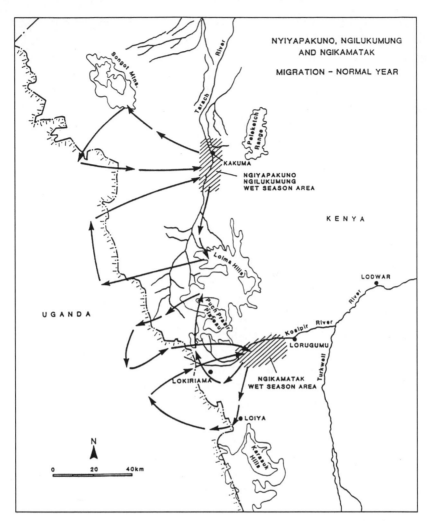

Figure 22.3 Generalized migration cycles of Ngiyapakuno, Ngilukumung, and Ngikamatak sections during a normal year.

son range (not counting the area in Uganda[2]), herbaceous production is estimated at 2,750 kg/ha, and woody and shrub production is estimated at 290 kg/ha.

Ngilukumong and Ngiyapakuno Movement Patterns

The drainage of the Tarach River is occupied by two sections of the Turkana, the Ngilukumong and the Ngiyapakuno. The Tarach rangelands include high volcanic plateaus, the southern extension of the vast Lotikippi Plains, and the Rift escarpment in the west.

J. Terrence McCabe

The higher elevations receive up to 600 mm of rainfall each year, while the lower elevations remain arid. Approximately one-half of the lowlands are dominated by annual grasses. This area has the potential for great productivity in good years, but because the vegetation is primarily grasses, productivity drops drastically during dry years. The other half of the lowlands are characterized by bushy or woody grasslands.

Like the Ngikamatak, the people of the Tarach region utilize the lowlands during the wet season and gradually move into the highlands as the dry season progresses. However, because the Ngiyapakuno and the Ngilukumong do not have friendly relations with their Ugandan neighbors (the Jie and Dodoth), they move above the escarpment only during times of drought (fig. 22.4). It should be noted that when people from the Tarach region move into Uganda, it is done by force.

Our estimates of primary productivity for the wet season range are 2,100 kg/ha for the grasslands during good years and 90 kg/ha for woody foliage production. Primary productivity for the dry season range is estimated at 2,700 kg/ha for the grasslands and 260 kg/ha for the woodlands.

Comparison of Movement Patterns
From the description presented above, it is clear that there is a pattern of mobility common to all three groups. The people in each group conceptualize their territories as consisting of wet season home areas, a dry season area, and a drought reserve. Movement into the drought reserve areas is usually accompanied by significant risk (in this case, disease or the presence of enemies), but these areas are critical resources for the pastoral population.

The data on pastoral mobility demonstrate that each group utilizes a rather unproductive area during the wet season, at the height of its productivity. Each group gradually moves into more productive areas as the vegetal production is dropping off, and each group only occasionally utilizes the most productive section of its territory during times of stress (see table 22.3).

Discussion
The pattern of land use described above is not limited to livestock-keeping peoples, but is similar to that described for many wild ungulate species. Michael Coughenour, an ecologist on the Turkana project, noted that wildebeest in the Serengeti (L. Pennycuik 1975;

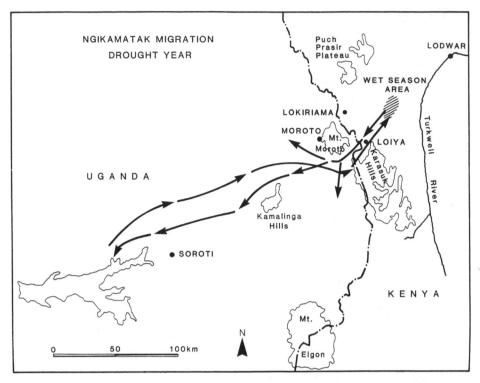

Figure 22.4 Generalized migration cycle of the Ngikamatak section during a drought year.

Fryxell 1995) and white-eared kob in the southern Sudan (Fryxell and Sinclair 1988) each move long distances to areas of low primary productivity during the rainy season. Coughenour and his colleagues, discussing the relevance of this observation to Turkana land use, feel that for herbivores not to utilize the forage in areas of low productivity would represent a significant "lost opportunity," which might result in decreased animal numbers (Coughenour, Coppock, and Ellis 1990).

Sinclair and Fryxell have also noted that pastoral peoples in the Sahel follow migration routes similar to those of the populations of wild ungulates that live there. These migration routes also utilize areas of low productivity in the wet season and areas of higher productivity in the dry season.

Although the explanations for migratory movement by wild ungulates vary (access to forage quantity: Fryxell 1995; access to forage quality: Murray 1995), it is fairly clear that the migratory pattern described above results in an increase in animal numbers

J. Terrence McCabe

Table 22.3 Range characteristics by season for three sections of the Turkana

Section	Wet	Dry	Drought
Ngisonyoka			
Herbaceous	1,800 kg/ha	2,300 kg/ha	2,800 kg/ha
Woody	100 kg/ha	200 kg/ha	200 kg/ha
Ngikamatak			
Herbaceous	1,300 kg/ha	2,750 kg/ha	>[a]
Woody	120 kg/ha	290 kg/ha	>[a]
Tarach (Ngilukumonq and Ngiyapakuno)			
Herbaceous	2,100 kg/ha[b]	2,700 kg/ha	>[a]
Woody	90 kg/ha	260 kg/ha	>[a]

[a]The drought reserve areas for these sections are located above the escarpment in Uganda. Biomass estimates are higher for this area, but because no ground truthing was possible, no calculations were made.

[b]This area is subject to extreme fluctuations in productivity.

compared with what would be possible if the animals remained resident in any one area. The same applies to the Turkana and their livestock. I am using the term "increase" here rather than employing the concept of maximization or optimization. I really do not have access to data that would allow for an evaluation of the maximization or optimization concepts, and thus feel uncomfortable using them. For example, to really assess whether the Turkana are optimizing land use, each movement decision would have to be evaluated with respect to all the available alternatives. In addition, all constraints on movement would also have to be included in the evaluation. Even with all the ecological data available from the South Turkana Ecosystem Project, this type of analysis remains problematic.

Complexity is added to the Turkana situation by, among other things, the predictions of emerons, varying relations among the Turkana and their neighbors, and the "appropriation" of natural resources into a cultural system. It is tempting to explain the predictions of emerons as having a "latent function"[3] of influencing large groups of people to do the right thing from an ecological perspective, and often these predictions do have a beneficial effect, such as setting aside areas for later use. However, the extent to which the beneficial effects of these predictions are functionally related or an unintended consequence are impossible to determine.

Although a case can be made that the Turkana mobility pattern is designed to increase livestock numbers, other aspects of herd management are best seen as attempts to reduce risk. The keeping of multiple-species herds is certainly one component of a risk reduction strategy. Another way to reduce risk is to avoid areas that are dangerous and to move away from areas as risk in those areas increases. In the case study presented above, both mechanisms are employed. In table 22.4, I summarize the benefits, costs, and risks of the most important Turkana decisions related to mobility and herd management.

The degree of variation from the "typical" pattern must also be considered. During the years 1982, 1993, and 1994 (20% of the total annual cycles studied) the pattern of mobility differed significantly from the long general pattern. In addition, a new form of social organization has emerged in southern Turkana—the arum rum mentioned above—and it will have an influence on the pattern of group mobility. It is certainly possible to explain why these variations occurred, but it would be impossible to predict their occurrence.

The case study also demonstrates the difficulty of categorizing a particular people's movement pattern. The Ngisonyoka's movement pattern generally fits Ingold's category of "full nomadism," but became "tethered nomadism" with the introduction of famine relief camps, and then returned to "full nomadism." Even within the "full nomadism" of the Ngisonyoka, their cattle follow a migratory cycle that more closely resembles "transhuman migration" than it does "full nomadism." For an analysis of the effect of famine relief camps among the Turkana, see McCabe 1990.

Conclusions

The Turkana case study demonstrates that pastoral mobility is strongly influenced by the environment, and that even though there is individual variation, there is also significant patterning in migratory cycles. The case study also demonstrates that a full appreciation of Turkana mobility is impossible without an understanding of how decisions ranging from reproduction to herd management are embedded within the cultural system.

For many anthropologists who have studied pastoral nomads, the influence of the environment seems so obvious that pastoral mobility is explained by references to pasture quality or resource availability without attempting to go into further detail. Examples of this type of explanation can be found in Ekvall's account of Tibetan pastoralism, whereby people and herds "follow the growth of grass" (Ekvall 1968, 34), in Capot-Rey's statement concerning the

J. Terrence McCabe

Table 22.4 Costs, benefits, and risks of management decisions

Management decision	Benefit	Cost	Risk
Move into drought reserve	Gain access to abundant forage	Reduce access to forage resources	Large-scale loss of livestock due to raiding or disease
Move away from dangerous areas	Reduce risk of livestock loss		
Divide herds and manage separately	More efficient use of environment	Loss of livestock products to the human population	Dependence on young herd managers
Multiple-species herds	Reduce risk of losses due to disease in one species	More labor needed	
Lend livestock to others	Strengthen ties between families	Loss of livestock products to the human population	Possible loss of livestock if one family becomes destitute

nomadic pastoralists of the Sahara: "The first rule of the game—to tell the truth—the only rule—is that the nomad follows the rain" (Capot-Rey 1953, 251), and in Croze's observation that pastoralists in East Africa "chase ephemeral protein and water around the ecosystem" (Croze and Gwynne 1981, 350).

Although I agree with the ecological explanation of pastoral mobility, the Turkana case study demonstrates that the quotes provided above present a far too simplistic explanation for group migration. Differential access to resources, reduction of risk, and annual variation in migratory cycles must all be accounted for.

The cultural context within which Turkana decisions concerning mobility and herd management are made is also critical in understanding why the Turkana do what they do. Natural resources do not exist in isolation; they are brought into the cultural system. For example, grazing resources are corporately held by the members of the section, while many water sources are owned by individuals. Those individuals who do not own wells in a certain area cannot use the grazing resources in that area during the dry season. Individuals accumulate livestock through inheritance, and the inheritance pattern is culturally determined. Likewise, a man's ability to marry is determined by his ability to make bridewealth payments. Thus his reproductive success is strongly influenced by the cultural institution of marriage.

Finally, it should also be noted that the Turkana are one of the most isolated pastoral peoples, in terms of integration into a national economy, found anywhere in the world. Of course, there are really no truly isolated peoples. Wilmsen (1989) has shown that the hunting and foraging peoples in the Kalahari were, to some extent, connected to the world economic system 5,000 years ago. The relative isolation of the Turkana may limit the generalizability of the Turkana case study, especially to pastoral groups fully integrated into a market economy. However, the need to understand the interplay of ecological and cultural variables will remain critical regardless of the human group being considered.

Acknowledgments
I would like to thank the Turkana people with whom I have worked for their friendship and willingness to let me participate in their lives. They taught me much more than how they use the land or manage their livestock. I would also like to thank the government of Kenya for granting permission to undertake the years of field-

work that these research findings are based upon. I could not have conducted fieldwork without the financial support of the National Science Foundation and the Norwegian Agency for International Development, and for that I thank them. Finally, I would also like all the members of the South Turkana Ecosystem Project to know how much they influenced my thoughts and contributed to my understanding of human behavior and arid land ecosystems.

Notes

1. Dyson-Hudson (1972) actually advocated the use of two continua in the analysis of pastoral nomads. The first was a continuum stretching from completely nomadic to completely sedentary. The other was aligned along an axis connecting total dependence on livestock to no dependence on livestock.

2. It was impossible for the study team to visit the areas within Uganda that are utilized in time of drought. Although they are highly productive, the ecologists felt that it was inappropriate to estimate their productivity without ground truthing, and they are therefore not included in the analysis.

3. The use of the term "latent function" for certain aspects of human behavior that communicate ecological information is associated with the ecological anthropology of the 1960s, especially Rappaport's *Pigs for the Ancestors* (1968). For a critique, see Friedman 1974.

New Directions for Group Movement

SUE BOINSKI AND PAUL A. GARBER

> "Cheshire puss," she began, rather timidly, as she did not at all know whether it would like the name: however, it only grinned a little wider. "Come, it's pleased so far," thought Alice, and she went on. "Would you tell me, please, which way I ought to go from here?"
>
> "That depends a good deal on where you want to get to." said the Cat.
>
> "I don't much care where—" said Alice.
>
> "Then it doesn't much matter which way you go," said the Cat.
>
> "—so long as I get *somewhere*," Alice added as an explanation.
>
> "Oh, you're sure to do that ," said the Cat, "if you only walk long enough."
>
> —Lewis Carroll, *Alice's Adventures in Wonderland,* 1866, 88

The perambulations of Alice and the inhabitants of Wonderland represent the many strategies available to individuals in goal-oriented travel: follow apparently knowledgeable individuals, use predetermined travel rules and routes, seek recognizable landscape features, employ sundry social gambits to obtain useful information, cooperation, and protection, and avoid risky habitats and beasts. Contributors to this volume have focused their attention on a special form of goal-directed travel, the movement of individuals within social groups. In part, our editorial motivation was the conviction that group movement has been caricatured in recent publications with a much heavier and simpler hand than Lewis Carroll would have ever wielded. In primates, for example, reports commonly dichotomize group movement into the intricate and longer travel routes of large-brained, finicky frugivores and the more circumscribed, mowerlike swaths of dim-witted grazing folivores. The reality we have observed in our fieldwork over the years is far less tidy. At times travel in monkey troops was highly coordinated, at other times somewhat chaotic, and at still other times seemed to represent a negotiation between conflicting individual interests. We have seen troops travel to distant, highly productive feeding sites as well as to distant feeding sites just denuded of fruits by neighboring groups. We have followed troops traveling beyond their normal range that became bewildered, even panicky, in their apparent efforts to select a return route to more familiar areas. We have also seen troops "stalled" for over an hour while one faction of troop members attempted to lead the troop on a trajectory that intersected with a cool, shady resting site, while another faction vigorously indicated an alternative travel route leading to a fruit patch. Eventually, the troop traveled in the direction indicated by the faction calling last and loudest.

This richness in form and outcome made it clear to us that group movement is a complex phenomenon contingent upon sensory perceptions, psychological responses, and morphology. It is also dependent upon processes of learning, motivation, individual experience, and decision making within a broader spatial, social, and ecological context. Unfortunately, a significant portion of recent research has been narrowly framed to the point that group movement is approximated by a single quantitative measure, such as daily path length or home range area. Therefore we decided to embrace fully the multiple parameters and messy interactions inherent in the coordinated travel of social groups. Not only have we encouraged contributors to address processes and cognitive abilities in addition to patterns, but these chapters are nearly as diverse in characters as those that animate the episodes in Lewis Carroll's classic; creatures from anthropoid and prosimian primates to ants, birds, cetaceans, and carnivores are among those considered.

Several major propositions emerge from the preceding chapters. The first is contrary to our expectation when we initially organized this volume. There is a lack of behavioral evidence, especially from field observations, to indicate that significant taxonomic differences exist in the abilities of primate species to locate feeding sites and navigate in space effectively, or between those of primates and many other social mammals and birds. Although species may use different social and environmental information in the decision-making process, and may exploit home ranges that vary greatly in size and structural complexity, there are no data suggesting that Old World monkeys are more efficient foragers than New World monkeys, or that apes are more efficient foragers than lemurs, food-caching birds, or bottlenose dolphins. Moreover, regarding abilities cut from the same cognitive cloth as travel efficiency in foraging, primates and other vertebrate taxa cannot yet be qualitatively distinguished in regard to their sensitivity to the density and distribution of predators or resource competitors. In short, the available data are still inadequate to discriminate between the two major explanations for taxonomic variation in travel decisions: divergent behavioral and ecological strategies versus differences in cognitive abilities. This volume also does not present evidence of distinctions among non-human primates, at least among the monkeys and apes, in the complexity of or demands on cognition required for group movement. We anticipate that the spatial and ecological skills evident in the group movements of gregarious prosimians will be found indistinguishable from those of anthropoid primates, although different types of sensory information may be involved. If differences do exist between prosimians and anthropoids, they will probably lie in the realm of social cognition (Byrne and Whiten 1988; Harcourt and deWaal 1992).

Perhaps the best substantiated conclusion from this volume is the pervasive influence of sociality on group movement. Group movement is as much a social behavior as it is an ecological response to the distribution and availability of resources and risks. Incorporation of social behavior into group movement decisions, in turn, is directly pertinent to the two prevalent, albeit divergent, paradigms in animal behavior that relate decision making and communication to group movement. Under what conditions are the behaviors required for coordinating group movement best explained by simple stimulus-response patterns or basic rules of thumb (a series of if-then statements), and under what conditions are they more flexible and contingent on both the immediate context and anticipation of subsequent decisions (a series of if A, then B, C, D depending on E, F, and G statements)? In the latter case, complex decision making includes elements of planning. If social and ecological factors are concurrently weighed in group movement decisions, as many chapters in this volume strongly indicate, the latter, more complex and flexible decision-making model is more appropriate. Even if simple rules of thumb and similar rudimentary cognitive tools form the basis of much decision making across taxa, however, many different rules of thumb must often be operating simultaneously and may be applied in anticipation of future decisions. Although any one decision may appear to be simple and straightforward, and could be modeled in a similar fashion across many taxa, the number of concurrent decisions made by social animals during group movement appears to be great.

Consideration of the decision processes involved in group move-

S. Boinski and P. A. Garber

ment leads directly to another controversial topic: Do cognitive abilities differ between primates and other social mammals? Is there any empirical basis for the popular perception of "clever" primates? Unique circumstances related to group movement in primates appear to have made flexibility, learning, expanded memory, and foresight beyond that which can be accorded to rule-based decision making advantageous. Moreover, the complexity of decision making, at least in terms of the number of choices that must be simultaneously made, appears at least quantitatively, and perhaps qualitatively, greater in primates. Compared with social carnivores, cetaceans, birds, ungulates, and other animals commonly living social groups, only primates (at least the anthropoids) exhibit the following constellation of traits: often intense within-and between-group food competition for discrete, patchy food resources; infrequent partitioning of troops into independent subgroups; the high travel costs of terrestrial locomotion as compared with flight and swimming; numerous opportunities for individuals to extract ecological benefits at a cost to social companions; attainment of group travel decisions by processes that can be considered a form of negotiation, and, possibly, planning of extended travel routes. To the best that we can determine, other social taxa exhibit only subsets of these traits.

- Extensive within- and between-group competition exists among many primate species for access to food patches. Whether folivorous or frugivorous, primates can be characterized as having a highly selective diet, with food items commonly distributed in discrete and dispersed patches. Individuals within a primate troop often feed concurrently on dissimilar items, some of markedly better or worse quality. Furthermore, even when all members of a primate troop are feeding on the same item, extensive individual variation in feeding rates occurs that cannot be accorded directly to individual variation in skill or body size.
- Most primates live in permanent social groups, albeit with a cohesion that can stretch like rubber on occasion. In contrast to social carnivores, cetaceans, and birds, reduction of within-group food competition by individual departure is a strategy available to only a few primate species. Moreover, the outcome of competition for resources between primate groups is frequently dependent on relative group size and cooperation, with larger, more cooperative groups often the more successful.
- Compared with birds, cetaceans, and many terrestrial mammals, primates have reduced travel efficiency. In at least some primate species the apparent consequential hesitancy to travel to more distant and

less familiar food patches may also be confounded with strategies to reduce predation risk. In any case, as a result of the disinclination of individual primates to travel to distant food patches, within- and between-group food competition is exacerbated, especially as fission-fusion strategies are seldom an option. The potential benefits obtained from social strategies that enhance foraging success and predation avoidance are thereby increased relative to taxa in less complex competitive regimes.

- The social processes underlying long-distance travel decisions in some primate species may indicate greater social complexity and cognitive demands than those on nonprimate social mammals. Multiple travel trajectories and (probably) travel goals can be simultaneously indicated and apparently resolved through negotiation-like processes, as well as dominance interactions, alliances, and, possibly, deceit. There is also evidence suggesting that appropriate usage of signals initiating and leading group travel in primates requires practice and learning. This interpretation is consistent with two other recent conclusions that reinforce the view that the cognitive abilities of nonhuman primates are relatively sophisticated. First, social alliances and manipulations are employed to gain access to food and other resources to an extent yet to be documented in social animals other than primates (Harcourt and deWaal 1992). Second, laboratory studies indicate that primates can comprehend "tertiary relationships," interactions and relationships among objects and individuals that do not involve themselves, also an ability not yet shown for nonprimates (Tomasello and Call 1997).

- Travel route planning and resource monitoring are cognitive aspects of group movement for which few quantitative data currently exist. By planning, we refer to mental abilities in which present decisions reflect not only immediate individual or group goals, but are part of a more complex effort to attain some future goal. Thus, decisions about which direction to travel in when leaving a sleeping site and which set of feeding sites to visit first may be based on where the individual or group wants to be at some later time during the day, possibly to monitor current food availability. This changes decision making from a step-by-step process into a one, two, three or more-step-ahead cognitive process. Data on baboons traveling to water holes, tamarins visiting nearest-neighbor fruiting trees, and chimpanzees selecting hammer stones before reaching the sites where these tools will be used are provocative, hinting at elements of a decision-making process that was planned for hours earlier. Future studies of primate foraging and movement patterns designed to test hypotheses of planning and forethought directly are certainly warranted.

As a consequence of these factors, primates are faced with a dynamic gauntlet of decisions that are probably more numerous and complex than those facing nonprimates. It is advantageous for individual primates not only to track where they are spatially within

S. Boinski and P. A. Garber

the large-scale environment of their range and the small-scale environment of the troop dispersion, but also to track and respond to the positions of other individual group members. Individual primates that successfully anticipate the future spatial, social, and ecological behavior of group members and modify their own behavior accordingly accrue social and ecological benefits. The success of an individual within a primate group is predicated to a great extent on the interaction of diverse frequency-dependent behaviors of its groupmates and the individual's own strategic responses. In effect, the game theory (Maynard Smith 1974; Parker and Hammerstein 1985) situations that primates are documented to solve commonly exceed the intricacy of those encountered by most other social taxa.

Yet another issue relevant to multifactorial decision making by primates and other social taxa concerns cognitive models premised on the existence of discrete cognitive domains. These domains have been described as genetically specified, functionally distinct brain modules that input and process information and output decisions (Fodor 1983; Karmiloff-Smith 1992). This modular concept of brain function was originally a reaction to primary reliance on one psychological process, associative learning (based on presumably general principles of neural organization and function) to explain the varied mental abilities of vertebrates. Associative learning and other general cognitive rules (Wasserman 1993) are clearly inadequate as a complete explanation for the intelligence of rats, cats, monkeys, and humans given the robust evidence for the narrow aspects of information that can be harvested and processed by vertebrate taxa, as exemplified by bat sonar and bird orientation during migration. Yet the contrary, extremely modular perspective on cognitive function may be equally unrealistic as currently derived and presented in discussions of the evolution of intelligence. While these discussions have certainly facilitated current consideration of brain function and evolution, no cognitive domain, such as those postulated for foraging or deception, has yet been identified within a human or other animal brain. In particular, the cognitive and social domains are argued to be separate in nonhuman primates and early humans, joining together only in modern humans (i.e., Tooby and Cosmides 1991; Mithen 1996). Researchers invoking the existence of these brain modules explicitly treat the presumed inner workings as black boxes that are only minimally connected or integrated with one another. As the chapters in this volume repeatedly document, however, social, ecological, and spatial information processing and decision making are so intertwined as to be effectively inseparable in primates and many other group-living animals. The concept of discrete cognitive domains is clearly untenable when confronted with the reality of group movement.

More specific examples of contradictions to a modular domain model provided by group movement are easy to identify. First, although some cognitive abilities may have evolved to solve restricted problems, they can be useful in other domains as well. For example, foraging and social domains would presumably summon the same spatial skills when recalling the locations and relative profitabilities of unseen nearby foraging sites and social companions, respectively. Second, on many occasions in the course of group movement, individuals are faced with multiple, simultaneous problems to which they have to find a balanced solution. In other words, behaviors are often the result of finding a single solution to a combination of problems referring putatively to separate domains. Several domains would be called upon, for instance, when males in a troop had to decide whether to concur with or oppose a travel initiative by female troop members that would lead the troop to a potential interaction with males of another troop.

We end this volume with the question posed by Alice to the Cheshire Cat. Which way ought we to go from here? The Cheshire Cat's response is also apropos: with group movement as a starting point, we are certain to connect with a rich literature and unresolved questions no matter which trajectory is selected. Group movement is an underexploited phenomenon from which to synthesize across multiple levels of biology, behavior, cognition, and ecology. We exhort our readers to be more ambitious in the scope and methodology of their research. This volume proves the value of anecdotes and impressions as inspiration and clues for future researchers; nevertheless, unexploited opportunities for quantitative documentation, prediction testing, and field experiments are almost dizzying in their vastness. Although recording the outcome of group behavior is important, the most challenging task will be explaining the underlying processes. In our fin-de-siècle era of behavioral and cognitive science, this will depend, in part, on the cleverness of investigators in designing their research. To navigate the rest of the route, researchers will need to emulate the collaboration and communication of their study subjects.

S. Boinski and P. A. Garber

Classification of living primates

Class	Mammalia	
Order	Primates	
Superfamily	Lemuroidea	
Family	Cheirogaleidae	*Allocebus*
		Cheirogaleus
		Microcebus
		Phaner
		Mirza
	Lemuridae	*Eulemur*
		Lemur
		Hapalemur
		Varecia
	Megaladapidae	*Lepilemur*
	Indriidae	*Avahi*
		Indri
		Propithecus
	Daubentoniidae	*Daubentonia*
Superfamily	Lorisoidea	
Family	Lorisidae	*Arctocebus*
		Euoticus
		Galago
		Galagoides
		Otolemur
Superfamily	Tarsioidea	
Family	Tarsiidae	*Tarsius*
Superfamily	Ceboidea	
Subfamily	Aotinae	*Aotus*
Subfamily	Atelinae	*Alouatta*
		Ateles
		Brachyteles
		Lagothrix

Subfamily	Callitrichinae	*Callimico* *Callithrix* (including *Cebuella*) *Leontopithecus* *Saguinus*
Subfamily	Cebinae	*Cebus* *Saimiri*
Subfamily	Pitheciinae	*Callicebus* *Cacajao* *Pithecia* *Chiropotes*

Superfamily
Family
Subfamily
	Cercopithecoidea
	Cercopithecidae
	Cercopithecinae
		
		Allenopithecus
		Cercocebus
		Cercopithecus
		Chlorocebus (formerly
		  Cercopithecus aethiops)
		Erythrocebus
		Lophocebus (formerly
		  Cercocebus albigena)
		Macaca
		Mandrillus
		Miopithecus
		Papio
		Theropithecus

Subfamily	Colobinae	*Colobus*
		Procolobus (formerly *Colobus*)
		Nasalis
		Presbytis
		Pygathrix (includes
		  Rhinopithecus)
		Semnopithecus (formerly
		  Presbytis entellus)
		Trachypithecus (formerly
		  Presbytis)

Superfamily Family	Hominoidea Hylobatidae	 *Hylobates* (includes *Symphalangus*)
Family Subfamily Subfamily	Hominidae Ponginae Homininae	 *Pongo* *Gorilla* *Homo* *Pan*

Robert A. Barton
Evolutionary Anthropology
Research Group
University of Durham
Durham DH1 3HN
United Kingdom

Benjamin B. Beck
Department of Mammals and
Division of Zoological
Research
National Zoological Park
Smithsonian Institution
Washington, D.C. 20008
U.S.A.

Sue Boinski
Department of Anthropology
and Division of Comparative
Medicine
University of Florida
Gainesville, FL. 32611-7305
U.S.A.

Erin E. Boydston
Department of Zoology
Michigan State University
East Lansing, MI 48824-1115
U.S.A.

Richard W. Byrne
Scottish Primate Research
Group
School of Psychology
University of St. Andrews
St. Andrews
Fife KY16 9JU
Scotland

Colin A. Chapman
Department of Zoology
University of Florida
Gainesville, FL 32611
U.S.A.

Lauren J. Chapman
Department of Zoology
University of Florida
Gainesville, FL 32611
U.S.A.

Marina Cords
Department of Anthropology
Columbia University
New York, NY 10027
U.S.A.

Fred C. Dyer
Department of Zoology
Michigan State University
East Lansing, MI 48824
U.S.A.

Paul A. Garber
Department of Anthropology
University of Illinois
Urbana, IL 61801 U.S.A.

Russell Greenberg
Smithsonian Migratory Bird
Center
National Zoological Park
Washington, D.C. 20008
U.S.A.

Kay E. Holekamp
Department of Zoology
Michigan State University
East Lansing, MI 48824-1115
U.S.A.

Charles Janson
Department of Ecology and
Evolution
State University of New York
Stony Brook, NY 11794-5245
U.S.A.

Peter M. Kappeler
Deutsches Primatenzentrum
Abt. Verhaltenforschung/
Ökologie
Kellnerweg 4
37077 Göttingen
Germany

Margaret F. Kinnaird
Wildlife Conservation Society
2300 Southern Blvd.
Bronx, NY 10460
U.S.A.

William R. Leonard
Department of Anthropology
Northwestern University
Evanston, IL 60208
U.S.A.

J. Terrence McCabe
Department of Anthropology
University of Colorado
Boulder, CO 80309
U.S.A.

Charles R. Menzel
Language Research Center
Georgia State University
3401 Panthersville Road
Decatur, GA 30034
U.S.A.

Katharine Milton
Department of Environmental
Science, Policy & Management
Division of Insect Biology
University of California
Berkeley, CA 94720-3112
U.S.A.

Timothy G. O'Brien
Wildlife Conservation Society
2300 Southern Blvd.
Bronx, NY 10460
U.S.A.

Carlos A. Peres
School of Environmental
Sciences
University of East Anglia
Norwich NR4 7TJ
United Kingdom

Marcia L. Robertson
Department of Anthropology
Northwestern University
Evanston, IL 60208
U.S.A.

Laura Smale
Department of Psychology
Michigan State University
East Lansing, MI 48824-1117
U.S.A.

Rachel Smolker
Biology Department
University of Vermont
Burlington, VT 05405-0086
U.S.A.

Karen Steudel
Department of Zoology
University of Wisconsin-
Madison
Madison, WI 53706
U.S.A.

Adrian Treves
Department of Psychology
University of Wisconsin-
Madison
Madison, WI 53706
U.S.A.

David P. Watts
Department of Anthropology
Yale University
New Haven, CT 06520-8277
U.S.A.

David Sloan Wilson
Department of Biological
Sciences
Binghamton University
Binghamton, NY 13902-6000
U.S.A.

REFERENCES

Able, K. P. 1996. The debate over olfactory navigation by homing pigeons. *J. Exp. Biol.* 199:121–24.

Able, K. P., and M. A. Able. 1996. The flexible migratory orientation system of the savannah sparrow *(Passerculus sanwichensis)*. *J. Exp. Biol.* 199:3–8.

Abrams, P. A. 1994. Should prey overestimate the risk of predation? *Am. Nat.* 144:317–28.

Addicott, J. F., J. M. Aho, M. F. Antolin, D. K. Padilla, J. S. Richardson, and D. A. Soluk. 1987. Ecological neighborhoods: Scaling environmental patterns. *Oikos* 49:340–46.

Aiello, L. C. 1996. Terrestriality, bipedalism, and the origin of language. *Proc. Brit. Acad.* 88:269–89.

Aiello, L. C., and Wheeler, P. 1995. The expensive tissue hypothesis: The brain and the digestive system in human and primate evolution. *Curr. Anthropol.* 36:199–221.

Aitkin, L. M., M. M. Merzenich, D. R. F. Irvine, J. C. Clarey, and J. E. Nelson. 1986. Frequency representation in auditory cortex of the common marmoset *(Callithrix jacchus jacchus)*. *J. Comp. Neurol.* 252:2:175–85.

Alatalo, R. V. 1981. Interspecific competition in tits *Parus* spp. and the goldcrest *Regulus regulus:* Foraging shifts in multi-specific flocks. *Oikos* 37:335–44.

Albernaz, A. L. K. M. 1993. Área de uso de *Callithrix argentata:* Habitats, disponibilidade de alimento e variação entre grupos. M. Sc. thesis, Fundação Universidade do Amazonas/INPA, Manaus.

Alberts, S. C. 1994. Vigilance in young baboons: Effects of habitat, age, sex and maternal rank on glance rate. *Anim. Behav.* 47:749–55.

Alberts, S. C., J. Altmann, and M. L. Wilson. 1996. Mate guarding constrains foraging activity of male baboons. *Anim. Behav.* 51:1269–77.

Aldrich-Blake, F. P. G. 1970. Problems of social structure in forest monkeys. In *Social behaviour in birds and mammals,* edited by J. H. Crook, 79–101. London: Academic Press.

Alevizon, W. S. 1976. Mixed schooling and its possible significance in a tropical western Atlantic parrotfish and sturgeon fish. *Copeia* 1976:769–98.

Allman, J. 1987. Primates: Evolution of the brain. In *The Oxford companion to the mind,* edited by R. L. Gregory, 633–39. Oxford: Oxford University Press.

———. 1990. The origin of the neocortex. *Semin. Neurosci.* 2:257–62.

Allman, J., and E. McGuinness. 1988. Visual cortex in primates. In *Comparative primate biology,* vol. 4, edited by H. D. Steklis and J. Erwin, 279–326. New York: Alan R. Liss.

Allman, J., T. McLaughlin, and A. Hakeem. 1993. Brain weight and life-span in primate species. *Proc. Natl. Acad. Sci. USA* 90:118–22.

Alonso, C., and A. Langguth. 1989. Ecologia e comportamento de *Callithrix jacchus* (Primates: Callitrichidae) numa ilha de floresta atlântica. *Revista Nordestina de Biologia* 6:105–37.

Altmann, J. 1974. Observational study of behaviour: Sampling methods. *Behaviour* 49:227–65.

———. 1980. *Baboon mothers and infants.* Cambridge, MA: Harvard University Press.

Altmann, J., and A. Samuels. 1992. Costs of maternal care: Infant-carrying in baboons. *Behav. Ecol. Sociobiol.* 29:391–98.

Altmann, S. A. 1959. Field observations on a howling monkey society. *J. Mammal.* 40:317–30.

———. 1974. Baboons: Space, time, and energy. *Am. Zool.* 14:221–48.

———. 1979. Baboon progressions: Order or chaos? A study of one-dimensional group geometry. *Anim. Behav.* 27:46–80.

———. 1987. The impact of locomotor energetics on mammalian foraging. *J. Zool.* 211:215–25.

———. 1998. *Foraging for survival: Yearling baboons in Africa.* Chicago: University of Chicago Press.

Altmann, S. A., and J. Altmann. 1970. *Baboon ecology.* Chicago: University of Chicago Press.

Amundin, M., and T. Cranford. 1990. Forehead anatomy of *Phocoena phocoena* and *Cephalorhynchus commersonii:* Three-dimensional computer reconstructions with and emphasis on nasal diverticula. In *Sensory abilities of cetaceans: Laboratory and field evidence,* edited by J. A. Thomas and R. A. Kastelein, 1–18. New York: Plenum Press.

Andeldt, W. F. 1985. Behavioral ecology of coyotes in south Texas. *Wildl. Monogr.* 94:1–45.

Andeldt, W. F., D. P. Althoff, and P. S. Gipson. 1979. Movements of breeding coyotes with emphasis on den site relationships. *J. Mammal.* 60:568–75.

Anderson, C. M. 1981. Intertroop relations of chacma baboons *(Papio ursinus).* *Int. J. Primatol.* 2:285–310.

———. 1983. Levels of social organization and male-female bonding in the genus *Papio. Am. J. Phys. Anthropol.* 60:15–22.

———. 1986. Predation and primate evolution. *Primates* 27:15–39.

Anderson, J. D. 1983. Optimal foraging and the traveling salesman. *Theor. Popul. Biol.* 24:145–59.

Anderson, J. R. 1984. Ethology and ecology of sleeping in monkeys and apes. *Adv. Stud. Behav.* 14:165–228.

———. 1990. Use of objects as hammers to open nuts by capuchin monkeys. *Folia Primatol.* 54:138–45.

Anderson, L. E., and W. K. Balzer. 1991. The effects of timing of leaders' opinions on problem-solving groups: A field experiment. *Group Org. Stud.* 16:86–101.

Andriatsarafara, F. R. 1988. Etude ecoethologique de deux lemuriens sypatriques de la foret seche caducifoliee d'Ampijoroa: *Lemur fulvus fulvus et Lemur mongoz.* Ph.D. thesis, University of Madagascar, Antananarivo.

Antón, S. C. 1997. Developmental age and taxonomic affinity of the Mojokerto Child, Java, Indonesia. *Am. J. Phys. Anthropol.* 102:497–514.

Arditi, R., and B. Dacorogna. 1988. Optimal foraging on arbitrary food distributions and the definition of habitat patches. *Am. Nat.* 131:837–46.

Areval, J. E., and A. G. Gosler. 1994. The behavior of treecreepers *Certhia familiaris* in mixed species flocks in winter. *Bird Stud.* 46:1–6.

Arhem, K. 1985. *Pastoral man in the Garden of Eden: The Maasai of the Ngorongoro Conservation Area, Tanzania.* Uppsala: Research Reports in Cultural Anthropology.

Armstrong, E. 1983. Relative brain size and metabolism in mammals. *Science* 220:1302–4.

Armstrong, R. B., M. H. Laughlin, L. Rome, and C. R. Taylor. 1983. Metabolism of rats running up and down an incline. *J. Appl. Physiol.* 55:518–21.

Austin, G. T., and E. L. Smith. 1972. Winter foraging ecology of mixed insectivorous bird flocks in oak woodlands in southern Arizona. *Condor* 74:17–24.

Avery, M. L. 1994. Finding good food and avoiding bad food: Does it help to associate with experienced flockmates? *Anim. Behav.* 48:1371–78.

Avitabile, A., R. A. Morse, and R. Boch. 1975. Swarming honey bees guided by pheromones. *Ann. Entomol. Soc. Am.* 68:1079–82.

Baenninger, R., R. D. Estes, and S. Baldwin. 1977. Anti-predator behaviour of baboons and impalas toward a cheetah. *E. Afr. Wildl. J.* 15:327–29.

Baerends, G. P. 1941. Fortpflanzungsverhalten und Orientierung der Grabwaspe *Ammophila campestris* Jur. *Tijdschr. Ent. Deel* 84:68–275.

Baird, R. W., and L. M. Dill. 1995. Occurrence and behaviour of transient killer whales: Seasonal and pod-specific variability, foraging behaviour and prey handling. *Can. J. Zool.* 73:1300–1311.

Baird, R. W., and P. J. Stacey. 1988. Foraging and feeding behavior of transient killer whales. *Whalewatcher J.* 22:11–15.

Baker, A. J., J. M. Dietz, and D. G. Kleiman. 1993. Behavioural evidence for monopolization of paternity in multi-male groups of golden lion tamarins. *Anim. Behav.* 46:1091–1103.

Baker, K. C., and F. Aureli. 1996. The neighbor effect: Other groups influence intragroup agonistic behavior in captive chimpanzees. *Am. J. Primatol.* 40:283–91.

Baker, R. G. V. 1982. Place utility fields. *Geogr. Anal.* 14:10–28.

Baker, R. R. 1978. *The evolutionary ecology of animal migration.* New York: Holmes and Meier.

———. 1980. Goal orientation by blindfolded humans after long-distance displacement: Possible involvement of a magnetic sense. *Science* 210:555–57.

———. 1982. *Migration: Paths through time and space.* London: Hodder and Stoughton.

Balas, E., and P. Toth. 1985. Branch and bound methods. In *The traveling salesman problem,* edited by E. L. Lawler, J. K. Lenstra, A. H. G. Rinooy Kan, and D. B. Shmoys. J. Wiley and Sons, Chichester.

Balda, R. P., and A. C. Kamil. 1988. The spatial memory of Clark's nutcrackers

(Nucifraga columbiana) in an analogue of the radial arm maze. *Anim. Learn. Behav.* 16:116–22.

———. 1989. A comparative study of cache recovery by three corvid species. *Anim. Behav.* 38:486–95.

Baldwin, J. D., and J. I. Baldwin. 1976. The vocalizations of howler monkeys *(Alouatta palliata)* in southwestern Panama. *Folia Primatol.* 26:81–108.

Balldellou, M., and S. P. Henzi. 1992. Vigilance, predator detection and the presence of supernumerary males in vervet monkey troops. *Anim. Behav.* 43:451–61.

Banfield, A. W. F. 1951. Populations and movements of the Saskatchewan timber wolf *(Canis lupus knightii)* in the Prince Albert National Park, Saskatchewan. Wildlife Management Bulletin 4, Canadian Wildlife Service.

Barbujani, G. 1991. What do languages tell us about human microevolution? *Trends Ecol. Evol.* 6:150–55.

Bardin, A. 1984. Structure of mixed-species flocks of tits. *Acta Congressus Intern. Ornithol.,* XVIII:1078.

Barkow, J. H., L. Cosmides, and J. Tooby, ed. 1992. *The adapted mind: Evolutionary psychology and the generation of culture.* Oxford: Oxford University Press.

Barnard, C. J., and D. B. A. Thompson. 1985. *Gulls and plovers: The ecology and behavior of mixed-species feeding groups.* Croom Helm. London.

Barnett, A. 1976. On searching for events of limited duration. *Oper. Res.* 24:438–51.

Baron-Cohen, S. 1994. A model of the mindreading system: Neuropsychological and neurobiological perspectives. In *Origins of an understanding of mind,* edited by P. Mitchell and C. Lewis. Hillsdale, NJ: Lawrence Erlbaum Associates.

Barth, F. 1961. *Nomads of South Persia: The Basseri Tribe of the Khamseh Confederacy.* Boston: Little Brown.

Bartholomew, G. A., and J. B. Birdsell. 1953. Ecology and the protohominids. *Am. Anthropol.* 55:481–98.

Barton, R. A. 1996a. Independent contrasts analysis of neocortical size and socio-ecology in primates. *Behav. Brain Sci.* 16:694–95.

———. 1996b. Neocortex size and behavioral ecology in primates. *Proc. R. Soc. Lond.* B 263:173–77.

———. 1998. Visual specialisation and brain evolution in primates. *Proc. R. Soc. Lond.* B 265:1933–37.

———. In press. The evolutionary ecology of the primate brain. In *Comparative primate socioecology,* edited by P. C. Lee. Cambridge: Cambridge University Press.

Barton, R. A., R. W. Byrne, and A. Whiten. 1996. Ecology, feeding competition and female bonding in baboons. *Behav. Ecol. Sociobiol.* 38:321–29.

Barton, R. A., and Dean, P. 1993. Comparative evidence indicating neural specialisation for predatory behaviour in mammals. *Proc. R. Soc. Lond.* B 254:63–68.

Barton, R., and Dunbar, R. I. M. 1997. Evolution of the social brain. In *Machiavellian Intelligence II,* edited by A. Whiten and R. Byrne, 240–63. Cambridge: Cambridge University Press.

Barton, R. A., and A. J. Purvis. 1994. Primate brains and ecology: Looking beneath the surface. In *Current primatology,* vol. 1, edited by J. R. Anderson and N. Herrenschmidt, 1–10. Strasbourg University Press.

Barton, R. A., A. Purvis, and P. H. Harvey. 1995. Evolutionary radiation of visual and olfactory brain systems in primates, bats and insectivores. *Phil. Trans. R. Soc. Lond.* B 348:381–92.

Barton, R. A., and A. Whiten. 1993. Feeding competition among female olive baboons. *Anim. Behav.* 46:777–89.

———. 1994. Reducing complex diets to simple rules: Food selection by olive baboons. *Behav. Ecol. Sociobiol.* 35:283–93.

Barton, R. A., A. Whiten, S. C. Strum, R. W. Byrne, and A. J. Simpson. 1992. Habitat use and resource availability in baboons. *Anim. Behav.* 43:831–44.

Bass, B. M. 1990. *Bass and Stogdill's handbook of leadership.* 3d ed. New York: Free Press.

Bassett, T. J. 1986. Fulani herd movement. *Geogr. J.* 76(3):233–48.

Bates, D. 1971. The role of the state in peasant-nomad mutualism. *Anthropol. Q.* 44:109–31.

Bateson, M., and A. Kacelnik. 1996. Rate currencies and the foraging starling: The fallacy of the averages revisited. *Behav. Ecol.* 7:341–52.

Bauchop, T., and R. W. Martucci. 1968. Ruminant-like digestion of the langur monkey. *Science* 161:698–700.

Baudinette, R. V. 1980. Physiological responses to locomotion in marsupials. In *Comparative physiology: Primitive mammals,* edited by K. Schmidt-Nielsen, L. Bolis, and C. R. Taylor, 200–212. New York: Cambridge University Press.

Bearder, S. K. 1986. Lorises, bushbabies, and tarsiers: Diverse societies in solitary foragers. In *Primate societies,* edited by B. B. Smuts, D. L. Cheney, R. M. Seyfarth, R. W. Wrangham, and T. T. Struhsaker, 11–24. Chicago: University of Chicago Press.

Beardwood, J., J. H. Halton, and J. M. Hammersley. 1959. The shortest path through many points. *Proc. Cambridge Phil. Soc.* 55:299–327.

Beck, B. B. 1995. Reintroduction, zoos, conservation, and animal welfare. In *Ethics on the Ark: Zoos, animal welfare, and wildlife conservation,* edited by B. G. Norton, M. Hutchins, E. F. Stevens, and T. L. Maple, 155–63. Washington, DC: Smithsonian Institution Press.

Beck, B. B., D. G. Kleiman, J. M. Dietz, I. Castro, C. Carvalho, A. Martins, and B. Rettberg-Beck. 1991. Losses and reproduction in reintroduced golden lion tamarins *Leontopithecus rosalia. Dodo* 27:50–61.

Becker, L. 1958. Untersuchungen über das Heimfindevermögen. der Bienen. *Z. vergl. Physiol.* 41:1–25.

Begon, M., J. L. Harper, and C. R. Townsend. 1990. *Ecology: Individuals, populations and communities.* 2d ed. Boston: Blackwell Scientific.

Behrensmeyer, A. K., and H. B. S. Cooke. 1985. Paleoenvironments, stratigraphy, and taphonomy in the African Pliocene and early Pleistocene. In *Ancestors: The hard evidence,* edited by E. Delson, 60–62. New York: Alan R. Liss.

Bekoff, M. 1975. Social behavior and ecology of the African Canidae: A review. In *The Wild Canids,* edited by M. W. Fox, 120–42. New York: Van Nostrand Reinhold.

Bekoff, M., T. J. Daniels, and J. L. Gittleman. 1984. Life history patterns and the comparative social ecology of carnivores. *Annu. Rev. Ecol. Syst.* 15:191–232.

Bekoff, M., and M. C. Wells. 1982. Behavioral ecology of coyotes: Social organization, rearing patterns, space use, and resource defense. *Z. Tierpsychol.* 60: 281–305.

————. 1986. Social ecology and behavior of coyotes. *Adv. Stud. Behav.* 16:251–338.

Bell, H. L. 1980. Composition and seasonality of mixed species feeding flocks of insectivorous birds in the Australian Capital area. *Emu* 80:227–32.

————. 1983. A bird community of lowland rainforest in New Guinea: 5. Mixed-species feeding flocks. *Emu* 82:256–75.

Bell, W. J. 1991. *Searching behaviour: The behavioural ecology of finding resources.* London: Chapman and Hall.

Belovsky, G. E. 1981. Diet optimization in a generalist herbivore: The moose. *Theor. Popul. Biol.* 14:105–34.

Benhamou, S. 1994. Spatial memory and searching efficiency. *Anim. Behav.* 47:1423–33.

————. 1996. No evidence for cognitive mapping in rats. *Anim. Behav.* 52:201–12.

Benhamou, S., J. Sauve, and P. Bovet. 1990. Spatial memory in large-scale movements: Efficiency and limitations of the egocentric coding process. *J. Theor. Biol.* 145:1–12.

Bennett, A. T. D. 1993. Remembering landmarks. *Nature* 364:293–94.

————. 1996. Do animals have cognitive maps? *J. Exp. Biol.* 199:219–24.

Bennett, E. L. 1986. Environmental correlates of ranging behavior in the banded langur, *Presbytis melalophos. Folia Primatol.* 47:26–38.

Bennett, E. L., and A. G. Davies. 1994. The ecology of Asian colobines. In *Colobine monkeys: Their ecology, behaviour, and evolution,* edited by A. G. Davies and J. F. Oates, 129–72. Cambridge: Cambridge University Press.

Berger, J. 1991. Pregnancy incentives, predation constraints and habitat shifts: Experimental and field evidence for wild bighorn sheep. *Anim. Behav.* 41:61–77.

Berner, T. O., and T. C. Grubb. 1985. An experimental analysis of mixed species flocking in birds of deciduous woodland. *Ecology* 66:1229–36.

Bertram, B. C. R. 1975. Social factors influencing reproduction in wild lions. *J. Zool.* 177:463–82.

————. 1978. Living in groups: Predators and prey. In *Behavioural ecology: An evolutionary approach.* edited by J. R. Krebs and N. B. Davies, 64–96. Oxford: Blackwell Scientific Publications.

————. 1979. Serengeti predators and their social systems. In *Serengeti: Dynamics of an ecosystem,* edited by A. R. E. Sinclair and M. Norton-Griffiths, 159–79. Chicago: University of Chicago Press.

Beyon, P., and O. A. E. Rasa. 1989. Do dwarf mongooses have a language?: Warning vocalizations transmit complex information. *Suid-Afrikaanse Tydskrif vir Wetenskop* 85:447–50.

Bhide, S., N. John, and M. R. Kabuka. 1993. A Boolean neural network approach for the Traveling Salesman Problem. *IEEE Trans. Comput.* 42:1271–78.

Biebach, H., M. Gordijn, and J. R. Krebs. 1989. Time-and-place learning by garden warblers, *Sylvia borin. Anim. Behav.* 37:353–60.

Bigg, M. A., G. E. Ellis, J. K. B. Ford, and K. A. Balcomb. 1987. *Killer whales: A study of their identification, geneology, and natural history in British Columbia and Washington State.* Nanaimo, British Columbia: Phantom Press.

Bigg, M. A., P. F. Olesiuk, G. M. Ellis, J. K. B. Ford, and K. C. Balcomb. 1990. Social organization and genealogy of resident killer whales *(Orcinus orca)* in the coastal waters of British Columbia and Washington State. International Whaling Commission Special Issue 12.

Bishop, N., S. B. Hardy, J. Teas, and J. Moore. 1981. Measures of human influence in habitats of South Asian Monkeys. *Int. J. Primatol.* 2:153–67.

Bitterman, M. E. 1996. Comparative analysis of learning in honeybees. *Anim. Learn. Behav.* 24:123–41.

Blickhan, R. 1989. The spring-mass model for running and hopping. *J. Biomech.* 22:1217–27.

Blumstein, D. T., and K. B. Armitage. 1997. Does sociality drive the evolution of communicative complexity? A comparative test with ground-dwelling sciurid alarm calls. *Am. Nat.* 150:179–200.

Blurton Jones, N. G., Smith, L. C., O'Connell, J. F., Hawkes, K., and C. L. Kamuzora. 1992. Demography of the Hadza, an increasing and high density population of savanna foragers. *Am. J. Phys. Anthropol.* 89:159–81.

Boake, R. B. 1986. A method for testing adaptive hypotheses of mate choice. *Am. Nat.* 127:654–66.

Boaz, N. T. 1979. Early hominid population densities: New estimates. *Science* 206:592–95.

Bock, C. 1969. Intra- versus interspecific aggression in Pigmy Nuthatch flocks. *Ecology* 50:903–5.

Boehm, C. 1982. The evolutionary development of morality as an effect of dominance behavior and conflict interference. *J. Soc. Biol. Struct.* 5:413–22.

———. 1992. Segmentary 'warfare' and the management of conflict: Comparison of East African chimpanzees and patrilineal-patrilocal humans. In *Coalitions and alliances in humans and other animals,* edited by A. H. Harcourt and F. B. M. de Waal, 137–73. Oxford: Oxford University Press.

———. 1993. Egalitarian behavior and reverse dominance hierarchy. *Curr. Anthropol.* 34:227–54.

———. 1997. Egalitarian behaviour and the evolution of political intelligence. In *Machiavellian Intelligence II,* edited by A. Whiten and R. Byrne, 341–64. Cambridge: Cambridge University Press.

Boesch, C. 1991a. The effects of leopard predation on grouping patterns in chimpanzees. *Behaviour* 117:220–42.

———. 1991b. Symbolic communication in wild chimpanzees? *Hum. Evol.* 6:81–90.

———. 1994. Hunting strategies of Gombe and Tai chimpanzees. In *Chimpanzee cultures,* edited by R. W. Wrangham, W. C. McGrew, F. B. M. de Waal, and P. G. Heltne, 77–91. Cambridge, MA: Harvard University Press.

Boesch, C., and H. Boesch. 1984. Mental map in wild chimpanzees: An analysis of hammer transports for nut cracking. *Primates* 25:160–70.

———. 1989. Hunting behaviour of wild chimpanzees in the Tai National Park. *Am. J. Phys. Anthropol.* 78:547–73.

Boese, G. 1975. Social behaviour and ecological considerations of West African baboons *(Papio papio).* In *Socioecology and ecology of primates,* edited by R. H. Tuttle. The Hague: Mouton.

Boinski, S. 1987a. Birth synchrony in squirrel monkeys *(Saimiri oerstedi):* A strategy to reduce neonatal predation. *Behav. Ecol. Sociobiol.* 21:393–400.

———. 1987b. Habitat use by squirrel monkeys *(Saimiri oerstedi)* in Costa Rica. *Folia Primatol.* 49:151–67.

———. 1988a. Sex differences in the foraging behavior of squirrel monkeys in a seasonal habitat. *Behav. Ecol. Sociobiol.* 23:177–86.

————. 1988b. Use of a club by a white-faced capuchin *(Cebus capucinus)* to attack a venomous snake *(Bothrops asper). Am. J. Primatol.* 14:177–79.

————. 1989. Why don't *Saimiri oerstedii* and *Cebus capucinus* form mixed-species groups? *Int. J. Primatol.* 10:103–14.

————. 1991. The coordination of spatial position: A field study of the vocal behaviour of adult female squirrel monkeys. *Anim. Behav.* 41:89–102.

————. 1993. Vocal coordination of group movement among white-faced capuchin monkeys, *Cebus capucinus. Am. J. Primatol.* 30:85–100.

————. 1996. Vocal coordination of troop movement in squirrel monkeys (*Saimiri oerstedi* and *S. sciureus*) and white-faced capuchins *(Cebus capucinus).* In *Adaptive radiations of Neotropical primates,* edited by M. Norconk, A. L. Rosenberger, and P. A. Garber, 251–69. New York: Plenum Press.

Boinski, S., and Campbell, A. F. 1995. Use of trill vocalizations to coordinate troop movement among white-faced capuchins: A second field test. *Behaviour* 132:875–901.

————. 1996. The huh vocalization of the white-faced capuchins: A spacing call disguised as a food call? *Ethology* 102:826–40.

Boinski, S., and Chapman, C. A. 1995. Predation in primates: Where are we and what next? *Evol. Anthropol.* 4:1–3.

Boinski, S., and S. J. Cropp. 1999. Disparate data sets resolve squirrel monkey *(Saimiri)* taxonomy: Implications for behavioral ecology and biomedical usage. *Int. J. Primatol.* 20:237–56.

Boinski, S., and D. M. Fragaszy. 1989. The ontogeny of foraging in squirrel monkeys, *Saimiri oerstedi. Anim. Behav.* 37:415–28.

Boinski, S., T. S. Gross, and K. J. Davis. In press. Terrestrial predator alarm vocalizations are a valid monitor of stress in captive brown capuchins *(Cebus apella). Zoo Biology.* In press.

Boinski, S., and C. L. Mitchell. 1992. The ecological and social factors affecting adult female squirrel monkey vocal behavior. *Ethology* 92:316–30.

————. 1994. Male residence and association patterns in Costa Rican squirrel monkeys *(Saimiri oerstedi). Am. J. Primatol.* 35:129–38.

————. 1995. Wild squirrel monkey *(Saimiri sciureus)* "caregiver" calls: Contexts and acoustic structure. *Am. J. Primatol.* 35:129–38.

————. 1997. Chuck vocalizations of wild female squirrel monkeys *(Saimiri sciureus)* contain information on caller identity and foraging activity. *Int. J. Primatol.* 18:975–93.

Boinski, S., E. Moraes, D. G. Kleiman, J. M. Dietz, and A. J. Baker. 1994. Intragroup vocal behaviour in wild golden lion tamarins, *Leontopithecus rosalia:* Honest communication of individual activity. *Behaviour* 130:53–76.

Boinski, S., and Newman, J. D. 1988. Preliminary observations on squirrel monkey *(Saimiri oerstedi)* vocalizations in Costa Rica. *Am. J. Primatol.* 14:329–43.

Borgerhoff Mulder, M. 1988. Kipsigis Bridewealth Payments. In *Human reproductive behavior,* edited by L. L. Betzig, M. Borgerhoff Mulder, and P. W. Turke, 65–82. Cambridge: Cambridge University Press.

————. 1992. *Demography of pastoralists: Preliminary data on the Datoga of Tanzania. Hum. Ecol.* 20:1–23.

Borgerhoff Mulder, M., and D. Sellen. 1994. *Pastoralist decision-making: A behavioral ecological perspective.* In *African pastoralist systems,* edited by E. Fratkin, E. Roth, and K. Galvin, 205–29. Boulder, CO: Lynne Reiner Publishers.

Borgers, A., and H. Timmermans. 1986. A model of pedestrian route choice and demand for retail facilities within inner-city shopping areas. *Geogr. Anal.* 18:115–28.

Boshoff, A. F., N. G. Palmer, G. Avery, R. A. G. Davies, and M. J. F. Jarvis. 1991. Biogeographical and topographical variation in the prey of the black eagle in the Cape Province, South Africa. *Ostrich* 62:59–72.

Boubli, J. In press. Feeding ecology of black-headed uakaris *(Cacajo melanocephalus melanocephalus)* in Pico da Neblina National Park, Brazil. *Int. J. Primatol.*

Bouskila, A., and D. T. Blumstein. 1992. Rules of thumb for predation hazard assessment: predictions from a dynamic model. *Am. Nat.* 139:161–76.

Bovet, J. 1984. Strategies of homing behavior in the red squirrel, *Tamiasciurus hudsonicus. Behav. Ecol. Sociobiol.* 16:81–88.

———. 1992. Mammals. In *Animal homing,* edited by F. Papi, 321–61. London: Chapman and Hall.

Bowen, W. D. 1981. Variation in coyote social organization: The influence of prey size. *Can. J. Zool.* 59:639–52.

———. 1982. Home range and spatial organization of coyotes in Jasper National Park, Alberta. *J. Wildl. Mgmt.* 46:201–16.

Bowyer, R. T. 1987. Coyote group size relative to predation on mule deer. *Mammalia* 51:515–26.

Boyd, R., and P. J. Richerson. 1990. Group selection among alternative evolutionarily stable strategies. *J. Theor. Biol.* 145:331–42.

———. 1992. Punishment allows the evolution of cooperation (or anything else) in sizable groups. *Ethol. Sociobiol.* 13:171–95.

Bradbury, J. W., and S. L. Vehrencamp. 1976. Social organization and foraging in emballonurid bats. II. A model for the determination of group size. *Behav. Ecol. Sociobiol.* 1:383–404.

Bradshaw, J. W. S. 1992. Behavioural biology. In *Waltham book of dog and cat behaviour,* edited by C. J. Thorne, 31–52. Oxford: Pergamon Press.

Brakke, K. E., and E. S. Savage-Rumbaugh. 1995. The development of language skills in bonobo and chimpanzee. I. Comprehension. *Lang. Communic.* 15:121–48.

———. 1996. The development of language skills in *Pan.* II. Production. *Lang. Communic.* 16:361–80.

Braslau-Shneck, S. L., and L. M. Herman. 1993. Synchronous coordination of creative behaviors by bottlenose dolphins. Abstract: Tenth Biennial Conference on the Biology of Marine Mammals, Galveston, TX.

Breder, C. M. J. 1954. Equations descriptive of fish schools and other animal aggregations. *Ecology* 35:361–70.

Bridge, B. 1951. Animal strategy. *Afr. Wildl.* 5:121–25.

Brill, R. L., M. L. Sevenich, T. J. Sullivan, J. D. Sustman, and R. E. Witt. 1988. Behavioral evidence for hearing through the lower jaw by an echolocating dolphin *(Tursiops truncatus). Mar. Mammal Sci.* 4:223.

Brizzee, K. R., and W. P. Dunlap. 1986. Growth. In *Comparative primate biology,* vol. 3, *Reproduction and development,* edited by W. R. Dukelow and J. Erwin, 363–413. New York: Alan R. Liss.

Brodie, E. D., D. R. Formanowicz, and E. D. Brodie. 1991. Predator avoidance and antipredator mechanisms: Distinct pathways to survival. *Ethol. Ecol. Evol.* 3:73–77.

Bronikowski, A. M., and J. Altmann. 1996. Foraging in a variable environment: Weather patterns and the behavioral ecology of baboons. *Behav. Ecol. Sociobiol.* 39:11–25.

Brothers, L. 1990. The social brain: A project for integrating primate behavior and neurophysiology in a new domain. *Concepts Neurosci.* 1:27–51.

Brown, C. H. 1982. Auditory localization and primate vocal behavior. In *Primate communication,* edited by C. T. Snowdon, C. H. Brown, and M. Peterson, 144–64. Cambridge: Cambridge University Press.

————. 1989. The acoustic ecology of east African primates and the perception of vocal signals by grey-cheeked mangabeys and blue monkeys. In *The comparative psychology of audition: Perceiving complex sounds,* edited by R. J. Dooling and S. H. Hulse, 201–39. Hillsdale, NJ: Lawrence Erlbaum Associates.

Brown, C. S., and C. L. Gass. 1993. Spatial association learning by hummingbirds. *Anim. Behav.* 46:487–97.

Brown, D. H., and K. S. Norris. 1956. Observations of captive and wild cetaceans. *J. Mammal.* 37 (3):311–26.

Brown, J. L. 1964. The evolution and diversity of avian territorial systems. Wilson Bull. 6:160–69.

Brown, J. L., and G. H. Orians. 1970. Spacing patterns in mobile animals. *Annu. Rev. Ecol. Syst.* 1:239–62.

Brown, L. H. 1966. Observations on some Kenya eagles. *Ibis* 108:531–72.

Bshary, R., and R. Noë. 1997. Red colobus and Diana monkeys provide mutual protection against predators. *Anim. Behav.* 54:1461–74.

Buchanan-Smith, H. M. 1990. Polyspecific association of two tamarin species, *Saguinus labiatus* and *Saguinus fuscicollis,* in Bolivia. *Am. J. Primatol.* 22:205–14.

————. 1991. A field study on the red-bellied tamarin, *Saguinus l. labiatus,* in Bolivia. *Int. J. Primatol.* 12:259–76.

Buchanan-Smith, H. M., and S. M. Hardie. 1997. Tamarin mixed-species groups: An evaluation of a combined captive and field approach. *Folia Primatol.* 68:272–86.

Budnitz, N., and K. Dainis. 1975. *Lemur catta:* Ecology and behavior. In *Lemur biology,* edited by I. Tattersall and R. W. Sussman, 219–35. New York: Plenum Press.

Bunn, H. T., and E. M. Kroll. 1986. Systematic butchery by Plio/Pleistocene hominids at Olduvai Gorge, Tanzania. *Curr. Anthropol.* 27:431–52.

Burger, J., and M. Gochfeld. 1992. Effect of group size on vigilance while drinking in the coati, *Nasua narica,* in Costa Rica. *Anim. Behav.* 44:1053–57.

Burkholder, B. L. 1959. Movements and behavior of a wolf pack in Alaska. *J. Wildl. Mgmt.* 23:1–11.

Burt, W. H. 1943. Territoriality and home range concepts as applied to mammals. *J. Mammal.* 24:346–52.

Buskirk, W. H. 1976. Social systems in a tropical forest avifauna. *Am. Nat.* 110:293–310.

Buskirk, W. H., R. E. Buskirk, and W. J. Hamilton II. 1974. Troop-mobilizing behavior of adult male chacma baboons. *Folia Primatol.* 22:9–18.

Buskirk, W. H., G. V. N. Powell, J. Wittenberger, R. E. Buskirk, and T. U. Powell. 1972. Interspecific bird flocks in tropical highland Panama. *Auk* 89:612–24.

Busse, C. D. 1978. Do chimpanzees hunt cooperatively? *Am. Nat.* 112:767–70.

———. 1980. Leopard and lion predation upon chacma baboons living in the Moremi Wildlife Reserve. *Botswana Notes and Records* 12:15–21.

———. 1984. Spatial structure of chacma baboon groups. *Int. J. Primatol.* 5:247–62.

Butynski, T. M. 1982a. Harem-male replacement and infanticide in the blue monkey *(Cercopithecus mitis stuhlmanni)* in Kibale Forest, Uganda. *Am. J. Primatol.* 3:1–22.

———. 1982b. Vertebrate predation by primates: A review of hunting patterns and prey. *J. Hum. Evol.* 11:421–30.

———. 1990. Comparative ecology of blue monkeys *(Cercopithecus mitis)* in high- and low-density subpopulations. *Ecol. Monogr.* 60:1–26.

Butynski, T. M., C. A. Chapman, L. J. Chapman, and D. M. Weary. 1992. Use of male blue monkey "pyow" calls for long-term individual identification. *Am. J. Primatol.* 28:183–89.

Byers, J. A. 1997. *American pronghorns: Social adaptations and the ghosts of predators past.* University of Chicago Press, Chicago.

Bygott, J. D., B. C. R. Bertram, and J. P. Hanby. 1979. Male lions in large coalitions gain reproductive advantages. *Nature* 282:839–41.

Byrne, R. W. 1979. Memory for urban geography. *Q. J. Exp. Psychol.* 31:147–54.

———. 1981. Distance vocalizations of Guinea baboons *(Papio papio):* An analysis of function. *Behaviour* 78:283–312.

———. 1982. Geographical knowledge and orientation. In *Normality and pathology of cognitive function,* edited by A. Ellis, 239–64. London: Academic Press.

———. 1989. Social relationships of mountain baboons: Leadership and affiliation in a non-female bonded monkey. *Am. J. Primatol.* 18:191–207.

———. 1993. Do larger brains mean greater intelligence? *Behav. Brain Sci.* 16:696–97.

Byrne, R. W. 1994. The evolution of intelligence. In *Behaviour and evolution,* edited by P. J. B. Slater and T. R. Halliday, 223–65. Cambridge: Cambridge University Press.

———. 1995a. Primate cognition: Comparing problems and skills. *Am. J. Primatol.* 37:127–41.

———. 1995b. *The thinking ape: Evolutionary origins of intelligence.* Oxford: Oxford University Press.

Byrne, R., and A. Whiten, eds. 1988. *Machiavellian intelligence: Social expertise and the evolution of intellect in monkeys, apes, and humans.* Oxford: Clarendon Press.

———. 1991. Computation and mindreading in primate tactical deception. In *Natural theories of mind,* edited by A. Whiten, 127–41. Oxford: Basil Blackwell.

———. 1992. Cognitive evolution in primates: Evidence from tactical deception. *Man* 27:609–27.

Byrne, R. W., A. Whiten, and S. P. Henzi. 1987. One-male groups and intergroup interactions of mountain baboons *(Papio ursinus). Int. J. Primatol.* 8:615–33.

———. 1990. Social relationships in mountain baboons: Leadership and affiliation in a non-female bonded monkey. *Am. J. Primatol.* 20:313–29.

Byrne, R. W., A. Whiten, S. P. Henzi, and F. M. McCulloch. 1993. Nutritional constraints on mountain baboons *(Papio ursinus):* Implications for baboon socio-ecology. *Behav. Ecol. Sociobiol.* 33:233–46.

Caine, N. G. 1986. Vigilance, vocalizations, and cryptic behaviour at retirement in captive groups of red-bellied tamarins *(Saguinus labiatus)*. *Am. J. Primatol.* 12:241–50.

———. 1993. Flexibility and cooperation as unifying themes in *Saguinus* social organization and behaviour: The role of predation pressures. In *Marmosets and tamarins: Systematics, behavior, and ecology*, edited by A. B. Rylands, 200–219. Oxford: Oxford University Press.

Caine, N. G., and C. Stevens. 1990. Evidence for a "monitoring" call in captive red-bellied tamarins. *Am. J. Primatol.* 22:251–62.

Caldwell, G. S. 1981. Attraction to tropical mixed-species heron flocks: Proximate mechanisms and consequences. *Behav. Ecol. Sociobiol.* 8:99–103.

Caldwell, M. C., and D. K. Caldwell. 1965. Individual whistle contours in dolphins *(Tursiops truncatus)*. *Nature* 207:434–35.

———. 1968. Vocalization of naive captive dolphins in small groups. *Science* 159:1121–23.

———. 1972. Vocal mimicry in the whistle mode by an Atlantic bottlenose dolphin. *Cetology* 9:1–8.

———. 1977. Cetaceans. In *How animals communicate*, edited by T. A. Sebeok, 794–808. Bloomington: Indiana University Press.

Caldwell, M. C., D. K. Caldwell, and N. R. Hall. 1973. Ability of an Atlantic bottlenosed dolphin *(Tursiops truncatus)* to discriminate between, and potentially identify to individual, the whistles of another species, the common dolphin *(Delphinus delphis)*. *Cetology* 14:1–7.

Caldwell, M. C., D. K. Caldwell, and P. L. Tyack. 1990. Review of the signature-whistle hypothesis for the Atlantic bottlenose dolphin. In *The Bottlenose dolphin*, edited by S. Leatherwood and R. R. Reeves, 199–234. San Diego, CA: Academic Press.

Camenzind, F. J. 1978. Behavioral ecology of coyotes in the National Elk Refuge, Jackson Hole, Wyoming. In *Coyotes: Biology, behavior and management*, edited by M. Bekoff, 267–94. New York: Academic Press.

Camp, G. D. 1956. A look ahead: New developments and applications. In *Operations research: A basic approach*, 99–111. American Management Association Special Report no. 13. New York: American Management Association.

Campos, F., S. de la Torre, and T. de Vries. 1992. Territorial behavior and home range establishment of *Callicebus torquatus* (Primates: Cebidae) in Amazonian Ecuador. Paper presented at the XIVth Congress of the International Primatological Society, Strasbourg.

Cannon, C. H., and M. Leighton. 1994. Comparative locomotor ecology of gibbons and macaques: Selection of canopy elements for crossing gaps. *Am. J. Phys. Anthropol.* 93:505–24.

Cant, J. G. H. 1977. Ecology, locomotion and social organization of spider monkeys *(Ateles geofffroyi)*. Ph.D. dissertation, University of California, Davis.

Capaldi, E. A., and F. C. Dyer. 1999. The role of orientation flights on homing performance in honey bees. *J. Exp. Biol.* In press.

Capot-Rey, R. 1953. *Le Sahara francais*. Paris. Presses Universitaires de France.

Caraco, T., and L. L. Wolf. 1975. Ecological determinants of group size in foraging lions. *Am. Nat.* 109:343–52.

Caro, T. M. 1976. Observations on the ranging behaviour and daily activities of lone silverback mountain gorillas *(Gorilla gorilla beringei). Anim. Behav.* 24:889–97.

———. 1994. *Cheetahs of the Serengeti plain.* Chicago: University of Chicago Press.

Carpenter, C. R. 1934. A field study of the behavior and social relations of howling monkeys. *Comp. Psychol. Monogr.* 10, 48:1–168.

———. 1964. *Naturalistic behavior of nonhuman primates.* University Park: Pennsylvania State University Press.

Carr, H., and J. B. Watson. 1908. Orientation of the white rat. *J. Comp. Neurol. Psychol.* 18:27–44.

Cartmill, M. 1974. Rethinking primate origins. *Science* 184:436–43.

Cartwright, B. A., and T. S. Collett. 1983. Landmark learning in bees. *J. Comp. Physiol.* 151:521–43.

———. 1987. Landmark maps for honeybees. *Biol. Cybern.* 57:85–93.

Carvalho, C. T., and C. F. Carvalho. 1989. A organização social dos sauís-pretos *(Leontopithecus chrysopygus* Mikan) na reserva em Teodoro Sampaio, São Paulo. *Revista Brasileira de Zoologia* 6:707–17.

Cashdan, E., ed. 1990. *Risk and uncertainty in tribal and peasant economies.* Boulder, CO: Westview Press.

Casimir, M. J. 1975. Feeding ecology and nutrition of an eastern gorilla group in the Mt. Kahuzi region (Republique du Zaire). *Folia Primatol.* 24:81–136.

Casimir, M. J., and R. E. Butenandt. 1973. Migration and core area shifting in relation to some ecological factors in a mountain gorilla group *(Gorilla gorilla beringei)* in the Mt. Kahuzi region (Republique du Zaire). *Z. Tierpsychol.* 33:514–22.

Casimir, M. J., and A. Rao, eds. 1992. *Mobility and territoriality: Social and spatial boundaries among foragers, fishers, pastoralists and peripatetics.* New York: Berg Publishers.

Castro, N. R. 1991. Behavioral ecology of two coexisting tamarin species (*Saguinus fuscicollis nigrifrons* and *Saguinus mystax mystax,* Callitrichidae, Primates) in Amazonian Peru. Ph.D. thesis, Washington University, St. Louis.

Castro, R., and P. Soini. 1977. Field studies on *Saguinus mystax* and other callitrichids in Amazonian Peru. In *The biology and conservation of the Callitrichidae,* edited by D. G. Kleiman, 73–78. Washington, DC: Smithsonian Institution Press.

Chance, M. R. A. 1955. The sociability of monkeys. *Man* 55:162–65.

———. 1967. Attention structure as a basis for primate rank orders. *Man* 2:503–18.

Chance, M. R. A., and C. J. Jolly. 1970. *Social groups of monkeys, apes and man.* London: Jonathan Cape.

Chapman, C. A. 1986. Boa constrictor predation and group response in white-faced cebus monkeys. *Biotropica* 18:171–72.

———. 1988a. Patch use and patch depletion by the spider and howling monkeys of Santa Rosa National Park, Costa Rica. *Behaviour* 105:88–116.

———. 1988b. Patterns of foraging and range use by three species of Neotropical primates. *Primates* 29:177–94.

———. 1989a. Ecological constraints on group size in three species of Neotropical primates. *Folia Primatol.* 73:1–9.

———. 1989b. Spider monkey sleeping sites: Implications for primate group structure. *Am. J. Primatol.* 18:53–60.

———. 1990a. Association patterns of spider monkeys: The influence of ecology and sex on social organization. *Behav. Ecol. Sociobiol.* 26:409–14.

———. 1990b. Ecological constraints on group size in three species of Neotropical primates. *Folia Primatol.* 55:1–9.

Chapman, C. A., and L. J. Chapman. 1990. Reproductive biology of captive and free-ranging spider monkeys. *Zoo Biol.* 9:1–10.

———. 1996. Mixed species primate groups in the Kibale Forest: Ecological constraints on association. *Int. J. Primatol.* 17:31–50.

Chapman, C. A., L. J. Chapman, and R. L. McLaughlan. 1989. Multiple central place foraging by spider monkeys: Travel consequences of using many sleeping sites. *Oecologia* 79:506–11.

Chapman, C. A., and L. M. Fedigan. 1984. Territoriality in the St. Kitts Vervet, *Cercopithecus aethiops. J. Hum. Evol.* 13:677–786.

———. 1990. Dietary differences between neighboring *Cebus capucinus* groups: Local traditions, food availability, or responses to food profitability. *Folia Primatol.* 54:177–86.

Chapman, C. A., and L. Lefebvre. 1990. Manipulating foraging group size: Spider monkey food calls at fruiting trees. *Anim. Behav.* 39:891–96.

Chapman, C. A., and D. Onderdonk. 1998. Forests without primates: Primate/plant codependency. *Am. J. Primatol.* 45:127–41.

Chapman, C. A., and D. M. Weary. 1990. Variability in spider monkey vocalizations may provide basis for individual recognition. *Am. J. Primatol.* 22:279–84.

Chapman, C. A., F. J. White, and R. W. Wrangham. 1994. Party size in chimpanzees and bonobos: A reevaluation of theory based on two similarly forested sites. In *Chimpanzee culture,* edited by R. W. Wrangham, W. C. McGrew, F. B. de Waal, and P. G. Heltne, 41–58. Cambridge, MA: Harvard University Press.

Chapman, C. A., R. W. Wrangham, and L. J. Chapman. 1995. Ecological constraints on group size: An analysis of spider monkey and chimpanzee subgroups. *Behav. Ecol. Sociobiol.* 36:59–70.

Chapuis, N. 1987. Detour and shortcut abilities in several species of mammals. In *Cognitive processes and spatial orientation in animal and man,* edited by P. Ellen and C. Thinus-Blanc, 97–106. Dordrecht: Martinus Nijhoff Publishers.

Chapuis, N., and C. Varlet. 1987. Short-cuts by dogs in natural surroundings. *Q. J. Exp. Psychol.* 39B:49–64.

Charles-Dominique, P. 1977a. *Ecology and behavior of nocturnal primates.* New York: Columbia University Press.

———. 1977b. Urine marking and territoriality in *Galago alleni* (Waterhouse, 1837, Lorisoidea, Primates): A field-study by radio-telemetry. *Z. Tierpsychol.* 43:113–48.

Charles-Dominique, P., and Hladik, M. 1971. Le lepilemur du sud de Madagascar: Ecologie, alimentation, et vie social. *Terre et Vie* 25:3–66.

Charles-Dominique, P., and J. J. Petter. 1980. Ecology and social life of *Phaner furcifer.* In *Nocturnal Malagasy primates,* edited by P. Charles-Dominique, H. M. Cooper, A. Hladik, C. M. Hladik, E. Pages, G. F. Pariente, A. Petter-Rousseaux, A. Schilling, and J. Petter, 75–96. New York: Academic Press.

Charnov, E. L. 1976. Optimal foraging: The marginal value theorum. *Theor. Popul. Biol.* 9:129–36.

Cheney, D. L. 1986. Interactions and relationships between groups. In *Primate societies,* edited by B. B. Smuts, D. L. Cheney, R. M. Seyfarth, R. W. Wrangham, and T. T. Struhsaker, 267–81. Chicago: University of Chicago Press.

———. 1992. Intragroup cohesion and intergroup hostility: The relation between grooming distributions and intergroup competition among female primates. *Behav. Ecol.* 3:334–45.

Cheney, D. L., P. C. Lee, and R. M. Seyfarth. 1981. Behavioral correlates of nonrandom mortality among free-ranging adult female vervet monkeys. *Behav. Ecol. Sociobiol.* 9:153–61.

Cheney, D. L., and R. M. Seyfarth. 1981. Selective forces affecting the predator alarm calls of vervet monkeys. *Behaviour* 76:25–61.

———. 1982. Recognition of individuals within and between groups of free-ranging vervet monkeys. *Am. Zool.* 22:519–29.

———. 1985. Vervet monkey alarm calls: Manipulation through shared information. *Behaviour* 93:150–66.

———. 1990. *How monkeys see the world: Inside the mind of another species.* Chicago: University of Chicago Press.

———. 1996. Function and intention in the calls of non-human primates. *Proc. Brit. Acad.* 88:59–76.

Cheney, D. L., R. M. Seyfarth, and R. Palombit. 1996. The function and mechanisms underlying baboon "contact" barks. *Anim. Behav.* 52:507–18.

Cheney, D. L., and R. W. Wrangham. 1986. Predation. In *Primate societies,* edited by B. B. Smuts, D. L. Cheney, R. M. Seyfarth, R. W. Wrangham, and T. T. Struhsaker, 227–239. Chicago: University of Chicago Press.

Cheverud, J. M., M. M. Dow, and W. Leutenegger. 1985. The quantitative assessment of phylogenetic constraints in comparative analyses: Sexual dimorphism in body weights among primates. *Evolution* 39:1335–51.

Chism, J., D. K. Olson, and T. E. Rowell. 1983. Diurnal births and perinatal behavior among wild patas monkeys: Evidence of an adaptive pattern. *Int. J. Primatol.* 4:167–84.

Chivers, D. J. 1969. On the daily behavior and spacing of howling monkey groups. *Folia Primatol.* 10:48–102.

———. 1974. *The siamang in Malaya: A field study of a primate in tropical rain forest.* Karger, Basel.

———. 1977. The feeding behaviour of Siamang *(Symphalangus syndactylus).* In *Primate ecology,* edited by T. H. Clutton-Brock and P. H. Harvey, 355–82. New York: Academic Press.

Christen, A., and T. Geissmann. 1994. A primate survey in northern Bolivia, with special reference to Goeldi's monkey, *Callimico goeldii. Int. J. Primatol.* 15:239–74.

Christenson, B., and L. Persson. 1993. Species-specific antipredatory behaviours: Effects on prey choice in different habitats. *Behav. Ecol. Sociobiol.* 32:1–9.

Christophides, N. 1985. Vehicle routing. In *The traveling salesman problem,* edited by E. L. Lawler, J. K. Lenstra, A. H. G. Rinnooy Kan, and D. B. Shmoys. Chichester: J. Wiley and Sons.

Clark, A. B. 1978. Olfactory communication in *Galago crassicaudatus,* and the social life of prosimians. In *Recent advances in primatology,* vol. 3, edited by D. S. Chivers and J. S. Joysey, 109–17. New York: Academic Press.

Clark, A. P., and R. W. Wrangham. 1993. Acoustic analysis of wild chimpanzee pant hoots: Do Kibale Forest chimpanzees have an acoustically distinct food arrival pant hoot? *Am. J. Primatol.* 31:99–110.

Clark, C. W. 1987. The lazy, adaptable lion: A Markovian model of group foraging. *Anim. Behav.* 35:361–68.

———. 1989. Call tracks of bowhead whales based on call characteristics as an independent means of determining tracking parameters. *Reports to the International Whaling Commission* 39:111–12.

———. 1990. Acoustic behavior of mysticete whales. In *Sensory abilities of cetaceans: Laboratory and field evidence,* edited by J. Thomas and R. A. Kastelein, 571–84. New York: Plenum Press.

Clark, C. W., C. J. Gagnon, and D. K. Mellinger. 1993. Whales '93: The application of the Navy IUSS for low-frequency marine mammal research. Abstract. Tenth Biennial Conference on the Biology of Marine Mammals, Galveston, TX.

Clark, C. W., and M. Mangel. 1984. Foraging and flocking strategies: Information in an uncertain environment. *Am. Nat.* 123:626–41.

———. 1986. The evolutionary advantages of group foraging. *Theor. Popul. Biol.* 30:45–75.

Clarke, A. S., and S. Boinski. Temperament in nonhuman primates. *Am. J. Primatol.* 37:103–26.

Clastres, P. 1972. The Guayaki. In *Hunters and gatherers today,* edited by M. G. Bicchieri, 138–74. New York: Collins.

Clayton, N. S., and A. Dickinson. 1998. Episodic-like memory during cache recovery by scrub jays. *Nature* 395:272–74.

Clifton, K. E. 1991. Subordinate group members act as food-finders within striped parrotfish territories. *J. Exp. Mar. Biol. Ecol.* 145:141–48.

Clutton-Brock, T. H. 1975. Ranging behavior of red colobus *(Colobus badius tephrosceles)* in the Gombe National Park. *Anim. Behav.* 23:706–22.

———. 1989. Mammalian mating systems. *Proc. R. Soc. Lond.* B 236:339–72.

Clutton-Brock, T. H., and S. D. Albon. 1979. The roaring of red deer and the evolution of honest advertisement. *Behaviour* 69:145–70.

Clutton-Brock, T. H., F. E. Guiness, and S. D. Albon. 1982. *The red deer: Behavior and ecology of two sexes.* Chicago: University of Chicago Press.

Clutton-Brock, T. H., and P. H. Harvey. 1977a. Primate ecology and social organization. *J. Zool.* 183:1–39.

———. 1977b. Species differences in feeding and ranging behaviour in primates. In *Primate ecology,* edited by T. H. Clutton-Brock, 557–84. London: Academic Press.

———. 1980. Primates, brains and ecology. *J. Zool.* 190:309–323.

Clutton-Brock, T. H., and P. H. Harvey. 1983. The functional significance of variation in body size among mammals. In *Advances in the study of mammalian behavior,* edited by J. F. Eisenberg and D. Kleiman, 632–59. Special Publication no. 7. Lawrence, KS: American Society of Mammalogists.

Clutton-Brock, T. H., and G. A. Parker. 1995. Punishment in animal societies. *Nature* 373:209–16.

Cody, M. L. 1971. Finch flocks in the Mojave desert. *Theor. Popul. Biol.* 2:141–58.

Coelho, A. M. Jr. 1985. Baboon dimorphism: Growth in weight, length and adiposity from birth to 8 years of age. In *Nonhuman primate models for human growth and development,* edited by E. S. Watts, 125–59. New York: Alan R. Liss.

Coimbra-Filho, A. F. 1977. Natural shelters of *Leontopithecus rosalia* and some ecological implications (Callitrichidae: Primates). In *The biology and conservation of the Callitrichidae,* edited by D. G. Kleiman, 79–89. Washington, DC: Smithsonian Institution Press.

Collett, T. S. 1995. Making learning easy: The acquisition of visual information during the orientation flights of social wasps. *J. Comp. Physiol.* A 177:737–47.

———. 1996. Insect navigation en route to the goal: Multiple strategies for the use of landmarks. *J. Exp. Biol.* 199:227–35.

Collett, T. S., and J. Baron. 1994. Biological compasses and the coordinate frame of landmark memories in honeybees. *Nature* 368:137–40.

Collett, T. S., B. A. Cartwright, and B. A. Smith. 1986. Landmark learning and visual-spatial memories in gerbils. *J. Comp. Physiol.* A 158:835–51.

Collett, T. S., E. Dilmann, A. Giger, and R. Wehner. 1992. Visual landmarks and route following in desert ants. *J. Comp. Physiol.* A 170:435–42.

Collett, T. S., and M. Lehrer. 1993. Looking and learning: A spatial pattern in the orientation flight of *Vespula vulgaris. Phil. Trans. R. Soc. Lond.* B 252:129–34.

Collins, B. E., and B. H. Raven. 1969. Group structure: Attraction, coalitions, communication and power. In *The handbook of social psychology,* edited by G. Lindzey and E. Aronson, 102–204. Reading, MA: Addison-Wesley.

Collins, D. A. 1984. Spatial pattern in a troop of yellow baboons *(Papio cynocephalus)* in Tanzania. *Anim. Behav.* 32:536–53.

Collins, N., and C. Southwick. 1952. A field study of population density and social organization in howling monkeys. *Proc. Am. Phil. Soc.* 96:143–56.

Condy, P. R., J. J. Van Aarde, and M. N. Bester. 1978. The seasonal occurrence and behavior of killer whales, *Orcinus orca,* at Marion Island. *J. Zool.* 184:449–64.

Connor, R. C., and R. Smolker. 1985. Habituated dolphins (*Tursiops* sp.) in Western Australia. *J. Mammal.* 66(2):398–400.

Connor, R. C., R. Smolker, and A. F. Richards. 1992a. Dolphin alliances and coalitions. In *Coalitions and alliances in humans and other animals,* edited by A. H. Harcourt and F. B. M. de Waal, 415–42. Oxford University Press, New York.

———. 1992b. Two levels of alliance formation among male bottlenose dolphins (*Tursiops* sp.). *Proc. Natl. Acad. Sci. USA* 89:987–90.

Cooper, H. M., M. Herbin, and E. Nevo. 1993. Ocular regression conceals adaptive progression of the visual system in a blind subterranean mammal. *Nature* 361:156–59.

Cooper, S. M. 1990. The hunting behaviour of spotted hyaenas *(Crocuta crocuta)* in a region containing both sedentary and migratory populations of herbivores. *Afr. J. Ecol.* 28:131–41.

Corbett, L. K. 1995. *The dingo in Australia and Asia.* Ithaca, NY: Cornell University Press.

Cords, M. 1986a. Forest guenons and patas monkeys: Male-male competition in one male groups. In *Primate societies,* edited by B. B. Smuts, D. L. Cheney,

R. M. Seyfarth, R. W. Wrangham, and T. T. Struhsaker, 98–111. Chicago: University of Chicago Press.

———. 1986b. Interspecific and intraspecific variation in diet of two forest guenons, *Cercopithecus ascanius* and *C. mitis. J. Anim. Ecol.* 55:811–27.

———. 1987. Mixed-species associations of *Cercopithecus* monkeys in the Kakamega Forest, Kenya. *Univ. Calif. Publ. Zool.* 117:1–109.

———. 1990a. Mixed-species association of East African guenons: General patterns or specific examples? *Am. J. Primatol.* 21:101–14.

———. 1990b. Vigilance and mixed-species associations of some East African forest monkeys. *Behav. Ecol. Sociobiol.* 26:297–300.

Cords, M., and Killen, M. 1998. Conflict resolution in human and nonhuman primates. In *Piaget, evolution, and development,* edited by J. Langer and M. Killen, 193–219. Mahwah, NJ: Lawrence Erlbaum Associates.

Cotes, J. E., and F. Meade. 1960. The energy expenditure and mechanical energy demand in walking. *Ergonomics* 3:97–120.

Coughenour, M. B., D. L. Coppock, and J. E. Ellis. 1990. Herbaceous forage variability in an arid pastoral region of Kenya: Importance of topographic and rainfall gradients. *J. Arid Environ.* 19:147–59.

Cowlishaw, G. 1992. Song function in gibbons. *Behaviour* 121:131–53.

———. 1994. Vulnerability to predation in baboon populations. *Behaviour* 131:293–304.

———. 1996. Sexual selection and information content in gibbon song bouts. *Ethology* 102:272–84.

———. 1997. Trade-offs between foraging and predation risk determine habitat use in a desert baboon population. *Anim. Behav.* 53:667–86.

Cramer, A. E. 1995. Computation on metric cognitive maps: How vervet monkeys solve the traveling salesman problem. Ph.D. thesis, University of California, Los Angeles.

Crandlemire-Sacco, J. L. 1986. The ecology of the saddle-back tamarin, *Saguinus fuscicollis,* of southeastern Peru. Ph.D. thesis, University of Pittsburgh.

———. 1988. An ecological comparison of two sympatric primates: *Saguinus fuscicollis* and *Callicebus moloch* of Amazonian Peru. *Primates* 29:465–75.

Cranford, T. W. 1988. The anatomy of acoustic structures in the spinner dolphin forehead as shown by X-ray computer tomography and computer graphics. In *Animal sonar: Processes and performance,* edited by P. E. Nachtigall and P. W. B. Moore, 67–78. New York: Plenum Press.

Creel, S. R. 1996. Behavioral endocrinology and social organization in dwarf mongooses. In *Carnivore behavior, ecology, and evolution,* edited by J. L. Gittleman, 46–77. Ithaca, NY: Cornell University Press.

———. 1997. Cooperative hunting and group size: Assumptions and currencies. *Anim. Behav.* 54:1319–24.

Creel, S. R., and N. M. Creel. 1991. Energetics, reproductive suppression, and obligate communal breeding in carnivores. *Behav. Ecol. Sociobiol.* 28:263–70.

———. 1995. Communal hunting and pack size in African wild dogs, *Lycaon pictus. Anim. Behav.* 50:1325–39.

Creel, S. R., S. L. Monfort, D. E. Wildt, and P. M. Waser. 1992. Behavioral and endocrine mechanisms of reproductive suppression in Serengeti dwarf mongooses. *Anim. Behav.* 43:231–45.

Creel, S. R., and P. M. Waser. 1994. Inclusive fitness and reproductive strategies in dwarf mongooses. *Behav. Ecol.* 5:339–48.

Crimprich, D. A., and T. C. Grubb. 1994. Consequences for Carolina Chickadees of foraging with Tufted Titmice in winter. *Ecology* 75:1615–25.

Crockett, C. M., and J. F. Eisenberg. 1986. Howlers: Variation in group size and demography. In Primate societies, edited by B. B. Smuts, D. L. Cheney, R. M. Seyfarth, R. W. Wrangham, and T. T. Struhsaker, 54–68. Chicago: University of Chicago Press.

Cronk, L. 1989. From hunters to herders: Subsistence change as a reproductive strategy among the Mukogodo. *Curr. Anthropol.* 30(2):224–34.

———. 1993. Wealth, status, and reproductive success among the Mukogodo of Kenya. *Am. Anthropol.* 93:345–58.

———. 1995. Is there a role for culture in human behavioral ecology. *Ethol. Sociobiol.* 16:185–205.

Crook, J. H., and J. S. Gartlan. 1996. Evolution of primate societies. *Nature* 210:1200–1203.

Croxall, J. P. 1976. The composition and behavior of some mixed species bird flocks in Sarawak. *Ibis* 118:333–46.

Croze, H., and G. Michael. 1981. A methodology for the inventory and monitoring of pastoral ecosystem processes. In *The future of pastoral peoples,* edited by J. Galaty, D. Aronson, P. Salzman, and A. Chouinard, 340–52. Ottawa: International Development Research Center.

Csermely, D. 1996. Antipredator behavior in lemurs: Evidence of an extinct eagle on Madagascar or something else? *Int. J. Primatol.* 17:349–54.

Curio, E. 1978. The adaptive significance of avian mobbing. *Z. Tierpsychol.* 48:175–83.

Curry, B. E., M. Milinkovitch, J. Smith, and A. E. Dizon. 1995. Stock structure of bottlenose dolphins, *Tursiops truncatus.* Abstract. Eleventh Biennial Conference on the Biology of Marine Mammals, Orlando, FL.

Curtis, D. J. 1997. The mongoose lemur *(Eulemur mongoz)*: A study in behavior and ecology. Ph.D. thesis, University of Zürich.

Daniels, T. J., and M. Bekoff. 1989a. Population and social biology of free-ranging dogs, *Canis familiaris. J. Mammal.* 70:754–62.

———. 1989b. Spatial and temporal resource use by feral and abandoned dogs. *Ethology* 81:300–312.

Darling, J. D., K. M. Gibson, and G. K. Silber. 1983. Observations on the abundance and behavior of humpback whales *(Megaptera novaeangliae)* off West Maui, Hawaii. In *Communication and behavior of whales,* edited by R. D. Payne, 201–21. Boulder, CO: Westview Press.

Darling, J. D., C. Nicklin, K. S. Norris, H. Whitehead, and B. Würsig. 1995. *Whales, dolphins and porpoises.* Washington, DC: National Geographic Society.

Darwin, C. 1859. *On the origin of species.* London: John Murray.

———. 1871. *The descent of man and selection in relation to sex.* London: John Murray.

Dasilva, G. L. 1992. The western black-and-white colobus monkey as a low-energy strategist: Activity budgets, energy expenditure and energy intake. *J. Anim. Ecol.* 61:79–91.

Davidar, E. R. C. 1975. Ecology and behavior of the dhole or Indian wild dog

Cuon alpinus (Pallas). In *The wild canids,* edited by M. W. Fox, 109–19. New York: Van Nostrand Reinhold.

Davies, N. B. 1991. Mating systems. In *Behavioural ecology,* 3d ed., edited by J. R. Krebs and N. B. Davies, eds. 263–94. Oxford: Blackwell.

Davies, N. B., and A. I. Houston. 1981. Owners and satellites: The economics of territorial defense in the pied wagtail, *Motacilla alba. J. Anim. Ecol.* 50:157–80.

———. 1984. Territory economics. In *Behavioural ecology: An evolutionary approach,* 2d ed., edited by J. R. Krebs and N. B. Davies, 148–69. Oxford: Blackwell Scientific Publications.

Davis, D. E. 1946. A seasonal analysis of mixed flocks of birds in Brazil. *Ecology* 27:168–81.

Dawkins, R. 1976. *The selfish gene.* Oxford: Oxford University Press.

———. 1982. *The extended phenotype.* Oxford: Oxford University Press.

Dawson, G. A. 1979. The use of time and space by the Panamanian Tamarin, *Saguinus oedipus. Folia Primatol.* 31:253–84.

Dawson, T. J., and C. R. Taylor. 1973. Energetic cost of locomotion in kangaroos. *Nature* 246:313–14.

Dawson, W. W. 1980. The cetacean eye. In *Cetacean behavior: Mechanisms and functions,* edited by L. M. Herman, 53–100. New York: J. Wiley and Sons.

Deacon, T. W. 1990. Fallacies of progression in theories of brain size evolution. *Int. J. Primatol.* 11:193–236.

Deaton, A., and J. Muellbauer. 1980. *Economics and consumer behavior.* Cambridge: Cambridge University Press.

DeBoer, W. F., and H. H. Prins. 1989. Decisions of cattle herdsmen in Burkina Faso and optimal foraging models. *Hum. Ecol.* 17(4):445–64.

Defler, T. R. 1983. Some population characteristics of *Callicebus torquatus lugens* (Humbolt, 1812) (Primates: Cebidae) in eastern Colombia. *Lozania (Acta Zoologica Colombiana)* 38:1–9.

de la Torre, S., F. Campos, and T. de Vries. 1995. Home range and birth seasonality of *Saguinus nigricollis graellsi* in Ecuadorian Amazonia. *Am. J. Primatol.* 37:39–56.

DeLillo, C., E. Visalberghi, and M. Aversano. 1997. The organization of exhaustive searches in a "patchy" space by capuchin monkeys *(Cebus apella). J. Comp. Psychol.* 111:82–90.

Dennett, D.C. 1983. Intentional systems in cognitive ethology: The "Panglossian paradigm" defended. *Behav. Brain Sci.* 6:343–55.

Descola, P. 1994. *In the society of nature: A native ecology of Amazonia.* Cambridge: Cambridge University Press.

Descola, P., and G. Palsson. 1996. Introduction. In *Nature and society: Anthropological perspectives,* 1–21. New York: Routledge.

Devoogd, T. J., J. R. Krebs, S. D. Healy, and A. Purvis. 1993. Relations between song repertoire size and the volume of brain nuclei related to song: Comparative evolutionary analyses amongst oscine birds. *Proc. R. Soc. Lond.* B 254:75–82.

DeVore, I. ed. 1965. *Primate behavior: Field studies of monkeys and apes.* New York: Holt, Rinehart & Winston.

DeVore, I., and K. R. L. Hall. 1965. Baboon ecology and human evolution. In *Primate behavior: Field studies of monkeys and apes,* edited by I. DeVore, 20–52. New York: Holt, Rinehart & Winston.

DeVore, I., and S. L. Washburn. 1963. Baboon ecology and human evolution. In *African ecology and human evolution,* edited by F. C. Howell and F. Bourliere, 335–67. Chicago: Aldine.

Diamond, J. 1987. Flocks of brown and black New Guinea birds: A bicoloured mixed species foraging association. *Emu* 87:201–11.

Diamond, J., and J. Terborgh. 1967. Observation of bird distribution and feeding assemblages along the Rio Galleria, Department of Loreta, Peru. *Wilson Bull.* 79:273–82.

Dickinson, J. A. 1994. Bees link local landmarks with celestial compass cues. *Naturwissenschaften* 81:465–67.

———. 1996. How do honey bees learn the sun's course? Alternative representations. Ph.D. dissertation, Michigan State University.

Dickinson, J. A., and F. C. Dyer. 1996. How insects learn about the sun's course: Alternative modeling approaches. In *From animals to animats 4: Proceedings of the Fourth International Conference on Simulation of Adaptive Behavior,* edited by P. Maes, M. Mataric, J.-A. Meyer, J. Pollack, and S. W. Wilson. Cambridge, MA: MIT Press.

Dietz, J. M., and A. J. Baker. 1993. Polygny and female reproductive success in golden lion tamarins *(Leontopithecus rosalia). Anim. Behav.* 46:1067–78.

Dietz, J. M., C. A. Peres, and L. Pinder. 1997. Foraging ecology and use of space in wild golden lion tamarins *(Leontopithecus rosalia). Am. J. Primatol.* 41:289–306.

Dietz, J. M., S. N. F. Sousa, and J. R. Silva. 1994. Population structure and territory size in golden-headed lion tamarins, *Leontopithecus chrysomelas. Neotropical Primates* 2(suppl.):21–23.

Digby, L. J., and C. E. Barreto. 1996. Activity patterns in common marmosets *(Callithrix jacchus)*: Implications for reproductive strategies. In *Adaptive radiations in Neotropical primates,* edited by M. Norconk, A. Rosenberger, and P. Garber, 173–85. New York: Plenum Press.

Dill, D. B. 1965. Oxygen used in horizontal and grade walking and running on the treadmill. *J. Appl. Physiol.* 20:19–22.

Dittus, W. P. J. 1979. The evolution of behaviour regulating density and age-specific sex ratios in a primate population. *Behaviour* 69:265–302.

———. 1984. Toque macaque food calls: Semantic communication concerning food distribution in the environment. *Anim. Behav.* 32:470–77.

Dolins, F. L. 1993. Spatial relational learning and foraging in cotton-top tamarins. Ph.D. thesis, University of Stirling, Scotland.

Doran, D. M., and A. J. McNeilage. 1998. Gorilla ecology and behavior. *Evol. Anthropol.* 6:120–31.

Drury, H. A., D.C. van Essen, C. H. Anderson, C. W. Lee, T. A. Coogan, and J. W. Lewis. 1996. Computerized mappings of the cerebral cortex: A multiresolution flattening method and a surface-based coordinate system. *J. Cognitive Neurosci.* 8:1–28.

Dugatkin, L. A., and H. K. Reeve. 1994. Behavioral ecology and levels of selection: Dissolving the group selection controversy. *Adv. Stud. Behav.* 23:101–33.

Dumbacher, J. P., B. M. Beehler, T. F. Spande, H. F. Garraffo, and J. W. Daly. 1992. Homobatrachotoxin in the genus *Pitohui:* Chemical defense in birds. *Science* 258:799–801.

Dunbar, R. I. M. 1983. Structure of gelada baboon reproductive units. IV. Integration at group level. *Z. Tierpsychol.* 63:265–82.

———. 1988. *Primate social systems.* Ithaca: Cornell University Press.

———. 1992. Neocortex size as a constraint on group size in primates. *J. Hum. Evol.* 20:469–93.

———. 1993. Coevolution of neocortical size, group size and language in humans. *Behav. Brain Sci.* 16:681–735.

———. 1995. Neocortex size and group size in primates: A test of the hypothesis. *J. Hum. Evol.* 28:287–96.

———. 1997. The monkeys, defense alliance. *Nature* 386:555–57.

Dunbar, R. I. M., and J. Bever. 1998. Neocortex size predicts group size in insectivores and some carnivores. *Ethology* 104:695–708.

Dunbar, R. I. M., and P. Dunbar. 1988. Maternal time budgets of gelada baboons. *Anim. Behav.* 36:970–80.

Duplaix, N. 1980. Observations on the ecology and behavior of the giant river otter *Pteronura brasiliensis* in Surinam. *Revue d'ecologies: La terre et la vie* 34:495–620.

D'Vincent, C. G., R. M., Nilson, and R. E. Hanna. 1985. Vocalizations and coordinated feeding behavior of the humpback whale in southeastern Alaska. *Sci. Rep. Whales Res. Inst.* 36:41–48.

Dyer, F. C. 1985. Nocturnal orientation by the Asian honey bee, *Apis dorsata. Anim. Behav.* 33:769–74.

———. 1987. Memory and sun compensation in honey bees. *J. Comp. Physiol.* A 160:621–33.

———. 1991. Bees acquire route-based memories but not cognitive maps in a familiar landscape. *Anim. Behav.* 41:239–46.

———. 1993. How honey bees find familiar feeding sites after changing nesting sites with a swarm. *Anim. Behav.* 46:813–16.

———. 1994. Spatial cognition and navigation in insects. In *Behavioral mechanisms in evolutionary ecology,* edited by L. A. Real, 66–98. Chicago: University of Chicago Press.

———. 1996. Spatial memory and navigation by honeybees on the scale of the foraging range. *J. Exp. Biol.* 199:147–54.

———. 1998. Cognitive ecology of navigation. In *Cognitive ecology,* edited by R. Dukas. Chicago: University of Chicago Press.

Dyer, F. C., N. A. Berry, and A. S. Richard. 1993. Honey bee spatial memory: Use of route-based memories after displacement. *Anim. Behav.* 45:1028–30.

Dyer, F. C., and J. A. Dickinson. 1994. Development of sun compensation by honeybees: How partially experienced bees estimate the sun's course. *Proc. Natl. Acad. Sci. USA* 91:4471–74.

———. 1996. Sun-compass learning in insects: Representation in a simple mind. *Curr. Dir. Psychol. Sci.* 5:67–72.

Dyer, F. C., and J. L. Gould. 1981. Honey bee orientation: A backup system for cloudy days. *Science* 214:1041–42.

Dyer, F. C., and T. D. Seeley. 1991. Dance dialects and foraging range in three Asian honey bee species. *Behav. Ecol. Sociobiol.* 28:227–33.

———. 1994. Colony migration in the tropical honey bee *Apis dorsata* F. (Hymenoptera: Apidae). *Ins. Soc.* 41:129–40.

Dyson-Hudson, N. 1966. *Karimojong politics.* Oxford: Clarendon Press.

―――. 1972. The study of nomads. In *Perspectives on nomadism,* edited by W. Irons and N. Dyson-Hudson. Leiden: E. J. Brill.

Dyson-Hudson, N., and R. Dyson-Hudson. 1969. Subsistence herding in Uganda. *Sci. Am.* 220:(2):76–89.

Dyson-Hudson, R., and J. T. McCabe. 1985. *South Turkana nomadism: Coping with an unpredictably varying environment.* Ethnography Series FL17–001. New Haven, CT: HRAFlex Books.

Easley, S. P. 1982. Ecology and behavior of *Callicebus torquatus:* Cebidae, Primates. Ph.D. thesis, Washington University, St. Louis, MO.

Easley, S. P., and W. G. Kinkey. 1986. Territorial shift in the yellow-handed titi monkey *(Callicebus torquatus). Am. J. Primatol.* 11:307–18.

Eason, P. 1989. Harpy eagle attempts predation on adult howler monkey. *Condor* 91:469–70.

East, M. L., and H. Hofer. 1991a. Loud calling in a female-dominated mammalian society: II. Behavioural contexts and functions of whooping of spotted hyaenas, *Crocuta crocuta. Anim. Behav.* 42:651–70.

―――. 1991b. Loud calling in a female-dominated mammalian society: I. Structure and composition of whooping bouts of spotted hyaenas, *Crocuta crocuta. Anim. Behav.* 42:637–50.

East, M. L., H. Hofer, and A. Turk. 1989. Functions of birth dens in spotted hyaenas *(Crocuta crocuta). J. Zool.* 219:690–97.

East, M. L., H. Hofer, and W. Wickler. 1993. The erect "penis" is a flag of submission in a female-dominated society: Greetings in Serengeti spotted hyenas. *Behav. Ecol. Sociobiol.* 33:355–70.

Eaton, R. L. 1969. Co-operative hunting by cheetah and jackal and a theory of domestication of the dog. *Mammalia* 33:87–92.

Eckert, R. 1988. *Animal physiology: Mechanisms and adaptations.* 3d ed. New York: W. H. Freeman.

Economos, A. C. 1980. Brain-life span conjecture: A re-evaluation of the evidence. *Gerontology* 26:82–89.

Eggeling, W. J. 1947. Observations on the ecology of the Budongo rain forest, Uganda. *J. Ecol.* 34:20–87.

Egler, S. G. 1992. Feeding ecology of *Saguinus bicolor bicolor* (Callitrichidae: Primates) in a relict forest in Manaus, Brazilian Amazonia. *Folia Primatol.* 59:61–76.

Ehrlich, P., and Erlich, A. 1973. Coevolution: Heterotypic schooling in Caribbean reef fish. *Am. Nat.* 107:157–60.

Eichinger, J., and D. J. Moriarty. 1985. Movement of Mojave Desert sparrow flocks. *Wilson Bull.* 97:511–16.

Eisenberg, J. F. 1966. The social organization of mammals. *Handbook of Zoology* VIII (1017), leiferung 39, 92 pages.

―――. 1973. Mammalian social systems: Are primate social systems unique? In *Precultural primate behavior,* edited by E. W. Menzel Jr., 232–49. Basel: S. Karger.

―――. 1977. The evolution of the reproductive unit in the Class Mammalia. In *Reproductive behavior and evolution,* edited by J. S. Rosenblatt and B. R. Komisaruk, 39–71. New York: Plenum.

————. 1981. *The mammalian radiations.* Chicago: University of Chicago Press.

Eisenberg, J. F., and R. Kuehn. 1976. The behavior of *Ateles geoffroyi* and related species. Smithsonian Miscellaneous Collections 151 (8):1–63.

Eisenberg, J. F., N. Muckenhirn, and R. Rudran. 1972. The relationship between ecology and social structure in primates. *Science* 176:863–74.

Eisenberg, J. F., and D. E. Wilson. 1978. Relative brain size and feeding strategies in the Chiroptera. *Am. Nat.* 32:740–51.

Ekman, J. 1979. Coherence, composition and territories of winter social groups of the Willow Tit *Parus montanus* and the Crested Tit *P. cristatus. Ornis Scand.* 10:56–68.

Ekvall, R. 1968. *Fields on the hoof: Nexus of Tibetan nomadic pastoralism.* New York: Holt, Rinehart & Winston.

Elam, Y. 1979. Nomadism in Ankole as a substitute for rebellion. *Africa* 49(2):147–58.

Elgar, M. A. 1986. House sparrows establish foraging flocks by giving chirrup calls if the resources are divisible. *Anim. Behav.* 34:169–74.

Elgar, M. A., and P. H. Harvey. 1987. Basal metabolic rates in mammals: Allometry, phylogeny and ecology. *Funct. Ecol.* 1:25–36.

Ellefson, J. O. 1968. Territorial behavior in the common white-handed gibbon, *Hylobates lar* Linn. In *Primates: Studies in adaptation and variability,* edited by P. C. Jay, 180–99. New York: Holt, Rinehart & Winston.

Ellen, R. 1982. *Environment, subsistence and system: The ecology of small-scale social formations.* Cambridge: Cambridge University Press.

Ellis, J., K. Galvin, J. T. McCabe, and D. Swift. 1987. Pastoralism and drought in Turkana District: Kenya. Report submitted to the Norwegian Agency for International Development, Oslo and Nairobi.

Ellis, J., and D. Swift. 1988. Stability of African pastoral ecosystems: Alternate paradigms and implications for development. *J. Range Mgmt.* 41:450–59.

Ellison, W. T., C. W. Clark, and G. C. Bishop. 1987. Potential use of surface reverberation by bowhead whales, *Balaena mysticetus,* in under-ice navigation: Preliminary considerations. Reports to the International Whaling Commission, 37.

Emlen, J. M. 1966. The role of time and energy in food preferences. *Am. Nat.* 100:611–17.

Emlen, S. T., and L. W. Oring. 1977. Ecology, sexual selection and the evolution of mating systems. *Science* 197:215–23.

Emmons, L. J. 1987. Comparative feeding ecology of felids in a Neotropical rainforest. *Behav. Ecol. Sociobiol.* 20:271–83.

Endler, J. A. 1985. *Natural selection in the wild.* Princeton Monographs in Population Biology no. 35. Princeton, NJ: Princeton University Press.

————. 1991. Interactions between predators and prey. In *Behavioural ecology: An evolutionary approach,* 3d ed., edited by J. R. Krebs and N. B. Davies, 169–96. Oxford: Blackwell.

Enquist, M., and O. Leiman. 1993. The evolution of cooperation in mobile organisms. *Anim. Behav.* 45:747–57.

Essen, D.C. van, C. H. Anderson, and D. J. Felleman. 1992. Information processing in the primate visual system: An integrated systems perspective. *Science* 255:419–23.

Estes, R. D., and J. Goddard. 1967. Prey selection and hunting behavior of the African wild dog. *J. Wildl. Mgmt.* 32:52–76.

Estrada, A., and R. Coates-Estrada. 1986. Frugivory by howling monkeys *(Alouatta palliata)* at Los Tuxlas, Mexico: Dispersal and fate of seeds. In *Frugivores and seed dispersal,* edited by A. Estrada and T. H. Fleming, 93–104. The Hague: W. Junk.

Etienne, A. S., R. Maurer, and F. Saucy. 1988. Limitations in the assessment of path dependent information. *Behaviour* 106:81–111.

Etienne, A. S., R. Maurer, and V. Séguinot. 1996. Path integration in mammals and its interactions with visual landmarks. *J. Exp. Biol.* 199:201–9.

Ettema, G. J. C., P. A. Huijing, G. J. Van Ingen Schenau, and A. De Hann. 1990. Effects of prestretch at the onset of stimulation on mechanical work output of rat medial gastrocnemius muscle-tendon complex. *J. Exp. Biol.* 152:333–51.

Evans-Pritchard, E. E. 1940. *The Nuer: A description of the modes of livelihood and political institutions of a Nilotic people.* Oxford: Oxford University Press.

Ewer, R. F. 1973. *The carnivores.* London: Weidenfield and Nicholson.

Fanshawe, J. H., and C. D. Fitzgibbon. 1993. Factors influencing the hunting success of an African wild dog pack. *Anim. Behav.* 45:479–90.

Fantino, E., and N. Abarca. 1985. Choice, optimal foraging, and delay-reduction hypothesis. *Behav. Brain Sci.* 8;315–30.

Farabaugh, S. M., A. Linzenbold, and R. J. Dooling 1994. Vocal plasticity in budgerigars *Melospittacus undulatus:* Evidence for social factors in the copying of contact calls. *J. Comp. Psychol.* 108:81–92.

Faria, D. S. 1989. O estudo de campo com o mico-estrela do planalto central brasileiro. In *Etologia de animais e de homens,* edited by C. Ades, 109–21. São Paulo: EDICON/EDUSP.

Farley, C. T., J. Glasheen, and T. A. McMahon. 1993. Running springs: Speed and animal size. *J. Exp. Biol.* 185:71–86.

Fay, J. M., R. Carroll, J. Peterhans, and D. Harris. 1995. Leopard attack on and consumption of gorillas in the Central African Republic. *J. Hum. Evol.* 29:93–99.

Fedak, M. A., L. Rome, and H. J. Seeherman. 1981. One-step N2-dilution technique for calibrating open-circuit Vo2 measuring systems. *J. Appl. Physiol: Respir. Environ. Exercise Physiol.* 51(3):772–76.

Fedigan, L. M. 1990. Vertebrate predation in *Cebus capucinus:* Meat-eating in a Neotropical monkey. *Folia Primatol.* 54:196–205.

———. 1993. Sex differences and intersexual relations in adult white-faced capuchins, *Cebus capucinus. Int. J. Primatol.* 14:853–77.

Fedigan, L. M., L. M. Rose, and R. M. Avila. 1996. See how they grow: Tracking capuchin monkeys *(Cebus capucinus)* populations in a regenerating Costa Rican dry forest. In *Adaptive radiation of Neotropical primates,* edited by M. Norconk, A. L. Rosenberger, and P. A. Garber, 289–307. New York: Plenum Press.

Felleman, F. L., J. R. Heimlich-Boran, and R. W. Osborne. 1991. The feeding ecology of killer whales *(Orcinus orca)* in the Pacific Northwest. In *Dolphin societies,* edited by K. Pryor and K. S. Norris, 113–48. University of California Press, Los Angeles.

Felsenstein, J. 1985. Phylogenies and the comparative method. *Am. Nat.* 125:1–15.

Ferguson, S. H., A. T. Bergerud, and R. Ferguson. 1988. Predation risk and habitat selection in the persistence of a remnant caribou population. *Oecologia* 76:236–45.

Ferrari, S. F. 1988. The behaviour and ecology of the buffy-headed marmoset, *Callithrix flaviceps* (O. Thomas, 1903). Ph.D. thesis, University College London.

Ferrari, S. F., H. K. M. Corrêa, and P. E. G. Coutinho. 1996. Ecology of the "southern" marmosets (*Callithrix aurita* and *Callithrix flaviceps*). In *Adaptive radiations in Neotropical primates,* edited by M. Norconk, A. Rosenberger, and P. Garber, 157–71. New York: Plenum Press.

Ferrari, S. F., and M. A. Lopes-Ferrari. 1989. A reevaluation of the social organisation of the Callitrichidae, with reference to the ecological differences between genera. *Folia Primatol.* 52:132–47.

———. 1990. Predator avoidance behaviour in the buffy-headed marmoset, *Callithrix flaviceps. Primates* 31:323–38.

Finlay, B. L., and R. B. Darlington. 1995. Linked regularities in the development and evolution of mammalian brains. *Science* 268:1578–84.

Fisher, R. A. 1958. *The genetical theory of natural selection.* 2d rev. ed. New York: Dover.

Fittinghoff, N. A., and D. G. Lindburg. 1980. Riverine refuging in East Bornean *Macaca fascicularis.* In *The macaques: Studies in ecology, behavior, and evolution,* edited by D. G. Lindburg, 182–214. New York: Van Nostrand Reinhold.

Fitzgibbon, C. D. 1990. Mixed species grouping in Thomson's and Grant's gazelles: The antipredator benefits. *Anim. Behav.* 39:1116–26.

Fleming, T. H. 1988. *The short-tailed fruit bat: A study in plant-animal interactions.* Chicago: University of Chicago Press.

Fleming, T. H., E. R. Heithaus, and W. B. Sawyer. 1977. An experimental analysis of the food location behavior of frugivorous bats. *Ecology* 58:619–27.

Fleury, M. C., and A. Gautier-Hion. 1997. Better to live with allogenerics than to live alone? The case of single male *Cercopithecus pogonias* in troops of *Colobus satanas. Int. J. Primatol.* 18:967–74.

Fodor, J. A. 1983. *The modularity of mind.* Cambridge, MA: MIT Press.

Foley, R., and P. C. Lee. 1989. Finite social space, evolutionary pathways, and reconstructing hominid behavior. *Science* 243:901–6.

Fonseca, G. A. B., and T. Lacher Jr. 1984. Exudate-feeding by *Callithrix jacchus penicillata* in semi-deciduous woodland (cerrado) in central Brazil. *Primates* 25:441–50.

Ford, J. K. B. 1989. Acoustic behaviour of resident killer whales *(Orcinus orca)* off Vancouver Island, British Columbia. *Can. J. Zool.* 67:727–45.

———. 1991. Vocal traditions among resident killer whales *(Orcinus orca)* in coastal waters of British Columbia. *Can. J. Zool.* 69:1454–83.

Fossey, D. 1974. Observations on the home range of one group of mountain gorillas *(Gorilla gorilla beringei). Anim. Behav.* 22:568–81.

Fossey, D., and A. H. Harcourt. 1977. Feeding ecology of free-ranging mountain gorilla *(Gorilla gorilla beringei).* In *Primate ecology,* edited by T. H. Clutton-Brock, 415–77. London: Academic Press.

Fox, M. W. 1972. *The behaviour of wolves, dogs, and related canids.* New York: Harper and Row.

———. 1975. Evolution of social behavior in canids. In *The wild canids,* edited by M. W. Fox, 429–60. New York: Van Nostrand Reinhold.

Fragaszy, D. M. 1980. Comparative studies of squirrel monkeys *(Saimiri)* and titi monkeys *(Callicebus)* in travel tasks. *Z. Tierpsychol.* 54:1–36.

Fragaszy, D. M., and S. Boinski. 1995. Patterns of individual choice and efficiency of foraging and diet in the wedge-capped capuchin, *Cebus olivaceus. J. Comp. Psychol.* 109:339–48.

Fragaszy, D. M., and E. Visalberghi. 1989. Social influences on the acquisition of tool-using behaviors in tufted capuchin monkeys *(Cebus apella). J. Comp. Psychol.* 103:159–70.

Frame, L. H., J. R. Malcolm, G. W. Frame, and H. van Lawick. 1979. Social organization of African wild dogs *(Lycaon pictus)* on the Serengeti plains, Tanzania, 1967–78. *Z. Tierpsychol.* 50:225–49.

Frank, L. G. 1986. Social organisation of the spotted hyaena *(Crocuta crocuta).* II. Dominance and reproduction. *Anim. Behav.* 35:1510–27.

Frank, L. G., K. E. Holekamp, and L. Smale. 1995. Dominance, demography, and reproductive success of female spotted hyenas. In *Serengeti II: Dynamics, management, and conservation of an ecosystem,* edited by A. R. E. Sinclair and P. Arcese, 364–84. Chicago: University of Chicago Press.

Frankel, A. S., C. W. Clark, L. M. Herman, T. R. Freeman, C. M. Gabriele, and M. A. Hoffhines. 1991. The spacing function of humpback whale song. Abstract. Ninth Biennial Conference on the Biology of Marine Mammals, Chicago, Illinois.

Franks, N. R. 1985. Reproduction, foraging efficiency and worker polymorphism in army ants. In *Experimental behavioral ecology and sociobiology,* edited by B. Hölldobler and M. Lindauer, 91–107. Stuttgart: Gustav Fischer Verlag.

———. 1989. Army ants: A collective intelligence. *Am. Sci.* 77:138–45.

Franks, N. R., and C. R. Fletcher. 1983. Spatial patterns in army ant foraging and migration: *Eciton burchelli* on Barro Colorado Island, Panama. *Behav. Ecol. Sociobiol.* 12:261–70.

Freeland, W. J., and D. H. Janzen. 1974. Strategies of herbivory in mammals: The role of plant secondary compounds. *Am. Nat.* 108:269–89.

Freese, C. H. 1978. The behavior of white-faced capuchins *(Cebus capucinus)* at a dry season waterhole. *Primates* 19:275–86.

Freese, C. H., and J. R. Oppenheimer. 1981. The capuchin monkeys, genus *Cebus.* In *Ecology and behavior of Neotropical primates,* vol. 1, edited by A. F. Coimbra-Filho and R. A. Mittermeier, 331–90. Rio de Janeiro: Academia Brasileira de Ciências.

Frid, A. 1997. Vigilance by female Dall's sheep: Interactions between predation risk factors. *Anim. Behav.* 53:799–808.

Friedl, W. A., P. E. Nachtigall, P. W. B. Moore, N. K. W. Chun, J. E. Haun, R. W. Hall, and J. L. Richards. 1990. Taste reception in the Pacific bottlenose dolphin *(Tursiops truncatus gilli)* and the California sea lion *(Zalophus californianus).* In *Sensory abilities of cetaceans: Laboratory and field evidence,* edited by J. Thomas and R. A. Kastelein, 1447–54. New York: Plenum Press.

Friedman, J. 1974. Marxism, structuralism and vulgar materialism. *Man* 9(3):444–69.

Frisch, K. von. 1967. *The dance language and orientation of bees.* Cambridge, MA: Harvard University Press.

Frisch, K. von, and M. Lindauer. 1954. Himmel und Erde in Konkurrenz bei der Orientierung der Bienen. *Naturwissenschaften* 41:245–53.

Fryxell, J. M. 1995. Aggregation and migration by grazing ungulates in relation to resources and predators. In *Serengeti II: Dynamics, management, and conservation of an ecosystem,* edited by A. R. E. Sinclair and P. Arcese, 257–73. Chicago: University of Chicago Press.

Fryxell, J. M., J. Greever, and A. R. E. Sinclair. 1988. Why are migratory ungulates so abundant? *Am. Nat.* 131:781–98.

Fryxell, J. M., and A. R. E. Sinclair. 1988. Seasonal migration of the white eared kob in relation to resources. *Afr. J. Ecol.* 26:17–31.

Fujita, N., R. L. Klatsky, J. M. Loomis, and R. G. Golledge. 1993. The encoding-error model of pathway completion without vision. *Geogr. Anal.* 25:295–314.

Fuller, T. K., and P. W. Kat. 1990. Movements, activity and prey relationships of African wild dogs *(Lycaon pictus)* near Aitong, southwestern Kenya. *Afr. J. Ecol.* 28:330–50.

———. 1993. Hunting success of African wild dogs in southwestern Kenya. *J. Mammal.* 74:464–67.

Gadagkar, R. 1991. *Belonogaster, Mischocyttarus, Parapolybia,* and independent-founding *Ropalidia.* In *The social biology of wasps,* edited by K. G. Ross and R. W. Matthews, 149–90. Ithaca, NY: Comstock.

Gaddis, P. 1980. Mixed flocks, accipiters and antipredator behavior. *Condor* 82:348–49.

Gagnon, G. J., and C. W. Clark. 1993. The use of U.S. Navy IUSS passive sonar to monitor the movement of blue whales. Abstract, Tenth Biennial Conference on the Biology of Marine Mammals, Galveston, TX.

Galdikas, B., and C. P. Yeager. 1984. Crocodile predation on a crab-eating macaque. *Am. J. Primatol.* 6:49–51.

Galef, B. G. Jr. 1983. Utilization by Norway rats *(Rattus norvegicus)* of multiple messages concerning distant foods. *J. Comp. Psychol.* 97:364–71.

———. 1988. Imitation in animals: History, definitions and interpretation of data from the psychological laboratory. In *Social learning,* edited by T. R. Zentall and B. G. Galef, 3–28. Hillsdale, NJ: Lawrence Erlbaum Associates.

Galef, B. G. Jr., and L. L. Buckley. 1996. Use of foraging trails by Norway rats. *Anim. Behav.* 51:765–71.

Galef, B. G. Jr., R. A. Mittermeier, and R. C. Bailey. 1976. Predation by the tayra *(Eira barbara).* *J. Mammal.* 57:760–61.

Gallistel, C. R. 1989. *Animal cognition: The representation of space, time, and number. Annu. Rev. Psychol.* 40:155–89.

———. 1990. *The organization of learning.* Cambridge, MA: MIT Press.

———. 1996. Brains as symbol-processors: The case of insect navigation. In *An invitation to cognitive science,* vol. 4, *Conceptual and methodological foundations,* edited by S. Sternberg and D. Osherson. Cambridge, MA: MIT Press.

Gallistel, C. R., and A. E. Cramer. 1996. Computations on metric maps in mammals: Getting oriented and choosing a multi-destination route. *J. Exp. Biol.* 199:211–17.

Garber, P. A. 1984. Use of time and positional behavior in a Neotropical primate, *Saguinus oedipus.* In *Adaptations for foraging in nonhuman primates,* edited by P. S. Rodman and J. G. Cant, 112–33. New York: Columbia University Press.

———. 1988a. Diet, foraging patterns, and resource defense in a mixed species troop of *Saguinus mystax* and *Saguinus fuscicollis* in Amazonian Peru. *Behaviour* 105:18–34.

————. 1988b. Foraging decisions during nectar feeding by tamarin monkeys (*Saguinus mystax* and *Saguinus fuscicollis*, Callitrichidae, Primates) in Amazonian Peru. *Biotropica* 20:100–106.

————. 1989. Role of spatial memory in primate foraging patterns: *Saguinus mystax* and *Saguinus fuscicollis*. *Am. J. Primatol.* 19:203–16.

————. 1993a. Feeding ecology and behaviour of the genus *Saguinus*. In *Marmosets and tamarins: Systematics, ecology and behaviour*, edited by A. B. Rylands, 273–95. Oxford: Oxford University Press.

————. 1993b. Seasonal patterns of diet and ranging in two species of tamarin monkeys: Stability versus variability. *Int. J. Primatol.* 14:145–66.

————. 1994. Phylogenetic approach to the study of tamarin and marmoset social systems. *Am. J. Primatol.* 34:199–219.

————. 1997. One for all and breeding for one: Cooperation and competition as a tamarin reproductive strategy. *Evol. Anthropol.* 5:135–47.

Garber, P. A., and F. L. Dolins. 1996. Testing learning paradigms in the field: Evidence for use of spatial and perceptual information and rule-based foraging in wild moustached tamarins. In *Adaptive radiation of Neotropical primates*, edited by M. Norconk, A. L. Rosenberger, and P. A. Garber, 201–16. New York: Plenum Press.

Garber, P. A., F. Encarnación, L. Moya, and J. D. Pruetz. 1993. Demographic and reproductive patterns of mating in moustached tamarin monkeys *(Saguinus mystax)*: Implications for reconstructing platyrrhine mating systems. *Am. J. Primatol.* 29:235–54.

Garber, P. A., and B. Hannon. 1993. Modeling monkeys: A comparison of computer generated and naturally occurring foraging patterns in 2 species of Neotropical primates. *Int. J. Primatol.* 14:827–52.

Garber, P. A., and L. M. Paciulli. 1997. Experimental field study of spatial memory and learning in wild capuchin monkeys *(Cebus capucinus)*. *Folia Primatol.* 68:236–54.

Garber, P. A., J. D. Pruetz, and J. Isaacson. 1993. Patterns of range use, range defense, and intergroup spacing in moustached tamarin monkeys *(Saguinus mystax)*. *Primates* 34:11–25.

Garber, P. A., and M. F. Teaford. 1986. Body weights in mixed species troops of *Saguinus mystax mystax* and *Saguinus fuscicollis nigrifrons* in Amazonian Peru. *Am. J. Phys. Anthropol.* 71:331–36.

Garcia, J. E., and F. Braza. 1987. Activity rhythms and use of space of a group of *Aotus azarae* in Bolivia during the rainy season. *Primates* 28:337–42.

Garland, T. Jr. 1983. Scaling the ecological cost of transport to body mass in terrestrial mammals. *Am. Nat.* 121:571–87.

Gartlan, J. S., and C. K. Brain. 1968. Ecology and social variability in *Cercopithecus aethiops* and *C. mitis*. In *Primates*, edited by P. Jay, 253–92. New York: Holt, Rinehart & Winston.

Gartlan, J. S., and T. T. Struhsaker. 1972. Polyspecific associations and niche separation of rainforest anthropoids in Cameroon, West Africa. *J. Zool.* 168:221–26.

Gasaway, W. C., K. T. Mossestad, and P. E. Stander. 1989. Demography of spotted hyaenas in an arid savanna, Etosha National Park, South West Africa/Namibia. *Madoqua* 16:121–27.

————. 1991. Food acquisition by spotted hyaenas in Etosha National Park, Namibia: Predation versus scavenging. *Afr. J. Ecol.* 29:64–75.

Gautier, J. P., and A. Gautier-Hion. 1969. Les associations polyspecifiques chez les Cercopithecidae de Gabon. *La Terre et Vie* 2:164–201.

————. 1977. Communication in Old World monkeys. In *How animals communicate,* edited by T. Sebeok, 890–964. Bloomington: Indiana University Press.

————. 1983. Comportement vocal des males adultes et organisation supraspecifique dans les troupes polyspecifiques de cercopitheques. *Folia Primatol.* 40:161–74.

Gautier-Hion, A. 1973. Social and ecological features of talapoin monkeys: Comparisons with sympatric cercopithecines. In *Comparative ecology and behaviour of primates,* edited by R. P. Michael and J. H. Crook, 148–70. New York: Academic Press.

Gautier-Hion, A. 1988. Polyspecific associations among forest guenons: Ecological, behavioural and evolutionary aspects. In *A primate radiation: Evolutionary biology of the African guenons,* edited by A. Gautier-Hion, F. Bourlière, J. P. Gautier, and J. Kingdon, 452–76. Cambridge: Cambridge University Press.

Gautier-Hion, A., and J. P. Gautier. 1974. Les associations polyspécifiques de Cercopithèues du Plateau de M'passa (Gabon). *Folia Primatol.* 22:134–77.

Gautier-Hion, A., J P. Gautier, and R. Quris. 1981. Forest structure and fruit availability as complementary factors influencing use by a troop of monkeys *(Cercopiticus cephus).* *Revue d'ecologies: La terre et la vie* 35:511–36.

Gautier-Hion, A., and F. Maisels. 1994. Mutualism between a leguminous tree and large African monkeys as pollinators. *Behav. Ecol. Sociobiol.* 34:203–10.

Gautier-Hion, A., R. Quris, and J. P. Gautier. 1983. Monospecific vs polyspecific life: A comparative study of foraging and antipredatory tactics in a community of *Cercopithecus* monkeys. *Behav. Ecol. Sociobiol.* 12:325–35.

Gautier-Hion, A., and C. E. G. Tutin. 1988. Simultaneous attack by adult males of a polyspecific troop of monkeys against a crowned hawk eagle. *Folia Primatol.* 51:149–51.

Gebo, D. L., C. A. Chapman, L. J. Chapman, and J. Lambert. 1994. Locomotory response to predator threat in red colobus. *Primates* 35:219–23.

Geffen, E., and D. W. Macdonald. 1992. Small size and monogamy: Spatial organization of Blanford's foxes, *Vulpes cana. Anim. Behav.* 44:1123–30.

Geist, V. 1971. *Mountain sheep: A study in behavior and evolution.* Chicago: University of Chicago Press.

George, J. C., C. W. Clark, G. M. Carroll, and W. T. Ellison. 1989. Observations on the ice-breaking and ice navigation behavior of migrating bowhead whales *(Balaena mysticetus)* near Point Barrow, Alaska, spring 1985. *Arctic* 42:24–30.

Gese, E. M., D. R. Rongstad, and W. R. Mytton. 1988. Relationship between coyote group size and diet in southeastern Colorado. *J. Wildl. Mgmt.* 52:647–53.

Getty, T., and H. R. Pulliam. 1993. Search and prey detection by foraging sparrows. *Ecology* 74:734–42.

Ghiglieri, M. P. 1984. *The chimpanzees of Kibale Forest.* New York: Columbia University Press.

Ghosh, A. 1982. A model of periodic marketing. *Geogr. Anal.* 14:155–66.

Ghosh, A., and S. McLafferty. 1984. A model of consumer propensity for multipurpose shopping. *Geogr. Anal.* 16:244–49.

Gibb, J. A. 1954. Feeding ecology of tits, with notes of Treecreeper and Goldcrest. *Ibis* 96:513–43.

Gibbons, E. F. Jr, and E. W. Menzel Jr. 1980. Rank orders of arising, eating, and retiring in a family group of tamarins *(Saguinus fuscicollis)*. *Primates* 21:44–52.

Gibson, K. R. 1990. New perspectives on instincts and intelligence: Brain size and the emergence of hierarchical mental constructional skills. In *Language and intelligence in monkeys and apes,* edited by S. T. Parker and K. R. Gibson, 97–128. Cambridge: Cambridge University Press.

Gill, F. B. 1988. Trapline foraging by hermit hummingbirds: Competition for an undefended, renewable resource. *Ecology* 69:1933–42.

Giraldeau, L-A. 1984. Group foraging: The skill pool effect and frequency dependent learning. *Am. Nat.* 124:72–79.

Gish, S. 1979. A quantitative description of two-way acoustic communication between captive Atlantic bottlenosed dolphins *(Tursiops truncatus)*. Ph.D. thesis, University of California, Santa Cruz.

Gittins, S. P. 1984. Territorial advertisement and defense in gibbons. In *The lesser apes: Evolutionary and behavioral theory,* edited by H. Preuschoft, D. J. Chivers, W. Y. Brockelman, and N. Creel, 420–24. Edinburgh: Edinburgh University Press.

Gittleman, J. L. 1984. The behavioural ecology of carnivores. Ph.D. thesis, University of Sussex, Brighton, England.

———. 1986. Carnivore brain size, behavioral ecology, and phylogeny. *J. Mammal.* 67:23–36.

———. 1989. Carnivore group living: Comparative trends. In *Carnivore behavior, ecology and evolution,* edited by J. L. Gittleman, 183–207. Ithaca, NY: Cornell University Press.

Gittleman, J. L., and P. H. Harvey. 1982. Carnivore home range size, metabolic needs and ecology. *Behav. Ecol. Sociobiol.* 10:57–63.

Gladwin, T. 1970. *East is a big bird.* Cambridge, MA: Harvard University Press.

Glander, K. E. 1978. Howling monkey feeding behavior and plant secondary compounds: A study of strategies. In *The ecology of arboreal folivores,* edited by G. G. Montgomery, 561–73. Washington, DC: Smithsonian Institution Press.

Glander, K., R. J. Tapia, and F. T. Augusto. 1984. Impact of cropping on wild populations of *Saguinus mystax* and *Saguinus fuscicollis* in Peru. *Am. J. Primatol.* 7:89–97.

Glickman, S. E., C. J. Zabel, S. I. Yoerg, M. L. Weldele, C. M. Drea, and L. G. Frank. 1997. Social facilitation, affiliation, and dominance in the social life of spotted hyenas. *Ann. NY Acad. Sci.* 807:175–84.

Golani, I., and A. Keller. 1975. A longitudinal field study of the behaviour of a pair of golden jackals. In *The wild canids,* edited by M. W. Fox, 303–35. New York: Van Nostrand Reinhold.

Golden, B. L., L. Levy, and R. Vohra. 1987. The orienteering problem. *Naval Res. Logistics* 34:307–18.

Golden, B. L., and W. R. Stewart. 1985. Empirical analysis of heuristics. In *The traveling salesman problem,* edited by E. L. Lawler, J. K. Lenstra, A. H. G. Rinooy Kan, and D. B. Shmoys. Chichester: J. Wiley and Sons.

Golden, B. L., Q. Wang, and L. Liu. 1988. A multifaceted heuristic for the orienteering problem. *Naval Res. Logistics* 35:359–66.

Goldizen, A. W. 1986. Tamarins and marmosets: Communal care of offspring. In *Primate societies,* edited by B. B. Smuts, D. L. Cheney, R. M. Seyfarth, R. W. Wrangham, and T. T. Struhsaker, 34–43. Chicago: University of Chicago Press.

Goldizen, A. W., J. Mendelson, M. van Vlaardingen, and J. Terborgh. 1996. Saddle-back tamarin *(Saguinus fuscicollis)* reproductive strategies: Evidence from a thirteen-year study of a marked population. *Am. J. Primatol.* 38:57–83.

Goldman, C. A. 1987. *Crossarchus obscurus. Mammal. Spec.* 290:1–5.

Goldman-Rakic, P. S. 1996. The prefrontal landscape: Implications of functional architecture for understanding human mentation and the central executive. *Phil. Trans. R. Soc. Lond.* B 346:1445–53.

Gompper, M. E. 1996. Foraging costs and benefits of coati *(Nasua narica)* sociality and asociality. *Behav. Ecol.* 7:254–63.

Gompper, M. E., J. L. Gittleman, and R. K. Wayne. 1997. Genetic relatedness, coalitions and social behaviour of white-nosed coatis, *Nasua narica. Anim. Behav.* 53:781–97.

Goodall, A. 1977. Feeding and ranging behavior of a mountain gorilla group *(Gorilla gorilla beringei)* in the Tshibinda-Kahuzi region, Zaire. In *Primate ecology,* edited by T. H. Clutton-Brock, 449–79. London: Academic Press.

Goodall, J. 1968. Behaviour of free-living chimpanzees of the Gombe stream area. *Anim. Behav. Monogr.* 1:163–311.

———. 1986. *The chimpanzees of Gombe: Patterns of behavior.* Cambridge, MA: The Belknap Press of Harvard University.

Goodall, J., A. Bandora, E. Bergmann, C. Busse, H. Matama, E. Mpongo, A. Pierce, and D. Riss. 1979. Intercommunity interactions in the chimpanzee population of the Gombe National Park. In *The great apes,* edited by D. A. Hamburg and E. R. McCown, 13–54. Menlo Park, CA: Benjamin Cummings.

Goodman, S. M. 1994. The enigma of antipredator behaviour in lemurs: Evidence of a large extinct eagle on Madagascar. *Int. J. Primatol.* 15:129–34.

Goodman, S. M., and O. Langrand. 1996. A high mountain population of the ring-tailed lemur *Lemur catta* on the Andringitra Massif, Madagascar. *Oryx* 30:259–68.

Goodman, S. M., S. O'Connor, and O. Langrand. 1993. A review of predation on lemurs: Implications for the evolution of social behaviour in small nocturnal primates. In *Lemur social systems and their ecological basis,* edited by P. M. Kappeler and J. U. Ganzhorn, 51–66. New York: Plenum.

Goodman, S. M., and L. Rakotozafy. 1995. Evidence for the existence of two species of *Aquila* on Madagascar during the Quarternary. *Geobios* 28:241–46.

Gordon, D. M. 1995. The expandable network of ant exploration. *Anim. Behav.* 50:995–1007.

Gould, J. L. 1986. The locale map of honeybees: Do insects have cognitive maps? *Science* 232:861–63.

Gould, J. L., and K. P. Able. 1981. Human homing: An elusive phenomenon. *Science* 212:1061–63.

Gould, L. 1996. Vigilance behavior during the birth and lactation season in naturally occurring ring-tailed lemurs *(Lemur catta)* at the Beza-Mahafaly Reserve, Madagascar. *Int. J. Primatol.* 17:331–47.

Gould, L., L. M. Fedigan, and L. M. Rose. 1997. Why be vigilant? The case of the alpha animal. *Int. J. Primatol.* 18:401–14.

Gouzoules, H., and S. Gouzoules. 1989. Design features and developmental modification of pigtail macaque, *Macaca nemestrina,* agonistic screams. *Anim. Behav.* 37:383–401.

———. 1995. Recruitment screams of pigtail monkeys *(Macaca nemestrina):* Ontogenetic perspectives. *Behaviour* 132:431–50.

Gouzoules, H., S. Gouzoules, and J. Ashley. 1995. Representational signaling in non-human primate vocal communication. In *Current topics in primate vocal communication,* edited by E. Zimmermann, J. D. Newman, and U. Jürgens, 235–52. New York: Plenum Press.

Gouzoules, H., S. Gouzoules, and P. Marler. 1984. Rhesus monkey *(Macaca mulatta)* screams: Representational signalling in the recruitment of agonistic aid. *Anim. Behav.* 32:182–93.

Gradwohl, J., and R. Greenberg. 1980. The formation of antwren flocks on Barro Colorado Island, Panama. *Auk* 97:385–95.

———. 1982. The effect of a single species of avian predator on the arthropods of aerial leaf litter. *Ecology* 6:581–83.

Grant, J. W. A., C. A. Chapman, and K. S. Richardson. 1992. Defended versus undefended home range size of carnivores, ungulates and primates. *Behav. Ecol. Sociobiol.* 31:149–61.

Graves, G. R., and N.J. Gotelli. 1993. Assembly of avian mixed-species feeding flocks in Amazonia. *Proc. Natl. Acad. Sci. USA* 90:1388–91.

Green, S. 1975. Dialects in Japanese monkeys: Vocal learning and cultural transmission of locale-specific vocal behaviors? *Z. Tierpsychol.* 38:304–14.

Greenberg, R. 1984. The winter exploitation system of Bay-breasted and Chestnut-sided warblers in Panama. *Univ. Calif. Publ. Zool.* 116:1–107.

———. 1987a. Development of dead leaf foraging in a tropical migrant warbler. *Ecology* 68:130–41.

———. 1987b. Social facilitation does not reduce neophobia in Chestnut-sided Warblers (Parulinae: *Dendroica pensylvanica*). *J. Ethol.* 5:7–10.

Greenberg, R., and J. Gradwohl. 1985. A comparative study of the social behavior of antwrens on Barro Colorado Island, Panama. *Ornith. Monogr.* 845–55.

———. 1986. Constant density and stable territoriality in some tropical forest insectivorous birds. *Oecologia* 69:618–25.

———. 1997. Territoriality, adult survival, and dispersal in the Checker-Throated Antwren in Panama. *J. Avian Biol.* 28:103–10.

Greig-Smith, P. W. 1978. The formation, structure and function of mixed-species insectivorous bird flocks in West African savanna woodlands. *Ibis* 120:284–97.

Griffin, D. R. 1984. *Animal thinking.* Cambridge MA: Harvard University Press.

———. 1992. *Animal minds.* Chicago: University of Chicago Press.

Grinnell, J., C. Packer, and A. E. Pusey. 1995. Cooperation in male lions: Kinship, reciprocity or mutualism? *Anim. Behav.* 49:95–105.

Guetzkow, H., and H. A. Simon. 1955. Communication patterns in task-oriented groups. *Mgmt. Sci.* 1:233–50.

Guggisberg, C. A. W. 1962. *Simba.* London: Bailey Brothers and Swinfen.

Guinet, C. 1991. Intentional stranding apprenticeship and social play in killer whales *(Orcinus orca). Can. J. Zool.* 69:2712–16.

Gulliver, P. H. 1951. A preliminary survey of the Turkana. New Series no. 26. Commonwealth School of African Studies, Capetown.

————. 1955. *The family herds: A study of two pastoral tribes.* London: Routledge and Kegan Paul.

————. 1975. Nomadic movements: Causes and implications. In *Pastoralism in tropical Africa,* edited by T. Monod, 369–86. Oxford: Oxford University Press.

Hain, J. H. W., G. R. Carter, S. D. Kraus, C. A. Mayo, and H. E. Winn. 1982. Feeding behavior of the humpback whale, *Megaptera novaengliae,* in the western North Atlantic. *U.S. Fish Wildl. Serv. Fishery Bull.* 80(2):259–68.

Haldane, J. B. S. 1932. The causes of evolution. London: Longmans Green & Co.

Hall, C. L., and L. M. Fedigan. 1997. Spatial benefits afforded by high rank in white-faced capuchins. *Anim. Behav.* 53:1069–82.

Hall, J. D. 1986. Notes on the distribution and feeding behavior of killer whales in Prince William Sound, Alaska. In *The behavioural biology of killer whales,* edited by B. C. Kirkevold and J. S. Lockard, 69–83. New York: Alan R. Liss.

Hall, K. R. L., and I. DeVore. 1965. Baboon social behavior. In *Primate behavior: Field studies of monkeys and apes,* edited by I. DeVore, 53–110. New York: Holt, Rinehart & Winston.

Halloy, M., and D. G. Kleiman. 1994. Acoustic structure of long calls in free-ranging groups of golden lion tamarins, *Leontopithecus rosalia. Am. J. Primatol.* 32:303–10.

Hamilton, T. H., and R. H. Barth Jr. 1962. The biological significance of season change in male plumage appearance in some New World migratory bird species. *Am. Nat.* 96:129–44.

Hamilton, W. D. 1964. The genetical evolution of social behaviour. *J. Theor. Biol.* 7:1–51.

Hamilton, W. D. 1971. Geometry for the selfish herd. *J. Theor. Biol.* 31:295–311.

Hamilton, W. D. 1975. Innate social aptitudes in man, an approach from evolutionary genetics. In *Biosocial anthropology,* edited by R. Fox. London: Malaby Press.

Hamilton, W. D. 1982. Baboon sleeping site preferences and relationships to primate grouping patterns. *Am. J. Primatol.* 3:41–53.

Hamilton, W. D., R. Buskirk, and W. Buskirk. 1975. Defensive stoning by baboons. *Nature* 256:488–89.

Hamilton, W. D., and K. E. F. Watt. 1970. Refuging. *Annu. Rev. Ecol. Syst.* 1:263–86.

Hansen, L. J. 1990. California coastal bottlenose dolphins. In *The bottlenose dolphin,* edited by S. Leatherwood and R. R. Reeves, 403–20. San Diego, CA: Academic Press.

Haraiwa-Hasegawa, M., R. W. Byrne, H. Takasaki, and J. M. E. Byrne. 1986. Aggression toward large carnivores by wild chimpanzees of Mahale Mountains National Park, Tanzania. *Folia Primatol.* 47:8–13.

Harcourt, A. H., and de Waal, F. B. M., eds. 1992. *Coalitions and alliances in humans and other animals.* Oxford: Oxford University Press.

Harcourt, A. H., and K. J. Stewart. 1994. Gorillas' vocalizations during rest periods: Signals of impending departure? *Behaviour* 130:29–40.

Harcourt, A. H., K. J. Stewart, and M. Hauser. 1993. Functions of wild gorilla "close" calls. I. Repertoire, context, and interspecific comparison. *Behaviour* 124:89–122.

Harcourt, C. S. 1981. An examination of the function of urine washing in *Galago senegalensis. Z. Tierpsychol.* 55:119–28.

Harcourt, C. S. 1991. Diet and behaviour of a nocturnal lemur, *Avahi laniger,* in the wild. *J. Zool.* 223:667–74.

Hardie, S. M., and H. M. Buchanan-Smith. 1997. Vigilance in single- and mixed-species groups of tamarins *(Saguinus labiatus* and *Saguinus fuscicollis).* *Int. J. Primatol.* 18:217–34.

Harding, R. S. O. 1973. Predation by a troop of olive baboons *(Papio anubis).* *Am. J. Phys. Anthropol.* 38:587–92.

Harding, R. S. O. 1977. Patterns of movement in open country baboons. *Am. J. Phys. Anthropol.* 47:349–54.

Hardy, J. W. 1966. Physical and behavioral factors in sociality and evolution of certain parrots *(Aratinga).* *Auk* 83:66–83.

Harestad, A. S., and F. L. Bunnell. 1979. Home range and body weight: A re-evaluation. *Ecology* 60:389–402.

Harlow, H. F. 1949. The formation of learning sets. *Psychol. Rev.* 56:51–65.

Harrington, F. H. 1987. Aggressive howling in wolves. *Anim. Behav.* 35:7–12.

Harrington, F. H., and L. D. Mech. 1979. Wolf howling and its role in territory maintenance. *Behaviour* 68:207–49.

———. 1983. Wolf pack spacing: Howling as a territory-independent spacing mechanism in a territorial population. *Behav. Ecol. Sociobiol.* 12:161–68.

Harris, J. M. 1985. Age and paleoecology of the upper Laetolil beds, Laetoli, Tanzania. In *Ancestors: The hard evidence,* edited by E. Delson, 76–81. New York: Alan R. Liss.

Hartmann, G., and Wehner, R. 1995. The ant's path integration system: A neural architecture. *Biol. Cybern.* 73:483–97.

Harvey, P H., and T. H. Clutton-Brock 1981. Primate home-range size and metabolic needs. *Behav. Ecol. Sociobiol.* 8:151–55.

Harvey, P. H., and J. R. Krebs. 1990. Comparing brains. *Science* 249:140–46.

Harvey, P. H., R. D. Martin, and T. H. Clutton-Brock. 1986. Life histories in comparative perspective. In *Primate societies,* edited by B. B. Smuts, D. L. Cheney, R. M. Seyfarth, R. W. Wrangham, and T. T. Struhsaker, 181–96. Chicago: University of Chicago Press.

Harvey, P. H., and M. D. Pagel. 1991. *The comparative method in evolutionary biology.* Oxford: Oxford University Press.

Hassell, M. P., and Southwood, T. R. E. 1978. Foraging strategies in insects. *Annu. Rev. Ecol. Syst.* 9:75–98.

Hauser, M. D. 1996. *The evolution of communication.* Cambridge, MA: MIT Press.

Hauser, M. D., and R. Wrangham. 1987. Manipulation of food calls in captive chimpanzees: A preliminary report. *Folia Primatol.* 48:24–35.

Hausfater, G. 1976. Predatory behavior of yellow baboons. *Behaviour* 56:44–68.

Haven Wiley, R. 1983. The evolution of communication: Information and manipulation. In *Animal Behaviour,* vol. 2, *Communication,* edited by T. R. Halliday and P. J. B. Slater. Oxford: Blackwell Scientific Press.

Hawkes, K., K. Hill, and J. F. O'Connell. 1982. Why hunters gather: Optimal foraging and the Ache of eastern Paraguay. *Am. Ethnol.* 9:379–98.

Haynes, K. E., and A. S. Fotheringham. 1984. *Gravity and spatial interaction models.* Newbury Park, CA: Sage.

Heal, O. W., and S. F. McLean. 1975. Comparative productivity in ecosystems: Secondary productivity. In *Unifying concepts in ecology,* edited by W. H. van Dobben and R. H. Lowe-McConnell, 89–108. The Hague: W. Junk.

Healy, S., and T. Guilford. 1990. Olfactory bulb size and nocturnality in birds. *Evolution* 44:339–46.

Heglund, N. C., and C. R. Taylor. 1988. Speed, stride frequency and energy cost per stride: How do they change with body size and gait? *J. Exp. Biol.* 138:301–18.

Heimlich-Boran, J. R. 1986. Fishery correlations with the occurrence of *Orcinus orca* in Greater Puget Sound. In *The behavioral biology of killer whales,* edited by B. C. Kirkevold and J. S. Lockard, 113–131. New York: Alan R. Liss.

Heinrich, B. 1979a. *Bumblebee economics.* Cambridge, MA: Harvard University Press.

———. 1979b. Resource heterogeneity and patterns of movement in foraging bumblebees. *Oecologia* 140:235–45.

Heinsohn, R. 1997. Group territoriality in two populations of African lions. *Anim. Behav.* 53:1143–47.

Heinsohn, R., and C. Packer. 1995. Complex cooperative strategies in group-territorial African lions. *Science* 269:1260–62.

Helweg, D. A., A. S. Frankel, J. R. Mobley Jr., and L. M. Herman. 1992. Humpback whale song: Our current understanding. In *Marine mammal sensory systems,* edited by J. Thomas, 459–83. New York: Plenum Press.

Helweg, D. A., and L. M. Herman. 1994. Diurnal patterns of behaviour and group membership of humpback whales *(Megaptera novaengliae)* wintering in Hawaiian waters. *Ethology* 98:298–311.

Hemmer, H. 1978. Socialization by intelligence: Social behavior as a function of relative brain size and environment. *Carnivore* 1:102–5.

Henderson, I. G. 1989. The exploitation of tits *Parus* species, Long-tailed Tit *Aegithos caudatus* and Goldcrests *Regulus regulus* by tree creeper: A behavioral study. *Bird Stud.* 36:99–104.

Henley, P. 1982. *The Panare: Tradition and change on the Amazonian frontier.* New Haven, CT: Yale University Press.

Henschel, J. R., and J. D. Skinner. 1987. Social relationships and dispersal patterns in a clan of spotted hyenas *Crocuta crocuta* in the Kruger National Park. *S. Afr. J. Zool.* 22:18–24.

———. 1991. Territorial behaviour by a clan of spotted hyaenas *Crocuta crocuta. Ethology* 88:223–35.

Henzi, S. P., R. W. Byrne, and A. Whitten. 1992. Patterns of movement by baboons in the Drakesnberg Mountains: Primary responses to the environment. *Int. J. Primatol.* 13:601–29.

Henzi, S. P., and J. W. Lycett. 1995. Population structure, demography and dynamics of mountain baboons. *Am. J. Primatol.* 35:155–63.

Herman, L. M., ed. 1980. *Cetacean behavior: Mechanisms and functions.* New York: J. Wiley and Sons.

Herman, L. M., and W. N. Tavolga. 1980. The communication systems of cetaceans. In *Cetacean behavior: Mechanisms and functions,* edited by L. M. Herman, 149–210. New York: J. Wiley and Sons.

Hershkovitz, P. 1977. *Living New World Monkeys* (Platyrrhini). Vol 1. Chicago: University of Chicago Press.

Heymann, E. W. 1987. A field observation of predation on a mustached tamarin *(Saguinus mystax)* by an anaconda. *Int. J. Primatol.* 8:193–95.

————. 1990a. Interspecific relations in a mixed-species troop of moustached tamarins, *Saguinus mystax,* and saddle-back tamarins, *Saguinus fuscicollis* (Platyrrhini: Callitrichidae), at the Rio Blanco, Peruvian Amazonia. *Am. J. Primatol.* 21:115–27.

————. 1990b. Reactions of wild tamarins, *Saguinus mystax* and *Saguinus fuscicollis,* to avian predators. *Int. J. Primatol.* 11:327–37.

————. 1995. Sleeping habits of tamarins, *Saguinus mystax* and *Saguinus fuscicollis* (Mammalia; Primates; Callitrichidae), in North-Eastern Peru. *J. Zool.* 237:211–26.

————. 1997. The relationship between body size and mixed-species troops of tamarins (*Saguinus* spp.). *Folia Primatol.* 68:287–95.

Hill, C. M. 1994. The role of female diana monkeys, *Cercopithecus diana,* in territorial defence. *Anim. Behav.* 47:425–31.

Hill, K. 1982. Hunting and human evolution. *J. Hum. Evol.* 11:521–44.

Hill, K., and K. Hawkes. 1983. Neotropical hunting among the Aché of eastern Paraguay. In *Adaptive responses of Native Amazonians,* edited by R. B. Hames and W. T. Vickers, 139–88. New York: Academic Press.

Hill, K., K. Hawkes, M. Hurtado, and H. Kaplan. 1984. Seasonal variance in the diet of Ache hunter-gatherers in Eastern Paraguay. *Hum. Ecol.* 12:101–35.

Hill, K. R., and A. M. Hurtado. 1996. *Aché life history: The ecology and demography of a foraging people.* New York: Aldine-DeGruyter.

Hillier, F. S., and G. J. Lieberman. 1967. *Introduction to operations research.* San Francisco: Holden-Day, Inc.

Hilton, H. 1978. Systematics and ecology of the eastern coyote. In *Coyotes: Biology, behavior and management,* edited by M. Bekoff, 210–28. New York: Academic Press.

Hilton, S. C., and J. R. Krebs. 1990. Spatial memory of four species of *Parus:* Performance in an open-field analogue of a radial maze. *Q. J. Exp. Psychol.* (B) 42:345–68.

Hinde, R. A. 1983. *Primate social relationships: An integrated approach.* Oxford: Blackwell.

Hladik, C. M., P., Charles-Dominique, and J. J. Petter. 1980. Feeding strategies of five nocturnal prosimians in the dry forest of the west coast of Madagascar. In *Nocturnal Malagasy primates: Ecology, physiology and behavior,* edited by P. Charles-Dominique, H. M. Cooper, A. Hladik, C. M. Hladik, E. Pages, G. F. Pariente, A., Petter Rousseaux, A. Schilling, and J. J. Petter, 41–74. New York: Academic Press.

Hoelzel, A. R. 1991. Killer whale predation on marine mammals at Punta Norte, Argentina: Food sharing, provisioning and foraging strategy. *Behav. Ecol. Sociobiol.* 29:197–204.

Hoese, H. D. 1991. Dolphin feeding out of water in a salt marsh. *J. Mammal.* 52:222–23.

Hofer, H., and East, M. L. 1993. How predators cope with migratory prey: The commuting system of spotted hyaenas in the Serengeti. I. Social organization. *Anim. Behav.* 46:547–57.

Hoffman, A. J., and P. Wolfe. 1985. History. In *The traveling salesman problem. A guided tour of combinatorial optimization,* edited by E. L. Lawler, J. K. Lenstra, A. H. G. Rinnoy Kan, and D. B. Shmoys. Chichester: John Wiley and Sons.

Hogan, R., G. J. Curphy, and J. Hogan. 1994. What we know about leadership: Effectiveness and personality. *Am. Psychol.* 49:493–504.

Hogsted, O. 1988. Advantages to social foraging in Willow tits *Parus montanus. Ibis* 130:275–83.

Hohmann, G. 1991. Comparative analyses of age- and sex-specific patterns of vocal behavior in four species of Old World monkey. *Folia Primatol.* 36:133–56.

Hohmann, G., and B. Fruth. 1994. Structure and use of distance calls in wild bonobos *(Pan paniscus). Am. J. Primatol.* 15(5):767–82.

Holekamp, K. E., E. E. Boydston, M. Szykman, I. Graham, K. Nutt, S. Birch, A. Piskiel, and M. Singh. 1999. Vocal recognition in the spotted hyaena and its possible implications regarding the evolution of intelligence. *Anim. Behav.* In press.

Holekamp, K. E., S. M. Cooper, C. I. Katona, N. A. Berry, L. G. Frank, and L. Smale. 1997. Patterns of association among female spotted hyenas *(Crocuta crocuta). J. Mammal.* 78:55–64.

Holekamp, K. E., J. O. Ogutu, H. T. Dublin, L. G. Frank, and L. Smale. 1993. Fission of a spotted hyena clan: Consequences of prolonged female absenteeism and causes of female emigration. *Ethology* 93:285–99.

Holekamp, K. E., and Smale, L. 1990. Provisioning and food-sharing by lactating spotted hyenas *(Crocuta crocuta). Ethology* 86:191–202.

———. 1991. Rank acquisition during mammalian social development: The "inheritance" of maternal rank. *Am. Zool.* 31:306–17.

———. 1993. Ontogeny of dominance in free-living spotted hyaenas: Juvenile rank relations with other immature individuals. *Anim. Behav.* 46:451–66.

———. 1995. Rapid change in offspring sex ratios after clan fission in the spotted hyaena. *Am. Nat.* 145:261–77.

Holekamp, K. E., L. Smale, R. Berg, and S. M. Cooper. 1997. Hunting rates and hunting success in the spotted hyena. *J. Zool.* 241:1–15.

Holekamp, K. E., L. Smale, and M. Szykman. 1996. Rank and reproduction in the female spotted hyaena. *J. Reprod. Fertil.* 108:229–37.

Holenweg, A., R. Noë, and M. Schabel. 1996. Waser's gas model applied to associations between red colobus and diana monkeys in the Taï National Park, Ivory Coast. *Folia Primatol.* 67:125–36.

Hölldobler, B., and E. O. Wilson. 1990. *The ants.* Cambridge, MA: Harvard University Press.

Holling, C. S. 1966. The functional response of invertebrate predators to prey density. *Mem. Entomol. Soc. Can.* 48:1–86.

Homewood, K. M. 1976. Ecology and behaviour of the Tana River mangabey *(Cercocebus galeritus galeritus).* Ph.D. dissertation, University of London.

Homscher, C., W. F. H. M. Mommaerts, N. V. Ricchiutti, and A. Wallner. 1972. Activation heat, activation metabolism, and tension related heat in frog semitendinosus muscles. *J. Physiol* 220:601–25.

Höner, O. P., L. Leumann, and R. Nöe. 1997. Dyadic associations of red colobus and diana monkey groups in the Taï National Park, Ivory Coast. *Primates* 38:281–91.

Hopfield, J. J., and D. W. Tank. 1985. "Neural" computation of decisions in optimization problems. *Biol. Cybern.* 52:141–52.

Horn, A. G., M. L. Leonard, and D. M. Weary. 1995. Oxygen consumption during crowing by roosters: Talk is cheap. *Anim. Behav.* 50:1171–75.

Houston, A., C. Clark, J. McNamara, and M. Mangel. 1988. Dynamic models in behavioural and evolutionary ecology. *Nature* 332:29–34.

Hoyt, D. F., and C. R. Taylor. 1981. Gait and the energetics of locomotion in horses. *Nature* 292:239–40.

Hreljac, A. 1993. Preferred and energetically optimal gait transition speeds in human locomotion. *Med. Sci. Sports Exerc.* 25:1158–62.

Huang, W., R. Ciochon, G. Yumin, R. Larick, F. Qiren, H. Schwarcz, C. Yonge, J. deVos, and W. Rink. 1995. Early *Homo* and associated artifacts from Asia. *Nature* 378:275–78.

Hubbell, S. P. 1979. Tree dispersion, abundance, and diversity in a tropical dry forest. *Science* 203:1299–1309.

Hubbell, S. P., and R. B. Foster. 1990. Structure, dynamics and equilibrium status of old-growth forest on Barro Colorado Island. In *Four Neotropical rainforests,* edited by A. Gentry, 522–41. New Haven, CT: Yale University Press.

Hubrecht, R. C. 1985. Home-range and use and territorial behavior in the common marmoset, *Callithrix jacchus jacchus,* at Tapacura Field Station, Recife, Brazil. *Int. J. Primatol.* 6:533–50.

Humphrey, N. K. 1976. The social function of intellect. In *Growing points in ethology,* edited by 303–25. Cambridge: Cambridge University Press.

Hutchins, E. 1995. *Cognition in the wild.* Cambridge, MA: MIT Press.

Hutto, R. L. 1988. Foraging behavior patterns suggest a possible cost associated with mixed-species bird flocks. *Oikos* 51:79–83.

———. 1994. The composition and social organization of mixed species flocks in a tropical deciduous forest in Mexico. *Condor* 96:105–18.

Ingmanson, E. J. 1996. Tool-using behavior in wild *Pan paniscus:* Social and ecological considerations. In *Reaching into thought: The minds of the great apes,* edited by A. E. Russon, K. Bard, and S. Taylor, 190–210. Cambridge: Cambridge University Press.

Ingmanson, E. J., and T. Kano. 1993. Waging peace. *Int. Wildl,* Nov./Dec., 31–37.

Ingold, T. 1987. *The appropriation of nature: Essays on human ecology and social relations.* Iowa City: University of Iowa Press.

Ingold, T. 1996. The optimal forager and economic man. In *Nature and society: Anthropological perspectives,* 25–44. New York: Routledge.

Irons, W. 1979. Natural selection, adaptation, and human social behavior. In *Evolutionary biology and human social behavior: An anthropological perspective,* edited by N. Chagnon and W. Irons, 4–39. North Scituate, MA: Duxbury Press.

Isaac, G. L. 1978. Food sharing and human evolution: Archaeological evidence from the Plio-Pleistocene of East Africa. *J. Anthropol. Res.* 34:311–25.

Isack, H. A., and H. U. Reyer. 1989. Honeyguides and honeygatherers: Interspecific communication in a symbiotic relationship. *Science* 243:1343–46.

Isbell, L. A. 1983. Daily ranging behavior of red colobus *(Colobus badius tephrosceles)* in Kibale Forest, Uganda. *Folia Primatol.* 41:34–49.

Isbell, L. A. 1990. Sudden short-term increase in mortality of vervet monkeys *(Cercopithecus aethiops)* due to leopard predation in Amboseli National Park, Kenya. *Am. J. Primatol.* 21:41–52.

Isbell, L. A. 1991. Contest and scramble competition: Patterns of female aggression and ranging behaviour among primates. *Behav. Ecol.* 2:143–55.

Isbell, L. A. 1994. Predation on primates: Ecological patterns and evolutionary consequences. *Evol. Anthropol.* 3:61–71.

Isbell, L. A., and D. Van Vuren. 1996. Differential costs of locational and social dispersal and their consequences for female group-living primates. *Behaviour* 133:1–36.

Isbell, L. A., and T. P. Young. 1993. Social and ecological influences on activity budgets of vervet monkeys, and their implications for group living. *Behav. Ecol. Sociobiol.* 32:377–85.

Isbell, L. A., J. D. Pruetz, M. Lewis, and T. Young. 1998. Locomotor activity differences between sympatric patas monkeys *(Erythrocebus patas)* and vervet monkeys *(Cercopithecus aethiops):* Implications for the evolution of long hindlimb length in *Homo. Am. J. Phys. Anthropol.* 105:199–207.

Isbell, L. A., D. L. Cheney, and R. M. Seyfarth. 1990. Costs and benefits of home range shifts among vervet monkeys *(Cercopithecus aethiops)* in Amboseli National Park, Kenya. *Behav. Ecol. Sociobiol.* 27:351–58.

Itani, J. 1963. Vocal communication of the wild Japanese monkey. *Primates* 4:11–66.

Ivins, B. L., and A. T. Smith. 1983. Responses of pikas (*Ochotona princeps,* Lagomorpha) to naturally occurring terrestrial predators. *Behav. Ecol. Sociobiol.* 13:277–85.

Iwamoto, T., A. Mori, M. Kawai, and A. Bekele. 1996. Anti-predator behavior of gelada baboons. *Primates* 37:389–97.

Iyengar, S. S., and R. L. Kashyap. 1991. Neural networks: A computational perspective. In *Neural networks: Concepts, applications, and implementations,* edited by P. Antognetti and V. Milutinovic. Englewood Cliffs, NJ: Prentice Hall.

Izawa, K. 1978. A field study of the ecology and behavior of the black-mantle tamarin *(Saguinus nigricollis). Primates* 19:241–74.

Izawa, K., and M. Yoneda. 1981. Habitat utilization of nonhuman primates in a forest of western Pando, Bolivia. *Kyoto University Overseas Research Reports on New World Monkeys* (1981):13–22.

Izor, R. J. 1985. Sloths and other mammalian prey of the harpy eagle. In *The evolution and ecology of armadillos, sloths and vermilinguas,* edited by G. G. Montgomery, 343–46. Washington, DC: Smithsonian Institution Press.

Jackendoff, R. 1992. *Consciousness and the computational mind.* Cambridge, MA: MIT Press.

Jacobs, A. H. 1965. African pastoralism: Some general remarks. *Anthropol. Q.* 38:144–54.

Jacobs, G. H. 1993. The distribution and nature of colour vision among the mammals. *Biol. Rev. Cambridge Phil. Soc.* 68:413–71.

Jacobs, G. H. 1995. Variations in primate colour vision: Mechanisms and utility. *Evol. Anthropol.* 3:196–205.

Jacobs, L. F., S. J. C. Gaulin, D. F. Sherry, and G. E. Hoffman. 1990. Evolution of spatial cognition: Sex specific patterns of spatial behavior predict hippocampal size. *Proc. Natl. Acad. Sci. USA* 87:6349–52.

Jacobs, M. S., J. Morgane, and W. McFarland. 1971. The anatomy of the brain of the bottlenose dolphin *(Tursiops truncatus),* rhinic lobe (rhinencephalon). 1. The paleocortex. *J. Comp. Neurol.* 141:205–72.

Jacobsen, J. K. 1986. The behavior of *Orcinus orca* in the Johnstone Strait, British Columbia. In *The behavioral biology of killer whales,* edited by B. C. Kirkevold and J. S. Lockard, 135–85. New York: Alan R. Liss.

James, W. P. T., and E. C. Schofield. 1990. *Human energy requirements: A manual for planners and nutritionists.* Oxford: Oxford Medical Publications.

Janetos, A. C., and B. J. Cole. 1981. Imperfectly optimal animals. *Behav. Ecol. Sociobiol.* 9:203–10.

Janik, V. M. 1997. Food-related calling in bottlenose dolphins, *Tursiops truncatus.* Advances in Ethology 32: Contributions to the XXV International Ethological Conference, 20–27 August 1997, edited by M. Taborsky and B. Taborsky.

Janson, C. H. 1990a. Ecological consequences of individual spatial choice in foraging groups of brown capuchin monkeys. *Cebus apella. Anim. Behav.* 40:922–34.

Janson, C. H. 1985. Aggressive competition and individual food consumption in wild brown capuchin monkeys *(Cebus apella). Behav. Ecol. Sociobiol.* 18:125–38.

Janson, C. H. 1988a. Food competition in brown capuchin monkeys *(Cebus apella):* Quantitative effects of group size and tree productivity. *Behaviour* 105:53–76.

Janson, C. H. 1988b. Intra-specific food competition and primate social structure: A synthesis. *Behaviour* 105:1–17.

Janson, C. H. 1990b. Social correlates of individual spatial choice in foraging groups of brown capuchin monkeys, *Cebus apella. Anim. Behav.* 40:910–21.

Janson, C. H. 1992. Evolutionary ecology of primate social structure. In *Evolutionary ecology and human behavior,* edited by E. A. Smith and B. Winterhalder, 95–130. New York: Aldine.

Janson, C. H. 1998. Experimental evidence for spatial memory in foraging wild capuchin monkeys, *Cebus apella. Anim. Behav.* 55:1129–43.

Janson, C. H., and S. Boinski. 1992. Morphological and behavioral adaptations for foraging in generalist primates: The case of the cebines. *Am. J. Phys. Anthropol.* 88:483–98.

Janson, C. H., and M. S. Di Bitetti. 1997. Experimental analysis of food detection in capuchin monkeys: Effects of distance, travel speed, and resource size. *Behav. Ecol. Sociobiol.* 41:17–24.

Janson, C. H., and M. L. Goldsmith. 1995. Predicting group size in primates: Foraging costs and predation risks. *Behav. Ecol.* 6:326–36.

Janson, C. H., and C. P. van Schaik. 1988. Recognizing the many faces of primate food competition: Methods. *Behaviour* 105:165–86.

Janson, C. H., E. W. Stiles, and D. W. White. 1986. Selection on plant fruiting traits by brown capuchin monkeys: A multivariate approach. In *Frugivores and seed dispersal,* edited by A. Estrada and T. Fleming. The Hague: Junk.

Janson, C. H., J. Terborgh, and L. E. Emmons. 1981. Nonflying mammals as pollinating agents in the Amazonian forest. *Biotropica* 13:1–6.

Janson, C. H. 1996. Towards an experimental socioecology of primates: Examples for Argentine brown capuchin monkeys *(Cebus apella nigritus).* In *Adaptive radiation of Neotropical primates,* edited by M. Norconk, A. L. Rosenberger, and P. A. Garber, 309–25. New York: Plenum Press.

Jay, P. C. 1968. *Primates: Studies in adaptation and variability.* New York: Holt.

Jeanne, R. L. 1991. The swarm-founding Polistinae. In *The social biology of wasps,* edited by K. G. Ross and R. W. Matthews, 191–231. Ithaca, NY: Comstock.

Jefferson, T. A., P. J. Stacey, and R. W. Baird. 1991. A review of killer whale interactions with other marine mammals. *Mammal Rev.* 21:151–80.

Jerison, H. J. 1973. *Evolution of the brain and intelligence.* New York: Academic Press.

Jerison, H. J. 1991. *Brain size and the evolution of mind.* New York: American Museum of Natural History.

Joffe, T. H., and R. I. M. Dunbar. 1997. Visual and socio-cognitive information processing in primate brain evolution. *Proc. R. Soc. Lond.* B 264:1303–7.

Johnsingh, A. J. T. 1982. Reproductive and social behaviour of the dhole, *Cuon alpinus* (Canidae). *J. Zool.* 198:443–63.

Johnson, D. 1969. *The nature of nomadism: A comparative study of pastoral migrations in Southwestern Asia and Northern Africa.* Chicago: Department of Geography Research Paper no. 118.

Johnson, D. S., and C. H. Papadimitriou. 1985. Computational complexity. In *The traveling salesman problem. A guided tour of combinatorial optimization,* edited by E. L. Lawler, J. K. Lenstra, A. H. G. Rinnoy Kan, and D. B. Shmoys. Chichester: John Wiley and Sons.

Johnson, G. S. 1986. Dolphin audition and echolocation capabilities. In: *Dolphin cognition and behavior: A comparative approach,* edited by R. J. Schusterman, J. A. Thomas, and F. G. Wood, 115–36. Hillsdale, NJ: Lawrence Erlbaum Associates.

Johnstone, R. A., and A. Grafen. 1993. Dishonesty and the handicap principle. *Anim. Behav.* 46:759–64.

Jolly, A., R. Hantanirina, H. R. Rasamimanana, M. F. Kinnaird, T. G. O'Brien, H. M. Crowley, C. S. Harcourt, S. Gardner, and J. M. Davidson. 1993. Territoriality in *Lemur catta* groups during the birth season at Berenty, Madagascar. In *Lemur social systems and their ecological basis,* edited by P. M. Kappeler and J. U. Ganzhorn, 85–109. New York: Plenum Press.

Jolly, A. 1966. *Lemur behavior.* Chicago: University of Chicago Press.

Jolly, A. 1972. Troop continuity and troop spacing in *Propithecus verreauxi* and *Lemur catta* at Berenty (Madagascar). *Folia Primatol.* 17:335–62.

Jolly, A. 1998. Pair-bonding, female aggression, and the evolution of lemur societies. *Folia Primatol.* In press.

Jones, E. E. 1955. Ecological studies on the rainforest of southern Nigeria. *J. Ecol.* 43:564–94.

Jordan, P. A., P. C. Shelton, and D. L. Allen. 1967. Numbers, turnover and social structure in the Isle Royale wolf population. *Am. Zool.* 7:233–52.

Joslin, P. W. B. 1967. Movements and homesites of timber wolves in Algonquin Park. *Am. Zool.* 7:279–88.

Jouventin, P. 1975. Observations sur la socioecologie du mandrill. *Terre et Vie* 29:493–532.

Judd, T. M., and P. W. Sherman. 1996. Naked mole-rats recruit colony mates to food sources. *Anim. Behav.* 52:957–69.

Julliot, C. 1994. Predation of a young spider monkey *(Ateles paniscus)* by a crested eagle *(Morphnus guianensis). Folia Primatol.* 63:75–77.

Jungers, W. L. 1988. New estimates of body size in australopithecines. In *Evolutionary history of the "robust" australopithecines,* edited by F. Grine, 115–25. New York: Aldine de Gruyter.

Jurasz, C. M., and V. P. Jurasz. 1979. Feeding modes of the humpback whale

(Megaptera novaeangliae) in southeast Alaska. *Sci. Rep. Whales Res. Inst.* 31:69–83.

Kaack, B., L. Walker, and K. R. Brizee. 1979. The growth and development of squirrel monkeys *(Saimiri sciureus)*. *Growth* 43:116–35.

Kaas, J. H. 1995. The evolution of isocortex. *Brain Behav. Evol.* 46:187–96.

Kamil, A. C. 1978. Systematic foraging by a nectar-feeding bird, the Amakihi *(Loxops virens)*. *J. Comp. Physiol. Psychol.* 92:388–96.

Kamil, A. C. 1994. A synthetic approach to the study of animal intelligence. In *Behavioral mechanisms in evolutionary ecology,* edited by L. A. Real, 11–45. Chicago, University of Chicago Press.

Kaplan, H., and K. Hill. 1992. The Evolutionary Ecology of Food Acquisition. In *Evolutionary ecology and human behavior,* edited by E. A. Smith and B. Winterhalder, 167–201. New York: Aldine De Gruyter.

Kappeler, P. M. 1993a. Female dominance in primates and other mammals. In *Perspectives in ethology,* vol. 10, *Behaviour and evolution,* edited by P. P. G. Bateson, P. H. Klopfer, and N. S. Thompson, 143–58. New York: Plenum Press.

Kappeler, P. M. 1993b. Variation in social structure: The effects of sex and kinship on social interactions in three lemur species. *Ethology* 93:125–45.

Kappeler, P. M. 1997a. Determinants of primate social organization: Comparative evidence and new insights from Malagasy lemurs. *Biol. Rev.* 72:111–51.

Kappeler, P. M. 1997b. Intrasexual selection in *Mirza coquereli:* Evidence for scramble competition polygyny in a solitary primate. *Behav. Ecol. Sociobiol.* 41:115–28.

Kappeler, P. M. 1998. Convergence and divergence in primate social systems. In *Primate communities,* edited by J. Fleagle, C. Janson, and K. Reed. Cambridge: Cambridge University Press. In press.

Kappeler, P. M., and Heymann, E. W. 1996. Non-convergence in the evolution of primate life history and socio-ecology. *Biol. J. Linn. Soc.* 59:297–326.

Karmiloff-Smith, A. 1992. *Beyond modularity: A developmental perspective on cognitive science.* Cambridge, MA: MIT Press.

Katzmarzyk, P. T., and Leonard, W. R. 1998. Climatic influences on body size and proportions: Ecological adaptations and secular trends. *Am. J. Phys. Anthropol.* 106:483–503.

Kaufmann, J. H. 1962. Ecology and social behavior of the coati, *Nasua narica,* on Barro Colorado Island, Panama. *U. Calif. Pub. Zool.* 60:95–222.

Kaufmann, J. H. 1983. On the definitions and functions of dominance and territoriality. *Biol. Rev.* 58:1–20.

Kaufmann, J. H., D. V. Lanning, and S. E. Poole. 1976. Current status and distribution of the coati in the United States. *J. Mammal.* 57:621–37.

Kavanagh, M. 1981. Variable territoriality among tantalus monkeys in Cameroon. *Folia Primatol.* 36:76–98.

Keller, C. P., and M. F. Goodchild. 1988. The multiobjective vending problem: A generalization of the travelling salesman problem. *Environ. Plan. B* 15:447–60.

Kelly, R. 1995. *The foraging spectrum: Diversity in hunter-gatherer lifeways.* Washington, DC: Smithsonian Institution Press.

Kennedy, J. S. 1992. *The new anthropomorphism.* Cambridge: Cambridge University Press.

Kennedy, M., C. R. Shave, H. G. Spencer, and R. D. Gray. 1994. Quantifying the effect of predation risk on foraging bullies: No need to assume an Ideal Free Distribution. *Ecology* 75:2220–26.

Kennedy, R. S. 1977. Notes on the biology and population status of the monkey-eating eagle of the Philippines. *Wilson Bull.* 89:1–20.

Kerfoot, W. C., and A. Sih. 1987. Introduction. In *Predation,* edited by W. C. Kerfort and A. Sih, vii–viii. Hanover, NH: University Press of New England.

Kessler, P. 1995. Preliminary field study of the red-handed tamarin, *Saguinus midas,* in French Guiana. *Neotropical Primates* 3:184–85.

Keuroghlian, A. 1990. Observations on the behavioral ecology of the black lion tamarin *(Leontopithecus chrysopygus)* at Caetetus Reserve, São Paulo, Brazil. M.Sc. thesis, West Virginia University, Morgantown.

Khazanov, A. M. 1994. *Nomads and the outside world.* 2d ed. Madison: University of Wisconsin Press.

Kiester, A. R., and M. Slatkin. 1974. A strategy of movement and resource utilization. *Theor. Popul. Biol.* 6:1–20.

Kils, U. 1986. Verhaltenphysiologie Untersuchungen an pelagischen Schwarmen Schwarmbuildung als Strategie zur Orientierung in Umwelt-Gradienten Bedeutung der Schwarmbuildung in der Aquakultur. Habilitationsschrift, Institut für Meereskunde. Mathematisch-Naturwissenschaftliche Fakultat, Christian-Albrechts-Universistat Kiel, Germany, 1–168.

King, J. E., R. F. Becker, and J. E. Markee. 1964. Studies on olfactory discrimination in dogs: 3. Ability to detect human odour trace. *Anim. Behav.* 12:311–15.

Kinnaird, M. F. 1990. Behavioral and demographic responses to habitat change by the Tana River crested mangabey *(Cercocebus galeritus galeritus).* Ph.D. dissertation, University of Florida, Gainesville.

Kinnaird, M. F. 1992a. Phenology of flowering and fruiting in an East African riverine forest ecosystem. *Biotropica* 24:187–94.

Kinnaird, M. F. 1992b. Variable resource defense by the Tana river crested managbey. *Behav. Ecol. Sociobiol.* 31:115–22.

Kinnaird, M. F., and O'Brien, T. G. 1995. Contextual analysis of the loud call of the Sulawesi black macaque. In *Variation in fruit resources and effects on vertebrate frugivores: The role of disturbance regimes in Sulawesi rainforests.* Final Report submitted to Indonesian Institute of Science, Jl. J. Sudirman, Jakarta, Indonesia.

Kinnaird, M. F., and T. G. O'Brien. In press. Contextual analysis of the loud call of the Sulawesi crested black macaque. *Trop. Biodiversity.*

Kinzey, W. G. 1977. Diet and feeding behaviour of *Callicebus torquatus.* In *Primate Ecology,* edited by T. H. Clutton-Brock, 127–51. London: Academic Press.

Kinzey, W. G. 1981. The titi monkeys, genus *Callicebus.* In *Ecology and behavior of Neotropical primates,* edited by A. F. Coimbra-Filho and R. A. Mittermeier, 241–76. Rio de Janeiro: Academia Brasileira de Ciências.

Kinzey, W. G., and M. Becker. 1983. Activity pattern of the masked titi monkey, *Callicebus personatus. Primates* 24:337–43.

Kinzey, W. G., and J. G. Robinson. 1983. Intergroup loud calls, range size, and spacing in *Callicebus torquatus. Am. J. Phys. Anthropol.* 60:539–44.

Kirchner, W. H., and U. Braun. 1994. Dancing honey bees indicate the location of food sources using path integration rather than cognitive maps. *Anim. Behav.* 48, 1437–41.

Kirkpatrick, M., and M. J. Ryan. 1991. The evolution of mating preferences and the paradox of the lek. *Nature* 350:33–38.

Kleiber, M. 1961. *The fire of life: An introduction to animal energetics.* Huntington, NY: Kreiger.

Kleiman, D. G., and J. F. Eisenberg. 1973. Comparisons of canid and felid social systems from an evolutionary perspective. *Anim. Behav.* 21:637–59.

Kleiman, D., B. Beck, J. Dietz, L. Dietz, J. Ballou, and A. Coimbra-Filho. 1986. Conservation program for the golden lion tamarin: Captive research and management, ecological studies, educational strategies, and reintroduction. In *Primates: The road to self-sustaining populations,* edited by K. Benirschke, 959–79. New York: Springer-Verlag.

Klein, L. L., and D. J. Klein. 1973. Observations on two types of Neotropical primate intertaxa associations. *Am. J. Phys. Anthropol.* 38:649–54.

Klein, L. L., and D. J. Klein. 1977. Feeding behaviour of the Columbian spider monkey *Ateles belzebuth.* In *Primate ecology,* edited by T. H. Clutton-Brock, 153–81. London: Academic Press.

Klinowska, M. 1991. *Dolphins, porpoises and whales of the world.* IUCN Red Data Book. Gland, Switzerland: IUCN.

Klopfer, P. H., and J. Dugard. 1976. Patterns of maternal care in lemurs. III. *Lemur variegatus. Z. Tierpsychol.* 40:210–20.

Klopfer, P. H., and A. Jolly. 1970. The stability of territorial boundaries in a lemur troop. *Folia Primatol.* 12:199–208.

Klump, G. M., and M. D. Shalter. 1984. Acoustic behaviour of birds and mammals in the predator context. *Z. Tierpsychol.* 66:189–226.

Koeniger, N., and G. Koeniger. 1980. Observations and experiments on migration and dance communication of *Apis dorsata* in Sri Lanka. *J. Apic. Res.* 19:21–34.

Köhler, W. 1925. *The mentality of apes.* New York: Harcourt Brace.

Kortlandt, A. 1963. Bipedal armed fighting in chimpanzees. Proceedings of the 16th International Congress on Zoology 3:64.

Kortlandt, A. 1980. How might early hominids have defended themselves against large predators and food competitors? *J. Hum. Evol.* 9:79–112.

Kosslyn, S. M. 1988. Aspects of a cognitive neuroscience of mental imagery. *Science* 240:1621–26.

Krader, L. 1959. The ecology of pastoral nomadism. *Int. Soc. Sci. J.* 11:499–510.

Kram, R., and C. R. Taylor. 1990. Energetics of running: A new perspective. *Nature* 346:265–67.

Krause, J. 1993. Differential fitness returns in relation to spatial position in groups. *Biol. Rev.* 69:187–206.

Krebs, J. R. 1971. Territory and breeding density in the Great Tit, *Parus major. Ecology* 52:2–22.

Krebs, J. R. 1973. Social learning and the significance of mixed-species flocks of chickadees (*Parus* spp.). *Can. J. Zool.* 51:1275–88.

Krebs, J. R. 1978. Optimal foraging decision rules for predators. In *Behavioural ecology: An evolutionary approach,* edited by J. R. Krebs and N. B. Davies, 23–63. Oxford: Blackwell Scientific.

Krebs, J. R. 1990. Food-storing birds: Adaptive specialization in brain and behaviour? *Phil. Trans. R. Soc. Lond.* 329:153–60.

Krebs, J. R., and N. B. Davies. 1992. *Behavioural ecology.* 3d ed. Oxford: Blackwell Scientific.

Krebs, J. R., and R. Dawkins. 1984. Animal signals: Mind reading and manipulation. In *Behavioural ecology: An evolutionary approach,* edited by J. R. Krebs and N. B. Davies, 380–401. Oxford: Blackwell.

Krebs, J. R., and A. J. Inman. 1992. Learning and foraging: Individuals, groups, and populations. *Am. Nat.* 240:S63–S84.

Krebs, J. R. et al. 1996. The ecology of the avian brain: Food storing memory and the hippocampus. *Ibis* 138:34–46.

Krebs, J. R., S. D. Healy, and S. J. Shettleworth. 1990. Spatial memory of Paridae: Comparison of a storing and a non-storing species, the coal tit, *Parus ater,* and the great tit, *P. major. Anim. Behav.* 39:1127–37.

Krebs, J. R., A. Kacelnik, and P. Taylor. 1978. Tests of optimal sampling by foraging great tits. *Nature* 275:27–31.

Krebs, J. R., M. H. MacRoberts, and J. M. Cullen. 1972. Flocking and feeding in the great tit *Parus major:* An experimental study. *Ibis* 114:507–30.

Kruglanski, A. W., and D. M. Webster. 1991. Group member's reactions to opinion deviates and conformists at varying degrees of proximity to decision deadline and of environmental noise. *J. Pers. Soc. Psychol.* 1:212–25.

Kruuk, H. 1972. *The spotted hyena: A study of predation and social behavior.* Chicago: University of Chicago Press.

Kruuk, H. 1975. Functional aspects of social hunting by carnivores. In *Function and evolution in behavior,* edited by G. Baerends, C. Beer, and A. Manning, 119–41. London: Clarendon Press.

Kruuk, H. 1989. *The social badger.* Oxford: Oxford University Press.

Kruuk, H., and M. Turner. 1967. Comparative notes on predation by lion, leopard, cheetah, and wild dog in the Serengeti area, East Africa. *Mammalia* 31:1–27.

Kubzdela, K., A. F. Richard, and M. E. Pereira. 1992. Social relations in semi-free-ranging sifakas *(Propithecus verreauxi coquereli)* and the question of female dominance. *Am. J. Primatol.* 28:139–45.

Kudo, H. 1987. The study of vocal communication of wild mandrills in Cameroon in relation to their social structure. *Primates* 28:289–308.

Kuhme, W. 1965a. Communal food distribution and division of labor in African hunting dogs. *Nature* 205:443–44.

Kuhme, W. 1965b. Freilandstudien zur soziologie des hyaenahundes (*Lycaon pictus lupinus* Thomas 1902). *Z. Tierpsychol.* 22:495–551.

Kummer, H. 1968a. *Social organisation of hamadryas baboons.* Chicago: University of Chicago Press.

Kummer, H. 1968b. Two variations in the social organization of baboons. In *Primates: Studies in adaptation and behavior,* edited by P. J. Jay, 293–312. Chicago: University of Chicago Press.

Kummer, H. 1971. *Primate societies: Group techniques of ecological adaptation.* Chicago: Aldine.

Kummer, H. 1995. *In quest of the sacred baboon: A scientist's journey.* Princeton, NJ: Princeton University Press.

Kummer, J. 1967. Tripartite relations in hamadryas baboons. In *Social communica-*

tion among primates, edited by S. A. Altmann, 63–71. Chicago: University of Chicago Press.

Kurland, J. A., and J. D. Pearson. 1986. Ecological significance of hypometabolism in nonhuman primates: Allometry, adaptation and deviant diets. *Am. J. Phys. Anthropol.* 71:445–57.

Kuroda, S., T. Nishihara, S. Suzuki, and R. A. Oko. 1996. Sympatric chimpanzees and gorillas in the Ndoki Forest. In *Great ape societies,* edited by W. C. McGrew; L. F. Marchant, and T. Nishida, 71–81. Cambridge, Cambridge University Press.

Kurta, A., G. P. Bell, K. A. Nagy, and T. H. Kunz. 1989. Energetics of pregnancy and lactation in free-ranging little brown bats *(Myotis lucifugus). Physiol. Zool.* 62:804–18.

Kuznetzov, V. B. 1990. Chemical sense of dolphin: Quasi-olfaction. In *Sensory abilities of cetaceans: Laboratory and field evidence,* edited by J. Thomas and R. A. Kastelein, 481–504. New York: Plenum Press.

Laman, T. G. 1992. Composition of mixed species foraging flocks in a Bornean rainforest. *Malayan Nature J.* 46:131–44.

Lamphere, J. 1992. *The scattering time: Turkana responses to colonial rule.* Oxford: Clarendon Press.

Lamprecht, J. 1978. On diet, foraging behaviour and interspecific food competition of jackals in the Serengeti National Park, East Africa. *Z. Saugetierkunde* 43:210–23.

———. 1981. The function of social hunting in larger terrestrial carnivores. *Mammal Rev.* 11:169–79.

———. 1991. Factors influencing leadership: A study of goose families *(Anser indicus). Ethology* 89:265–74.

Lancaster, J. 1968. Primate communication systems and the emergence of human language. In *Primates: Studies in adaptation and variability,* edited by P. Jay, 439–57. New York: Holt, Rinehart & Winston.

———. 1978. Carrying and sharing in human evolution. *Hum. Nature,* Feb., 42–48.

Langdale-Brown, I., H. A. Osmaston, and J. G. Wilson. 1964. *The vegetation of Uganda and its bearing on land use.* Entebbe: Government of Uganda.

Lanning, D. V. 1976. Density and movements of the coati in Arizona. *J. Mammal.* 57:609–11.

Lattimore, O. 1940. *Inner Asian frontiers of China.* Research Series no. 21. New York: American Geographical Society.

Laws, M. J., and P. Henzi. 1995. Inter-group encounters in blue monkeys: How territorial must a territorial species be? *Anim. Behav.* 49:240–43.

Laws, R. M. 1970. Elephants as agents of habitat and landscape change in East Africa. *Oikos* 21:1–15.

Laws, R. M., I. S. C. Parker, and R. C. B. Johnstone. 1975. *Elephants and their habitats: The ecology of elephants in North Bunyoro, Uganda.* Oxford: Clarendon Press.

Leatherwood, S., R. R. Reeves, and L. Foster. 1983. *The Sierra Club handbook of whales and dolphins.* San Francisco: Sierra Club Books.

Leck, C. F. 1971. Measurement of social attractions between tropical passerine birds. *Wilson Bull.* 83:278–83.

Lee, R. B. 1968. What hunters do for a living, or how to make out on scarce resources. In *Man the hunter,* edited by R. B. Lee and I. Devore, 30–48. Chicago: Aldine.

——. 1979. *The !Kung San: Men, women and work in a foraging society.* Cambridge: Cambridge University Press.

Lee, S., and S. Chang. 1993. Neural networks for routing of communication networks with unreliable components. *IEEE Trans. Neural Networks* 4:854–63.

Lehner, P. N. 1978. Coyote vocalizations: A lexicon and comparison with other canids. *Anim. Behav.* 26:712–22.

——. 1982. Differential vocal response of coyotes to "group howl" and "group yip-howl" playbacks. *J. Mammal.* 63:675–79.

Lehrer, M. 1991. Bees which turn back and look. *Naturwissenschaften* 78:274–76.

——. 1996. Small-scale navigation in the honey bee: Active acquisition of information about the goal. *J. Exp. Biol.* 199:253–61.

Leighton, D. R. 1986. Gibbons: Territoriality and monogamy. In *Primate societies,* edited by B. B. Smuts, D. L. Cheney, R. M. Seyfarth, R. W. Wrangham, and T. T. Struhsaker, 134–45. Chicago: University of Chicago Press.

Leighton, M. 1993. Modeling dietary selectivity by Bornean orangutans: Evidence for integration of multiple criteria in fruit selection. *Int. J. Primatol.* 14:257–314.

Leighton, M., and D. R. Leighton. 1982. The relationship of size and feeding aggregate to size of food patch: Howler monkeys *(Alouatta palliata)* feeding in *Trichilia cipo* fruit trees on Barro Colorado Island. *Biotropica* 14:81–90.

Leith, H. 1975. Primary productivity of ecosystems: Comparative analyses of global patterns. In *Unifying concepts in ecology,* edited by W. H. van Dobben and Lowe-McConnell, 67–88. The Hague: W Junk.

Lentnek, S., M. Harwitz, and S. C. Narula. 1981. Spatial choice in consumer behavior: Towards a contextual theory of demand. *Econ. Geogr.* 57:362–72.

Leonard, B., and B. L. McNaughton. 1990. Spatial representation in the rat: Conceptual, behavioural and neurophysiological perspectives. In *Neurobiology of comparative cognition,* edited by R. P. Kesner and D. S. Olton. 363–422. London: Lawrence Erlbaum.

Leonard, W. R., and M. L. Robertson. 1992. Nutritional requirements and human evolution: A bioenergetics model. *Am. J. Hum. Biol.* 4:179–95.

——. 1994. Evolutionary perspectives on human nutrition: The influence of brain and body size on diet and metabolism. *Am. J. Hum. Biol.* 6:77–88.

——. 1997. Comparative primate energetics and hominid evolution. *Am. J. Phys. Anthropol.* 102:265–81.

Leutenegger, W. 1973. Maternal-fetal weight relationships in primates. *Folia Primatol.* 20:280–93.

Lewis, A. C. 1986. Memory constraints and flower choice in *Pieris rapae. Science* 232:863–65.

Lieberman, P. 1992. On Neanderthal speech and Neanderthal extinction. *Curr. Anthropol.* 33:409–10.

Lima, S. L. 1992. Strong preferences for apparently dangerous habitats? A consequence of differential escape form predators. *Oikos* 64:597–600.

——. 1993. Ecological and evolutionary perspectives on escape from predatory attack: A survey of North American Birds. *Wilson Bull.* 105:1–47.

————. 1995. Collective detection of predatory attack by social foragers: Fraught with ambiguity. *Anim. Behav.* 50:1097–1108.

Lima, S. L., and L. M. Dill. 1990. Behavioural decisions under the risk of predation: A review and prospectus. *Can. J. Zool.* 68:619–40.

Lima, S. L., and T. J. Valone. 1991. Predators and avian community organization: An experiment in semi-desert grassland. *Oecologia* 86:105–12.

Lindauer, M. 1955. Schwarmbienen auf Wohnungsucche. *Z. vergl. Physiol.* 37:263–324.

————. 1957. Sonnenorientierung der Bienen unter der Aequatorsonne und zur Nachtzeit. *Naturwissenschaften* 44:1–6.

————. 1959. Angeborene und erlernte Komponenten in der Sonnenorientierung der Bienen. *Z. vergl. Physiol.* 42:43–62.

————. 1961. *Communication among social bees.* Cambridge, MA: Harvard University Press.

Livingstone, M. S., and D. H. Hubel. 1988. Segregation of form, color, movement and depth: Anatomy, physiology and perception. *Science* 240:740–49.

Ljungblad, D. K., and S. E. Moore. 1983. Killer whales *(Orcinus orca)* chasing grey whales *(Eschrichtius robustus)* in the northern Bering Sea. *Arctic* 36:361–64.

Ljungblad, D. K., P. O. Thompson, and S. E. Moore. 1982. Underwater sounds recorded from migrating bowhead whales, *Balaena mysticetus,* in 1979. *J. Acoust. Soc. Am.* 71:477–82.

Loomis, J. M., R. L. Klatzky, R. G. Golledge, J. G. Cicinelli, J. W. Pellegrino, and P. A. Fry. Nonvisual navigation by blind and sighted: Assessment of path integration ability. *J. Exp. Psychol. Gen.* 122:73–91.

Lopes, M. A., and S. F. Ferrari. 1994. Foraging behavior of a tamarin group *(Saguinus fuscicollis weddelli)* and interactions with marmosets *(Callithrix emiliae).* *Int. J. Primatol.* 15:373–87.

Lopez, J. C., and Lopez, D. 1985. Killer whales *(Orcinus orca)* of Patagonia and their behavior of intentional stranding while hunting nearshore. *J. Mammal.* 66:181–83.

Lowen, C., and R. I. M. Dunbar. 1994. Territory size and defendability in primates. *Behav. Ecol. Sociobiol.* 35:347–54.

Lucas, J. R., and P. M. Waser. 1989. Defense through exploitation: A Skinner box for tropical rain forests. *Trends Ecol. Evol.* 4:62–63.

Ludescher, F. B. 1973. Sumpfmeise *(Parus p. palustris* L.) and Weidenmeise *(P. montanus saolicarius)* als sympatrische Zwillingaarten. *J. Ornithol.* 114:3–56.

MacArthur, R., and R. Pianka. E. 1966. On optimal use of a patchy environment. *Am. Nat.* 100:603–9.

Macdonald, D. W. 1979a. The flexible social system of the golden jackal, *Canis aureus. Behav. Ecol. Sociobiol.* 5:17–38.

————. 1979b. Helpers in fox society. *Nature* 282:69–71.

————. 1981. Resource dispersion and the social organization of the red fox, *Vulpes vulpes.* In *Proceedings of the Worldwide Furbearers Conference,* Frostburg, MD, edited by J. A. Chapman and D. Ureley, 918–49.

————. 1983. The ecology of carnivore social behavior. *Nature* 301:379–84.

MacDonald, S. E., and D. M. Wilkie. 1990. Yellow-nosed monkeys *(Cercopithecus ascanius whitesidei):* Spatial memory in a simulated foraging environment. *J. Comp. Psychol.* 104:382–87.

Mace, G. M., and J. F. Eisenberg. 1982. Competition, niche specialisation and the evolution of brain size in the genus *Peromyscus. Biol. J. Linn. Soc.* 17:243–57.

Mace, G. M., P. H. Harvey, and T. H. Clutton-Brock. 1980. Is brain size an ecological variable? *Trends Neurosci.* 3:193–96.

———. 1983. Vertebrate home range size and energetic requirements. In *The ecology of animal movement,* edited by I. R. Swingland and P. J. Greenwood, 32–53. Oxford: Clarendon Press.

Mace, R. 1990. Pastoralist herd compositions in unpredictable environments: A comparison of model predictions and data from camel keeping groups. *Agric. Syst.* 33:1–11.

———. 1993. Transitions between cultivation and pastoralism in sub-Saharan Africa. *Curr. Anthropol.* 34(4):363–82.

Mace, R., and Houston, A. 1989. Pastoralist strategies for survival in unpredictable environments: A model of herd composition that maximizes household viability. *Agric. Syst.* 31:185–204.

Macedonia, J. M. 1994. The vocal repertoire of the ringtailed lemur *(Lemur catta). Folia Primatol.* 61:186–217.

Macedonia, J. M., and C. S. Evans. 1992. Variation among mammalian alarm call systems and the problem of meaning in animal signals. *Ethology* 93:177–97.

Macedonia, J. M., and D. Evans. 1993. Adaptation and phylogenetic constraints in the antipredator behavior of ringtailed and ruffed lemurs. In *Lemur social systems and their ecological basis,* edited by P. M. Kappeler and J. U. Ganzhorn, 67–84. New York: Plenum Press.

Macedonia, J. M., and L. L. Taylor. 1985. Subspecific divergence in a loud call of the ruffed lemur *(Varecia variegata). Am. J. Primatol.* 9:295–304.

Macedonia, J. M., and P. L. Young. 1991. Auditory assessment of avian predator threat in semicaptive ringtailed lemurs *(Lemur catta). Primates* 32:169–82.

MacKinnon, J. 1974. The behaviour and ecology of wild orang-utans *(Pongo pygmaeus). Anim. Behav.* 22:3–74.

———. 1978. *The ape within us.* London: Collins.

Macphail, E. M. 1982. *Brain and intelligence in vertebrates.* Oxford: Clarendon Press.

Madsen, C. J., and L. M. Herman. 1980. Social and ecological correlates of cetacean vision and visual appearance. In *Cetacean behavior: Mechanisms and functions,* edited by L. M. Herman, 101–48. New York: J. Wiley and Sons.

Maestripieri, D. 1996. Gestural communication and its cognitive implications in pigtail macaques *(Macaca nemestrina). Behaviour* 133:997–1022.

Maestripieri, D., G. Schino, F. Aureli, and A. Troisi. 1992. A modest proposal: Displacement activities as an indicator of emotions in primates. *Anim. Behav.* 44:967–79.

Maier, W., C. Alonso, and A. Langguth. 1982. Field observations on *Callithrix jacchus jacchus. Z. Saugetierkunde* 47:334–46.

Maisels, F. G., J. P. Gautier, A. Cruickshank, and J. P. Bosefe. 1994. Attacks by crowned hawk eagles *(Stephanoaetus coronatus)* on monkeys in Zaire. *Folia Primatol.* 61:157–59.

Malcolm, J. R. 1979. Social organization and the communal rearing of pups in African wild dogs *(Lycaon pictus).* Ph.D. thesis, Harvard University.

Malcolm, J. R., and K. Marten. 1982. Natural selection and the communal rearing of pups in the African wild dog *(Lycaon pictus). Behav. Ecol. Sociobiol.* 10:1–13.

Malcolm, J. R., and H. van Lawick. 1975. Notes on wild dogs *(Lycaon pictus)* hunting zebras. *Mammalia* 39:231–40.

Mann, J., R. Smolker, and B. B. Smuts. 1995. Responses to calf entanglement in free-ranging bottlenose dolphins. *Mar. Mammal Sci.* 11:100–106.

Mann, J., and B. B. Smuts. 1998. Natal attraction: Allomaternal care and mother-infant separations in wild bottlenose dolphins. *Anim. Behav.* 55:1–17.

Margaria, R., R. P. Cerretelli, P. Aghemo, and G. Sassi. 1963. Energy cost of running. *J. Appl. Physiol.* 18:367–70.

Marler, P. 1957. Species distinctiveness in the communication signals of birds. *Behaviour* 11:13–39.

———. 1965. Communication in monkeys and apes. In *Primate behavior,* edited by I. DeVore, 544–84. New York: Holt, Rinehart & Winston.

———. 1970. Vocalizations of East African monkeys. *Folia Primatol.* 13:8191.

———. 1976a. Sensory templates in species-specific behavior. In *Simpler networks and behavior,* edited by J. Fentress, 314–29. Sunderland, MA: Sinauer Associates.

———. 1976b. Social organization, communication and graded signals: The chimpanzee and gorilla. In *Growing points in ethology,* editcd by P. P. P. Bateson and R. A. Hinde, 239–80. Cambridge: Cambridge University Press.

Marler, P., and C. S. Evans. 1996. Bird calls: Just emotional displays or something more? *Ibis* 138:26–33.

Marler, P., C. S. Evans, and M. D. Hauser. 1992. Animal signals: Motivational, referential, or both? In *Nonverbal communication: Comparative and developmental approaches,* edited by H. Papousek, U. Jurgens, and M. Papousek, 66–86. Cambridge: Cambridge University Press.

Marler, P., and L. Hobbett. 1975. Individuality in a long-range vocalization of wild chimpanzees. *Z. Tierpsychol.* 38:97–109.

Marquardt, F. R. 1976. Zur akustischen Kommunikation beim Zwwergmungo *(Helogale undulata rufula).* Diplomarbeit, University of Marburg.

Marsh, C. W. 1981a. Ranging behaviour and its relation to diet selection in the Tana River red colobus *(Colobus badius rufomitratus).* *J. Zool.* 195:473–92.

———. 1981b. Time budget of the Tana River Red Cobobus. *Folia Primatol.* 35:30–50.

Martin, R. A. 1981. On extinct hominid population densities. *J. Hum. Evol.* 10:427–28.

Martin, R. D. 1972. Adaptive radiation and behaviour of the Malagasy lemurs. *Phil. Trans. R. Soc. Lond.* 264:320–52.

———. 1981a. Field studies of primate behaviour. *Symp. Zool. Soc. Lond.* 46:287–336.

———. 1981b. Relative brain size and basal metabolic rate in terrestrial vertebrates. *Nature* 293:57–60.

———. 1990. *Primate origins and evolution: A phylogenetic reconstruction.* London: Chapman and Hall.

———. 1996. Scaling of the mammalian brain: The maternal energy hypothesis. *News Physiol. Sci.* 11:149–56.

Martinez, D. R., and E. Klinghammer. 1970. The behavior of the whale *Orcinus orca:* A review of the literature. *Z. Tierpsychol.* 27:828–39.

Martins, E. P., and T. F. Hansen. 1996. The statistical analysis of interspecific data: A review and evaluation of the phylogenetic comparative methods. In *Phyloge-*

nies and the comparative methods in animal behavior, edited by E. P. Martins, 22–75. Oxford: Oxford University Press.

Marx, E. 1967. *Bedouin of the Negev.* Manchester: Manchester University Press.

Masataka, N. 1986. Rudimentary representational vocal signalling of fellow group members in spider monkeys. *Behaviour* 96:49–61.

Mason, W. A. 1968. Use of space by *Callicebus* groups. In *Primates: Studies in adaptation and variability,* edited by P. C. Jay, 200–216. New York: Holt, Rinehart & Winston.

Mason, W. A., and S. P. Mendoza. 1993. *Primate social conflict.* Albany, NY: State University of New York Press.

Matano, S. 1985. Volume comparisons in the cerebellar complex of primates. II. Cerebellar nuclei. *Folia Primatol.* 44:182–203.

Mate, B. R., S. Nieukirk, and S. D. Kraus. 1997. Satellite monitored movements of the Northern Right Whale. *J. Wildl. Mgmt.* 61(4):1393–1405.

Maurer, R., and V. Séguinot. 1995. What is modeling for? A critical review of the models of path integration. *J. Theor. Biol.* 175:457–75.

Maynard Smith, J. 1964. Group selection and kin selection. *Nature* 201:1145–46.

———. 1974. The theory of games and the evolution of animal conflicts. *J. Theor. Biol.* 47:209–21.

McArdle, W. D., F. I. Katch, and V. I. Katch. 1991. *Exercise physiology: Energy, nutrition, and human performance.* 3d ed. Philadelphia: Lea and Febiger.

McBride, A. F., and H. Kritzler. 1951. Observations on pregnancy, parturition, and post-natal behavior in the bottlenose dolphin. *J. Mammal.* 32 (3):251–66.

McCabe, J. T. 1984. Livestock management among the Turkana: A social and ecological analysis of herding in an East African pastoral population. Ph.D dissertation, State University of New York at Binghamton.

———. 1990. Success and failure: The breakdown of traditional drought coping institutions among the Turkana of Kenya. *J. Asian Afr. Stud.* 25, 3–4:146–60.

———. 1994. Pastoral mobility: New interpretations to old conceptual problems. In *African pastoralist systems,* edited by E. Fratkin, E. Roth, and K. Galvin, 69–89. Boulder, CO: Lynne Reiner Publishers.

McClure, E. H. 1967. The composition of mixed species flocks in lowland and sub-montane forest of Malaya. *Wilson Bull.* 79:131–55.

McComb, K. 1987. Roaring by red deer stags advances the date of oestrus in hinds. *Nature* 330:648–49.

———. 1991. Female choice for high roaring rates in red deer, *Cervus elaphus. Anim. Behav.* 41:79–88.

McComb, K., C. Packer, and A. E. Pusey. 1994. Roaring and numerical assessment in contests between groups of female lions, *Panthera leo. Anim. Behav.* 47:379–87.

McComb, K., A. E. Pusey, C. Packer, and J. Grinnell. 1993. Female lions can identify potentially infanticidal males from their roars. *Proc. R. Soc. Lond.* B 252:59–64.

McCowan, B., and D. Reiss. 1995a. Quantitative comparison of whistle repetoires from captive adult bottlenose dolphins (Delphinidae, *Tursiops truncatus*): A reevaluation of the signal whistle hypothesis. *Ethology* 100:194–209.

———. 1995b. Whistle contour development in captive-born infant bottlenose dolphins *(Tursiops truncatus):* The role of learning. *J. Comp. Psychol.* 109 (3):247–60.

McDonald, D. W., and D. G. Henderson. 1972. Aspects of the behavior and ecology of mixed species bird flocks in Kashmir. *Ibis* 119:481–91.

McDonald, M. V. 1989. Function of song in Scott's Seaside Sparrow *Amodramus maritima peninsulae. Anim. Behav.* 38:468–85.

McFarland, D. J. 1971. *Feedback mechanisms in animal behaviour.* London: Academic Press.

———. 1977. Decision making in animals. *Nature* 269:15–21.

McGraw, W. S. 1996. Cercopithecid locomotion, support use, and support availability in the Taï forest, Ivory Coast. *Am. J. Phys. Anthropol.* 100:507–22.

McHenry, H. M. 1988. New estimates of body weight in early hominids and their significance to encephalization and megadontia in "robust" australopithecines. In *Evolutionary history of the "robust" australopithecine,* edited by F. Grine, 133–48. New York: Aldine de Gruyter.

———. 1992. Body size and proportions in early hominids. *Am. J. Phys. Anthropol.* 87:407–31.

———. 1994. Behavioral ecological implications of early hominid body size. J. Hum. Evol. 27:77–87.

McNab, B. K. 1963. Bioenergetics and the determination of home range size. *Am. Nat.* 97:133–40.

———. 1995. Energy-expenditure and conservation in frugivorous and mixed-diet carnivorans. *J. Mammal.* 76:206–22.

McNab, B. K., and J. F. Eisenberg. 1989. Brain size and its relation to the rate of metabolism in mammals. *Am. Nat.* 133:157–67.

McNamara, J. M., and A. I. Houston. 1986. The common currency for behavioral decisions. *Am. Nat.* 127:358–78.

McNaughton, B. L., C. A. Barnes, J. L. Gerrard, K. Gothard, M. W. Jung, J. J. Knierem, H. Kudrimoti, Y. Qin, W. E. Skaggs, M. Suster, and K. L. Weaver. 1996. Deciphering the hippocampal polyglot: The hippocampus as a path integration system. *J. Exp. Biol.* 199:173–85.

McNaughton, B. L., L. L. Chen, and E. J. Markus. 1991. "Dead reckoning," landmark learning, and sense of direction: A neurophysiological and computational hypothesis. *J. Cog. Neurosci.* 3:190–202.

McNaughton, B. L., B. Leonard, and L. L. Chen. 1989. Cortical-hippocampal interactions and cognitive mapping: A hypothesis based on reintegration of the parietal and inferotemporal pathways for visual processing. *Psychobiology* 17:236–46.

McNaughton, B. L., and R. G. M. Morris. 1987. Hippocampal synaptic enhancement and information storage within a distributed memory system. *Trends Neurosci.* 10:408–15.

McNaughton, S. J. 1985. Ecology of a grazing ecosystem: The Serengeti. *Ecol. Monogr.* 55:259–94.

McNaughton, S. J., and F. F. Banyikwa. 1995. Plant communities and herbivory. In *Serengeti II: Dynamics, management, and conservation of an ecosystem,* edited by A. R. E. Sinclair and P. Arcese, 49–70. Chicago: University of Chicago Press.

McNeilage, A. R. 1995. Mountain gorillas in the Virunga Volcanos: Feeding ecology and carrying capacity. Ph.D. thesis, University of Bristol.

McNutt, J. W. 1996. Sex biased dispersal in African wild dogs, *Lycaon pictus. Anim. Behav.* 52:1067–77.

Mech, L. D. 1966. The wolves of Isle Royale. *U.S. Natl. Park Serv. Fauna* 7:1–210.

————. 1970. *The wolf: Ecology and behaviour of an endangered species.* Garden City, NY: The Natural History Press.

Medawar, P. B. 1976. Does ethology throw any light on human behaviour? In *Growing points in ethology,* edited by P. P. P. Bateson and R. A. Hinde, 497–506. Cambridge: Cambridge University Press.

Medley, K. E. 1990. Forest ecology and conservation in the Tana River National Primate Reserve, Kenya. Ph.D. dissertation, Michigan State University, East Lansing, MI.

Mehlman, P. T. 1984. Aspects of the ecology and conservation of the Barbary macaque in fir forest habitat of the Moroccan Rif mountains. In *The Barbary macaque: A case study in conservation,* edited by J. E. Fa, 165–99. New York: Plenum Press.

————. 1996. Branch shaking and related displays in wild Barbary macaques. In *Evolution and ecology of macaque societies,* edited by J. E. Fa and D. G. Lindburg, 503–26. Cambridge: Cambridge University Press.

Mehmet Ali, M. K., and F. Kamoun. 1993. Neural networks for shortest path computation and routing in computer networks. *IEEE Trans. Neural Networks* 4:941.

Meier, V. O., A. E. Rasa, and H. Scheich. 1983. Call system similarity in a ground-living social bird and a mammal in the bush habitat. *Behav. Ecol. Sociobiol.* 12:5–9.

Melnick, D. J., and M. C. Pearl. 1986. Cercopithecines in multimale groups: Genetic diversity and population structure. In *Primate societies,* edited by B. B. Smuts, D. L. Cheney, R. M. Seyfarth, R. W. Wrangham, and T. T. Struhsaker, 121–34. Chicago: University of Chicago Press.

Meltzoff, A. N. 1988. The human infant as *Homo imitans.* In *Social learning,* edited by T. R. Zentall and B. G. Galef, 319–41. Hillsdale, NJ: Lawrence Erlbaum Associates.

Menzel, C. R. 1986. Structural aspects of arboreality in titi monkeys *(Callicebus moloch). Am. J. Phys. Anthropol.* 70:167–76.

————. 1991. Cognitive aspects of foraging in Japanese monkeys. *Anim. Behav.* 41:397–402.

————. 1993. Coordination and conflict in *Callicebus* social groups. In *Primate social conflict,* edited by W. A. Mason and S. P. Mendoza, 253–90. Albany: State University of New York Press.

————. 1996. Structure-guided foraging in long-tailed macaques. *Am. J. Primatol.* 38:117–32.

————. 1997. Primates' knowledge of their natural habitats: As indicated in foraging. In *Machiavellian intelligence II: Extensions and valuations,* edited by A. Whiten and R. W. Byrne, 207–39. Cambridge: Cambridge University Press.

Menzel, C. R. In press. Unprompted recall and reporting of hidden objects by a chimpanzee *(Pan troglodytes)* after extended delays. *J. Comp. Psychol.*

————. 1997. Chimpanzee *(Pan paniscus)* spatial memory and communication in a 20 hectare forest. *Am. J. Primatol.* 42:134.

Menzel, E. W. Jr. 1971a. Communication about the environment in a group of young chimpanzees. *Folia Primatol.* 15:220–32.

————. 1971b. Group behavior in young chimpanzees: Responsiveness to cumulative novel changes in a large outdoor enclosure. *J. Comp. Physiol. Psychol.* 74:46–51.

————. 1973a. Chimpanzee spatial memory organization. *Science* 182:943–45.

————. 1973b. Leadership and communication in young chimpanzees. In: *Precultural primate behavior,* edited by E. W. Menzel, 192–225. Symposia of the Fourth International Congress of Primatology. vol. 1. Basel: Karger.

————. 1974. A group of young chimpanzees in a one-acre field. In *Behavior of nonhuman primates,* vol. 5, edited by A. M. Schrier and F. Stollnitz, 83–153. New York: Academic Press.

————. 1978. Cognitive mapping in chimpanzees. In *Cognitive processes in animal behavior,* edited by S. Hulse, H. Fowler and W. K. Honig, 375–422. Hillsdale, NJ: Lawrence Erlbaum Associates.

————. 1979a. Communication of object-locations in a group of young chimpanzees. In *The great apes,* edited by D. A. Hamburg and E. R. McGown, 359–71. Menlo Park, CA: Benjamin/Cummings.

————. 1979b. General discussion of the methodological problems involved in the study of social interaction. In *Social interaction analysis: Methodological issues,* edited by M. E. Lamb, S. J. Suomi, and G. R. Stephenson, 291–309. Madison: University of Wisconsin Press.

————. 1987. Behavior as a locationist views it. In *Cognitive processes and spatial orientation in animals and man,* vol. 1, edited by P. Ellen and C. Thinus-Blanc, 55–72. NATO ASI Series. Dordrecht, Holland: Martin Nijhoff.

Menzel, E. W. Jr., and M. K. Johnson. 1976. Communication and cognitive organization in humans and other animals. *Ann. NY Acad. Sci.* 280:131–42.

Menzel, E. W. Jr., and C. Juno. 1982. Marmosets *(Saguinus fuscicollis):* Are learning sets learned? *Science* 217:750–52.

————. 1985. Social foraging in marmoset monkeys and the question of intelligence. *Phil. Trans. R. Soc. Lond.* B 308:145–58.

Menzel, R., L. Chittka, S. Eichmüller, K. Geiger, D. Peitsch, and P. Knoll. 1990. Dominance of celestial cues over landmarks disproves map-like orientation in honey bees. *Z. Naturforschung* C 45:723–26.

Merigan, W. H., and J. H. R. Maunsell. 1993. How parallel are the primate visual pathways? *Annu. Rev. Neurosci.* 16:369–402.

Mertl-Millhollen, A. S. 1988. Olfactory demarcation of territorial but not home range boundaries by *Lemur catta. Folia Primatol.* 50:175–87.

Messeri, P., E. Masi, R. Piazza, and Dessi'-Fulgheri, F. 1987. A study of the vocal repertoire of the banded mongoose *(Mungos mungos). Monitore Zoologico Italiano* (N.S.) Suppl. 22:341–73.

Messier, F. 1985. Social organization, spatial distribution, and population density of wolves in relation to moose density. *Can. J. Zool.* 63:1068–77.

Michaelsen, L. K., W. E. Watson, and R. H. Black. 1989. A realistic test of individual versus group consensus decision making. *J. Appl. Psychol.* 74:834–39.

Michener, C. D. 1974. *The social behavior of the bees.* Cambridge, MA: Harvard University Press.

Mills, M. G. L. 1985. Related spotted hyaenas forage together but do not cooperate in rearing young. *Nature* 316:61–62.

————. 1990. *Kalahari hyaenas: The behavioural ecology of two species.* London: Unwin Hyman.

Milton, K. 1977. The foraging strategy of mantled howler monkeys in the tropical forest on Barro Colorado Island. Ph.D. dissertation, New York University.

———. 1979a. Factors influencing leaf choice by howler monkeys: A test of some hypotheses of food selection by generalist herbivores. *Am. Nat.* 117:476–95.

———. 1979b. Spatial and temporal patterns of plant food in tropical forests as a stimulus to intellectual development in primates. *Am. J. Phys. Anthropol.* 50:464–65.

———. 1980. *The foraging strategy of howler monkeys: A study in primate economics.* New York: Columbia University Press.

———. 1981a. Diversity of plant foods in tropical forests as a stimulus to mental development in primates. *Am. Anthropol.* 83:534–48.

———. 1981b. Food choice and digestive strategies of two sympatric primate species. *Am. Nat.* 117:476–95.

———. 1982. The role of resource seasonality in density regulation of a wild primate population. In *Ecology of a tropical forest,* edited by E. G. Leigh, A. S. Rand, and D. W. Windsor, 273–89. Washington, DC: Smithsonian Institution Press.

———. 1984. Habitat, diet, and activity patterns of free-ranging woolly spider monkeys (*Brachyteles arachnoides* E. Geoffroyi 1806). *Int. J. Primatol.* 5:491–514.

———. 1987. Primate diets and gut morphology. In *Food and evolution: Toward a theory of human food habits,* edited by M. Harris and E. B. Ross, 93–116. Philadelphia: Temple University Press.

———. 1988. Foraging behaviour and the evolution of primate cognition. In *Machiavellian intelligence: Social expertise and the evolution of intellect in monkeys, apes and humans,* edited by R. W. Byrne and A. Whiten, 285–305. Oxford: Clarendon Press.

———. 1991. Annual patterns of leaf and fruit production by six Neotropical Moraceae species. *J. Ecol.* 79:1–26.

———. 1992. Comparative aspects of diet in Amazonian forest-dwellers. In *Foraging strategies and natural diets of monkeys, apes and humans,* edited by A. Whiten and E. M. Widdowson, 93–103. Oxford: Clarendon Press.

———. 1993a. Diet and social behavior of a free-ranging spider monkey population: The development of species-typical behaviors in the absence of adults. In *Juvenile primates: Life history, development and behavior,* edited by M. E. Pereira and L. A. Fairbanks, 173–81. Oxford: Oxford University Press.

———. 1993b. Diet and primate evolution. *Sci. Am.* 269(2):86–93.

Milton, K., and M. May. 1976. Body weight, diet and home range in primates. *Nature* 259:459–62.

Minetti, A. E., C. Capelli, P. Zamparo, P. E. di Prampero, and F. Saibene. 1995. Effects of stride frequency on mechanical power and energy expenditure of walking. *Med. Sci. Sports Exerc.* 27:1194–1202.

Mitani, J. C. 1985a. Gibbon song duets and intergroup spacing. *Behaviour* 92:59–96.

———. 1985b. Sexual selection and adult male orangutan long calls. *Anim. Behav.* 33:272–83.

———. 1988. Male gibbon *(Hylobates agilis)* singing behavior: Natural history, song variations and function. *Ethology* 79:177–94.

———. 1996. Comparative studies of African ape vocal behavior. In *Great ape societies,* edited by W. C. McGrew, L. F. Marchant, and T. Nishida, 241–54. Cambridge: Cambridge University Press.

Mitani, J. C., T. Hasegawa, J. Gros-Louis, P. Marler, and R. Byrne. 1992. Dialects in wild chimpanzees? *Am. J. Primatol.* 27:233–43.

Mitani, J. C., and Nishida, T. 1993. Contexts and social correlates of long-distance calling by male chimpanzees. *Anim. Behav.* 45:735–46.

Mitani, J. C., and P. S. Rodman. 1979. Territoriality: The relation of ranging pattern and home range size to defendability, with an analysis of territoriality among primate species. *Behav. Ecol. Sociobiol.* 5:241–51.

Mitchell, C. L. 1990. The ecological basis for female social dominance: A behavioral study of the squirrel monkey *(Saimiri sciureus)* in the wild. Ph.D. thesis, Princeton University, Princeton, NJ.

Mitchell, C. L., S. Boinski, and C. P. van Schaik. 1991. Competitive regimes and female bonding in two species of squirrel monkey *(Saimiri oerstedi* and *S. sciureus).* *Behav. Ecol. Sociobiol.* 28:55–60.

Mithen, S. 1996. The early prehistory of human social behaviour: Issues of archaeological inference and cognitive evolution. *Proc. Brit. Acad.* 88:145–77.

Mittermeier, R. A. 1977. The distribution, synecology and conservation of Surinam monkeys. Ph.D. thesis, Harvard University, Cambridge, MA.

Mittermeier, R. A., and M. G. M. van Roosmalen. 1981. Preliminary observations on habitat utilization and diet in eight Surinam monkeys. *Folia Primatol.* 36:1–39.

Mobley, J. R. Jr., and D. A. Helweg. 1990. Visual ecology and cognition in cetaceans. In *Sensory abilities of cetaceans: Laboratory and field evidence,* edited by J. Thomas and R. A. Kastelein, 519–35. Plenum Press, New York.

Mobley, J. R. Jr., and L. M. Herman. 1995. Transience of social affiliations among humpback whales *(Megaptera novaeangliae)* on the Hawaiian wintering grounds. *Can. J. Zool.* 63:762–72.

Moehlman, P. D. 1979. Jackal helpers and pup survival. *Nature* 277:382–83.

———. 1983. Socioecology of silver-backed and golden jackals *(Canis mesomelas, C. aureus).* In *Recent advances in the study of mammalian behavior,* edited by J. F. Eisenerg and D. G. Kleiman, 423–53. Special publication no. 7. Lawrence, KS: American Society of Mammalogists.

———. 1989. Intraspecific variation in canid social systems. In *Carnivore behavior, ecology and evolution,* edited by J. L. Gittleman, 143–63. Ithaca, NY: Cornell University Press.

Møller, A. P. 1988. False alarms as a means of resource usurpation in the Great Tit Parus major. *Ethology* 798:25–30.

Møller, K.-H. 1995. Ranging in masked titi monkeys *(Callicebus personatus)* in Brazil. *Folia Primatol.* 65:224–28.

Mollon, J. D. 1989. Tho she kneeld in that place where they grew: The uses and origins of primate color-vision. *J. Exp. Biol.* 146:21–38.

Møkkønen, M., J. T. Forsman, and P. Hella. 1996. Multi-species foraging aggregations and heterospecific attraction in boreal bird communities. *Oikos* 77:127–36.

Moody, M. L., and E. W. Menzel. 1976. Vocalizations and their behavioral contexts in the tamarin, *Saguinus fuscicollis. Folia Primatol.* 25:73–94.

Moore, J. A. 1981. The effect of information networks in hunter-gatherer societies. In *Hunter-gatherer foraging strategies,* edited by B. Winterhalder and E. A. Smith, 194–217. Chicago: University of Chicago Press.

Moore, K. E., W. A. Watkins, and P. L. Tyack. 1993. Pattern similarity in shared codas from sperm whales *(Physeter catodon). Mar. Mammal Sci.* 9:1–9.

Morgane, P. J., M. S. Jacobs, and A. Galaburda. 1986. Evolutionary morphology of the dolphin brain. In *Dolphin cognition and behavior: A comparative approach,* edited by R. J. Schusterman, J. A. Thomas, and F. G. Wood, 5–30. Hillsdale, NJ: Lawrence Erlbaum Associates.

Mori, A. 1983. Comparison of the communicative vocalizations and behaviors of group ranging in eastern gorillas, chimpanzees and pygmy chimpanzees. *Primates* 24:486–500.

Morland, H. S. 1990. Parental behavior and infant development in ruffed lemurs *(Varecia variegata)* in a northeast Madagascar rainforest. *Am. J. Primatol.* 20:253–65.

———. 1991. Preliminary report on the social organization of ruffed lemurs *(Varecia variegata variegata)* in a northeast Madagascar rain forest. *Folia Primatol.* 56:157–61.

Morris, R. G. 1981. Spatial localization does not require the presence of local cues. *Learn. Motiv.* 12:239–60.

Morris, R. G. M., E. Anderson, G. S. Lynch, and M. Baudry. 1986. Selective impairment of learning and blockade of long-term potentiation by an N-methyl-D-aspartate receptor agonist, AP5. *Nature* 319:774–76.

Morrison, D. W. 1975. The foraging behavior and feeding ecology of a Neotropical fruit bat, *Artibeus jamaicensis.* Ph.D. dissertation, Cornell University, Ithaca, NY.

Morse, D. H. 1970. Ecological aspects of some mixed-species foraging flocks of birds. *Ecol. Monogr.* 40:119–68.

———. 1977. Feeding behavior and predator avoidance in heterospecific groups. *Bioscience* 27:332–39.

———. 1978. Structure and foraging patterns of flocks of tits and associated species in an English woodland during winter. *Ibis* 120:298–312.

Morton, A. 1990. A quantitative comparison of the behaviour of resident and transient forms of the killer whale off the Central British Columbia coast. International Whaling Commission Special Report no. 12.

Morton, C. S. 1980. Adaptations to seasonal changes by migrant land birds in the Panama Canal Zone. In *Migrant birds in the Neotropics: Behavior, ecology, distribution, and conservation,* edited by A. Keast and E. S. Morton, 437–53. Washington, DC: Smithsonian Institution Press.

Mostrom, A. 1993. The social organization of Carolina Chickadee *(Parus carolinensis)* flocks in the non-breeding season. Ph.D. dissertation, University of Pennsylvania, Philadelphia.

Moutsouka, S. 1980. Pseudo-warning calls in titmice. *Tori* 29:97–90.

Moynihan, M. 1962. The organization and probable evolution of some mixed species flocks of Neotropical birds. *Smithsonian Miscellaneous Collections* 143:1–140.

———. 1968. Social mimicry: Character convergence versus character displacement. *Evolution* 22:315–31.

———. 1976. *The New World primates.* Princeton: Princeton University Press.

Müller, M., and R. Wehner. 1988. Path integration in desert ants, *Cataglyphis fortis. Proc. Natl. Acad. Sci. USA* 85:5287–90.

Mundinger, P. C. 1970. Vocal imitation and individual recognition of finch calls. *Science* 168:480–82.

Munn, C. A. 1985. Permanent canopy and understory flocks in Amazonia: Species composition and population density. *Ornithol. Monogr.* 36:683–710.

———. 1986. Birds that "cry wolf." *Nature* 319:143–45.

Munn, C. A., and J. Terborgh. 1979. Multispecies territoriality in Neotropical foraging flocks. *Condor* 81:338–47.

Murray, M. G. 1995. Specific nutrient requirements and migration of wildebeest. In *Serengeti II: Dynamics, management, and conservation of an ecosystem,* edited by A. R. E. Sinclair and P. Arcese, 231–56. Chicago: University of Chicago Press.

Muskin, A. 1984. Field notes and geographic distribution of *Callithrix aurita* in eastern Brazil. *Am. J. Primatol.* 7:377–80.

Mwanza, N.J., J. Yamagiwa, T. Yumoto, and T. Maruhashi. 1992. Distribution and range utilization of eastern lowland gorillas. In *Topics in primatology,* vol. 2, *Ecology, behavior, and conservation,* edited by N. Itoigawa, Y. Sugiyama, G. P. Sackett, and R. K. K. Thompson, 283–300. Tokyo: University of Tokyo Press.

Myers, M. J., K. L. Steudel, and S. C. White. 1993. Uncoupling the correlates of locomotor costs: A factorial approach. *J. Exp. Zool.* 265:211–23.

Nachtigall, P. 1986. Vision, audition and chemoreception in dolphins and other marine mammals. In *Dolphin cognition and behavior: A comparative approach,* edited by R. J. Schusterman, J. A. Thomas, and F. G. Wood, 79–114. Hillsdale, NJ: Lawrence Erlbaum Associates.

Nadel, L. 1990. Varieties of spatial cognition: Psychobiological considerations. *Ann. NY Acad. Sci.* 608:613–36.

Nagamachi, C. Y., J. C. Pieczarka, M. Schwarz, R. M. S. Barros, and M. Mattevi. 1997. Chromosomal similarities and differences between tamarins, *Leontopithecus* and *Saguinus* (Platyrrhini, Primates). *Am. J. Primatol.* 43:265–76.

Nagy, K. A. 1987. Field metabolic rate and food requirement scaling in mammals and birds. *Ecol. Monogr.* 57:111–28.

———. 1994. Field bioenergetics of mammals: What determines field metabolic rates? *Aust. J. Zool.* 42:43–53.

Nelson, R. K. 1973. *Hunters of the Northern Forest.* Chicago: University of Chicago Press.

New, D. A. T., and J. K. New. 1962. The dances of honey-bees at small zenith distances of the sun. *J. Exp. Biol.* 39:279–91.

Newell, A., J. C. Shaw, and H. A. Simon. 1958. Elements of a theory of human problem solving. *Psychol. Rev.* 65:151–66.

Newell, A., and H. A. Simon. 1972. *Human problem solving.* New York: Prentice-Hall.

Newton, P. N. 1992. Feeding and ranging patterns of forest Hanuman langurs *(Presbytis entellus). Int. J. Primatol.* 13:245–82.

Neyman, P. F. 1977. Aspects of the ecology and social organization of the free-ranging cotton-top tamarins *(Saguinus oedipus)* and the conservation status of the species. In *The biology and conservation of the Callitrichidae,* edited by D. G. Kleiman, 39–69. Washington, DC: Smithsonian Institution Press.

Nicholson, A. J. 1954. An outline of the dynamics of animal populations. *Aust. J. Zool.* 2:9–65.

Nishida, T. 1968. The social group of wild chimpanzees in the Mahali Mountains. *Primates* 9:167–224.

Noda, M., K. Gushima, and S. Kakuda. 1994. Local prey search based on spatial memory and expectation in the planktivorous reef fish, *Chromis chrysurus.* *Anim. Behav.* 47:1413–22.

Noë, R., and R. Bshary. 1997. The formation of red colobus-diana monkey associations under predation pressure from chimpanzees. *Proc. R. Soc. Lond.* B 264:253–59.

Noordwijk, M. A. van, C. Hemelrijk, L. Herremans, and E. Sterck. 1993. Spatial position and behavioral sex differences in juvenile long-tailed macaques. In *Juvenile primates: Life history, development and behavior,* edited by M. E. Pereira and L. A. Fairbanks, 77–85. New York: Oxford University Press.

Noordwijk, M. A. van, and C. P. van Schaik. 1986. Competition among female long-tailed macaques, *Macaca fascicularis. Anim. Behav.* 35:577–589.

Norconk, M. A. 1986. Interactions between primate species in a Neotropical forest: Mixed-species troops of *Saguinus mystax* and *S. fuscicollis* (Callitrichidae). Ph.D. thesis, University of California, Los Angeles.

———. 1990a. Introductory remarks: Ecological and behavioural correlates of polyspecific primate troops. *Am. J. Primatol.* 21:81–85.

———. 1990b. Mechanisms promoting stability in mixed *Saguinus mystax* and *S. fuscicollis* troops. *Am. J. Primatol.* 21:159–70.

Norconk, M. A., and W. G. Kinzey. 1994. Challenge of Neotropical frugivory: Travel patterns of spider monkeys and bearded sakis. *Am. J. Primatol.* 34: 171–83.

Norris, K. S. 1968. The evolution of acoustic mechanisms in odontocete cetaceans. In *Evolution and environment,* edited by E. T. Drake. New Haven, CT: Yale University Press.

Norris, K. S., and T. P. Dohl. 1980a. Behavior of the Hawaiian spinner dolphin, *Stenella longirostris. Fish. Bull.* 77 (4):821–49.

———. 1980b. The structure and functions of cetacean schools. In *Cetacean behavior: Mechanisms and functions,* edited by L. M. Herman. 211–61. New York: John Wiley & Sons.

Norris, K. S., and Harvey, G. W. 1974. Sound transmission in the porpoise head. *J. Acoust. Soc. Am.* 56:659–94.

Norris, K. S., and Schilt, C. R. 1988. Cooperative societies in three-dimensional space: On the origins of aggregations, flocks, and schools, with special reference to dolphins and fish. *Ethol. Sociobiol.* 9:149–79.

Norris, K. S., B. Würsig, R. S. Wells, M. Würsig, S. M. Brownlee, C. M. Johnson, and J. Solow. 1985. The behavior of the Hawaiian spinner dolphin, *Stenella longirostris.* Rept. LJ-85–06C. National Marine Fisheries Service.

Norton, G. W. 1986. Leadership decision processes of group movement in yellow baboons. In *Primate ecology and conservation,* edited by J. G. Else and P. C. Lee, 145–56. Cambridge: Cambridge University Press.

Norton, K. I., M. T. Jones, and R. B. Armstrong. 1990. Oxygen consumption and distribution of blood flow in rats climbing a laddermill. *J. Appl. Physiol.* 68:241–47.

Nottebohm, F. 1970. Ontogeny of bird song. *Science* 167:950–56.

Nowicki, S. 1983. Flock-specific recognition of chickadee calls. *Behav. Ecol. Sociobiol.* 12:317–20.

———. 1989. Vocal plasticity in captive Black-capped Chickadees: The acoustic basis and rate of call convergence. *Anim. Behav.* 37:64–73.

Noy-Meir, I. 1978. Stability in simple grazing models: Effects of explicit function. *J. Theor. Biol.* 71:347–80.

Nunney, L. 1985. Group selection, altruism, and structured-deme models. *Am. Nat.* 126:212–30.

Oates, J. F. 1977. The guereza and its food. In *Primate Ecology: Studies of feeding and ranging behaviour in lemurs, monkeys, and apes,* edited by T. H. Clutton-Brock, 276–321. London: Academic Press.

———. 1986. Food distribution and foraging behavior. In *Primate societies,* edited by B. B. Smuts, D. L. Cheney, R. M. Seyfarth, R. W. Wrangham, and T. T. Struhsaker, 197–209. Chicago: University Chicago Press.

Oates, J. F., and G. H. Whitesides. 1990. Association between olive colobus *(Procolobus verus),* diana guenons *(Cercopithecus diana),* and other forest monkeys in Sierra Leone. *Am. J. Primatol.* 21:129–46.

O'Brien, T. G., and M. F. Kinnaird. 1996. Changing populations of birds and mammals in North Sulawesi. *Oryx* 30(2):150–56.

———. 1997. Behavior, diet and movements of the Sulawesi crested black macaque, *Macaca nigra. Int. J. Primatol.* 18:321–52.

Oda, R. 1996. Effects of contextual and social variables on contact call production in free-ranging ringtailed lemurs *(Lemur catta). Int. J. Primatol.* 17:191–205.

O'Dea, K. 1992. Traditional diet and food preferences of Australian aboriginal hunter-gatherers. In *Foraging strategies and natural diets of monkeys, apes and humans,* edited by A. Whiten and E. M. Widdowson, 73–81. Oxford: Clarendon Press.

Odum, E. P. 1942. Annual cycle of the Black-capped Chickadee. 3. *Auk* 59:499–531.

———. 1971. *Fundamentals of ecology.* 3d ed. Philadelphia: W. B. Saunders.

O'Keefe, J., and L. Nadel. 1978. *The hippocampus as a cognitive map.* Oxford: Oxford University Press.

Olejniczak, C. 1994. Report on a pilot study of western lowland gorillas at Mbeli Bai, Nouabala-Ndoki Reserve, northern Congo. *Gorilla Conserv. News* 8:9–11.

Olesiuk, P. F., M. Bigg, and G. M. Ellis. 1990. Life history and population dynamics of resident killer whales *(Orcinus orca)* in the coastal waters of British Columbia and Washington State. International Whaling Commission Special Report 12.

Oliveira, A. C. M. 1996. Ecologia e comportamento alimentar de um grupo de *Saguinus midas niger* (Callitrichidae, Primates) na Amazonia oriental. M.Sc. thesis, Universidade Federal do Pará, Belem, Brazil.

Olmos, F. 1994. Jaguar predation on muriqui *Brachyteles arachnoides. Neotropical Primates* 2:16.

Olupot, W., C. A. Chapman, C. Brown, and P. Waser. 1994. Mangabey *(Cercocebus albigena)* population density, group size, and ranging: A twenty-year comparison. *Am. J. Primatol.* 32:197–205.

Opfinger, E. 1931. Über die Orientierung der Biene an der Futterquelle. *Z. vergl. Physiol.* 15:431–87.

Oppenheimer, J. R. 1982. Home range, population dynamics, and interspecific relationships. In *The ecology of a tropical forest: Seasonal rhythms and long-terms changes,* edited by E. G. Leigh, A. S. Rand, and D. M. Windsor, 253–72. Washington, DC: Smithsonian Institution Press.

Osorio, D., and M. Vorobyev. 1996. Colour vision as an adaptation to frugivory in primates. *Proc. R. Soc. Lond.* B 263:593–99.

Oster, G. F., and E. O. Wilson. 1978. *Caste and ecology in social insects.* Princeton, NJ: Princeton University Press.

Overdorff, D. J. 1988. Preliminary report on the activity cycle and diet of the red-bellied lemur *(Lemur rubriventer)* in Madagascar. *Am. J. Primatol.* 16:143–54.

————. 1993a. Ecological and reproductive correlates to range use in red-bellied lemurs *(Eulemur rubriventer)* and rufous lemurs *(Eulemur fulvus rufus).* In *Lemur social systems and their ecological basis,* edited by P. M. Kappeler and J. U. Ganzhorn, 167–78. New York: Plenum Press.

————. 1993b. Similarities, differences, and seasonal patterns in the diets of *Eulemur rubriventer* and *Eulemur fulvus rufus* in the Ranomafana National Park, Madagascar. *Int. J. Primatol.* 14:721–53.

————. 1996. Ecological correlates to social structure in two lemur species in Madagascar. *Am. J. Phys. Anthropol.* 100:487–506.

Overdorff, D. J., and Rasmussen, M. 1995. Determinants of nighttime activity in "diurnal" lemurid primates. In *Creatures of the dark: The nocturnal prosimians,* edited by L. Alterman, G. Doyle, and M. Izard, 61–74. New York: Plenum Press.

Owings, D. H., and D. W. Leger. 1980. Chatter vocalizations of California ground squirrels: Predator-and social-role specificity. *Z. Tierpsychol.* 54:163–84.

Packer, C. 1986. The ecology of sociality in felids. In *Ecological aspects of social evolution,* edited by D. I. Rubenstein and R. W. Wrangham, 429–51. Princeton, NJ: Princeton University Press.

Packer, C., and T. M. Caro. 1997. Foraging costs in social carnivores. *Anim. Behav.* 54:1317–18.

Packer, C., D. A. Gilbert, A. E. Pusey, and S. J. O'Brien. 1991. A molecular genetic analysis of kinship and cooperation in African lions. *Nature* 351:562–65.

Packer, C., L. Herbst, A. E. Pusey, J. D. Bygott, S. J. Cairns, J. P. Hanby, and M. Borgerhoff-Mulder. 1988. Reproductive success of lions. In *Reproductive success,* edited by T. Clutton-Brock, 363–83. Chicago: University of Chicago Press.

Packer, C., and A. E. Pusey. 1983. Adaptations of female lions to infanticide by incoming males. *Am. Nat.* 121:716–28.

————. 1984. Infanticide in carnivores. In *Infanticide: Comparative and evolutionary perspectives,* edited by G. Hausfater and S. Blaffer-Hrdy, 31–42. Hawthorne, NY: Aldine.

Packer, C., and L. Ruttan. 1988. The evolution of cooperative hunting. *Am. Nat.* 132:159–98.

Packer, C., D. Scheel, and A. E. Pusey. 1990. Why lions form groups: Food is not enough. *Am. Nat.* 136:1–19.

Packer, C., and A. E. Pusey. 1982. Cooperation and competition within coalitions of male lions: Kin selection or game theory? *Nature* 296:740–42.

Pagel, M. D., and P. H. Harvey. 1988. How mammals produce large-brained offspring. *Evolution* 42:948–57.

————. 1989. Comparative methods for examining adaptation depend on evolutionary models. *Folia Primatol.* 53:203–20.

Pailhous, J. 1984. The representation of urban space: Its development and its role in the organisation of journeys. In *Social representations,* edited by R. Raff and S. Moscovi, 311–27. Cambridge: Cambridge University Press.

Paine, R. 1971. Animals as capital: Comparisons among northern nomadic herders and hunters. *Am. Anthropol.* 44(3):157–72.

Park, T. 1954. Experimental studies of interspecific competition. II. Temperature, humidity and competition in two species of *Tribolium. Physiol. Zool.* 27:177–238.

Parker, G. A. 1974. Assessment strategy and the evolution of fighting behaviour. *J. Theor. Biol.* 47:223–43.

Parker, G. A., and P. Hammerstein. 1985. Game theory and animal behaviour. In *Evolution: Essays in honor of John Maynard Smith,* edited by P. J. Greenwood, P. H. Harvey, and M. Slatkin, 73–94. Cambridge: Cambridge University Press.

Parker, I. 1996. Richard Dawkins's evolution. *New Yorker,* September 9, 41–45.

Parker, S. T., and K. R. Gibson. 1977. Object manipulation, tool use and sensorimotor intelligence as feeding adaptations in great apes and cebus monkeys. *J. Hum. Evol.* 6:623–41.

Parker, S. T., and P. Poti. 1990. The role of innate motor patterns in ontogenetic and experimental development of intelligent use of sticks in cebus monkeys. In *Language and intelligence in monkeys and apes,* edited by S. T. Parker and K. R. Gibson, 219–43. Cambridge: Cambridge University Press.

Parsons, P. E., and Taylor, C. R. 1977. Energetics of brachiation versus walking: A comparison of a suspended and inverted pendulum mechanism. *Physiol. Zool.* 50:182–88.

Passamani, M. 1995. Field observations of a group of Geoffroy's marmosets mobbing a margay cat. *Folia Primatol.* 64:163–66.

Passos, F. C. 1992. Hábito alimentar do mico-leão-preto, *Leontopithecus chrysopygus* (Mikan, 1823) (Callithrichidae, Primates) na Estação Ecológica dos Caetetus, Município de Gália, SP. MSc thesis, Universidade Estadual de Campinas, Campinas.

———. 1997. Padrão de atividades, dieta e uso do espaço em um grupo de mico-leão-preto *(Leontopithecus chrysopygus)* na Estação Ecológica dos Caetetus, SP. Ph.D. thesis, Universidade de São Carlos, São Carlos, Brazil.

Paton, D.C., and F. L. Carpenter. 1984. Peripheral foraging by territorial rufous hummingbirds: Defense by exploitation. *Ecology* 65:1808–19.

Payne, K. B., W. R. Langbauer Jr., and E. M. Thomas. 1986. Infrasonic calls of the Asian elephant *(Elephas maximus). Behav. Ecol. Sociobiol.* 18:297–301.

Payne, R. S., and S. McVay. 1971. Songs of humpback whales. *Science* 173:585–97.

Payne, R. S., and D. Webb. 1971. Orientation by means of long range acoustic signalling in baleen whales. *Ann. NY Acad. Sci.* 188:110–41.

Pearson, R. 1972. *The avian brain.* London: Academic Press.

Peetz, A., M. A. Norconk, and W. G. Kinzey. 1992. Predation by jaguar on howler monkeys *(Alouatta seniculous)* in Venezuela. *Am. J. Primatol.* 28:223–28.

Pennycuick, C. J. 1975. On the running of the gnu *(Connochaetes taurinus)* and other animals. *J. Exp. Biol.* 63:775–800.

Pennycuick, L. 1975. Movements of the migratory wildebeest population in the Serengeti area between 1969 and 1973. *E. Afr. Wildl. J.* 13:65–87.

Pereira, M. E. 1995. Development and dominance among group-living primates. *Am. J. Primatol.* 37:143–75.

Pereira, M. E., and P. M. Kappeler. 1997. Divergent systems of agonistic behaviour in lemurid primates. *Behaviour* 134:225–74.

Pereira, M. E., A. Klepper, and E. L. Simons. 1987. Tactics of care for young infants by forest-living ruffed lemurs *(Varecia variegata variegata):* Ground nests, parking, and biparental guarding. *Am. J. Primatol.* 13:129–44.

Pereira, M. E., and J. M. Macedonia. 1991. Ringtailed lemur anti-predator calls denote predator class, not response urgency. *Anim. Behav.* 41:543–44.

Pereira, M. E., M. L. Seeligson, and J. M. Macedonia. 1989. The behavioral repertoire of the black and white ruffed lemur, *Varecia variegata variegata* (Primates, Lemuridae). *Folia Primatol.* 51:1–32.

Peres, C. A. 1986a. Costs and benefits of territorial defense in golden lion tamarins, *Leontopithecus rosalia.* M.Sc. thesis, University of Florida, Gainesville.

———. 1986b. Ranging patterns and habitat selection in golden lion tamarins, *Leontopithecus rosalia* (L., 1766). In *A primatologia no Brasil,* vol. 2., edited by M. Thiago de Mello, 223–33. Belo Horizonte: Universidade Federal de Minas Gerais.

———. 1989. Costs and benefits of territorial defense in golden lion tamarins *(Leontopithecus rosalia). Behav. Ecol. Sociobiol.* 25:227–33.

———. 1990. Effects of hunting on western Amazonian primate communities. *Biol. Conserv.* 54:47–59.

———. 1991a. Ecology of mixed-species groups of tamarins in Amazonian terra firme forests. Ph.D. thesis, University of Cambridge.

———. 1991b. Intergroup interactions, movements, and use of space in wild golden lion tamarins *(Leontopithecus rosalia).* In *A primatologia no Brasil,* vol. 3, edited by A. B. Rylands, C. Valle, and G. B. Fosenca, 173–89. Belo Horizonte: Universidade Federal de Minas Gerais.

———. 1992a. Consequences of joint territoriality in a mixed-species group of tamarin monkeys. *Behaviour* 123:220–46.

———. 1992b. Prey capture benefits in a mixed-species group of Amazonian tamarins, *Saguinus fuscicollis* and *S. mystax. Behav. Ecol. Sociobiol.* 31:339–47.

———. 1993a. Antipredation benefits in mixed-species groups of Amazonian tamarins. *Folia Primatol.* 61:61–76.

———. 1993b. Diet and feeding ecology of saddle-back and moustached tamarins in an Amazonian terra firme forest. *J. Zool.,* 230:567–92.

———. 1993c. Structure and spatial organization of an Amazonian terra firme forest primate community. *J. Trop. Ecol.* 9:259–76.

———. 1994. Primate responses to phenological changes in an Amazonian terra firme forest. *Biotropica* 26:98–112.

———. 1996a. Food patch structure and plant resource partitioning in interspecific associations of Amazonian tamarins. *Int. J. Primatol.* 17:695–724.

———. 1996b. Use of space, spatial group structure, and foraging group size of Gray Woolly Monkeys *(Lagothrix lagotricha cana)* at Urucu, Brazil: A review of the Atelinae. In *Adaptive radiation of Neotropical primates,* edited by M. Norconk, A. L. Rosenberger, and P. A. Garber, 467–88. New York: Plenum Press.

———. 1997. Primate community structure at twenty Amazonian flooded and unflooded forests. *J. Trop. Ecol.* 13:381–405.

Perrett, D. I., J. K. Hietanen, M. W. Oram, and P. J. Benson. 1992. Organization and function of cells responsive to faces in the temporal cortex. *Phil. Trans. R. Soc. Lond.* B 335:23–30.

Peters, G., and W. C. Wozencraft. 1989. Acoustic communication by fissiped carnivores. In *Carnivore behavior, ecology, and evolution,* edited by J. L. Gittleman, 14–56, Ithaca. NY: Cornell University Press.

Peters, R. H. 1983. *The ecological implications of body size.* New York: Cambridge University Press.

Peterson, R. O. 1977. Wolf ecology and prey relationships on Isle Royale. National Park Service Monograph, Serial no. 11.

Petter, J. J. 1962. Ecological and behavioral studies of Madagascar lemurs in the field. Ann. NY. Acad. Sci. 102:206–82.

Petter, J. J., A. Schilling, and G. Pariente. 1975. Observations on behavior and ecology of *Phaner furcifer.* In *Lemur biology,* edited by I. Tattersall and R. W. Sussman, 209–18. New York: Plenum Press.

Phillips, K. A. 1995. Resource patch size and flexible foraging in white-face capuchins *(Cebus capucinus). Int. J. Primatol.* 16:509–21.

Phillips-Conroy, J. E., and C. J. Jolly. 1981. Sexual dimorphism in two subspecies of Ethiopian baboons *(Papio hamadryas)* and their hybrids. *Am. J. Phys. Anthropol.* 56:115–30.

Piaget, J., and B. Inhelder. 1956. *The child's conception of space.* London: Routledge and Kegan Paul.

Picardi, A., and W. Siefert. 1976. A tragedy of the commons in the Sahel. *Tech. Rev.* 78:42–51.

Pitcher, T. J., and J. K. Parrish. 1993. Functions of shoaling behaviour in teleosts. In *Behaviour of teleost fishes,* edited by T. J. Pitcher, 363–439. London: Chapman and Hall.

Plumptre, A. J. 1993. The effects of trampling damage by herbivores on the vegetation of the Parc National des Volcans, Rwanda. *Afr. J. Ecol.* 32:115–29.

———. 1995. The chemical composition of montane plants and its influence on the diet of large mammalian herbivores in the Parc National des Volcans, Rwanda. *J. Zool.* 235:323–37.

Pocock, J. W. 1956. Operations research: A challenge to management. In *Operations research: A basic approach,* 7–21. Special Report no. 13. New York: American Management Association.

Podolsky, R. D. 1990. Effects of mixed-species associations on resource use by *Saimiri sciureus* and *Cebus apella. Am. J. Primatol.* 21:147–58.

Pollock, J. I. 1975. Field observations on *Indri indri,* a preliminary report. In *Lemur biology,* edited by I. Tattersall and R. W. Sussman, 287–311. New York: Plenum Press.

———. 1979a. Female dominance in *Indri indri. Folia Primatol.* 31:143–64.

———. 1979b. Spatial distribution and ranging behavior in lemurs. In *The study of prosimian behavior,* edited by G. A. Doyle and R. D. Martin, 359–409. New York: Academic Press.

———. 1986. The song of the Indris (*Indri indri,* Primates: Lemuroidea): Natural history, form and function. *Int. J. Primatol.* 7:225–64.

Pook, A. G., and G. Pook. 1979. The conservation status of the Goeldi's monkey, *Callimico goeldii,* in Bolivia. *J. Jersey Wildl. Presev. Trust* 16:40–45.

———. 1981. A field study of the socio-ecology of the Goeldi's monkey *(Callimico goeldii)* in Northern Bolivia. *Folia Primatol.* 35:288–312.

————. 1982. Polyspecific association between *Saguinus fuscicollis, Saguinus labiatus,* and *Callimico goeldii* and other primates in north-western Bolivia. *Folia Primatol.* 38:196–216.

Poole, J. H., K. Payne, W. R. Langbauer Jr., and C. Moss. 1988. The social contexts of some very low frequency calls of African elephants. *Behav. Ecol. Sociobiol.* 22:385–92.

Popper, A. N. 1980. Sound emission and detection by delphinids. In *Cetacean behavior: Mechanisms and functions,* edited by L. M. Herman, 1–52. New York: J. Wiley & Sons.

Possingham, H. P. 1989. The distribution and abundance of resources encountered by a forager. *Am. Nat.* 133:42–60.

Post, D. G. 1978. Feeding and ranging behaviour of the yellow baboon. Ph.D. thesis, Yale University, New Haven, CT.

————. 1981. Activity patterns of yellow baboons *(Papio cynocephalus)* in the Amboseli National Park, Kenya. *Anim. Behav.* 29:357–74.

Potts, R. 1988. *Early hominid activities at Olduvai.* New York: Aldine.

Poucet, B. 1993. Spatial cognitive maps in animals: New hypotheses on their structure and neural mechanisms. *Psychol. Rev.* 100:163–82.

Poucet, B., C. Thinus-Blanc, and N. Chapuis. 1983. Route planning in cats, in relation to the visibility of the goal. *Anim. Behav.* 31:594–99.

Poulsen, B. O. 1994. Movements of single birds and mixed flocks between isolated fragments of cloud forest in Ecuador. *Stud. Neotrop. Fauna Environ.* 29:149–60.

Powell, G. V. N. 1974. Experimental analysis of the social value of flocking by Starlings *(Sturnus vulgaris)* in relation to predation and foraging. *Anim. Behav.* 22:508–18.

————. 1979. Structure and dynamics of interspecific flocks in a mid-elevational Neotropical forest. *Auk* 96:375–90.

————. 1980. Migrant participation in Neotropical mixed species flocks. In *Migrant birds in the Neotropics: Behavior, ecology, distribution, and conservation,* edited by A. Keast and E. S. Morton, 477–84. Washington, DC: Smithsonian Institution Press.

————. 1985. Sociobiology and adaptive significance of interspecific foraging flocks in the Neotropics. *Ornithol. Monogr.* 36:713–32.

————. 1989. On the possible contribution of mixed species flocks to species richness in Neotropical avifaunas. *Behav. Ecol. Sociobiol.* 24:387–94.

Pravosudov, V. V. 1987. Ecology of two closely related species of tits (*Parus cinctus* and *Parus montanus*) in the north western part of the USSR. *Ornitologia* (Moscow) 22:68–75.

Premack, D. 1976. *Intelligence in ape and man.* Hillsdale, NJ: Lawrence Erlbaum Associates.

Premack, D., and G. Woodruff. 1978. Does the chimpanzee have a theory of mind? *Behav. Brain Sci.* 1:515–26.

Preuss, T. M. 1993. The role of the neurosciences in primate evolutionary biology. In *Primates and their relatives in phylogenetic perspective,* edited by R. S. D. E. Macphee, 333–62. New York: Plenum Press.

Price, E. C. 1992. Adaptation of captive-bred cotton-top tamarins *(Saguinus oedipus)* to a natural environment. *Zoo Biol.* 11:107–20.

Prins, H. H. T. 1996. *Ecology and behaviour of the African buffalo.* London: Chapman & Hall.

Provenza, F. D., and R. P. Cincotta. 1993. Foraging as a self-organizational learning process: Accepting adaptability at the expense of predictability. In *Diet selection: An interdisciplinary approach to foraging behaviour,* edited by R. N. Hughes, 78–101. London: Blackwell Scientific Publications.

Pryor, K., and I. K. Kang. 1991. Social structure in spotted dolphins *(Stenella attenuata)* in the tuna purse-seine fishery in the Eastern Tropical Pacific. In *Dolphin societies: Discoveries and puzzles,* edited by K. Pryor and K. S. Norris, 161–98. Los Angeles: University of California Press.

Pulliam, H. R. 1973. On the advantages of flocking. *J. Theor. Biol.* 38:419–22.

———. 1975. On the theory of optimal diets. *Am. Nat.* 108:59–74.

Pulliam, H. R., and T. Caraco. 1984. Living in groups: Is there an optimal group size? In *Behavioural ecology,* edited by J. R. Krebs and N. Davies, 122–47. Oxford: Blackwell Scientific.

Purvis, A., and A. Rambaut. 1995. Comparative analysis by independent contrasts (CAIC): An Apple Macintosh application for analysing comparative data. *Comp. Appl. Biosci.* 11:247–51.

Pusey, A. E., and C. Packer. 1994. Non-offspring nursing in social carnivores. *Behav. Ecol.* 4:362–74.

Pyke, G. H. 1978. Optimal foraging: Movement patterns of bumblebees between inflorescences. *Theor. Popul. Biol.* 13:72–97.

———. 1983. Animal movements: An optimal foraging approach. In *The ecology of animal movement,* edited by I. R. Swingland and P. J. Greenwood. Oxford: Clarendon.

Pyke, G. H., H. R. Pulliam, and E. L. Charnov. 1977. Optimal foraging theory: A selective review of theory and tests. *Q. Rev. Biol.* 52:137–54.

Quek, V. S. H., and P. Trayhurn. 1990. Calorimetric study of the energetics of pregnancy in golden hamsters. *Am. J. Physiol.* 259:R807–R812.

Rabb, G. B., J. H. Woolpy, and B. E. Ginsburg. 1967. Social relationships in a group of captive wolves. *Am. Zool.* 7:305–11.

Radakov, D. V. 1973. *Schooling in the ecology of fish.* New York: Wiley.

Rajpurohit, L., and Sommer, V. 1991. Sex differences in mortality among langurs *(Presbytis entellus)* of Jodhpur, Rajasthan. *Folia Primatol.* 56:17–27.

Rall, J. A. 1986. Energetic aspects of skeletal muscle contraction: Implications of fiber types. In *Exercise and sports sciences reviews* 13, edited by R. L. Terjung, 33–74. New York: Macmillan.

Ramirez, M. 1984. Population recovery in the moustached tamarin *(Saguinus mystax):* Management strategies and mechanisms of recovery. *Am. J. Primatol.* 7:245–59.

———. 1986. Feeding ecology of the moustached tamarin, *Saguinus mystax.* In *A primatologia no Brasil,* vol. 2, edited by M. Thiago de Mello, 211–12. Belo Horizonte: Universidade Federal de Minas Gerais.

Ramirez, M., C. H. Freese, and J. Revilla. 1977. Feeding ecology of the pygmy marmoset, *Cebuella pygmaea.* In *The biology and conservation of the Callitrichidae,* edited by D. G. Kleiman, 91–104. Washington, DC: Smithsonian Institution Press.

Ranta, E., R. Hannu, and K. Lindström. 1993. Competition versus cooperation: Success of individuals foraging alone and in groups. *Am. Nat.* 142:42–58.

Rappaport, R. 1968. *Pigs for the ancestors: Ritual in the ecology of a New Guinea people.* New Haven, CT: Yale University Press.

———. 1990. Ecosystems, populations and people. In *The ecosystem approach in anthropology: From concept to practice,* 41–72. Ann Arbor: University of Michigan Press.

Rasa, O. A. E. 1977a. Differences in group member response to intruding conspecifics and potentially dangerous stimuli in dwarf mongooses *(Helogale parvula rufula).* Z. *Saugetierkunde* 42:108–12.

———. 1977b. The ethology and sociology of the dwarf mongoose *(Helogale parvula rufula).* Z. *Tierpsychol.* 43:337–406.

———. 1983. Dwarf mongoose and hornbill mutualism in the Taru Desert, Kenya. *Behav. Ecol. Sociobiol.* 12:181–90.

———. 1986a. Coordinated vigilance in dwarf mongoose family groups: The "watchman's song" hypothesis and the costs of guarding. Z. *Tierpsychol.* 71:340–44.

———. 1986b. *Mongoose watch.* New York: Doubleday.

———. 1987. The dwarf mongoose: A study of behavior and social structure in relation to ecology in a small social carnivore. *Adv. Stud. Behav.* 17:121–63.

Rasmussen, D. R. 1979. Correlates of patterns of range use of a troop of yellow baboons *(Papio cynocephalus).* I: Sleeping sites, impregnable females, births, and male emigrations and immigrations. *Anim. Behav.* 27:1098–1112.

———. 1983. Correlates of patterns of range use of a troop of yellow baboons *(Papio cynocephalus).* II. Spatial structure, cover density, food gathering, and individual behaviour patterns. *Anim. Behav.* 31:834–57.

Rasmussen, M., C. A. Barnes, and B. L. McNaughton. 1989. A systematic test of cognitive mapping, working memory, and temporal discontiguity theories of hippocampal function. *Psychobiology* 17:335–48.

Rasoloarison, R. M., B. P. N. Raslonadrasana, J. U. Ganzhorn, and S. M. Goodman. 1995. Predation on vertebrates in the Kirindy Forest, Western Madagascar. *Ecotropica* 1:59–65.

Ratnayeke, S., A. Bixler, and J. L. Gittleman. 1994. Home range movements of solitary, reproductive female coatis, *Nasua narica,* in southeastern Arizona. *J. Zool.* 233:322–26.

Ratnieks, F. L. W. 1991. Africanized bees: Natural selection for colonizing ability. In *The "African" honey bee,* edited by M. Spivak, D. J. C. Fletcher, and M. D. Breed, 119–35. Boulder, CO: Westview Press.

Real, L. A. 1994. Information processing and the evolutionary ecology of cognitive architecture. In *Behavioural mechanisms in evolutionary ecology,* edited by L. A. Real, 99–132. Chicago: University of Chicago Press.

Reed, C. R., T. G. O'Brien, and M. F. Kinnaird. 1997. Male social behavior and dominance hierarchy in the Sulawesi crested black macaque, *Macaca nigra. Int. J. Primatol.* 18:247–60.

Reiss, D., and B. McCowan. 1993. Spontaneous vocal mimicry and production by bottlenose dolphins *(Tursiops truncatus):* Evidence for vocal learning. *J. Comp. Psychol.* 107 (3):301–12.

Remis, M. 1997a. Ranging and grouping patterns of a lowland group *(Gorilla*

gorilla gorilla) at Bai Hokou, Central African Republic. *Am. J. Primatol.* 43:111–33.

———. 1997b. Western lowland gorillas *(Gorilla gorilla gorilla)* as seasonal frugivores: Use of variable resources. *Am. J. Primatol.* 43:87–109.

Rettig, N. L. 1978. Breeding behavior of the harpy eagle *(Harpia harpyja)*. *Auk* 95:629–43.

Rhine, R. J. 1975. The order of movement of yellow baboons. *(Papio cynocephalus)*. *Folia Primatol.* 23:72–104.

Rhine, R. J., and B. J. Westlund. 1981. Adult male positioning in baboon progressions: Order and chaos revisited. *Folia Primatol.* 35:77–116.

Richard, A. F. 1978. *Behavioral variation: Case study of a Malagasy lemur.* Lewisburg: Bucknell University Press.

———. 1985a. *Primates in nature.* New York: W. H. Freeman.

———. 1985b. Social boundaries in a Malagasy prosimian, the sifaka *(Propithecus verreauxi)*. *Int. J. Primatol.* 6:553–68.

———. 1986. Malagasy prosimians: Female dominance. In *Primate societies,* edited by B. B. Smuts, D. L. Cheney, R. M. Seyfarth, R. W. Wrangham and T. T. Struhsaker, 25–33. Chicago: University of Chicago Press.

Richard, A. F., P. Rakotomanga, and M. Schwartz. 1991. Demography of *Propithecus verreauxi* at Beza Mahafali, Madagascar: Sex ratio, survival, and fertility. *Am. J. Phys. Anthropol.* 84:307–22.

———. 1993. Dispersal by *Propithecus verreauxi* at Beza Mahafaly, Madagascar: 1984–1991. *Am. J. Primatol.* 30:1–20.

Richard, K. R., M. C. Dillon, H. Whitehead, and J. M. Wright. 1996. Patterns of kinship in groups of free-living sperm whales *(Physeter macrocephalus)* revealed by molecular genetic analyses. *Proc. Natl. Acad. Sci. USA* 93:8792–95.

Richards, A. F. 1996. Life history and behavior of female dolphins (*Tursiops* sp.) in Shark Bay, Western Australia. Ph.D. thesis, University of Michigan, Ann Arbor.

Richards, D. G., J. P. Wolz, and L. M. Herman. 1984. Vocal mimicry of computer-generated sounds and vocal labeling of objects by a bottlenosed dolphin, *Tursiops truncatus. J. Comp. Psychol.* 98:10–28.

Richerson, P., and R. Boyd. 1992. Cultural inheritance and evolutionary ecology. In *Evolutionary ecology and human behavior,* edited by E. A. Smith and B. Winterhalder, 61–92. New York: Aldine De Gruyter.

Ridgely, R. 1982. The distribution, status and conservation of Neotropical mainland parrots. Ph.D. dissertation, Yale University, New Haven, CT.

Ridgway, S. H. 1986. Physiological observations on dolphin brains. In *Dolphin cognition and behavior: A comparative approach,* edited by R. J. Schusterman, J. A. Thomas, and F. G. Wood, 31–60. Hillsdale, NJ: Lawrence Erlbaum Associates.

Rigamonti, M. M. 1993. Home range and diet in red ruffed lemurs *(Varecia variegata rubra)* on the Masoala peninsula, Madagascar. In *Lemur social systems and their ecological basis,* edited by P. M. Kappeler and J. U. Ganzhorn, 25–40. New York: Plenum Press.

Rigley, L. 1983. Dolphins feeding in a North Carolina salt marsh. *Whalewatcher* 17:3–5.

Rijksen, H. D. 1978. A field study on Sumatran orang-utans (*Pongo pygmaeus abelii,* Lesson 1827): Ecology, behaviour, and conservation. Wageningen, The Netherlands: H. Veenman and Zonen.

Robbins, M. L. 1995. A demographic analysis of male life history and social structure of mountain gorillas. *Behaviour* 132:21–47.

Robinson, G. E., and C. Dyer. F. 1993. Plasticity of spatial memory in honey bees: Reorientation following colony fission. *Anim. Behav.* 46:311–20.

Robinson, J. G. 1979. Vocal regulation of use of space by groups of titi monkeys *Callicebus moloch. Behav. Ecol. Sociobiol.* 5:1–15.

———. 1981a. Spatial structure in foraging groups of wedge-capped capuchin monkeys. *Cebus nigrivittatus. Anim. Behav.* 29:1036–56.

———. 1981b. Vocal regulation of inter- and intra-group spacing during boundary encounters in the titi monkey, *Callicebus moloch. Primates* 22:161–72.

———. 1982. Vocal systems regulating within-group spacing. In *Primate communication,* edited by C. T. Snowdon, C. H. Brown, and M. Peterson, 94–116. Cambridge: Cambridge University Press.

———. 1986. Seasonal variations in use of time and space by wedge-capped capuchin monkeys, *Cebus olivaceus:* Implications for foraging theory. Smithsonian Contributions to Zoology 431:1–60.

———. 1988. Demography and group structure in wedge-capped capuchin monkeys, *Cebus olivaceus. Behaviour* 104:202–32.

Robinson, J. G., and C. H. Janson. 1986. Capuchins, squirrel monkeys and atelines: Socioecological convergence with Old World primates. In *Primate societies,* edited by B. B. Smuts, D. L. Cheney, R. M. Seyfarth, R. W. Wrangham, and T. T. Struhsaker, 69–82. Chicago: University of Chicago Press.

Robinson, J. G., P. C. Wright, and W. G. Kinzey. 1986. Monogamous cebids and their relatives: Intergroup calls and spacing. In *Primate societies,* edited by B. B. Smuts, D. L. Cheney, R. M. Seyfarth, R. W. Wrangham, and T. T. Struhsaker, 44–53. Chicago: University of Chicago Press.

Rodman, P. S. 1973a. Population composition and adaptive organization among orang-utans of the Kutai reserve. In *Comparative ecology and behaviour of primates,* edited by R. P. Michael and J. H. Crook, 171–209. London: Academic Press.

———. 1973b. Synecology of Bornean primates: I. A test for interspecific interactions in spatial distribution of five species. *Am. J. Phys. Anthropol.* 38:655–60.

———. 1988. Resources and group sizes of forest primates. In *The ecology of social behavior,* edited by C. N. Slobodchikoff, 83–108. New York: Academic Press.

———. 1991. Structural differentiation of microhabitats of sympatric *Macaca fascicularis* and *M. nemistrina* in East Kalimantan, Indonesia. *Int. J. Primatol.* 12(4):357–75.

Rodman, P. S., and J. C. Mitani. 1986. Orangutans: Sexual dimorphism in a solitary species. In *Primate societies,* edited by B. B. Smuts, D. L. Cheney, R. M. Seyfarth, R. W. Wrangham, and T. T. Struhsaker, 146–54. Chicago: University of Chicago Press.

Rogers, M. E., E. L. Williamson, C. E. G. Tutin, and M. Fernandez. 1988. Effects of the dry season on gorilla diet in Gabon. *Primate Rep.* 22:25–33.

Roitblat, H. L. 1987. *Introduction to comparative cognition.* New York: W. H. Freeman.

Ron, T. 1996. Who is responsible for fission in a free-ranging troop of baboons? *Ethology* 102:128–33.

Ron, T., S. P. Henzi, and U. Motro. 1996. Do female chacma baboons compete for a safe spatial position in a southern woodland habitat? *Behaviour* 133:475–90.

Rood, J. P. 1978. Dwarf mongoose helpers at the den. *Z. Tierpsychol.* 48:277–87.

———. 1983. The social system of the dwarf mongoose. In *Advances in the study of mammalian behavior,* edited by J. F. Eisenberg and D. Kleiman, 454–88. Special Publication no. 7. Lawrence, KS: American Society of Mammalogists.

———. 1986. Ecology and social evolution in the mongooses. In *Ecological aspects of social evolution,* edited by D. I. Rubenstein and R. W. Wrangham, 131–52. Princeton, NJ: Princeton University Press.

———. 1990. Group size, survival, reproduction and routes to breeding in dwarf mongoose. *Anim. Behav.* 39:566–72.

Roosevelt, T., and E. Heller. 1915. *Life histories of African game animals.* London: John Murray.

Roosmalen, M. G. M. van 1985. Habitat preferences, diet, feeding strategy and social organization of the black spider monkey (*Ateles paniscus paniscus* Linnaeus 1758) in Surinam. *Acta Amazonia* 15: 3/4c suplemento.

Rose, L. M. 1994a. Benefits and costs of resident males to females in white-faced capuchins, Cebus capucinus. *Am. J. Primatol.* 32:235–48.

———. 1994b. Sex differences in diet and foraging behaviour in white-faced capuchins *(Cebus capucinus). Int. J. Primatol.* 15:95–114.

———. 1997. Vertebrate predation and food-sharing in *Cebus* and *Pan. Int. J. Primatol.* 18:727–65.

Rose, L. M., and L. M. Fedigan. 1995. Vigilance in white-faced capuchins *(Cebus capucinus)* in Costa Rica. *Anim. Behav.* 49:63–70.

Rosenberger, A. L. 1984. Fossil New World monkeys dispute the molecular clock. *J. Hum. Evol.* 13:737–42.

Ross, C. 1992. Basal metabolic rate, body weight and diet in primates: An evaluation of the evidence. *Folia Primatol.* 58:7–23.

———. 1993. Predator mobbing by an all-male band of Hanuman langurs *(Presbytis entellus). Primates* 34:105–7.

Ross, G. J. B., and V. G. Cockroft. 1990. Comments on Australian bottlenose dolphins and the taxonomic status of *Tursiops aduncus* (Ehrenberg, 1832). In *The bottlenose dolphin,* edited by S. Leatherwood and R. R. Reeves, 101–28. Academic Press, San Diego, CA.

Roubik, D. W. 1989. *Ecology and natural history of tropical bees.* Cambridge: Cambridge University Press.

Roughgarden, P. 1997. *Principles of population ecology.* New York: McGraw Hill.

Rowell, T. 1972a. *The social behaviour of monkeys.* Harmondsworth: Penguin Books.

———. 1972b. Female reproductive cycles and social behaviour in primates. *Adv. Stud. Behav.* 4:69–105.

Rowell, T. E., and D. K. Olson. 1983. Alternative mechanisms of social organisation in monkeys. *Behaviour* 86:31–54.

Rowley, I., and G. Chapman. 1991. The breeding biology, food, social organization, demography and conservation of the Major Mitchell or pink cockatoo, *Cacatua leadbeateri,* on the margin of the western Australian wheatbelt. *Aust. J. Zool.* 39:211–61.

Rubenstein, R. I. 1986. Ecology and sociality in horses and zebras. In *Ecological aspects of social evolution,* edited by D. I. Rubenstein and R. W. Wrangham, 282–302. Princeton: Princeton University Press.

———. 1994. Ecology of female social behaviour in horses, zebras and asses. In *Animal societies: Individuals, interactions and organizations,* edited by P. J. Jarman and A. Rossiter, 13–28. Kyoto: Kyoto University Press.

Rudnai, J. A. 1973. *The social life of the lion.* Lancaster, UK: Medical and Technical Publishing

Ruiter, J. R. de 1986. The influence of group size on predator scanning and foraging behaviour of wedge-capped capuchin monkeys *(Cebus olivaceus). Behaviour* 98:240–58.

Russell, J. K. 1981. Exclusion of adult male coatis from social groups: Protection from predation. *J. Mammal.* 62:206–8.

———. 1983. Altruism in coati bands: Nepotism or reciprocity? In *Social behavior of female vertebrates,* edited by S. K. Wasser, 263–90. New York: Academic Press.

Russell, R. J. 1977. The behavior, ecology, and environmental physiology of a nocturnal primate, *Lepilemur mustelinus.* Ph.D. thesis, Duke University.

Ryan, M. J. 1990. Sexual selection, sensory systems, and sensory exploitation. *Oxford Surv. Evol. Biol.* 7:158–95.

———. 1994. Mechanisms underlying sexual selection. In *Behavioral mechanisms in evolutionary ecology,* edited by L. A. Real, 190–215. Chicago: University of Chicago Press.

Ryden, H. 1975. *God's dog.* New York: Coward, McCann, and Geoghegan.

Rylands, A. B. 1981. Preliminary field observations on the marmoset, *Callithrix humeralifer intermedius* (Hershkovitz, 1977) at Dardanelos, Rio Aripuanã, Mato Grosso. *Primates* 22:46–59.

———. 1982. The behaviour and ecology of three species of marmosets and tamarins (Callitrichidae, Primates) in Brazil. Ph.D. thesis, University of Cambridge, Cambridge.

———. 1986. Ranging behaviour and habitat preference of a wild marmoset group, *Callithrix humeralifer* (Callitrichidae, Primates). *J. Zool.* A 210:489–514.

———. 1989. Sympatric Brazilian callitrichids: The black tufted-ear marmoset, *Callithrix kuhli,* and the golden-headed lion tamarin, *Leontopithecus chrysomelas. J. Hum. Evol.* 18:679–95.

———. 1993. The ecology of the lion tamarins, *Leontopithecus:* Some intrageneric differences and comparisons with other callitrichids. In *Marmosets and tamarins: Systematics, behaviour, and ecology,* edited by A. B. Rylands, 296–313. Oxford: Oxford University Press.

Ryon, J. 1986. Den digging and pup care in captive coyotes *(Canis latrans). Can. J. Zool.* 64:1582–85.

Sacher, G. A. 1959. Relationship of lifespan to brain weight and body weight in mammals. In *CIBA Foundation Symposium on the Lifespan of Animals,* edited by G. E. W. Wolstenholme and M. O'Connor. Boston: Little Brown.

Sailer, L. D., S. J. C. Gaulin, J. S. Boster, and J. A. Kurland. 1985. Measuring the relationship between dietary quality and body size in primates. *Primates* 26:14–27.

Saisa, J., and Garling, T. 1987. Sequential spatial choices in the large-scale environment. *Environ. Behav.* 19:614–35.

Salzman, P. C. 1969. Multi-resource nomadism in Iranian Baluchistan. Paper presented at the Annual Meetings of the American Anthropological Association, New Orleans.

———. 1971. Movement and resource extraction among pastoral nomads: The case of the Shah Nawazi Baluch. *Anthropol. Q.* 44:185–97.

Sapolsky, R. M., and J. C. Ray. 1989. Styles of dominance and their endocrine correlates among wild olive baboons *(Papio anubis). Am. J. Primatol.* 18:1–13.

Saunders, D. A. 1983. Vocal repertoire and individual vocal recognition in the short-billed white-tailed black cockatoo, *Calyptorhynchus funereus latirostris* Carnaby. *Aust. J. Wildl. Res.* 10:527–36.

Sauther, M. L., and R. W. Sussman. 1993. A new interpretation of the social organization and mating system of the ringtailed lemur *(Lemur catta).* In *Lemur social systems and their ecological basis,* edited by P. M. Kappeler and J. U. Ganzhorn, 111–21. New York: Plenum Press.

Savage-Rumbaugh, E. S., K. McDonald, R. A. Sevcik, W. D. Hopkins, and E. Rubert. 1986. Spontaneous symbol acquisition and communicative use by pygmy chimpanzees *(Pan paniscus). J. Exp. Psychol. Gen.* 115:211–35.

Savage-Rumbaugh, E. S., J. Murphy, R. Sevcik, K. E. Brakke, S. L. Williams, and D. M. Rumbaugh. 1993. *Language comprehension in ape and child.* Monographs of the Society for Research on Child Development 58(3–4, Serial no. 233).

Savage-Rumbaugh, E. S., S. L. Williams, T. Furuichi, and T. Kano. 1996. Language perceived: *Paniscus* branches out. In *Great ape societies,* edited by W. C. McGrew, L. F. Marchant, and T. Nishida, 173–84. Cambridge: Cambridge University Press.

Sawaguchi, T. 1992. The size of the neocortex in relation to ecology and social structure in monkeys and apes *Folia Primatol.* 58:131–45.

Sawaguchi, T., and H. Kudo. 1990. Neocortical development and social structure in primates. *Primates* 31:283–89.

Sayigh, L. S. 1992. Development and functions of signature whistles of free-ranging bottlenose dolphins, *Tursiops truncatus.* Ph.D. thesis, Woods Hole Oceanographic Institution and Massachusetts Institute of Technology.

Sayigh, L. S., P. L. Tyack, and R. S. Wells. 1994. Recording underwater sounds of free-ranging dolphins while underway in a small boat. *Mar. Mammal Sci.* 9 (2):209–12.

Sayigh, L. S., P. L. Tyack, R. S. Wells, and M. D. Scott. 1990. Signature whistles of free-ranging bottlenose dolphins, *Tursiops truncatus:* Stability and mother-offspring comparisons. *Behav. Ecol. Sociobiol.* 26:247–60.

Scanlon, C. E., N. R. Chalmers, and M. A. O. Monteiro da Cruz. 1989. Home range use and exploitation of gum in the marmoset *Callithrix jacchus jacchus. Int. J. Primatol.* 10:123–36.

Schaik, C. P. van. 1983. Why are diurnal primates living in groups? *Behaviour* 87:120–44.

———. 1989. The ecology of social relationships amongst female primates. In *Comparative socioecology,* edited by V. Standen and R. A. Foley, 195–218. Cambridge: Blackwell.

———. 1996. Social evolution in primates: The role of ecological factors and male behaviour. *Proc. Brit. Acad.* 88:9–31.

Schaik, C. P. van, A. van Amerongen, and M. A. van Noordwijk. 1996. Riverine refuging by wild Sumatran long-tailed macaques. In *Evolution and ecology of macaque societies,* edited by J. E. Fa and D. G. Lindburg, 160–81. Cambridge: Cambridge University Press.

Schaik, C. P. van, P. R. Assink, and N. Salafsky. 1992. Territorial behavior in southeast Asian langurs: Resource defense or mate defense? *Am. J. Primatol.* 26:233–42.

Schaik, C. P. van, and R. I. M. Dunbar. 1990. The evolution of monogamy in large primates: A new hypothesis and some crucial tests. *Behaviour* 115:30–61.

Schaik, C. P. van, and J. A. R. A. M. van Hooff. 1983. On the ultimate causes of primate social systems. *Behaviour* 85:91–117.

———. 1996. Towards an understanding of the orangutan's social system. In *Great Ape Societies,* edited by W. C. McGrew, L. F. Marchant, and T. Nishida, 3–15. Cambridge: Cambridge University Press.

Schaik, C. P. van, and M. Hörstermann. 1994. Predation risk and the number of adult males in a primate group: A comparative test. *Behav. Ecol. Sociobiol.* 35:261–72.

Schaik, C. P. van, and P. M. Kappeler. 1993. Life history, activity period and lemur social systems. In *Lemur social systems and their ecological basis,* edited by P. M. Kappeler and J. U. Ganzhorn, 241–60. New York: Plenum Press.

———. 1996. The social systems of gregarious lemurs: Lack of convergence with anthropoids due to evolutionary equilibria? *Ethology* 102:915–41.

Schaik, C. P. van, and T. Mitrasetia. 1990. Changes in the behaviour of wild long-tailed macaques *(Macaca fascicularis)* after encounters with a model python. *Folia Primatol.* 55:104–8.

Schaik, C. P. van, and M. A. van Noordwijk. 1985. Evolutionary effect of the absence of felids on the social organization of the macaques on the island of Simeulue (*Macaca fascicularis,* Miller 1903). *Folia Primatol.* 44:138–47.

———. 1986. The hidden costs of sociality: Intra-group variation in feeding strategies in sumatran long-tailed macaques *(Macaca fascicularis). Behaviour* 99:296–315.

———. 1988. Scramble and contest in feeding competition among female long-tailed macaques *(Macaca fascicularis). Behaviour* 105:77–98.

———. 1989. The special role of male *Cebus* monkeys in predation avoidance and its effects on group composition. *Behav. Ecol. Sociobiol.* 24:265–76.

Schaik, C. P. van, M. A. van Noordwijk, R. J. Boer, and I. Den Tonkelaar. 1983. The effect of group size on time budgets and social behaviour in wild long-tailed macaques *(Macaca fascicularis). Behav. Ecol. Sociobiol.* 13:173–81.

Schaller, G. 1963. *The mountain gorilla: Ecology and behavior.* Chicago University of Chicago Press.

———. 1965. The behavior of the mountain gorilla. In *Primate behavior: Field studies of monkeys and apes,* edited by I. DeVore, 324–67. New York: Holt, Rinehart & Winston.

———. 1972. *The Serengeti lion: A study of predator-prey relations.* Chicago: University of Chicago Press.

———. 1983. Mammals and their biomass on a Brazilian Ranch. *Arquivos de Zoologia* 31:1–36.

Schaller, G., and G. Lowther. 1969. The relevance of carnivore behavior to the study of early hominids. *Southwest J. Anthropol.* 25:307–41.

Scheel, D., and C. Packer. 1991. Group hunting behavior of lions: A search for cooperation. *Anim. Behav.* 41:697–709.

Schenkel, R. 1947. Expression studies of wolves. *Behaviour* 1:81–129.

Scherer, K. 1992. Vocal affect expression as symptom, symbol, and appeal. In *Nonverbal communication: Comparative and developmental approaches,* edited by H. Papousek, U. Jurgens, and M. Papousek, 43–60. Cambridge: Cambridge University Press.

Schilling, A. 1979. Olfactory communication in prosimians. In *The study of prosimian behavior,* edited by G. A. Doyle and R. D. Martin, 461–542. New York: Academic Press.

———. 1980. The possible role of urine in territoriality of some nocturnal prosimians. *Symp. Zool. Soc. Lond.* 45:166–93.

Schmidt-Koenig, K., J. U. Ganzhorn, and R. Ranvaud. 1991. The sun compass. In *Orientation in birds,* edited by P. Berthold, 1–15. Basel: Birkhauser.

Schmidt-Nielsen, K. 1972. Locomotion: The energy cost of swimming, flying, and running. *Science* 177:222–28.

Schneider, H., M. I. C. Schneider, M. L. Harada, M. Stanhope, and M. Goodman. 1993. Molecular phylogeny of the New World monkeys (Platyrrhini, Primates). *Mol. Phylog. Evol.* 2:225–42.

Schneider, S. S., and L. C. McNally. 1994. Waggle dance behavior associated with seasonal absconding in colonies of the African honey bee, *Apis mellifera scutellata. Ins. Soc.* 41:115–27.

Schneirla, T. C. 1971. *The army ants: A study in social organization.* San Francisco: W. H. Freeman.

Schoener, T. W. 1971. Theory of feeding strategies. *Annu. Rev. Ecol. Syst.* 2:369–404.

———. 1974. The compression hypothesis and temporal resource partitioning. *Proc. Natl. Acad. Sci. USA* 71:4169–4172.

———. 1983. Simple models of optimal feeding-territory size: A reconciliation. *Am. Nat.* 121:608–29.

———. 1987. Time budgets and territory size: Some simultaneous optimization models for energy maximizers. *Am. Zool.* 27:259–91.

Schöne, H. 1984. *Spatial orientation.* Princeton, NJ: Princeton University Press.

Schultz, A. H. 1969. *The life of primates.* New York: Universe Books.

Scott, J. P. 1965. *Dog behavior.* Chicago: University of Chicago Press.

Scott, M. P. 1996. Communal breeding in burying beetles. *Am. Sci.* 84:376–82.

Seeley, T. D. 1985. *Honeybee ecology.* Princeton, NJ: Princeton University Press.

———. 1989. The honey bee colony as a superorganism. *Am. Sci.* 77:546–53.

———. 1995. *The wisdom of the hive: The social physiology of honey bee colonies.* Cambridge, MA: Harvard University Press.

———. 1997. Honey-bee colonies are group-level adaptive units. *Am. Nat.* 150: S22–S41.

Seeley, T. D., and S. C. Buhrman. 1999. Group decision making in swarms of honey bees. *Behav. Ecol. Sociobiol.* 45:19–31.

Seeley, T. D., and R. A. Morse. 1977. Dispersal behavior of honey bee swarms. *Psyche* 83:199–209.

———. 1978. Nest site selection by the honey bee, *Apis mellifera. Insectes Sociaux* 25:323–37.

Seeley, T. D., R. A. Morse, and P. K. Visscher. 1979. The natural history of the flight of honey bee swarms. *Psyche* 86:103–13.

Seeley, T. D., R. H. Seeley, and P. Akratanakul. 1982. Colony defense strategies of the honeybees in Thailand. *Ecol. Monogr.* 52:43–63.

Seger, J. 1991. Cooperation and conflict in social insects. In *Behavioural ecology: An evolutionary approach,* 3d ed., edited by J. R. Krebs and N. B. Davies, 338–73. Oxford: Blackwell.

Seidensticker, J. 1983. Predation by *Panthera* cats and measures of human influence in habitats of South Asian monkeys. *Am. J. Primatol.* 4:323–26.

Sekulic, R. 1981. The significance of howling in the red howler monkey, *Alouatta seniculus.* Ph.D. dissertation, Department of Zoology, University of Maryland.

———. 1982. The function of howling in red howler monkeys *(Alouatta seniculous). Behaviour* 81:38–54.

Seyfarth, R. M., and D. L. Cheney. 1980. The ontogeny of vervet monkey alarm-calling behavior: A preliminary report. *Z. Tierpsychol.* 54:37–56.

———. 1984. Grooming, alliances, and reciprocal altruism in vervet monkeys. *Nature* 308:541–43.

Sharpe, F., L. Dill, B. Spellman, and V. Beaver. 1998. Acoustic studies of bubblenet feeding groups of Alaskan humpback whales. Abstract. Twelfth Biennial Conference on the Biology of Marine Mammals, Monaco.

Shaw, E. 1962. The schooling of fishes. *Sci. Am.* 206:128–38.

Shea, B. T. 1983. Phyletic size change and brain/body allometry: A consideration based on the African pongids and other primates. *Int. J. Primatol.* 4:33–60.

Shemyakin, F. N. 1962. Orientation in space. In *Psychological science in the U.S.S.R.,* vol. 1, part 1 (Report No. 11466), edited by B. G. Anan'yer, 186–255. Washington, DC: U.S. Office of Technical Reports.

Sheppard, E. S. 1978. Theoretical underpinnings of the gravity hypothesis. *Geogr. Anal.* 10:386–402.

Sherman, P. T. 1991. Harpy eagle predation on a red howler monkey. *Folia Primatol.* 56:53–56.

Sherman, P. W. 1977. Nepotism and the evolution of alarm calls. *Science* 197:1246–53.

Sherman, P. W., J. U. M. Jarvis, and R. D. Alexander. 1991. *The biology of the naked mole-rat.* Princeton, NJ: Princeton University Press.

Sherry, D. F. 1989. Food storage in the Paridae. *Wilson Bull.* 101:289–304.

———. 1996. Middle-scale navigation: The vertebrate case. *J. Exp. Biol.* 199:163–64.

Sherry, D. F., and S. J. Duff. 1996. Behavioural and neural bases of orientation in food-storing birds. *J. Exp. Biol.* 199:165–72.

Sherry, D. F, J. R. Krebs, and R. J. Cowie. 1981. Memory for the location of stored food in marsh tits. *Anim. Behav.* 29:1260–66.

Sherry, D. F., and D. L. Schacter. 1987. The evolution of multiple memory systems. *Psychol. Rev.* 94:439–54.

Sherry, D. F., A. L. Vaccarino, K. Buckenham, and R. S. Herz. 1989. The hippocampal complex of food-storing birds. *Brain Behav. Evol.* 34:308–17.

Shettleworth, S. J. 1993. Varieties of learning and memory in animals. *J. Exp. Psychol.: Anim. Behav. Proc.* 19:5–14.

Shettleworth, S. J., and J. R. Krebs. 1986. Stored and encountered seeds: A comparison of two spatial memory tasks in marsh tits and chickadees. *J. Exp. Psych: Anim. Behav. Proc.* 12:248–57.

Shine, R., P. S. Harlow, J. S. Keogh, and Boeadi. 1998. The influence of sex and body size on food habits of a giant tropical snake, *Python reticulatus. Funct. Ecol.* 12:248–58.

Shulz, H.-D. 1967. Metrische untersuchungen an den schichten des corpus geniculatum laterale tag- und nachtaktiver primaten. Doctoral dissertation, Johann Wolfgang Geothe-Universitt Frankfurt.

Sicotte, P. 1993. Inter-group encounters and female transfer in mountain gorillas: Influence of group composition on male behavior. *Am. J. Primatol.* 30:21–36.

Siegel, A. W., and S. H. White. 1975. The development of spatial representation of large-scale environments. In *Advances in child development and behavior,* edited by H. W. Reese. New York: Academic Press.

Siegfried, W. R., and L. G. Underhill. 1975. Flocking as an antipredator strategy in doves. *Anim. Behav.* 23:504–8.

Sigg, J., and A. Stolba. 1981. Home range and daily march in a hamadryas baboon troop. *Folia Primatol.* 36:40–75.

Sih, A. 1987. Predators and prey lifestyles: An evolutionary and ecological overview. In *Predation,* edited by W. C. Kerfort and A. Sih, 203–24. Hanover: University Press of New England.

———. 1992. Prey uncertainty and the balancing of antipredator and feeding needs. *Am. Nat.* 139(5):1052–69.

Sih, A., and K. Milton. 1985. Optimal diet theory: Should the !Kung eat mongongos? *Am. Anthropol.* 87:396–401.

Silber, G. K., M. W. Newcomer, and M. H. Perez-Cortez. 1990. Killer whales *(Orcinus orca)* attack and kill a Bryde's whale *(Balaenoptera edeni). Can. J. Zool.* 68:1603–6.

Simila, T., and F. Ugarte. 1993. Surface and underwater observations of cooperatively feeding killer whales in northern Norway. *Can. J. Zool.* 71:1494–99.

Simon, H. A. 1981. *The sciences of the artificial.* 2nd ed. Cambridge, MA: MIT Press.

Sinclair, A. R. E., and J. M. Fryxell. 1985. The Sahel of Africa: Ecology of a disaster. *Can. J. Zool.* 63:987–94.

Skiena, S. 1997. *The algorithm design manual.* New York: Springer Verlag.

Skorupa, J. P. 1989. Crowned eagles *Stephanoaetus coronatus* in rainforest: Observations on breeding chronology and diet at a nest in Uganda. *Ibis* 131:294–98.

Smale, L., L. G. Frank, and K. E. Holekamp. 1993. Ontogeny of dominance in free-living spotted hyenas: Juvenile rank relations with adults. *Anim. Behav.* 46:467–77.

Smith, C. C. 1977. Feeding behaviour and social organization in howling monkeys. In *Primate societies,* edited by T. H. Clutton-Brock, 97–126. London: Academic Press.

Smith, E. A. 1991. *Inujjuamiut foraging strategies: Evolutionary ecology of an Arctic hunting economy.* New York: Aldine De Gruyter.

Smith, E. A., and B. Winterhalder. 1992. Natural selection and decision making: Some fundamental principles. In *Evolution, ecology and human behavior,* edited by E. A. Smith and B. Winterhalder, 25–60. New York: Walter de Gruyter.

Smith, J. M. N. 1974. The food searching behavior of two European thrushes. II. The adaptiveness of search patterns. *Behaviour* 49:1–61.

Smith, R. J. 1996. Biology and body size in human evolution: Statistical inference misapplied. *Curr. Anthropol.* 37:451–81.

Smith, S. M. 1991. *The Black-capped Chickadee: Behavioral ecology and natural history.* Ithaca, NY: Cornell University Press.

Smith, T. G., D. B. Siniff, R. Reichle, and S. Stone. 1981. Coordinated behavior of killer whales, *Orcinus orca,* hunting a crabeater seal, *Lobodon carcinophagus. Can. J. Zool.* 59:1185–89.

Smith, W. J. 1969. Messages of vertebrate communication. *Science* 165:145–50.

Smolker, R., J. Mann, and B. B. Smuts. 1993. Use of signature whistles during separations and reunions by wild bottlenose dolphin (*Tursiops* spp.) mothers and infants. *Behav. Ecol. Sociobiol.* 33:393–402.

Smolker, R., and J. W. Pepper. In press. Whistle convergence among allied male bottlenose dolphins (Delphinidae, *Tursiops* sp.). *Ethology* 105.

Smolker, R., and A. F. Richards. 1988. Loud sounds during feeding in Indian Ocean bottlenose dolphins. In *Animal sonar: Processes and performance,* edited by P. Nachtigall and P. W. B. Moore, 703–6. New York: Plenum Press.

Smolker, R., A. F. Richards, R. C. Connor, and J. W. Pepper. 1992. Patterns of association among Indian Ocean Bottlenose Dolphins. *Behaviour* 123:38–69.

Smuts, B. B. 1985. *Sex and friendship in baboons.* New York: Aldine de Gruyter.

Smuts, B. B., D. L. Cheney, R. M. Seyfarth, R. W. Wrangham, and T. T. Struhsaker (eds.). 1986. *Primate societies.* Chicago: Chicago University Press.

Smuts, B. B., and R. W. Smuts. 1993. Male aggression and sexual coercion of females in nonhuman primates and other mammals: Evidence and theoretical implications. *Adv. Stud. Behav.* 22:1–63.

Smythe, N. 1970. The adaptive value of the social organization of the coati *(Nasua narica). J. Mammal.* 51:818–20.

Snowdon, C. T., J. Cleveland, and J. A. French. 1983. Responses to context- and individual-specific cues in cotton-top tamarin long calls. *Anim. Behav.* 31:92–101.

Snowdon, C. T., and A. Hodun. 1981. Acoustic adaptations in pygmy marmoset contact calls: Locational cues vary with distance between conspecifics. *Behav. Ecol. Sociobiol.* 9:295–300.

Snowdon, C. T., A. Hodun, A. L. Rosenberger, and A. F. Coimbra-Filho. 1986. Long-call structure and its relation to taxonomy in lion tamarins. *Am. J. Primatol.* 11:253–61.

Snowdon, C. T., and P. Soini. 1988. The tamarins, genus *Saguinus.* In *Ecology and behavior of Neotropical primates,* vol. 2, edited by R. A. Mittermeier, A. F. Coimbra-Filho, and G. A. B. da Fonseca, 223–98. Washington, DC: World Wildlife Fund.

Sober, E., and D. S. Wilson. 1998. *Unto others: The evolution and psychology of unselfish behavior.* Harvard University Press.

Soini, P. 1981. The pygmy marmoset, genus *Cebuella.* In *Ecology and behavior of Neotropical primates,* edited by R. A. Mittermeier, A. B. Rylands, and A. F. Coimbra-Filho, 79–129. Rio de Janeiro: Academia Brasileira de Ciencias.

Soini, P. 1982. Ecology and population dynamics of the pygmy marmoset, *Cebuella pygmaea. Folia Primatol.* 39:1–21.

Soini, P. 1987. Ecology of the saddle-back tamarins *Saguinus fuscicollis illigeri* on the Rio Pacaya, northeastern Peru. *Folia Primatol.* 49:11–32.

Soini, P., and M. Cûppula. 1981. Ecologia y dinamica poblacional del pichico, *Saguinus fuscicollis* (Primates: Callitrichidae). Informe de Pacaya, No. 4.

Solano, C. 1996. Activity patterns and action area of the night monkey *(Aotus brumbacki)* Hershkovitz 1993 (Primates: Cebidae), Tinigua Natural National Park, Colombia. Paper presented at the XVIth Congress of the International Primatological Society, Madison, WI.

Soltz, R. L. 1986. Foraging path selection in bumblebees: Hindsight or foresight. *Behaviour* 99:1–21.

Soule, R. G., and R. F. Goldman. 1969. Energy cost of loads carried on the head, hands, or feet. *J. Appl. Physiol.* 27:687–90.

Soule, R. G., K. B. Pandolf, and R. F. Goldman. 1978. Energy expenditure of heavy load carriage. *Ergonomics* 21:373–81.

Srikosamatara, S. 1986. Group size in Wedge-capped Capuchin monkeys *(Cebus olivaceus):* Vulnerability to predators, intragroup and intergroup feeding competition. Ph.D. dissertation, University Florida, Gainesville, FL.

Srivastava, A. 1991. Cultural transmission of snake-mobbing in free-ranging Hanuman langurs, *Folia Primatol.* 56:117–20.

Stafford, B. J., and F. M. Ferreira. 1995. Predation attempts on Callitrichids in the Atlantic Coastal Rain Forest of Brazil. *Folia Primatol.* 65:229–33.

Stammbach, E. 1986. Desert forest and montane baboons: Multi-level societies. In *Primate societies,* edited by B. B. Smuts, D. L. Cheney, R. M. Seyfarth, R. W. Wrangham, and T. T. Struhsaker, 112–20. Chicago: University of Chicago Press.

Stamps, J. A. 1986. Conspecifics as cues to territory quality: A preference of juvenile lizards *(Anolis aeneus)* for previously used territories. *Am. Nat.* 129:629–42.

Stander, P. E. 1992a. Cooperative hunting in lions: The role of the individual. *Behav. Ecol. Sociobiol.* 29:445–54.

———. 1992b. Foraging dynamics of lions in a semi-arid environment. *Can. J. Zool.* 70:8–21.

Stander, P. E., and S. D. Albon. 1993. Hunting success of lions in a semi-arid environment. In *Mammals as predators,* edited by N. Dunstone and M. L. Gorman, 127–43. Zoological Society of London Symposium no 65. Oxford: Oxford Scientific Publications.

Stanford, C. B. 1989. Predation by jackals *(Canis aureus)* on capped langurs *(Presbytis pileata)* in Bangladesh. *Am. J. Primatol.* 26:53–56.

———. 1990. Colobine socioecology and female-bonded models of primate social structure. *Kroeber Anthropol. Soc. Pap.* 71–72:21–28.

———. 1991. *The capped langur in Bangladesh: Behavioral ecology and reproductive tactics.* Basel: Karger.

———. 1993. Mixed species flock composition in two forest types in Bangladesh. *J. Bombay Nat. Hist. Soc.* 90:99–103.

———. 1995a. Chimpanzee hunting behavior and human evolution. *Am. Sci.* 83:256–61.

———. 1995b. The influence of chimpanzee predation on group size and antipredator behavior in red colobus monkeys. *Anim. Behav.* 49:577–87.

———. 1998. Predation and male bonds in primate societies. *Behaviour* 135:513–33.

Stanford, C. B., J. Wallis, H. Matama, and J. Goodall. 1994. Patterns of predation by chimpanzees on red colobus monkeys in Gombe National Park, Tanzania 1982–1991. *Am. J. Phys. Anthropol.* 94:213–28.

Stanger, K. 1993. Structure and function of the vocalizations of nocturnal prosimians (Cheirogeleidae). Ph.D. thesis, University Tübingen, Germany.

Starin, D. 1993. The kindness of strangers. *Nat. Hist.* 10:44–49.

Steiner, W. J., J. H. Hain, H. E. Winn, and P. J. Perkins. 1979. Vocalizations and feeding behavior of the killer whale *(Orcinus orca). J. Mammal.* 60 (4):823–27.

Steltner, H., S. Steltner, and D. E. Sargent. 1984. Killer whales *(Orcinus orca)* prey on narwhals *(Monodon monoceros):* An eyewitness account. *Can. Field Nat.* 98 (4):458–62.

Stenning, D. 1957. Transhumance, migratory drift, migration: Patterns of pastoral Fulani nomadism. *J. R. Anthropol. Inst.* 57–73.

———. 1959. *Savanna nomads: A study of the Wodaabe pastoral Fulani of Western Bornu Province, Northern Region, Nigeria.* London: Oxford University Press.

Stephan, H., G. Baron, and H. Frahm. 1988. Comparative size of brains and brain components. In *Comparative primate biology,* vol. 4, edited by H. D. Steklis, and J. Erwin, 1–37. New York: Alan R. Liss.

Stephan, H., H. Frahm, and G. Baron. 1981. New and revised data on the brain structures in insectivores and primates. *Folia Primatol.* 35:1–29.

Stephens, D. W. 1993. Learning and behavioral ecology: Incomplete information and environmental unpredictability. In *Insect learning: Ecological and evolutionary perspectives,* edited by D. R. Papaj and A. C. Lewis, 195–218. New York: Chapman and Hall.

Stephens, D. W., and J. R. Krebs. 1986. *Foraging theory.* Princeton, NJ: Princeton University Press.

Sterck, E. A. 1997. Determinants of female dispersal in Thomas langurs. *Am. J. Primatol.* 42:179–98.

Sterck, E. A., and R. Steenbeck. 1997. Female dominance relations and food competition in the sympatric Thomas langur and longtail macaques. *Behaviour* 134:749–74.

Sterck, E. A., D. P. Watts, and C. P. van Schaik. 1997. The evolution of female social relationships in primates. *Behav. Ecol. Sociobiol.* 41:291–310.

Sterling, E. 1993. Patterns of range use and social organization in ayes-ayes *(Daubentonia madagascariensis)* on Nosy Mangabe. In *Lemur social systems and their ecological basis,* edited by P. M. Kappeler and J. U. Ganzhorn, 1–10. New York: Plenum Press.

Sterling, E., and A. Richard. 1995. Social organization in the aye-aye *(Daubentonia madagascariensis)* and the perceived distinctiveness of nocturnal primates. In *Creatures of the dark: The nocturnal prosimians,* edited by L. Alterman, G. Doyle and M. Izard, 439–51. New York: Plenum Press.

Steudel, K. 1990. The work and energetic cost of locomotion: 1. The effects of limb mass distribution in quadrupeds. *J. Exp. Biol.* 154:273–85.

Steudel, K., and J. Beattie. 1995. Does limb length predict the relative energetic cost of locomotion in mammals? *J. Zool.* 235:501–14.

Stevenson, M. F., and A. B. Rylands. 1988. The marmosets, genus *Callithrix.* In *Ecology and behavior of Neotropical primates,* vol. 2, edited by R. A. Mittermeier, A. B. Rylands, A. F. Coimbra-Filho and G. A. B. Fonseca, 131–222. Washington, DC: World Wildlife Fund.

Stevenson-Hamilton, J. 1954. *Wildlife in South Africa.* London: Cassell.

Stewart, K. J., and A. H. Harcourt. 1986. Gorillas: Variation in female relationship. In *Primate societies,* edited by B. B. Smuts, D. L. Cheney, R. M. Seyfarth, R. W. Wrangham, and T. T. Struhsaker, 155–64. Chicago: University of Chicago Press.

———. 1994. Gorillas' vocalizations during rest periods: Signals of impending departure? *Behaviour* 130:29–40.

Steyn, P. 1982. *Birds of prey of southern Africa.* Dover: Tanager Books.

Stillman, R. A., and W. J. Sutherland. 1990. The optimal search path in a patchy environment. *J. Theor. Biol.* 145:177–82.

Stolba, A. 1979. Entscheidungsfindung in Verbanden von *Papio hamadryas.* Dissertation, University of Zurich.

Stolz, L., and G. Saayman. 1970. Ecology and behavior of baboons in the Northern Transvaal. *Ann. Transvaal Mus.* 26:99–143.

Storer, J. A., and J. H. Reif. 1994. Shortest paths in the plane with polygonal obstacles. *J. Assoc. Comput. Mach.* 41:982–1012.

Strang, K., and K. Steudel. 1990. Explaining the scaling of transport costs: The role of stride frequency and stride length. *J. Zool.* 221:343–58.

Strier, K. B. 1987. Ranging behavior of woolly spider monkeys. *Int. J. Primatol.* 8:575–91.

———. 1989. Effects of patch size on feeding associations in muriquis *(Brachyteles arachnoides). Folia Primatol.* 52:70–77.

Struhsaker, T. T. 1967a. Auditory communication among vervet monkeys *(Cercopithecus aethiops).* In *Social communication among primates,* edited by S. A. Altmann, 281–324. Chicago: University of Chicago Press.

———. 1967b. Social structure among vervet monkeys *(Cercopithecus aethiops). Behaviour* 29:83–121.

———. 1969. Correlates of ecology and social organization among African cercopithecines. *Folia Primatol.* 11:80–118.

———. 1970. Phylogenetic implications of some vocalizations of *Cercopithecus* monkeys. In *Old World monkeys: Evolution, systematics and behavior,* edited by J. R. Napier and P. H. Napier, 365–444. New York: Academic Press.

———. 1975. *The red colobus monkey.* Chicago: University of Chicago Press.

———. 1980. Comparison of the behavior and ecology of red colobus and redtail monkeys in the Kibale Forest, Uganda. *Afr. J. Ecol.* 18:33–51.

———. 1981. Polyspecific associations among tropical rain-forest primates. *Z. Tierpsychol.* 57:268–304.

Struhsaker, T. T., and M. Leakey. 1990. Prey selectivity by crowned hawk-eagles on monkeys in the Kibale forest, Uganda. *Behav. Ecol. Sociobiol.* 26:435–43.

Struhsaker, T. T., and L. Leland. 1979. Socioecology of five sympatric monkey species in the Kibale Forest, Uganda. *Adv. Stud. Behav.* 9:159–227.

———. 1986. Colobines: Infanticide by adult males. In *Primate societies,* edited by B. B. Smuts, D. L. Cheney, R. M. Seyfarth, R. W. Wrangham, and T. T. Struhsaker, 83–97. Chicago: University of Chicago Press.

Struhsaker, T. T., J. S. Lwanga, and J. M. Kasenene. 1996. Elephants, selective logging, and forest regeneration in the Kibale Forest, Uganda. *J. Trop. Ecol.* 12:45–64.

Strum, S. C. 1981. Processes and products of change: Baboon predatory behavior at Gilgil, Kenya. In *Omnivorous primates,* edited by R. S. O. Harding and G. Teleki, 255–302. New York: Columbia University Press.

Sugardjito, J., C. H. Southwick, J. Supriatna, A. Kolhaas, S. Baker, J. Erwin, K. Froehlich, and N. Lerche. 1989. Population survey of macaques in Northern Sulawesi. *Am. J. Primatol.* 18:285–301.

Sugiyama, Y. 1976. Characteristics of the ecology of the Himalayan langurs. *J. Hum. Evol.* 5:249–77.

Suhonen, J. 1993. Risk of predation and foraging sites of individuals in mixed-species tit flocks. *Anim. Behav.* 45:1193–98.

Suhonen, J., M. Halone, and T. Mappes. 1992. Predation risk and the organization of the *Parus* guild. *Oikos* 66:94–100.

Suhonen, J., and K. Iniki. 1992. Recovery of willow tit food caches by other willow tits and great tits. *Anim. Behav.* 44:180–81.

Sullivan, K. 1984. The advantages of social foraging in Downy Woodpeckers *Picoides pubescens. Anim. Behav.* 32:16–22.

Sunquist, M. E. 1981. The social organization of tigers *(Panthera tigris)* in Royal Chitawan National Park, Nepal. *Smithsonian Contributions to Zoology* 336:1–92.

Sussman, R. W. 1974. Ecological distinctions of sympatric species of Lemur. In *Prosimian biology,* edited by R. D. Martin, G. A. Doyle and A. C. Walker, 75–108. London: Duckworth.

———. 1977a. Feeding behaviour of *Lemur catta* and *Lemur fulvus.* In *Primate ecology,* edited by T. H. Clutton-Brock, 1–36. London: Academic Press.

———. 1977b. Socialization, social structure and ecology of two sympatric species of Lemur. In *Primate bio-social development: Biological, social, and ecological determinants,* edited by S. Chevalier-Skolnikoff and F. E. Poirier, 515–28. New York: Garland.

———. 1991a. Demography and social organization of free-ranging *Lemur catta* in the Beza Mahafaly Reserve, Madagascar. *Am. J. Phys. Anthropol.* 84:43–58.

———. 1991b. Primate origins and the evolution of angiosperms. *Am. J. Primatol.* 23:209–23.

———. 1992. Male life history and intergroup mobility among ringtailed lemurs *(Lemur catta). Int. J. Primatol.* 13:395–414.

Sussman, R. W., and W. G. Kinzey. 1984. The ecological role of the Callitrichidae: A review. *Am. J. Phys. Anthropol.* 64:419–49.

Sussman, R. W., and P. H. Raven. 1978. Pollination by lemurs: An archaic coevolutionary system. *Science* 200:731–36.

Swidler, W. 1969. Some demographic factors regulating the formation of flocks and camps among the Brahui. Paper presented at the Annual Meetings of the American Anthropological Association, New Orleans.

Swisher, C. C., G. H. Curtis, T. Jacob, A. G. Getty, and A. Suprijo. 1994. Age of earliest known hominids in Java, Indonesia. *Science* 263:1118–21.

Symington, M. M. 1987. Ecological and social correlates of party size in the black spider monkey *Ateles paniscus chamek.* Ph.D. dissertation, Princeton University, Princeton, NJ.

————. 1988a. Food competition and foraging party size in the black spider monkey *(Ateles paniscus chamek)*. *Behaviour* 105:117–34.

————. 1988b. Demography, ranging patterns and activity budgets of black spider monkeys *(Ateles paniscus chamek)* in the Manu National Park, Peru. *Am. J. Primatol.* 15:45–67.

————. 1990. Fission-fusion social organization in *Ateles* and *Pan*. *Int. J. Primatol.* 11:47–61.

Symmes, D., and M. Biben. 1992. Vocal development of nonhuman primates. In *Nonverbal vocal communication: Comparative and developmental approaches,* edited by H. Papousek, U. Jurgens, and M. Papousek, 123–40. Cambridge: Cambridge University Press.

Takahashi, T. T., and C. H. Keller. 1994. Representation of multiple sound sources in the owl's auditory space map. *J. Neurosci.* 14:4780–93.

Tandon, M., P. T. Cummings, and M. D. LeVan. 1995. Scheduling of multiple products on parallel units with tardiness penalties using simulated annealing. *Comput. Chem. Engin.* 19:1069–76.

Taneyhill, D. E. 1994. Evolution of complex foraging behavior in bumble bees. Ph.D. dissertation, State University of New York at Stony Brook.

Tardif, S. D., M. L. Harrison, and M. A. Simek. 1993. Communal infant care in marmosets and tamarins: Relation to energetics, ecology, and social organization. In *Marmosets and tamarins: Systematics, behaviour, and ecology,* edited by A. B. Rylands, 220–34. Oxford: Oxford Science Publications.

Tarpy, C. 1979. Killer whale attack. *Nat. Geogr.* 155:542–45.

Tattersall, I. 1977. Ecology and behavior of *Lemur fulvus mayottensis* (Primates, Lemuriformes). *Anthropol. Pap. Am. Mus. Nat. Hist.* 54:425–82.

————. 1987. Cathemeral activity in primates: A definition. *Folia Primatol.* 49:200–202.

————. 1993. Madagascar's lemurs. *Sci. Am.* Jan., 90–97.

Tattersall, I., and R. W. Sussman. 1975. Observations on the ecology and behavior of the mongoose lemur *Lemur mongoz mongoz* Linnaeus (Primates, Lemuriformes), at Ampijoroa, Madagascar. *Anthropol. Pap. Am. Mus. Nat. Hist.* 52:195–216.

————. 1985. Homing behavior in an artificially released female Mauritian long-tailed macaque. *Mammalia* 49:323–28.

Tayler, C. K., and G. S. Saayman. 1972. The social organisation and behaviour of dolphins *(Tursiops aduncus)* and baboons *(Papio ursinus):* Some comparisons and assessments. *Ann. Cape Prov. Mus.* (Nat. Hist.) 9:11–49.

Taylor, C. R. 1977. The energetics of terrestrial locomotion and body size in vertebrates. In *Scale effects in animal locomotion,* edited by T. J. Pedley, 127–41. Academic Press, New York.

————. 1980. Energetics of locomotion: Primitive and advanced mammals. In *Comparative physiology: Primitive mammals,* edited by K. Schmidt-Nielsen, L. Bolis, and C. R. Taylor, 192–99. Cambridge: Cambridge University Press.

Taylor, C. R., S. L. Caldwell, and V. J. Rowntree. 1972. Running up and down hills: Some consequences of size. *Science* 178:1096–97.

Taylor, C. R., N. C. Heglund, and G. M. O. Maloiy. 1982. Energetics and mechanics of terrestrial locomotion. I. Metabolic energy consumption as a function of speed and body size in birds and mammals. *J. Exp. Biol.* 97:1–21.

Taylor, C. R., N. C. Heglund, T. A. McMahon, and T. R. Looney. 1980. Energetic cost of generating muscular force during running: A comparison of large and small mammals. *J. Exp. Biol.* 86:9–18.

Taylor, C. R., K. Schmidt-Nielsen, and J. L. Raab. 1970. Scaling of energetic cost of running to body size in mammals. *Am. J. Physiol.* 219:1104–7.

Taylor, H. A., and B. Tversky. 1992a. Descriptions and depictions of environments. *Mem. Cog.* 20:483–96.

———. 1992b. Spatial mental models derived from survey and route descriptions. *J. Mem. Lang.* 31:261–92.

Teleki, G. 1973. *The predatory behaviour of wild chimpanzees.* Lewisburg, PA: Bucknell University Press.

Templeton, J. T., and L. A. Giraldeau. 1995. Public information cues affect the scrounging decisions of starlings. *Anim. Behav.* 49:1617–26.

Tenaza, R. R. 1975. Territory and monogamy among Kloss' gibbons *(Hylobates klossii)* in Siberut Island, Indonesia. *Folia Primatol.* 24:68–80.

———. 1976. Songs and related behavior of Kloss' gibbon *(Hylobates klossi)* in Siberut Island, Indonesia. *Z. Tierpsychol.* 40:37–52.

Terborgh, J. 1983. *Five New World primates: A study of comparative ecology.* Princeton, NJ: Princeton University Press.

———. 1990. Mixed flock and polyspecific associations: Costs and benefits of mixed groups to birds and monkeys. *Am. J. Primatol.* 21:87–100.

Terborgh, J., and C. H. Janson. 1986. The socioecology of primate groups. *Annu. Rev. Ecol. Syst.* 17:111–35.

Terborgh, J., and M. Stern. 1987. The surreptitious life of the saddle-backed tamarin. *Am. Sci.* 75:260–69.

Thinus-Blanc, C. 1989. Animal spatial cognition. In *Thought without language,* edited by L. Weiskranz, 371–95. Oxford: Clarendon Press.

Thomas, B. 1976. Energy flow at high altitude. In *Man in the Andes: A multidisciplinary study of high altitude Quechua,* edited by P. T. Baker and M. Little. Stroudsburg, PA: Dowden, Huchinson.

Thomas, J. A., and R. A. Kastelein (eds.). 1990. *Sensory abilities of cetaceans: Laboratory and field evidence.* New York: Plenum Press.

Thompson, S. D., R. E. MacMillen, E. M. Burke, and C. R. Taylor. 1980. The energetic cost of bipedal hopping in small mammals. *Nature* 287:223–24.

Thomson, J. D., M. Slatkin, and B. Thomson. 1997. Trapline foraging in bumble bees: II. Definition and detection from sequence data. *Behav. Ecol.* 8:199–210.

Thomson, P. C. 1992. The behavioural ecology of dingoes in north-western Australia: IV. Social and spatial organization, and movements. *Wildl. Res.* 41:543–63.

Thomson, T. J., H. E. Winn, and P. J. Perkins. 1979. Mysticete sounds. In *Behavior of marine animals: Current perspectives in research,* edited by W. E. Winn and B. L. Olla, 403–28. New York: Plenum Press.

Thorpe, W. H. 1956. *Learning and instinct in animals.* London: Methuen.

Tiebout, H. M. 1996. Costs and benefits of interspecific dominance rank: Are subordinates better at finding novel food locations? *Anim. Behav.* 51:1375–81.

Tilson, R. L., and W. J. Hamilton. 1984. Social dominance and feeding patterns of spotted hyaenas. *Anim. Behav.* 32:715–24.

Tilson, R. L., and J. R. Henschel. 1986. Spatial arrangement of spotted hyaena groups in a desert environment, Namibia. *Afr. J. Ecol.* 24:173–180.

Timm, R. M., D. E. Wilson, B. L. Clavson, R. K. LaVal, and C. S. Vaughan. 1989. *Mammals of the La Selva-Bracilio Carrille Complex, Costa Rica.* North American Fauna, no. 75. Washington, DC: U.S. Fish and Wildlife Service.

Tinbergen, N., and W. van Kruyt. 1938. Über die Orientierung des Bienenwolfes (*Philanthus triangulum* Fabr.). III. Die Bevorzugung bestimmter Wegmarken. *Z. vergl. Physiol.* 25:292–334.

Tinkelpaugh, O. L. 1932. Multiple delayed reaction with chimpanzee and monkeys. *J. Comp. Psychol.* 13:207–43.

Tobin, H., and A. W. Logue. 1994. Self-control across species (*Columba livia, Homo sapiens,* and *Rattus norvegicus*). *J. Comp. Psychol.* 108:126–33.

Tobin, H., A. W. Logue, J. J. Chelonis, and K. T. Ackerman. 1996. Self-control in the monkey *Macaca fascicularis. Anim. Learn. Behav.* 24:168–74.

Tolman, E. C. 1948. Cognitive maps in rats and men. *Psychol. Rev.* 55:189–208.

Tomasello, M., and J. Call. 1997. *Primate cognition.* New York: Oxford University Press.

Tooby, J., and L. Cosmides. 1991. The psychological foundation of culture. In *The adapted mind: Evolutionary psychology and the generation of culture,* edited by J. H. Barkow, L. Cosmides, and J. Tooby, 19–136, New York: Oxford University Press.

Tooby, J., and I. DeVore. 1987. The reconstruction of hominid behavioral evolution through strategic modeling. In *The evolution of human behavior: Primate models,* edited by W. G. Kinzey, 183–238. Albany: State University of New York Press.

Topkis, D. M. 1988. A k shortest path algorithm for adaptive routing in communications networks. *IEEE Trans. Communic.* 36:855–59.

Torigoe, T. 1985. Comparison of object manipulation among 74 species of nonhuman primates. *Primates* 26:182–94.

Torres de Assumpção, C. 1983. An ecological study of primates in southern Brazil, with a reappraisal of *Cebus apella* races. Ph.D. dissertation, University of Edinburgh.

Tovée, M. H. 1996. *An introduction to the visual system.* Cambridge: Cambridge University Press.

Treves, A. 1997a. Self-protection in primates. Ph.D. dissertation, Harvard University, Cambridge, MA.

———. 1997b. Vigilance and use of micro-habitat in solitary rainforest mammals. *Mammàlia* 61(4):511–25.

———. 1998. The influence of group size and near neighbors on vigilance in two species of arboreal primates. *Behaviour* 135(4):453–82.

Treves, A., and C. A. Chapman. 1996. Conspecific threat, predation avoidance, and resource defense: Implications for grouping and alliances in langurs. *Behav. Ecol. Sociobiol.* 39:43–53.

Treves, A., and L. Naughton-Treves. 1999. Risk and opportunity for humans coexisting with large carnivores. *J. Hum. Evol.* In press.

Trivers, R. L. 1971. The evolution of reciprocal altruism. *Q. Rev. Biol.* 46:35–57.

Tsinglia, H. M., and T. E. Rowell. 1984. The behaviour of adult male blue monkeys. *Z. Tierpsychol.* 64:253–68.

Tucker, V. A. 1975. The energetic cost of moving about. *Am. Sci.* 63:413–19.

Tullock, G. 1971. The coal tit as a careful shopper. *Am. Nat.* 105:77–80.

Tutin, C. E. G. 1996. Ranging and social structure of lowland gorillas in the Lopé Reserve, Gabon. In *Great ape societies,* edited by W. C. McGrew, L. F. Marchant, and T. Nishida, 58–70. Cambridge: Cambridge University Press.

Tutin, C. E. G., and M. Fernandez. 1993. Composition of the diet of chimpanzees and comparison with that of sympatric lowland gorillas in the Lopé Reserve, Gabon. *Int. J. Primatol.* 30:195–211.

Tutin, C. E. G., W. C. McGrew, and P. J. Baldwin. 1981. Social organization of savanna-dwelling chimpanzees, *Pan troglodytes verus,* at Mt. Assirik, Senegal. *Primates* 24:154–73.

———. 1983. Responses of wild chimpanzees to potential predators. In *Primate behaviour and sociobiology,* edited by B. Chiarelli and R. Corrucini, 136–41. Berlin: Springer-Verlag.

Tyack, P. 1981. Interactions between singing Hawaiian humpback whales and conspecifics nearby. *Behav. Ecol. Sociobiol.* 8:105–16.

———. 1983. Differential response of humpback whales, *Megaptera novaeangliae,* to playback of song or social sounds. *Behav. Ecol. Sociobiol.* 13:49–55.

———. 1986. Whistle repertoires of two bottlenosed dolphins, *Tursiops truncatus:* Mimicry of signature whistles? *Behav. Ecol. Sociobiol.* 18:251–57.

———. 1993. Animal language research needs a broader comparative and evolutionary framework. In *Language and communication: Comparative perspectives,* edited by H. L. Roitblat, L. M. Herman, and P. E. Nachtigall, 115–52. Hillsdale, NJ: Lawrence Erlbaum Associates.

Tyack, P., and H. Whitehead. 1983. Male competition in large groups of wintering humpback whales. *Behaviour* 83:132–54.

Uehara, S., T. Nishida, M. Hamai, T. Hasegawa, H. Hayaki, M. A. Huffman, K. Kawanaka, S. Kobayashi, J. C. Mitani, Y. Takahata, H. Takasaki, and T. Tsukahara. 1992. Characteristics of predation by the chimpanzees in the Mahale Mountains National Park, Tanzania. In *Topics in primatology: Behavior, ecology, and conservation,* edited by N. Itoigawa, Y. Sugiyama, G. P. Sackett, and R. K. R. Thompson, 143–58. Tokyo: University of Tokyo Press.

Uganda News. 1997. Crocodiles kill 32 on Victoria shores. Features African Network News Bulletin, Oct. 15, 1997.

Ullmann-Margalit, E. 1977. *The emergence of norms.* Oxford: Oxford University Press.

Underwood, B. A. 1990. Seasonal nesting cycle and migration patterns of the Himalayan honey bee *Apis laboriosa. Nat. Geogr. Res.* 6:276–90.

Utami, S. S., and J. A. R. A. M. van Hooff. 1997. Meat-eating by adult female Sumatran orangutans *(Pongo pygmaeus abelii). Am. J. Primatol.* 43:159–65.

Valburg, L. K. 1992. Flocking and frugivory: The effect of social groupings on resource use in the Common Bush-Tanager. *Condor* 94:358–63.

Valladares-Padua, C. 1993. The ecology, behavior and conservation of the black lion tamarins (*Leontopithecus chrysopygus,* Mikan, 1823). Ph.D. thesis, University of Florida, Gainesville.

Valone, T. J. 1989. Group foraging, public information, and patch estimation. *Oikos* 56:357–63.

Valone, T. J., and L. A. Giraldeau. 1993. Patch estimation by group foragers: What information is used. *Anim. Behav.* 45:721–28.

Vanbuskirk, J., and D. C. Smith. 1989. Individual variation in winter foraging of Black-capped Chickadees. *Behav. Ecol. Sociobiol.* 24:257–63.

Vander Wall, S. B. 1990. *Food hoarding in animals.* Chicago: University of Chicago Press.

VanKrunkelsven, E., J. Dupain, L. Van Elsacker, and R. F. Verheyen. 1996. Food calling by captive bonobos *(Pan paniscus):* An experiment. *Int. J. Primatol.* 17:207–17.

Vayda, A. P. 1995. Failures of explanation in Darwinian ecological anthropology. *Phil. Soc. Sci.* 25(2):219–49, 25(3):360–75.

Vedder, A. L. 1984. Movement patterns of a group of free-ranging mountain gorillas *(Gorilla gorilla beringei)* and their relation to food availability. *Am. J. Primatol.* 7:73–88.

Venkataraman, A. B., R. Arumugan, and R. Sukumar. 1995. The foraging ecology of dhole *(Cuon alpinus)* in Mudumalai Sanctuary, southern India. *J. Zool.* 237:543–61.

Veracini, C. 1996. Preliminary observations on the ecology of *Callithrix argentata* and its relationships with *Saguinus midas niger.* Paper presented at the XVIth Congress of the International Primatological Society, Madison, WI.

Vick, L. G., and M. E. Pereira. 1989. Episodic targeting aggression and the histories of Lemur social groups. *Behav. Ecol. Sociobiol.* 25:3–12.

Vickery, W. I., L. A. Giraldeau, J. J. Templeton, D. Dramer, and C. Chapman. 1991. Producers, scroungers, and group foragers. *Am. Nat.* 137:847–63.

Vine, I. 1971. Risk of visual detection and pursuit by a predator and the selective advantage of flocking behaviour. *J. Theor. Biol.* 30:405–22.

Visalberghi, E. 1986. Aspects of space representation in an infant gorilla. In *Current perspectives in primate social dynamics,* edited by D. M. Taub and F. A. King, 445–52. New York: Van Nostrand Reinhold.

Visalberghi, E., and C. DeLillo. 1995. Understanding primate behaviour: A cooperative effort of field and laboratory research. In *Behavioural brain research in naturalistic and semi-naturalistic settings,* edited by E. Alleva et al., 413–24. The Netherlands: Kluwer Academic Publishers.

Visalberghi, E., and D. Fragaszy. 1990. Do monkeys ape? In *Language and intelligence in monkeys and apes,* edited by S. Parker and K. Gibson, 247–74. Cambridge: Cambridge University Press.

———. 1995. The behavior of capuchin monkeys, *Cebus apella,* with novel foods: The role of social context. *Anim. Behav.* 49:1089–95.

Visscher, P. K., and S. Camazine. 1999. Collective decisions and cognition in bees. *Nature* 397:400.

Visscher, P. K., and T. D. Seeley. 1982. Foraging strategy of honeybee colonies in a temperate deciduous forest. *Ecology* 63:1790–1801.

Vogel, C. 1973. Acoustical communication among-free-ranging common Indian langurs *(Presbytis entellus)* in two different habitats of North India. *Am. J. Phys. Anthropol.* 38:469–80.

Vogel, C., and H. Loch. 1984. Reproductive parameters, adult-male replacements, and infanticide among free-ranging langurs *(Presbytis entellus)* at Jodhpur (Rajasthan), India. In *Infanticide: Comparative and evolutionary perspectives,* edited by G. Hausfater and S. B. Hardy, 237–56. Hawthorne, NY: Aldine.

Vollbehr, J. 1975. Zur Orientierung junger Honigbienen bei ihrem ersten Orientierungsflug. *Zool. Jb. allg. Zool. Physiol.* 79:33–69.

von Hippel, F. A. 1996. Interactions between overlapping multimale troops of black and white colobus monkeys *(Colobus guereza)* in the Kakamega Forest, Kenya. *Am. J. Primatol.* 38:193–209.

Vrba, E. S. 1985. Ecological and adaptive changes associated with early hominid evolution. In *Ancestors: The hard evidence,* edited by E. Delson, 63–71. New York: Alan R. Liss.

———. 1988. Late Pliocene climatic events and hominid evolution. In *Evolutionary history of the "robust" australopithecines,* edited by F. E. Grine, 405–26. Hawthorne, NY: Aldine.

———. 1993. The pulse that produced us. *Nat. Hist.* 102 (May):47–51.

———. 1995. The fossil record of African antelopes (Mammalia: Bovidae) in relation to human evolution and paleoclimate. In *Paleoclimate and evolution, with emphasis on human origins,* edited by E. S. Vrba, G. H. Denton, T. C. Partridge, and L. H. Burckle, 385–424. New Haven, CT: Yale University Press.

Waal, F. B. M. de. 1989. Dominance "style" and primate social organization. In *Comparative socioecology,* edited by V. Standen and R. A. Foley, 243–63. Oxford: Blackwell.

———. 1992. Coalitions as part of reciprocal relations in the Arnhem chimpanzee colony. In *Coalitions and alliances in humans and other animals,* edited by A. H. Harcourt and F. B. M. de Waal, 233–57. Oxford: Oxford University Press.

———. 1995. *Good natured.* Cambridge, MA: Harvard University Press.

Waal, F. B. M. de, and L. M. Luttrell. 1989. Toward a comparative socioecology of the genus *Macaca:* Different dominance styles in rhesus and stumptail monkeys. *Am. J. Primatol.* 19:83–110.

Wachter, B., M. Schabel, and R. Noë. 1997. Diet overlap and polyspecific associations of red colobus and Diana monkeys in the Taï National Park, Ivory Coast. *Ethology* 103:514–26.

Wahome, J. M., T. E. Rowell, and H. M. Tsinglia. 1993. The natural history of the de Brazza's monkey in Kenya. *Int. J. Primatol.* 14:445–66.

Waite, T. A., and T. C. Grubb. 1988. Copying of foraging locations in mixed species flocks of temperate deciduous woodland: An experimental study. *Condor* 90:132–40.

Walker, S. E. 1996. The evolution of positional behavior in the saki-uacaris *(Pithecia, Chiroptes,* and *Cacajao).* In *Adaptive Radiations of Neotropical Primates,* edited by M. Norconk, A. L. Rosenberger, and P. A. Garber, 335–68. New York: Plenum Press.

Walters, J. R., and R. M. Seyfarth. 1986. Conflict and cooperation. In *Primate societies,* edited by B. B. Smuts, D. L. Cheney, R. M. Seyfarth, R. W. Wrangham, and T. T. Struhsaker, 358–69. Chicago: University of Chicago Press.

Ward, P., and A. Zahavi. 1973. The importance of certain assemblages of birds as "information centres" for food-finding. *Ibis* 115:517–34.

Warren, R. 1994. Lazy leapers: A study of the locomotor ecology of two species of saltatory nocturnal lemur in sympatry at Ampijoroa, Madagascar. Ph.D. thesis, University of Liverpool.

Warren, R., and R. H. Crompton. 1997. A comparative study of the ranging behaviour, activity rhythms and sociality of *Lepilemur edwardsi* (Primates, Lepilemuridae) and *Avahi occidentalis* (Primates, Indriidae) at Ampijoroa, Madagascar. *J. Zool.* 243:397–415.

Waser, P. M. 1974. Inter-group interactions in a forest monkey, the mangabey *Cercocebus albigena*. Ph.D. dissertation, Rockefeller University, New York.

———. 1975. Experimental playbacks show vocal mediation of intergroup avoidance in a forest monkey. *Nature* 255:56–58.

———. 1976. *Cercocebus albigena:* Site attachment, avoidance and intergroup spacing. *Am. Nat.* 110:911–35.

———. 1977a. Feeding, ranging and group size in the mangabey *Cercocebus albigena*. In *Primate ecology,* edited by T. H. Clutton-Brock, 182–222. London: Academic Press.

———. 1977b. Individual recognition, intragroup cohesion and intergroup spacing: Evidence from sound playback to forest monkeys. *Behaviour* 60:28–74.

———. 1981. Sociality or territorial defense? The influence of resource renewal. *Behav. Ecol. Sociobiol.* 8:231–37.

———. 1982a. The evolution of male loud calls among mangabeys and baboons. In *Primate communication,* edited by C. T. Snowdon, C. H. Brown, and M. R. Petersen, 117–43. Cambridge: Cambridge University Press.

———. 1982b. Polyspecific associations: Do they occur by chance? *Anim. Behav.* 30:1–8.

———. 1984a. "Chance" and mixed-species associations. *Behav. Ecol. Sociobiol.* 15:197–202.

———. 1984b. Ecological differences and behavioral contrasts between two mangabey species. In *Adaptations for foraging in nonhuman primates,* edited by P. S. Rodman and J. G. H. Cant, 195–216. New York: Columbia University Press.

Waser, P. M., and C. H. Brown. 1986. Habitat acoustics and primate communication. *Am. J. Primatol.* 10:135–54.

Waser, P. W., and O. Floody. 1974. Ranging patterns of the mangabey, *Cercocebus albigina,* in the Kibale Forest, Uganda. *Z. Tierpsychol.* 35:85–101.

Waser, P. M., and K. Homewood. 1979. Cost-benefit approaches to territoriality: A test with forest primates. *Behav. Ecol. Sociobiol.* 6:115–19.

Waser, P. M., and R. H. Wiley. 1979. Mechanisms and evolution of spacing in animals. In *Handbook of behavioral neurobiology,* vol. 3, edited by P. Marler and J. Vandenburgh, 159–223. New York: Plenum Press.

Washburn, S. L., and I. DeVore. 1961. The social life of baboons. *Sci. Am.* 204:62–71.

Washburn, S. L., and J. B. Lancaster. 1968. The evolution of hunting. In *Man the hunter,* edited by R. Lee and I. DeVore, 293–303. Chicago: Aldine.

Wasserman, E. A. 1993. Comparative cognition: Toward a general understanding of cognition in behavior. *Psychol. Sci.* 4:156–61.

Watanabe, K. 1981. Variation in group composition and population density of two sympatric Metawaian leaf monkeys. *Primates* 22:145–60.

Waterman, P. G., and K. M. Kool, 1995. Colobine food selection and food chemistry. In *Colobine monkeys: Their behavior, ecology, and evolution,* edited by A. G. Davies and J. F. Oates, 251–84. Cambridge: Cambridge University Press.

Waterman, P. G., G. M. Choo, A. L. Vedder, and D. P. Watts. 1983. Digestibility, digestion inhibitors, and nutrients of herbaceous foliage and green stems from an African montane flora and comparison with other tropical flora. *Oecologia* 60:244–49.

Watkins, W. A. 1993. Sperm whale tracking underwater and at the surface. Abstract. Tenth Biennial Conference on the Biology of Marine Mammals, Galveston, Texas.

Watkins, W. A., and W. E. Schevill. 1977. Sperm whale codas. *J. Acoust. Soc. Am.* 62:1485–90.

Watkins, W. A., P. Tyack, K. E. Moore, and J. E. Bird. 1987. The 20 hz signals of finback whales *(Balaenoptera physalis). J. Acoust. Soc. Am.* 82:1901–12.

Watts, D. P. 1984. Composition and variability of mountain gorilla diets in the central Virungas. *Am. J. Primatol.* 7:323–56.

———. 1985. Relationships between group size and composition and feeding competition in mountain gorilla groups. *Anim. Behav.* 33:72–85.

———. 1987. The influence of mountain gorilla foraging activities on the productivity of their food species. *Afr. J. Ecol.* 25:155–63.

———. 1988. Environmental influences on mountain gorilla time budgets. *Am. J. Primatol.* 15:295–312.

———. 1989. Infanticide in mountain gorillas: New cases and a reconsideration of the evidence. *Ethology* 81:1–18.

———. 1990. Ecology of gorillas and its relationship to female transfer in mountain gorillas. *Int. J. Primatol.* 11:21–45.

———. 1991. Strategies of habitat use by mountain gorillas. *Folia Primatol.* 56:1–16.

———. 1992. Social relationships of immigrant and resident female mountain gorillas. I. Male-female relationships. *Am. J. Primatol.* 28:159–81.

———. 1994. The influence of male mating strategies on habitat use in mountain gorillas *(Gorilla gorilla beringei). Primates* 35:35–47.

———. 1996. Comparative socioecology of gorillas. In *Great ape societies,* edited by W. C. McGrew, L. F. Marchant, and T. Nishida, 16–28. Cambridge: Cambridge University Press.

———. 1998a. Long-term habitat use by mountain gorillas *(Gorilla gorilla beringei).* 1. Consistency, variation, and home range size and stability. *Int. J. Primatol.* 19:651–80.

———. 1998b. Long-term habitat use by mountain gorillas *(Gorilla gorilla beringei).* 2. Re-use of foraging areas in relation to resource abundance, quality, and depletion, *Int. J. Primatol.* 19:681–702.

———. 1998c. Seasonality in the ecology and life histories of mountain gorillas *(Gorilla gorilla beringei). Int. J. Primatol.* 19:929–48.

Weber, A. W., and A. L. Vedder. 1983. Population dynamics of the Virunga gorillas, 1959–1978. *Biol. Conserv.* 26:341–66.

Wehner, R. 1981. Spatial vision in arthropods. In *Handbook of sensory physiology,* vol. VII/6C, edited by H. Autrum, 287–616. Berlin-Heidelberg-New York: Springer.

———. 1982. Himmelsnavigation bei Insekten. *Neurophysiologie und Verhalten. Vierteljahrsschr. Naturforsch. Ges. Zürich* 5:1–132.

———. 1984. Astronavigation in insects. *Annu. Rev. Entomol.* 29:277–98.

———. 1991. Visuelle Navigation: Kleinstgehirn-Strategien. *Verh. Dt. Zool. Ges.* 84:89–104.

Wehner, R., S. Bleuler, C. Nievergelt, and D. Shah. 1990. Bees navigate by using vectors and routes rather than maps. *Naturwissenschaften* 77:479–82.

Wehner, R., R. D. Harkness, and P. Schmid-Hempel. 1983. Foraging strategies in individually searching ants, *Cataglyphis bicolor* (Hymenoptera: Formicidae). Stuttgart-New York: Gustav Fischer Verlag.

Wehner, R., M. Lehrer, and W. R. Harvey (eds.). 1996. Navigation. *J. Exp. Biol.* 199(1):1–261.

Wehner, R., B. Michel, and P. Antonsen. 1996. Visual navigation in insects: Coupling of egocentric and geocentric information. *J. Exp. Biol.* 199:129–40.

Wehner, R., and M. Müller. 1993. How do ants acquire their celestial ephemeris function? *Naturwissenschaften* 80:331–33.

Wehner, R., and S. Wehner. 1990. Insect navigation: Use of maps or Ariadne's thread? *Ethol. Ecol. Evol.* 2:27–48.

Weilgart, L. S., and H. Whitehead. 1988. Distinctive vocalizations from mature male sperm whales *(Physeter macrocephalus)*. *Can. J. Zool.* 66:1931–37.

———. 1997. Group-specific dialects and geographical variation in coda repertoire in South Pacific sperm whales. *Behav. Ecol. Sociobiol.* 40:277–85.

Weilgart, L. S., H. Whitehead, and K. Payne. 1996. A collossal convergence. *Am. Sci.* 84:278–87.

Weinrich, M. 1991. Stable social associations among humpback whales *(Megaptera novaeangliae)* in the southern Gulf of Maine. *Can. J. Zool.* 69:3012–18.

Weiskrantz, L., L. Willner, and R. D. Martin. 1985. Some basic principles of mammalian sexual dimorphism. In *Human sexual dimorphism,* edited by J. Ghesquiere, R. D. Martin, and F. Newcombe, 1–19. London: Taylor and Francis.

Weller, D. W. 1991. The social ecology of Pacific coast bottlenose dolphins. Masters' thesis, San Diego State University.

Wells, R. S., L. J. Hansen, A. Baldridge, T. P. Dohl, D. L. Kelly, and R. H. Defran. 1990. Northward extension of the range of bottlenose dolphins along the California coast. In *The bottlenose dolphin,* edited by S. Leatherwood and R. R. Reeves, 421–31. San Diego, CA: Academic Press.

Wells, R. S., M. D. Scott, and B. Irvine A. 1987. The social structure of free-ranging bottlenose dolphins. In *Current mammalogy,* vol. 1, edited by H. Genoways, 247–305. New York: Plenum Press.

West-Eberhard, M. J. 1978. Temporary queens in Matapolybia wasps: Non-reproductive helpers without altruism? *Science* 200:441–43.

Western, D. 1982. The environment and ecology of pastoralists in arid savannas. *Devel. Change* 13:183–211.

Westoby, M. 1974. An analysis of diet selection by large generalist herbivores. *Am. Nat.* 108:290–304.

Whately, A., and P. M. Brooks. 1978. Numbers and movements of spotted hyaenas in Hluhluwe Game Reserve. *Lammergeyer* 26:44–52.

Wheatley, B. P. 1980. Feeding and ranging of East Borneo *Macaca fascicularis.* In *The macaques,* edited by D. G. Lindburg, 215–46. New York: Van Nostrand Reinhold.

Wheeler, P. E. 1991. Thermoregulatory advantages of hominid bipedalism in open equatorial environments: The contribution of increased heat loss and cutaneous evaporative cooling. *J. Hum. Evol.* 21:107–15.

White, F. J. 1991. Social organization, feeding ecology, and reproductive strategy of ruffed lemurs, *Varecia variegata.* In *Proceedings of the XIII Congress of the International Primatological Society,* edited by A. Ehara, T. Kimura, O. Takenaka, and M. Iwamoto, 81–84. Amsterdam: Elsevier.

White, F. J., and R. W. Wrangham. 1988. Feeding competition and patch size in the chimpanzee species *Pan paniscus* and *Pan troglodytes. Behaviour* 105:148–64.

White, L. J. T. 1994. Biomass of rain forest mammals in the Lopé Reserve, Gabon. *J. Anim. Ecol.* 63:499–512.

White, L. J. T., M. E. Rogers, C. E. G. Tutin, E. A. Williamson, and J. M. Fernandez. 1995. Herbaceous vegetation in different forest types in the Lopé Reserve, Gabon: Implications for keystone food availability. *Afr. J. Ecol.* 33:124–41.

White, L. J. T., C. E. G. Tutin, and J. M. Fernandez. 1993. Group composition and diet of forest elephants, *Loxodonta africana cyclotis* Matschie 1900, in the Lopé Reserve, Gabon. *Afr. J. Ecol.* 31:181–99.

White, T. D., G. Suwa, and B. Asfaw. 1994. *Australopithecus ramidus,* a new species of early hominid from Aramis, Ethiopia. *Nature* 371:306–12.

Whitehead, H. 1983. Structure and stability of humpback whale groups off Newfoundland. *Can. J. Zool.* 61:1391–97.

———. 1989. Formations of foraging sperm whales, *Physeter macrocephalus,* off the Galapagos Islands. *Can. J. Zool.* 67:2131–39.

———. 1993. The behavior of mature male sperm whales on the Galapagos breeding grounds. *Can. J. Zool.* 71:689–99.

———. 1996. Babysitting, dive synchrony and indications of alloparental care in sperm whales. *Behav. Ecol. Sociobiol.* 38:237–44.

Whitehead, H., S. Waters, and T. Lyrholm. 1991. Social organization of female sperm whales and their offspring: Constant companions and casual acquaintances. *Behav. Ecol. Sociobiol.* 29:385–89.

Whitehead, J. M. 1987. Vocally mediated reciprocity between neighboring groups of mantled howling monkeys, *Alouatta palliata palliata. Anim. Behav.* 35:1615–27.

———. 1989. The effect of the location of a simulated intruder on responses to long-distance vocalizations of mantled howling monkeys, *Alouatta palliata palliata. Behaviour* 108:73–103.

———. 1995. Vox Alouattinae: A preliminary survey of the acoustic characteristics of long-distance calls of howling monkeys. *Int. J. Primatol.* 16:121–144.

Whiten, A., and R. Byrne. 1986. The St. Andrews catalogue of tactical deception in primates. St. Andrews Psychological Reports, no. 10.

———. 1988. The manipulation of attention in primate tactical deception. In *Machiavellian intelligence: Social expertise and the evolution of intellect on monkeys, apes, and humans,* edited by R. Byrne and A. Whiten, 211–23. Oxford: Clarendon Press.

Whiten, A., R. W. Byrne, R. A. Barton, P. G. Waterman, and S. P. Henzi, 1991. Dietary and foraging strategies of baboons. *Phil. Trans. R. Soc. Lond.* B 334:187–97.

Whiten, A., R. W. Byrne, and S. P. Henzi. 1987. The behavioural ecology of mountain baboons. *Int. J. Primatol.* 8:367–88.

Whiten, A., and R. Ham. 1992. On the nature and evolution of imitation in the animal kingdom: Reappraisal of a century of research. *Adv. Stud. Behav.* 21:239–83.

Whitesides, G. H. 1989. Interspecific associations of Diana monkeys, *Cercopithecus diana,* in Sierra Leone, West Africa: Biological significance or chance? *Anim. Behav.* 37:760–76.

Whitten, A. J. 1982. Home range use by Kloss' gibbons *(Hylobates klossii)* on Siberut Island, Indonesia. *Anim. Behav.* 30:182–98.

Whitten, P. 1983. Diet and dominance among female vervet monkeys *(Cercopithecus aethiops). Am. J. Primatol.* 5:139–59.

———. 1988. Effects of patch quality and feeding subgroup size on feeding success in vervet monkeys *(Cercopithecus aethiops). Behaviour* 105:35–52.

Wiley, R. H. 1971. Cooperative roles in mixed species flocks of antwrens (Formicariidae). *Auk* 88:881–92.

———. 1980. Multispecies antbird societies in lowland forests of Surinam and Ecuador: Stable membership and foraging differences. *J. Zool.* 191:127–41.

———. 1994. Errors, exaggeration, and deception in animal communication. In *Behavioral mechanisms in evolutionary ecology,* edited by L. A. Real, 157–89. Chicago: University of Chicago Press.

Wille, A. 1983. Biology of the stingless bees. *Annu. Rev. Ecol. Syst.* 28:41–64.

Williams, G. C. 1966. *Adaptation and natural selection: A critique of some current evolutionary thought.* Princeton, NJ: Princeton University Press.

———. 1992. *Natural selection: Domains, levels and challenges.* Oxford: Oxford University Press.

Williams, G. C., and D.C. Williams. 1957. Natural selection of individually harmful social adaptations among sibs with special reference to social insects. *Evolution* 11:32–39.

Williams, T. M., W. A. Friedl, M. L. Fong, R. M. Yamada, P. Dedivy, and J. Haun. 1992. Travel at low energetic cost by swimming and wave-riding bottlenose dolphins. *Nature* 355:821–23.

Williamson, E. A., C. E. G. Tutin, M. E. Rogers, and M. Fernandez. 1990. Composition of the diet of lowland gorillas at Lopé in Gabon. *Am. J. Primatol.* 21:265–77.

Willis, E. O. 1972. The behavior of spotted antbirds. Ornithological Monograph no. 10.

Willis, E. O., and Y. Oniki. 1978. Birds and army ants. *Annu. Rev. Ecol. Syst.* 9:243–63.

Wilmhurst, J. F., J. M. Fryxell, and R. J. Hudson. 1995. Forage quality and patch choice by wapiti *(Cervus elephus). Behav. Ecol.* 6:209–17.

Wilmsen, E. 1989. *Land filled with flies: A political economy of the Kalahari.* Chicago: University of Chicago Press.

Wilson, D. S. 1975. A theory of group selection. *Proc. Natl. Acad. Sci. USA* 72:143–46.

———. 1987. Altruism in Mendelian populations derived from sibgroups: The haystack model revisited. *Evolution* 41:1059–70.

———. 1990. Weak altruism, strong group selection. *Oikos* 59:135–40.

———. 1997a. Altruism and organism: Disentangling the themes of multilevel selection theory. *Am. Nat.* 150: S122–S134.

———. 1997b. Incorporating group selection into the adaptationist program: A case study involving human decision making. In *Evolutionary social psychology,* edited by J. Simpson and D. Kendrick. Mahwah, NJ: Lawrence Erlbaum Associates.

Wilson, D. S., and L. A. Dugatkin. 1997. Group selection and assortative interactions. *Am. Nat.* 149:336–51.

Wilson, D. S., and E. Sober. 1994. Reintroducing group selection to the human behavioral sciences. *Behav. Brain Sci.* 17:585–654.

Wilson, D. S., C. Wilczynski, A. Wells, and L. Weiser. 2000. Gossip and other aspects of language as group-level adaptations. In *Evolution of cognition,* edited by C. Heyes and L. Huber. MIT Press. In press.

Wilson, E. O. 1971. *The insect societies.* Cambridge, MA: Harvard University Press.

Wilson, J. M., P. D. Stewart, G. S. Ramangason, A. M. Denning, and M. S. Hutchings. 1989. Ecology and conservation of the crowned lemur, *Lemur coronatus,* at Ankarana, N. Madagascar. *Folia Primatol.* 52:1–26.

Wiltschko, R., and W. Wiltschko. 1981. The development of sun compass orientation in young homing pigeons. *Behav. Ecol. Sociobiol.* 9:135–41.

Wiltschko, W., and R. Wiltschko. 1996. Magnetic orientation in birds. *J. Exp. Biol.* 199:29–38.

Windfelder, T. 1997. Polyspecific association of saddleback tamarins *(Saguinus fuscicollis)* and emperor tamarins *(S. imperator).* Ph.D. dissertation, Duke University.

Winston, M. L. 1987. *The biology of the honey bee.* Cambridge, MA: Harvard University Press.

Winterhalder, B. 1981. Optimal foraging strategies and hunter-gatherer research in anthropology: Theory and models. In *Hunter-gatherer foraging strategies,* edited by B. Winterhalder and E. A. Smith, 13–35. Chicago: University of Chicago Press.

Winterhalder, B., and E. Smith. 1981. *Hunter gatherer foraging strategies.* Chicago: University of Chicago Press.

WoldeGabriel, G., T. D. White, G. Suwa, P. Renne, J. deHeinzelin, W. K. Hart, and G. Helken. 1994. Ecological and temporal placement of early Pliocene hominids at Aramis, Ethiopia. *Nature* 371:330–33.

Wolf, N. G. 1985. Odd fish abandon mixed-species groups when threatened. *Behav. Ecol. Sociobiol.* 17:47–52.

Wolpoff, M. H. 1980. *Paleoanthropology.* New York: Knopf.

Wood, F. G., and W. E. Evans. 1980. Adaptiveness and ecology of ecoloction in toothed whales. In *Animal sonar systems,* edited by R. G. Busnel and J. F. Fish, 381–425. New York: Plenum Press.

Woodburn, J. C. 1972. Ecology, nomadic movement and the composition of the local group among hunters and gatherers: An East African example and its implication. In *Man, settlement and urbanism,* edited by P. J. Ucko, R. Tringham, and G. W. Dimbleby. London: Duckworth.

Work, T. H., H. Trapido, D. P. Narasimba Murthy, R. Laxmana Rao, R. N. Bhatt, and K. G. Kulkarni. 1957. Kyasanur forest disease. III. A preliminary report on the nature of the infection and clinical manifestations in human beings. *Indian J. Med. Sci.* 11:619–45.

Wrangham, R. W. 1977. Feeding behavior of chimpanzees in Gombe National Park, Tanzania. In *Primate ecology,* edited by T. H. Clutton-Brock, 504–38. London: Academic Press.

———. 1979. On the evolution of ape social systems. Soc. Sci. Infor. 18:334–68.

———. 1980. An ecological model of female-bonded primate groups. *Behaviour* 75:262–300.

————. 1986. Evolution of social structure. In *Primate societies,* edited by B. B. Smuts, D. L. Cheney, R. M. Seyfarth, R. W. Wrangham and T. T. Struhsaker, 282–97. Chicago: University of Chicago Press.

Wrangham, R. W., J. Gittleman, and C. A. Chapman. 1993. Constraints on group size in primates and carnivores: Population density and day-range as assays of exploitation competition. *Behav. Ecol. Sociobiol.* 32:199–210.

Wrangham, R. W., and E. Z. B. Riss. 1990. Rates of predation on mammals by Gombe chimpanzees, 1972–1975. *Primates* 31:157–70.

Wright, P. C. 1978. Home range, activity pattern, and agonistic encounters of a group of night monkeys *(Aotus trivirgatus)* in Peru. *Folia Primatol.* 29:43–55.

————. 1981. The night monkeys, genus *Aotus.* In *Ecology and behaviour of Neotropical primates,* edited by A. F. Coimbra-Filho and R. A. Mittermeier. Rio de Janeiro: Academia Brasileira de Ciencias.

————. 1984. Biparental care in *Aotus trivirgatus* and *Callicebus moloch.* In *Female primates: Studies by women primatologists,* edited by M. F. Small, 59–85. New York: Alan R. Liss.

————. 1986. Costs and benefits of nocturnality in night monkeys, *Aotus.* Ph.D. thesis, City University of New York.

————. 1989. The nocturnal primate niche in the New World. *J. Hum. Evol.* 18:635–58.

————. 1994. The behavior and ecology of the owl monkey. In Aotus: *The owl monkey,* edited by J. F. Baer, R. E. Weller, and I. Kakoma, 97–112. New York: Academic Press.

————. 1995. Demography and life history of free-ranging *Propithecus diadema edwardsi* in Ranomafana National Park, Madagascar. *Int. J. Primatol.* 10:835–54.

————. 1998. Impact of predation risk on the behaviour of *Propithecus diadema edwardsi* in the rain forest of Madagascar. *Behaviour* 135:483–512.

Wright, P. C., S. K. Heckscher, and A. E. Dunham. 1997. Predation on Milne-Edward's sifaka *(Propithecus diadema edwardsi)* by the fossa *(Cryptoprocta ferox)* in the rain forest of southeastern Madagascar. *Folia Primatol.* 68:34–43.

Wright, P. C., and L. B. Martin. 1995. Predation, pollination and torpor in two nocturnal prosimians: *Cheirogaleus major* and *Microcebus rufus* in the rainforest of Madagascar. In *Creatures of the dark: The nocturnal prosimians,* edited by L. Alterman, G. A. Doyle, and M. K. Izard, 45–60. New York: Plenum Press.

Wright, S. 1945. Tempo and mode in evolution: A critical review. *Ecology* 26:415–19.

————. 1948. Genetics of populations. *Encyclopedia Britannica* (1961 printing), 10:111D–112.

Wright, S. J., M. E. Gompper, and B. DeLeon. 1994. Are large predators keystone species in Neotropical forests? The evidence from Barro Colorado Island. *Oikos* 71:279–94.

Wright, T. F. 1996. Regional dialects in the contact call of a parrot. *Proc. R. Soc. Lond.* B 263:867–72.

————. 1997. Vocal communication in the Yellow-naped Amazon, *Amazona auropalliata.* Ph.D. dissertation, University of California, San Diego.

Wunder, B. A., and P. R. Morrison. 1974. Red squirrel metabolism during incline running. *Comp. Biochem. Physiol.* 48A:153–61.

Würsig, B. 1978. Occurrence and group organization of Atlantic bottlenose porpoises *(Tursiops truncatus)* in an Argentine Bay. *Biol. Bull.* 154 (2):348–59.

———. 1986. Delphinid foraging strategies. In *Dolphin cognition and behavior: A comparative approach,* edited by R. J. Schusterman, J. A. Thomas, and F. G. Wood, 347–60. Hillsdale, NJ: Lawrence Erlbaum Associates.

Würsig, B., F. Cipriano, and M. Würsig. 1991. Dolphin movement patterns: Information from radio and theodolite tracking studies. In *Dolphin societies: Discoveries and puzzles,* edited by K. Pryor and K. S. Norris, 79–112. Los Angeles: University of California Press.

Würsig, B., and G. Harris. 1990. Site and association fidelity in bottlenose dolphins off Argentina. In *The bottlenose dolphin,* edited by S. Leatherwood and R. R. Reeves, 361–65. San Diego, CA: Academic Press.

Würsig, B., T. Keikhefer, and T. A. Jefferson. 1990. Visual displays for communication in cetaceans. In *Sensory abilities of cetaceans: Laboratory and field evidence,* edited by J. Thomas and R. A. Kastelein, 545–60. New York: Plenum Press.

Würsig, B., and M. Würsig. 1980. Behavior and ecology of the dusky dolphin *(Lagenorhynchus obscurus)* in the South Atlantic. *Fish. Bull.* 77 (4):871–90.

Wyman, J. 1967. The jackals of the Serengeti. *Animals* 10:79–83.

Yamagiwa, J. 1986. Activity rhythm and the ranging of a solitary male mountain gorilla. *Primates* 27:273–82.

Yamagiwa, J., T. Maruhashi, T. Yumoto, and N. Mwanza. 1996. Dietary and range overlap in sympatric chimpanzees and gorillas in Kahuzi-Biega National Park, Zaire. In *Great ape societies,* edited by W. C. McGrew, L. F. Marchant, and T. Nishida, 82–98. Cambridge: Cambridge University Press.

Yamagiwa, J., and N. Mwanza. 1994. Day journey length and daily diet of solitary male gorillas in lowland and highland habitats. *Int. J. Primatol.* 15:207–24.

Ydenberg, R. C., and M. Dill. L. 1986. The economics of fleeing from predators. *Adv. Stud. Behav.* 16:229–49.

Yip, P. P., and Y. Pao. 1995. Combinatorial optimization with use of guided evolutionary simulated annealing. *IEEE Trans. Neural Networks* 6:290–95.

Yoder, A. D., M. Cartmill, M. Ruvolo, K. Smith, and R. Vilgalys. 1996. Ancient single origin for Malagasy primates. *Proc. Natl. Acad. Sci. USA,* 93:5122–26.

Yoneda, M. 1984. Comparative studies on vertical separation, foraging behavior and traveling mode of saddle-backed tamarins *(Saguinus fuscicollis)* and red-chested moustached tamarins *(Saguninus labiatus)* in northern Bolivia. *Primates* 25:414–22.

Yost, J. A., and P. M. Kelley. 1983. Shotguns, blowguns and spears: The analysis of technological efficiency. In *Adaptive responses of native Amazonians,* edited by R. B. Hames and W. T. Vickers, 189–224. New York: Academic Press.

Young, M. P. 1993. The organization of neural systems in the primate cerebral-cortex. *Proc. R. S. Lond.* B 252:13–18.

Young, M. P., J. W. Scannell, G. A. P. C. Burns, and C. Blakemore. 1994. Analysis of connectivity: Neural systems in the cerebral cortex. *Rev. Neurosci.* 227–49.

Zabel, C. J., S. E. Glickman, L. G. Frank, K. B. Woodmansee, and G. Keppel. 1992. Coalition formation in a colony of prebubertal spotted hyaenas. In *Coalitions and alliances in humans and other animals,* edited by A. H. Harcourt and F. B. M. de Waal, 113–35. Oxford: Oxford University Press.

Zahavi, A. 1975. Mate selection: A selection for a handicap. *J. Theor. Biol.* 53:205–14.

Zajonc, R. B. 1965. Social facilitation. *Science* 149:269–74.

Zar, J. H. 1996. *Biostatistical analysis.* 3d ed. London: Prentice-Hall International.

Zarrugh, M. Y., and C. W. Radcliffe. 1978. Predicting metabolic cost of level walking. *Eur. J. Appl. Physiol.* 38:215–23.

Zarrugh, M. Y., F. N. Todd, and H. J. Ralston. 1974. Optimization of energy expenditure during level walking. *Eur. J. Appl. Physiol.* 33:293–306.

Zeki, S. M., and S. Shipp. 1988. The functional logic of cortical connections. *Nature* 335:311–17.

Zemel, A., and Y. Lubin. 1995. Inter-group competition and stable group sizes. *Anim. Behav.* 50:485–88.

Zhao, Q. -K. 1996. Etho-ecology of Tibetan macaques at Mount Emei, China. In *Evolution and ecology of macaque societies,* edited by J. E. Fa and D. G. Lindburg, 160–81. Cambridge: Cambridge University Press.

Zimen, E. 1975. Social dynamics of the wolf pack. In *The wild canids,* edited by M. W. Fox, 336–62. New York: Van Nostrand Reinhold.

———. 1976. On the regulation of pack size in wolves. *Z. Tierpsychol.* 40:300–341.

Zimmerman, M. 1979. Optimal foraging: A case for random movement. *Oecologia* 43:261–67.

Zimmermann, E. 1995a. Acoustic communication in nocturnal prosimians. In *Creatures of the dark: The nocturnal prosimians,* edited by L. Alterman, G. Doyle, and M. Izard, 311–30. New York: Plenum Press.

———. 1995b. Loud calls in nocturnal prosimians: Structure, evolution and ontogeny. In *Current topics in primate vocal communication,* edited by E. Zimmerman, J. D. Newman, and U. Jurgens, 47–72. New York: Plenum Press.

Zimmermann, E., and C. Lerch. 1993. The complex acoustic design of an advertisement call in male mouse lemurs (*Microcebus murinus,* Prosimii, Primates) and sources of its variation. *Ethology* 93:211–24.

Zuberbühler, K., R. Noë, and R. M. Seyfarth. 1997. Diana monkey long-distance calls: Messages for conspecifics and predators. *Anim. Behav.* 53:589–604.

birds (*continued*)

coordinated travel, 524, 532, 536, 539, 545, 549, 551, 554, 557

cuckoo-shrikes, 529

deception, 544

decision making, 554

degree/timing of association, 527, 530–32, 543, 552, 556–57

dominance, 540–41

drongos, 538, 552

egrets, 537

factors promoting interspecific gregariousness, 535–36, 539–41, 547–49, 553, 556–57

falcons, 550

feeding/foraging behavior, 524–25, 535, 537, 541–45, 547–48, 557–58

finches, 539, 544

flock organization, 521–22, 529, 536, 538, 540

flock size, 549

flycatchers, 523, 529, 535

greenlets, 529

herons, 537

honey creepers, 525, 535, 537

kleptoparasitism, 539, 541, 544

klingets, 541

leadership, 551–52, 554, 557–58

megaflocks, 525

movement strategies, 545–46, 549, 554

nuclear species, 529–30, 535–36, 540–41, 545–47, 549–51, 554

number of participating species, 525–27, 531

nutches, 541–42

predation, 541, 545–48, 550–52, 558

sentinel species, 551–52, 555

signal convergence, 536–40

social facilitation, 543

southern bentbills, 523

sparrows, 548

spatial memory, 546

tanagers, 525, 528–29, 536, 541, 544, 548

territoriality, 523, 530–32, 545, 552–56

thornbills, 528

titmice, 532, 540–41, 545–46, 548, 551

tits, 528–29, 532, 535–36, 541–44, 547–48, 555

treecreepers, 547

trogans, 523

troop cohesion, 527, 551

vigilance, 541, 547, 550

vocal signals, 528, 532–36, 538–39, 544–47, 550–52, 554, 557

warblers, 523, 529, 532, 535, 543, 555

woodcreepers, 524, 530

woodpeckers, 531, 541–42, 547

wrens, 528–29

bonobo

coordination of group movement, 440, 467

food long calls, 396–397

foraging patterns, 395

lexigrams and spatial information, 323

long distance calls, 399

brain

adult brain size-fetal brain size ratio, 208

brain size, 204–6, 208–9, 211–13, 215, 217, 220, 227, 491, 620–621

brain size and energetics, 205–206, 209, 225, 628, 646

brain size/body weight relationship, 207–209, 212–213, 224–225

brain size/gut size, 208

brain size/home range size, 635, 646–47

brain evolution, 204–205, 208–209, 211–12, 219, 223–24, 227, 234, 236, 628, 646

brain organization/structures, 210–11, 213, 215–19, 221, 223, 226, 231, 233–234

cerebellum, 214–16, 228

cerebral cortex, 408–9

dichromatic color vision, 221

foraging and diet, 205–6, 208–9, 220–22, 224–30, 233, 235, 458–59

group size/brain size, 201, 205, 220, 232–33; hand-eye coordination, 216

hippocampus, 210, 213–15, 226–27, 230, 546

information processing/cognition, 204, 209, 221

lateral geniculate nucleus, 214, 217–18, 221–25, 235

limbic system, 214, 223, 232–33

magnocellular layers, 217–18, 221–22, 227

modularity, 216, 227, 235

neocortex, 211–18, 220–24, 226, 229–30, 233–34, 366, 620–21

neural network, 182, 194, 207, 210, 230–31

olfactory bulb, 210, 213–15, 220

parvocellular layers, 217–18, 221–25, 227, 231

carnivores (*continued*)
division of labor, 614–15
domestic dogs, 600, 608
dominance hierarchy, 598, 600, 602, 605, 609, 623–24
dwarf mongoose, 592–93, 596–97, 605, 608–10, 621
effect of neonates/young on adult behavior, 590, 595–96, 600, 602, 609–10
effect of prey abundance on group movement, 587, 598, 601
energetic costs of travel, 16–17
food sharing, 619, 624
feeding competition, 37, 589–90, 602, 609
feral dogs, 597
fission-fusion societies, 590, 601–2, 609
foxes, 597
giant river otter, 593, 607–9
golden jackal, 593, 599–600, 607, 609
gregarious, 589, 598, 606
grooming, 596
group cohesion, 590–92, 595–98, 601, 608, 622, 625
group hunting, 601, 612
group living carnivores, 589
group movement/travel, 592, 596, 598, 604, 606, 608–10, 614–15, 618, 621–24
group size and foraging, 598, 602–3, 606, 612–13, 617–18, 621
group size and composition, 592–94, 598, 600, 602
group size and intergroup competition, 601
habitat, 593–94
home range overlap, 595, 598
home range size, 589, 593–94, 598, 604, 621–22
hunting strategies, 587, 589–90, 603, 612–18, 621, 624
hyena, 425, 503, 566, 617
initiation of group movement, 596, 599–601, 604, 613, 624
intergroup spacing, 607
jackals, 597–98
leadership, 595–97, 600–602, 604–6, 608, 614, 623–24
leopards, 432
lion, 425, 429, 503, 592, 594, 600–601, 604, 606–7, 609, 612, 614–16, 618, 624
matrilines, 602

meerkat, 593, 596, 608–9
mongooses, 589, 607, 609, 624
motivation, 615
olfaction, 595–98, 604–5, 607–8, 610
packs, 592, 597–98, 613
planning, 616–17
predation, 596, 601, 621
as a predator of primates, 503
prey selection, 587–92, 598, 612–13, 615, 617, 621
recruitment calls, 605
reproduction, 597, 600, 603, 605, 610, 622
resource defense, 598
resource distribution and abundance, 621
response facilitation, 607
rule of thumb, 611, 614–16, 619, 621, 623, 625–26
scent glands, 588, 596
scent marking, 604–8, 610
sex differences in behavior, 597, 605–6, 610, 623
social canids, 592, 597, 607, 609, 612–13
social cognition, 616
social facilitation, 607, 621–22, 624
social rallies (pep rallies), 599, 605–6, 622, 624
solitary hunting, 618
solitary individuals, 589, 592, 598, 615, 618, 620
spotted hyenas, 587, 591–92, 594, 602–5, 607, 609–10, 612–14, 616, 618
subgrouping, 595, 598, 601–2, 622, 624
territorial behavior, 592, 596, 598, 601–2, 604–6, 622
travel groups, 603–4
travel rate, 601
vocalizations, 425, 429, 595–97, 599–601, 605, 607–8
wolves, 425, 593, 597–600, 607–10, 612–15
cetacean
acoustics/vocalizations, 562–63, 573–74, 576–83, 585
aggression, 569, 576
alliances, 559, 565, 568, 570, 572, 584
baleen, 562
benefits/costs of group movement, 564–65
blue whales, 582
bottlenose dolphins, 212, 407–9, 559–60, 566–71, 573–75, 579, 581, 584
bowhead whales, 582

humans (*continued*)
 brain evolution, 628, 646
 celestial cues, 402
 cognition, 616
 cognitive maps, 402, 515
 cooperation, 251, 616, 619
 cooperative hunting, 619
 coordinated hunting, 616
 cultural, economic, and political factors,
  652–55, 665
 decision-making, 250–55, 650–51, 653–54
 dietary strategy, 630, 633, 642–44, 646
 division of labor, 254–55, 405–6, 616,
  619
 dynamic optimality model, 654–55
 early *Homo,* 629–30, 638–40, 643, 646,
  648
 ecology, 651–54
 energy demands, 630, 640, 642–44, 646
 energy efficiency, 662, 665
 food search, 402–5, 629
 food sharing, 616, 620
 geometric map, 155
 group-level cognition, 250–51, 255, 619
 group mobility (*see also* humans, Turkana), 650, 652–55
 group movement (*see also* humans, Turkana), 649, 650, 655
 Guayaki, 631, 633
 Hadza, 631, 633, 652
 home range/body size relationship, 635
 home range, body size, and diet quality,
  636–37, 644, 647–48.
 home range size, 629–30, 633, 642,
  645–46
 hominid evolution, 628–30, 633, 637–40,
  642, 645, 647–48
 Homo erectus, 629, 637–38, 640, 642–48
 Homo habilis, 638, 643
 Homo sapiens, 637–38, 643
 hunter-gatherers, 405, 415, 649–54
 hunting, 403–4, 406, 415, 616–17
 Karimojong, 653
 Kutchin hunting, 405, 417
 intelligent behavior, 616, 620
 language, 255, 415–16, 463
 leadership, 253–54
 locomotion, 12, 18
 Maasai, 653
 Matis, 403–4
 Moyoruna, 403
 Mubti, 633
 navigation, 269, 402
 nomads, 649–51, 653, 655

paleoenvironmental reconstruction, 640,
  645
 pastoralists, 649–54, 656
 planning, 189, 402, 616–17, 619, 625–26
 reproductive success, 620, 665
 social facilitation, 622–23
 social rallies, 622
 spatial routes, 166, 170, 193, 194, 199,
  402
 thermal stress, 629
 Turkana
  aggregations, 660
  annual range, 63–64
  bridewealth, 665, 676
  cattle, 651, 657
  climate/habitat, 657–58, 660, 666–73
  cooperation, 662
  coordination of group movement, 659
  culture, 664–65, 673, 675–76
  decision making, 650–51, 659–65,
   674–76
  ecology, 660–61, 666–69, 674–76
  group mobility, 651, 659–61, 669, 671,
   673–74
  group movement, 650, 662, 666–69,
   671, 673–74
  herd ownership, 663–64, 666
  herd use strategies, 664–67, 669, 672–75
  households, 659–61, 665–67, 669
  information, 661–62
  kinship, 659, 676
  land use patterns, 651, 669–73
  leadership, 659
  livestock, 657, 664
  neighborhoods, 659–60, 662–63, 673
  pastoralists, 649–50, 657, 674
  population size, 657
  primary productivity, 667
  raiding, 660, 665, 667–69
  rights to grazing and water resources,
   658, 661–63, 676
  seasonal movements, 659–61, 664,
   667–69, 671
  social organization, 658–60
  soothsayers, 660, 663–64
  upper Paleolithic, 616
  Waorani, 631, 633
  Yoruk, 652
hunting. *See* capuchins, carnivores, cetaceans; chimpanzees; humans

indri, visual cortex, 221
 insectivores, brain size, 212–13, 215–17,
  227

kangaroos
  energetic cost of travel, 11
  as a prey species, 612

langurs
  loud calls and intergroup encounters, 433
  travel decisions, 433
leaf monkeys
  initiation of group movement, 440
  intention movements, 440
  intergroup encounters, 432
learning, behavioral flexibility, 415
lemurs. *See also* prosimians; strepshirhines
  acoustic/vocal signals, 476, 478, 483–86
  adaptive radiation, 471
  aye-aye, 472, 476, 477
  bamboo lemur, 473
  body size, 473
  care of young, 474
  cathemeral, 473, 479, 481, 490
  coordinating individual and group move-
    ment, 477–79, 481, 486, 488
  dominance, 488
  dwarf lemur, 472, 476
  eulemurs, 482–83
  female dominance/leadership, 479, 483,
    485–86
  fork-marked lemur, 478
  gentle lemur, 473
  indri, 479–80, 481
  leadership, 457, 485
  mating patterns, 472
  mouse lemur, 472, 476
  nocturnality, 472–74
  predation, 488
  range size, 482
  ranging patterns and group movement,
    475
  red-bellied lemurs, 479
  resource distribution, 487
  ring-tailed lemur, 484–85, 488
  roaming behavior, 476, 485
  ruffed lemur, 484
  scentmarking, 474, 476, 486
  sifaka, 485, 488
  social evolution, 472
  social organization, 472–73, 481
  solitary living, 473, 476
  subgroups, 483, 485, 487
  territorial behavior, 484
  woolly lemurs, 478–79
lion tamarins
  coordination of group movement, 299,
    318–22, 440

day range, 107, 113, 302
description of study site, 104
experimental conditions, 304–7
group size and composition, 302
home range, 107, 113
homing, 190, 301, 304, 307–17
intergroup encounters, 109, 302
loud calls, 100, 430
ranging behavior, 106–7, 115, 302
reintroduction, 303–4, 320–21, 323–24
travel calls, 440, 443–47, 456, 461, 466
loris, energetic cost of travel, 13

macaques
  agonistic screams, 463
  coordination of group movement, 435,
    440, 467
  foraging strategy, 200, 273–74
  group size and day range, 38
  location of sleeping sites, 52
  predation, 53
  spatial memory, 401
Madagascar, primate predators, 52–53
manabeys
  intergroup encounters, 432
  movement patterns, 330
  loud calls and territoriality, 430–31
marmosets
  day range, 115
  home range, 115
mixed-species associations. *See also* birds;
    capuchins; guenons; tamarins; squir-
    rel monkeys
  aggregations, 528
  aggression, 525
  alarm calls/responses, 84, 528, 547,
    551–52
  attendant species, 528–30
  attraction, 74, 86, 527
  benefits/costs, 76–77, 85, 540–43, 548
  cohesion, 73, 87, 91–92, 97, 105, 551–52
  competition, 542–43
  core species, 528, 555–56
  deception, 544
  degree of association, 75, 521, 525, 527
  disbanding, 95
  effects on group movement and ranging,
    74–75, 78–86, 93–94, 97, 527
  feeding/foraging efficiency, 85–87, 524–
    25, 548
  forest flocks, 521–22
  gregariousness, 527, 529–30, 535
  habitat utilization, 83–84, 521
  kleptoparasitism, 539, 541, 544

group size/brain size, 205
first order intentionality, 460
foraging, 262
honest communication, 422, 423
individual recognition, 492
leadership, 253, 319, 322, 368, 380, 394–
    95, 449–52, 457, 479, 482, 485, 489,
    492, 495–503, 517, 551–52, 554, 557–
    58, 595–97, 600–602, 604–6, 608,
    614, 623–24, 678
manipulation, 423, 468–69, 582
memory, 458
negotiation, 517
reconciliation, 489
social cohesion, 471
social evolution, 472
social intelligence, 229–30, 232, 414
social insects
    ants, 129, 134–35, 145, 148, 150–51, 188,
        550
    army ants, 132, 134, 137, 524
    bumble bees, 129, 187
    carpenter bees, 128
    chemical cues, 134, 139, 145–48, 150
    colony fission/dispersal, 135, 138–39,
        142, 144, 157, 161
    colony organization, 129–30
    communication, 131, 138–44, 146–47,
        246–47
    coordination of activities, 127, 131–32,
        134–38, 147, 163, 246–47
    defense of colony/nest, 130–34, 138, 149
    division of labor, 130–31, 137–38
    feeding behavior, 130, 132, 142, 149, 152,
        157, 161
    honey bees, 128–29, 131–39, 142, 144–45,
        148–50, 158, 160, 246
    initiation of movement, 144, 146
    landmarks, 150–57, 161–62
    metric map, 156, 228
    migration, 143–44, 146, 157
    navigation, 150–51, 155
    navigation using the sun, 151–52, 157
    recruitment pheromones, 146–48
    route map, 155, 157, 162
    scouts, 139–40, 142
    spatial cognition, 150–51, 153–56
    stingless bees, 135–36, 138–39, 149–50
    swarms, 129, 139, 144, 146–48, 158–61
    termites, 129, 150
    wasps, 128–29, 135, 138, 149, 155, 158
    weaver ants, 138
social organization

aggression, 482
breeding pair, 596
cohesion, 470, 482
fission-fusion, 34, 36, 68, 164, 205, 236,
    393, 407, 411, 414–15, 566–67
genetic structure, 472, 481
group-living, 471–73, 481
group size, 472, 484
grouping pattern, 470, 482, 484, 494
mate guarding, 495
monogamous pair, 597
neighborhoods, 481
pair-living, 472–73, 477, 481, 484
social cohesion, 470, 494
socioecological theory, 470
solitary primates, 472–73
subgrouping, 393, 482, 484, 494, 498,
    503–4
socio-spatial cognition
    social cognition hypothesis, 206, 234
    social information and foraging, 204–5,
        233, 236, 262, 269, 271–73, 288, 296,
        366, 680, 682
spatial representations
    cognitive map, 149–50, 153–54, 156, 168,
        205, 209, 213, 228–30, 234, 264–65,
        269, 288–90, 390, 402, 509–12, 517,
        678
    dead-reckoning, 389
    Euclidean map, 168, 171, 175, 230, 265,
        267, 288–89, 291, 295, 510–11, 514,
        516
    food caching/storage, 170
    geometric map (see also spatial represen-
        tations: Euclidean map; metric map;
        vector map), 155–56, 265, 267,
        389–90
    homing, 152, 158
    landmarks, 150–54, 156, 162, 189, 263,
        265–66, 285, 288–89, 291, 294–95,
        297, 316, 389, 390, 402, 515
    large-scale space, 153, 155–56, 190, 263,
        265–66, 285, 288–89, 291, 294–95,
        297, 492, 509–10, 512
    Manhattan metric, 171
    mental map (see also spatial representa-
        tions, cognitive map), 152, 264
    metric map, 156, 511
    navigation, 190, 228, 265–66, 269, 290,
        316–17, 389, 491, 511–12, 514–15
    network map (see also spatial representa-
        tions, topological map), 182, 512–16
    path integration (see also spatial represen-

spatial representations (*continued*)
tations, dead-reckoning), 152–53,
267–68, 292, 316, 389
route map, 155, 162, 189, 154
shortest path problem, 167–69, 511, 516
small-scale space, 153, 263, 265–66, 269,
277, 287–89, 291, 295, 297
string of commands, 513
topological map (*see also* spatial representations: network map; route map),
267–69, 288–89, 294, 317, 366, 514,
545
traveling salesman problem, 165–72,
175–77, 182–86, 194, 197–98, 262;
utility, 168–69, 180
vector map (*see also* spatial representations, geometric map), 366, 511–12,
515–16
vehicle routing problem, 172–73
spider monkey
Barro Colorado Island, 392
brain size and foraging, 205
day range, 392
diet, 392
energetic costs of travel, 12
food calls, 396–97
foraging pattern, 393, 400
group size and composition, 394
group size and patch size, 37, 393–94,
400
home range, 394
leading females, 395
patch depletion, 31–32
patch size, density, and distribution, 34,
392
ranging patterns, 263
sex differences in ranging, 394–95
sleeping sites, 63
subgrouping, 392, 296, 400
squirrel monkey
alarm calls, 43
communication, 1
contact calls, 71, 448
energetic costs of travel, 19–20, 22
foraging routes, 192, 236
home range overlap, 120
mixed species troops
coordinating troop movement, 90, 93,
448
leading species, 447–48
leading troop movement, 94
predation, 46, 50, 453
travel calls, 440, 443–46, 450, 453, 461

strepsirhines, brain size, 212–13, 217
Sulawesi crested black macaque
avoidance of neighboring groups, 341–42
day range, 333, 336
diet, 333–35, 338–40
group size and composition, 333
home range size, 333
intergroup encounters, 347–49
patch size, 338
range boundaries, 341, 345
range use, 335, 337, 339–40, 342–46, 349
ranging behavior, 331
travel calls, 440, 443–47

talapoin monkey, brain size, 217
tamarins
alarm calls, 70
effects of predation on group movement,
62, 66
goal directed travel, 377, 516
least-effort travel routes, 516
mixed species troops
coordination of troop movement, 90,
93
day range, 107, 113, 115
description of study site, 104
general, 75, 79
home range size, 107, 115
intergroup encounters, 109
leading troop movement, 94
loud calls, 100
ranging behavior, 106–7, 119
studies in captivity, 98
troop cohesion, 105
rule guided foraging, 274–80, 287, 292
spatial memory, 190–92, 200, 293–95,
516
wah-wah calls, 466
Tana river crested mangabey
avoidance of neighboring groups, 341
day range, 333, 336
diet, 333–35, 338–40
group size and composition, 333
home range size, 333
intergroup encounters, 347–49
Kenya, 331
patch size, 338
range boundaries, 341, 345
range use, 335, 337, 339–40, 342–46, 349
resource defense, 331, 345
territoriality. *See also* birds; mixed-species
associations
cost of defense, 102, 330, 553

SPECIES INDEX

Bold italicized page numbers indicate a figure and/or table.

animalia (*continued*)

Parus major (*see* great tit), 544, 555

Parus montanus (*see* willowtit), 535, 547

Parus rufescens (*see* chestnut-backed chickadee), 542

Parus spp. (*see* tit), 192

Picoides pubescens (*see* downy woodpecker), 531

Phylloscopus (*see* warbler), *526*

Phylloscopus occipitalis (*see* western crown warbler), 532

Stephanoaetus coronatus (*see* crowned hawk eagle), 64, 432

Tachyphonus (*see* tanager), *526, 528*

Tachyphonus luctuosus (*see* tanager), 530, 536

Tangara (*see* tanager), 528

Tangara inornata (*see* plain-colored tanager), 536

Terenotriccus erythrurus (*see* ruddtailed flycatcher), 523

Thamnomanes (*see* antshrike), 532, *533,* 541, 544, 552

Thamnophilus puncttatus (*see* slaty antshrike), 523

Trogon rufus (*see* green and yellow black-throated trogon), 523

Xiphorhynchus, *554*

Xiphorhynchus guttatus (*see* buff-billed woodcreeper), 530

Hexapoda, Hymenoptera

Apis, spp. (*see* honeybee), 128, 135

Apis cerana (*see* Asian hive bee), 138

Apis dorsata (*see* Asian "rock bee"), 133–34, 142, *143,* 157

Apis florea (*see* dwarf honeybee), 132, 134, 137, 149, 160

Apis laboriosa (*see* rock bee), 133, 137

Apis mellifera (*see* European honeybee), 139, 246

Apis mellifera scutellata (*see* African hive bee), 142, 143, 157

Bembix (*see* digger wasp), 128

Bombus spp. (*see* bumblebee), 187

Eciton spp. (*see* army ant), 132, 135

Eciton burchelli (*see* army ant), *133,* 134

Lapidus praedator (*see* army ant), 145

Melipona spp. (*see* stingless bee), 135

Oecophylla smaradina (*see* weaver ant), 138

Philanthus (*see* digger wasp), 128

Trigona spp. (*see* stingless bee), 135

Xylocopa (*see* carpenter bee), 128

Mammalia, Eutheria, Artiodactylia

Alces alces (*see* moose), *627*

Cervus elephas (*see* red deer), 433, *627*

Connochaetes taurinus (*see* wildebeest), 18, *627*

Gazella spp. (*see* gazelle), *627*

Odocoileus spp. (*see* deer), *627*

Ovis dalli (*see* mountain sheep), 59, 457

Phacochoerus aethiopicus (*see* warthog), *627*

Syncerus caffer (*see* African buffalo), 248, *627*

Mammalia, Eutheria, Carnivora

Canis aureus (*see* golden jackal), *627*

Canis familiaris (*see* dog; dingo), *627*

Canis latrans (*see* coyote), 425, *627*

Canis lupus (*see* wolf), 425, *627*

Canis mesomelas (*see* black-backed jackal), *627*

Crocuta crocuta (*see* spotted hyena), 425, *627*

Crossarchus obscurus (*see* common cusimanse), *627*

Cuon alpinus (*see* dhole), *627*

Helogale parvula (*see* dwarf mongoose), *627*

Lobodon carcinophagus (*see* seal), 575

Lycaon pictus (*see* cape hunting dog), *627*

Meles meles (*see* European badger), *627*

Mirounga leonina (*see* pinniped), 575

Mungos mungo (*see* banded mongoose), *627*

Nasua narica (*see* coati), 272, *627*

Otaria flavescens (*see* pinniped), 575

Panthera leo (*see* lion), 425, 429, *627*

Panthera pardus (*see* leopard), 432

Pteronura brasiliensis (*see* giant river otter), *627*

Suricata suricata (*see* meerkat), *627*

Vulpes spp. (*see* fox), *627*

Mammalia, Eutheria, Cetacea

Balaena mysticetus (*see* bowhead whale), 563

Balaenoptera physalus (*see* fin whale), 563

Delphinus delphis (*see* common dolphin), 565

Eschrichtius robustus (*see* grey whale), 582

animalia (*continued*)